AF598227

Two

THE
INTERNATIONAL SERIES
OF
MONOGRAPHS ON PHYSICS

INTERNATIONAL SERIES OF MONOGRAPHS ON PHYSICS

99. L. Mestel: *Stellar magnetism*
98. K. H. Bennemann: *Nonlinear optics in metals*
97. D. Salzmann: *Atomic physics in hot plasmas*
96. M. Brambilla: *Kinetic theory of plasma waves*
95. M. Wakatani: *Stellarator and heliotron devices*
94. S. Chikazumi: *Physics of ferromagnetism*
93. A. Aharoni: *Introduction to the theory of ferromagnetism*
92. J. Zinn-Justin: *Quantum field theory and critical phenomena*
91. R. A. Bertlmann: *Anomalies in quantum field theory*
90. P. K. Gosh: *Ion traps*
89. E. Simánek: *Inhomogeneous superconductors*
88. S. L. Adler: *Quaternionic quantum mechanics and quantum fields*
87. P. S. Joshi: *Global aspects in gravitation and cosmology*
86. E. R. Pike, S. Sarkar: *The quantum theory of radiation*
84. V. Z. Kresin, H. Morawitz, S. A. Wolf: *Mechanisms of conventional and high* T_c *superconductivity*
83. P. G. de Gennes, J. Prost: *The physics of liquid crystals*
82. B. H. Bransden, M. R. C. McDowell: *Charge exchange and the theory of ion–atom collision*
81. J. Jensen, A. R. Mackintosh: *Rare earth magnetism*
80. R. Gastmans, T. T. Wu: *The ubiquitous photon*
79. P. Luchini, H. Motz: *Undulators and free-electron lasers*
78. P. Weinberger: *Electron scattering theory*
76. H. Aoki, H. Kamimura: *The physics of interacting electrons in disordered systems*
75. J. D. Lawson: *The physics of charged particle beams*
73. M. Doi, S. F. Edwards: *The theory of polymer dynamics*
71. E. L. Wolf: *Principles of electron tunneling spectroscopy*
70. H. K. Henisch: *Semiconductor contacts*
69. S. Chandrasekhar: *The mathematical theory of black holes*
68. G. R. Satchler: *Direct nuclear reactions*
51. C. Møller: *The theory of relativity*
46. H. E. Stanley: *Introduction to phase transitions and critical phenomena*
32. A. Abragam: *Principles of nuclear magnetism*
27. P. A. M. Dirac: *Principles of quantum mechanics*
23. R. E. Peierls: *Quantum theory of solids*

F. P. Bowden, D. Tabor: *The friction and lubrication of solids*
J. M. Ziman: *Electrons and phonons*
M. Born, K. Huang: *Dynamical theory of crystal lattices*
M. E. Lines, A. M. Glass: *Principles and applications of ferroelectrics and related materials*

Stellar Magnetism

LEON MESTEL

Emeritus Professor of Astronomy
University of Sussex

CLARENDON PRESS • OXFORD

1999

Oxford University Press, Great Clarendon Street, Oxford OX2 6DP
Oxford New York
Athens Auckland Bangkok Bogota Buenos Aires Calcutta
Cape Town Chennai Dar es Salaam Delhi Florence Hong Kong Istanbul
Karachi Kuala Lumpur Madrid Melbourne Mexico City Mumbai
Nairobi Paris São Paolo Singapore Taipei Tokyo Toronto Warsaw
and associated companies in
Berlin Ibadan

Oxford is a registered trade mark of Oxford University Press

Published in the United States
by Oxford University Press Inc., New York

A catalogue record for this book is available from the British Library

Library of Congress Cataloging in Publication Data
(Data available)

ISBN 0 19 851761 0

Typeset by the author using LaTeX

Printed at Thomson Press (India) Ltd

In Memoriam

SOLOMON MESTEL 1886 – 1966

RACHEL MESTEL (née BRODETSKY) 1891 – 1974

'IT IS NOT THY DUTY TO COMPLETE THE LABOUR,

BUT NEITHER ART THOU FREE TO DESIST THEREFROM'

Ethics of the Fathers, II, 21.

PREFACE

This book has emerged from lectures, papers and review articles written over the decades. A lifelong interest in stellar structure began when I was Fred Hoyle's Ph.D. student at Cambridge University in 1948–51, and in cosmical electrodynamics from three years as an I.C.I. research fellow at Leeds University in 1951–54, working with the late Thomas G. Cowling. Before returning to Cambridge as a junior faculty member, I had a further rewarding year as Commonwealth Fund Fellow at Princeton University Observatory, working with the late Martin Schwarzschild and Lyman Spitzer Jr. Among others of my seniors from whom I recall learning – whether from personal discussion or from their published work – I mention especially the late Ludwig Biermann, Subrahmanyan Chandrasekhar, Vincenzo Ferraro and William B. Thompson, and also Hermann Bondi, Thomas Gold, C.C. Lin, William H. McCrea and Maurice Pryce. My debt – whether explicit or implicit – to all of them, and also to many colleagues of my own and younger generations, is gratefully acknowledged.

The book begins with the introductory Chapter 1, containing a brief historical survey and a summary of the observational data. Chapter 2 presents the basic physics, primarily for non-relativistic applications, in particular the two-fluid and three-fluid models respectively for fully ionized and partially ionized gases. Chapter 3 consists of applications of the basic theory, introducing concepts that are household words for the worker in cosmical electrodynamics.

The choice of astronomical topics for detailed study, though clearly influenced strongly by those areas in which I have been active, has been motivated by the hope that the book will complement other recent additions to the literature. Thus with excellent volumes available such as Eric Priest's 'Solar magnetohydrodynamics', Michael Stix's 'The Sun' and Jack Thomas and Nigel Weiss's 'Sunspots: observations and theory', it appears sufficient to restrict the treatment of 'solar activity' to fundamentals. Chapter 4 on the interaction of magnetism and convection attempts to set the scene.

The dynamo problem (Chapters 6 and 8) continues to be a challenge, both in spite of and because of the plethora of models in the literature. There is now a belated recognition that some of the euphoria following the genuinely remarkable developments of the last two or three decades should have been tempered by an explicit recognition that much dirt has been swept under the carpet, especially by the introduction of a blanket 'turbulent resistivity'. As noted in Chapter 4, the escalation in computing power is already leading to much more sophisticated descriptions of magnetoconvection, and one looks forward to further enlightenment on the fundamentals of dynamo theory. However, I hazard a guess that something like the phenomenological 'standard dynamo equations' will survive as encapsulating much of the essence of what is going on, especially in solar-type stars.

Interaction between rotation and magnetism is a recurrent theme in the book. The theory of the braking of rotating stars through magnetic coupling with a stellar wind (Chapter 7) is conceptually clear, and when combined with a phenomenological dynamo dependence on rotation leads to a satisfying if provisional account of the 'solar–stellar connection', in particular of the rotational history of late-type and pre-main sequence stars, with suggestive hints towards the understanding of planetary system formation (Chapters 8 and 10). Closer inspection. however, shows that a complete understanding of the rotating magnetic wind and of its instabilities still faces subtle physical and mathematical difficulties.

The studies in Chapter 5 of axisymmetric rotating magnetic stars are a necessary preliminary to the attempt in Chapter 9 to understand the early-type magnetic stars in terms of the essentially non-axisymmetric oblique rotator. This phenomenological model has turned out to be remarkably successful for the interpretation of the observations. To one trained in the traditions of the Cambridge Mathematical Tripos, there is an irresistible imperative to understand in depth the mutual interaction of the thermo-gravitational field and the perturbing centrifugo-magnetic field in a structure of such low symmetry – to some extent analogous to the geophysicist's studies of polar wandering and latitude variation. Although changes in the observational situation have lessened the chances of even indirect vindication of theoretical predictions, the analysis retains at least its pedagogical interest.

The problems of star formation are a fruitful field for the application of magnetohydrodynamics, especially in the form appropriate for lightly ionized gases (Chapters 11 and 12). However, studies that instead follow on from the pioneering work of the days before magnetohydrodynamics may yet turn out to be relevant to the earliest epochs, before galactic dynamos had time to operate; and in any case, such studies are of value in showing how radically the introduction of magnetic coupling transforms the discussion. It is surely premature (not to say foolhardy) to claim either that star formation cannot occur at all in the absence of a magnetic field, or that the presence of the field has only a marginal effect on the initial mass function – a primary goal of any theory of star formation. It may be that the initial mass function will be determined in the later stages of star formation: beyond the coverage of Chapters 11 and 12, but earlier than the epoch covered by Chapter 10 on the pre-main sequence stars, which is devoted mainly to accretion discs, Alfvén-wave driven winds, and the problem of jet collimation.

Chapters 13 and 14 deal with the formidable and still unresolved pulsar magnetosphere problem. It is emphasized that the pulsed radio emission is in fact a diagnostic of a much more powerful process that carries off far more energy and angular momentum, as is readily inferred from the observed spin-down of the star. The motivation behind the work of the Sussex group has been largely the feeling that it is just intolerable that we should not understand how a rotating magnetic neutron star – even in the aligned, 'non-pulsar' case – is able to come to terms with its environment, whether in the context of classical or quantum physics. Both the similarities to and the differences from the analogous non-relativistic stellar spin-down problem (Chapter 7) justify a reappraisal of the approximations adopted in standard plasma physics (Chapter 2). Incoherent radiation, in particular gamma-ray emission that can lead to electron–positron pair production, is an essential feature of the magnetosphere problem, but the reader interested in the weaker but much more complicated problem of coherent radio emission is referred to excellent reviews, especially by Don Melrose. Pulsar wind theory faces subtle mathematical difficulties analogous to those found in the non-relativistic problem.

The theory of X-ray binary sources appear to be much closer than that of radio pulsars to classical magnetohydrodynamics. However, to keep an upper bound on an already rather long book, it was decided there was no need to compete with works such as Juhan Frank *et al.*'s 'Accretion power in astrophysics' and Christopher Campbell's 'Magnetohydrodynamics in binary stars'.

The process of writing has been both greatly influenced and facilitated by the technological revolution of the latter half of the century. The personal computer used as a word-processor has largely compensated for the decline in the level of secretarial support that universities can now afford to offer their faculty members. In particular, the mastering of LaTeX has been a most profitable investment of time and effort. My thanks are due to my colleagues Robert Smith, Peter Thomas and Andrew Liddle, to Stuart Keir, manager of the Sussex Starlink Node, and to the many Sussex graduate students and research fellows for their kindness and patience in helping an old dog to learn new tricks.

The ease with which LaTeX files can be propagated around the globe would be of little consequence were it not for the readiness of colleagues, erstwhile students and collaborators to read preliminary drafts of chapters, not only to check that their work is being put into proper context, but also to offer constructive suggestions on improving the balance of space devoted to different topics – perhaps the most difficult hurdle to negotiate. I am particularly grateful to my late colleague Roger Tayler, the late Lyman Spitzer Jr., and to Jacob Bekenstein, Axel Brandenburg, Andrew Cameron, Christopher Campbell, Cathie Clarke, Jean-Francois Donati, Douglas Gough, Alan Hood, David Hughes, Jan Kuijpers, Sanjiv Kumar, John Landstreet, Thierry Lanz, Donald Lynden-Bell, Don Melrose, Chris McKee, David Moss, Aake Nordlund, Colin Norman, John Papaloizou, Richard Paris, Eric Priest, Jim Pringle, Gunther Rüdiger, Manfred Schüssler, Shinpei Shibata, Sami Solanki, Henk Spruit, Kandu Subramanian, Kanaris Tsinganos, Yi-Ming Wang, Nigel Weiss and Ellen Zweibel.

Much of the final writing and editing was done when I used the text as the basis for a lecture course delivered as visiting professor at the Racah Institute of Theoretical Physics, Hebrew University of Jerusalem in 1995–96. The final polishing of the text took place when I was a visitor at the Institute of Astronomy of the University of Utrecht during the autumn of 1997. The support from the Royal Society/Israel Academy programme and the Netherlands Organization for Scientific Research (NWO) is gratefully acknowledged.

Thanks are due also to Denys Wilkinson, a Delegate of the Oxford University Press, and to Martin Rees, one of the series editors, for the encouragement I have received over the years, and to the OUP staff for their help during both the gestation and production periods. And finally to my family, especially to my wife Louise for tolerating the amount of time I have taken up paying court to her rival Urania.

Falmer, E.Sussex L.M.
March 1998

CONTENTS

The plates may be found between pages 12 and 13.

FIGURE ACKNOWLEDGEMENTS

Plates 1(a,b) are reproduced with permission from L. Golub.
Plate 2 is reproduced with permission from L. Tao, N.O. Weiss, D.P. Brownjohn and M.R.E. Proctor.

The following figures are reproduced with the permission of the American Astronomical Society and of the authors:

Fig. 4.3 from D.J. Galloway and N.O. Weiss (1981), *Astrophysical Journal*, **243**, 945.
Fig. 4.6 from N. Hulburt and J. Toomre (1988), *Astrophysical Journal*, **327**, 920.
Fig. 6.4(a,b) from E.N. Parker (1993), *Astrophysical Journal*, **408**, 707.
Fig. 6.7 from S.I. Vainshtein, E.N. Parker and R. Rosner (1993), *Astrophysical Journal*, **404**, 773.
Fig. 8.1 from R.W. Noyes *et al.* (1984), *Astrophysical Journal*, **279**, 763.
Fig. 8.10 from P.A. Gilman (1983), *Astrophysical Journal Supplement*, **33**, 243.
Fig. 10.2 from J. Li (1996), *Astrophysical Journal*, **456**, 696.
Fig. 10.3 from S. Edwards *et al.* (1993), *Astronomical Journal*, **106**, 1, 372.
Fig. 10.9 from A. Brandenburg *et al.* (1995), *Astrophysical Journal*, **446**, 741.

The following figures are reproduced with the permission of the Royal Astronomical Society:

Fig. 3.7 from D. Lynden-Bell and C. Boily (1994), *Monthly Notices Royal Astronomical Society*, **267**, 146.
Figs. 3.15(a,b,c) from L. Mestel and P.A. Strittmatter (1967), *Monthly Notices Royal Astronomical Society*, **137**, 95.
Fig. 8.4 from A. Collier Cameron and J. Li (1994), *Monthly Notices Royal Astronomical Society*, **269**, 1099.
Fig. 8.5 from R.N. Bracewell (1988), *Quarterly Journal Royal Astronomical Society*, **29**, 119.
Figs. 9.4(a,b) from D. Moss, L. Mestel and R.J. Tayler (1990), *Monthly Notices Royal Astronomical Society*, **245**, 550.
Figs. 9.5(a,b,c) from D. Moss (1984), *Monthly Notices Royal Astronomical Society*, **209**, 607.
Fig. 10.5 from P.J. Armitage and C. Clarke (1996), *Monthly Notices Royal Astronomical Society*, **280**, 458.

Fig. 10.6 from P.J. Armitage (1995), *Monthly Notices Royal Astronomical Society*, **274**, 1242.
Fig. 10.13 from H.C. Spruit, T. Foglizzo and R. Stehle (1997), *Monthly Notices Royal Astronomical Society*, **288**, 333.
Fig. 11.8 from L. Mestel and R.B. Paris (1979), *Monthly Notices Royal Astronomical Society*, **187**, 337.
Fig. 12.1 from D.M. Barker and L. Mestel (1990), *Monthly Notices Royal Astronomical Society*, **245**, 147.
Figs. 12.2, 12.3 and 12.4 from D.M. Barker and L. Mestel (1996), *Monthly Notices Royal Astronomical Society*, **282**, 317.
Figs. 12.5, 12.6, 12.7, 12.8 and 12.9 from L. Mestel and T.P. Ray (1985), *Monthly Notices Royal Astronomical Society*, **212**, 275.
Fig. 13.4 from L. Mestel and M.H.L. Pryce (1992), *Monthly Notices Royal Astronomical Society*, **254**, 355.
Figs. 14.1, 14.2, 14.3 and 14.4 from S. Shibata (1997), *Monthly Notices Royal Astronomical Society*, **287**, 262.
Figs. 14.8, 14.9, 14.10 and 14.11 from L. Mestel and S. Shibata (1994), *Monthly Notices Royal Astronomical Society*, **271**, 621.

The following figures are reproduced with the permission of *Astronomy and Astrophysics*:

Fig. 4.2(a,b) from M. Küker, G. Rüdiger and L.L. Kitchatinov (1993), *Astronomy and Astrophysics*, **279**, L1.
Fig. 8.11 from S.M. Tobias (1996), *Astronomy and Astrophysics*, **307**, L21.
Fig. 10.4 from A. Collier Cameron *et al.* (1995), *Astronomy and Astrophysics*, **298**, 133.
Figs. 10.7 and 10.8 from U. Torkelsson and A. Brandenburg (1994), *Astronomy and Astrophysics*, **283**, 677.
Figs. 11.6 and 11.7 from L. Mestel and R.B. Paris (1984), *Astronomy and Astrophysics*, **136**, 98.

The following figures are reproduced with kind permission from Kluwer Academic Publishers and from the authors.

Figs. 2.2, 3.1, 3.2, 3.3 and 3.4 from figs. 2.4, 4.7, 5.4, 5.5 and 5.6 in E.R. Priest (1982), *Solar magnetohydrodynamics*.
From K.C. Tsinganos (ed.) (1996), *Solar and astrophysical magnetohydrodynamic flows*:
Figs. 4.7, 4.8 and 4.9 from figs. 3, 6 and 8 in M. Schüssler, *Magnetic flux tubes and the solar dynamo*, p. 17; and

Figs. 4.10, 4.11 and 4.12 from figs. 2.3 and 4 in B.C. Low, *Spontaneous formation of current sheets*, p. 109.
Fig. 1.3 from fig. 3.5 in Ye Shi-Hui (1994), *Magnetic fields of celestial bodies.*
Fig. 3.5 from fig. 3 in J. Kuijpers, *Physics of flares in stars and accretion discs*, in J.T. Schmelz and J.C. Brown (eds.) (1992), *The Sun – a laboratory for astrophysics.*
Fig. 4.5 from B.W. Lites, A. Nordlund and G.S. Scharmer, *Comparison of computed spectral lines with observation*, in R.J. Rutten and G. Severino (eds.) (1989), *Solar and stellar granulation*, NATO 263.
Fig. 8.6(b) from M. Stuiver and T.F. Braziunas, in F.R. Stephenson and A.W. Wolfendale (eds.) (1988), *Secular solar and geomagnetic variations in the last 10,000 years.*
Figs. 10.10 and 10.12 from figs. 13.10 and 13.12 in H.C. Spruit, *MHD jets and winds*, in R. Wijers *et al.* (eds) (1996), *Evolutionary processes in binary stars.*
Fig. 10.11 from fig. 1 in H.C. Spruit, *Magnetic winds from stars and disks*, in D. Lynden-Bell (ed.) (1994), *Cosmical magnetism.*

Fig. 1.1 appears courtesy of S.M. Tobias and of D.N. Hathaway (NASA/Marshall Flight Center).
Fig. 1.4 is reproduced from A. Collier Cameron in *Surface inhomogeneities in late-type stars* (eds P.B. Byrne and D.J. Mullan), (1992), by permission of Springer, Berlin.
Figs. 2.4 and 2.5 are reproduced from *Plasma physics* (ed. B.E. Keen) (1974), by permission of Institute of Physics.
Figs. 3.11, 3.12 and 3.13 are reproduced from G. Bateman, *MHD instabilities* (1978), by permission of MIT Press.
Fig. 3.18 is reproduced from E.R. Priest and T.G. Forbes, *Journal Geophysical Research*, **97**, 16,757 (1992), published by American Geophysical Union.
Fig. 4.1 is reproduced from G. Rüdiger and L.L. Kitchatinov in *The solar engine* (ed. E. Ribes) (1994), by permission of Springer, Berlin.
Fig. 4.4 is reproduced from *Plasma instabilities in astrophysics* (eds D.G. Wentzel and D.A. Tidman) (1968), by permission of Gordon and Breach Science Publishers.
Figs. 6.1 and 6.2 are reproduced from figs. 4.2 and 4.5 in Ya. B. Zeldovich *et al.*, *Magnetic fields in astrophysics* (1983), by permission of Gordon and Breach Science Publishers.
Figs. 6.3(a,b,c) are reproduced from fig. 2.4 in D. Moss, *Magnetic fields in stars* (1981), *Physics Reports.* **140**, 1, by kind permission of Elsevier Science-NL, Sara Burgerhartstraat 25, 1055 KV Amsterdam, The Netherlands.
Fig. 6.5 is reproduced from figs. 4 and 6 in N.O. Weiss, F. Cattaneo and C.A. Jones (1984), *Geophysical Astrophysical Fluid Dynamics*, **30**, 305, by permission of Gordon and Breach Science Publishers.
Figs. 6.8 and 6.9 are reproduced from figs. 1 and 8 in A. Brandenburg, R.L. Jen-

nings, A. Nordlund, M. Rieutord, R.F. Stein and I. Tuominen (1996), *Magnetic structures in a dynamo simulation, Journal Fluid Mechanics*, **306**, 325. Cambridge University Press.

Figs. 8.2 and 8.7 are reproduced with permission from *Nature*, **348**, 520, S. Baliunas and R. Jastrow, *Evidence for long-term brightness changes of solar-type stars.* Copyright 1990, Macmillan Magazines Ltd.

Fig. 8.3 is reproduced from fig. 7 in S.H. Saar and S.L. Baliunas, in *The solar cycle* (ed. K. Harvey), PASP Conference Series **27**. Copyright 1992, *Astronomical Society of the Pacific*; reproduced with permission.

Fig. 8.6(a) is reproduced from G.M. Raisbeck *et al.* (1990), *Philosophical Transactions Royal Society A*, **330**, 463.

Fig. 8.8 is reproduced from M.J. Thompson *et al.* (1996), *Science*, **272**, 1300. Copyright 1996, American Association for the Advancement of Science.

Fig. 8.9 appears courtesy of Y.-M. Wang.

Fig. 8.12 is reproduced from D.F. Gray in *The Sun and cool stars* (eds I. Tuominen, D. Moss and G. Rüdiger), (1991), by permission of Springer, Berlin.

Fig. 9.1 is reproduced from G. Preston (1971), *PASP*, **83**, 571. Copyright 1971, *Astronomical Society of the Pacific*; reproduced with permission.

Fig. 9.2 is reproduced from fig. 8 in H. Gollnow, *Observational Results*, in *Stellar and solar magnetic fields* (1965), by permission of the editor R. Lüst.

Figs. 11.1(a,b) and 11.2 are reproduced by permission of R.B. Paris from his 1971 Ph.D. Dissertation, University of Manchester.

Figs. 14.5 and 14.6 are reproduced from L. Mestel in *Pulsars: problems and progress* (eds S. Johnston, M.A. Walker and M. Bailes), PASP Conference Series **105**. Copyright 1996, *Astronomical Society of the Pacific*; reproduced with permission.

Every effort has been made to obtain permission to reproduce copyright figures. If notified, the publisher will be pleased to rectify any errors or omissions at the earliest opportunity.

1

INTRODUCTION

1.1 Historical survey

Recognition that the Earth itself is a gigantic magnet goes back to the sixteenth century, as recorded by William Gilbert in his treatise *De magnete*. Not long after, it was established that the Earth's magnetic properties slowly vary, in particular showing a westward drift. Edmund Halley's extensive oceanic survey yielded the first world magnetic chart. Halley was writing before the discoveries of Oersted, Ampère and Faraday, and so he naturally interpreted the terrestrial field as due to permanently magnetized matter. He pictured the Earth as having two massive blocks of lodestone, an outer shell with two magnetic poles and an inner nucleus, concentric with the shell and with two poles of its own (Armitage 1966). To account for the variations, the magnetic axes of shell and nucleus must be inclined both to each other and to the axis of the Earth's diurnal rotation, about which the two components turn at slightly different rates, yielding a slow relative motion of the magnetic poles. Thus he inferred that the Earth's interior must be partially fluid, so allowing the two giant magnets, 'frozen' into the two solid components, to yield the observed magnetic variations. The existence of the seismologically-inferred fluid domain which is also conducting is indeed built into modern discussions of terrestrial magnetism, based on magnetohydrodynamics.

The study of extraterrestrial magnetism began with Hale's (1908) paper 'On the possible existence of a magnetic field in sunspots'. In his own words, 'Thanks to Zeeman's discovery of the effect of magnetism on radiation, it appeared that the detection of such a magnetic field should offer no great difficulty, provided that it were sufficiently intense': and in fact in that paper he reported on the detection of field strengths of several kilogauss in the dark patches on the solar surface known as 'sunspots'. Some twenty years earlier, impressed by the similarity of solar coronal plumes as seen during a total eclipse to the field lines of a magnetized sphere, Bigelow had speculated that the Sun might possess a general magnetic field like that of the Earth. In 1913 and again in 1918, Hale and colleagues reported more tentatively on the likely presence of a general solar field, though much weaker than the fields discovered in sunspots. Over the years there has followed a steady extension of the list of cosmical bodies in which magnetic fields are observed, directly or indirectly, and of cosmical phenomena in which magnetism plays a crucial role.

Sunspots are reported to have been seen with the naked eye in 350 BC by Theophrastus, a pupil of Aristotle. They were observed systematically by the Chinese from 23 BC onwards, and were independently rediscovered by Galileo and others in the seventeenth century (Wilson 1994). Johann Goldsmid (known as Fabricius) was the first actually to publish his observations and to infer that the Sun must rotate. Galileo identified the solar rotation period as close to a lunar month, and remarked also on the relative motion of the spots within a single group. Both Galileo and his antagonist Christopher Scheiner noted that the spots occurred in zones of low latitude but never near the poles. It was accepted that the spots were indeed solar surface phenomena after Galileo had shown that Scheiner's suggestion that they might be small planets was incompatible with their observed changes in size and shape. Scheiner appears to have been the first to record that the solar surface rotation varies with latitude and that the rotation axis is inclined to the ecliptic, and his drawings of sunspots distinguished clearly umbra and penumbra.

It was not till the nineteenth century that the 11-year cycle was discovered by Heinrich Schwabe and the associated latitude drift of sunspot pairs by Richard Carrington and by Spörer. Perhaps the first hint of a magnetic connection came with the discovery in 1852 that the cycle is related to the occurrence of geomagnetic storms. Hale's observations implied that the 11-year cycle is in fact one half of a basic 22-year cycle, for the leading spot of a pair is found to have a magnetic polarity which reverses in the following 11-year half-cycle. And a crucial later discovery (Babcock 1959) was that the weak, large-scale solar field appears to reverse along with the sunspot cycle. An updated version of the celebrated 'butterfly diagram' (Maunder 1913) is given in Fig. 1.1. Earlier, Babcock and Babcock (1955) had confirmed the estimate by Kiepenheuer and others, reported in Kuiper (1953) p. 361 *et seq.*, that the general magnetic field of the Sun has the strength ≈ 1 G rather than Hale's value of 25–30 G.

Parallel observations, especially over the nineteenth and early twentieth centuries, yielded a wide variety of solar phenomena, and solar physics became a major part of astrophysics. The Sun's visible surface (the photosphere) has a fine structure, the 'granulation', later put down as a manifestation of turbulent convection. Systematic observation during solar eclipses revealed clearly the structure of the outer layers of the solar atmosphere, in particular the chromosphere and the corona with the dramatic outward increase of temperature. Observations of solar prominences and filaments, and the recognition that the form of the corona changes during a solar cycle, point to a magnetic contribution to the mechanical support of the solar atmosphere. Likewise, the interpretation of observed mass motions – local, such as the Evershed flow in sunspot penumbrae, or global, such as the solar wind – must be through the dynamics of conducting magnetized gases rather than through classical gas dynamics. And the recognition that both the solar and terrestrial fields show reversals gave an enormous fillip to the cosmical dynamo problem.

The pioneering work of Wilson (1978) brought out that one of the best-known diagnostics of the solar cycle – the intensity of the chromospheric CaI and CaII lines – shows a similar periodic behaviour in other slowly rotating, late-type stars. For higher rotations, the behaviour is again non-steady but chaotic rather than periodic. The link with stellar rotation – of great theoretical importance – is now shown through the direct photometric observation of 'starspots', which give

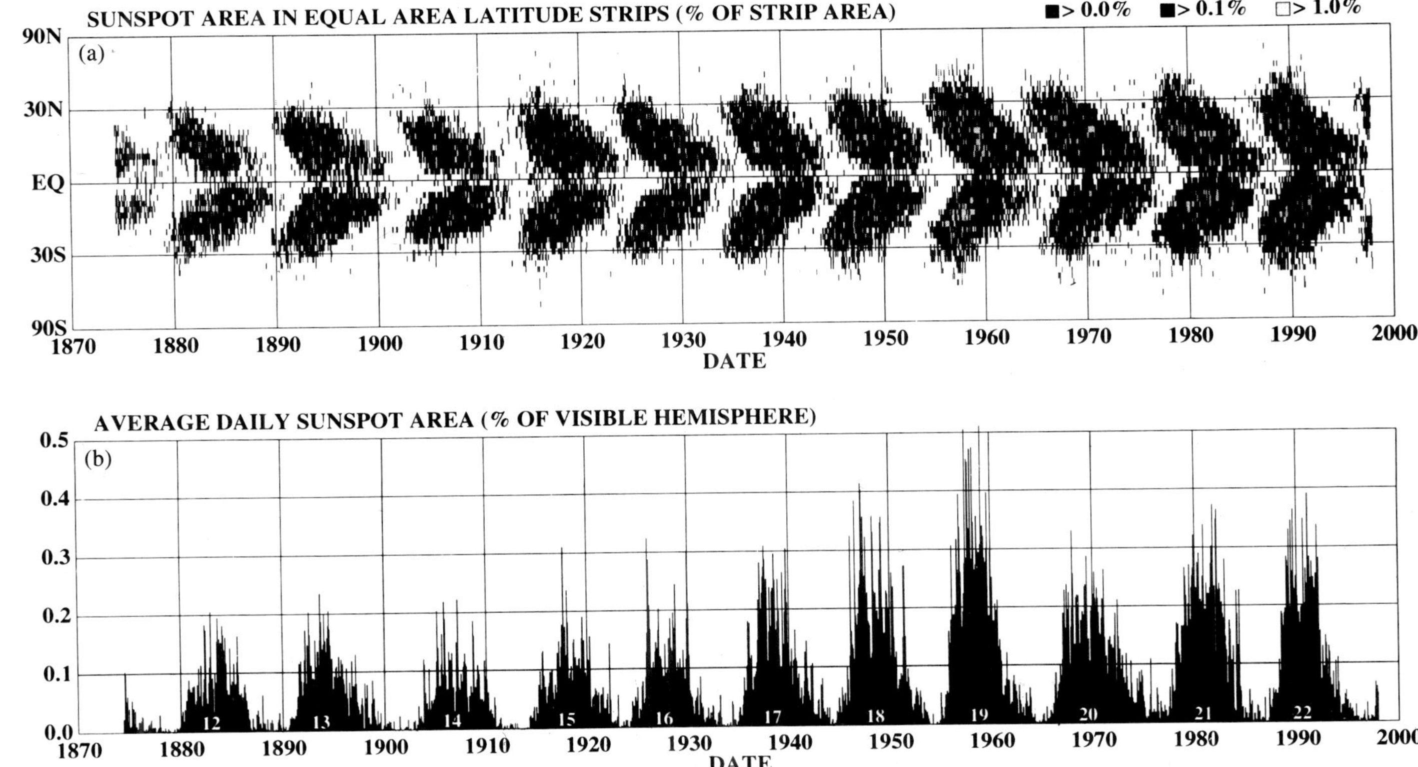

FIG. 1.1. (a) The solar 'butterfly diagram', showing the incidence of sunspots as a function of colatitude from 1875 to 1998. (b) Average daily sunspot area as a function of time. The irregular 11-year cycle and the equatorward drift of the sunspot zones are manifest. (Courtesy of S.M. Tobias and D.N. Hathaway, NASA/Marshall Flight Center.)

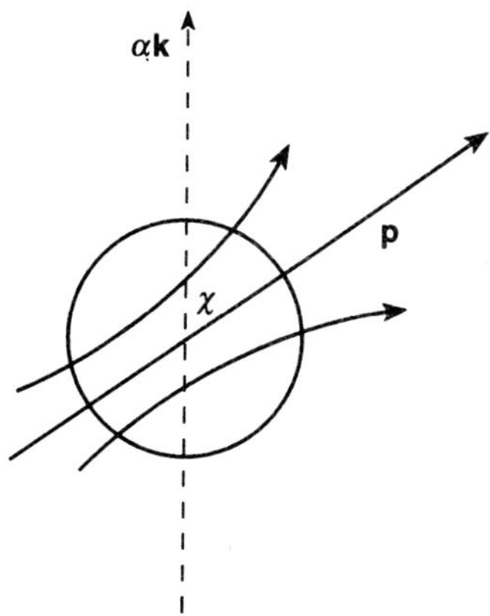

FIG. 1.2. The oblique rotator model: a non-axially symmetric magnetic structure and associated element distribution, shown up by the star's rotation.

the surface a non-axisymmetric structure, so that periodic variations are picked up. The systematic extension of the 'solar–stellar connection' is of no surprise either to observers or to theorists, but is none the less gratifying. Other late-type stars – evolved giants, and pre-main sequence stars such as the T Tauri stars – are likewise the subject of active research in the optical, UV and X-ray domains.

The appearance of solar-type activity amply justifies the inferred presence of magnetic fields. Direct observation of a field on a late-type star other than the Sun was first reported by Robinson *et al.* (1980). There followed detection of a field on a cool subgiant by Giampapa *et al.* (1983) and on a T Tauri star by Basri *et al.* (1992).

Early-type stars with surface temperatures above 10^4 K do not have extensive sub-photospheric convective zones, so one would expect any magnetic phenomena to be of a different form. The first detection of an early-type magnetic star was by Babcock (1947). Since then, a significant and growing fraction shows strong and apparently stable fields, observed through the Zeeman effect. A typical field strength is about 10^3 G, with the strongest known being about 3×10^4 G. The fields show systematic periodic variation, often reversing. Some version of the *oblique rotator* model (Fig. 1.2) yields a satisfactory phenomenological description of the observations. In the simplest picture, a dipolar magnetic field of moment $\mathbf{p}$ rotates with angular velocity α about the axis defined by the unit vector $\mathbf{k}$, with $\mathbf{p}$ and $\mathbf{k}$ mutually inclined at the obliquity angle χ. Again, an essentially non-axisymmetric structure is picked up through rotation. The original magnetic variables were all a subclass of the 'Ap' stars, with anomalous spectra varying with the same period as the magnetic field and the associated small light variations. The class of magnetic variables has now been extended up the main sequence.

The oblique rotator is the paradigm for the study also of condensed stars with correspondingly strong magnetic fields – some 40 white dwarfs, and many more magnetic neutron stars, manifesting themselves as X-ray and radio pulsars. A reliable diagnostic for white dwarf magnetic fields is continuum circular polarization (Kemp 1970; Kemp *et al.* 1970), due to the differing dependence of the opacity on B and on the angle between $\mathbf{B}$ and the line of sight for the ordinary and extraordinary rays respectively. This yielded 10^7 G in the first definitive case, and subsequently fields between 10^6 G and 5×10^8 G in others (Chanmugan 1992; Landstreet 1994). Inference of a field of $(3-5) \times 10^{12}$ G on the neutron star Her X1 came from the detection in the pulsed X-ray spectrum of a cyclotron line, resulting from a transition between Landau levels. Optical cyclotron lines are clearly seen in some Am Her binaries. For radio pulsars, the field has to be inferred indirectly from spin-down rates, yielding 10^{11}–10^{13} G for some 95% and 10^8–10^{11} G in the rest.

Over the last half-century, evidence has steadily built up for interstellar magnetic fields – both large-scale and small-scale – in both our own and other galaxies. Exercised by the need to generate cosmic rays, Fermi pictured an interaction between massive gas clouds, moving randomly with non-relativistic velocities, and charged particles, initially above an energy threshold, which attempt the impossible task of achieving equipartition of energy with the clouds, but which do succeed in building a high-energy tail in their distribution function. The interaction is mediated by a magnetic field that is dragged about by the clouds, yielding a net input of energy into the particles via the magnetic mirror effect. Simultaneously, the discovery of polarization of starlight led to the same magnetic field being called upon to align non-spherical dust grains, so that absorption of starlight would be accompanied by polarization with the electric vector indicating the local direction of the galactic $\mathbf{B}$.

Fifty years on, most workers favour cosmic ray acceleration by relativistic shock waves, but still with the galactic magnetic field an essential ingredient. Direct evidence for galactic fields comes from several physical processes. Faraday rotation of waves from radio galaxies is recognizable by its characteristic λ^2-dependence. Synchrotron radiation, emitted by relativistic particles gyrating in a magnetic field, is picked up from localized sources such as the Crab nebula, from the galactic background and from other galaxies. When the magnetic field is large-scale, the radiation is polarized with the electric vector perpendicular to $\mathbf{B}$. The Zeeman effect, for example on the 21-cm line from atomic hydrogen, is detected in gas clouds, especially those massive enough for gravitational amplification of the field strength. Difficulties in achieving mutually consistent estimates from the different methods seem to have been resolved, especially through improved measurements of the cosmic ray electron density, plus the recognition that a fluctuating component to the galactic field will contribute to the synchrotron radiation but not to the Faraday rotation measurements. A consensus has been reached, putting the large-scale field at $\approx 3 \times 10^{-6}$ G in regions where the density is $\approx 1 m_{\mathrm{H}}\ \mathrm{cm}^{-3}$. Such a field – a compromise between the original 'strong' and 'weak' fields – is dynamically significant on the scale of the massive gas clouds which appear to be the loci of star formation.

There was initially some resistance to the introduction of macroscopic electromagnetism into the physics of the interstellar medium. One of the arguments was that a field of the strength required for Fermi's mechanism to work, and also able to align the dust grains responsible for optical polarization, would make star formation impossible. This objection was useful in forcing a reappraisal of both the basic physics of the problem and of the dynamics of gravitational collapse. Another argument – still heard – is that impossibly large voltages would be required to build up a large-scale galactic field even over a Hubble time. The question is part and parcel of the dynamo problem in general, which is undergoing close and critical scrutiny.

It is now recognized that cosmical magnetic fields manifest themselves over a wide range of scales. The appearance of the jet phenomenon both in active galactic nuclei and in young stellar objects is perhaps the most striking example. For a recent review of galactic magnetism in general, see Beck *et al.* 1996, and for a discussion of magnetic fields in high redshift objects, see Perry *et al.* 1993 and Perry 1993.

1.2 Stellar magnetic fields

For all but the degenerate stars, the Zeeman effect is the principal reliable diagnostic of photospheric magnetic fields. The simplest example is of a magnetic field aligned along the line of sight. In the 'normal' Zeeman effect, the line has only two components, with opposite circular polarizations, and with the wavelength difference $2\Delta\lambda_B$ given in standard notation by

$$2\Delta\lambda_B = (e/2\pi m_e c^2) g\lambda^2 B = 9.34 \times 10^{-13} g\lambda^2 B, \tag{1.1}$$

with λ the wavelength in Angstroms, g the quantum Landé factor, measuring the line's magnetic sensitivity, and B the field strength in Gauss. When the field is directed perpendicular to the line of sight, one sees both the unshifted line, plane polarized along the field, and a shifted component on either side, plane polarized across the field and displaced again by $\Delta\lambda_B$.

Direct measurement of B is possible only if $\Delta\lambda_B$ exceeds the half-width $\Delta\lambda_D$ of the unsplit line. In the visible part of the spectrum, natural broadening and turbulent broadening ensure that the typical width of a Zeeman-sensitive spectral line is $\simeq$ 0.1 Angstrom, so $\Delta\lambda_B < \Delta\lambda_D$ unless B exceeds 1500 G. Thus on the Sun, Zeeman splitting is detectable in sunspots, whereas local fields outside spots, of strength 1–10^2 G, yield only Zeeman broadening.

When interpreting observations of a spatially unresolved stellar photosphere, some explicit or implicit model is adopted. The simplest description of the inhomogeneous field of a late-type star has two components: a field of strength B covering the fraction f (the filling factor) of the stellar surface, with zero field in the remaining faction $(1-f)$. The measured light intensity I can then be written roughly as

$$I = fI_m + (1-f)I_{nm}, \tag{1.2}$$

where I_m, I_{nm} are respectively the magnetic and non-magnetic intensity components, and effects due to the spherical shape of the surface and the related projection and field line inclination effects are temporarily neglected. Thus if f is small, the observed line profile will be only slightly affected by the field.

Further information can often be gained from the polarization of the spectrum. The net circular polarization (the Stokes parameter V) gives a measure of the line-of-sight component of $\mathbf{B}$. Light coming from a field-free region of the stellar atmosphere has zero net circular polarization, so $V = 0$; hence measurement of V yields information exclusively on the magnetic features, even though they are spatially unresolved.

More precisely: Zeeman broadening (or splitting, when observable) yields the mean surface field

$$B_s = \frac{\int |\mathbf{B}| I \, \mathrm{d}S}{\int I \, \mathrm{d}S} \tag{1.3}$$

where $|\mathbf{B}|$ is the local scalar field strength; whereas measurement of the circular polarization yields the mean line-of-sight or effective field

$$B_{\mathrm{eff}} = \frac{\int B \cos\gamma \, I \, \mathrm{d}S}{\int I \, \mathrm{d}S}, \tag{1.4}$$

where γ is the angle between the line of sight and the direction of $\mathbf{B}$. However, whereas all local fields contribute positively to B_s, the contributions to B_{eff} change sign with the local line-of-sight component, so that oppositely-directed fields within a spatial resolution element lead to partial or complete cancellation of the V signal.

A revolution in magnetic field observations was achieved by H.D. and H.W. Babcock with the photoelectric magnetograph (Babcock and Babcock 1952; Babcock 1953). The principle of the device is illustrated in Fig. 1.3. When observing parallel to the field, one sees two circularly polarized components. If the field is weak, the two components are close to each other: their profiles are shown in Fig. 1.3 by the full and dotted lines. In the focal plane of a spectrograph, two identical slits S_1 and S_2 are placed at symmetrical positions in the wings of a magnetosensitive spectral line. In front of the slit of the spectrograph there is an automatic device for admitting left- and right-hand circular polarizations, with a frequency of more than a 100 times per second. Thus the solid and dotted profiles in Fig. 1.3 appear in turn, and the light flux passing through S_1 and S_2 is constantly changing:

$$\delta F = 9.34 \times 10^{-5} I_c d \frac{\partial r_\lambda}{\partial \lambda} g \lambda^2 B \cos\gamma, \tag{1.5}$$

where d is the slit width, I_c the intensity of the neighbouring continuum, and r_λ the residual intensity. Hence measurement of δF yields $B \cos\gamma$ where γ is an average over a spatial resolution element along the slit.

Plate 1 shows a *magnetogram* of the full solar disc – a map of V in the wing of a spectral line, corresponding to a map of the line-of-sight $\mathbf{B}$-component. It

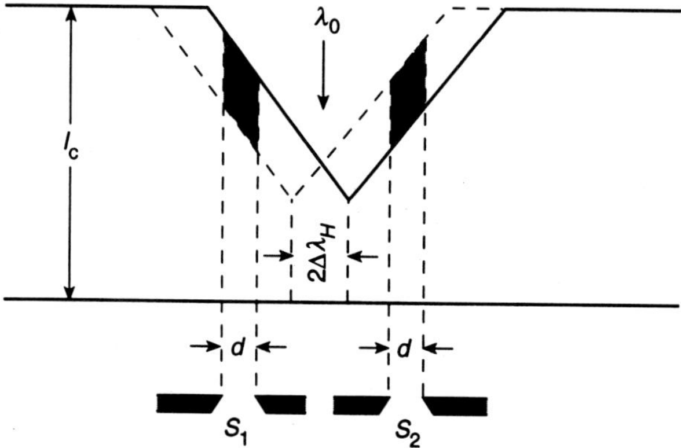

FIG. 1.3. The photoelectric magnetograph.

is clear that the solar field has a complex spatial structure, consisting of bipolar active regions and a quiet network. Much of the structure is at small scales; the smallest spatial scales currently resolvable being $\simeq$ 200–300 km. Further, almost equal amounts of opposite polarity fluxes are present, so that the average of V over the solar surface is very small, well below the detectable level of stellar polarimetry. This implies that if other solar-type stars have fields similar to that of the Sun, direct detection of $\mathbf{B}$ by the circular polarization technique will be unlikely for slow rotators. After many frustrating years of negative results, net circular polarization has been detected in just a handful of slowly rotating late-type stars. In rapid rotators, however, regions of opposite polarity yield circular polarization signatures that are separated in velocity space, so that cancellation no longer occurs. In Donati *et al.* (1997), details are given of detected circular polarization signatures in the photospheric absorption lines of 14 rapidly rotating, late-type stars. From these and other observations, both earlier and later, it can now be stated (Donati, private communication) that fields are detected in every late-type object studied, provided it is rapidly rotating and bright enough.

The situation appears to be radically different for the early-type magnetic stars. The circular polarization measurements readily yield values for B_{eff}, showing that these stars do have a large-scale field, and with the periodic variations which are convincingly described by the oblique rotator (Fig. 1.2) (e.g. Borra *et al.* 1982; Landstreet 1992). The factor $\cos\gamma$ in (1.4) is now an average over the whole unresolved disc, so the inferred B_{eff} is much more dependent on the field geometry assumed than is B_s in the limited number of cases when Zeeman splitting is observable (strong fields, narrow lines).

The example of the Sun – a small magnetic filling factor, and the mixture of polarities on the surface – is an incentive to the search for a detection technique that uses unpolarized light (Robinson *et al.* 1980). Two spectral lines are selected which are as close as possible in their properties – ideally, belonging to the same ion, with the same oscillator strength and excitation potential and similar wavelengths – but with differing Zeeman sensitivity as shown by the Landé factor. In the non-magnetic part of the stellar atmosphere the line profiles are then almost identical; but in the magnetic regions, the Zeeman effect changes the shapes of the lines by differing amounts. Interpretation of the observations, for example with the two-component model, allows determination of B and the filling factor f. Although the method is beset with snags (cf. Solanki 1991, 1992), the number of reliable detections in late-type stars is now about 30. In particular, the plot of fB against rotation Ω shows a striking positive correlation, though admittedly with a large scatter: a least-squares fit yields the tentative relation $fB \propto \Omega^{1.3}$ (Saar 1991).

The name 'Doppler imaging' is given to the process by which maps or images of stellar surface features are obtained by analysing the time-varying profiles of rotating stars (e.g. Collier Cameron 1992; Rice 1996 and references therein). The process is illustrated in Fig. 1.4 for the simplest case of a star with a single small cool spot. Figure 1.4(a) illustrates the spectrum near a rotationally-broadened line λ_0 of an unspotted, 'immaculate' star, and also the contribution made to that spectrum of the region A if it has the same temperature as the rest of the star, showing the narrow absorption feature Doppler-shifted by rotation. Now replace the region A by a cool spot, making virtually no contribution at all wavelengths (Fig. 1.4(b)). The new line spectrum is now measured against a continuum that is reduced by the contribution from A, yielding the profile with a *bright* bump. The argument clearly assumes that the star's rotation is so rapid that the rotational broadening is significantly larger than the local line profile at a single point on the surface. The bump moves across the profile as the star rotates, and from observations at different phases, the longitude and latitude of the bump can be inferred. The same general principle applies to the Ap-stars, where the surface 'spots' are regions of anomalously high element abundances, so the 'bump' is a local valley of increased line depth. Pioneering work on surface element mapping began with Deutsch (1958).

A related procedure is Zeeman–Doppler imaging, already referred to. The main difference is the use of sets of circularly polarized line profiles, measured by V, to recover a vector image of the surface magnetic topology. Although, as stated, direct measurement of B requires that Zeeman splitting exceed the line half-width, the weak-field expression (1.5) gives the average flux density over a spatial resolution element of scale d. The method is less sensitive than normal Doppler imaging to errors in stellar parameters and models of local line profiles, again because zero field yields null polarized profiles, so that the unperturbed profile is unambiguously known. As before, the use of circularly polarized profiles means that one is measuring flux, and deduced field strengths can be considerable underestimates. For details, see Shi-hui (1994), Donati *et al.* (1997), Donati and Brown (1997) and references therein.

Attention has turned recently to the Hanle effect as a diagnostic of weak solar magnetic fields, especially through the development of highly sensitive po-

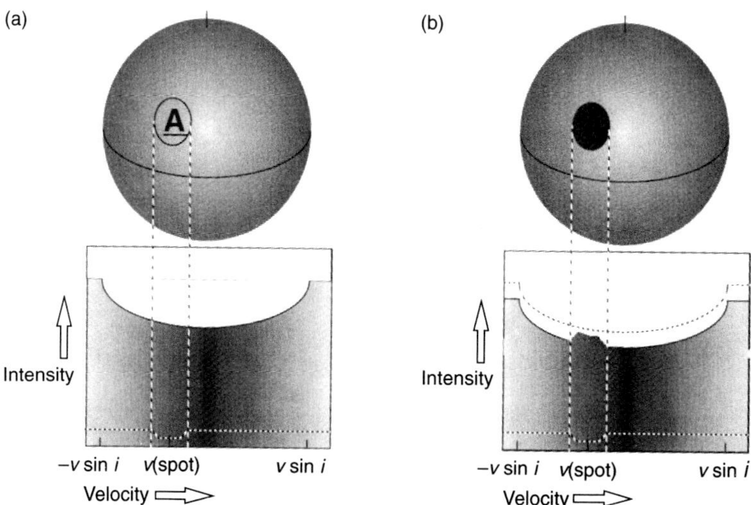

FIG. 1.4. A schematic picture of the formation of a bright bump in the rotationally-broadened spectral line of central wavelength λ_0, in a star with a dark starspot. The significance of the figures (a) and (b) is explained in the text. (After Collier Cameron 1992.)

larimeters (Zirin 1988; Solanki 1993; Stenflo 1994; Bianda *et al.* 1998 and 1998, in press; and references therein). A magnetic field modifies the intrinsic polarization produced by coherent scattering in spectral lines. For fields above 15-20 G, Zeeman splitting exceeds spin-orbit splitting, and the J-levels are mixed: the light is re-emitted in a transition involving different J-levels and forgets its original polarization. For lower fields the polarization persists but is reduced, and also the plane of linear polarization is rotated because of mixing of the different m_J-levels. By comparison of both the angle of rotation and the degree of polarization with theoretical prediction, fields of moderate strength can be measured. In the words of Bianda *et al.*: 'The diagnostic possibilities with the Hanle effect are based on complex physical processes with subtle observational effects, which have only begun to be explored.'

Observations with the *Einstein* and *ROSAT* satellites showed the ubiquity of stellar X-ray emission, indicating the presence of hot coronae. As in the solar corona, a magnetic field will impose much detailed structure on the corona, as revealed especially by solar observations from the more recent *Skylab* and *YOHKOH* satellites (Plate 1(b)). Current X-ray telescopes cannot spatially resolve stellar coronae, but again structure can be inferred from rotational modulation in single stars and from eclipses in binary systems. The difficulties in the way of inferring in general the three-dimensional coronal structure are brought out by Schmitt (1996), but in at least one case the observations are convincingly described in terms of a simple magnetosphere model (cf. Chapter 8).

References

Armitage, A. (1966). *Edmund Halley*. Nelson, London.

Babcock, H.W. (1947). *Astrophysical Journal*, **105**, 105.

Babcock, H.W. (1953). *Astrophysical Journal*, **118**, 387.

Babcock, H.D. (1959). *Astrophysical Journal*, **130**, 364.

Babcock, H.D. and Babcock, H.W. (1952). *Publications Astronomical Society Pacific*, **64**, 282.

Babcock, H.W. and Babcock, H.D. (1955). *Astrophysical Journal*, **121**, 349.

Basri, G., Marcy, G.W. and Valenti, J.A. (1992). *Astrophysical Journal*, **390**, 622.

Beck, R., Brandenburg, A., Moss, D., Shukurov, A. and Sokoloff, D. (1996). *Annual Review Astronomy Astrophysics*, **34**, 155.

Bianda, M., Solanki, S.K. and Stenflo, J.O. (1998). *Astronomy and Astrophysics*, **331**, 760.

Bianda, M., Stenflo, J.O. and Solanki, S.K. (1998). *Astronomy and Astrophysics*, in press.

Borra, E.F., Landstreet, J.D. and Mestel, L. (1982). *Annual Review Astronomy Astrophysics*, **20**, 191.

Chanmugam, G. (1992). *Annual Review Astronomy Astrophysics*, **30**, 143.

Collier Cameron, A. (1992). In *Surface inhomogeneities on late-type stars* (eds P.B. Byrne and D.J. Mullan), p. 33. Springer, Berlin.

Deutsch, A.J. (1958). In *Electromagnetic phenomena in cosmical physics* (ed. B. Lehnert), p. 209. Cambridge University Press.

Donati, J.-F. and Brown, S.F. (1997). *Astronomy and Astrophysics*, **326**, 1135.

Donati, J.-F., Semel, M., Carter, B.D., Rees, D.E. and Collier Cameron, A. (1997). *Monthly Notices Royal Astronomical Society*, **291**, 658.

Giampapa, M.S., Golub, L. and Worden, S.P. (1983). *Astrophysical Journal*, **268**, L121.

Hale, G.E. (1908). *Astrophysical Journal*, **28**, 315.

Kemp, J.C. (1970). *Astrophysical Journal*, **162**, 169.

Kemp. J.C., Swedlund, J.B., Landstreet, J.D. and Angel, J.R.P. (1970). *Astrophysical Journal*, **161**, L77.

Kuiper, G.P. (ed.) (1953). *The Sun*. Chicago University Press.

Landstreet, J.D. (1992). *Astronomy Astrophysics Reviews*, **4**, 25.

Landstreet, J.D. (1994). In *Cosmical magnetism* (ed. D. Lynden-Bell), p. 55. Kluwer, Dordrecht.

Maunder, E.W. (1913). *Monthly Notices Royal Astronomical Society*, **64**, 747.

Perry, J.J. (1993). In *Cosmical magnetism, contributed papers* (ed. D. Lynden-Bell), p. 144. Institute of Astronomy, Cambridge.

Perry, J.J., Watson, A.M. and Kronberg, P.P. (1993). *Astrophysical Journal*, **406**, 407.

Rice, J.B. (1996). In *Stellar surface structure* (eds K.G. Strassmeier and J.L. Linsky), p. 19. Kluwer, Dordrecht.

Robinson, R.D., Worden, S.P. and Harvey, J.W. (1980). *Astrophysical Journal*, **236**, L155.

Saar, S.H. (1991). In *The Sun and cool stars* (eds I. Tuominen, D. Moss and G. Rüdiger), p. 389. Springer, Berlin.

Schmitt, J.H.M.M. (1996). In *Stellar surface structure* (eds K.G. Strassmeier and J.L. Linsky), p. 85. Kluwer, Dordrecht.

Shi-hui, Y. (1994). *Magnetic fields of celestial bodies.* Kluwer, Dordrecht.

Solanki, S.K. (1991). In *Reviews modern astronomy*, Vol. 4 (ed. G. Klare), p. 208. Springer, Berlin.

Solanki, S.K. (1992). In *Cool stars, stellar systems and the Sun VII* (eds J. Bookbinder and M. Giampapa). Publications Astronomical Society Pacific, Conference Series, p. 211.

Solanki, S.K. (1993). *Space Science Reviews*, **65**, 1.

Stenflo, J.O. (1994). *Solar magnetic fields – polarized radiation diagnostics.* Kluwer, Dordrecht.

Wilson, O.C. (1978). *Astrophysical Journal*, **226**, 379.

Wilson, P.R. (1994). *Solar and stellar activity cycles.* Cambridge University Press.

Zirin, H. (1988). *Astrophysics of the Sun.* Cambridge University Press.

Magnetogram (NSO/Kitt Peak)

— 1 arcmin

(a)

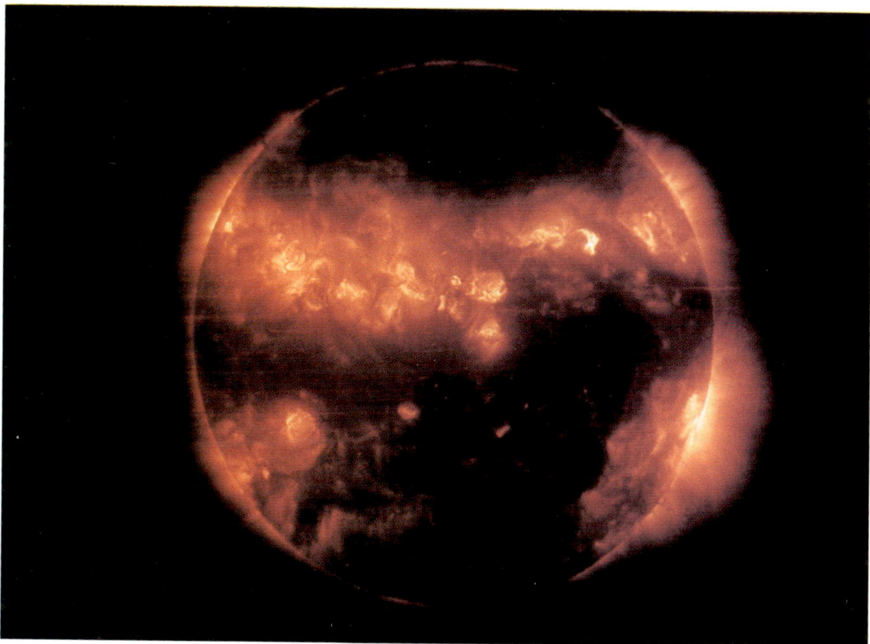

(b)

Plate 1. (a) A magnetogram of the solar disc. (Courtesy Kitt Peak National Observatory) (b) A simultaneous photograph of the Sun taken in X-rays. (Courtesy IBM Research and SAO; L. Golub and N. O. Weiss)

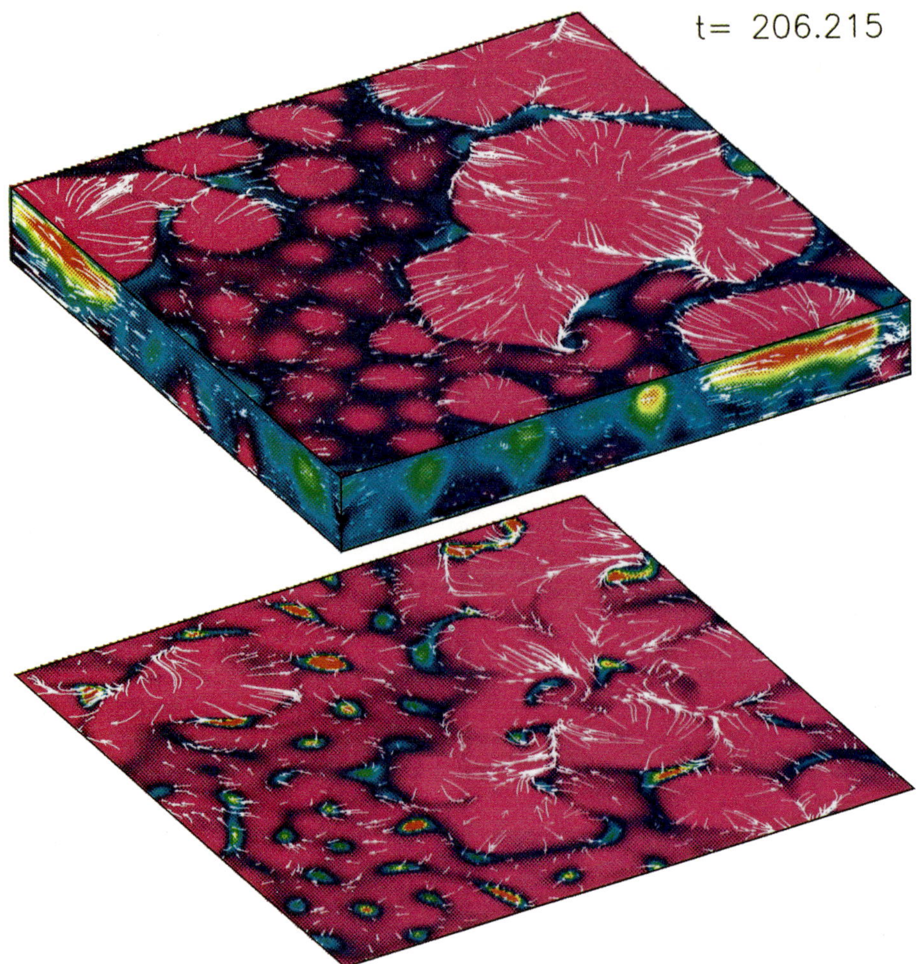

Plate 2. Flux separation for non-linear convection in an imposed vertical magnetic field. The Rayleigh number $\hat{R}$, defined at the middle of the layer, has the value 10^5, the value at the midpoint of the ratio $\hat{\zeta}$ of the magnetic to thermal diffusivity is 1.2, the Chandrasekhar number $Q = 10^3$, and the aspect ratio $\lambda = 8$. The colour coding from violet (low) to red (high) indicates the value of $|\mathbf{B}|^2$ at the top and the bottom of the box and of the temperature fluctuation on its sidewalls. The arrows represent the component of velocity parallel to the wall of the box. (After Tao *et al.* 1998.)

2

THEORETICAL BASIS

2.1 Maxwell's equations and the magnetohydrodynamic approximation

In a given frame of reference S, the vectors $\mathbf{E}$ and $\mathbf{B}$ satisfy Maxwell's equations in the form appropriate for fully or partially ionized gases in vacuum:

$$\begin{aligned} \nabla \cdot \mathbf{E} &= 4\pi\rho_e, & \nabla \cdot \mathbf{B} &= 0, \\ \nabla\times\mathbf{E} &= -\frac{1}{c}\frac{\partial \mathbf{B}}{\partial t}, & \nabla\times\mathbf{B} &= \frac{4\pi}{c}\mathbf{j} + \frac{1}{c}\frac{\partial \mathbf{E}}{\partial t}, \end{aligned} \tag{2.1}$$

where we use Gaussian units, with charge and current densities ρ_e, $\mathbf{j}$ both measured in electrostatic units. The physical quantities are all 'macro'-scalars and vectors, averaged over regions large compared with the scale of thermal fluctuations. Under a Lorentz transformation to a frame S' moving with velocity $\mathbf{u}$ relative to S, the components of $\mathbf{E}$ and $\mathbf{B}$ parallel to $\mathbf{u}$ are unchanged, but the perpendicular components transform according to

$$\mathbf{E}' = \gamma(\mathbf{E} + \mathbf{u} \times \mathbf{B}/c), \quad \mathbf{B}' = \gamma(\mathbf{B} - \mathbf{u} \times \mathbf{E}/c). \tag{2.2}$$

where $\gamma = (1 - (\mathbf{u}/c)^2)^{-1/2}$ in standard notation. The components of the current perpendicular to $\mathbf{u}$ are unchanged, but the charge density and the parallel component of $\mathbf{j}$ transform according to

$$\rho_e' = \gamma(\rho_e - \mathbf{u} \cdot \mathbf{j}/c^2), \quad \mathbf{j}' = \gamma(\mathbf{j} - \rho_e \mathbf{u}). \tag{2.3}$$

In most applications – with the important exception of pulsar electrodynamics, discussed in Chapters 13 and 14 – we are concerned with non-relativistic speeds. Terms of order $(u/c)^2$ can then be dropped, $\gamma \approx 1$, and the Lorentz transformation reduces to the Galilean transformation. Further, we anticipate that in many problems the 'high-conductivity approximation' is valid – i.e. at each point there exists a frame moving with a non-relativistic velocity $\mathbf{v}$ in which the electric field is 'small' :

$$\mathbf{E} + \frac{\mathbf{v} \times \mathbf{B}}{c} \simeq 0, \tag{2.4}$$

with the terms on the right of magnitude small compared with E. In a fully ionized gas, $\mathbf{v}$ is close to the bulk velocity (Section 2.3), whereas in a lightly ionized gas it is close to the velocity with which the ionized component moves with respect to the neutral bulk (Section 2.7). With these approximations, eqns (2.2) reduce to

$$\mathbf{E}' = \mathbf{E} + \frac{\mathbf{u} \times \mathbf{B}}{c}, \quad \mathbf{B}' = \mathbf{B}. \tag{2.5}$$

Thus in the non-relativistic, high conductivity approximation, one may speak of 'the magnetic field' without specifying the frame in which it is measured, whereas the electric field is not invariant but transforms according to (2.5) (Alfvén 1950).

If D is a characteristic scale of variation of the field quantities, from (2.1) and (2.4)

$$|\rho_e| \simeq \left|\nabla \cdot \left(\frac{\mathbf{v} \times \mathbf{B}}{4\pi c}\right)\right| \simeq \frac{v}{c}\left(\frac{B}{4\pi D}\right). \tag{2.6}$$

The convection of this charge density by the fluid yields a contribution to the total current-density $\mathbf{j}$:

$$|\rho_e \mathbf{v}| \simeq \left(\frac{v}{c}\right)^2 \left(\frac{cB}{4\pi D}\right) \tag{2.7}$$

$$\simeq \left(\frac{v}{c}\right)^2 \left(\frac{c|\nabla \times \mathbf{B}|}{4\pi}\right) \tag{2.8}$$

if $|\nabla \times \mathbf{B}|$ is of order B/D. If also the displacement current can be dropped from (2.1), then the total current $\mathbf{j}$ is given by $(c/4\pi)\nabla \times \mathbf{B}$ and in magnitude is of order $cB/4\pi D$. Thus (2.8) implies that the convection of the net charge density makes a negligible contribution to the total current, which will be due mainly to a 'conduction current' – the drift of electrons relative to the ions with velocity $\mathbf{V}$ (see below). Further,

$$\frac{|\rho_e|}{n_e e} \simeq \frac{vB/c}{4\pi D n_e e} \simeq \left(\frac{v}{c^2}\right)\frac{j}{n_e e} = \left(\frac{v}{c}\right)\left(\frac{V}{c}\right). \tag{2.9}$$

In most cases we shall find $V \ll v$, so that in a non-relativistic theory the charge separation is small. It will be seen below that the electric force density is smaller than the magnetic by a similar factor $\mathrm{O}(v/c)^2$. The neglect of the displacement current is easily justified. If τ is a typical time of variation of the field quantities, this requires

$$|\partial \mathbf{E}/c\,\partial t| \simeq (E/c\tau) \ll |\nabla \times \mathbf{B}| \simeq B/D, \tag{2.10}$$

or by (2.4)

$$\tau \gg (v/c)(D/c). \tag{2.11}$$

The usual condition for the 'quasi-static' approximation is that $\tau \gg D/c$ – electromagnetic waves cross a region of scale D in a time short compared with the time of variation of the field quantities. In the present scheme of approximation, the condition (2.11) introduces an extra factor (v/c). (See Section 13.1.2 for the modifications to the arguments when they are applied to a single-sign domain.)

The truncated Maxwell equations are: the Ampère–Maxwell equation

$$\nabla \times \mathbf{B} = \frac{4\pi}{c}\mathbf{j}; \tag{2.12}$$

the Faraday–Neumann equation,

$$\nabla \times \mathbf{E} = -\frac{1}{c}\frac{\partial \mathbf{B}}{\partial t}, \tag{2.13}$$

supplemented by

$$\nabla \cdot \mathbf{B} = 0; \tag{2.14}$$

and the Poisson–Maxwell equation, conveniently written

$$\rho_e = \frac{\nabla \cdot \mathbf{E}}{4\pi}. \tag{2.15}$$

These equations are invariant to the Galilean transformation

$$\mathbf{r}' = \mathbf{r} - \mathbf{u}t, \quad t' = t, \tag{2.16}$$

provided $\mathbf{E}$, $\mathbf{B}$ transform according to (2.5), and by (2.3), $\rho_e, \mathbf{j}$ according to

$$\rho_e' = \rho_e - \frac{\mathbf{u} \cdot \mathbf{j}}{c^2} \tag{2.17}$$

and

$$\mathbf{j}' = \mathbf{j}. \tag{2.18}$$

The dropping of the term $\rho_e \mathbf{u}$ compared with $\mathbf{j}$ in (2.18) is mandatory, because of the neglect of all other terms of order $(v/c)^2$. However, as noted by Schlüter (1961), $\mathbf{u} \cdot \mathbf{j}/c^2$ is of the same order as ρ_e and so must be retained in (2.17): since by (2.9) ρ_e is relativistically small, for consistency this essentially non-relativistic treatment must nevertheless retain this curious relic of relativistic theory.

The truncated Maxwell equations contain just the one time-derivative $\partial \mathbf{B}/\partial t$. If all the field quantities are known at any time, then the Faraday–Neumann equation yields $\mathbf{B}$ an instant later, and the Ampère equation gives the associated current density. Only one more equation is needed for the computation of the electromagnetic quantities to proceed.

2.2 Properties of cosmical plasmas

The electric and magnetic fields that appear in (2.12) et seq. are 'collective', as compared with the small-scale random fields present in a thermal plasma. If the gas is near thermodynamic equilibrium, the maximum length-scale over which thermal fluctuations cause significant separation of positive and negative charges is estimated by supposing that all the ions are removed from a sphere of radius λ, leaving an electron sphere with electric potential energy $\simeq (3/5)(4\pi\lambda^3 n_e e/3)^2/\lambda$ of the same order as the thermal energy $(3kT/2)(4\pi n_e \lambda^3/3)$. A more precise treatment (e.g. Spitzer 1956) yields the same value (apart from a numerical factor) for this *Debye shielding length*:

$$\lambda_D = \left(\frac{kT}{4\pi n_e e^2}\right)^{1/2} \simeq 7\left(\frac{T}{n_e}\right)^{1/2} \text{ cm.} \tag{2.19}$$

Provided the macroscopic length-scales $\gg \lambda_D$, the electromagnetic field may be divided into two parts. The large-scale collective field appearing explicitly in Maxwell's equations is maintained by the macroscopic charge-current field. As shown above, in non-relativistic problems the magnetic energy dominates over the electric. The small-scale random field is essentially the sum of the unshielded Coulomb fields of individual particles; it acts so as to restore a general particle distribution function to the Maxwellian form, and to inhibit the drift of electrons relative to ions.

Inside a star, λ_D varies between 10^{-8} and 10^{-5} cm and so is far less than any macroscopic length. At the base of the solar corona $\lambda_D \simeq 1$ cm; in the solar wind near the earth $\lambda_D \simeq 10^3$ cm. It appears that in most problems of interest the basic plasma condition of a 'small' Debye length will be easily satisfied.

In discussing the effect of random, small-scale electric fields on the particle distribution function, it is customary to introduce a mean-free-path

$$\lambda = \frac{1}{n\pi b^2} \tag{2.20}$$

where n is the number density of the scattering particles and b the 'collision radius', by analogy with elementary kinetic theory. For elastic scattering, e.g. of electrons by ions, a first estimate for b is given by equating the mutual Coulomb energy to the kinetic energy relative to the mass-centre: $Ze^2/b \approx 3kT/2$ where Ze is the ionic charge. However, this 'billiard-ball' picture, appropriate for short-range interparticle forces, is not obviously applicable for long-range Coulomb scattering. The differential Rutherford cross-section for scattering of electrons of mass m by massive ions is (e.g. Thompson 1962; Sturrock 1994)

$$\sigma(\theta, v)\,\mathrm{d}\Omega = \frac{Z^2 e^4}{4(mv^2)^2}\operatorname{cosec}^4(\theta/2)\,\mathrm{d}\Omega \tag{2.21}$$

where θ is the scattering angle, v the asymptotic velocity of the electron, and $\mathrm{d}\Omega = 2\pi \sin\theta\,\mathrm{d}\theta$ the element of solid angle (cf. Fig. 2.1). The scattered particle loses momentum $mv(1 - \cos\theta)$ in the direction of its initial motion; hence the momentum transfer cross-section $Q = \pi b^2$ for scattering over all angles between a lower limit θ_m and π is

$$Q = \int \sigma(\theta, v)(1 - \cos\theta)\,\mathrm{d}\Omega = 4\pi\frac{Z^2 e^4}{m^2 v^4}\log\operatorname{cosec}(\theta_m/2). \tag{2.22}$$

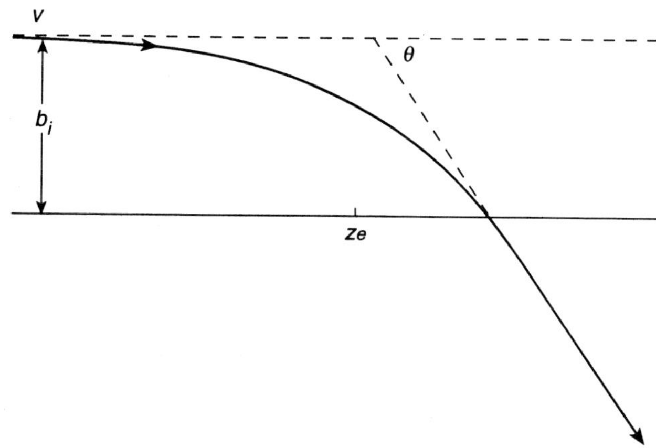

FIG. 2.1. Coulomb scattering.

The angle θ is related to the impact parameter b_i by $\tan^2\theta/2 = \left(Ze^2/mv^2b_i\right)^2$, and if b_i is allowed to have all values up to infinity, $\theta_m \to 0$ and (2.22) diverges. A plausible restriction imposes an upper limit $b_i = \lambda_D$, given by (2.19), for beyond this distance the Coulomb field of the scattering ion is effectively screened. This assumption typically yields for the effective collision radius

$$b \approx 3Ze^2/2kT, \tag{2.23}$$

a factor $\approx 9/4$ greater than the first estimate. In a plasma that behaves like a nearly perfect gas, the Coulomb coupling energy over the mean interparticle distance $\simeq n^{-1/3}$ must be small compared with the thermal energy, a condition equivalent to

$$\frac{b}{n^{-1/3}} \ll 1. \tag{2.24}$$

The ratio of the interparticle distance to the Debye length is also $\simeq (bn^{1/3})^{1/2} \ll 1$.

With this value for b, the mean-free-path λ_{ei} for the scattering of electrons by protons is

$$\lambda_{ei} \simeq 6 \times 10^4 \frac{T^2}{n}. \tag{2.25}$$

One can then conveniently define the mean collision time

$$\tau_{ei} = \frac{\lambda_{ei}}{(v_T)_e} \tag{2.26}$$

in which a sub-thermal electron drift velocity relative to the ions is randomized. Similar lengths can be defined for ion–ion and electron–electron encounters. The mean-free-paths are small compared with the macroscopic lengths in many problems of interest but not in all: e.g. in the solar wind, near the earth $\lambda_i \simeq 6\times 10^{13}$, greater than an astronomical unit.

The conceptually simplest plasma problems are those with imposed macroscopic scales L large compared with the mean-free-paths, and with associated time-scales long compared with the collision times λ/v_T. One can then adopt implicitly the Chapman–Enskog approach to kinetic theory, admirably set out in Chapman and Cowling (1970). Particle distribution functions are nearly Maxwellian, with deviations of order λ/L which yield transport phenomena – thermal conduction, viscosity, electrical resistivity. The macroscopic equations can, however, yield spontaneously structures with scales so small that the implicit expansion procedure fails: domains of nearly dissipation-free flow are then linked by a shock front of microscopic thickness, within which the particle distribution functions make rapid adjustments from one Maxwellian state to another.

In a fully ionized gas with a magnetic field, the picture is complicated by the anisotropy introduced by the field. A single particle of charge Ze and mass m gyrates in a uniform field $\mathbf{B}$ with angular frequency $\omega = ZeB/mc$. For ions of mass Am_H and for electrons,

$$\omega_i = ZeB/m_i c = 10^4 (Z/A)B, \quad \omega_e = eB/m_e c = 2\times 10^7 B \tag{2.27}$$

respectively. The mean gyration radius a_ω of a species is v_T/ω, and this must be compared with the mean-free-path λ; or equivalently, the time $1/\omega$ for gyration through one radian is to be compared with the appropriate mean collision time τ. If for example

$$\omega_e \tau_{ei} \simeq 1.75 B(T^{3/2}/n) \gg 1, \tag{2.28}$$

then the field introduces a microscopic anisotropy – an electron performs a number of turns of a spiral before having its motion randomized by 'collisions' with the ion gas. For ions there is an extra factor $(m_e/m_i)^{1/2}$: the anisotropy appears if

$$\omega_i \tau_i \simeq 4.2\times 10^{-2} B(T^{3/2}/n) \gg 1, \tag{2.29}$$

where for simplicity we have put $Z = A = 1$ (a pure hydrogen gas), so that λ is roughly the same for ion–ion, electron–electron and ion–electron encounters.

Inside most stars, with any plausibly estimated field B, both $\omega_i\tau_i$ and $\omega_e\tau_e$ are small, but they become large compared with unity, e.g. in low density magnetized stellar coronae and in diffuse interstellar clouds, whether an 'HII zone' of hot gas, ionized by nearby early-type stars, a lightly ionized zone of atomic hydrogen, or a cloud of molecular hydrogen H_2. The effect of this microscopic anisotropy on the analogues of Ohm's law, for both fully and partially ionized gases, is discussed in Sections 2.3 and 2.7. For the moment we note that in the collision-dominated limit $\omega\tau \ll 1$, the random velocity distribution will be kept Maxwellian, so that the partial pressures are kept isotropic, and both charge species have the same temperature. At the other extreme with $\omega\tau \gg 1$, one can expect close isotropy in the two dimensions normal to $\mathbf{B}$, but $(p_e)_\perp$, $(p_i)_\perp$ need not equal respectively

$(p_e)_{\|}$, $(p_i)_{\|}$, nor need the parallel and perpendicular temperatures of electrons and ions agree (Chew *et al.* 1956). This will not be so important in domains where $(B^2/8\pi p) \gg 1$, but is a serious complication if macroscopic length-scales are small compared with the local mean-free-path, as in the solar wind close to the Earth, or if externally-driven time variations are more rapid than $1/\tau$ but slower than ω.

2.3 Macroscopic equations for a fully ionized gas: the two-fluid model

One way of deriving macroscopic plasma quantities starts by solving the microscopic equations for individual particles, on the assumption that all macroscopic quantities vary slowly in both space and time over a gyration radius and in a gyration period. The motion of an individual charge q in a magnetized plasma can be described as the sum of the gyration (2.27) about a moving instantaneous centre of gyration. This 'guiding centre' has a motion parallel to $\mathbf{B}$, driven by the components $\mathbf{F}_{\|}$ of the non-magnetic forces, together with a set of drifts across $\mathbf{B}$. Each component $\mathbf{F}_{\perp}$ across $\mathbf{B}$, including the inertial force due to motion along a curved field line, yields the drift

$$\frac{c\mathbf{F}_{\perp} \times \mathbf{B}}{qB^2}. \tag{2.30}$$

Further drifts arise from inhomogeneities in $\mathbf{B}$ (Alfvén 1950). The macroscopic current density and velocity can then be derived by an averaging over all the relevant particles present, a procedure of some subtlety (cf. Schlüter 1952 and Spitzer 1952, 1956). In particular, the macroscopic mean velocity of a particle species need not agree with the mean drift velocity $\mathbf{w}_d$, defined as the mean velocity of the *guiding centres* in a volume element. If one considers *all* the particles gyrating about a given guiding centre, then the different particles will have random phases of gyration, and so the mean velocity $\mathbf{v}'$ will indeed be equal to $\mathbf{w}_d$. However, the macroscopic velocity $\mathbf{v}$ is by definition the mean velocity of all the particles in a given volume element, regardless of the location of their guiding centres: to obtain $\mathbf{v}$, one must omit from the averaging all particles that are outside the volume element but with their guiding centres within it, and include those particles inside the element but with external guiding centres.

When correctly performed, this procedure removes spurious contradictions between the microscopic and macroscopic treatments. A uniform electric field $\mathbf{E}$ in the stationary frame does in fact yield $\mathbf{v} = \mathbf{w}_d$ (equal by (2.5) and (2.30) to the velocity of the frame in which the electric field vanishes). A uniform field $\mathbf{B}$ by itself yields no microscopic drifts, but in the presence of a pressure gradient there result macroscopic currents across $\mathbf{B}$. On the other hand, a time-independent but inhomogeneous field $\mathbf{B}$ yields microscopic drift velocities, but with a vanishing macroscopic velocity.

It is pedagogically important to show that the microscopic and macroscopic approaches are consistent. For most purposes, however, it is simpler to derive macroscopic quantities such as $\mathbf{j}$ and $\mathbf{v}$ as solutions of *macroscopic* equations, constructed by summing over the corresponding microscopic equations.

2.3.1 *Equations to the flow of the whole gas*

For simplicity, we take the gas to consist of electrons of charge $-e$ and of one ionic species, of charge Ze; the results are easily generalized. The charge and current density entering Maxwell's equations are written in terms of $n_i, n_e, \mathbf{v}_i, \mathbf{v}_e$, the densities and mean drift velocities of the ions and electrons:

$$\rho_e = n_i Ze - n_e e, \tag{2.31}$$

$$\begin{aligned}\mathbf{j} &= n_i Ze\mathbf{v}_i - n_e e\mathbf{v}_e = \rho_e \mathbf{v}_i + n_e e(\mathbf{v}_i - \mathbf{v}_e) \\ &\simeq -n_e e\mathbf{V}\end{aligned} \tag{2.32}$$

where $\mathbf{V} = \mathbf{v}_e - \mathbf{v}_i$ is again the mean drift of electrons relative to ions. The ionic mass $m_i \gg m_e$, so the gas density $\rho \approx n_i m_i$, and the bulk velocity $\mathbf{v} \approx \mathbf{v}_i$.

The mutual friction between electrons and ions disappears when we consider the flow of the whole gas. The electromagnetic (Lorentz) force density is

$$n_i Ze\left(\mathbf{E} + \frac{\mathbf{v}_i \times \mathbf{B}}{c}\right) - n_e e\left(\mathbf{E} + \frac{\mathbf{v}_e \times \mathbf{B}}{c}\right) = \rho_e \mathbf{E} + \frac{\mathbf{j} \times \mathbf{B}}{c}. \tag{2.33}$$

From (2.4) and (2.6), the electrical part is of order

$$\left(\frac{vB}{c}\right)\left(\frac{vB}{4\pi cD}\right) \simeq \left(\frac{v}{c}\right)^2 \left(\frac{B^2}{4\pi D}\right) \simeq \left(\frac{v}{c}\right)^2 \left|\frac{\mathbf{j} \times \mathbf{B}}{c}\right| \tag{2.34}$$

and so is normally negligible in a non-relativistic approximation, again due to the smallness of the net charge density allowed by the Coulomb forces. (By contrast, in the essentially relativistic pulsar magnetosphere problem, the electric and magnetic force densities are comparable.) Thus the equation of motion of the gas is just the modified Navier–Stokes equation

$$\rho \frac{\mathrm{d}\mathbf{v}}{\mathrm{d}t} = \rho\left(\frac{\partial \mathbf{v}}{\partial t} + \nabla\left(\frac{\mathbf{v}^2}{2}\right) - \mathbf{v} \times (\nabla \times \mathbf{v})\right) = -\nabla p + \rho\mathbf{g} + \frac{\mathbf{j} \times \mathbf{B}}{c} + \mathbf{F}_\nu, \tag{2.35}$$

where the gravitational acceleration is derived from the potential V,

$$\mathbf{g} = \nabla V, \tag{2.36}$$

and $\mathbf{F}_\nu$ is the viscous body force. Centrifugal and Coriolis forces may be included if it is found convenient to use a rotating frame of reference. The scalar product of (2.35) with $\mathbf{v}$ is

$$\rho \frac{\mathrm{d}}{\mathrm{d}t}\left(\frac{\mathbf{v}^2}{2}\right) = -\mathbf{v} \cdot \nabla p + (\rho\mathbf{g} + \mathbf{F}_\nu) \cdot \mathbf{v} + \frac{\mathbf{j} \times \mathbf{B}}{c} \cdot \mathbf{v}, \tag{2.37}$$

showing how the Lorentz body force contributes to the changes in the bulk kinetic energy of the gas.

In a non-recombining plasma, the conservation of mass is as usual expressed by the continuity equation

$$\frac{\mathrm{d}\rho}{\mathrm{d}t} + \rho\nabla\cdot\mathbf{v} = \frac{\partial\rho}{\partial t} + \nabla\cdot(\rho\mathbf{v}) = 0. \tag{2.38}$$

The Lorentz force density $\mathbf{f}$ may be expanded in the form

$$\mathbf{f} = \frac{(\nabla\times\mathbf{B})\times\mathbf{B}}{4\pi} = \frac{(\mathbf{B}\cdot\nabla)\mathbf{B}}{4\pi} - \nabla\left(\frac{\mathbf{B}^2}{8\pi}\right). \tag{2.39}$$

The first term on the right is non-zero if $\mathbf{B}$ varies along the direction of $\mathbf{B}$: it represents the effect of a tension $\mathbf{B}^2/4\pi$ per unit area, and so contributes to $\mathbf{f}$ when the field lines are curved. If we write $\mathbf{B} = B\hat{\mathbf{s}}$ with $\hat{\mathbf{s}}$ the unit vector along the field, then this tension term can be written

$$\frac{B}{4\pi}\frac{\mathrm{d}(B\hat{\mathbf{s}})}{\mathrm{d}s} = \frac{\mathrm{d}}{\mathrm{d}s}\left(\frac{B^2}{8\pi}\right)\hat{\mathbf{s}} + \frac{B^2}{4\pi}\frac{\hat{\mathbf{n}}}{R_c} \tag{2.40}$$

where $\hat{\mathbf{n}}$ is the unit vector in the direction of the principal normal to the field line, and R_c is the local radius of curvature. The second term in (2.39) is the gradient of the isotropic pressure $B^2/8\pi$: its component along the field cancels the first term in (2.40), ensuring that the total magnetic force density is normal to $\mathbf{B}$, as is manifest from the original form $\mathbf{j}\times\mathbf{B}/c$.

In suffix notation,

$$f_i = -\frac{\partial}{\partial x_j}T_{ij} \tag{2.41}$$

with the magnetic part of the Maxwell stress tensor

$$T_{ij} = \frac{\mathbf{B}^2}{8\pi}\delta_{ij} - \frac{B_iB_j}{4\pi}. \tag{2.42}$$

From (2.41), the total magnetic force acting on the currents within a volume τ surrounded by a surface S is

$$F_i = \int_\tau f_i\,\mathrm{d}\tau = -\int_S T_{ij}n_j\,\mathrm{d}S \tag{2.43}$$

where n_j is the local unit vector out of the volume τ. Thus the force F_i can be described as equivalent to the integrated inflow of the i-component of linear momentum of amount $T_{ij}n_j$ per unit area across elements of area $n_j\,\mathrm{d}S$. Likewise, the magnetic torque density

$$l_i = (\mathbf{r}\times\mathbf{f})_i = \varepsilon_{ijk}x_j\left(-\frac{\partial}{\partial x_l}T_{kl}\right), \tag{2.44}$$

and because of the symmetry of T_{kl}, this can be written

$$l_i = -\frac{\partial}{\partial x_l} D_{il} \tag{2.45}$$

where

$$D_{il} = \varepsilon_{ijk} x_j T_{kl} = x_j T_{kl} - x_k T_{jl} \tag{2.46}$$

with the convention that (i, j, k) is a cyclic permutation of (1,2,3). Then

$$-\int_\tau l_i \, d\tau = \int_S D_{il} n_l \, dS, \tag{2.47}$$

so that by exact analogy, $D_{il} n_l$ represents the *outflow* per unit area of the i-component of angular momentum across the element $n_l \, dS$ (Lüst and Schlüter 1955; Mestel 1968). Further, even if l_i were to vanish everywhere within the domain τ, so that the the integral (2.47) also vanishes, D_{il} can still be thought of as measuring the transport of angular momentum per unit area, but for the special case in which outflow from τ happens to be balanced by inflow.

In a system in which the magnetic field is symmetric about the rotation axis, the net outward transport of angular momentum is carried just by the magnetic tensions, the magnetic pressures making no contribution. Thus if $i = 3$ refers to the axis of symmetry Oz and $k = 2$ to the azimuthal (toroidal) direction of the cylindrical polar coordinate system (ϖ, ϕ, z), then from (2.42) and (2.46), $D_{3l} = (x_1 T_{2l} - x_2 T_{1l}) = \varpi((\mathbf{B}^2/8\pi)\delta_{2l} - B_l B_\phi/4\pi)$. If n_l is in the direction of the meridional (poloidal) component $\mathbf{B}_\mathrm{p} = (B_1, 0, B_3)$, then

$$D_{3l} n_l = -\frac{\varpi B_\phi B_l}{4\pi} \tag{2.48}$$

is the flow of angular momentum per unit area along $\mathbf{B}_\mathrm{p}$. The $\varpi\mathbf{B}^2/8\pi$ term can be thought of as transporting angular momentum in the $l = 2$ (toroidal) direction, but because of the axial symmetry, neither the magnetic nor the gas pressures affect its net distribution.

In tensor notation, the equations of motion (2.35) and of continuity (2.38) become

$$\frac{\partial \rho}{\partial t} + \frac{\partial}{\partial x_j}(\rho v_j) = 0, \tag{2.49}$$

and

$$\begin{aligned} \rho \frac{\partial v_i}{\partial t} + \rho v_j \frac{\partial}{\partial x_j} v_i &= \frac{\partial}{\partial t}(\rho v_i) + \frac{\partial}{\partial x_j}(\rho v_i v_j) \\ &= -\frac{\partial}{\partial x_i} p + \rho \frac{\partial V}{\partial x_i} - \frac{\partial}{\partial x_j} T_{ij} + (F_\nu)_i. \end{aligned} \tag{2.50}$$

The viscous force is given the standard form

$$(F_\nu)_i = \frac{\partial}{\partial x_j}\left(2\rho\nu\left[e_{ij} - \frac{2}{3}\nabla\cdot\mathbf{v}\right]\right), \qquad e_{ij} = \frac{1}{2}\left(\frac{\partial v_i}{\partial x_j} + \frac{\partial v_j}{\partial x_i}\right) \tag{2.51}$$

with ν the kinematic viscosity. Equations (2.49) and (2.50) must be supplemented by the heat equation

$$\begin{aligned}\rho T\frac{\mathrm{d}s}{\mathrm{d}t} = \rho\left[\frac{\mathrm{d}u}{\mathrm{d}t} + p\frac{\mathrm{d}}{\mathrm{d}t}\left(\frac{1}{\rho}\right)\right] &= \rho\frac{\mathrm{d}u}{\mathrm{d}t} - \frac{p}{\rho}\frac{\mathrm{d}\rho}{\mathrm{d}t}\\ &= \rho\frac{\mathrm{d}u}{\mathrm{d}t} + p\nabla\cdot\mathbf{v} = -\mathcal{L},\end{aligned} \tag{2.52}$$

where u is the internal energy and s the entropy per gram, $\mathcal{L}$ the net loss of heat per cm^3 s, and use is made of (2.38). For an ideal gas with principal specific heats c_v, $c_p = \gamma c_v$ in standard notation,

$$u = \frac{p}{(\gamma-1)\rho} = \frac{\mathcal{R}T}{\mu(\gamma-1)} = c_v T, \tag{2.53}$$

and (2.52) may be written

$$\begin{aligned}\rho c_v T\frac{\mathrm{d}}{\mathrm{d}t}\log\frac{p}{\rho^\gamma} &= \frac{1}{\gamma-1}\left(\frac{\mathrm{d}p}{\mathrm{d}t} - \frac{\gamma p}{\rho}\frac{\mathrm{d}\rho}{\mathrm{d}t}\right)\\ &= \rho\frac{\mathrm{d}}{\mathrm{d}t}(c_p T) - \frac{\mathrm{d}p}{\mathrm{d}t} = -\mathcal{L},\end{aligned} \tag{2.54}$$

where $c_p T = \gamma p/(\gamma-1)\rho$ is the enthalpy per gram. If $\mathcal{L}$ is negligible, (2.52) yields constant specific entropy (adiabaticity) of the moving matter. We return to the expression for $\mathcal{L}$ in Section 2.4.

2.3.2 *The generalized Ohm's law*

To complete the system of equations, one must consider the drift of electrons relative to the ions, driven by the macroscopic fields **E**, **B** but impeded by collisions with the ions. The treatment follows closely that given by Cowling (1953, 1957), but with insights from other workers also.

As discussed above, the effect of the random electron velocities is again represented by an isotropic partial pressure p_e. The gravitational force on the electrons is negligible. The loss of electron drift momentum through collisions with the ions is equivalent to a frictional force $\mathbf{F}_{ei}$ per electron. It is idealized by supposing the relative velocity **V** to be completely randomized at intervals τ_{ei}, the 'collision time' introduced above. Thus $\mathbf{F}_{ei} = -m_e\mathbf{V}/\tau_{ei}$, and the equation of motion of the electrons becomes

$$-\nabla p_e - n_e e\left(\mathbf{E} + \frac{\mathbf{v}\times\mathbf{B}}{c}\right) - \frac{n_e e\mathbf{V}\times\mathbf{B}}{c} - \frac{n_e m_e\mathbf{V}}{\tau_{ei}} = 0. \tag{2.55}$$

By ignoring the electron inertia we assume implicitly that the macroscopic timescales are long compared with both the gyration period $2\pi/\omega_e$ and the plasma oscillation period (see below).

Equation (2.55) may be written

$$\frac{\mathbf{j}}{\sigma} = \mathbf{E} + \frac{\mathbf{v} \times \mathbf{B}}{c} + \frac{(\nabla p_e - \mathbf{j} \times \mathbf{B}/c)}{n_e e}, \tag{2.56}$$

where

$$\sigma = \frac{n_e e^2 \tau_{ei}}{m_e} \tag{2.57}$$

is the electrical conductivity (sometimes called the 'unreduced conductivity'). Equivalently,

$$\mathbf{E}' = \frac{\mathbf{j}}{\sigma} + \frac{\mathbf{j} \times \mathbf{B}}{c n_e e}, \tag{2.58}$$

where

$$\mathbf{E}' = \mathbf{E} + \frac{\mathbf{v} \times \mathbf{B}}{c} + \frac{\nabla p_e}{n_e e}. \tag{2.59}$$

The vector $\mathbf{E}'$ – sometimes called an 'equivalent electric field' – is the sum of the real electric field, as seen in the frame moving with the local bulk velocity of the gas, and the extra term $\nabla p_e / n_e e$, from now on referred to as the 'battery term'. From (2.58), $\mathbf{E}'$ is given by the sum of the usual 'Ohmic term' $\mathbf{j}/\sigma$ and the 'Hall term' $\mathbf{j} \times \mathbf{B}/c n_e e$, which arises because of the non-vanishing Lorentz force. The ratio of the Hall term to the Ohmic term is

$$\frac{\sigma B}{n_e e c} = \omega_e \tau_{ei} \tag{2.60}$$

from (2.27) and (2.57).

One can formally 'solve' (2.58) for the current components respectively parallel and perpendicular to $\mathbf{B}$:

$$\mathbf{j} = \sigma \mathbf{E}'_{\parallel} + \sigma_1 \mathbf{E}'_{\perp} + \sigma_2 \frac{\mathbf{B} \times \mathbf{E}'}{B}, \tag{2.61}$$

where

$$\sigma_1 = \frac{\sigma}{1 + (\omega_e \tau_{ei})^2}, \qquad \sigma_2 = \frac{(\omega_e \tau_{ei})\sigma}{1 + (\omega_e \tau_{ei})^2}. \tag{2.62}$$

If $\omega_e \tau_{ei} \ll 1$, then $\sigma_1 \approx \sigma$ and $\sigma_2 \ll \sigma$, so that (2.61) reduces to the case of isotropic conductivity $\mathbf{j} = \sigma \mathbf{E}'$. This is the case for high density and weak magnetic field: the electrons do not have time to take cognizance of the anisotropy caused by the presence of $\mathbf{B}$, because a collision with an ion occurs before the electron has travelled far along its circle of gyration. The presence of the field $\mathbf{B}$ is felt just through the definition of $\mathbf{E}'$. By contrast, when the Hall term dominates, so that $\omega_e \tau_{ei} \gg 1$, then $\sigma_1 \ll \sigma_2 \ll \sigma$. This is the limit with 'free spiralling between collisions': the magnetic field makes itself felt microscopically by impeding the perpendicular motion, whereas motion along $\mathbf{B}$ is restricted only by collisions with the ions. In the limit $\tau_{ei} \to \infty$, then $\sigma \to \infty$, $\sigma_1 \to 0$, and $\sigma_2 \to n_e ec/B$: the current across $\mathbf{B}$ becomes

$$\mathbf{j} = -n_e e \left(-c\frac{\mathbf{B} \times \mathbf{E}'}{B^2} \right) \tag{2.63}$$

– just the current due to electrons drifting across the field with the velocity $-c(\mathbf{B} \times \mathbf{E}'/B^2)$. In the absence of friction a component of $\mathbf{E}'$ parallel to $\mathbf{B}$ continually *accelerates* electrons, and yields an infinite parallel conductivity: whereas a component $\mathbf{E}'_\perp$ yields a finite drift *velocity* perpendicular to both $\mathbf{B}$ and $\mathbf{E}'$ (cf. (2.30)).

As emphasized by Schlüter (1950, 1958), this 'reduced conductivity' for currents flowing perpendicular to $\mathbf{B}$ does not imply increased dissipation. To the order of the present treatment, the number of ion–electron 'collisions' is unaffected by the presence of $\mathbf{B}$, and the volume rate of dissipation is still given by $\mathbf{j}^2/\sigma$, where $\mathbf{j} = (c/4\pi)\nabla \times \mathbf{B}$ and σ is the unreduced conductivity (2.57). A more accurate treatment (e.g. Spitzer 1956), which uses the Boltzmann equation to construct the distribution function, yields a modest increase in dissipation – by up to a factor 2 – for currents $\mathbf{j}_\perp$ as compared with $\mathbf{j}_\parallel$. This is of course far smaller than the enormous (spurious) increases that would result from the illicit use of σ_1 or σ_2 instead of σ in $\mathbf{j}^2/\sigma$.

For current flow perpendicular to $\mathbf{B}$ one can define another conductivity σ_3 by the ratio of $\mathbf{j}_\perp$ to the component of $\mathbf{E}'$ in the direction of $\mathbf{j}_\perp$ (which by (2.61) is inclined to $\mathbf{E}'$):

$$j_\perp = \sigma_3(\mathbf{j}_\perp \cdot \mathbf{E}'_\perp)/j_\perp. \tag{2.64}$$

Combining (2.61) and (2.64) yields

$$(\sigma_1^2 + \sigma_2^2)E'^2 = j_\perp^2 = \sigma_3\sigma_1 E'^2 \tag{2.65}$$

or

$$\sigma_3 = \sigma_1 + \sigma_2^2/\sigma_1 = \sigma \tag{2.66}$$

for the present case of a fully ionized gas. In fact, from (2.58), $j^2/\sigma = \mathbf{j} \cdot \mathbf{E}'$ with $\mathbf{E}'$ defined by (2.59); and just as by (2.61) $\mathbf{j}_\parallel \cdot \mathbf{E}'_\parallel = j_\parallel^2/\sigma$, by (2.64) and (2.66), $\mathbf{j}_\perp \cdot \mathbf{E}'_\perp = j_\perp^2/\sigma_3 = j_\perp^2/\sigma$ – the same effective *resistivity* for currents both parallel to and perpendicular to $\mathbf{B}$. (The contrasting results for a lightly ionized gas are brought out in Section 2.7.)

The anisotropy of the conductivity arises from the Hall term in (2.58). Since this is proportional to the Lorentz force density $\mathbf{j} \times \mathbf{B}/c$, it can be eliminated by use of the bulk equation of motion (2.35) (with the viscous term dropped), yielding 'Ohm's law' in the apparently more familiar form (e.g. Schlüter 1958)

$$\mathbf{j} = \sigma \mathbf{E}'', \tag{2.67}$$

where the new 'equivalent electric field'

$$\mathbf{E}'' = \mathbf{E} + \frac{\mathbf{v} \times \mathbf{B}}{c} + \frac{1}{n_e e}\left(\rho \mathbf{g} - \nabla p_i - \rho \frac{d\mathbf{v}}{dt}\right), \tag{2.68}$$

does not involve $\mathbf{j}$. This can sometimes be a more convenient form to use than (2.56); but as emphasized by Cowling (1953), as it is derived from (2.56) and (2.35), in any problem involving both the equation of motion and either form of 'Ohm's law', the results must be identical. Equation (2.67) involves the unreduced conductivity, even for components perpendicular to $\mathbf{B}$; hence $(\mathbf{E}'')_\perp$ must be less than $(\mathbf{E}')_\perp$, and rotated relative to it.

Equations (2.68) and (2.67), with $\mathbf{j}/\sigma$ replacing $\mathbf{E}''$, can then formally be rewritten so that $\rho d\mathbf{v}/dt$ is given in terms of field vectors. However, Schlüter (1958) points out that the equation cannot usefully be regarded as an alternative to the equation of motion (2.35). This is because the electric field $\mathbf{E}$ is not known a priori; it consists of an irrotational part, related to the distribution of space charges, and a solenoidal part, related to the change in the magnetic field. In a highly conducting medium one cannot prescribe a charge distribution and so determine the associated electric field from the Poisson–Maxwell equation; rather, the mobile charges distribute themselves so as to satisfy $\rho_e = \nabla \cdot \mathbf{E}/4\pi$, where $\mathbf{E}$ is given by the analogue of Ohm's law.

By neglecting the displacement current and electron inertia, we imply that the space–charge field associated with a varying $\mathbf{E}$ is built up instantaneously. If instead the linear electron inertia term is retained, Ohm's law (2.56) becomes

$$\frac{m_e}{n_e e^2}\frac{\partial \mathbf{j}}{\partial t} = \mathbf{E} + \mathbf{R} - \frac{\mathbf{j}}{\sigma} \tag{2.69}$$

where $\mathbf{R} \equiv (\mathbf{v} \times \mathbf{B}/c + \nabla p_e/n_e e - \mathbf{j} \times \mathbf{B}/cn_e e)$. For simplicity, take n_e and σ uniform. From the divergence of (2.69),

$$-\frac{m_e}{n_e e^2}\ddot{\rho}_e = 4\pi\rho_e + \frac{\dot{\rho}_e}{\sigma} + \nabla \cdot \mathbf{R}. \tag{2.70}$$

Equation (2.70) implies that the charge density oscillates with the plasma frequency $\omega_p = (4\pi n_e e^2/m_e)^{1/2}$ for a time of order $8\pi\sigma/\omega_p^2$ (σ in esu), after which the electron–ion collisions have damped out the oscillations, and the charge density is given by taking the divergence of the normal form (2.56) of 'Ohm's law'.

The 'unreduced conductivity' σ is by (2.57), (2.25) and (2.26)

$$\frac{n_e e^2}{m_e}\frac{1}{n_e \pi (3kT/m_e)^{1/2}(3Ze^2/2kT)^2} \simeq \frac{2 \times 10^7 T^{3/2}}{Z}. \tag{2.71}$$

The density disappears, as the number of carriers of current n_e increases proportionately to the number of scatterers n_i.

The ratio of the decay time of plasma oscillations to their period is

$$\frac{8\pi\sigma/\omega_p^2}{2\pi/\omega_p} = \frac{4\sigma}{\omega_p} \simeq \left(\frac{1}{n_e^{1/3} b}\right)^{3/2} \tag{2.72}$$

where b is again the collision radius $\simeq e^2/kT$. By (2.24), b is always much smaller than the interparticle distance $n_e^{-1/3}$ in an uncondensed plasma, so the description of the process in terms of 'slowly' damped plasma oscillations is seen to be valid; but the damping time is always far shorter than any macroscopic time-scales.

A more precise treatment of the cut-off in Rutherford scattering (Spitzer 1952) replaces (2.71) by

$$\sigma = \frac{1.4 \times 10^8 T^{3/2}}{Z \ln \Lambda}, \tag{2.73}$$

where the Coulomb logarithm ($\ln \Lambda$) is a slowly varying function of ρ and T, with typical values between 10 and 20 (Priest 1982).

2.4 The energy equation of a fully ionized gas

The scalar product of 'Ohm's law' in the form (2.56) with the current density $\mathbf{j}$ yields

$$\mathbf{j} \cdot \mathbf{E} = \frac{\mathbf{j}^2}{\sigma} - \frac{\nabla p_e}{n_e e} \cdot \mathbf{j} + \frac{\mathbf{j} \times \mathbf{B}}{c} \cdot \mathbf{v}. \tag{2.74}$$

The term $\mathbf{j} \cdot \mathbf{E} = [n_i Z e \mathbf{v} - n_e e(\mathbf{v} + \mathbf{V})] \cdot \mathbf{E}$ is the total rate of working per cm^3 on the charges of the electric force (the magnetic force is perpendicular to the velocity and so does no work). The truncated Maxwell equations (2.12)–(2.14) yield for Poynting's theorem

$$\int_\tau \mathbf{j} \cdot \mathbf{E}\, \mathrm{d}\tau = -\int_S \left(\frac{c}{4\pi} \mathbf{E} \times \mathbf{B}\right) \cdot \mathbf{n}\, \mathrm{d}\mathbf{S} - \frac{\mathrm{d}}{\mathrm{d}t} \int \frac{\mathbf{B}^2}{8\pi}\, \mathrm{d}\tau \tag{2.75}$$

where S is the surface bounding the volume τ and $\mathbf{n}$ is the outward-drawn normal. Because of our neglect of the displacement current the relativistically small term in $\mathbf{E}^2/8\pi$ is automatically missing. Also, the magnetic field due to a 'quasi-static' current density of finite dimensions falls off like the inverse square, so yielding a vanishing Poynting integral at infinity: the system does not lose any energy by radiation.

The first term on the right of (2.74) is the familiar Joule heat per $\mathrm{cm}^3\,\mathrm{s}$. By (2.32) and (2.57) it may be written:

$$\frac{\mathbf{j}^2}{\sigma} = -\mathbf{V} \cdot \left(\frac{-n_e m_e \mathbf{V}}{\tau_{ei}}\right) = -n_e \mathbf{V} \cdot \mathbf{F}_{ei}, \tag{2.76}$$

showing that it indeed represents the energy dissipated through randomization of the electron drift velocity $\mathbf{V}$, as described by the effective frictional force $\mathbf{F}_{ei}$. Note that σ is the 'unreduced' conductivity. The Hall term – which we saw is responsible for the anisotropy in the conductivity – has disappeared from (2.76), it being perpendicular to $\mathbf{j}$: as already noted, according to the simple mean-free-path, two-fluid model, the dissipation term is the same for currents along or across the field. The term $-(\nabla p_e/n_e e) \cdot \mathbf{j}$ is minus the rate of working per cm^3 of the 'battery field' on the currents. If $\nabla p_e/n_e e$ is essentially parallel to $\mathbf{j}$, it pumps energy into the magnetic field; if anti-parallel, magnetic energy is converted into thermal energy – the 'battery is being charged'. The third term appears in equation (2.37) for the bulk kinetic energy of the gas: it is the rate of working of the magnetic body force $\mathbf{j} \times \mathbf{B}/c$ on the fluid moving with bulk velocity $\mathbf{v}$. If the two vectors are essentially parallel this work is positive, as in an electric motor; if essentially anti-parallel, the work is negative, and the non-magnetic forces driving the flow against the opposition of the magnetic forces pump energy into the magnetic field, as in a dynamo.

The energy loss function $\mathcal{L}$ may now be written down:

$$\mathcal{L} = \nabla \cdot \mathbf{q} + L_{\text{rad}} - \frac{\mathbf{j}^2}{\sigma} + \mathbf{j} \cdot \frac{\nabla p_e}{n_e e} - H, \tag{2.77}$$

where the Ohmic and battery terms contribute, $\mathbf{q}$ is the heat flux due to heat conduction, L_{rad} the losses due to radiation, and H includes the viscous heating and heating from magnetosonic shock waves (Section 3.2). Inside a non-degenerate star, $\mathbf{q}$ is negligible, and heat transfer is by radiation in the small mean-free-path approximation: in standard notation,

$$L_{\text{rad}} = -\nabla \cdot \left(\frac{4acT^3}{3\kappa_r \rho} \nabla T \right) \tag{2.78}$$

where κ_r is the appropriate opacity. In an optically thin region such as a stellar corona or a diffuse gas cloud, L_{rad} is conveniently written as $n_e n_H Q(T)$, where the function Q has been constructed by a number of authors (cf. Priest 1982 for details). In a hot stellar corona, heat conduction is important, and becomes anisotropic in the presence of a strong magnetic field:

$$-\nabla \cdot \mathbf{q} = \nabla_{\parallel} \cdot (\kappa_{\parallel} \nabla_{\parallel} T) + \nabla_{\perp} \cdot (\kappa_{\perp} \nabla_{\perp} T), \tag{2.79}$$

where the κ values are conductivities respectively along and across the field. Conduction along the field is primarily by electrons, yielding for a fully ionized hydrogen plasma (Spitzer 1962)

$$\kappa_{\parallel} = 1.8 \times 10^{-5} \frac{T^{5/2}}{\ln \Lambda}, \tag{2.80}$$

with $\ln \Lambda$ again the Coulomb logarithm. Conduction across the field is mainly by ions. A genuine reduction of the trans-field conductivity occurs when the ions spiral many times before mutual collision:

$$\frac{\kappa_{\perp}}{\kappa_{\parallel}} = 2 \times 10^{-27} \frac{n^2}{T^3 B^2}, \tag{2.81}$$

where n is the number of particles in a hydrogen plasma.

The rate of viscous heating is

$$H_\nu = \rho\nu\left(2e_{ij}e_{ij} - \frac{2}{3}(\nabla\cdot\mathbf{v})^2\right). \tag{2.82}$$

For a hydrogen plasma

$$\rho\nu = 2.2\times 10^{-15}\frac{T^{5/2}}{\ln\Lambda} \tag{2.83}$$

(Spitzer 1962). However, when dealing with turbulent media, the microviscosity is often replaced by a much larger turbulent viscosity (cf. Section 4.2).

2.5 Kinematic coupling

For the moment, ignore the thermal and Hall fields, and use the simple (Galilean-invariant) form of Ohm's law

$$\frac{\mathbf{j}}{\sigma} = \mathbf{E}' = \mathbf{E} + \frac{\mathbf{v}\times\mathbf{B}}{c}, \tag{2.84}$$

where by (2.5) $\mathbf{E}'$ is the electric field as seen in the frame moving with the local bulk velocity $\mathbf{v}$ of the fluid. Substitution of $\mathbf{E}$ into Faraday's law (2.13) and replacing $\mathbf{j}$ from the truncated Ampère–Maxwell law (2.12) yields the basic equation of magnetokinematics:

$$\frac{\partial\mathbf{B}}{\partial t} = \nabla\times(\mathbf{v}\times\mathbf{B}) - \nabla\times(\lambda\nabla\times\mathbf{B}), \tag{2.85}$$

where λ is the resistivity or the magnetic diffusivity,

$$\lambda = c^2/4\pi\sigma, \tag{2.86}$$

with the same dimensions (length^2/time) as the kinematic viscosity. If λ is constant, the diffusive term in (2.85) can be written $\lambda\nabla^2\mathbf{B}$. Let D be a characteristic scale of variation of $\mathbf{B}$. The first term in (2.85), describing the effect of the bulk motion on the field, is typically of order vB/D. The second term represents diffusion of the field through the fluid and is typically of order $\lambda B/D^2$. The relative importance of the two terms is therefore measured by the *magnetic Reynolds number* (MRN)

$$\mathrm{R_m} = vD/\lambda = 4\pi\sigma vD/c^2 \tag{2.87}$$

by analogy with the viscous Reynolds number in fluid dynamics. When $\mathrm{R_m} \gg 1$, convective effects dominate the evolution of the field; when $\mathrm{R_m} \ll 1$, resistive diffusion dominates.

It is convenient to consider first the limit of infinite conductivity, for which (2.85) reduces to

$$\frac{\partial\mathbf{B}}{\partial t} = \nabla\times(\mathbf{v}\times\mathbf{B}). \tag{2.88}$$

As is well known, (2.88) implies that the magnetic flux through a surface moving with the fluid remains constant, and that if two material elements are initially

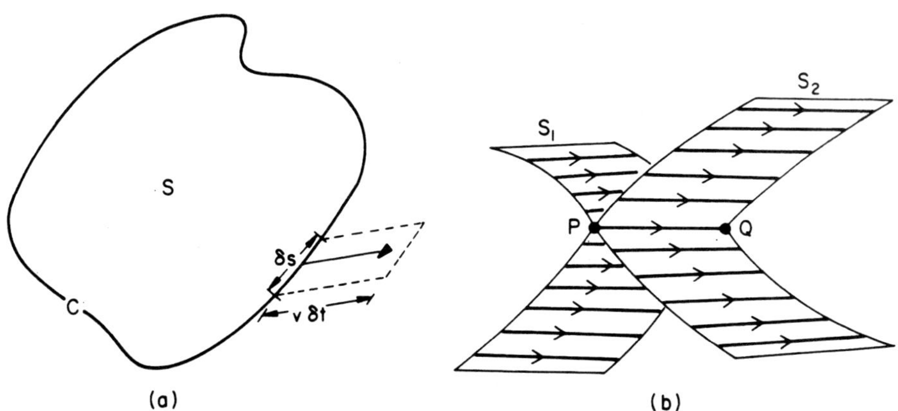

FIG. 2.2. Flux freezing. (a) A surface S bounded by a curve C that moves with the fluid. (b) Two magnetic flux surfaces S_1, S_2, intersecting in the field line PQ. (From Priest 1982.)

on the same line of force, then they will remain on that line for all time. In Fig. 2.2(a), the surface S bounded by the closed contour C moves with the fluid. The total rate of change of flux through S is a combination of the Eulerian time-derivative and the flux entering the surface as a result of its displacement: in general,

$$\begin{aligned}\frac{\mathrm{d}}{\mathrm{d}t}\int_S \mathbf{B}\cdot\mathrm{d}\mathbf{S} &= \int_S \frac{\partial \mathbf{B}}{\partial t}\cdot\mathrm{d}\mathbf{S} + \oint_C \mathbf{B}\cdot\mathbf{v}\times\mathrm{d}\mathbf{s} = \int_S \frac{\partial \mathbf{B}}{\partial t}\cdot\mathrm{d}\mathbf{S} - \oint_C \mathbf{v}\times\mathbf{B}\cdot\mathrm{d}\mathbf{s}\\ &= \int_S\left(\frac{\partial \mathbf{B}}{\partial t} - \nabla\times(\mathbf{v}\times\mathbf{B})\right)\cdot\mathrm{d}\mathbf{S}. \qquad (2.89)\end{aligned}$$

However, when (2.88) holds, (2.89) vanishes: the flux through S remains constant – the analogue of the Kelvin–Helmholtz vorticity theorem for an inviscid fluid.

To isolate a segment of a line of force, define it by the intersection of two surfaces. Thus in Fig. 2.2(b), let S_1 and S_2 be two surfaces made up of such segments, and intersecting in the segment PQ. The flux through each surface is initially zero and must remain so if the surfaces move with the fluid. Hence the surfaces are always made up of lines of force and so their intersection must remain a line of force PQ.

Qualitatively, we may say that whereas a flow with $\mathbf{v}$ parallel to $\mathbf{B}$ has no direct electromagnetic effect, because $c\mathbf{E} + \mathbf{v}\times\mathbf{B}$ vanishes in a system with $\mathrm{R_m}$ large, flow across the field drags the field with it. However, flow along and across the field are linked by the equation of continuity. Figure 2.3 shows a short segment of length s, part of a thin flux-tube of cross-section A. Conservation of flux defines A as satisfying BA constant, and flux-freezing requires that the

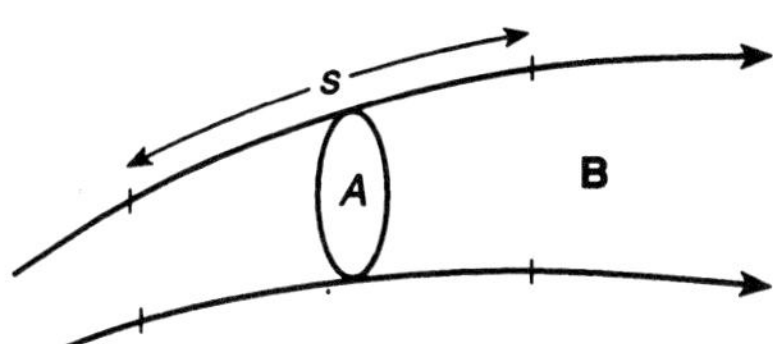

FIG. 2.3. Flux freezing. A segment of length s of a thin flux tube of local cross-section A.

segment contains the same fluid particles, so that ρAs is also constant, ρ being the local density. Thus (2.89) implies that

$$B \propto \rho s. \tag{2.90}$$

In an incompressible fluid, or in a flow with modest changes in ρ, one can think in terms of $B \propto s$: a shearing motion increases s, so reducing A and increasing B (the 'spaghetti process'). Examples that will recur are the distortion of a uniform field by convection (Chapter 4), and the generation of an azimuthal ('toroidal') field from a meridional ('poloidal') field by differential rotation (*passim*).

Formally, the right-hand side of (2.88) can be expanded to yield

$$\frac{\mathrm{d}\mathbf{B}}{\mathrm{d}t} \equiv \frac{\partial \mathbf{B}}{\partial t} + (\mathbf{v}\cdot\nabla)\mathbf{B} = (\mathbf{B}\cdot\nabla)\mathbf{v} - \mathbf{B}(\nabla\cdot\mathbf{v}). \tag{2.91}$$

The left-hand side of (2.91) is the Lagrangian derivative of $\mathbf{B}$, while the terms on the right represent the stretching and compression of the flux tubes. Combining (2.91) with the continuity equation (2.38) in the form

$$\frac{\mathrm{d}\rho}{\mathrm{d}t} = -\rho\nabla\cdot\mathbf{v} \tag{2.92}$$

yields

$$\frac{\mathrm{d}}{\mathrm{d}t}\left(\frac{\mathbf{B}}{\rho}\right) = \left[\left(\frac{\mathbf{B}}{\rho}\right)\cdot\nabla\right]\mathbf{v}. \tag{2.93}$$

This equation has the formal Lagrangian integral

$$\left(\frac{\mathbf{B}}{\rho}\right) = \left[\left(\frac{\mathbf{B}}{\rho}\right)_0\cdot\nabla_0\right]\mathbf{r}, \tag{2.94}$$

where $\mathbf{r} = \mathbf{r}(\mathbf{r}_0, t)$, ∇_0 is the gradient operator with respect to the Lagrangian variable $\mathbf{r}_0$ and $\rho = \rho(\mathbf{r}, t)$, $\rho_0 = \rho_0(\mathbf{r}_0, t_0)$, etc. (Roberts 1967).

At the other extreme, when $\mathrm{R_m} \ll 1$, (2.85) reduces to a diffusion equation:

$$\frac{\partial \mathbf{B}}{\partial t} = -\nabla\times(\lambda\nabla\times\mathbf{B}) = \lambda\nabla^2\mathbf{B} \tag{2.95}$$

when σ is uniform, implying that non-uniformities in the field drift so as mutually to cancel. A characteristic time for the decay of a field of scale D is

$$\tau_d = 4\pi(\sigma/c^2)D^2 = D^2/\lambda. \tag{2.96}$$

Division of the energy density $B^2/8\pi$ by the rate of Ohmic dissipation/cm^3, $j^2/\sigma \simeq [(c/4\pi)(B/D)]^2/\sigma$, yields a time-scale of the same order. The condition $\mathrm{R_m} \gg 1$ for the approximate freezing-in of the field can be rewritten as $\tau_d \gg D/v$ – the decay time long compared with the time of flow across D. For a large-scale field within a star, $\tau_d \simeq 5 \times 10^9$ yr; in a fairly small sunspot, $D \simeq 10^8$ cm and $\tau_d \simeq 50$ yr.

It is clearly the large length-scales in cosmical physics that suggest (2.88) as the appropriate paradigm rather than (2.95). However, recall that D is the characteristic length in which a field changes significantly, and this can become locally small (cf. Section 5.8). Also, a modest departure from strict flux freezing can cause a change in the topology of the field (Section 3.7).

Consider now the effect of the Hall and battery terms in (2.56). In the limit of infinite conductivity, (2.88) is replaced by

$$\frac{\partial \mathbf{B}}{\partial t} = \nabla \times [(\mathbf{v} + \mathbf{V}) \times \mathbf{B}] + \nabla \times \left(\frac{c\nabla p_e}{n_e e}\right), \tag{2.97}$$

where the Hall term has been written in its original form $\mathbf{V} \times \mathbf{B}/c$ with $\mathbf{V}$ the drift of the electrons relative to the ions. Without the battery term, this equation implies that flux is invariant through a circuit moving with the electron velocity $(\mathbf{v}+\mathbf{V})$ rather than the bulk velocity $\mathbf{v}$. As long as $V \ll v$ this correction is small (even in spontaneously forming current sheets, the development of an anomalous resistivity (Section 2.8) is likely to be a much more important effect). However, the battery term is the analogue of the Peltier effect and can indeed generate magnetic flux. From (2.89) and (2.56), the rate of change of flux across a surface S spanning a circuit C that moves with the *electrons* is

$$c\int_S \nabla \times \left(\frac{\nabla p_e}{n_e e} - \frac{\mathbf{j}}{\sigma}\right) \cdot \mathrm{d}\mathbf{S} = c\oint_C \left(\frac{\nabla p_e}{n_e e} - \frac{\mathbf{j}}{\sigma}\right) \cdot \mathrm{d}\mathbf{s} \tag{2.98}$$

The Ohmic term acts to destroy the flux through C, but a non-zero ‘emf’ of the battery field around C will generate flux.

The ratio $(|\nabla p_e|/n_e e)/(vB/c) \simeq (v_T/v)(v_T/\omega_e L)$, where v_T is the electron thermal speed and L a macroscopic length-scale, and this will normally be small. Thus the battery term – like the Hall term – will not affect much the freezing-in condition, but it can still be significant through having a non-zero component in the direction parallel to $\mathbf{B}$, along which the induction term vanishes. We return to this effect in Chapter 5.

2.6 Dynamical coupling

So far, the interaction between the magnetic and kinetic fields has been one-way: $\mathbf{v}$ has been prescribed and the consequences for $\mathbf{B}$ and $\mathbf{j}$ examined. The magnetic field in fact reacts back on the flow through the Lorentz force $\mathbf{f} = \mathbf{j} \times \mathbf{B}/c = (\nabla\times\mathbf{B}) \times \mathbf{B}/4\pi$. In the first example, the prescribed initial field $\mathbf{B}$ is supposed curl-free, maintained by currents external to the region studied. From Ohm's law (with the Hall and battery terms ignored), a flow of gas across $\mathbf{B}$ yields currents

$$\mathbf{j} = \sigma\left(\mathbf{E} + \frac{\mathbf{v} \times \mathbf{B}}{c}\right). \tag{2.99}$$

To bring out again the importance of electromagnetic induction, we begin by ignoring it, writing $\mathbf{E} = -\nabla\phi$ and by implication keeping $\mathbf{B}$ unchanged. For simplicity, σ is assumed uniform. If also $\nabla\times\mathbf{v} = 0$, $\nabla \cdot (\mathbf{v} \times \mathbf{B}) = \mathbf{B} \cdot \nabla\times\mathbf{v} - \mathbf{v} \cdot \nabla\times\mathbf{B} = 0$, so from (2.99) $\nabla^2\phi = 0$ everywhere; whence with appropriate boundary conditions at infinity, $\nabla\phi = 0$, and the whole of $\mathbf{v}\times\mathbf{B}/c$ is available to drive currents. More generally, the curl-free part will be cancelled by an electric field, but for order-of-magnitude estimates we may take $\mathbf{j} = \sigma(\mathbf{v} \times \mathbf{B}/c)$, with a consequent Lorentz force $-(\sigma B^2/c^2)\mathbf{v}_\perp$. The equation of motion across $\mathbf{B}$ is

$$\rho\frac{\mathrm{d}\mathbf{v}_\perp}{\mathrm{d}t} = \mathbf{P}_\perp - \frac{\sigma B^2}{c^2}\mathbf{v}_\perp, \tag{2.100}$$

where $\mathbf{P}$ is the density of the other forces acting on the fluid. Then (2.100) implies an exponential decay of $\mathbf{v}_\perp$, with a time-constant $\rho/(\sigma/c^2)B^2$, to the value $\mathbf{P}_\perp/(\sigma/c^2)B^2$, with the current density $\mathbf{j}$ approaching the value $c\mathbf{P}\times\mathbf{B}/B^2$ that yields the magnetostatic condition $\mathbf{P}_\perp + \mathbf{j} \times \mathbf{B}/c = 0$. In this equilibrium state, the current is independent of σ: Ohm's law (2.99) merely fixes the velocity of drift across the field required by the finite resistivity. Instead of Ohm's law determining the current and the equation of motion the velocity, the roles are reversed (Cowling 1953; Spitzer 1956).

It should be emphasized that this conclusion depends on the conductivity being high, so that $v_\perp \propto 1/\sigma B^2$ is small and the non-linear inertial term in (2.100) is negligible. Further, the approach to equilibrium depends on self-induction being negligible. The flow of current $j \simeq \sigma(vB/c)$ in a region of scale D generates by Ampère's law a field $B' \simeq (4\pi D/c)(\sigma vB/c)$ in a characteristic time D/v, so that by Faraday's law the induced electric field is $\simeq (-D/c)(4\pi D\sigma vB/c^2)/(D/v) = -(4\pi\sigma/c^3)Dv^2B$, and this is comparable with vB/c when $\mathrm{R_m} = (4\pi\sigma/c^2)vD \approx 1$ – just the magnetic Reynolds number condition again. When $\mathrm{R_m} \gg 1$ the zero-order approximation to be used is no longer $\mathbf{j} = \sigma(\mathbf{v} \times \mathbf{B})/c$ but the infinite conductivity limit $c\mathbf{E} + \mathbf{v} \times \mathbf{B} = 0$, so that at any time the field-freezing condition (2.88) holds. The current density is then fixed by Ampère's law and so the instantaneous Lorentz force density is known.

The dragging of the field by the moving fluid generates a force density which does not automatically vanish when $\mathbf{v} = 0$ as in the case above. As already noted,

f is the divergence of the Maxwell stress tensor, comprising an isotropic pressure $B^2/8\pi$ and a tension $B^2/4\pi$ along the direction of $\mathbf{B}$. With matter moving with the field, we may regard each field line of infinitesimal area A as an elastic string under a tensile force $(B^2/4\pi)A$ and with a line density ρA. A disturbance to a uniform field will not then be damped out but will generate an *Alfvén wave*, travelling with speed

$$v_{\mathrm{A}} = \frac{B}{(4\pi\rho)^{1/2}}, \tag{2.101}$$

and with a continuous interchange of energy between the kinetic and magnetic fields (Alfvén 1942, 1950).

In the simplest example, consider a perfectly conducting fluid, incompressible, inviscid and of uniform density ρ. There are no non-magnetic body forces. The medium is permeated by a uniform field $\mathbf{B}_0$ parallel to the z-axis. All quantities are assumed to depend only on z and t. The incompressibility condition

$$\nabla \cdot \mathbf{v} = 0 \tag{2.102}$$

yields $\partial v_z/\partial z = 0$, so there exists a frame with $v_z = 0$ in which the fluid motions are transverse. The field-freezing condition (2.88) together with (2.102) then yields

$$\frac{\partial \mathbf{B}}{\partial t} = \left(B_z \frac{\partial}{\partial z}\right) \mathbf{v}. \tag{2.103}$$

From $\nabla \cdot \mathbf{B} = 0$ and (2.103), B_z is independent of z and t, so we may write

$$\mathbf{B} = \mathbf{B}_0 + \mathbf{B}', \quad \mathbf{j} = (c/4\pi)\nabla\times\mathbf{B}', \tag{2.104}$$

with $\mathbf{B}'$ transverse ($\mathbf{B}_0 \cdot \mathbf{B}' = 0$), and satisfying

$$\frac{\partial \mathbf{B}'}{\partial t} = \left(\mathbf{B}_0 \frac{\partial}{\partial z}\right) \mathbf{v}. \tag{2.105}$$

The Lorentz force density is

$$\frac{1}{4\pi}\left(B_0\frac{\partial B'_x}{\partial z},\ B_0\frac{\partial B'_y}{\partial z},\ -\frac{\partial \mathbf{B}'^2/2}{\partial z}\right) \tag{2.106}$$

in Cartesians. The equation of motion (2.35) yields for the z-component

$$p + \frac{B'^2_x + B'^2_y}{8\pi} = p_0 \tag{2.107}$$

where p_0 is the fluid pressure far from the disturbance. The other components yield

$$\rho\frac{\partial v_x}{\partial t} = \frac{B_0}{4\pi}\frac{\partial B'_x}{\partial z}, \tag{2.108}$$

etc., so that v_x, v_y, B'_x, B'_y all satisfy the one-dimensional wave equation

$$\frac{\partial^2 B'_x}{\partial t^2} = v_{A0}^2 \frac{\partial^2 B'_x}{\partial z^2}, \quad v_{A0} = \frac{B_0}{(4\pi\rho)^{1/2}}, \tag{2.109}$$

etc. Plane-polarized transverse waves are propagated along the field without distortion. Note that no linearization approximation has been made in this treatment. Since $\mathbf{v}$ and $\mathbf{B}'$ vanish together outside the disturbance domain, (2.105) and (2.108) yield

$$\mathbf{v} = \pm \frac{\mathbf{B}'}{(4\pi\rho)^{1/2}}, \tag{2.110}$$

whence

$$\frac{1}{2}\rho \mathbf{v}^2 = \frac{(\mathbf{B}_0 + \mathbf{B}')^2 - \mathbf{B}_0^2}{8\pi} = \frac{\mathbf{B}'^2}{8\pi} \tag{2.111}$$

– equipartition of energy between the kinetic field and the disturbance to the magnetic field.

In fact, there is no need to require $\mathbf{B}'$ to have any particular spatial dependence (Walén 1944). With ρ constant and with (2.110) still holding, the terms $(\nabla\times\mathbf{v}) \times \mathbf{v}$ and $(\nabla\times\mathbf{B}) \times \mathbf{B}$ in the equation of motion mutually cancel, again leaving only a linear magnetic force term. If there is also a conservative body force $\mathbf{g} = -\nabla V$, then

$$p - \rho V + \frac{(\mathbf{B}_0 + \mathbf{B}')^2}{8\pi} = \text{constant} \tag{2.112}$$

is the analogue of (2.107).

Chandrasekhar (1957) showed that there exist solutions of the steady-state hydromagnetic equations for an infinitely conducting, inviscid medium such that

$$\mathbf{v} = \pm \frac{\mathbf{B}}{(4\pi\rho)^{1/2}}, \tag{2.113}$$

provided ρ is constant on the field-streamlines, so that $\nabla \cdot (\rho\mathbf{v}) = 0$ implies $\nabla \cdot \mathbf{v} = 0$, and also provided

$$\nabla \left(p + \frac{B^2}{8\pi} \right) = \rho \nabla V. \tag{2.114}$$

If ρ is uniform, then on writing $\mathbf{B} = \mathbf{B}_0 + \mathbf{B}'$ it follows that a non-steady solution of the equations is given by a Galilean transformation to a frame moving with velocity $\mathbf{B}_0/(4\pi\rho)^{1/2}$ relative to the original frame. In this frame

$$\mathbf{v}' = \mathbf{v} - \frac{\mathbf{B}_0}{(4\pi\rho)^{1/2}} = \frac{\mathbf{B}'}{(4\pi\rho)^{1/2}}, \tag{2.115}$$

i.e. we recover the Walén solution.

By including dissipative terms – the Ohmic field in Ohm's law, the viscous force in the equation of motion – one can demonstrate the progressive damping of Alfvén waves. Inclusion of the Hall field makes a further qualitative change, for as seen above, in a perfect conductor the field **B** moves strictly with the electrons rather than with the ions: consequently, modes $\propto \exp[i(\omega t - kz)]$ are circularly polarized, and have phase velocities differing slightly for right-handed and left-handed polarizations.

The Alfvén speed is less than c provided $B_0^2/4\pi < \rho c^2$, i.e. provided the magnetic energy density is less than one-half of the material energy density, a condition that holds in normal, non-relativistic systems but can be violated, e.g. in the pulsar magnetosphere (Chapters 13 and 14). As the two energies approach, the neglected 'inertia of the ether', as represented by the displacement current, becomes important. Provided the wave frequencies are much less than the plasma and gyro-frequencies, then one can generalize the wave treatment (Alfvén 1950), replacing $\mathbf{j} = (c/4\pi)\nabla\times\mathbf{B}$ by

$$\frac{c}{4\pi}\left[\nabla\times\mathbf{B} - \frac{1}{c}\frac{\partial\mathbf{E}}{\partial t}\right] = \frac{c}{4\pi}\left[\nabla\times\mathbf{B} + \frac{1}{c^2}\frac{\partial}{\partial t}(\mathbf{v}\times\mathbf{B})\right] \tag{2.116}$$

(substituting for **E** from (2.4)). Transverse waves are found to propagate with speed c', where

$$\begin{aligned}(c')^2 = \frac{c^2 v_\mathrm{A}^2}{c^2 + v_\mathrm{A}^2} &\approx c^2 \ \text{ if } v_\mathrm{A}^2 \gg c^2 \\ &\approx v_\mathrm{A}^2 \ \text{ if } c^2 \gg v_\mathrm{A}^2.\end{aligned} \tag{2.117}$$

The theory of small amplitude magnetohydrodynamic waves is developed further in Section 3.1. Magnetohydrodynamic (MHD) shock waves are discussed in Section 3.2.

2.7 The three-fluid model

In the outer regions and atmospheres of late-type stars, the temperature falls below the ionization temperature of hydrogen and neutral helium, so there is present a substantial fraction of neutral gas. This demands a generalization of the work of Sections 2.3 and 2.4 to the 'three-fluid model', discussed explicitly or implicitly by several authors (Schlüter 1951; Schlüter and Biermann, 1950; Cowling 1956, 1976; Piddington 1955; Mestel and Spitzer 1956). In the present section the general theory is presented, following closely Cowling's notation, which brings out both the similarities and the important contrasts with the two-fluid model of a fully ionized gas. The crucial importance of 'ambipolar diffusion' and the consequent greatly enhanced dissipation in a lightly ionized gas does not seem to have been recognized until about 1955. The principal application in this book is to the magnetohydrodynamics of cool gas clouds that are the observed loci of much contemporary star formation, where the fractional ionization is so low that a simplified treatment is appropriate (Chapters 11 and 12).

The three-fluid mixture consists of the two-fluid gas as discussed above, together with an interpenetrating neutral gas of mass m_n and number density n_n. As before, we anticipate that the degree of charge-separation allowed by the Poisson–Maxwell equation will turn out to be very small, so that again we can write $n_e = Zn_i \equiv n$. In the outer regions of a late-type star, the ions are nearly all of hydrogen and helium, and it is a good approximation to take m_i and m_n (which are in any case averages) as equal; however, that is not the case for the gas cloud case, and we shall leave them as distinct.

As for the two-fluid case, each gas is given a mean velocity, plus a random velocity distribution taken to be isotropic and so yielding the partial pressures p_e, p_i, p_n. The ion and electron fluids have mean velocities $\mathbf{v}+\mathbf{V}_i$ and $\mathbf{v}+\mathbf{V}_i+\mathbf{V}$, where $\mathbf{v}$ is the bulk velocity of the gas, whence the neutral gas has mean velocity $\mathbf{v}_n = \mathbf{v} - (F^{-1} - 1)\mathbf{V}_i$, where F is defined to be the fractional contribution of the non-ionized gas to the total density:

$$F \equiv \frac{n_n m_n}{n_n m_n + n_i m_i} \equiv \frac{\rho_n}{\rho_n + \rho_i} \tag{2.118}$$

(the electron mass again being negligible). The velocity of the ions relative to the neutrals is then $F^{-1}\mathbf{V}_i$, that of the electrons $\mathbf{V} + F^{-1}\mathbf{V}_i$. The relative motion of the three species is again inhibited by the friction due to mutual collisions.

Consider first the equation of motion of the electrons. At each scattering, the low-mass electrons are again well described as having their velocity relative to the neutrals completely randomized, so there is a new drag term $-nm_e(\mathbf{V} + f^{-1}\mathbf{V}_i)/\tau_e$, where τ_e is the mean time between successive collisions of an electron with neutral atoms. Because of the near charge-neutrality, the total current due to motion of both electrons and ions is again $\mathbf{j} = -ne\mathbf{V}$. It is convenient to define also the current

$$\mathbf{j}_i = n_i Ze\mathbf{V}_i = ne\mathbf{V}_i \tag{2.119}$$

due to the motion of the ions relative to the bulk, so that the corresponding electron current is $\mathbf{j} - \mathbf{j}_i$. The equation of motion of the electrons is then

$$-ne\left(\mathbf{E} + \frac{\mathbf{v}\times\mathbf{B}}{c}\right) - \nabla p_e + \frac{(\mathbf{j}-\mathbf{j}_i)\times\mathbf{B}}{c} + \frac{\kappa B\mathbf{j}}{c} + \frac{\kappa_e B}{c}(\mathbf{j} - F^{-1}\mathbf{j}_i) = 0 \tag{2.120}$$

where

$$\kappa^{-1} = \omega\tau, \; \kappa_e^{-1} = \omega\tau_e, \; \omega = eB/m_e c. \tag{2.121}$$

(Note that κB, $\kappa_e B$ are independent of B.) As for the two-fluid model, it is again convenient to use the joint equation of motion of the electrons and ions. The inertial term is taken to be $\rho_i \, \mathrm{d}\mathbf{v}_i/\mathrm{d}t$, i.e. it is assumed that the velocity $\mathbf{V}_i$ of the ions relative to the bulk is small. This will normally be a good approximation: when it does fail, the ion density will be exceptionally low and the inertial term is then in any case negligible (cf. Chapter 11). At each collision of an ion with a neutral atom, the fraction $m_n/(m_n + m_i)$ of its momentum $m_i\mathbf{V}_i/F$ relative to the ion is lost. Thus if we define τ_i to be $m_n/(m_n + m_i)$ times the ion–neutral collision interval, then by (2.119) the drag force per cm^3 on the ions may be written

$$-\kappa_i F^{-1} B \mathbf{j}_i / c \tag{2.122}$$

with

$$\kappa_i^{-1} = \omega_i \tau_i, \;\; \omega_i = ZeB/m_i c. \tag{2.123}$$

Hence since $\rho_i = (1 - F)\rho$, the joint ion–electron equation of motion is

$$(1-F)\rho\frac{\mathrm{d}\mathbf{v}}{\mathrm{d}t} = -\nabla(p_i + p_e) + \frac{\mathbf{j}\times\mathbf{B}}{c} + (1-F)\rho\nabla V + \frac{\kappa_e B}{c}\mathbf{j} - \frac{(\kappa_e + \kappa_i)B}{c}F^{-1}\mathbf{j}_i, \tag{2.124}$$

while the equation of motion of the gas as a whole is

$$\rho\frac{\mathrm{d}\mathbf{v}}{\mathrm{d}t} = -\nabla p + \rho\nabla V + \frac{\mathbf{j}\times\mathbf{B}}{c}. \tag{2.125}$$

As the results of greatest interest do not involve the pressures, for simplicity we write $(p_i + p_e) = (1 - f)p$, where f is the fractional contribution (assumed constant) of the neutral gas to the total pressure. Elimination of $\rho\,\mathrm{d}\mathbf{v}/\mathrm{d}t$ between (2.124) and (2.125) then yields an equation for $\mathbf{j}_i$:

$$(\kappa_e + \kappa_i)\frac{F^{-1}B}{c}\mathbf{j}_i = -(F - f)\nabla p + \frac{\kappa_e B}{c}\mathbf{j} + F\frac{\mathbf{j}\times\mathbf{B}}{c}. \tag{2.126}$$

Substitution into (2.120) then yields the analogue of (2.58) and (2.59):

$$\begin{aligned}\mathbf{E}' = \mathbf{E} + \frac{\mathbf{v}\times\mathbf{B}}{c} + \nabla p_e &= \frac{(\kappa_e + \kappa)B}{nec}\mathbf{j} + \frac{\mathbf{j}\times\mathbf{B}}{nec}\\ &+ \frac{F^2}{neB\kappa_i}\left((1 - fF^{-1})\nabla p - \frac{\mathbf{j}\times\mathbf{B}}{c}\right)\times\mathbf{B},\end{aligned} \tag{2.127}$$

where terms of order $(\kappa_e/\kappa_i) \simeq (m_e/m_i)^{1/2}$ have been dropped.

All the terms in (2.127) other than the last two are familiar. On the left is the 'equivalent electric field' (2.59). On the right, the second term is identical with the Hall term for a fully ionized gas. The first term on the right can be written as $\mathbf{j}/\sigma$, with $\sigma = (ne^2/m_e)(1/\tau + 1/\tau_e)$: it is just the ordinary Ohmic field but including electron collisions with neutrals as well as with ions. Because the Coulomb scattering cross-section is so much larger than that for electron scattering by a neutral particle (which depends on the induced polarization of the neutral particle), scattering by ions continues to dominate until $n_n/n \gg 1$ by several powers of 10. Equation (2.127) shows that for currents parallel to $\mathbf{B}$, the only change is this modification to the Ohmic term, which in any case remains small.

For currents perpendicular to $\mathbf{B}$, it is the extra term in (2.127) which dominates. In the most interesting applications the ∇p term is the smaller (in the gas cloud case it is quite negligible), and so from now on it is dropped. Recall that these extra terms arise after substitution for $\mathbf{j}_i$ from (2.126) into (2.120): in particular, the Lorentz force drives the ambipolar drift velocity $\mathbf{V}_i$ of the electrons and ions jointly relative to the neutrals, and there is a consequent dissipation due to the drag. From (2.126),

$$\mathbf{j}_i \approx \frac{f^2 c}{B\kappa_i}\left(\frac{\mathbf{j}\times\mathbf{B}}{c}\right), \tag{2.128}$$

showing that $j_i/j \gg 1$ when $\kappa_i \ll 1$ – again for 'free spiralling between collisions'. The dissipation per cm^3 s due to this dominant friction term is as usual the scalar product with the relative velocity:

$$\begin{aligned}(-\kappa_i F^{-1}B\mathbf{j}_i/c)\cdot(-\mathbf{V}_i) &= \frac{j_i^2\kappa_i B}{necF} = \frac{F^3 c}{ne\kappa_i B}\left(\mathbf{j}\times\mathbf{B}/c\right)^2 \\ &= \frac{F^3 cB}{ne\kappa_i c^2}j_\perp^2 = \frac{Z\tau_i B^2}{nm_i c^2}j_\perp^2\end{aligned} \tag{2.129}$$

when $F \simeq 1$.

Equivalently, one can follow Section 2.3.2 and write (2.127) for perpendicular currents $\mathbf{j}_\perp$ as

$$\mathbf{E}' = \mathbf{E} + \frac{\mathbf{v}\times\mathbf{B}}{c} \simeq \mathbf{j}_\perp\left(\frac{1}{\sigma} + \frac{F^3 Z\tau_i B^2}{nm_i c^2}\right) + \frac{\mathbf{j}\times\mathbf{B}}{cne} \tag{2.130}$$

The crucial difference from the fully ionized case, described by (2.58), is that whereas for parallel currents $\mathbf{E}' = \mathbf{j}_\parallel/\sigma$, for perpendicular currents the coefficient of $\mathbf{j}_\perp$ in (2.130) is dominated by the much larger second term. One can then treat this as in Section 2.3.2, first defining σ_1 and σ_2 by (2.61) and then σ_3 by (2.64). Again $\sigma_3 = \sigma_1 + \sigma_2^2/\sigma_1$, but now $1/\sigma_3$ is not $1/\sigma$ but essentially $F^3 Z\tau_i B^2/nm_i c^2$, yielding a much increased effective resistivity for perpendicular currents, due to ambipolar diffusion, driven by the Lorentz force.

2.8 'Anomalous' resistivity

The large length-scales of cosmical problems lead naturally to the $\mathrm{R_m} \to \infty$ limit as the zero-order approximation, with consequent freezing of the field into the flow. This can lead to strongly distorted field structures, including regions where field gradients are large and so the maintaining currents are concentrated locally into sheets. The question then arises whether the equations that have led to these structures will remain valid. There is the likelihood that instabilities in physical space will develop, yielding an effective macroscopic resistivity (cf. Sections 3.8, 7.4 and 14.5). Here we consider briefly the possible development of instability in momentum space. If a locally enhanced flow of electrons relative to ions leads to growing collective electrostatic modes, the randomization of the electron drift velocity can be much more effective than through binary collisions, yielding a very much higher resistivity (Buneman 1959).

We study the simplest model of the 'two-stream instability' (e.g. Thompson 1962; Wesson 1974; Sturrock 1994). An infinite cold plasma is assumed, uniform in the y and z directions. The zero-order state consists of a density n of electrons with charge $-e$ and of ions of charge e, each species moving in the x-direction with speeds v_e, v_i respectively. The uniform electric potential ϕ is normalized to zero. In the perturbed state, $n_e = n + n'_e$, etc., with all primed quantities having the factor $\exp[\mathrm{i}(kx - \omega t)]$. From the Poisson–Maxwell equation,

$$k^2\phi' = 4\pi e(n'_i - n'_e). \tag{2.131}$$

The equations of motion and of continuity for the electrons

$$m_e\left(\frac{\partial v_e}{\partial t} + v_e\frac{\partial v_e}{\partial x}\right) = e\frac{\partial \phi}{\partial x}, \tag{2.132}$$

$$\frac{\partial n_e}{\partial t} + \frac{\partial}{\partial x}(n_e v_e) = 0 \tag{2.133}$$

yield the linearized forms

$$m_e(-\omega + kv_e)v'_e = ek\phi', \tag{2.134}$$

$$(-\omega + kv_e)n'_e = -nkv'_e, \tag{2.135}$$

and similarly for the ions, with m_i, e replacing m_e, $-e$ respectively. For a non-vanishing solution to exist, the dispersion relation

$$G(\omega) \equiv -\frac{1}{\omega_p^2} + \frac{\mu}{(\omega - kv_i)^2} + \frac{1}{(\omega - kv_e^2)^2} = 0 \tag{2.136}$$

must hold, where $\omega_p^2 = 4\pi n_e e^2/m_e$, and $\mu = m_e/m_i \ll 1$.

Figure 2.4 illustrates the behaviour of $G(\omega)$. The wave of number k is stable provided the four roots of (2.136) are all real, which requires that the minimum of $G(\omega)$ is less than zero. This minimum occurs at

$$\omega = k(v_i + \mu^{1/3}v_e)/(1 + \mu^{1/3}), \tag{2.137}$$

yielding the stability criterion for the mode k

$$k^2v^2 > (1 + \mu^{1/3})\omega_p^2, \qquad V = (v_i - v_e). \tag{2.138}$$

If $V \neq 0$, this will be violated for sufficiently small k: in an infinite system, there are wavelengths $\propto 1/k$, sufficiently long to ensure instability. Surprisingly, the criterion yields instability for sufficiently *small* V, even though when $V = 0$ the plasma is stable; however, the growth rate – given by the imaginary part of ω – does go to zero as $kV \to 0$ (Fig. 2.5).

In reality, each component of the the plasma will have a thermal spread in velocity. As we are interested in a collective instability, with a growth time that is

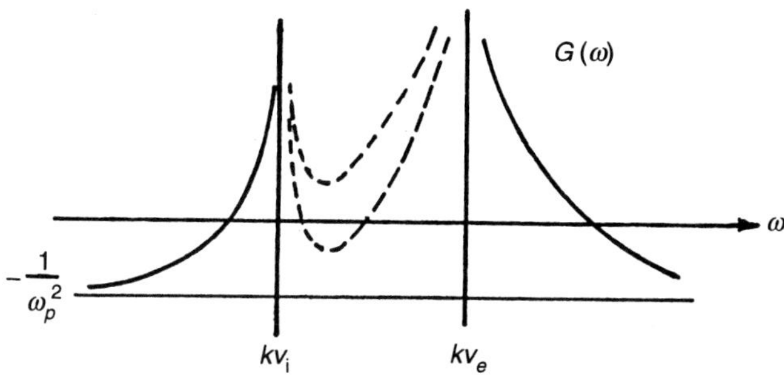

FIG. 2.4. Plot of $G(\omega) \equiv [1/(\omega - kv_e)^2 + \mu/(\omega - kv_i)^2 - 1/\omega_p^2]$ against ω.

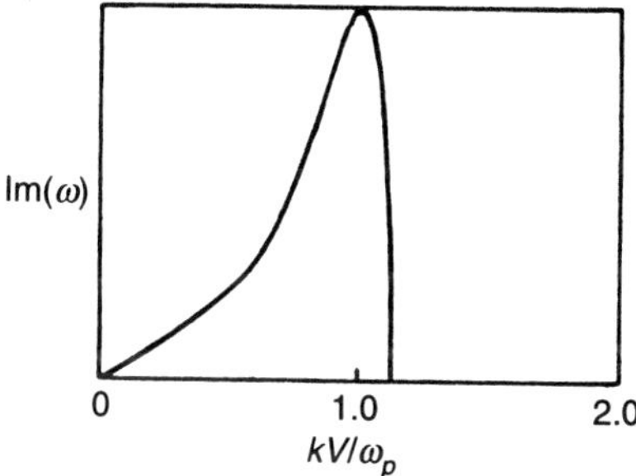

FIG. 2.5. Growth rate against reduced wave number (kV/ω_p) for the case $\mu = 1/1836$.

much shorter than the collision time (2.26), the problem should then be studied with the distribution functions f for the species $(q,\, m)$ satisfying the *collisionless* Boltzmann or Vlasov equation

$$\frac{\partial f}{\partial t} + v\frac{\partial f}{\partial x} + \frac{q}{m}E\frac{\partial f}{\partial v}, = 0. \tag{2.139}$$

and with the charge density appearing in Poisson's equation given by

$$\rho_e = \Sigma q \int f \mathrm{d}v. \tag{2.140}$$

Consider the simplest example of a gas with one, singly charged species of ion, and with an electron current streaming with velocity v_e as seen in the frame in which the ions are at rest. A simple distribution function that includes both a streaming velocity and a thermal spread is

$$f = \frac{1}{\pi} \frac{nv_T}{(v - v_e)^2 + (v_T)_e^2}. \tag{2.141}$$

It can then be shown (Wesson 1974) that (2.136) is replaced by

$$\frac{1}{\omega_p^2} = \frac{\mu}{(\omega + \mathrm{i}k(v_T)_i)^2} + \frac{1}{(\omega - kv_e + \mathrm{i}k(v_T)_e)^2}. \tag{2.142}$$

If the relative drift velocity v_e is sufficiently high, we recover essentially the two-stream instability. At lower v_e, when $1/\omega_p^2$ is negligible, the dispersion relation (2.142) becomes

$$\omega = \frac{k}{1+\mu}[\pm\mu^{1/2}((v_T)_e + (v_T)_i) + \mu v_e + \mathrm{i}(\pm\mu^{1/2}v_e - \mu(v_T)_e - (v_T)_i)]. \tag{2.143}$$

Instability occurs with the upper signs chosen and when

$$v_e > \mu^{1/2}(v_T)_e + \mu^{-1/2}(v_T)_i. \tag{2.144}$$

As $\mu \ll 1$, normally $(v_T)_i \ll (v_T)_e$; and if also $\mu^{1/2}|v_e| \ll (v_T)_e$, then from (2.143)

$$\mathrm{Re}(\omega) = k\mu^{1/2}(v_T)_e = kc_s \tag{2.145}$$

where c_s is the ion sound speed. For equal electron and ion temperatures, $(v_T)_i = \mu^{1/2}(v_T)_e$, and the instability criterion (2.144) becomes

$$v_e > (v_T)_e. \tag{2.146}$$

The overall conclusion is that instability occurs if the electron streaming velocity overtakes a suitably defined thermal speed. It may be that a system that finds the field-freezing condition too constraining may have the same total current break up into a filamentary structure, with v_e locally superthermal. The collective randomization of the energy of the current stream clearly corresponds to a greatly enhanced resistivity.

References

Alfvén, H. (1942). *Arkiv Matematik, Astronomi och Fysik*, **29B**, 2.

Alfvén, H. (1950). *Cosmical electrodynamics.* Clarendon Press, Oxford.

Buneman, O. (1959). *Physical Review*, **115**, 503.

Chandrasekhar, S. (1957). *Proceedings National Academy Sciences*, **43**, (1), 24.

Chapman, S. and Cowling, T.G. (1970). *The mathematical theory of non-uniform gases* (3rd edn). Cambridge University Press.

Chew, G.F., Goldberger, M.L. and Low, F.E. (1956). *Proceedings Royal Society* A, **236**, 112.

Cowling, T.G. (1953). *Solar electrodynamics.* In *The Sun* (ed. G.P. Kuiper), p. 532. Chicago University Press.

Cowling, T.G. (1956). *Monthly Notices Royal Astronomical Society*, **116**, 114.
Cowling, T.G. (1976). *Magnetohydrodynamics.* Adam Hilger, London. (Interscience edition 1957.)
Lüst, R. and Schlüter, A. (1955). *Zeitschrift Astrophysik*, **38**, 190.
Mestel, L. (1968). *Monthly Notices Royal Astronomical Society*, **140**, 177.
Mestel, L. and Spitzer Jr., L. (1956). *Monthly Notices Royal Astronomical Society*, **116**, 503.
Piddington, J.H. (1955). *Monthly Notices Royal Astronomical Society*, **114**, 638 and 651.
Priest, E.R. (1982). *Solar magnetohydrodynamics.* Reidel, Dordrecht.
Roberts, P.H. (1967). *An introduction to magnetohydrodynamics.* Longmans, London.
Schlüter, A. (1950). *Zeitschrift Naturforschung*, **5A**, 72.
Schlüter, A. (1951). *Zeitschrift Naturforschung*, **6A**, 73.
Schlüter, A. (1952). *Annalen Physik*, **6-10**, 422.
Schlüter, A. (1958). In *Electromagnetic phenomena in cosmical physics* (ed. B. Lehnert), p. 71. Cambridge University Press.
Schlüter, A. (1961). Lectures on plasma physics, Cambridge.
Schlüter, A. and Biermann, L. (1950). *Zeitschrift Naturforschung*, **5A**, 237.
Spitzer Jr., L. (1952). *Astrophysical Journal*, **116**, 299.
Spitzer Jr., L. (1956, 1962). *Physics of fully ionized gases.* Interscience, New York.
Sturrock, P.A. (1994). *Plasma physics.* Cambridge University Press.
Thompson, W.B. (1962). *An introduction to plasma physics.* Pergamon, Oxford.
Walén, C. (1944). *Arkiv Matematik, Astronomi och Fysik*, **30A**, No. 15; **31B**, No. 3.
Wesson, J.A. (1974). *Plasma stability theory*, in *Plasma physics* (ed. B.E. Keen), p. 93. Conference Series Number 20, Institute of Physics, London and Bristol.

3

APPLICATIONS

In this chapter, some of the topics introduced in Chapter 2 are discussed in greater detail.

3.1 Magnetosonic waves

The theory of hydromagnetic waves is now generalized to include the effects of compressibility. Again, a zero-order state is assumed with a perfectly conducting, inviscid medium of uniform density ρ, permeated by a uniform field $\mathbf{B}_0 \equiv B_0\mathbf{b}$, and subject to no non-magnetic body forces. For a non-uniform medium the results will be a tolerable approximation for disturbances of wavelength small compared with the scale of variation of ρ and $\mathbf{B}_0$. From (2.38) and (2.88), an infinitesimal displacement $\xi(\mathbf{r})$ yields perturbations

$$\rho' = -\rho\nabla\cdot\xi, \quad \mathbf{B}' = \nabla\times(\xi\times\mathbf{B}_0), \tag{3.1}$$

and the equation of motion (2.35) becomes

$$\ddot{\xi} = \frac{\partial\mathbf{v}}{\partial t} = -\frac{1}{\rho}\nabla p' + \frac{(\nabla\times\mathbf{B}')\times\mathbf{B}_0}{4\pi\rho} \tag{3.2}$$

The 'quasi-adiabatic' assumption $p = p(\rho)$ is made, with the functional form fixed by the particular problem under study. Then

$$p' = \left(\frac{\mathrm{d}p}{\mathrm{d}\rho}\rho'\right) \equiv c_s^2\rho' \tag{3.3}$$

with c_s the sound speed, and from (3.2),

$$\ddot{\xi} = c_s^2\nabla(\nabla\cdot\xi) + v_\mathrm{A}^2\left(\nabla\times\{\nabla\times(\zeta\times\mathbf{b})\}\right)\times\mathbf{b} \tag{3.4}$$

with again $v_\mathrm{A} = B_0/(4\pi\rho)^{1/2}$. For plane waves, $\xi \propto \exp\mathrm{i}(\mathbf{k}\cdot\mathbf{r}-\omega t)$, and so the gradient operator may be replaced by $\mathrm{i}\mathbf{k}$. From (3.1), $\mathbf{k}\cdot\mathbf{B}' = 0$, so the magnetic perturbation is normal to the direction of propagation. Equation (3.4) yields the dispersion relation

$$\omega^2\xi = [(c_s^2 + v_\mathrm{A}^2)\xi\cdot\mathbf{k} - v_\mathrm{A}^2(\mathbf{k}\cdot\mathbf{b})(\xi\cdot\mathbf{b})]\mathbf{k} + v_\mathrm{A}^2[(\mathbf{k}\cdot\mathbf{b})^2\xi - (\mathbf{k}\cdot\mathbf{b})(\xi\cdot\mathbf{k})\mathbf{b}]. \tag{3.5}$$

We look first for incompressible modes, with $0 = \nabla\cdot\xi \propto \mathbf{k}\cdot\xi$. The scalar product of (3.5) with $\mathbf{k}$ yields $\xi\cdot\mathbf{b} = 0$ – displacements normal to $\mathbf{B}_0$ – and

$$\omega = \pm v_{\rm A}(\mathbf{k}\cdot\mathbf{b}) = \pm(v_{\rm A}\cos\theta)k \tag{3.6}$$

where θ is the angle between $\mathbf{B}_0$ and $\mathbf{k}$. This mode, with phase velocity $\omega/k = v_{\rm A}\cos\theta$, corresponds to the Alfvén wave in incompressible media, discussed in Section 2.6. Equivalently, $\omega = v_{\rm A}k_Z$ where $k_Z = k\cos\theta$ and Z is the Cartesian axis along $\mathbf{b}$. As usual, the group velocity $\mathbf{v}_g = (\partial\omega/\partial k_X, \partial\omega/\partial k_Y, \partial\omega/\partial k_Z) = v_{\rm A}\mathbf{b}$: energy is carried at the rate $v_{\rm A}$ along $\mathbf{B}_0$, even though the individual waves can travel at any inclination to the field, but with the speed $v_{\rm A}\cos\theta$ that decreases to zero as $\theta \to \pi/2$. Returning to (3.1): for incompressible modes,

$$\mathbf{B}' = B_0(\mathbf{k}\cdot\mathbf{b})\xi, \tag{3.7}$$

so that

$$\mathbf{v}' = \dot{\xi} = \pm\mathbf{B}'/(4\pi\rho)^{1/2} \tag{3.8}$$

from (3.6) and (3.7). Hence $\mathbf{k}\cdot\mathbf{B}' \propto \mathbf{k}\cdot\xi = 0$ – both $\mathbf{B}'$ and ξ lie in a plane normal to the wave vector and so parallel to the wave-front; and since by (3.7), $\mathbf{B}'\cdot\mathbf{b} \propto \xi\cdot\mathbf{b} = 0$, $\mathbf{B}'$ is normal to $\mathbf{B}_0$.

The other solutions are found by taking the scalar products of (3.5) with $\mathbf{k}$ and $\mathbf{b}$ and eliminating $\xi\cdot\mathbf{b}$. There are two compressional modes, the fast and slow magnetosonic waves, with phase velocities given by

$$(v_f^2, v_s^2) = \frac{1}{2}\left((c_s^2 + v_{\rm A}^2) \pm [(c_s^2 + v_{\rm A}^2)^2 - 4c_s^2 v_{\rm A}^2\cos^2\theta]^{1/2}\right). \tag{3.9}$$

When $\theta = 0$ or $\pi/2$ the wave is purely longitudinal, but for intermediate directions of propagation the waves are neither longitudinal nor transverse. The polar diagrams (Fig. 3.1) are for the phase and group velocities. Note that if $c_s \ll v_{\rm A}$, then the fast wave travels nearly isotropically with $v \simeq v_g \simeq v_{\rm A}$, whereas the slow wave has the phase velocity $\simeq c_s\cos\theta$ and so the group velocity $\simeq c_s\mathbf{b}$. Thus energy can be transmitted isotropically with velocity $\simeq v_{\rm A}$ by the fast compressional wave, or anisotropically, strictly along the field at speed $v_{\rm A}$ by the torsional Alfvén wave, or nearly so but at speed c_s by the slow compressional wave.

It is instructive to return to the unapproximated forms (2.35) and (2.88) and look for (inviscid) *finite amplitude* Alfvén wave solutions of the form

$$\mathbf{v}_1 = \pm\frac{\mathbf{B}_1}{(4\pi\rho)^{1/2}}, \qquad \nabla\cdot\mathbf{v}_1 = 0, \tag{3.10}$$

superposed again on a medium with uniform ρ and $\mathbf{B}_0$. The terms $\mathbf{v}_1\times(\nabla\times\mathbf{v}_1) + [(\nabla\times\mathbf{B}_1)\times(\mathbf{B}_0+\mathbf{B}_1)]/(4\pi\rho)$ then leave just the linear term $(\nabla\times\mathbf{B}_1)\times\mathbf{B}_0/(4\pi\rho)$. The remaining non-linear term is $\nabla(v_1^2/2) = \nabla(B_1^2/8\pi\rho)$. In a single plane-polarized wave, B_1^2 fluctuates, causing pressure and so also density changes: the divergence-free condition (3.10) is valid only in the linear approximation. However, the superposition of two plane-polarized components, $\pi/2$ out of phase

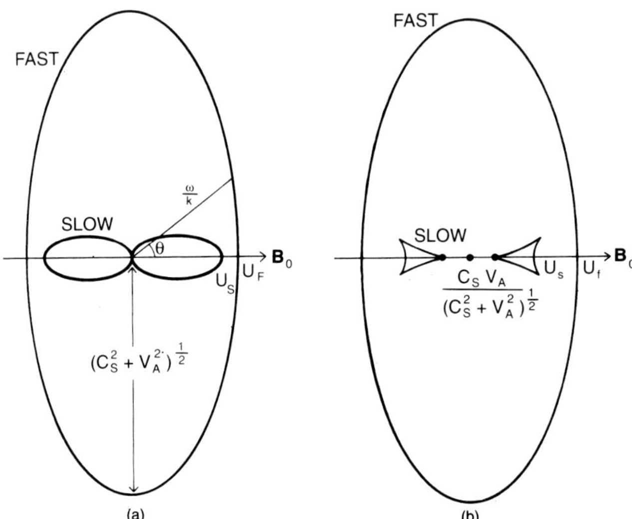

FIG. 3.1. Polar diagrams for fast and slow magnetoacoustic waves, propagating at an angle θ to the unperturbed field $\mathbf{B}_0$. The speeds u_s, u_f are the slower and faster, respectively, of the Alfvén speed v_A and the sound speed c_s: (a) shows the phase velocities, (b) the group velocities. (After Priest 1982.)

and of equal amplitude yields a circularly polarized wave with B_1^2 constant, so that the solution is exact for finite disturbances.

3.2 Magnetohydrodynamic shocks

Just as for sound waves in non-magnetic media, finite-amplitude compressional waves will spontaneously steepen to form shock waves. The theory has been developed by a number of authors (e.g. Jeffrey and Taniuti 1964). The following summary of the essentials follows closely that given in Priest (1982).

(a) Normal shocks

We consider two perfectly conducting media, separated by a plane shock front. The simplest generalization of the familiar hydrodynamic shock has the magnetic field parallel to the shock front in both domains and perpendicular to the velocity of both gas and shock (Fig. 3.2). It is convenient to use a reference frame stationary in the shock frame, and with suffices 1, 2 referring respectively to the unshocked and the shocked gas. Axes are chosen with the flow along the x-direction and the field along the y. From mass conservation across the shock,

$$\rho_1 v_1 = \rho_2 v_2. \tag{3.11}$$

The momentum equation balances the change across the shock of the momentum transport $(\rho v)v$ per $\mathrm{cm}^2\,\mathrm{s}$ against the joint action of the thermal pressure and

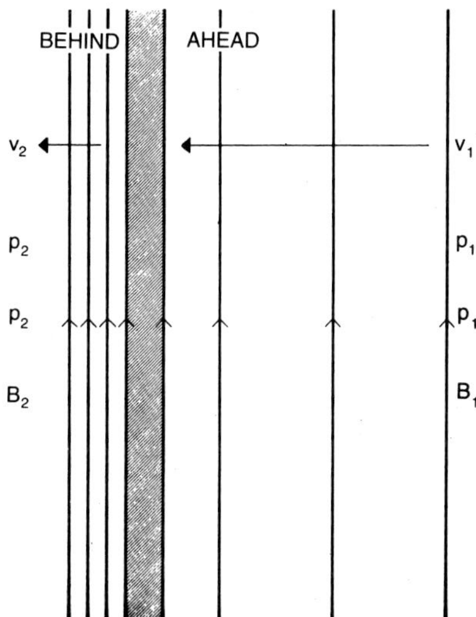

FIG. 3.2. The plane perpendicular shock, as seen in the shock frame.

the Maxwell stresses $(p\delta_{ij} + T_{ij})n_j = (p + B^2/8\pi)n_i$, since $B_j n_j = 0$ (cf. 2.42). Thus

$$p_1 + \frac{B_1^2}{8\pi} + \rho_1 v_1^2 = p_2 + \frac{B_2^2}{8\pi} + \rho_2 v_2^2. \tag{3.12}$$

The same total pressure does work $(p + B^2/8\pi)v$ per second on the gas crossing unit area with velocity v, while the energy transport per second is $(\rho u + \frac{1}{2}\rho v^2 + B^2/8\pi)v$, with the internal energy $u = p/(\gamma - 1)\rho$ per gram (cf. 2.53). The energy equation is thus

$$\left(p_1 + \frac{B_1^2}{8\pi}\right) v_1 + \left(\rho_1 u_1 + \frac{1}{2}\rho_1 v_1^2 + \frac{B_1^2}{8\pi}\right) v_1$$
$$= \left(p_2 + \frac{B_2^2}{8\pi}\right) v_2 + \left(\rho_2 u_2 + \frac{1}{2}\rho_2 v_2^2 + \frac{B_2^2}{8\pi}\right) v_2. \tag{3.13}$$

The last equation is the continuity of flux transport:

$$v_1 B_1 = v_2 B_2. \tag{3.14}$$

On either side of the shock, the perfect conductivity condition $\mathbf{E} = -\mathbf{v} \times \mathbf{B}/c = -(vB/c)\hat{\mathbf{y}}$. Thus $\mathbf{E}$ is also parallel to the shock, and so from the integral of the Faraday–Neumann equation (2.13) across an infinitely thin discontinuity, $\mathbf{E}$ is continuous and (3.14) follows. The Poynting flux $(c/4\pi)\mathbf{E} \times \mathbf{B}$ across unit area is $(vB^2/4\pi)\hat{\mathbf{x}}$, consistent with the sum of the two magnetic terms on each side of (3.13).

In terms of the density ratio $D \equiv \rho_2/\rho_1$, and the Mach number $M_1 = v_1/c_{s1}$ and the plasma $\beta_1 = p_1/(B_1^2/8\pi) = (2/\gamma)(c_s^2/v_A^2)_1$, both for the unshocked plasma, (3.11)–(3.14) reduce to

$$v_2/v_1 = D^{-1}, \tag{3.15}$$
$$B_2/B_1 = D, \tag{3.16}$$
$$p_2/p_1 = \gamma M_1^2(1 - D^{-1}) + \beta_1^{-1}(1 - D^2), \tag{3.17}$$

with D the positive root (unique, since $1 < \gamma < 2$) of

$$f(D) \equiv 2(2-\gamma)D^2 + (2\beta_1 + (\gamma-1)\beta_1 M_1^2 + 2)\gamma D - \gamma(\gamma+1)\beta_1 M_1^2 = 0. \tag{3.18}$$

The result $D = (\gamma+1)M_1^2/2 + (\gamma-1)M_1^2$ is recovered as $\beta_1 \to \infty$. The requirement that entropy increase across the shock restricts D to be greater than 1. In the range $1 \le D \le \infty$, $f(D)$ is monotonic increasing with D; hence for given parameters β_1, M_1, a solution $D > 1$ exists provided $f(1) \le 0$, implying $M_1^2 \ge 1 + (2/\gamma\beta_1)$, or

$$v_1^2 > c_{s1}^2 + v_{\mathrm{A}1}^2 \tag{3.19}$$

– ahead of the shock, the gas speed must exceed the fast magnetosonic speed given in (3.9), which therefore (unsurprisingly) plays the same role as the sound speed in standard gas dynamics. In the limit $M_1 \to \infty, D \to (\gamma+1)/(\gamma-1)$, as for a non-magnetic shock, and so the increase of B is limited by

$$1 < \frac{B_2}{B_1} < \frac{\gamma+1}{\gamma-1}. \tag{3.20}$$

(b) Oblique shocks

In general, the magnetic field contains components both parallel to and normal to the shock front. As before, axes are chosen moving with the shock; the axis Oy lies in the shock, with the magnetic and velocity vectors ahead of the shock in the plane Oxy, the jump conditions then ensuring the same holds for $\mathbf{B}$ and $\mathbf{v}$ behind the shock. Further simplification results if the origin of coordinates moves along the y-axis at the speed chosen so that (cf. Fig. 3.3)

$$v_{1y} = v_{1x}\frac{B_{1y}}{B_{1x}}. \tag{3.21}$$

(This transformation is clearly not possible for the perpendicular shock case with $B_{1x} = 0$). The choice (3.21) and the condition on $\mathbf{E}$ at the shock then ensure that $\mathbf{E} = -\mathbf{v} \times \mathbf{B}/c = 0$ and $\mathbf{v}$ is parallel to $\mathbf{B}$ behind as well as ahead of the shock. The Poynting vector $(c/4\pi)\mathbf{E} \times \mathbf{B}$ is now zero on both sides, so that in this frame the flow of magnetic energy is zero.

The condition $\nabla \cdot \mathbf{B} = 0$ requires

$$B_{1x} = B_{2x}, \tag{3.22}$$

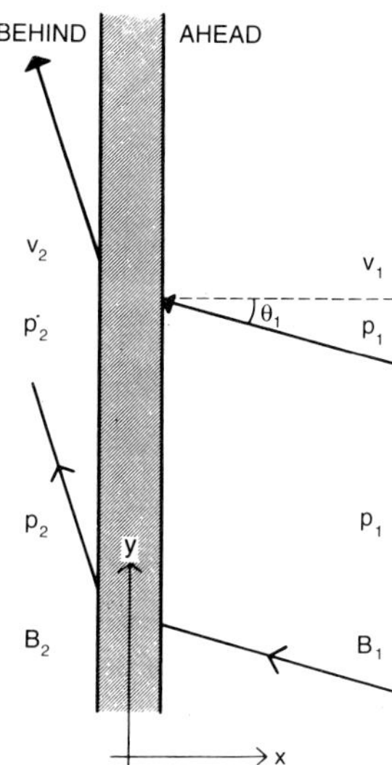

FIG. 3.3. The oblique shock wave, seen in the frame moving with the shock, and with the velocity along the shock chosen to make the plasma velocity **v** parallel to **B** both ahead of and behind the shock.

and mass conservation

$$\rho_1 v_{1x} = \rho_2 v_{2x}. \tag{3.23}$$

Momentum conservation – continuity of $(p\delta_{ij} + T_{ij} + \rho v_i v_j)n_j$ – now yields the two equations

$$p_1 + \frac{B_1^2}{8\pi} - \frac{B_{1x}^2}{4\pi} + \rho_1 v_{1x}^2 = p_2 + \frac{B_2^2}{8\pi} - \frac{B_{2x}^2}{4\pi} + \rho_2 v_{2x}^2, \tag{3.24}$$

$$\rho_1 v_{1x} v_{1y} - \frac{B_{1x}B_{1y}}{4\pi} = \rho_2 v_{2x} v_{2y} - \frac{B_{2x}B_{2y}}{4\pi}. \tag{3.25}$$

The vanishing of the Poynting flux in this frame ensures that no magnetic terms enter explicitly the energy equation:

$$(\gamma\rho_1 u_1)v_{1x} + \frac{1}{2}\rho_1(v_{1x}^2 + v_{1y}^2) = (\gamma\rho_2 u_2)v_{2x} = \frac{1}{2}(v_{2x}^2 + v_{2y}^2)v_{2x}, \tag{3.26}$$

or by use of (3.23),

$$\frac{\gamma p_1}{(\gamma - 1)\rho_1} + \frac{1}{2}v_1^2 = \frac{\gamma p_2}{(\gamma - 1)\rho_2}. \tag{3.27}$$

Again in terms of $D = \rho_2/\rho_1$, $c_{s1} = (\gamma p_1/\rho_1)^{1/2}$ and $v_{A1} = B_1/(4\pi\rho_1)^{1/2}$, (3.21), (3.23), (3.24) and (3.27) reduce to

$$\frac{v_{2x}}{v_{1x}} = D^{-1} \tag{3.28}$$

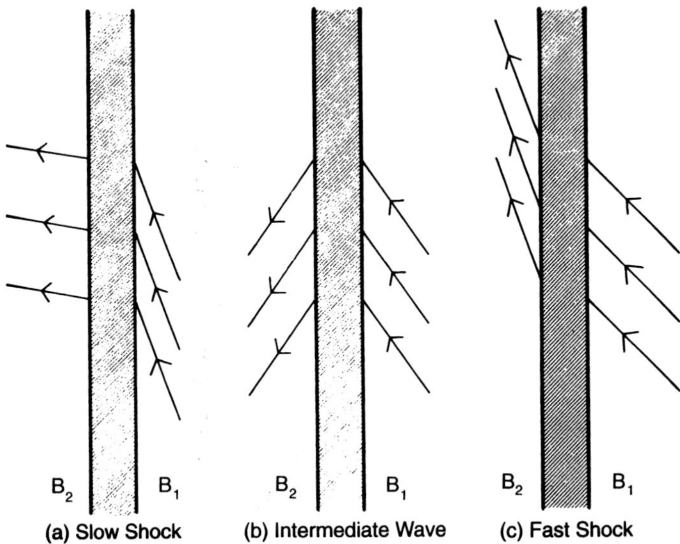

FIG. 3.4. The changes in the direction of **B** in the (a) slow, (b) Alfvén and (c) fast shocks.

$$\frac{v_{2y}}{v_{1y}} = \frac{v_1^2 - v_{A1}^2}{v_1^2 - Dv_{A1}^2}, \tag{3.29}$$

$$\frac{B_{2y}}{B_{1y}} = \frac{(v_1^2 - v_{A1}^2)D}{v_1^2 - Dv_{A1}^2}, \tag{3.30}$$

and

$$\frac{p_2}{p_1} = D + \frac{(\gamma-1)Dv_1^2}{2c_{s1}^2}\left(1 - \frac{v_2^2}{v_1^2}\right), \tag{3.31}$$

where X is a solution of

$$(v_1^2 - Dv_{A1}^2)^2 \left[Dc_{s1}^2 + \frac{1}{2}v_1^2\cos^2\theta(D(\gamma-1) - (\gamma+1))\right] \tag{3.32}$$
$$+ \frac{1}{2}v_{A1}^2 v_1^2 \sin^2\theta D\left[(\gamma + D(2-\gamma))v_1^2 - Dv_{A1}^2((\gamma+1) - D(\gamma-1))\right] = 0$$

with $v_{1x} = v_1\cos\theta$ (cf. Fig. 3.4).

There are three solutions of (3.32): a slow shock, an Alfvén wave, and a fast shock. In the infinitesimal limit $D \to 1$, these reduce to the three wave solutions of Section 3.1, with speeds v_A – the Alfvén wave – and the slow and fast wave speeds of (3.9).

For slow and fast shocks, the entropy increase condition $D = \rho_2/\rho_1 > 1$ applied to (3.31) yields $p_2/p_1 > 1$. There is in addition the *evolutionary condition* (Jeffrey and Taniuti 1964), which ensures that the perturbation caused by a small disturbance incident on a shock front is both small and unique. The most important consequence of this is to forbid a reversal of sign of the tangential field component across the shock. Thus $B_{2y}/B_{1y} > 0$ so that by (3.29), in a *slow shock*,

$$v_1^2 < v_{A1}^2 \ (< Dv_{A1}^2) \tag{3.33}$$

with $B_2 < B_1$ – the magnetic field is refracted towards the shock normal and so its strength decreases across the shock. By contrast, in a *fast shock*,

$$v_1^2 > Dv_{A1}^2 \ (> v_{A2}^2) \tag{3.34}$$

with $B_2 > B_1$ – the field is refracted away from the normal and the field strength increases. From (3.28), the flow speed is always decelerated in the direction normal to the shock. From (3.29) and (3.33), for a slow shock, $v_{2y} < v_{1y}$ – deceleration parallel to the shock, whereas from (3.34), for a fast shock, $v_{2y} > v_{1y}$ – acceleration parallel to the shock. From the evolutionary condition, for a (slow, fast) shock, v_{1x} is greater than the (slow, fast) magnetosonic speed of the shock ahead of the shock and is less than the (slow, fast) speed behind the shock, as one would expect.

When $v_1 = v_{A1}$, but $D > 1$, then by (3.30), $B_{2y} = 0$ – a 'switch-off shock'. Since $\mathbf{v}_1$ and $\mathbf{B}_1$ are parallel, the switch-off condition is equivalent to

$$v_{1x} = B_{1x}/(4\pi\rho_1)^{1/2} \tag{3.35}$$

– the inflow speed equals the Alfvén speed based on the normal field component. When $v_1 > v_{A1} > c_{s1}$, there exist also 'switch-on shocks', for which $B_{1y} = 0$, but $B_{2y} > 0$. When $v_1 = v_{A1}$, but $D = 1$, there exists the non-trivial solution with $B_{2y} = -B_{1y}$, $v_{2y} = -v_{1y}$, $B_{2x} = B_{1x}$, $v_{2x} = v_{1x}$ – just a finite-amplitude Alfvén wave.

3.3 Self-gravitating systems; the virial theorems

The discussion of Section 3.1 assumes implicitly that self-gravitation has no effect on wave-propagation. The classical attempt at generalization is Jeans's discussion of gravitationally-modified sound waves. Corresponding to the density perturbation ρ' of (3.1), there is a perturbation V' to the gravitational potential, satisfying the perturbed Poisson equation

$$\nabla^2 V' = -4\pi G\rho' = -\mathbf{k}^2 V' \tag{3.36}$$

for the plane wave $\propto \exp i(\mathbf{k}\cdot\mathbf{r} - \omega t)$. With the magnetic force dropped from (3.2) but the self-gravitation term $\rho\nabla V'$ included, the modified (3.4) yields longitudinal waves with the dispersion relation

$$\omega^2 = c_s^2\left(1 - \frac{4\pi G\rho}{c_s^2 k^2}\right). \tag{3.37}$$

Thus this treatment illustrates gravitational reduction of the elasticity of the medium, with ω becoming imaginary and so predicting *gravitational instability* for semi-wavelengths greater than the *Jeans length*

$$\lambda_{\mathrm{J}} = \frac{\pi^{1/2}}{2} \frac{c_s}{(G\rho)^{1/2}}. \tag{3.38}$$

On dimensional grounds, one expects the Jeans length and the associated Jeans mass $M_{\mathrm{J}} = 4\pi\rho\lambda_{\mathrm{J}}^3/3$ to be an essential part of the language in which the basic problems of self-gravitating media are discussed. The obvious objection to the treatment is that no account is taken of the zero-order gravitational field with its spatially-dependent potential V_0, which for equilibrium to hold requires a corresponding non-uniform pressure–density field, and with a scale-height of the order of the Jeans length. The implicit assumption of wavelengths small compared with the scale-height is valid only for sound waves slightly modified by self-gravitation, but not at the transition to instability. The Jeans treatment of gravitational instability is therefore sometimes referred to irreverently as the 'Jeans swindle'. However, rigorous treatments of the 'Jeans problem', i.e. the spontaneous gravitational instability of self-gravitating gaseous bodies *in equilibrium*, that do involve the necessarily non-uniform zero-order state have been given respectively by Ledoux (1951) for plane-parallel and Ebert (1957) for spherical geometry.

Now let there be a uniform magnetic field $B_0\mathbf{b}$ permeating the medium. Longitudinal plane waves travelling along $\mathbf{b}$ clearly do not interact with the field, so the original Jeans treatment (with the limitations just outlined) remains valid. For longitudinal waves perpendicular to $\mathbf{b}$, the analogous treatment yields the dispersion relation (3.5) (with $\mathbf{k} \cdot \mathbf{b} = 0$) and again with c_s^2 replaced by $(c_s^2 - 4\pi G\rho/k^2)$. Instability is now predicted for semi-wavelengths greater than the magnetic Jeans length λ_{B} defined by (3.38) but with c_s^2 replaced by $(c_s^2 + B_0^2/4\pi\rho)$ (Chandrasekhar 1961).

The effects of a large-scale magnetic field on the problems of gravitational collapse and fragmentation are discussed in depth in Chapters 11 and 12. The magnetic modification λ_{B} to λ_{J} will indeed appear, though not in the way that the linear Jeans-type treatment predicts: in fact, the exercise is an object lesson in the dangers sometimes hidden in extrapolation from linear analysis into the non-linear domain. It turns out that it is the integral *virial theorems* which can often give one a more reliable global view of the dynamics, in particular of the interaction of the magnetic and gravitational fields, in advance of a detailed analytical or numerical solution (Chandrasekhar and Fermi 1953; Chandrasekhar 1961). The equation of motion for an inviscid gas is written in tensor notation:

$$\rho \frac{\mathrm{d}v_i}{\mathrm{d}t} = -\frac{\partial}{\partial x_i}\left(p + \frac{\mathbf{B}^2}{8\pi}\right) + \rho g_i + \frac{\partial}{\partial x_j}\left(\frac{B_i B_j}{4\pi}\right), \tag{3.39}$$

where, as before, $\mathrm{d}/\mathrm{d}t \equiv \partial/\partial t + v_j\, \partial/\partial x_j$ is the total time derivative. (Note that with this definition, since x_i, t are independent variables, the velocity v_i of an element instantaneously at the point x_i is indeed $\mathrm{d}x_i/\mathrm{d}t = \partial x_i/\partial t + v_j\, \partial x_i/\partial x_j = v_i$.)

In the simplest case, the gravitational field $\mathbf{g}$ is due just to the gas in the volume τ over which the virial integrals below are constructed: $\mathbf{g} = \nabla V$, with

$$V(\mathbf{x}) = G\int_\tau \frac{\rho(\mathbf{x}')}{|\mathbf{x}-\mathbf{x}'|}\,\mathrm{d}\mathbf{x}'. \tag{3.40}$$

More generally, one can write $\mathbf{g} = \nabla V + \mathbf{g}_{\mathrm{ext}}$, where $\mathbf{g}_{\mathrm{ext}}$ is due to sources external to τ or to stars and proto-stars within.

We now define the following symmetric tensors:

(a)

$$V_{ik} = G\int_\tau \frac{\rho(\mathbf{x}')(x_i - x_i')(x_k - x_k')}{|\mathbf{x}-\mathbf{x}'|^3}\,\mathrm{d}\mathbf{x}' \tag{3.41}$$

– a generalization of (3.40), satisfying

$$V_{ii}(\mathbf{x}) \equiv V(\mathbf{x}). \tag{3.42}$$

(b)

$$\mathcal{V}_{ik} = -\frac{1}{2}G\int_\tau\int_\tau \rho(\mathbf{x})\rho(\mathbf{x}')\frac{(x_i - x_i')(x_k - x_k')}{|\mathbf{x}-\mathbf{x}'|^3}\,\mathrm{d}\mathbf{x}\,\mathrm{d}\mathbf{x}', \tag{3.43}$$

which reduces on contraction to the gravitational potential energy

$$\mathcal{V}_{ii} = \mathcal{V} = -\frac{1}{2}G\int \frac{\rho(\mathbf{x})\rho(\mathbf{x}')}{|\mathbf{x}-\mathbf{x}'|}\,\mathrm{d}\mathbf{x}\,\mathrm{d}\mathbf{x}'. \tag{3.44}$$

The identity

$$\mathcal{V}_{ik} = \int \rho(\mathbf{x})x_i\frac{\partial V}{\partial x_k}\,\mathrm{d}\mathbf{x} = \int \rho(\mathbf{x})x_k\frac{\partial V}{\partial x_i}\,\mathrm{d}\mathbf{x} \tag{3.45}$$

is readily established: by definition,

$$\begin{aligned}
&\int \rho(\mathbf{x})x_i\frac{\partial V}{\partial x_k}\,\mathrm{d}\mathbf{x} = G\int \mathrm{d}\mathbf{x}\rho(\mathbf{x})x_i\frac{\partial}{\partial x_k}\int \mathrm{d}\mathbf{x}'\frac{\rho(\mathbf{x}')}{|\mathbf{x}-\mathbf{x}'|}\\
&= -G\int\int \mathrm{d}\mathbf{x}\,\mathrm{d}\mathbf{x}'\rho(\mathbf{x})\rho(\mathbf{x}')\frac{x_i(x_k - x_k')}{|\mathbf{x}-\mathbf{x}'|^3}\\
&= (1/2)\,(\text{sum of expressions with primed and unprimed interchanged})\\
&= \mathcal{V}_{ik}
\end{aligned} \tag{3.46}$$

as required.

(c) The kinetic energy tensor

$$\mathcal{T}_{ik} = \frac{1}{2}\int \rho v_i v_k \, \mathrm{d}\mathbf{x}, \tag{3.47}$$

yielding

$$\mathcal{T} \equiv \mathcal{T}_{ii} = \frac{1}{2}\int \rho \mathbf{v}^2 \, \mathrm{d}\mathbf{x}. \tag{3.48}$$

(d) The magnetic energy tensor

$$\mathcal{M}_{ik} = \frac{1}{8\pi}\int B_i B_k \, \mathrm{d}\mathbf{x}, \tag{3.49}$$

yielding

$$\mathcal{M} \equiv \mathcal{M}_{ii} = \frac{1}{8\pi}\int \mathbf{B}^2 \, \mathrm{d}\mathbf{x}. \tag{3.50}$$

(e) The inertia tensor

$$I_{ik} = \int \rho x_i x_k \, \mathrm{d}\mathbf{x}, \tag{3.51}$$

yielding

$$I \equiv I_{ii} = \int \rho \mathbf{x}^2 \, \mathrm{d}\mathbf{x}. \tag{3.52}$$

Define also

$$P = \int p \, \mathrm{d}\mathbf{x}. \tag{3.53}$$

In the case of a perfect gas, $p = (\gamma - 1)u$, where u is the local internal energy density and γ the ratio of the two principal specific heats. If also γ is uniform, then

$$P - (\gamma - 1)\int u \, \mathrm{d}\mathbf{x} \equiv (\gamma - 1)U. \tag{3.54}$$

In the following the form (3.54) is used; if necessary, γ can be interpreted as an average.

Now multiply (3.39) by x_k and integrate over a volume τ bounded by a surface S. The left-hand side can be reduced to

$$\begin{aligned}\int \rho x_k \frac{\mathrm{d}v_i}{\mathrm{d}t} \, \mathrm{d}\mathbf{x} &= \int \rho x_k \frac{\mathrm{d}^2 x_i}{\mathrm{d}t^2} \, \mathrm{d}\mathbf{x} \\ &= \int \rho \frac{\mathrm{d}}{\mathrm{d}t}\left(x_k \frac{\mathrm{d}x_i}{\mathrm{d}t}\right) \mathrm{d}\mathbf{x} - \int \rho \frac{\mathrm{d}x_k}{\mathrm{d}t}\frac{\mathrm{d}x_i}{\mathrm{d}t} \, \mathrm{d}\mathbf{x}\end{aligned}$$

$$= \int \rho \frac{\mathrm{d}}{\mathrm{d}t}\left(x_k \frac{\mathrm{d}x_i}{\mathrm{d}t}\right) \mathrm{d}\mathbf{x} - 2\mathcal{T}_{ik}. \tag{3.55}$$

On the right-hand side, integration by parts yields

$$-\int x_k \frac{\partial}{\partial x_i}\left(p + \frac{\mathbf{B}^2}{8\pi}\right) \mathrm{d}\mathbf{x} = -\int_S \left(p + \frac{\mathbf{B}^2}{8\pi}\right) x_k \, \mathrm{d}S_i + \delta_{ik}\left\{(\gamma - 1)U + \mathcal{M}\right\}, \tag{3.56}$$

and likewise

$$\frac{1}{4\pi}\int x_k \frac{\partial}{\partial x_i}(B_i B_j)\, \mathrm{d}\mathbf{x} = \frac{1}{4\pi}\int x_k B_i B_j \, \mathrm{d}S_j - 2\mathcal{M}_{ik}. \tag{3.57}$$

Hence

$$\int \rho \frac{\mathrm{d}}{\mathrm{d}t}\left(x_k \frac{\mathrm{d}x_i}{\mathrm{d}t}\right) \mathrm{d}\mathbf{x} = 2\mathcal{T}_{ik} + \delta_{ik}\{(\gamma - 1)U + \mathcal{M}\} + \mathcal{V}_{ik} - 2\mathcal{M}_{ik} \\ - \int (p\delta_{il} + T_{il}) x_k n_l \, \mathrm{d}S \tag{3.58}$$

where $T_{ik} = (B^2/8\pi)\,\delta_{ik} - B_i B_k/4\pi$ is the Maxwell stress tensor (2.42). Making use of the symmetry of the various tensors, we find for the antisymmetric part of (3.58)

$$\int \rho \frac{\mathrm{d}}{\mathrm{d}t}\left(x_k \frac{\mathrm{d}x_i}{\mathrm{d}t} - x_i \frac{\mathrm{d}x_k}{\mathrm{d}t}\right) \mathrm{d}\mathbf{x} = -\int \left[x_k(p\delta_{ip} + T_{ip}) - x_i(p\delta_{kp} + T_{kp})\right] n_p \, \mathrm{d}S \\ = -\int (D_{jp} + \epsilon_{jkp} x_k p) n_p \, \mathrm{d}S \tag{3.59}$$

where D_{jp} is the Lüst–Schlüter tensor, defined in (2.46). The terms on the right of (3.59) represent the torque exerted over the surface S by the Maxwell stresses and the thermal pressure. In the simplest case the mass within the surface S is constant, and (3.59) can be written

$$\frac{\mathrm{d}}{\mathrm{d}t}\int \rho \left(x_k \frac{\mathrm{d}x_i}{\mathrm{d}t} - x_i \frac{\mathrm{d}x_k}{\mathrm{d}t}\right) \mathrm{d}\mathbf{x} = -\int (D_{jp} + \epsilon_{jkp} x_k p) n_p \, \mathrm{d}S : \tag{3.60}$$

the change in the angular momentum is due just to the moments of the surface stresses. If there is flow of gas across the surface S – whether S is fixed or moving – then there is an extra term due to the associated direct transport of angular momentum; and if there is an external gravitational field, it will also in general exert a torque.

We now construct the symmetric part of (3.58). The left-hand side is

$$\frac{1}{2}\int \rho \frac{\mathrm{d}}{\mathrm{d}t}\left(x_k \frac{\mathrm{d}x_i}{\mathrm{d}t} + x_i \frac{\mathrm{d}x_k}{\mathrm{d}t}\right) \mathrm{d}\mathbf{x} = \frac{1}{2}\int \rho \frac{\mathrm{d}^2}{\mathrm{d}t^2}(x_k x_i)\, \mathrm{d}\mathbf{x}. \tag{3.61}$$

Again in the simplest case with a fixed mass within S,

$$\frac{1}{2}\frac{\mathrm{d}^2}{\mathrm{d}t^2}I_{ik} = 2\mathcal{T}_{ik} - 2\mathcal{M}_{ik} + \mathcal{V}_{ik} + [(\gamma-1)U + \mathcal{M}]\delta_{ik}$$
$$- \frac{1}{2}\int (p\delta_{ij} + T_{ij})x_k n_j \,\mathrm{d}S - \frac{1}{2}\int (p\delta_{kj} + T_{kj})x_i n_j \,\mathrm{d}S. \quad (3.62)$$

The diagonal terms yield (e.g. for $i = k = 1$)

$$\frac{1}{2}\int \rho \frac{\mathrm{d}^2}{\mathrm{d}t^2}(x_1^2)\,\mathrm{d}\mathbf{x} = 2\mathcal{T}_{11} - \mathcal{M}_{11} + (\gamma-1)U + \mathcal{V}_{11}$$
$$- \int p x_1 n_1 \,\mathrm{d}S - \int x_1 T_{1j} n_j \,\mathrm{d}S. \quad (3.63)$$

A large-scale magnetic field inevitably introduces an essential anisotropy. The different diagonal components of the tensorial virial theorem, if used judiciously, enable one to take cognizance of this anisotropy in a simple way (cf. in particular Chapter 11). However, if one wants an estimate averaged over all three axial directions, contraction of (3.62) yields the scalar virial theorem (Chandrasekhar and Fermi 1953)

$$\frac{1}{2}\frac{\mathrm{d}^2 I}{\mathrm{d}t^2} = 2\mathcal{T} + \mathcal{M} + \mathcal{V} + 3(\gamma-1)U$$
$$- \int \left(p + \frac{\mathbf{B}^2}{8\pi}\right)(\mathbf{r}\cdot\mathbf{n})\,\mathrm{d}S + \int (\mathbf{B}\cdot\mathbf{r})(\mathbf{B}\cdot\mathbf{n})\,\mathrm{d}S. \quad (3.64)$$

The surface terms are often (but not always) small enough to be dropped (cf. Chapter 11).

3.4 Magnetostatic equilibrium; force-free fields

A necessary but clearly not a sufficient condition for a domain to be in magnetostatic equilibrium is that the right-hand side of (3.64) must vanish. Note that if the surface terms are negligible, then the magnetic field's contribution to equilibrium is through the essentially positive term $\mathcal{M}$: a localized magnetic field has a disruptive effect on the equilibrium, like the thermal and kinetic terms, requiring the negative gravitational term $\mathcal{V}$ to maintain equilibrium. However, a field B falling off like $1/r^{3/2}$ may in principle have the volume and surface magnetic terms in (3.64) cancelling (cf. below).

It will emerge that stellar magnetic fields are almost certainly too small to have much effect on the hydrostatic equilibrium of the dense bulk of the star. However, near and above the surface the density falls off exponentially, whereas the decline in B is limited by the divergence condition to be algebraic. Thus in low density regions located near a strong source of flux – near a sunspot in a solar-type star, in much of the magnetosphere of a strongly magnetic CP-star – the condition of equilibrium will approximate to the *force-free* condition

$$\mathbf{j} \times \mathbf{B} = 0, \tag{3.65}$$

which by (2.41) and (2.42) is equivalent to

$$\frac{\partial T_{ij}}{\partial x_j} = 0; \quad T_{ij} = \frac{1}{8\pi}\left(\mathbf{B}^2 \delta_{ij} - 2B_i B_j\right). \tag{3.66}$$

By Ampère's equation (2.12) and the divergence condition (2.14), (3.65) yields

$$\nabla \times \mathbf{B} = \alpha \mathbf{B}, \tag{3.67}$$

with the constraint

$$\mathbf{B} \cdot \nabla \alpha = 0. \tag{3.68}$$

The local currents must thus flow nearly parallel to the field, so that the Lorentz force vanishes, and the proportionality factor α must be constant on the identical field and current lines. In the degenerate case with $\alpha = 0$, the currents are zero and the force-free condition reduces to the curl-free condition.

Any force-free field must be 'twisted', and in particular an axisymmetric force-free field must have both a meridional (poloidal) part $\mathbf{B}_\mathrm{p}$ and an azimuthal (toroidal) part $\mathbf{B}_\mathrm{t}$.[1] In cylindrical polar coordinates (ϖ, ϕ, z), the divergence condition is automatically satisfied when

$$\mathbf{B} = \left(\frac{1}{\varpi}\frac{\partial P}{\partial z}, \quad -\frac{S}{\varpi}, \quad -\frac{1}{\varpi}\frac{\partial P}{\partial \varpi}\right) \tag{3.69}$$

with the scalar functions P and S depending on ϖ and z but not on ϕ. (No confusion should arise with the quantity P temporarily defined by (3.53) and (3.54).) Then

$$\nabla \times \mathbf{B} = \frac{1}{\varpi}\left(\frac{\partial S}{\partial z}, \quad \nabla^2 P - \frac{2P_\varpi}{\varpi}, \quad -\frac{\partial S}{\partial \varpi}\right), \tag{3.70}$$

and the force-free condition (3.67) yields

$$\alpha = \frac{\mathrm{d}S}{\mathrm{d}P}, \tag{3.71}$$

$$\nabla^2 P - \frac{2P_\varpi}{\varpi} = -S\frac{\mathrm{d}S}{\mathrm{d}P}. \tag{3.72}$$

The simplest force-free fields are cylindrical ($\partial/\partial z = 0$) and helical ($B_\varpi = 0$), and are generated from the function $F(\varpi)$ by the relations

$$B_\phi^2 = -\frac{1}{2}\varpi\frac{\mathrm{d}F}{\mathrm{d}\varpi}, \qquad B_z^2 = F(\varpi) - B_\phi^2 \tag{3.73}$$

[1]See Appendix for a more general definition of poloidal and toroidal fields.

(Parker 1974). An example is the uniform-twist field (Gold and Hoyle 1960; Priest 1982), generated by $F = B_0^2/[1+(\Phi\varpi/L)^2]$:

$$B_\phi = \frac{B_0(\Phi\varpi/L)}{1+(\Phi\varpi/L)^2}, \qquad B_z = \frac{B_0}{1+(\Phi\varpi/L)^2}, \tag{3.74}$$

for which the parameter Φ gives the angle of twist of a field line about the axis in going the distance L. In terms of the formalism of (3.69–3.72), this field is described by

$$P = -\frac{B_0}{2}\left(\frac{L}{\Phi}\right)^2 \ln\left[1+\left(\frac{\Phi\varpi}{L}\right)^2\right], \qquad S = -B_0\frac{L}{\Phi}\frac{(\Phi\varpi/L)^2}{1+(\Phi\varpi/L)^2}, \tag{3.75}$$

where

$$\alpha = \frac{\mathrm{d}S}{\mathrm{d}P} = \frac{2\Phi}{L}\frac{1}{1+(\Phi\varpi/L)^2} = \frac{2\Phi}{L}\exp\left[\left(\frac{2P}{B_0}\right)\left(\frac{\Phi}{L}\right)^2\right]. \tag{3.76}$$

Note that if the field (3.74) is limited to a finite cylindrical radius $\varpi = a$, then the total magnetic pressure $F(a)/8\pi$ on the boundary has to be balanced by an external pressure p_e, e.g. a thermal pressure combined with the pressure of an untwisted, curl-free magnetic field.

The Gold–Hoyle force-free field is a special case of the class of helical fields with

$$B_\phi = \frac{B_0(\Phi\varpi/L)}{1+(\Phi\varpi/L)^2}, \qquad B_z = \frac{B_0(\lambda\Phi/L)}{1+(\Phi\varpi/L)^2}, \tag{3.77}$$

with $\lambda\Phi/L \leq 1$, and with equilibrium maintained by the pressure

$$p = \frac{B_0^2}{8\pi}\frac{1-(\lambda\Phi/L)^2}{1+(\Phi\varpi/L)^2}. \tag{3.78}$$

The vector potential of the field (3.77) is

$$\mathbf{A} = \frac{B_0 L}{2\Phi}\ln\left[1+\left(\frac{\Phi\varpi}{L}\right)^2\right]\left(0, \frac{\lambda}{\varpi}, -1\right). \tag{3.79}$$

The field (3.74) is a simple example of a force-free field in which α varies from one field line to another. The helical, z-independent field in which α is the same constant on every field line is

$$\mathbf{B} = B_0\big[0,\ \mathrm{J}_1(\alpha\varpi),\ \mathrm{J}_0(\alpha\varpi)\big], \tag{3.80}$$

where J_0, J_1 are Bessel functions in standard notation. This field is a special case of the constant-α field which decays in z (Schatzman 1965):

$$\mathbf{B} = B_0 \exp(-lz)\big[(l/k)\mathrm{J}_1(k\varpi),\ (1 - l^2/k^2)^{1/2}\mathrm{J}_1(k\varpi),\ \mathrm{J}_0(k\varpi)\big], \tag{3.81}$$

with $\alpha = (k^2 - l^2)^{1/2}$. In general, the constant-α force-free fields yield linear differential equations and so are the simplest to construct. In spherical polar coordinates, the simplest global solutions of (3.71) and (3.72) for the axisymmetric linear force-free field are in terms of Legendre polynomials $\mathrm{P}_1(\cos\theta)$ and spherical Bessel functions of r of order $3/2$ (confirming the above inference from the virial theorem). The field breaks up into disjoint domains related to the zeros of the Bessel functions (Lüst and Schlüter 1954; Chandrasekhar 1956; Mestel 1959; Berger 1985). The non-axisymmetric generalizations involve surface harmonics in $(\theta,\ \phi)$ (Chandrasekhar and Kendall 1957; Priest 1982). Often of greater interest are the global force-free fields with poloidal parts that retain the topology of a vacuum dipole; they must have α varying between field lines, so that (3.67) is non-linear (Milsom and Wright 1976; Aly 1981, 1995; and below).

If a field is force-free within a volume τ surrounded by a closed surface S, then only the magnetic terms within the scalar virial theorem (3.64) survive, and there results

$$\begin{aligned}\int_\tau \frac{\mathbf{B}^2}{8\pi}\,\mathrm{d}\tau &= \int_S x_i T_{ij} n_j\,\mathrm{d}S \\ &= \int \frac{\mathbf{B}^2}{8\pi}(\mathbf{n}\cdot\mathbf{r})\,\mathrm{d}S - \frac{1}{4\pi}\int (\mathbf{B}\cdot\mathbf{n})(\mathbf{B}\cdot\mathbf{r})\,\mathrm{d}S. \end{aligned} \tag{3.82}$$

Thus non-vanishing Maxwell stresses are required over at least some of the surface S, and the surface integral fixes the energy of the force-free field within V (Aly 1984, 1985; Low 1986). As an example (Kuijpers 1992), consider a force-free field in the volume τ between two concentric spheres, one of radius R, near the base of the solar corona, just above the photosphere, and the other far from the star (Fig. 3.5). Provided the strength of the currents, flowing along the field lines and maintaining the field, falls off fast enough for the currents to be effectively limited to a finite domain, then over the distant sphere, $B \propto 1/r^2$ at the most, whence $B^2 r^3 \propto 1/r$ and the surface integrals go to zero. Then by (3.82), the magnetic energy within τ reduces to

$$(R^3/2)\left(\overline{B_n^2} - \overline{B_t^2}\right), \tag{3.83}$$

where B_n, B_t are respectively normal and tangential components and the bars refer to averaging over the inner sphere. Since magnetic energy is an essentially positive quantity, the force-free condition requires that $\overline{B_n^2}$ exceed $\overline{B_t^2}$. The work done on the field by sub-photospheric convective motions is almost certainly the source of energy for the coronal fields, but the ratio B_t/B_n cannot be imposed arbitrarily but is determined by the force-free condition. Thus if the ends of loops such as PP' are pulled in opposite senses, the field reacts by also increasing the mean normal component.

In general, with the normal field B_n on the bounding surface prescribed, the

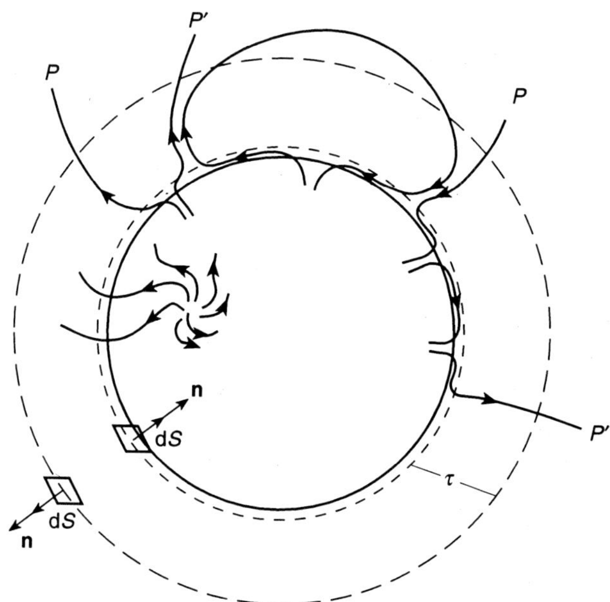

FIG. 3.5. A force-free field in the domain within a sphere just above the photosphere and a distant sphere.

magnetic energy within is a minimum if the field is *curl-free*. The result (3.83) shows that the maximum energy excess of a force-free field over the curl-free (vacuum) field with the same B_n-distribution is of the order of the curl-free field energy. Now consider the problem of a strictly force-free field within the domain limited by the finite boundary S_0, immersed in an ambient medium of constant thermal pressure p_e but with negligibly small magnetic fields (Lynden-Bell 1996). A surface current on S_0 cancels the normal component of the surface field; for equilibrium to hold, the magnetic pressure $\mathbf{B}^2/8\pi$ just within S_0 balances p_e. The magnetic field lines are taken to emerge from and return to the plane $z = 0$, where the flux is frozen and so B_z prescribed (Fig. 3.6). For simplicity, the system is from the start assumed symmetric about the z-axis.

A toroidal component is generated from an initially purely poloidal field by differential twisting at $z = 0$. By application of the 33-component of the tensorial virial theorem (3.63) to the volume $V(z)$ within S_0 above the plane $z = 0$,

$$\frac{1}{8\pi}\int_{V(z)} (B_\varpi^2 + B_\phi^2 - B_z^2)\,\mathrm{d}\tau = p_e V(z). \tag{3.84}$$

If $p_e = 0$, and the field dies away as required at infinity, then (3.84) becomes

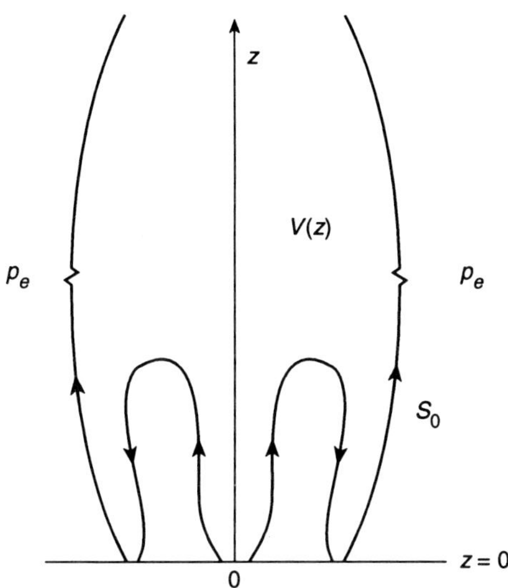

FIG. 3.6. An axisymmetric force-free field within a finite domain. The medium outside the boundary S_0 is field-free but exerts a thermal pressure p_e.

$$\int_{V(z)} (B_\varpi^2 + B_\phi^2)\,\mathrm{d}\tau = \int_{V(z)} B_z^2\,\mathrm{d}\tau. \tag{3.85}$$

This follows from a stronger result, given by integrating the z-component of (3.66) over the volume $V(z)$, and then transforming by Gauss's theorem to yield the vanishing of an integral over the plane z (Aly 1984):

$$0 = \int_z 8\pi T_{zk} n_k\,\mathrm{d}S = \int_z (B_z^2 - B_\varpi^2 + B_\phi^2)\,\mathrm{d}S. \tag{3.86}$$

Integration of (3.86) from z to ∞ then yields the volume integral (3.85).

Likewise, the scalar virial theorem (3.64) yields

$$\frac{1}{8\pi}\int_{V(z)} (B_\varpi^2 + B_\phi^2 + B_z^2)\,\mathrm{d}\tau = 3p_e V(z) + \frac{1}{4\pi}\int B_\varpi B_z \varpi\,\mathrm{d}S, \tag{3.87}$$

where the last term results from integrating the Maxwell stress surface term in (3.63) over z. Thus when applied to the domain $z > 0$, (3.84) and (3.87) combine to yield

$$\int (B_\varpi^2 + B_\phi^2)\,\mathrm{d}\tau = \int_{z=0} B_\varpi B_z \varpi\,\mathrm{d}S + 16\pi p_e V_0, \tag{3.88}$$

$$\int B_z^2\,\mathrm{d}\tau = \int_{z=0} B_\varpi B_z \varpi\,\mathrm{d}S + 8\pi p_e V_0. \tag{3.89}$$

In the problem without the p_e term, the input of energy accompanying the generation of a toroidal component by the steady twisting of the poloidal field would be expected to cause an expansion of the field. The condition (3.85) must however continue to hold, and with B_z kept fixed at $z = 0$, the field has no option but to occupy a growing volume. This effect, together with the associated opening up of the field lines, is demonstrated dramatically in work by Aly (1985, 1986, 1994, 1995), Van Ballegooijen (1994), Zylstra *et al.* (1989) and Lynden-Bell and Boily (1994). The latter have constructed a class of self-similar, axisymmetric force-free fields with the flux function $P \propto r^{-p}F(\mu)$, resulting from the systematic winding of an initially quadrupolar field, rooted in a massive conductor. The spherical polar components are

$$B_r = F_m r^{-(p+2)} f'(\mu), \tag{3.90}$$
$$B_\theta = -pF_m r^{-(p+2)} f(\mu)/(1-\mu^2)^{1/2}, \tag{3.91}$$
$$B_\phi = \left(3^{1/2}K/(p+1)\right) F_m r^{-(p+2)} f(\mu)/(1-\mu^2)^{1/2}, \tag{3.92}$$

where $\mu = \cos\theta$, $K = p\cosh^{-1}[\exp(1+1/p)]$ and F_m is an amplitude. The index p is small compared with the value $p = 2$ for the vacuum quadrupole field, so that $K \propto 1 + p(1+\ln 2)$, and $f(\mu)$ is well approximated by

$$f = 1 - \frac{p}{p+1}\ln\left\{\cosh\left[\frac{K}{p}(1-2\mu)\right]\right\}. \tag{3.93}$$

The associated current density is

$$\mathbf{j} = -\frac{3^{1/2}}{4\pi}\frac{K}{p}\frac{f^{1/p}}{r}\mathbf{B}. \tag{3.94}$$

As the system is wound up, the field configurations expand along a cone of semi-angle $\pi/3$. After just $3^{-1/2}$ turns, toroidal flux is released to infinity by a discharge around this cone. The case with $p = 1/3$ is shown in Fig. 3.7. Note that B_r changes sign on the line $\mu = 1/2$, and as $p \to 0$, the field on either side of this line approaches the structure of a split monopole, with all the flux extending to infinity. It is remarkable that this structure, strikingly different from that in the untwisted, vacuum quadrupolar field, is reached when the feet of the outgoing field lines are twisted relative to the incoming flux by no more than $3^{-1/2}$ turns.

If the initial field is dipolar rather than quadrupolar, there exist analogous solutions, differing primarily in having the equator $\mu = 0$ rather than the cone $\mu = \pi/3$ as the direction in which toroidal flux is released to infinity (Lynden-Bell 1997, personal communication). The fields have the form (3.90)–(3.92) but without the factor $3^{1/2}$ in B_ϕ, with K as before, and with

$$f(\mu) = 1 - \frac{p}{p+1}\ln\left\{\cosh\left[\frac{K}{p}\mu\right]\right\}. \tag{3.95}$$

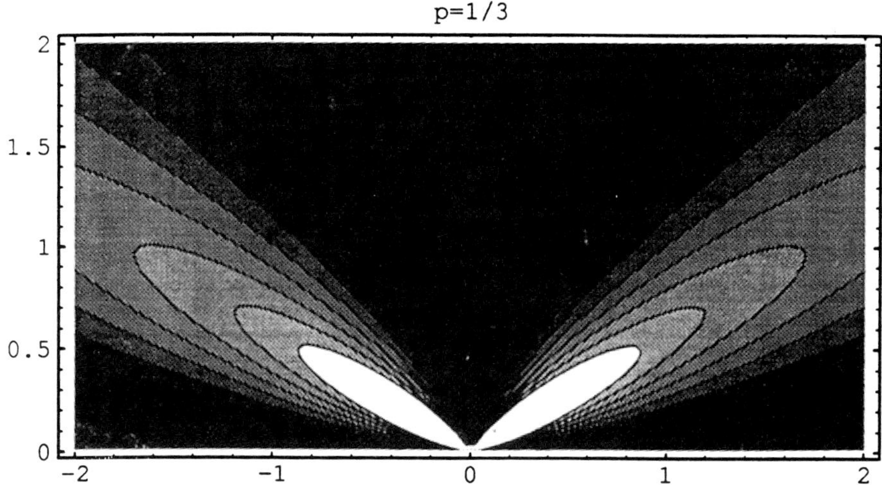

FIG. 3.7. The force-free field $P = r^{-1/3}f(\mu)$ with f given by (3.93).

The integral relations (3.88) and (3.89) show that in the general problem, with the surface pressure $p_e \neq 0$, the *net* pinching effect of the force-free field is still measured by $\int B_\varpi B_z \varpi \, \mathrm{d}S$ over the surface $z = 0$, where B_z is prescribed. The example just cited shows in detail how the mutual coupling of the field components through conditions (3.66) strongly controls the strength of the allowed B_ϕ that the twisting on $z = 0$ is trying to generate. The problem will clearly be changed when $p_e \neq 0$. We return to these questions when discussing collimation in Section 10.9. For the moment, we just note that p_e can be active or passive, depending on the problem under study: the Maxwell stresses may compress the non-magnetic external gas, building up p_e, or a strong external compression can modify profoundly the field structure within the force-free domain.

3.5 Magnetic helicity

It should be noted that if field-freezing holds strictly, then the twisting of an initially purely poloidal field, e.g. by shearing on the plane $z = 0$, does not change the topology of the field as a whole. Initially, a complete loop – extending into the domains both above and below $z = 0$ – of a purely poloidal field has zero linking (toroidal) flux. After shearing, there is in general a toroidal field component at each point, but in the absence of any diffusion, closed loops remain closed loops: a traveller starting from a point with azimuthal angle ϕ_0 and moving in the direction of $\mathbf{B}$ will return to the same value ϕ_0 (*without* any $\pm 2n\pi$ addition). Equally, if one starts, for example, with an axisymmetric field with a net toroidal flux linking each closed poloidal loop, then azimuthal shearing under perfect conductivity will change B_ϕ, but the linking toroidal flux is unaltered (cf. Section

5.5).

A quantity that is a measure of a topological property of the field and so is of particular interest for the magnetostatic problem is the *magnetic helicity* K (Woltjer 1958; Moffatt 1969, 1978; Berger 1984, 1985, 1994; Berger and Field 1984; Field 1986; and references therein). For a volume V that is simply-connected and *magnetically-closed*, i.e. enclosed by a surface S on which $\mathbf{B} \cdot \mathbf{n} = 0$, K is defined in terms of $\mathbf{B}$ and its vector potential $\mathbf{A}$:

$$K \equiv \int_V \mathbf{A} \cdot \mathbf{B} \, \mathrm{d}\tau. \tag{3.96}$$

Under a gauge transformation that replaces $\mathbf{A}$ by $\mathbf{A} + \nabla\chi$, the change in K is

$$\int_V (\mathbf{B} \cdot \nabla\chi) \, \mathrm{d}\tau = \int_V \nabla \cdot (\chi\mathbf{B}) \, \mathrm{d}\tau = \int_S \chi(\mathbf{B} \cdot \mathbf{n}) \, \mathrm{d}S = 0. \tag{3.97}$$

so the definition (3.96) is unique.

The helicity is a measure of the linkage of the field lines. Consider two thin, untwisted, mutually-linked flux-tubes, of volumes V_1, V_2 and strengths $\Phi_1 = B_1\sigma_1$, $\Phi_2 = B_2\sigma_2$, as in Fig. 3.8. They contribute $K_1 + K_2$ to the total helicity K, where

$$K_{1,2} = \int_{1,2} (\mathbf{A} \cdot \mathbf{B})\sigma \, \mathrm{d}s \tag{3.98}$$

respectively. Since $\mathbf{B} = B\hat{\mathbf{s}}$, $\mathrm{d}\mathbf{s} = (\mathrm{d}s)\hat{\mathbf{s}}$, then

$$K_1 = \int_1 (B\sigma)\mathbf{A} \cdot \mathrm{d}\mathbf{s} = \Phi_1 \int_1 \mathbf{A} \cdot \mathrm{d}\mathbf{s} = \Phi_1 \int_1 \mathbf{B} \cdot \mathrm{d}\mathbf{S} = \Phi_1\Phi_2. \tag{3.99}$$

Thus the total contribution of the two tubes is

$$K_{12} = 2\Phi_1\Phi_2, \tag{3.100}$$

a result that is easily generalized.

As emphasized by Berger and Field (1984), and is manifest from many studies of solar magnetic activity, it is often convenient to separate space into magnetic surfaces, e.g. nested toroidal surfaces with mutually linking poloidal and toroidal fields. With a slight adaptation of their terminology, it is seen that one such surface is identified by the toroidal flux ψ_{t} within it, and a neighbouring surface by $\psi_{\mathrm{t}} + \mathrm{d}\psi_{\mathrm{t}}$. The hole created by the surface is threaded by poloidal flux ψ_{p}, which reaches its maximum at the limiting surface that encloses zero volume and so zero ψ_{t}. For the infinitesimal annular volume containing flux $(\mathrm{d}\psi_{\mathrm{t}}, -\mathrm{d}\psi_{\mathrm{p}})$, the toroidal flux $\mathrm{d}\psi_{\mathrm{t}}$ links the poloidal flux ψ_{p}, while the poloidal flux $-\mathrm{d}\psi_{\mathrm{p}}$ links the toroidal flux ψ_{t}. Thus the helicity of this annular volume is

$$\mathrm{d}H = \psi_{\mathrm{p}} \, \mathrm{d}\psi_{\mathrm{t}} - \psi_{\mathrm{t}} \, \mathrm{d}\psi_{\mathrm{p}} \tag{3.101}$$

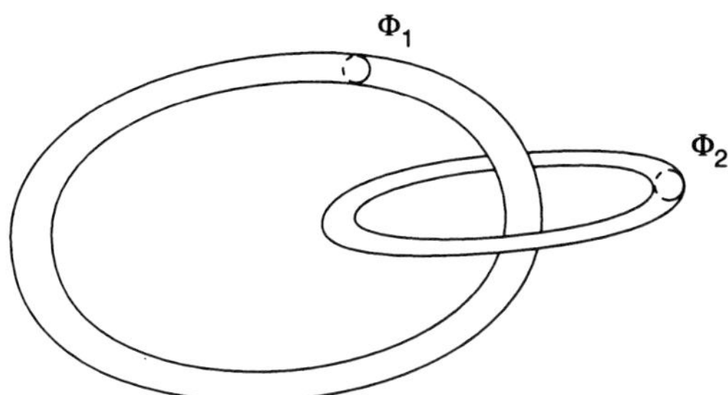

FIG. 3.8. The total helicity due to two thin mutually-linked flux-tubes.

(Kruskal and Kulsrud 1958). After integration by parts of the first term,

$$H = 2\int_0^{\Psi} T\psi_{\rm t}\, {\rm d}\psi_{\rm t} \tag{3.102}$$

where the twist

$$T \equiv -\frac{{\rm d}\psi_{\rm p}}{{\rm d}\psi_{\rm t}}, \tag{3.103}$$

and Ψ is the total toroidal flux. T represents the number of times a field line winds around the torus in the poloidal direction for one circuit in the toroidal direction.

The quantity (3.102) can be used to describe the total helicity of the field. However, it is often convenient to work in terms of thin but not infinitesimal tubes which contain linked flux. The total helicity is thought of as the sum of internal helicity – some combination of twist and kink helicity – and of external helicity, due to mutual linkages and knotting of the tubes:

$$H = H_{\rm T} + H_{\rm K}. \tag{3.104}$$

Only H is topologically invariant, as twists and kinks can convert into one another without any discontinuities being invoked. Within each tube, a central closed line is identified as the 'axis', with respect to which the twisting of each of the other lines is referred. For a single tube, $H_{\rm K}$ represents kinking of the axis and is written

$$H_{\rm K} = Wr\Psi^2, \tag{3.105}$$

where Wr is the *writhing number*, while

$$H_{\rm T} = \int_0^{\Psi} Tw(2\psi_{\rm t})\, {\rm d}\psi_{\rm t} \tag{3.106}$$

where Tw is the *twist number*. For further details, the reader is referred to Berger and Field (1984) and the references therein.

In ideal MHD, with flux frozen into a fluid satisfying the mass conservation law, helicity is conserved. The following formal proof is essentially as in Field (1986). In a general (ϕ, $\mathbf{A}$) gauge,

$$\frac{\partial}{\partial t}(\mathbf{A}\cdot\mathbf{B}) = \frac{\partial \mathbf{A}}{\partial t}\cdot\mathbf{B} + \mathbf{A}\cdot\frac{\partial \mathbf{B}}{\partial t} = -(c\mathbf{E} + c\nabla\phi)\cdot\mathbf{B} - c\mathbf{A}\cdot\nabla\times\mathbf{E}$$
$$= -c\nabla\cdot(\mathbf{E}\times\mathbf{A} + \phi\mathbf{B}) - 2c\mathbf{E}\cdot\mathbf{B}, \quad (3.107)$$

whence

$$\rho\frac{\mathrm{d}}{\mathrm{d}t}\left(\frac{\mathbf{A}\cdot\mathbf{B}}{\rho}\right) = \rho\left[\frac{\partial}{\partial t} + \mathbf{v}\cdot\nabla\right]\left(\frac{\mathbf{A}\cdot\mathbf{B}}{\rho}\right) = -\nabla\cdot[(c\phi - \mathbf{A}\cdot\mathbf{v})\mathbf{B}] \quad (3.108)$$

on use of the flux-freezing condition $c\mathbf{E} = -\mathbf{v}\times\mathbf{B}$ and the continuity condition $\mathrm{d}\rho/\mathrm{d}t = -\rho\nabla\cdot\mathbf{v}$. On integration over a volume bounded by a magnetic surface that moves with the fluid, the divergence will vanish, and

$$\frac{\mathrm{d}}{\mathrm{d}t}\int(\mathbf{A}\cdot\mathbf{B})\,\mathrm{d}\tau = \int\rho\frac{\mathrm{d}}{\mathrm{d}t}\left(\frac{\mathbf{A}\cdot\mathbf{B}}{\rho}\right)\mathrm{d}\tau = 0, \quad (3.109)$$

since

$$\frac{\mathrm{d}}{\mathrm{d}t}(\rho\,\mathrm{d}\tau) = 0; \quad (3.110)$$

under flux-freezing, the helicity within a volume bounded by a magnetic surface is indeed a constant of the motion.

This conservation of total helicity is particularly useful when considering the evolution of perfectly conducting flux tubes: a change in the external helicity, e.g. due to kinking, must be compensated by an equal and opposite change in the internal twist helicity (cf. Section 10.9).

Woltjer (1958) pointed out that helicity invariance for each flux-tube is a necessary constraint to be imposed in a Lagrangian formulation of the magnetostatic problem for a perfectly conducting fluid. Thus in addition to the magnetic energy and thermal energy contributions to the total field Lagrangian, there is the term

$$L_c = -\int_V \lambda(\mathbf{A}\cdot\mathbf{B})\,\mathrm{d}\tau. \quad (3.111)$$

The Lagrange multiplier λ is constant on each field loop, i.e.

$$\mathbf{B}\cdot\nabla\lambda = 0, \quad (3.112)$$

ensuring that (3.111) is gauge-invariant (Field 1986). In his original discussion, Woltjer (1958) ignored the internal energy and wrote down just the magnetic contribution $\int(\mathbf{B}^2/8\pi)\,\mathrm{d}\tau$ to the Lagrangian. He then inferred from the variational principle not only that the field must satisfy the force-free condition (3.67), but, surprisingly, that the function α should be constant everywhere, i.e. that the field be a linear force-free field. However, Field (1986) shows that with retention of the internal energy, the principle yields the usual balance of thermal pressure gradient and Lorentz force. Subsequent approach to the zero-pressure limit yields the force-free condition (subject to the imposed helicity constraints) but with no restriction beyond (3.68).

Strict conservation of helicity clearly holds only under perfect conductivity. Taylor (1974) conjectured that in a dissipative plasma, although the helicity of an individual loop is not conserved, the total helicity might be approximately conserved. Later, Field (1986) argued that helicity is sometimes dissipated much more slowly than the magnetic energy. Although as seen, Woltjer's argument is unacceptable, his conclusion that the system may prefer to relax towards a constant α force-free field may sometimes hold. Recent calculations by Hornig and Rastätter (1997) confirm the approximate conservation of magnetic helicity during reconnection in a nearly ideal plasma (cf. Section 3.7.2).

In solar physics one is interested in applying the helicity concept just to the optically thin regions, excluding the parts of the field in the unobservable regions below the photosphere, even though gauge-invariance requires that $\mathbf{B} \cdot \mathbf{n} = 0$ on the bounding surface. Berger and Field (1984) define a *relative helicity* which is again gauge-invariant. They consider fields which are identical below the surface R and have the same normal component at the surface, but which differ above the surface. They show that the difference in the helicities of the two fields is independent of the sub-surface field. In particular, they take a standard field above the surface which is the potential (current-free) field $\mathbf{P}_a$, continuous in its normal component with the sub-surface field, so that

$$H_R = H[\mathbf{B}_a, \mathbf{B}_b] - H[\mathbf{P}_a, \mathbf{B}_b] \tag{3.113}$$

(the suffices standing for 'above' and 'below' respectively). This is gauge-invariant, completely determined by the observable field $\mathbf{B}_a$ and its normal component on R, and is independent of the surface currents that maintain the discontinuity in the horizontal components. The procedure is generalized slightly by Berger (1985). The concept of relative helicity is useful in solar physics because one can often consider the field structure in the dense sub-photospheric regions as determined essentially by the local dynamics, supplying a boundary condition for the coronal field through the surface normal component, but being relatively unaffected by changes above the surface.

3.6 Stability

The stability of magnetohydrostatic equilibria is a crucial problem, of which our understanding is still limited. Only for the simplest equilibria can one hope to construct analytically the full spectrum of small-amplitude eigen-modes. For a fuller discussion of MHD instabilities in general, the reader is referred to Bateman (1978), and for an astrophysically slanted discussion to Priest (1982), Chapter 7.

3.6.1 *The MHD energy principle*

As for rigid body mechanics, an energy principle is available to study the stability of a wide class of dissipation-free systems (Bernstein *et al.* 1958; Schmidt 1966). Consider an equilibrium state with no zero-order motions, perturbed by an infinitesimal displacement field $\xi(\mathbf{r})$. Take the scalar product with ξ of the equation of motion (2.35), including the gravitational term (2.36) but ignoring the viscous term (2.37), and integrate over the volume occupied by fluid. Then with $\xi \propto \mathrm{e}^{\mathrm{i}\omega t}$,

$$\begin{aligned} -\frac{1}{2}\int \rho\xi\cdot\ddot{\xi}\,\mathrm{d}\tau &= \omega^2\int \frac{1}{2}\rho\xi^2\,\mathrm{d}\tau \\ &= \frac{1}{2}\left[\int \xi\cdot\nabla p\,\mathrm{d}\tau - \int \xi\cdot(\mathbf{j}\times\mathbf{B}/c)\,\mathrm{d}\tau - \int \rho\xi\cdot\nabla V\,\mathrm{d}\tau\right] \\ &\equiv \delta W \end{aligned} \tag{3.114}$$

by analogy with the one-dimensional oscillator. The contribution to δW must be evaluated to second order in ξ. The perturbed magnetic field $\delta\mathbf{B} = \nabla\times(\xi\times\mathbf{B})$ (cf. 2.88), and so the magnetic contribution to δW is

$$\begin{aligned} &\int \xi\cdot(\mathbf{j}\times\mathbf{B}/c)\,\mathrm{d}\tau = \int \xi\cdot(\delta\mathbf{j}\times\mathbf{B}/c)\,\mathrm{d}\tau + \int \xi\cdot(\mathbf{j}\times\delta\mathbf{B}/c)\,\mathrm{d}\tau \\ &= -\frac{1}{4\pi}\int(\xi\times\mathbf{B})\cdot\nabla\times\delta\mathbf{B}\,\mathrm{d}\tau + \int\xi\cdot(\mathbf{j}\times\delta\mathbf{B}/c)\,\mathrm{d}\tau \\ &= -\frac{1}{4\pi}\int\nabla\cdot[(\xi\times\mathbf{B})\times\delta\mathbf{B}]\,\mathrm{d}\tau - \frac{1}{4\pi}\int\nabla\times(\xi\times\mathbf{B})\cdot\delta\mathbf{B}\,\mathrm{d}\tau + \int\xi\cdot(\mathbf{j}\times\delta\mathbf{B}/c)\,\mathrm{d}\tau \\ &= -\frac{1}{4\pi}\int_S(\xi\times\delta\mathbf{B})\cdot\mathbf{n}\,\mathrm{d}S - \frac{1}{4\pi}\int\delta\mathbf{B}^2\,\mathrm{d}\tau + \int[\mathbf{j}\cdot(\delta\mathbf{B}\times\xi)/c]\,\mathrm{d}\tau. \end{aligned} \tag{3.115}$$

where S is a bounding surface. After similar transformations for the pressure and gravity terms, we obtain for δW the sum of the integral δW_{F} over the plasma volume:

$$\begin{aligned} \delta W_{\mathrm{F}} = \frac{1}{2}\int\Bigg(&\frac{\delta\mathbf{B}^2}{4\pi} - \frac{\mathbf{j}\cdot(\delta\mathbf{B}\times\xi)}{c} + \gamma p(\nabla\cdot\xi)^2 + (\xi\cdot\nabla p)\nabla\cdot\xi \\ &+(\xi\cdot\nabla V)\nabla\cdot(\rho\xi)\Bigg)\mathrm{d}\tau \end{aligned} \tag{3.116}$$

and the surface term

$$\frac{1}{2}\int \mathrm{d}S\,\mathbf{n}\cdot\xi(\delta p + \mathbf{B}\cdot\delta\mathbf{B}/4\pi). \tag{3.117}$$

If the displacement ξ vanishes on the boundary, the term (3.117) also vanishes. More realistically for astrophysical problems, the current-carrying plasma may be of finite extent but bounded by a low-density domain V with a curl-free magnetic field. The surface term (3.117) can then be integrated by parts to yield two extra terms to be added to (3.116):

$$\frac{1}{2}\int_V \mathrm{d}\tau \, \frac{(\delta \mathbf{B})^2}{4\pi} \tag{3.118}$$

and the surface integral

$$\delta W_{\mathrm{S}} = \frac{1}{2}\int (\mathbf{n}\cdot\xi)^2 \mathbf{n} \cdot \left\| \left\{ \nabla \left[p + \frac{\mathbf{B}^2}{8\pi} \right] \right\} \right\| \mathrm{d}S, \tag{3.119}$$

where the double vertical line represents the discontinuity on crossing the plasma surface outwards. The term (3.119) survives only if the plasma is maintained in equilibrium by surface currents (Bernstein *et al.* 1958; Schmidt 1966). A more general treatment that not only allows for finite contributions from the surface terms but also includes self-gravitation has been given by Kovetz (1966).

The thermal and magnetic terms in the integrand in (3.116) may alternatively be written (Furth *et al.* 1963)

$$\frac{|\delta \mathbf{B}_\perp|^2}{4\pi} + \frac{1}{4\pi}\left| \delta \mathbf{B}_\parallel - (4\pi)^2 \frac{\xi\cdot\nabla p}{\mathbf{B}^2}\mathbf{B} \right|^2 + \gamma p |\nabla\cdot\xi|^2 + \frac{\mathbf{j}\cdot\mathbf{B}}{c\mathbf{B}^2}(\mathbf{B}\times\xi)\cdot\delta\mathbf{B} - 2(\xi\cdot\nabla p)(\xi\cdot\kappa) \tag{3.120}$$

where κ is the normal curvature of the equilibrium magnetic field. Each term has its physical significance, as follows. The first term in (3.120) is the magnetic energy in the shear Alfvén wave; the second is the potential energy of the fast magnetosonic wave; the third the potential energy associated with ordinary sound waves. All three are stabilizing terms. Both the other two terms can destabilize: the fourth can drive kink (current-driven) instabilities and the fifth interchange or ballooning instabilities, analogues of Rayleigh–Taylor instability driven by pressure gradient and curvature. For a detailed treatment of examples, the reader is referred especially to Bateman (1978).

A *sufficient* condition for instability is that there should exist some possible $\xi(\mathbf{r})$ for which $\delta W < 0$. However, to prove stability it would be necessary to show that $\delta W > 0$ for all conceivable displacements, something that is usually too difficult.

3.6.2 *Illustrative examples*

(a) Consider first a simple non-magnetic problem, that of a spherical star in hydrostatic equilibrium. Stability against arbitrary displacements is then determined by the behaviour of

$$\begin{aligned} &\gamma p(\nabla\cdot\xi)^2 + (\xi\cdot\nabla p)\nabla\cdot\xi + (\xi\cdot\nabla p)(\nabla\cdot\xi + \xi\cdot\nabla\rho/\rho) \\ &\quad = \gamma p \xi_r^2 \left[\left(\frac{\nabla\cdot\xi}{\xi_r} + \frac{p'}{\gamma p} \right)^2 + \left(\frac{p'}{\gamma p} \right)\left(\frac{\rho'}{\rho} - \frac{p'}{\gamma p} \right) \right], \end{aligned} \tag{3.121}$$

on use of $\nabla p = \rho \nabla V$ and of the zero-order spherical symmetry which yields $\nabla p = p'\hat{\mathbf{r}}$, $\nabla \rho = \rho'\hat{\mathbf{r}}$. In the simplest standard treatment, a displaced element of gas is supposed to move with its pressure change Δp related adiabatically to its density change $\Delta\rho$, but with pressure equilibrium with the ambient gas being maintained:

$$\frac{\gamma p}{\rho}\Delta\rho = \Delta p = p'\xi_r. \tag{3.122}$$

By continuity,

$$0 = \delta\rho + \nabla\cdot(\rho\xi) = (\delta\rho + \xi\cdot\nabla\rho) + \rho\nabla\cdot\xi = \Delta\rho + \rho\nabla\cdot\xi, \tag{3.123}$$

whence

$$p'\xi_r = -\gamma p\nabla\cdot\xi \tag{3.124}$$

and the first bracket in (3.121) vanishes. Since $p' < 0$, the domain is clearly locally stable provided

$$\frac{\rho'}{\rho} < \frac{p'}{\gamma p}. \tag{3.125}$$

A disturbed element will then oscillate; by (3.114), the medium admits *internal gravity waves* with the characteristic *Brunt–Väisälä frequency* N, given by

$$N^2 = g\left(\frac{p'}{\gamma p} - \frac{\rho'}{\rho}\right) \tag{3.126}$$

(cf. e.g. Lighthill 1978, and Section 4.4 below). When the inequality sign in (3.125) is reversed, we arrive at the *Schwarzschild convective instability criterion.* This can be written

$$n < \frac{1}{\gamma - 1}, \tag{3.127}$$

where n is the local polytropic index, defined by

$$1 + \frac{1}{n} = \frac{\mathrm{d}\ln p}{\mathrm{d}\ln\rho}. \tag{3.128}$$

Equivalently, spontaneous instability occurs when

$$-\frac{\mathrm{d}T}{\mathrm{d}r} > \left(\frac{\mathrm{d}T}{\mathrm{d}r}\right)_{\mathrm{ad}} = \frac{(\gamma-1)}{\gamma}T\left(\frac{-p'}{p}\right), \tag{3.129}$$

where $(\mathrm{d}T/\mathrm{d}r)_{\mathrm{ad}}$ is the adiabatic temperature gradient associated with the given pressure gradient. The exponential rate of growth of a disturbance is then

$$\omega = |N| \tag{3.130}$$

with N^2 the negative quantity given by (3.126).

(b) Consider now a magnetic example (Hughes and Cattaneo 1987). A layer of gas is confined in the vertical direction by rigid surfaces at $z = d$ (top) and

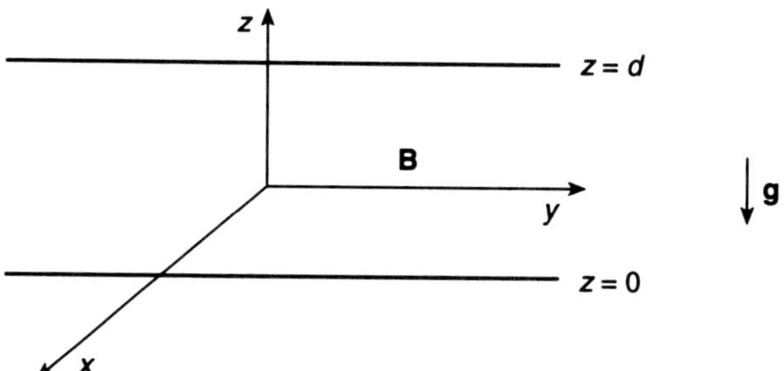

FIG. 3.9. **Instability of a fluid with partial support against gravity of a horizontal** magnetic field.

$z = 0$ (bottom), with the external gravity field acting in the negative z-direction (Fig. 3.9). The equilibrium is a balance between gravity, gas pressure and the Lorentz force due to a horizontal magnetic field $\mathbf{B} = B\hat{\mathbf{y}}$, with pressure, density, temperature and field strength all dependent just on z:

$$-\frac{\mathrm{d}p}{\mathrm{d}z} - \frac{\mathrm{d}}{\mathrm{d}z}\left(\frac{B^2}{8\pi}\right) = \rho g. \tag{3.131}$$

We investigate the stability of the equilibrium to undulatory disturbances of the form

$$\xi_x \propto \sin lx \sin my, \quad \xi_y \propto \cos lx \cos my, \quad \xi_z \propto \cos lx \sin my, \tag{3.132}$$

applied to a box of height d and suitably chosen length and breadth in the x- and y-dimensions. With the help of (3.131), (3.116) becomes

$$\delta W = \frac{1}{2}\int_0^d \left(\frac{B^2}{4\pi}[m^2(\xi_z^2 + \xi_x^2) + (\xi_z' + l\xi_x)^2] + \gamma p(\xi_z' - m\xi_y + l\xi_x)^2 \right.$$
$$\left. + 2\rho g\xi_z(\xi_z' - m\xi_y + l\xi_x) - g\xi_z^2\rho'\right) \mathrm{d}z, \tag{3.133}$$

where the prime denotes derivation with respect to z.

The simplest case is the interchange mode, with no motion ξ_y along the zero-order field, and no fluting in the y-direction: i.e. both $\xi_y = 0$ and $m \to 0$, but the coefficients of ξ_x, ξ_y becoming large like $1/m$ so as to yield finite, y-independent displacements in the x- and z-directions. A formal minimization of the integrand in the resulting form of (3.133) then yields

$$\delta W_{\min} = \frac{1}{2}\int_0^d \left[\left(-g\rho' - \frac{\rho^2 g^2}{\gamma p + B^2/4\pi}\right)\xi_z^2\right] \mathrm{d}z. \tag{3.134}$$

If $\rho' > 0$, then (3.134) implies instability – the standard Rayleigh–Taylor instability of a denser fluid resting on a lighter one, whether or not there is a magnetic field. In a normally stratified domain with $\rho' < 0$, the magnetic interchange Rayleigh–Taylor instability results if

$$-\rho' < \frac{\rho^2 g}{\gamma p + B^2/4\pi}, \tag{3.135}$$

(a result that is most easily obtained by actually interchanging two horizontal flux tubes, without any bending and with mass conservation – see Tayler 1973). Condition (3.135) can be written

$$\frac{-v_{\mathrm{A}}^2}{\gamma H_p} \frac{\mathrm{d}}{\mathrm{d}z} \ln \frac{B}{\rho} > N^2, \tag{3.136}$$

where $H_p \equiv (p/g\rho)$ is the pressure scale-height, $v_{\mathrm{A}} = B/(4\pi\rho)^{1/2}$ is again the Alfvén speed, and N is now the notional Brunt–Väisälä frequency, *defined* for the unperturbed $p - \rho$ field by (3.126).

For two-dimensional undular modes, l is identically zero. Minimizing (3.133) with respect to ξ_y, and then letting $m \to 0$, eliminates the work done against magnetic tension (Hughes and Cattaneo 1987) and so relaxes somewhat the instability condition. However, the instability becomes maximally efficient for three-dimensional modes. Minimization of (3.133) with respect to ξ_x and ξ_y, and then letting $m \to 0$ and $l \to \infty$, now yields instability if, somewhere in the fluid, $-\rho' < (\rho^2 g^2/\gamma p)$, equivalent to

$$-\frac{v_{\mathrm{A}}^2}{\gamma H_p} \frac{\mathrm{d}}{\mathrm{d}z} \ln B > N^2. \tag{3.137}$$

Comparison of (3.136) and (3.137) shows that whereas the interchange modes require an outward decrease in B/ρ, undular modes may be unstable if B alone decreases. However, the example of the purely horizontal field – in some ways a simulation of a toroidal field – is chosen just as the simplest illustration of the energy principle. Realistic studies will involve mixed poloidal–toroidal fields, and in the case of unstable systems, will emphasize the need for numerical studies of the non-linear developments (Chapters 4, 5 and 8).

3.6.3 *The pinched cylindrical discharge*

This is another classical example from laboratory studies of astrophysical relevance (already touched on in Sections 3.4 and 3.5). In its idealized form, a cylindrical column of plasma of radius a extends to infinity in the $\pm z$-directions, and contains the field $(0, B_{0\phi}, B_{0z})$, where (ϖ, ϕ, z) are cylindrical polar coordinates. All equilibrium quantities (with suffix zero) are functions just of the radial distance ϖ from the cylindrical axis. The balance of pressure gradient and Lorentz force yields

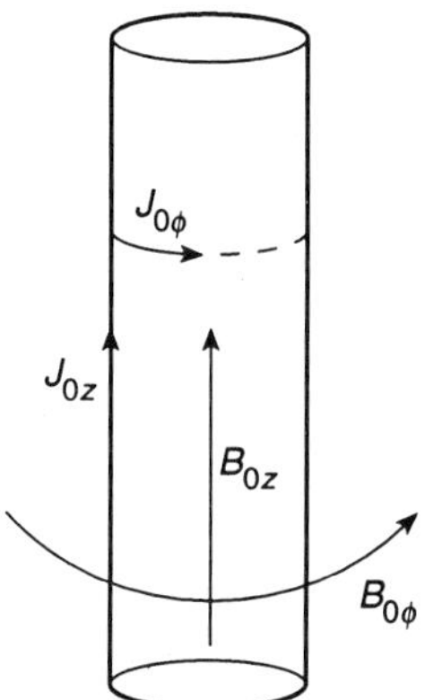

FIG. 3.10. An infinite cylindrical column of plasma in magnetothermal equilibrium: the special case with just surface currents ($J_{0\phi}$, J_{0z}).

$$\frac{\mathrm{d}}{\mathrm{d}\varpi}\left(p_e + \frac{B_{0z}^2}{8\pi}\right) + \frac{1}{4\pi\varpi}B_{0\phi}\frac{\mathrm{d}}{\mathrm{d}\varpi}(\varpi B_{0\phi}) = 0. \tag{3.138}$$

For each radius ϖ one can define the quantity (Bateman 1978; Priest 1982)

$$\Phi = \frac{2LB_{0\phi}}{\varpi B_{0z}}, \tag{3.139}$$

measuring the the amount of twist that a field line undergoes over the distance $2L$ parallel to the z-axis. The Gold–Hoyle fields discussed above are special in having the quantity Φ independent of ϖ.

In a particularly simple case (Fig. 3.10), the currents flow just on the surface of the cylinder – an azimuthal component $J_{0\phi}$ and a longitudinal component J_{0z} maintain B_{0z} and $B_{0\phi}$ respectively. The external field is purely toroidal,

$$B_{0\phi}(\varpi) = \frac{4\pi}{c}J_{0z}\frac{a}{\varpi}, \tag{3.140}$$

and the internal field is purely longitudinal and uniform:

$$B_{0z} = \frac{4\pi}{c}J_{0\phi}. \tag{3.141}$$

The plasma is of uniform density ρ_0 and exerts a uniform pressure p_e, satisfying the surface boundary condition of stress continuity:

$$p_e + \frac{(B_{0z})^2}{8\pi} = \frac{(B_{0\phi}(a))^2}{8\pi}. \tag{3.142}$$

If $B_{0z} = 0$, the discharge is subject to the 'sausage instability' (Fig. 3.11), growing at the rate $(2p_e k/\rho_0 a)^{1/2} \approx (2p_e/\rho_0)^{1/2}/a$ for the most disastrous modes with $ka \approx 1$. When $B_{0z} \neq 0$ and is large enough, it can be shown (e.g. Priest 1982, Section 7.5) that Alfvén waves propagating in the z-direction have the (non)-dispersion relation

$$\omega^2 = -\frac{2p_e}{\rho_0 a^2} + \frac{(B_{0z})^2}{4\pi\rho_0 a^2} = \frac{1}{4\pi\rho_0 a^2}[2(B_{0z})^2 - (B_{0\phi})^2] \tag{3.143}$$

on use of (3.142). Thus the sausage mode is stabilized by the superposition of a longitudinal field of strength greater than $|B_{0\phi}|/2^{1/2}$.

The pinched equilibrium with $B_{0z} = 0$ is subject also to the 'kink instability' (Lundquist 1951; Kruskal and Schwarzschild 1954; Bateman 1978; Priest 1982). The helical kink perturbation of the form $\xi(\varpi)\exp[\mathrm{i}(\phi - kz) + \mathrm{i}\omega t]$ gives instability for all wavenumbers k; hence a lateral kink $\propto \exp(i\phi)\cos kz$ and obtained by the superposition of the two oppositely twisted helical perturbations $\exp[\mathrm{i}(\phi \pm kz)]$ is also unstable. However, unlike the sausage mode, the kink instability cannot be stabilized for arbitrary wavelengths by the superposition of an axial, B_{0z}-component. For the above helical perturbation, the transition from stable short to unstable long wavelengths occurs when the wave-number vector $(0, \varpi^{-1}, k)$ is perpendicular to the unperturbed magnetic field – $\mathbf{k} \cdot \mathbf{B} = 0$; equivalently, for a *prescribed twist* Φ along a tube of length $2L$, defined by (3.139), instability has set in when the helical kink has wave-number k satisfies the cylindrical Kruskal–Shafranov condition

$$k \leq \frac{\Phi}{2L}. \tag{3.144}$$

A general physical discussion of these current-driven instabilities is instructive (e.g. Bateman 1978, p. 104 et seq.).[2] The sausage instability for the case with a pure B_ϕ-field (illustrated in Fig. 3.11) results from an axisymmetric, $m = 0$ perturbation under which the column is constricted at A and bulges at B. Since the same total current continues to flow through each area, the B_ϕ-field at the plasma surface increases at A and decreases at B, and so the changed magnetic forces act to accelerate the instability. As the plasma can flow along the tube, the plasma pressure does not change significantly, and so it cannot prevent the instability. As seen above, stabilization of this mode is possible if the unperturbed state includes a longitudinal component B_z, for then the compression and rarefaction at A and B respectively increase/decrease the contribution of $B_z^2/8\pi$ to the magnetic pressure.

However, the same longitudinal field introduces a possible new instability – the above-noted $m = 1$ kink mode, in which the perturbation is shaped like a corkscrew. In Bateman's first, illustrative example, the plasma column is pictured as a thin wire carrying current $\mathbf{I}$, embedded in a longitudinal field $B_z\hat{\mathbf{z}}$. In the

[2] Bateman's use of the terms 'poloidal' and 'toroidal' is the opposite to that among astrophysicists, as his work is primarily aimed at laboratory plasma physicists, involved in the thermonuclear fusion programme.

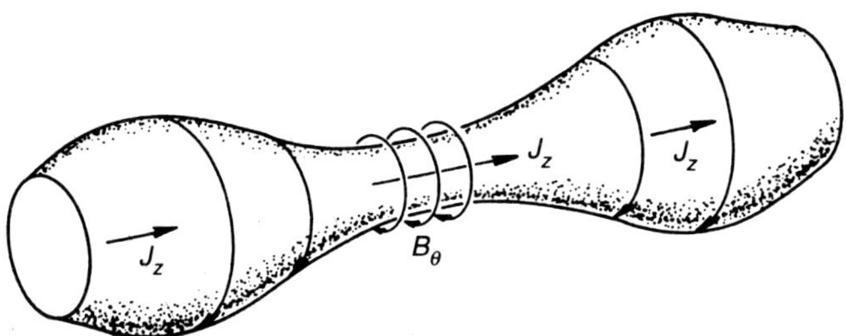

FIG. **3.11. The sausage instability of a purely toroidal field (after Bateman** 1978).

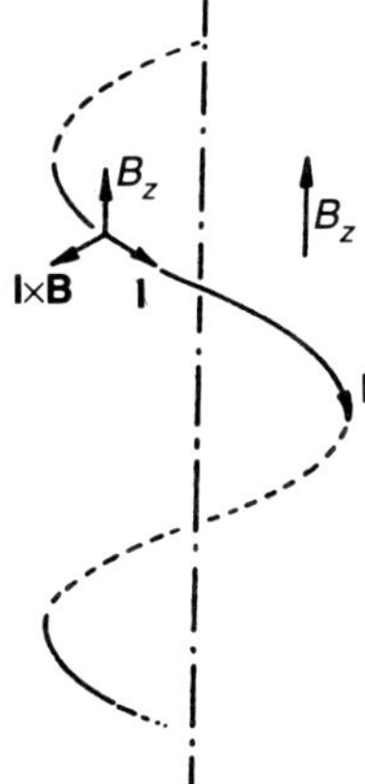

FIG. 3.12. **The $m = 1$ kink instability in a thin plasma column (aft**er Bateman 1978).

unperturbed, equilibrium state, $\mathbf{I} \times B_z\hat{\mathbf{z}} = 0$; but when the wire is given a helical perturbation, the same force – due to the perturbed current now crossing the unperturbed field – acts in the direction of the instability and so accelerates it (Fig. 3.12).

A rather more realistic case of the $m = 1$ kink instability is illustrated in Fig. 3.13 (given originally by Johnson *et al.* 1958). A fat plasma column is again given a helical distortion, but with the cross-section remaining unchanged. The sections A and B are separated by $\lambda/4 = \pi/2k$, where λ is the wavelength and k the corresponding wavenumber of the perturbation helix. Displayed also is a line of the unperturbed helical field, and also the same line of the perturbed field $\mathbf{B}_0 + \mathbf{B}_1$ that results from the displacement field ξ. As one moves from A to B,

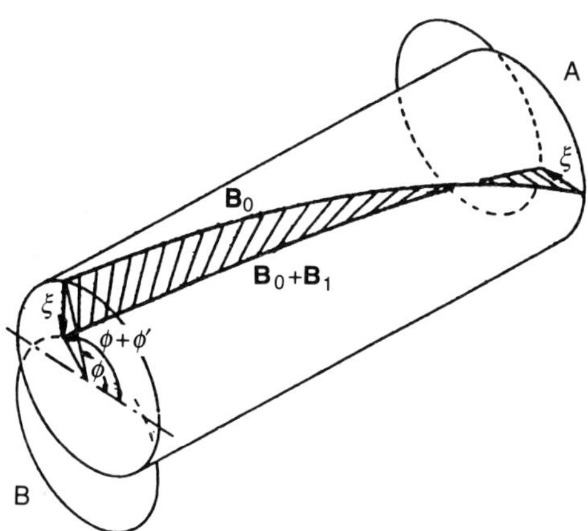

FIG. 3.13. Helical distortion of a fat plasma column (after Bateman 1978).

the angle ϕ subtended at the magnetic axis increases from 0 to ϕ^0. For the line drawn, $\phi^0 > \pi/2$. In the helical disturbance, the ξ vector, taken as horizontally inwards at A, will be vertically inwards at B; hence the perturbed field line will subtend a larger angle $\phi = \phi^0 + \phi^1$ at the original magnetic axis, and on the side of the column where the perturbation is pushing inwards. To traverse this larger angle, the field component in the cross-section must be increased by the perturbation, and the increased magnetic pressure again acts inwards in the direction of ξ, so driving instability. Equally, if $\phi^0 < \pi/2$, the perturbations act so as to decrease ϕ, so that the inward magnetic pressure is reduced, and the system is stable against this helical mode. Identifying ϕ_0 with Φ of (3.139) yields the criterion cited above.

Such a cylindrical column arises when discussing winds and jets (Chapters 7 and 10). It can also be regarded as an approximate model of a torus, with the two points $z = 0$, $z = 2L$ to be identified. The above theory can then be applied, but the length $2L$ must now be identified with a whole wavelength $2\pi/k$ rather than a fraction of a wavelength. Helical kink instability develops in such a torus if

$$\Phi > 2\pi \tag{3.145}$$

(Kruskal 1954; Shafranov 1957; Tayler 1957). But to apply the theory, for example, to coronal flux loops, the problem must be generalized to allow for stabilization due to anchoring of field lines in the photosphere (Priest 1982). Inside stars, the density is high and gravity must be taken into account (Section 5.3).

In his discussion of the stability of cylindrical structures, Parker (1979) lays stress on the question of whether the cylinder is under longitudinal tension or compression. In cylindrical polars, an axisymmetric field $(0, B_\phi(\varpi), B_z(\varpi))$ exerts over the cylindrical cross-section $0 < \varpi < a$ a total stress

$$Q(a) = \int T_{33}\,\mathrm{d}S = 2\pi \int_0^a \frac{B_z^2 - B_\phi^2}{8\pi} \varpi\,\mathrm{d}\varpi \tag{3.146}$$

where the Maxwell stress tensor T_{ij} is given by (2.42). A cylinder that is twisted too much will have $B_\phi^2 > B_z^2$ on average, so that $Q(a)$ will be negative: the cylinder is under compression rather than tension and so intuitively one expects it to be liable to 'buckle'. As an example, Parker considers the force-free magnetic rope, generated by the function $F(\varpi) = \mathbf{B}^2(\varpi)$, as in (3.73). This yields

$$Q(a) = 2\pi a^2 \left[\frac{B^2(a)}{8\pi} - \frac{1}{2\pi a^2}\left(2\pi \int_0^a \frac{B^2}{8\pi} \varpi\,\mathrm{d}\varpi\right)\right]. \tag{3.147}$$

We have seen that a force-free cylinder of finite radius requires an external pressure p_e (thermal and/or magnetic); then $Q(a)$ becomes negative if $p_e = B^2(a)/8\pi$ is less than half the mean magnetic energy over the cross-section.

The most obvious example of a cylinder under longitudinal compression is the pinched discharge of Fig. 3.10 but with a purely toroidal field B_ϕ, and this is unstable to the *axisymmetric* sausage mode of Fig. 3.11. Although one might expect non-axisymmetric modes to be more violently unstable, it is certainly not surprising that a cylinder under compression should be subject also to an axisymmetric instability. We have seen that the introduction of a sufficiently strong longitudinal component B_z suppresses the sausage mode, but now introduces unstable non-axisymmetric helical 'kink' modes. However, there appears to be no obvious link-up of this transition with a change in the sign of the integrated Maxwell stress $Q(a)$. Equally, one of the earliest studies of the equilibrium and stability of a (non-force-free) cylinder (Roberts 1956) shows instability setting in first via a long wavelength *axisymmetric* mode but with the longitudinal integrated tension positive. Hood (1994) defines 'buckling' as a lateral as opposed to a helical kink mode. He finds that for a force-free field, the first mode to go unstable is always the $m = 1$ helical kink mode. The $m = 0$ sausage mode always has a higher threshold, even when $B_\phi > B_z$ (de Bruyne and Hood 1992). Hood (personal communication) emphasizes that the instability criterion – as determined rigorously either by normal mode analysis or the energy principle – is not identical with the negative tension condition: there are cases with the net tension positive (or negative) and with the tube stable or unstable.

In his 1979 monograph (p. 172 et seq.), Parker uses the term 'buckling' to mean a more extreme form of helical kinking. It is recognized that kink instability sets in when the cylindrical Kruskal–Shafranov criterion (3.144) is satisfied, and for a long tube this can correspond to a very small twist, and with the mean tension still positive; however, it is argued that for small twisting, in the neighbouring non-linear domain there is a stable equilibrium with just a slightly corkscrew form. 'The criterion that the twisting be small enough for the tube to remain in tension is equivalent to the statement that the straight tube relapses into only a moderate corkscrew form.' Other workers (e.g. Spruit, cf. Section 10.9) assume rather that violent kinking sets in as soon as $B_\phi > B_z$, even if the criterion (3.144) is not satisfied. The time appears to be ripe for a systematic study of the non-linear development of the instabilities. One hazards a guess that Parker's negative tension criterion may turn out to be a non-rigorous but rough description of the demarcation between mild and violent instability.

3.6.4 *The Kelvin–Helmholtz instability*

The normal mode method is well illustrated by the magnetic Kelvin–Helmholtz instability. The classical K–H instability can arise when a fluid of density ρ_1 streams with velocity $\mathbf{V}$ relative to a lower fluid of density ρ_2 with $\rho_2 > \rho_1$ so that the Rayleigh–Taylor instability does not arise. Viscosity and surface tension are assumed negligible. For a disturbance of wave-number k, the stabilizing effect of gravity g is overcome if

$$\mathbf{V}^2 > \frac{(\rho_2^2 - \rho_1^2)g}{\rho_1 \rho_2 k}. \tag{3.148}$$

If instead, both the density $\rho(z)$ and the streaming velocity $V(z)$ are not discontinuous but vary smoothly with the vertical coordinate z, then instability occurs if the Richardson number Ri is not too large, i.e. if

$$\mathrm{Ri} \equiv \frac{-g\mathrm{d}\rho/\mathrm{d}z}{\rho(\mathrm{d}V/\mathrm{d}z)^2} < 0.25. \tag{3.149}$$

Now suppose there are uniform horizontal magnetic fields $\mathbf{B}_1$, $\mathbf{B}_2$ present in the two fluids. To isolate the stabilizing role of the magnetic field, we ignore gravity and study the purely magnetic K–H instability (e.g. Cowling 1976). This approximation is not normally acceptable within a star but may be relevant, for example, to the solar wind. To preserve algebraic symmetry, we use an inertial frame in which both undisturbed fluids move, with respective uniform velocities $\mathbf{V}_1$ and $\mathbf{V}_2$. Perfect conductivity is assumed. The perturbations $\mathbf{b}$, $\mathbf{v}$ in each fluid satisfy

$$\rho\left(\frac{\partial \mathbf{v}}{\partial t} + (\mathbf{V}\cdot\nabla)\mathbf{v}\right) = -\nabla\Pi + \mathbf{B}\cdot\nabla\mathbf{b}, \tag{3.150}$$

where

$$\Pi = p + \frac{\mathbf{B}^2}{8\pi}, \tag{3.151}$$

and

$$\frac{\partial \mathbf{b}}{\partial t} + (\mathbf{V}\cdot\nabla)\mathbf{b} = (\mathbf{B}\cdot\nabla)\mathbf{v}, \quad \nabla\cdot\mathbf{b} = 0. \tag{3.152}$$

For simplicity, incompressibility is also assumed, so that

$$\nabla\cdot\mathbf{v} = 0. \tag{3.153}$$

As well as proper behaviour at $z = \pm\infty$, the boundary conditions to be imposed are continuity on the plane $z = 0$ separating the two fluids of the total pressure Π

and the z-component of the fluid displacement. The divergence of (3.150) yields $\nabla^2\Pi = 0$, so that a wave solution propagating in the (x, y) planes has the form

$$\Pi = \Pi_0 \exp\{\mathrm{i}\omega t - \mathrm{i}(k_x x + k_y y) \pm kz\}, \quad k^2 = k_x^2 + k_y^2, \tag{3.154}$$

with conditions at infinity requiring that $+k$ correspond to $z < 0$ and $-k$ to $z > 0$. The variables $\mathbf{v}$ and $\mathbf{b}$ have the same space–time dependence.

In terms of the z-displacement ζ of the common surface,

$$v_z = \frac{\partial \zeta}{\partial t} + \mathbf{V} \cdot \nabla\zeta = \mathrm{i}(\omega - \mathbf{k} \cdot \mathbf{V})\zeta \tag{3.155}$$

so that $v_z/(\omega - \mathbf{k} \cdot \mathbf{V})$ is continuous. The z-components of (3.150) and (3.152) yield jointly

$$\rho(\omega - \mathbf{k} \cdot \mathbf{V})^2 v_z = \pm \mathrm{i}k(\omega - \mathbf{k} \cdot \mathbf{V})\Pi + (\mathbf{k} \cdot \mathbf{B})v_z/4\pi. \tag{3.156}$$

With continuity of Π and of $v_z/(\omega - \mathbf{k} \cdot \mathbf{V})$ imposed, (3.156) then yields the dispersion relation

$$\rho_1(\omega - \mathbf{k} \cdot \mathbf{V}_1)^2 + \rho_2(\omega - \mathbf{k} \cdot \mathbf{V}_2)^2 = [(\mathbf{k} \cdot \mathbf{B}_1)^2 + (\mathbf{k} \cdot \mathbf{B}_2)^2]/4\pi. \tag{3.157}$$

With $\mathbf{k}$ prescribed to be real, the real roots of (3.157) correspond to pure oscillatory (stable) solutions. When ω is real, the left-hand side of (3.157) has the minimum value

$$\frac{\rho_1\rho_2}{\rho_1 + \rho_2}[\mathbf{k} \cdot (\mathbf{V}_1 - \mathbf{V}_2)]^2, \tag{3.158}$$

attained for

$$\omega = \mathbf{k} \cdot \frac{\rho_1 \mathbf{V}_1 + \rho_2 \mathbf{V}_2}{\rho_1 + \rho_2}; \tag{3.159}$$

hence from (3.157), the condition for stability (ω real) is

$$\frac{\rho_1\rho_2}{\rho_1 + \rho_2}[\mathbf{k} \cdot (\mathbf{V}_1 - \mathbf{V}_2)]^2 < \frac{(\mathbf{k} \cdot \mathbf{B}_1)^2 + (\mathbf{k} \cdot \mathbf{B}_2)^2}{4\pi}. \tag{3.160}$$

This result is the analogue of (3.148) (with the inequality sign reversed). It depends not on the magnitude but on the direction of $\mathbf{k}$. For absolute stability, (3.160) must hold for all $\mathbf{k}$. Take the x-axis parallel to the relative velocity $\mathbf{V} = \mathbf{V}_1 - \mathbf{V}_2$. Then if B_{1y} and B_{2y} are not both zero, the ratio of the right-hand side of (3.160) to the left-hand side has a maximum at

$$\frac{k_x}{k_y} = -\frac{B_{1x}B_{1y} + B_{2x}B_{2y}}{B_{1y}^2 + B_{2y}^2}, \tag{3.161}$$

and yielding as a condition of absolute stability

$$\frac{\rho_1 \rho_2 V^2}{\rho_1 + \rho_2} < \frac{(B_{1x}B_{1y} - B_{2x}B_{2y})^2}{B_{1y}^2 + B_{2y}^2}. \tag{3.162}$$

This shows that that absolute stability can occur only if the fields $\mathbf{B}_1$, $\mathbf{B}_2$ are not parallel but are mutually sheared. If B_{1x}, B_{2x} are both zero, then disturbances with wave-number $(k_x, 0)$ exchange field lines without distorting them, hence the field has no stabilizing effect. At the other limit, with B_{1y}, B_{2y} both zero, then (3.160) does not depend on k_y but reduces to

$$\frac{\rho_1 \rho_2}{\rho_1 + \rho_2} V^2 < \frac{B_{1x}^2 + B_{2x}^2}{4\pi}. \tag{3.163}$$

In this case, a disturbance with $k_x \neq 0$ does bend the field lines, so the field has a stabilizing effect. In the limit as B_{1y}, B_{2y} go simultaneously to zero, (3.161) yields $k_x \to 0$, for which (3.157) yields $\omega = 0$: a wave in the y-direction merely exchanges field lines without bending them.

3.6.5 *Stability of rotating systems*

One is often interested in the stability of states with a zero-order velocity field, for example a star or a circumstellar disc with a non-uniform rotation field (cf. Sections 5.3 and 10.5). Frieman and Rotenberg (1960) discussed hydromagnetic systems without gravitational force; Lynden-Bell and Ostriker (1967) treated non-magnetic, differentially rotating stars. Following Chanmugam (1978), we consider differentially rotating, magnetic systems with an externally maintained gravitational field and with symmetry about the rotation axis.

The problem is treated in the inertial frame. Initially, the zero-order, steady velocity field $\mathbf{v}$ is not specialized as a rotation field. A displacement field $\xi(\mathbf{r}, t) = \xi(\mathbf{r}) \exp \mathrm{i}\omega t$, substituted into the equation of motion, now yields

$$-\omega^2 \rho \xi + 2\mathrm{i}\omega\rho(\mathbf{v} \cdot \nabla)\xi - \mathbf{F}(\xi) = 0, \tag{3.164}$$

where

$$\begin{aligned} F_i(\xi) = {} & \frac{\partial}{\partial x_i}(\gamma p \nabla \cdot \xi + (\xi \cdot \nabla)p - \mathbf{B} \cdot \mathbf{Q}) + (\mathbf{B} \cdot \nabla)Q_i + (\mathbf{Q} \cdot \nabla)B_i - \nabla \cdot (\rho\xi)\frac{\partial V}{\partial x_i} \\ & + \frac{\partial}{\partial x_j}\left[\rho \xi_j (\mathbf{v} \cdot \nabla)v_i - \rho v_j (\mathbf{v} \cdot \nabla)\xi_i\right], \end{aligned} \tag{3.165}$$

and

$$\mathbf{Q} = \nabla \times (\xi \times B). \tag{3.166}$$

The zero-order $\mathbf{v}$-field field shows up in the middle term in (3.164) and in the second block in (3.165).

Multiply (3.165) on the left by ξ^* and integrate over the whole body to obtain

$$-\omega^2 a + 2\omega b + c = 0, \tag{3.167}$$

where

$$a = \int \rho \xi^* \xi \, \mathrm{d}\tau, \tag{3.168}$$

$$b = \mathrm{i} \int \rho \xi^* (\mathbf{v} \cdot \nabla) \xi \, \mathrm{d}\tau, \tag{3.169}$$

$$c = -\int \xi^* \mathbf{F}(\xi) \, \mathrm{d}\tau. \tag{3.170}$$

The quantity a is manifestly real (and positive). It can be shown (Frieman and Rotenberg 1960; Lynden-Bell and Ostriker 1967) that the operators $\mathrm{i}\rho(\mathbf{v} \cdot \nabla)\xi$, and the modified $\mathbf{F}(\xi)$ entering into (3.164) are Hermitian, so that b, c are also real. For example, take two field vectors ξ, η, everywhere non-singular and with continuous first derivatives, and construct the 'matrix element'

$$\begin{aligned} \int \eta^* \cdot (\mathrm{i}\rho \mathbf{v} \cdot \nabla)\xi \, \mathrm{d}\tau &= \mathrm{i} \int \{\nabla \cdot [\rho \mathbf{v}(\eta^* \cdot \xi)] - \rho \xi \cdot (\mathbf{v} \cdot \nabla)\eta^*\} \, \mathrm{d}\tau \\ &= -\mathrm{i} \int \xi \cdot (\mathbf{v} \cdot \nabla)\eta^* \rho \, \mathrm{d}\tau \\ &= (\mathrm{i}/2) \int [\eta^* \cdot (\mathbf{v} \cdot \nabla)\xi - \xi \cdot (\mathbf{v} \cdot \nabla)\eta^*]\rho \, \mathrm{d}\tau, \\ &= \left(\int \xi^* \cdot (\mathrm{i}\rho \mathbf{v} \cdot \nabla)\eta \, \mathrm{d}\tau \right)^* . \end{aligned} \tag{3.171}$$

The divergence term disappears after transformation into an integral over the surface of the unperturbed system, which vanishes because the normal component of the unperturbed flow vanishes there. Thus the Hermitian property is seen to hold, and b is real; a similar proof shows that c is also real (see the references cited).

From (3.167),

$$\omega = (b \pm \sqrt{b^2 + ac})/a. \tag{3.172}$$

Since $a > 0$, ω real is ensured if $c > 0$, which is therefore a sufficient condition for stability; while a stronger condition is $b^2 + ac > 0$.

Consider now a star or a disc rotating with angular velocity $\Omega(\varpi, z)$ in cylindrical polar coordinates (ϖ, ϕ, z), so that $\mathbf{v} = (0, \Omega\varpi, 0)$ in the inertial frame. Hence

$$(\mathbf{v} \cdot \nabla)\mathbf{v} = -\Omega^2 \varpi(1, 0, 0), \tag{3.173}$$

$$(\mathbf{v} \cdot \nabla)\xi = \Omega(\varpi) \left(\frac{\partial \xi_\varpi}{\partial \phi} - \xi_\phi, \frac{\partial \xi_\phi}{\partial \phi} + \xi_\varpi, \frac{\partial \xi_z}{\partial \phi} \right). \tag{3.174}$$

Now write $\xi = (\xi_\varpi, \xi_\phi, \xi_z) \exp im\phi$, where ξ_ϖ, ξ_z are real and ξ_ϕ pure imaginary. After some algebra, we arrive at

$$b = -2\mathrm{i}\int \Omega\xi_\varpi\xi_\phi\rho\,\mathrm{d}\tau - m\int \rho\Omega\xi^*\cdot\xi\,\mathrm{d}\tau, \tag{3.175}$$

$$c = h + 2\delta W - m\int \rho\Omega^2[m\xi^*\cdot\xi + 4\mathrm{i}\xi_\varpi\xi_\phi]\,\mathrm{d}\tau, \tag{3.176}$$

where

$$h = \int \Omega^2\xi_\varpi\varpi\left[\frac{\partial\rho}{\partial\varpi}\xi_\varpi + \frac{\partial\rho}{\partial z}\xi_z + 2\varpi\Omega\rho\left(\varpi_r\frac{\partial\Omega}{\partial\varpi} + \xi_z\frac{\partial\Omega}{\partial z}\right) + \rho\nabla\cdot\xi\right]\mathrm{d}\tau, \tag{3.177}$$

and δW is the expression (3.134), valid in the absence of rotation. (The effect of rotation is, however, implicit in δW through its effect on the ∇p term through the equation of motion.)

To illustrate the effect of rotation, consider the simplest example: a system rotating with uniform angular velocity Ω, and subject to axisymmetric perturbations. Then the bracketed term in (3.177) is just $\nabla\cdot(\rho\xi)$, and so from (3.176), c reduces to the same expression δW but with gravity $\mathbf{g}\equiv\nabla V$ replaced by effective gravity $g_e \equiv \nabla V + \Omega^2\varpi$. (In fact, for this case the results are reached more immediately when one uses the frame rotating with angular velocity Ω; the unperturbed velocities are zero in this frame, but the centrifugal and Coriolis forces yield immediately the above modification to δW and the term $2\omega b$ in (3.167).) If $\Omega = 0$, we recover the standard energy principle discussed in Section 3.6.1. With $\Omega \neq 0$, then since a, b, c are all real, a *sufficient* condition for stability is that $\delta W > 0$ for all possible perturbations – identical with that for non-rotating systems, except that the zero-order effects of rotation are included in p and g_e, as noted. However, the Coriolis terms have a stabilizing effect, so $\delta W < 0$ is not sufficient to make the expression under the radical in (3.172) negative. Thus the modified energy criterion is sufficient but not necessary for stability (cf. Chandrasekhar 1987; Lebovitz 1966). The stability of differentially rotating systems is discussed below in the context of accretion disc theory (Section 10.5).

3.7 Effects of dissipation; reconnection

The various dissipative processes – finite Ohmic resistivity, viscosity, ambipolar diffusion – will have some anticipated effects, such as the damping of wave motions, of likely importance for the heating of solar and stellar chromospheres and coronae (Chapter 4). It is also known that in non-magnetic problems, departure from strict adiabaticity can lead to destabilization of modes that are otherwise dynamically stable (Goldreich and Schubert 1967). In hydromagnetic problems, a similar effect can also follow from *reconnection* – the relaxation of the severe topological constraint of strict flux-freezing. Thus consider a domain with a sheet current $\mathbf{j} = cB_0/4\pi l(0,\,1,\,0)$ of thickness l, maintaining the field $\mathbf{B} = B_0(z/l,\,0,\,0)$ in Cartesian coordinates (Fig. 3.14(a)). If gravity is ignorable, *dynamically* stable equilibrium is maintained by a balance of thermal and magnetic pressures $p + \mathbf{B}^2/8\pi = \text{constant}$. In the absence of mass motions, (2.95) predicts that the oppositely-directed field lines mutually annihilate by

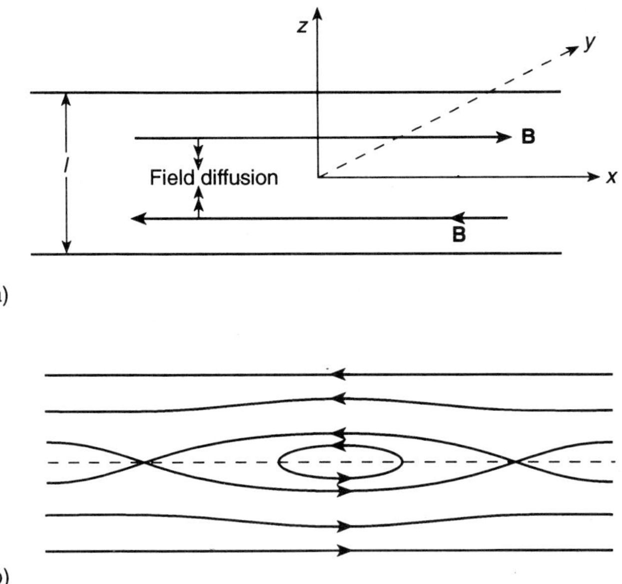

FIG. 3.14. (a) Unperturbed plane-parallel pinched zone, undergoing slow mutual annihilation of oppositely directed field lines. (b) The tearing mode instability.

slow diffusion in a characteristic time $\tau_d \simeq 4\pi\sigma l^2/c^2$. Linear perturbation analysis (Furth *et al.* 1963) yields in particular the *tearing mode* instability for modes $\propto \exp \mathrm{i}(kx - \omega t)$ with k in the approximate range $(\tau_\mathrm{A}/\tau_d)^{1/4} < kl < 1$, where $\tau_\mathrm{A} = l/[B_0/(4\pi\rho)^{1/2}]$. The instability is driven by the magnetic force arising from the changed topology of the field due to finite resistivity (Fig. 3.14(b)). A typical growth rate is

$$[\tau_d^3 \tau_\mathrm{A}^2 (kl)^2]^{-1/5}, \tag{3.178}$$

with a maximum (for $kl = (\tau_\mathrm{A}/\tau_d)^{1/4}$) of

$$(\tau_\mathrm{A}\tau_d)^{-1/2} \tag{3.179}$$

the geometric mean of the diffusive and dynamical rates. Numerical simulations into the non-linear domain show how the mutual annihilation process is indeed speeded up.

Reconnection of field lines is of great importance in a variety of problems. Sometimes it is the actual change in the field topology that is of interest, e.g. in theories of star formation that rely on magnetic field lines linking different domains to transfer angular momentum (Sections 11.6 and 12.3). An illustrative example is given in Section 3.7.1. In flare theory, one focuses rather on the explosive release of energy accompanying reconnection that occurs sufficiently rapidly. There the problem is to some extent analogous to the linear resistive instability: one looks for ways in which accompanying bulk motions can greatly accelerate the diffusion process (see Section 3.7.2).

3.7.1 *Reconnection in a medium at rest*

We recall that in the infinite conductivity approximation, the Galilean-invariant form of the simple Ohm's law predicts that the field moves with the bulk motion of the gas. Retention of the Hall term changes this to motion of the field with the electrons, but the difference is small except in regions of locally high current density where neglect of dissipative terms becomes questionable, especially if the normal resistivity is replaced by a collective response (Section 2.8). Likewise, in a lightly ionized gas (Section 2.7), the magnetokinematics is well approximated by having the field tied to the ionized component, with the joint ion–electron motion ('ambipolar diffusion') given by the balance between the Lorentz force and the ion–neutral friction. As long as the ionized component satisfies a continuity equation, without sources and sinks, there is no change in the topology of the field. For example, a field that is a local distortion of a large-scale background field, with 'infinite field lines', will remain so, even if there has been leakage of field lines plus charges by ambipolar diffusion (cf. Sections 11.7 and 12.4). Ohmic diffusion, however, relaxes the constraint of strict flux freezing and this can change the field topology, in particular leading to a domain with closed field lines nesting about an *O*-type neutral point, and linking with the background field at an *X*-type neutral point.

The process is most easily illustrated when the plasma has a non-anomalous, Ohmic resistivity $\lambda = c^2/4\pi\sigma$ which may depend on position through the temperature but does not depend on the field strength (Mestel and Strittmatter 1967). Suppose the gas is at rest, so that $\mathbf{j} = \sigma\mathbf{E}$, and the field evolves according to (2.95):

$$\frac{\partial \mathbf{B}}{\partial t} = -c\nabla\times\mathbf{E} = -\nabla\times(\lambda\nabla\times\mathbf{B}). \tag{3.180}$$

Let the field consist of a uniform background field $\mathbf{B}_0$ with a vanishing curl, plus the local distortion $\Delta\mathbf{B}$, symmetric about the direction of $\mathbf{B}_0$ (e.g. as in Fig. 11.1(a)). At time $t = t_0$ the total field $\mathbf{B}$ is as in Fig. 3.15(a), while $\Delta\mathbf{B}(t_0)$ is as in Fig. 3.15(b). Because (3.180) is linear it is satisfied by $\Delta\mathbf{B}$ alone. One can then solve (3.180) for $\Delta\mathbf{B}(t)$, starting from $\Delta\mathbf{B}(t_0)$: to view the total field at any subsequent epoch, one needs just to construct $\Delta\mathbf{B}(t) + \mathbf{B}_0$.

As pointed out by Sweet (1950), when currents flow perpendicular to the field, so that $\mathbf{j}\cdot\mathbf{B} = 0$ as for an axisymmetric 'poloidal' field, with lines in meridian planes, then Ohm's law $\mathbf{E} = \mathbf{j}/\sigma$ can be written

$$\mathbf{E} + \frac{\mathbf{v}_d \times \mathbf{B}}{c} = 0, \tag{3.181}$$

where

$$\mathbf{v}_d = \frac{c}{\sigma}\frac{\mathbf{j}\times\mathbf{B}}{B^2} = \lambda\frac{(\nabla\times\mathbf{B})\times\mathbf{B}}{B^2}. \tag{3.182}$$

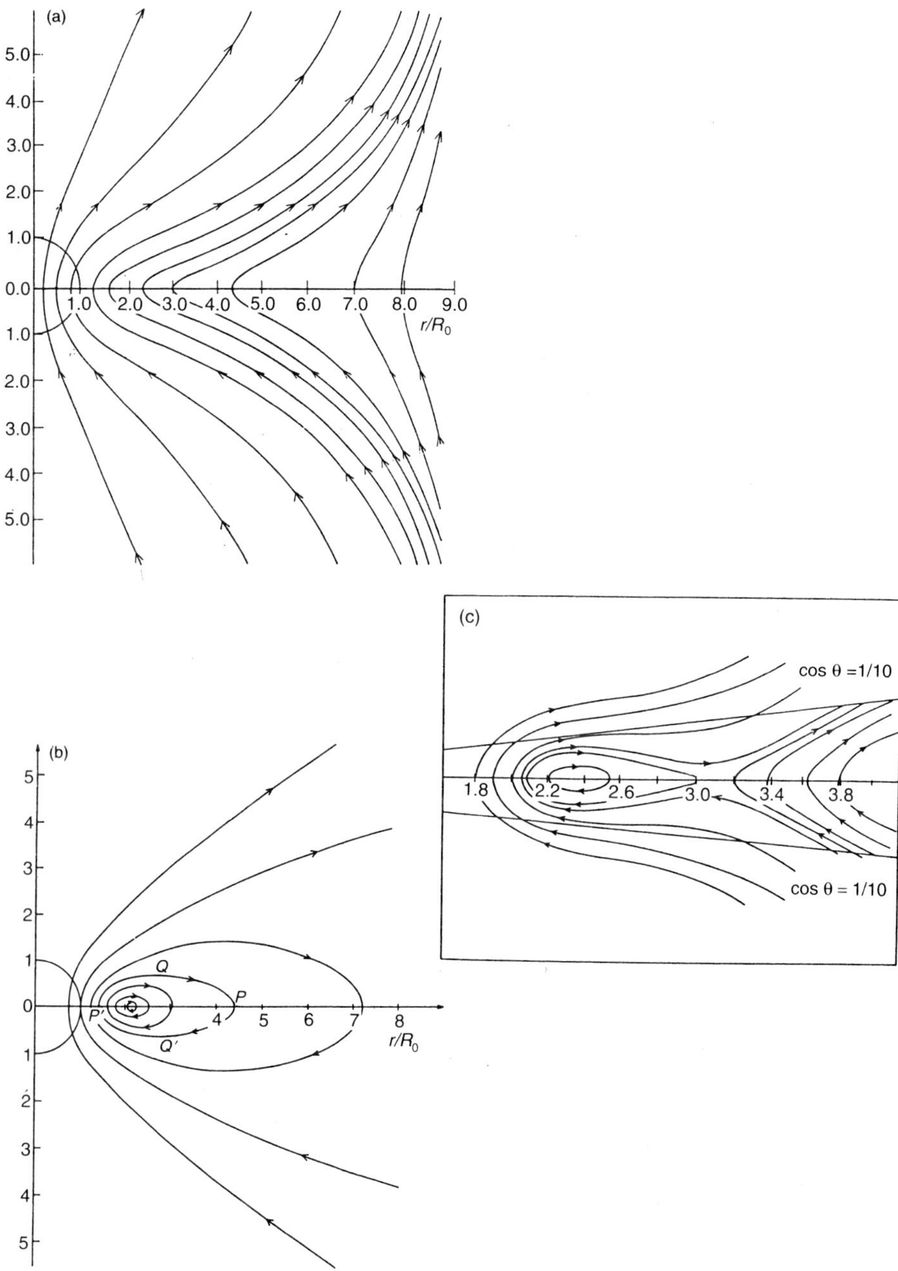

FIG. 3.15. Diffusion leading to field line detachment. (a) Field at time $t = 0$. (b) $\Delta\mathbf{B}$, the same field but with the uniform background field $\mathbf{B}_0$ subtracted out. (c) After some time, the total field has the configuration with O- and X-type neutral points and associated detached field lines.

Thus in this simple case, the diffusion can be treated as if the field lines move with speed $\mathbf{v}_d$. The inexorable decay of the field $\Delta\mathbf{B}$ through motion of the field lines into the neutral point O will be discussed further in Section 5.7. For the moment we suppose that λ is sufficiently non-uniform, being low near O, so that flux-loss is negligible over the times of interest, but high enough away from O for the changing shape of the loops to be important. It is anticipated that in an elongated loop such as L, on the equator $|\nabla\times\mathbf{B}|/B$ will be larger at P than at P' because of the larger curvature, and correspondingly smaller at off-equator points Q, Q'. Thus for a given value of λ, the diffusion is expected to make such loops more nearly circular, with points like P moving inwards. At $t=0$, at all points P, $\Delta B_z < 0$, but $|\Delta B_z| < B_0$, so that the total field is as in Fig. 3.15(a). However, as the loop L evolves, with P moving in, then $|\Delta B_z|$ will grow. At some epoch t, $\Delta B_z + B_0$ will be positive except at a unique point C where it vanishes: the total field has developed a cusp-like neutral point at C. At later epochs there is an interval on the equator in which $\Delta B_z + B_0$ is negative, extending from the O-type neutral point O to the X-type point X (Fig. 3.15(c)): a domain of detached field lines has spontaneously developed by diffusion.

Figure 3.15 shows a specific example worked out by Mestel and Strittmatter (1967) for uniform λ. In spherical polar coordinates based on the direction of $\mathbf{B}_0$, the field is written in terms of the flux function $P(r,\mu)$:

$$\mathbf{B} = \nabla\times(0,0,-P/r\sin\theta) \tag{3.183}$$

where $\mu = \cos\theta$. The solution for the total field is then found to be of the form

$$P = \sum\nolimits_k \left[\mathrm{P}_{2k+2}(\mu) - \mathrm{P}_{2k}(\mu)\right]\left(\frac{r^{1/2}}{2\lambda t}\right) \times \int_0^\infty z^{1/2}\mathrm{F}_{2k+1}(z)\mathrm{I}_{2k+3/2}\left(\frac{rz}{2\lambda t}\right)\exp\left[-\frac{(r^2+z^2)}{4\lambda t}\right]\mathrm{d}z, \tag{3.184}$$

where P_{2k} is a Legendre function, and the functions F_{2k+1} are defined in terms of the initial flux function $\bar{P}(r,\mu)$ by

$$\mathrm{F}_{2k+1}(r) = \int_{-1}^{1}\mathrm{P}_{2k+1}(\partial\bar{P}/\partial\mu)\,\mathrm{d}\mu. \tag{3.185}$$

The function P can thus be computed from (3.184) at all subsequent times. For the strongly distorted field adopted, even with constant λ, detachment of field lines is found.

3.7.2 *Reconnection in general*

The above example is merely illustrative of how relaxation of strict flux-freezing through the retention of finite diffusion can yield a qualitatively different picture. To achieve the much faster reconnection rate required to understand for example the solar flare phenomenon, one must allow for plasma motion. The early two-dimensional model of Sweet (1958) and Parker (1963) is as in Fig. 3.16. The system is mirror-symmetric in the y-axis. The diffusion region is that of a current sheet in the $-z$-direction, extending between $x = \pm L$ and $y = \pm l$. Inflowing gas drags in oppositely directed magnetic field lines from either side, to be mutually annihilated, while the gas flows out along the sheet in the $\pm x$-directions. In a steady state,

$$v_i = \lambda / l, \tag{3.186}$$

for the inward advection of flux at the speed v_i must equal the speed λ/l with which the oppositely directed field lines within $|y| \approx l$ diffuse towards each other, as given by (3.182). The gas within the diffusion zone has a thermal pressure equal to the pinching pressure $B_i^2/8\pi$ of the external magnetic field. For simplicity, consider the incompressible limit; then by Bernoulli's equation the gas squeezed out by this pressure acquires the speed $v_{\mathrm{A}i} = B_i/(4\pi\rho)^{1/2}$. By conservation of gas flowing into and out of the diffusion region,

$$L v_i = l v_{\mathrm{A}i}; \tag{3.187}$$

elimination of l then yields the inflow Alfvénic Mach number $(v/v_{\mathrm{A}})_i$ in terms of the inflow MRN $\mathrm{R}_{\mathrm{m}i} = L v_{\mathrm{A}i}/\lambda$:

$$M_i = \mathrm{R}_{\mathrm{m}i}^{-1/2}. \tag{3.188}$$

With values for $\mathrm{R_m} \simeq 10^6 - 10^{12}$, based on the microresistivity, $M_i \approx 10^{-3} - 10^{-6}$. This rate of reconnection is relevant to some recent versions of the dynamo problem (cf. Section 6.8), but is still many orders of magnitude below that required to explain rapid release of energy in a flare.

'Slow' reconnection as in the Sweet–Parker picture occurs at a rate determined by the value of the micro-diffusivity and the macro-length-scale, combining with $v_{\mathrm{A}i}$ to yield $\mathrm{R_m}$. The first 'fast' mechanism (Petschek 1964) limits diffusion to a small central region, of scale that adjusts so as to allow a much faster imposed reconnection rate. Most of the energy conversion occurs at slow shocks, which accelerate and heat the plasma to form two hot, fast, outflowing jets (Fig. 3.17). Petschek estimated an *upper limit* to the reconnection rate, given by

$$M_e^* = \frac{\pi}{8 \log \mathrm{R}_{\mathrm{m}e}} \tag{3.189}$$

with $\mathrm{R}_{\mathrm{m}e} = L v_{\mathrm{A}e}/\lambda$, the MRN defined by the inflow quantities. In practice, $M_e^* \approx 0.01$, much greater than the Sweet–Parker rate. Petschek's model in fact incorporates the Sweet–Parker picture, which describes the plasma and field in the diffusion region. The crucial advantage is the (admittedly non-rigorous) prediction of 'fast' reconnection, dependent essentially on the macro-quantities and only very weakly on the micro-resistivity.

Priest and Forbes (1986, 1992) have generalized the Petschek model, partly in order to resolve the apparent conflict with some numerical simulations, so bringing out the crucial role of the imposed boundary conditions in the different

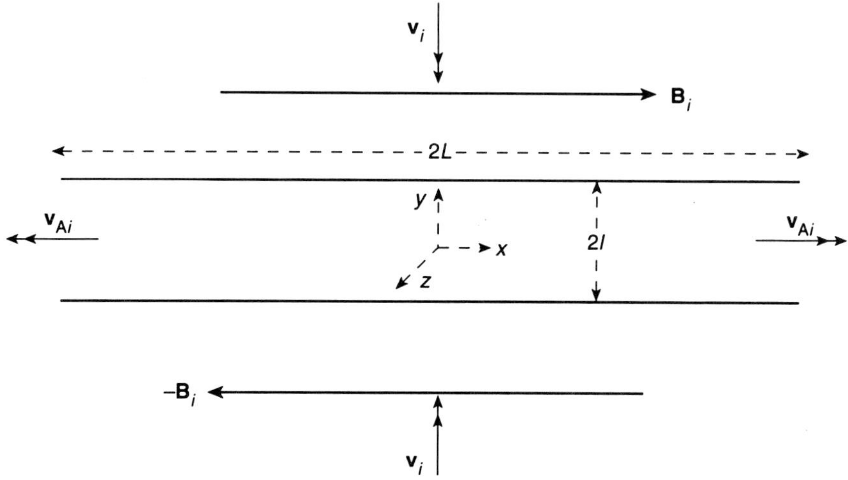

FIG. 3.16. The Sweet–Parker reconnection model.

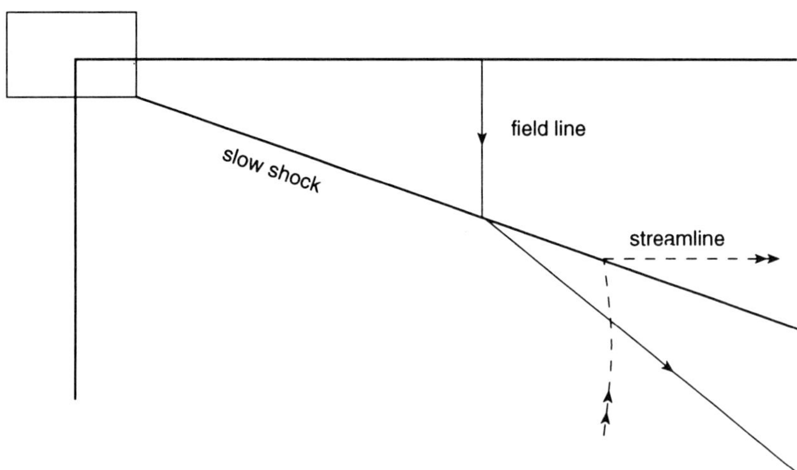

FIG. 3.17. The Petschek fast reconnection model. The diffusion region occupies only a small central location, while most of the energy conversion occurs at stationary slow-mode shocks, which accelerate and heat the plasma to form two hot fast jets.

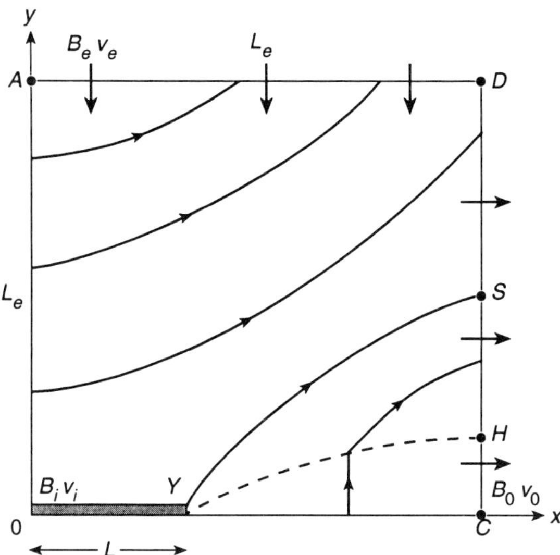

FIG. 3.18. Details of the reconnection configuration (after Priest and Forbes 1986, 1992). OY is the central current sheet of thickness l, YH the shock, YS the separatrix. The fluid flow speed and field strength change from the external values v_e, B_e to v_i, B_i at the entrance to the diffusion region, then to v_0, B_0 at the outflow point.

models. Their work links up also with studies by Sonnerup (1970) and Vasyliunas (1975). We have already emphasized that field-freezing is often an excellent first-order approximation, but also that systems will often try to find ways of violating the consequent severe topological constraints. The Priest–Forbes model is therefore presented in some detail, again as an illustration of principles. With the dynamo problem, for example, demanding increasingly sophisticated treatment (cf. Chapter 6), we anticipate growing effort to elucidate further the subtleties hidden in the equations of MHD at high MRN.

Figure 3.18 represents a quadrant of the reconnecting domain. Like Petschek's, the models studied are all 'nearly uniform', with the inflow magnetic field lines only slightly curved from the uniform background field $B_0\hat{\mathbf{x}}$. For simplicity, the flow is again assumed incompressible. The equations for steady, incompressible, two-dimensional flow

$$\rho(\mathbf{v}\cdot\nabla\mathbf{v}) = -\nabla p + \frac{(\nabla\times\mathbf{B})\times\mathbf{B}}{4\pi}, \tag{3.190}$$

$$c\mathbf{E} + \mathbf{v}\times\mathbf{B} = 0, \tag{3.191}$$

$$\nabla\cdot\mathbf{v} = 0, \quad \nabla\cdot\mathbf{B} = 0, \tag{3.192}$$

are solved by perturbing about a uniform field:

$$\mathbf{B} = B_0\hat{\mathbf{x}} + \mathbf{B}_1 + ..., \quad \mathbf{v} = \mathbf{v}_1 + ... \tag{3.193}$$

The vector potential $A_1\hat{\mathbf{z}}$ satisfies

$$\nabla^2 A_1 = -\frac{4\pi}{c} j_1, \tag{3.194}$$

where $B_{1x} = \partial A_1/\partial y$, $B_{1y} = -\partial A_1/\partial x$, and

$$j_1 = \frac{1}{B_0}\frac{\mathrm{d}p_1}{\mathrm{d}y}. \tag{3.195}$$

The boundary condition $B_{1y}(0, y) = 0$ follows from symmetry. The conditions $B_{1x}(x, L_e) = 0$, $\partial B_{1y}/\partial x = 0$ on $x = L_e$ serve to make the solutions unique in terms of the prescribed parameters. On $y = 0$ the boundary conditions

$$\begin{aligned} B_{1y} &= 2B_N x/L, \qquad 0 \le |x| \le L, \\ B_{1y} &= 2B_N \qquad\qquad L \le |x| \le L_e \end{aligned} \tag{3.196}$$

are imposed. The domain $|x| \le L$ is essentially the Sweet–Parker diffusion region. The quantity B_N is the magnetic field component normal to each of the Petschek slow-mode shocks. As the present model is almost uniform, each shock is inclined at a small angle to the x-axis and so the field behind is nearly normal: the shocks 'switch off' the transverse **B**-component (cf. Section 3.2). It is likewise consistent (as in (3.196)) to compute the first-order deviations from the uniform field $\mathbf{B}_0$ by applying the shock conditions on the line segments $L \le |x| \le L_e$, $y = 0$ instead of on the shocks. The factor 2 in (3.196) ensures that the shock surface currents in both domains – $y < 0$ as well as $y > 0$ – contribute to the field $\mathbf{B}_1$.

Note also that in the incompressible limit, the shocks are effectively finite-amplitude Alfvén waves: the switch-off condition (3.35) is applied first to the jump conditions (3.28) and (3.31) and then the limit $D \to 1$ is taken. In the present problem, since the flow ahead of the shocks is uniform to the first order, condition (3.35) can be written approximately

$$B_N = M_e B_e, \tag{3.197}$$

where M_e is the Alfvénic Mach number of the inflowing gas (i.e. at large $|y|$).

Subject to these conditions, (3.194) has the solution

$$A_1 = -\sum_0^\infty \frac{a_n}{(n+1/2)\pi}\left(b - \cos\left[(n+1/2)\pi\frac{x}{L_e}\right]\right)\cosh\left[(n+1/2)\pi\left(1 - \frac{y}{L_e}\right)\right], \tag{3.198}$$

where

$$a_n = \frac{4B_N L_e^2 \sin[(n+1/2)\pi L/L_e]}{L(n+1/2)^2\pi^2 \cosh[(n+1/2)\pi]}; \tag{3.199}$$

whence

$$B_{1x} = \sum_0^\infty a_n \left(b - \cos\left[(n+1/2)\,\pi\frac{x}{L_e}\right]\right) \sinh\left[(n+1/2)\,\pi\left(1-\frac{y}{L_e}\right)\right], \tag{3.200}$$

$$B_{1y} = \sum_0^\infty \frac{a_n}{L_e} \sin\left[(n+1/2)\pi\frac{x}{L_e}\right] \cosh\left[(n+1/2)\pi\left(1-\frac{y}{L_e}\right)\right], \tag{3.201}$$

and

$$\frac{4\pi}{c} j_1 = \sum_0^\infty a_n (n+1/2)\pi b \cosh\left[(n+1/2)\,\pi\left(1-\frac{y}{L_e}\right)\right]. \tag{3.202}$$

Equation (3.195) then yields the associated pressure p_1; (3.191) yields for the first-order flow the familiar uniform drift

$$\mathbf{v}_1 = c(E/B_0)\hat{\mathbf{y}}, \tag{3.203}$$

but for the second-order flow

$$\mathbf{v}_2 = -v_1(B_{1x}/B_0)\hat{\mathbf{y}} + (v_1/B_0)(B_{1x} - 4\pi x j_1/c)\hat{\mathbf{x}}. \tag{3.204}$$

The solution depends on the parameter b, which by (3.202) determines the strength of the volume currents j_1 in the inflow region. When $b = 0$ the modification to the external field B_0 is due just to the current sheets in the shocks. The model represents a Petschek-type solution, with the plasma and magnetic pressures decreasing together and the flow converging as the magnetic field is carried in – a 'fast-mode expansion'. As b increases, the inflow changes to being divergent (with strong slow-mode expansions). These modes differ from the Petschek-type modes in having long diffusion regions. Rapid reconnection occurs through a pile-up of magnetic flux at the entrance to the diffusion region, and now the diffusion region lengthens as the rate of reconnection goes up.

For each solution one is interested especially in the relation between M_i and M_e, respectively the Alfvén Mach numbers at the entrance to the diffusion region and in the inflow domain at $|y| = L_e$, so one needs to use the above solution to compute $B_i \equiv B_x(0,0) = B(0,0)$ and $B_e \equiv B_x(0,L_e) = B(0,L_e)$. The result is

$$M_e^{1/2} M_i^{-1/2} = 1 - 4M_e\pi^{-1}(1-b)[0.834 - \ln\tan(\pi \mathrm{R}_{me}^{-1} M_e^{-1/2} M_i^{-3/2}/4)], \tag{3.205}$$

which has the property – predicted by Petschek in his less rigorous treatment – that there is a maximum value for M_e, found to be quite close to the original estimate. Jardine and Priest (1988*a*,*b*) have extended the Priest–Forbes theory to higher order, including compressibility and energetics.

As emphasized by Priest and Forbes, the overall conclusion is that the type of reconnection regime and the associated rate of reconnection depend sensitively on the inflow boundary conditions. In particular, the numerical experiments of Biskamp (1986) yielded only the slow Sweet–Parker reconnection rate, because he adopted a fixed normal component B_y superposed on the $B_0\hat{\mathbf{x}}$-field. The same conditions applied to the analytical treatment yield the same result. Although most of the models in the literature are undoubtedly too idealized, nevertheless it appears that the rate of reconnection will often be determined by the gross dynamics and electrodynamics.

3.8 Macroscopic dissipation

The gas-dynamic equations of Chapter 2 have been written in terms of a bulk velocity for each species of particle, plus random microscopic velocities that yield pressures, viscosity, resistivity, etc. An explicit assumption is that the macroscopic length-scales remain large compared with the mean-free-path. The steepening of compressional waves is a familiar example of how the dissipation-free equations spontaneously develop structures which locally violate this basic assumption. However, this is dealt with by dividing the flow into domains within which the equations of magneto-fluid dynamics hold and which are linked by discontinuities; the jump conditions follow from the conservation laws subject to the entropy increase constraint (Section 3.2). In magnetized plasmas that are so diffuse that the inequality (2.29) holds (ion mean-free-path λ_i longer than the gyration radius $r_g = (v_T/\omega)_i$), then r_g replaces λ_i as the thickness of the 'collision-free' shock. A plausible provisional picture of, for example, the expanding solar corona (the 'solar wind' – cf. Chapter 7) has a laminar outflow of gas heated by dissipation of magnetosonic shock waves emanating from the convective envelope. The heat input H in (2.77) is fixed by the macro-dynamics of the waves in the emitted spectrum.

Again, a familiar hydrodynamic phenomenon is the spontaneous development of 'turbulence', a field of random *macroscopic* velocities superposed on a mean laminar flow. If the velocities are subsonic, then the density fluctuations will be modest, and the Navier–Stokes equations remain valid everywhere. A standard picture of hydrodynamic turbulence has an 'inertial range', in which the non-linear inertial terms cause a dissipation-free cascade of energy from the larger to the smaller eddies, until scales small enough are reached for the small micro-viscosity to operate. The time-scale of the dissipative process is that of the cascade, typically a turn-over time of the large eddies, while the viscosity fixes the minimum scale on which the motions persist. A simple parametrization of the actual dissipation rate then replaces the micro-viscosity by a macro-eddy-viscosity acting on the mean flow, with a macro-scale and a turbulent velocity replacing respectively the mean-free-path and the thermal velocity. When the turbulence develops from convective instability, the macro-scale is identified with the 'mixing-length' (cf. Section 4.2).

Referring to Section 3.7, it is seen that the Petschek picture and its variants have the properties of a turbulent process: the gross dynamics and electrodynamics fix the rate of reconnection, while the actual micro-resistivity fixes the dimensions of the domain in which reconnection occurs. And in fact the analogous concept of a turbulent resistivity – of the same order as the eddy-viscosity – arises as soon as one combines the cascade picture of hydrodynamic turbulence with field freezing, *provided the field remains passive and does not react back on the dynamics of the turbulence*; a big proviso, as tangling of the field by the turbulent eddies will prima facie steadily increase the Lorentz force density and so interfere with the cascade. This will come up again in Chapters 4, 5, 6 and 8, especially in connection with the dynamo problem.

APPENDIX

Poloidal and Toroidal Fields

A solenoidal vector field may in general be written as the sum of poloidal and toroidal vector fields, defined by (Chandrasekhar 1961, Appendix III)

$$\mathbf{S} = \nabla \times \left[\nabla \times \left(\frac{\Phi}{r}\mathbf{r}\right)\right] \equiv \nabla \times \left[\nabla \left(\frac{\Phi}{r}\right) \times \mathbf{r}\right], \tag{3.206}$$

$$\mathbf{T} = \nabla \times \left(\frac{\Psi}{r}\mathbf{r}\right) \equiv \nabla \left(\frac{\Psi}{r}\right) \times \mathbf{r}, \tag{3.207}$$

where Φ and Ψ are arbitrary scalar functions of position.

Since $\mathbf{T}$ is solenoidal,

$$\begin{aligned}\nabla\times(\nabla\times\mathbf{T})_i &= -\frac{\partial^2 T_i}{\partial x_s \partial x_s} = \frac{\partial^2}{\partial x_s^2}\left[\varepsilon_{ijk} x_j \frac{\partial}{\partial x_k}\left(\frac{\Psi}{r}\right)\right] \\ &= \varepsilon_{ijk} x_j \frac{\partial}{\partial x_k}\left[\nabla^2\left(\frac{\Psi}{r}\right)\right] = -\nabla\left[\nabla^2\left(\frac{\Psi}{r}\right)\right] \times \mathbf{r}. \end{aligned} \tag{3.208}$$

Similarly,

$$\nabla\times\mathbf{S} = -\nabla\left[\nabla^2\left(\frac{\Phi}{r}\right)\right] \times \mathbf{r}. \tag{3.209}$$

Thus from (3.206) and (3.207), $\nabla\times\mathbf{T}$ is a poloidal field with the same defining scalar Ψ, while $\nabla\times(\nabla\times\mathbf{T})$ is a toroidal field with the defining scalar $\tilde{\Psi} = -r\nabla^2(\Psi/r)$. Likewise, $\nabla\times\mathbf{S}$ is a toroidal field with the defining scalar $\tilde{\Phi} = -r\nabla^2(\Phi/r)$, while $\nabla\times(\nabla\times\mathbf{S})$ is a poloidal field with the same defining scalar $\tilde{\Phi}$.

In spherical polar coordinates $(r,\ \theta,\ \phi)$,

$$S_r = \frac{1}{r^2}L^2\Phi, \quad S_\theta = \frac{1}{r}\frac{\partial^2\Phi}{\partial r\partial\theta}, \quad S_\phi = \frac{1}{r\sin\theta}\frac{\partial^2\Phi}{\partial r\partial\phi}, \tag{3.210}$$

$$T_r = 0, \quad T_\theta = \frac{1}{r\sin\theta}\frac{\partial\Psi}{\partial\phi}, \quad T_\phi = -\frac{1}{r}\frac{\partial\Psi}{\partial\theta}, \tag{3.211}$$

where

$$L^2 = -\frac{1}{\sin\theta}\frac{\partial}{\partial\theta}\left(\sin\theta\frac{\partial}{\partial\theta}\right) - \frac{1}{\sin^2\theta}\frac{\partial^2}{\partial\phi^2}. \tag{3.212}$$

A fundamental basis (on a sphere) for toroidal and poloidal fields is obtained by expressing Φ and Ψ in spherical harmonics with coefficients that are functions of r. For details, see Chandrasekhar 1961, Appendix III.

In the axisymmetric case, $T_\theta = 0$, $S_\phi = 0$: toroidal vectors then have only azimuthal components, and poloidal vectors lie in meridian planes. On writing

$$\sin\theta\frac{\partial\Phi}{\partial\theta} \equiv P(r,\theta), \qquad -\sin\theta\frac{\partial\Psi}{\partial\theta} \equiv T(r,\theta), \tag{3.213}$$

we arrive at

$$\mathbf{S} = \left(-\frac{1}{r^2\sin\theta}\frac{\partial P}{\partial\theta}, \frac{1}{r\sin\theta}\frac{\partial P}{\partial r}, 0\right) = \frac{\nabla P\times\mathbf{t}}{r\sin\theta}, \tag{3.214}$$

$$\mathbf{T} = \left(0, 0, \frac{T}{r\sin\theta}\right). \tag{3.215}$$

where $\mathbf{t}$ is the unit azimuthal vector. These forms for axisymmetric poloidal and toroidal field will be continually used, sometimes in vector form, or in spherical polar or cylindrical polar components.

References

Aly, J.J. (1984). *Astrophysical Journal*, **283**, 349.

Aly, J.J. (1985). *Astronomy and Astrophysics*, **143**, 19.

Aly, J.J. (1986). In *Magnetospheric processes in astrophysics* (eds R.I. Epstein and W.C. Feldman), CP144, p. 45. American Institute of Physics, New York.

Aly, J.J. (1994). *Astronomy and Astrophysics*, **288**, 1012.

Aly, J.J. (1995). *Astrophysical Journal*, **439**, L63.

Bateman, G. (1978). *MHD instabilities.* MIT Press, Cambridge, Massachusetts.

Berger, M.A. (1984). *Geophysical Astrophysical Fluid Dynamics*, **30**, 79.

Berger, M.A. (1985). *Astrophysical Journal Supplement Series*, **59**, 433.

Berger, M.A. (1994). *Space Science Reviews*, **68**, 3.

Berger, M.A. and Field, G.B. (1984). *Journal Fluid Mechanics*, **147**, 133.

Bernstein, I.B., Frieman, E.A., Kruskal, M.D. and Kulsrud, R.M. (1958). *Proceedings Royal Society A*, **244**, 17.

Biskamp, D. (1986). *Physics Fluids*, **29**, 1520.

Chandrasekhar, S. (1956). *Proceedings National Academy Sciences*, **42**, 1, 1.

Chandrasekhar, S. (1961). *Hydrodynamic and hydromagnetic stability.* Clarendon Press, Oxford.

Chandrasekhar, S. (1987). *Ellipsoidal figures of equilibrium.* Dover, New York.

Chandrasekhar, S. and Fermi, E. (1953). *Astrophysical Journal*, **118**, 113, 116.

Chandrasekhar, S. and Kendall, P.C. (1957). *Astrophysical Journal*, **126**, 457.

Chanmugam, G. (1978). *Astrophysical Journal*, **221**, 965.

Cowling, T.G. (1976). *Magnetohydrodynamics.* Adam Hilger, London.
de Bruyne, P. and Hood, A.W. (1992). *Solar Physics*, **142**, 87.
Ebert, R. (1957). *Zeitschrift für Astrophysik*, **42**, 263.
Field, G.B. (1986). In *Magnetospheric processes in astrophysics* (eds R.I. Epstein and W.C. Feldman), CP144, p. 324. American Institute of Physics, New York.
Frieman, E.A. and Rotenberg, M. (1960). *Reviews Modern Physics*, **32**, 898.
Furth, H.P., Killeen, J. and Rosenbluth, M.N. (1963). *Physics Fluids*, **6**, 459.
Gold, T. and Hoyle, F. (1960). *Monthly Notices Royal Astronomical Society*, **120**, 89.
Goldreich, P. and Schubert, G. (1967). *Astrophysical Journal*, **150**, 571.
Hood, A. (1994). *Solar Physics*, **150**, 99.
Hornig, G. and Rastätter, L. (1997). *Advances Space Research*, **19**, 1789.
Hughes, D.W. and Cattaneo, F. (1987). *Geophysical Astrophysical Fluid Dynamics*, **39**, 65.
Jardine, M. and Priest, E.R. (1988*a*). *Journal Plasma Physics*, **40**, 143.
Jardine, M. and Priest, E.R. (1988*b*). *Geophysical Astrophysical Fluid Dynamics*, **42**, 163.
Jeffrey, A. and Taniuti, T. (1964). *Nonlinear wave propagation.* Academic Press, New York.
Johnson, J.L., Oberman, C.R., Kulsrud, R.M. and Frieman, E.A. (1958). *UN Geneva Conference*, **31**, 198.
Kovetz, A. (1966). *Astrophysical Journal*, **146**, 462.
Kruskal, M.D. (1954). U.S. Atomic Energy Commission Report No. NYO-6015.
Kruskal, M.D. and Kulsrud, R.M. (1958). *Physics Fluids*, **1**, 265.
Kruskal, M.D. and Schwarzschild, M. (1954). *Proceedings Royal Society A*, **223**, 348.
Kuijpers, J. (1992). In *The Sun, a laboratory for astrophysics* (eds J.T. Schmelz and J.C. Brown), p. 535. Kluwer, Dordrecht.
Lebovitz, N.R. (1966). *Astrophysical Journal*, **145**, 878.
Ledoux, P. (1951). *Annales d'Astrophysique*, **14**, 438.
Lighthill, M.J. (1978). *Waves in fluids.* Cambridge University Press.
Low, B.C. (1986). *Astrophysical Journal*, **307**, 205.
Lundquist, S. (1951). *Physical Review*, **83**, 307.
Lüst, R. and Schlüter, A. (1954). *Zeitschrift Astrophysik*, **34**, 263.
Lynden-Bell, D. (1996). *Monthly Notices Royal Astronomical Society*, **279**, 389.
Lynden-Bell, D. and Boily, C. (1994). *Monthly Notices Royal Astronomical Society*, **267**, 146.
Lynden-Bell, D. and Ostriker, J.P. (1967). *Monthly Notices Royal Astronomical Society*, **136**, 293.
Mestel, L. (1959). *Monthly Notices Royal Astronomical Society*, **119**, 223.
Mestel, L. and Strittmatter, P.A. (1967). *Monthly Notices Royal Astronomical Society*, **137**, 95.
Milsom, F. and Wright, G.A.E. (1976). *Monthly Notices Royal Astronomical Society*, **174**, 307.

Moffatt, H.K. (1969). *Journal Fluid Mechanics*, **35**, 117.
Moffatt, H.K. (1978). *Magnetic field generation in electrically conducting fluids.* Cambridge University Press.
Parker, E.N. (1963). *Astrophysical Journal Supplement*, Series **8**, 177.
Parker, E.N. (1974). *Astrophysical Journal*, **191**, 245.
Parker, E.N. (1979). *Cosmical magnetic fields.* Clarendon Press, Oxford.
Petschek, H.E. (1964). AAS–NASA Symposium *Physics of solar flares.* NASA SP-50, p. 425.
Priest, E.R. (1982). *Solar magnetohydrodynamics.* Reidel, Dordrecht.
Priest, E.R. and Forbes, T.G. (1986). *Journal Geophysical Research*, **91**, 5,579.
Priest, E.R. and Forbes, T.G. (1992). *Journal Geophysical Research*, **97**, A11, 16,757.
Roberts, P.H. (1956). *Astrophysical Journal*, **124**, 430.
Schatzman, E. (1965). *IAU Symposium* **22**, 337.
Schmidt, G. (1966). *Physics of high temperature plasmas.* Academic Press, New York.
Shafranov, V.D. (1957). *Journal Nuclear Energy III*, **5**, 86.
Sonnerup, B.U.Ö. (1970). *Journal Plasma Physics*, **4**, 161.
Sweet, P.A. (1950). *Monthly Notices Royal Astronomical Society*, **116**, 69.
Sweet, P.A. (1958). In *Electromagnetic phenomena in cosmical physics* (ed. B. Lehnert), p. 123. Cambridge University Press.
Tayler, R.J. (1957). *Proceedings Physical Society*, **B70**, 1049.
Tayler, R.J. (1973). *Monthly Notices Royal Astronomical Society*, **161**, 365.
Taylor, J.B. (1974). (1974). *Physical Review Letters*, **33**, 1139.
van Ballegooijen, A.A. (1994). *Space Science Reviews*, **68**, 299.
Vasyliunas, V. (1975). *Reviews Geophysics*, **13**, 303.
Woltjer, L. (1958). *Proceedings National Academy Sciences*, **44**, 489.
Zylstra, G.J., Lamb, F.K. and Aly, J.J. (1989). *Bulletin American Physical Society*, **34**, 1289.

4

MAGNETISM AND CONVECTION

4.1 Introduction

Our Sun is a veritable laboratory for the study of magnetohydrodynamics. It exhibits a wide variety of phenomena – sunspots, flares, prominences – in which a magnetic field is clearly the crucial feature. Further, late-type stars such as the Sun develop spontaneously a subphotospheric convective zone, with motions observable most strikingly as the violent, small-scale granulation, as well as the larger-scale meso- and supergranulation. In this chapter, we are concerned primarily with some aspects of the interaction between turbulent convection and a magnetic field in the Sun, and so also in other late-type stars with their deep outer convection zones. For the moment the solar magnetic field is accepted as a datum from observation: the problem of the origin of the field and of its detailed structure is postponed to Chapter 5 and especially to Chapters 6 and 8.

The Schwarzschild criterion (derived in Section 3.6) shows that a domain with a superadiabatic temperature gradient is convectively unstable. The first problem is to follow at least semi-quantitatively the non-linear development and saturation of the buoyancy instability. We begin by summarizing the canonical model, essentially as formulated in the classical work of Biermann (1932) and Cowling (1935). In particular, Biermann took over from laboratory fluid mechanics Prandtl's concept of the 'mixing-length' l, the distance which a rising or sinking blob of fluid is pictured as travelling before being re-absorbed by the ambient medium. Over this 'macroscopic mean-free-path', the blob in a superadiabatic domain feels a net buoyancy force. The mean speed v_{t} acquired by a blob is estimated by

$$\left(\frac{v_{\mathrm{t}}}{c_s}\right)^2 \simeq \frac{\Delta(\nabla T)}{(T/H_{\mathrm{p}})}\left(\frac{l}{H_{\mathrm{p}}}\right)^2 \tag{4.1}$$

with c_s the local sound speed, $H_{\mathrm{p}} = c_s^2/g$ the pressure scale-height, and $\Delta(\nabla T)$ the superadiabatic gradient (i.e. the excess of the actual temperature gradient over the adiabatic gradient corresponding to the same pressure gradient). The associated energy flux is

$$F^{\mathrm{conv}} \simeq \rho v_{\mathrm{t}} c_{\mathrm{p}} \Delta(\nabla T) l \simeq \left[\frac{\Delta(\nabla T)}{(T/H_{\mathrm{p}})}\right]^{3/2}\left(\frac{l}{H_{\mathrm{p}}}\right)^2 (c_p \rho T)(g H_{\mathrm{p}})^{1/2} \tag{4.2}$$

where c_{p} is the specific heat at constant pressure. The mixing-length is usually taken as of the order of the local pressure scale-height H_{p}. (This can be justified intuitively as maximizing the heat transport for a given $\Delta(\nabla T)$: with moderate values of l, the convective cells can have similar horizontal and vertical dimensions; but with larger values, the continuity condition would enforce a large horizontal splaying out of the upward moving flow, so reducing the number of cells.) Deep within a star, $l \simeq 10^9$ cm, enormously greater than the photon mean-free-path $\simeq 1$ cm, so that from (4.2), to transport an energy flux of the order of a stellar luminosity L, only a very small superadiabatic excess is required. Typically, $\Delta(\nabla T)/(T/H_\mathrm{p}) \approx 10^{-6}$, the Mach number of the turbulence is $\approx 10^{-4}$, and the convective 'turnover time' $\tau_c = l/v_\mathrm{t} \approx 10^6$ s. Very roughly, with $l \approx H_\mathrm{p}$, (4.1) and (4.2) yield

$$\frac{\Delta(\nabla T)}{(T/H_\mathrm{p})} \simeq \left(\frac{\bar{\rho}R^{1/2}}{\rho H_\mathrm{p}^{1/2}}\frac{\tau_\mathrm{ff}}{\tau_\mathrm{KH}}\right)^{2/3}, \qquad \left(\frac{v_\mathrm{t}}{c_s}\right) \simeq \left[\frac{\Delta(\nabla T)}{(T/H_\mathrm{p})}\right]^{1/2}, \tag{4.3}$$

where $\bar{\rho}$ is the mean stellar density, τ_ff = time of free-fall $\simeq R/c_s \simeq (G\bar{\rho})^{-1/2}$ and τ_KH = Kelvin–Helmholtz time $\simeq Mc_s^2/L$.

In a convective envelope, $\Delta(\nabla T)/(T/H_\mathrm{p})$ increases as one approaches the surface, but is still small over the bulk of the zone. Near to the photosphere, where the temperatures are closer to 10^4 K, characteristic values are $l \approx 10^8$ cm and $v_\mathrm{t} \approx 10^5$ cm s^{-1}, typical of the observed granulation. The observationally inferred weak mesogranulation, with length-scales $\approx 8 \times 10^8$ cm and velocities $\approx 4 \times 10^4$ cm s^{-1}, and the supergranulation, with length-scales $\approx$ (2–5)$\times 10^9$ cm and velocities $\approx$ (1–4)$\times 10^4$ cm s^{-1}, have been cited as evidence for the larger-scale motions determined by ambient conditions deep down (Simon and Weiss 1968, 1991) – cf. Section 4.5 below.

The onset of convection in a nearly *incompressible* fluid depends on the value of the Rayleigh number,

$$\mathrm{Ra} = \frac{g\alpha\,\Delta(\nabla T)}{\kappa\nu}d^4, \tag{4.4}$$

where g is gravity, d the depth of the fluid layer, and α, κ, ν are the coefficients of volume expansion, thermometric conductivity and kinematic viscosity respectively. The fluid is convectively unstable when Ra exceeds a value of the order of 1–2000, depending on the precise boundary conditions adopted (e.g. Chandrasekhar 1961). With a scale-height inserted for d, and with microscopic coefficients, one finds that only very small temperature gradients are required. In a compressible stellar domain, the Rayleigh criterion may be applied to a medium with a temperature gradient satisfying the Schwarzschild criterion, and so on the verge of instability if it were dissipation-free. However, it is then found that the degree of superadiabaticity required to satisfy the Rayleigh criterion is very small, well below the (also very small) estimate (4.3) required to carry an energy flux of solar order.

The picture outlined above has been standard dogma for decades, if only because of the absence of a successful analytical-plus-numerical scheme to replace it. Even if uncomfortably aware of the crudeness of the arguments, workers in stellar structure and evolution could reasonably claim that for the construction of models of homogeneous, main sequence stars, it is sufficient to know that over a convective core and over the bulk of a convective envelope, the temperature gradient hardly deviates from the adiabatic value. However, for the outermost, subphotospheric layers of the solar convective zone, and for the extensive convective envelopes of cool giants, the departure from adiabaticity as estimated from mixing-length theory is non-trivial. In particular, in giant stars, the markedly non-adiabatic domain linking the photosphere with the nearly adiabatic bulk of the convective zone yields an effective jump in entropy which is crucial for the structure of the star (Schwarzschild 1958; Gough and Weiss 1976). There is no doubt that the absence of a satisfactory deductive theory of turbulent convection has been probably the most serious defect in the theory of stellar evolution.

Mixing-length theory is about the simplest attempt to construct a model of convective heat transport which depends just on the *local* values of $\Delta(\nabla T)$ and the thermodynamic variables, even though the macroscopic mean-free-path l is taken to be of the order of a scale-height, a distance over which by definition the values of some 'local' quantities must change significantly. There are several other such local models in the literature, from the developments of the the original Biermann–Cowling formulation, e.g. by Böhm-Vitense (1958) and Gough (1977), to the mathematically more sophisticated theoretical work by Canuto (1991, 1993) and Canuto and Mazitelli (1991, 1992). They clearly differ in detail, but they all arrive at rather similar-looking expressions for the heat flux, of the form $\mathbf{F}(\Delta(\nabla T))$. There is also a long list of papers that build on mixing-length theory, but take some account of 'non-local' corrections. The paper by Grossman (1996) contains both a useful list of references and a detailed comparison of different formulations, with numerical results.

All models of convection contain parameters, such as the ratio l/H_{p}, which have to be calibrated before application to a particular stellar model. Gough and Weiss (1976) have compared different mixing-length models by evolving $1M_\odot$ stellar models from the zero-age main sequence to fit the age and the present luminosity and effective temperature of the Sun. They find that the resulting solar models then become almost indistinguishable, in particular producing the same depth for the solar convection zone. This insensitivity to details of the convection model is fortunate for stellar structure theory, but equally means that comparison of main sequence stellar models with observation will not yield reasons for preferring one convection model over another.

The Sun is observed to rotate non-uniformly, showing in particular a marked equatorial acceleration. The strong interaction between a non-uniform rotation and a magnetic field, already noted in Chapter 2, will be a recurring theme throughout this book. In convective zones, the Reynolds stresses of the turbulence will play an important, in fact probably a dominant, role in determining the preferred distribution of angular momentum, and so of the angular velocity field which reacts on the magnetic field. An interesting development of mixing-length theory exploits both the expected deviation from isotropy as well as the inhomogeneity of the Reynolds stresses so as to infer the expected distribution of angular momentum in a stellar convective zone. As a number of dynamical dynamo models in the recent literature build on this theory, it is conveniently summarized in Section 4.2.

In fact, it is becoming clear, for both dwarf and giant stars, that for studies of magnetic activity – not least for the dynamo problem – one needs a much more detailed and precise understanding of the motions within a convective zone (Chapters 6 and 8). A reliable non-local theory is clearly required near the boundary of a convective zone, e.g. in order to estimate the degree of overshoot from the solar convective envelope into the radiative interior (Section 8.6.3).

From the amount of detailed information emerging from recent numerical studies of both the pure convection problem and of magnetoconvection, summarized in Section 4.5, one is led to hope that the new generation of supercomputers will ultimately remove all the ambiguities. But at the time of writing there has not yet arisen a consensus, due at least in part to differences in the input physics. Different groups, each with impressive numerical expertise, describe their qualitative results in terms that sometimes overlap, but also contain important divergences. The ultimate effect is likely to change to some extent the language used in qualitative discussion, especially for convective stellar envelopes. In the words of a partisan of one group (Spruit 1997), the emerging picture will enforce a 'changing paradigm' (Kuhn 1970), with the philosophy behind the mixing-length picture, along with the extensive derived modelling, due for the historical rubbish bin. Others, especially those who have treated the mixing-length model more as a convenient parametrization, may continue to find it useful as a calibrator of new results.

In the convective envelope of a late-type star, the driving instability is due to the large latent heat of hydrogen ionization and the subsequent high opacity (cf. Section 8.1), whereas in the core of an early-type star, the convection is driven by too great a central concentration of the energy sources – a much closer analogy to the standard fluid dynamical picture of instability arising through 'heating from below'. It may nevertheless turn out, as is implicit in standard theory, that the developed convection is insensitive to the physical origin of the superadiabaticity, depending far more on the degree of density stratification.

In summarizing what appear to be the relevant studies to date, we shall lean heavily on several excellent texts and reviews, in particular those by Cowling (1976), Priest (1982), Hughes and Proctor (1988), Rüdiger (1989), Hughes (1991), Weiss (1991) and Thomas and Weiss (1992), as well as the stream of papers and review articles emanating from Cambridge, Copenhagen and other centres of activity in this area. In a period of rapid development, it is inevitable that some of the discussion will be in terms that may ultimately sound dated.

Different aspects of the interaction between magnetism and convection are discussed in Sections 4.3–4.5, and Chapters 6, 8 and 10. The generation of locally strong magnetic fields in convection zones manifests itself in the appearance of sunspots (Section 4.6). The effect of the emanating magnetic flux on both the equilibrium and the energetics of the superphotospheric domains is treated in Section 4.8.2.

The effect of magnetic fields on stellar radiative zones – i.e. zones that have subadiabatic temperature gradients and so are not spontaneously unstable according to the Schwarzschild criterion – is studied in some detail in Chapters 5 and 9. However, the region just beneath the solar convective envelope, though slightly subadiabatic, appears to be coupled strongly with the envelope, both through convective overshoot and the downward pumping of magnetic flux; hence some of the discussion in Section 4.7 will be on the equilibrium and stability of the overshoot domain at the base of a convective envelope. Section 4.8.2 summarizes work on the eruption of flux from the overshoot domain to the photosphere.

4.2 The angular velocity distribution in a convective zone

Let us for the moment continue using the canonical picture of turbulent convection, with its prediction of a mixing-length $l \simeq 10^9$ cm as the macroscopic analogue of the mean-free-path in kinetic theory, and with the typical associated turbulent velocity $v_t \simeq 3 \times 10^3$ cm s^{-1} (Spruit 1974). The terminology suggests the introduction of an 'eddy-viscosity' $\nu_t \simeq l v_t/3 \simeq 10^{12}$ cm^2 s^{-1}, as compared with a microviscosity (kinetic or radiative) $\simeq 10^3$. If the analogy were perfect, with the turbulent viscosity *isotropic*, then the turbulence would act to smooth out a shear over a scale L in the time L^2/ν_t. To destroy the observed solar differential rotation, a microviscosity would require $5 \times 10^{21}/10^3 \times 3 \times 10^7 \approx (10^{11})$ yr – longer than the Hubble time, whereas the eddy viscosity would take about 200 yr. This would imply that the equatorial acceleration of the Sun could not be a 'fossil' but would need to be generated by some process – e.g. meridian circulation – with a characteristic time shorter than this turbulent diffusion time. However, Lebedinski (1941) and later Wasiutynski (1946) and Biermann (1951) argued that the turbulence would itself be a cause of non-uniform rotation through its essential *an*isotropy, due to l and v_t being different in different directions. In the simplest example, appropriate for a slow rotator, the anisotropy is due just to the gravitational field, which distinguishes between the vertical and horizontal directions. For the equatorial plane, Biermann wrote the turbulent flux of angular momentum as the sum of two terms: the isotropic part $-A_1 r^2\, \mathrm{d}\Omega/\mathrm{d}r$ and the monotropic part $-A_2\, \mathrm{d}(\Omega r^2)/\mathrm{d}r$ which transports angular momentum radially; A_1 is the traditional eddy viscosity and A_2 an analogous coefficient, also of order $l v_t$. Thus the total flux is

$$-(A_1 + A_2) r^2 \frac{\mathrm{d}\Omega}{\mathrm{d}r} - 2A_2 r\Omega. \tag{4.5}$$

The crucial new effect is that (4.5) does not vanish when Ω is uniform: rather, the vanishing of (4.5) yields the law $\Omega \propto r^{-2A_2/(A_1+A_2)}$ for the steady state determined by the action of eddy viscosity alone.

A more general treatment is given by Rüdiger (1989) and by Rüdiger and Kitchatinov (1994), following the methods introduced by Osborne Reynolds. The velocity field in a turbulent zone is written as the sum of mean and fluctuating parts:

$$\mathbf{v} = \bar{\mathbf{v}} + \mathbf{v}' = (\Omega\varpi\mathbf{t} + \mathbf{v}_\mathrm{p}) + \mathbf{v}', \tag{4.6}$$

where the mean velocity $\bar{\mathbf{v}}$ has a rotatory (*toroidal*) component $\Omega(\varpi, z)\varpi\mathbf{t}$ and a meridional (*poloidal*) circulation component $\mathbf{v}_\mathrm{P}$. If the velocities $\mathbf{v}'$ are very subsonic, the associated density fluctuations ρ' can be ignored. When (4.6) is substituted into the equation of motion (2.35), the mean of the non-linear inertial term $\partial(\rho v_i v_j)/\partial x_j$ yields

$$\frac{\partial}{\partial x_j}(\rho \bar{v}_i \bar{v}_j) + \frac{\partial}{\partial x_j}\rho(\overline{v_i' v_j'}). \tag{4.7}$$

The last term is taken on the other side of the equation of motion so as to become a new effective force $-\nabla \cdot (\rho \mathcal{Q})$ felt by the mean flow, with the *Reynolds stress tensor* $-\rho Q_{ij}$ depending on the turbulence through the symmetric one-point correlation tensor

$$Q_{ij} = \overline{v_i'(\mathbf{r}, t) v_j'(\mathbf{r}, t)}. \tag{4.8}$$

The traditional parametrization in terms of an isotropic eddy viscosity just writes

$$Q_{ij} = -\nu_\mathrm{t} \left(\partial \bar{v}_i / \partial x_j + \partial \bar{v}_j / \partial x_i\right). \tag{4.9}$$

The generalization writes

$$Q_{ij} = Q_{ij}^\nu + Q_{ij}^\Lambda. \tag{4.10}$$

Here the diffusive part Q^ν depends on the velocity gradients and so is the analogue of (4.9):

$$Q_{ij}^\nu = -\mathcal{N}_{ijkl} \frac{\partial}{\partial x_l} \bar{v}_k, \tag{4.11}$$

where $\mathcal{N}$ is the eddy viscosity tensor. The non-diffusive part Q_{ij}^Λ – the so-called 'Λ-effect'– is written

$$Q_{ij}^\Lambda = \Lambda_{ijk} \Omega_k. \tag{4.12}$$

It is the generalization of the Lebedinski–Biermann terms that depend on Ω but not on its derivatives.

In spherical polar coordinates, the terms in Q_{ij} that contribute to angular momentum transport are

$$\begin{aligned} Q_{r\phi} &= -\nu_{vv} r \sin\theta \, \partial\Omega/\partial r + Q_{r\phi}^\Lambda, \\ Q_{\theta\phi} &= -\nu_{hh} \sin\theta \, \partial\Omega/\partial\theta + Q_{\theta\phi}^\Lambda, \end{aligned} \tag{4.13}$$

where the differing coefficients ν again allow for anisotropy in the eddy viscosity.

The simplest expression for Λ_{ijk} that yields a symmetric, polar Q^Λ uses just the unit vector $\hat{\mathbf{r}}$ anti-parallel to gravity as its preferred direction:

$$\Lambda_{ijk} = \Lambda_V (\varepsilon_{ipk} \hat{r}_j + \varepsilon_{jpq} \hat{r}_i) \hat{r}_p. \tag{4.14}$$

This expression is effectively a slight generalization of that used by Wasiutynski (1946), Biermann (1951), Kippenhahn (1963) and Köhler (1970): in spherical coordinates the only surviving contribution is $Q_{r\phi} = \Lambda_V \sin\theta\Omega$. This is appropriate for a slow rotator. Equations (4.13) now become

$$\begin{aligned} Q_{r\phi} &= -\nu_{vv} r \sin\theta\, \partial\Omega/\partial r + \Lambda_V \sin\theta\Omega, \\ Q_{\theta\phi} &= -\nu_{hh} \sin\theta\, \partial\Omega/\partial\theta. \end{aligned} \tag{4.15}$$

The value of the coefficients in the Reynolds stresses can be calculated from an analytical model of the turbulence (e.g. Kitchatinov and Rüdiger 1993) or from numerical simulations (e.g. Pulkkinen *et al.* 1993).

In this section, magnetic torques are supposed negligible. In a steady state, angular momentum transport by the Reynolds stresses is balanced by the advection of angular momentum by a laminar meridian circulation $\mathbf{v}_{\mathrm{p}}$, satisfying the continuity equation $\nabla \cdot (\rho \mathbf{v}_{\mathrm{p}}) = 0$:

$$\nabla \cdot \left(\rho\Omega\varpi^2 \mathbf{v}_{\mathrm{p}} + \rho\varpi\overline{v'_\phi \mathbf{v}'} \right) = 0. \tag{4.16}$$

If the circulation is negligibly slow, the mean steady state is determined by the Reynolds stresses alone, and (4.16) becomes

$$r^{-2}\, \partial(\rho r^3 Q_{r\phi})/\partial r + (\rho/\sin^2\theta)\, \partial(\sin^2\theta\, Q_{\theta\phi})/\partial\theta = 0. \tag{4.17}$$

In the slow rotation limit, $\Lambda_V, \nu_{vv}, \nu_{hh}$ will all be independent of θ. With application of the stress-free boundary condition at the surface and either stress-free or rigid rotation at the lower boundary, it is found, not surprisingly, that the steady-state solution is stress-free everywhere, with Ω a function just of r: if we write $\Lambda_V = \nu_{vv} V^0(r)$, the solution is

$$\Omega = \Omega_s \exp \int_R^r \frac{V^0(r)}{r}\, \mathrm{d}r. \tag{4.18}$$

That this simplest model predicts a latitude-independent rotation law was noted by Biermann and in particular by Kippenhahn (1963), who looked for more general steady states which include advection of angular momentum by a meridian circulation. The circulation was not put in by fiat, but justified by considering the equation of hydrostatic equilibrium:

$$-\frac{\nabla p}{\rho} + \nabla V + \Omega^2 \varpi = 0 \tag{4.19}$$

with both the magnetic forces and also the Reynolds stresses of the very subsonic turbulence (the 'turbulent pressure') assumed negligible. The curl of this equation is

$$\frac{1}{\rho^2}(\nabla\rho \times \nabla p) + \varpi \frac{\partial\Omega^2}{\partial z}\mathbf{t} = 0. \tag{4.20}$$

Then if one assumes that adiabaticity holds to a high approximation, the surfaces of constant p and ρ coincide and the first term in (4.20) vanishes, but the second term is non-zero, since Ω as constructed is a function of r and not of ϖ. The non-irrotational centrifugal field cannot be balanced by a barotropic (p, ρ)-field, and a meridian circulation is set up, with velocities fixed by making the drag due to the eddy viscosity balance the centrifugal force and the pressure. Equatorial acceleration results when the flow is equatorwards at the surface; this requires the zero-order Ω to increase outwards, in turn requiring a larger horizontal than vertical mixing-length.

For realistic solar parameters, Kippenhahn's iterative scheme did not in fact converge. A subsequent numerical treatment of a constant density domain by Köhler (1970), with the same assumptions, showed that circulation speeds of a few metres/sec would produce the observed rotation law. But Köhler's models, like others that assume adiabaticity, produce nearly cylindrical surfaces of constant Ω, whereas current helioseismological observations (cf. Chapter 8) suggest rather that the surface distribution $\Omega(\theta)$ persists with depth through the convective envelope. There is also a discrepancy between one of the Wasiutynski–Kippenhahn equations and the above covariant formulation which does not appear to have been resolved (Rüdiger 1989).

There are in fact several reasons why the pioneering Lebedinski–Biermann–Kippenhahn approach has to be modified. First of all, the Sun is not a 'slow rotator': for large-scale solar convection, the Coriolis number $\Omega^* = 2\tau_c\Omega$ (equivalently $4\pi/\mathrm{Ro}$, where $\mathrm{Ro} = (2\pi/\Omega)/\tau_c$ is the Rossby number) is ≈ 1, becoming small only for angular velocities well below the solar value. As pointed out first by Iroshnikov (1969), terms of higher order in Ω should then be included in the Λ-effect, and in particular there is now a non-zero contribution to $Q_{\theta\phi}$, so that the Reynolds stresses, *acting alone*, will yield a θ-dependent Ω-law. Further, it appears that the inhomogeneity due to density stratification yields a more powerful Λ-effect than the anisotropy (Roberts and Soward 1975; Kitchatinov 1987).

The poloidal equation of balance must still be satisfied. Again, the non-irrotational centrifugal field generated by the generalized Λ-effect cannot be balanced by an assumed adiabatic (p, ρ) distribution, and so it would drive a meridian circulation. We may very well need a circulation field, not only from observation, but also in order to construct a fully satisfactory solar dynamo model (cf. Sections 8.5 and 8.6). The problem is that *with adiabaticity assumed*, whatever form for the Λ-effect is adopted, the combined effects of the toroidal and poloidal equations again enforce isorotation contours that are nearly cylindrical (e.g. Brandenburg *et al.* 1990; 1992*b*).

In fact, as early as 1965, Weiss had stressed that the effect of the rotation on convective heat flux need not be negligible, and would in particular differ at pole and equator. Durney and Roxburgh (1971) introduced a latitude-dependent turbulent thermal conductivity into a study of thin rotating shells. By a suitable choice of parameter they were able to reproduce the observed solar surface rotation, but with the equator predicted to be markedly warmer (by 70 K) than the poles. The turbulent Prandtl number, the ratio of eddy viscosity to turbulent thermal conductivity, was not discussed. Moss and Vilhu (1983) emphasized that the computational results would be sensitive to modest departures from adiabaticity.

The introduction of an anisotropic turbulent conductivity looks promising (Brandenburg *et al.* 1992a), and is in fact incorporated into a recent comprehensive study by Rüdiger and Kitchatinov (1994). They appeal to a mixing-length model of quasi-isotropic, inhomogeneous turbulence (Kitchatinov and Rüdiger 1993), which has density stratification as the main inhomogeneity. The two off-diagonal components of the Q^{Λ}_{ij}-tensor relevant to angular momentum transport are

$$\begin{aligned} Q^{\Lambda}_{r\phi} &= \nu_{\rm t}(V^{(0)} + \sin^2\theta V^{(1)})\Omega\sin\theta, \\ Q^{\Lambda}_{\theta\phi} &= \nu_{\rm t}V^{(1)}\Omega\sin^2\theta\cos\theta, \end{aligned} \tag{4.21}$$

where $\nu_{\rm t}$ is again the eddy viscosity defined by (4.9), and

$$\begin{aligned} V^{(0)} &= k[\mathcal{T}_0(\Omega^*) + \mathcal{T}_1(\Omega^*)], \\ V^{(1)} &= -k[\mathcal{T}_1(\Omega^*)]. \end{aligned} \tag{4.22}$$

The argument Ω^* is again the Coriolis number $2\tau_c\Omega$; the coefficient $k = \tau_c^2\overline{v'^2}/H_\rho^2$ where H_ρ is the density scale-height. For a nearly adiabatic zone, k is close to unity; the functions $\mathcal{T}_0$, $\mathcal{T}_1$ are computed from the turbulence model (Fig. 4.1). For $\Omega^* \ll 1$, $V^{(1)}$ is small as in the Biermann–Kippenhahn model, so that the Reynolds stresses acting alone yield a radial non-uniformity in Ω. There is a sharp transition at $\Omega^* \approx 1$ to the high Ω^* regime, in which $V^{(0)}$ becomes negative and $V^{(1)}$ positive and of the same order: the average of $Q_{r\phi}$ becomes small but $Q_{\theta\phi}$ remains significant. The anisotropy due to the rotation modifies also $Q^{\nu}_{r\phi}$, $Q^{\nu}_{\theta\phi}$, each depending on both $\partial\Omega/\partial r$ and $\partial\Omega/\partial\theta$.

Equation (4.17) involving just the Reynolds stress terms is solved for a convective shell (representing the solar convection zone), subject to stress-free boundary conditions. There are only two parameters in the problem, the ratio of the mixing-length to the pressure scale-height, and the mean angular velocity Ω_0 of the shell, and both are fixed for the Sun. The resulting Ω-contours are presented in Fig. 4.2. They agree tolerably – perhaps remarkably – well with the currently inferred solar data (Fig. 8.8) in being approximately radial over most of the shell.

This rotation law yields a non-irrotational centrifugal field which will again drive a meridian circulation $\mathbf{v}_{\rm p}$. For consistency, the speeds must be slow enough for the corrections from the neglected angular momentum advection term in (4.16) to be small. The curl of the poloidal equation of motion – replacing (4.20) – is

$$\mathcal{D}(\mathbf{v}_{\rm p}) = \frac{1}{\rho^2}(\nabla\rho \times \nabla p) + \varpi\frac{\partial\Omega^2}{\partial z}\mathbf{t} \tag{4.23}$$

where the term in $\mathbf{v}_{\rm p}$ is the curl of the frictional drag on the laminar meridian circulation $\mathbf{v}_{\rm p}$ and so can be written in terms of the viscosity tensor $\mathcal{N}$ of (4.11). After non-dimensionalization, the term in Ω^2 in (4.23) has as a coefficient the Taylor number

$$\mathrm{Ta} = \left(\frac{2\Omega R^2}{\nu_{\rm t}}\right)^2. \tag{4.24}$$

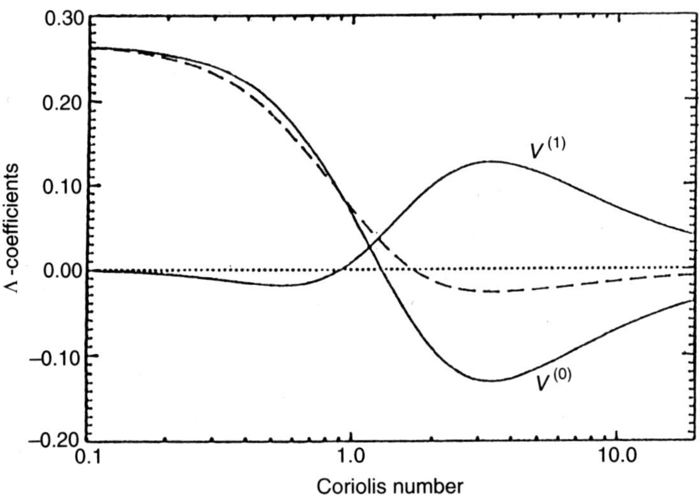

FIG. 4.1. The Rüdiger–Kitchatinov model: the quantities $V^{(0)}$, $V^{(1)}$ in (4.21) and (4.22) as functions of the Coriolis number Ω^*. Dashed line: $V^{(0)} + V^{(1)}$.

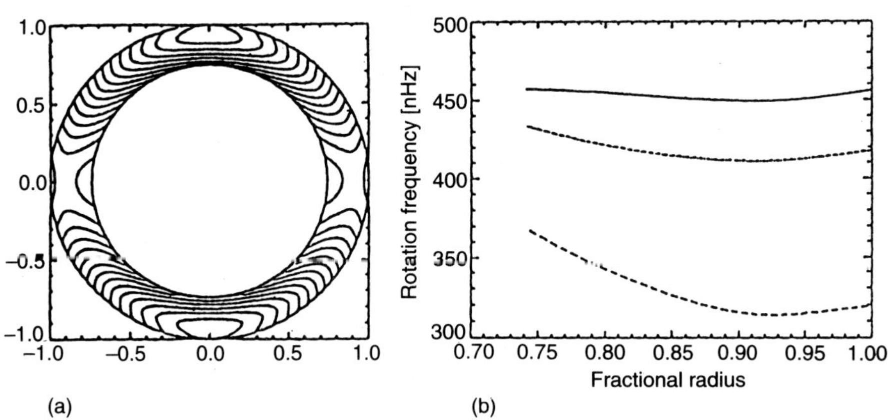

FIG. 4.2. The Rüdiger–Kitchatinov model: the results for $\Omega_0 = 2.7 \times 10^{-6}\ \mathrm{s}^{-1}$. (a) Isorotation contours. (b) Angular velocity for the equatorial plane (solid line), the polar axis (dashed), and at 30°-latitude (dashed–dotted). (Taken from Küker *et al.* 1993.)

In the absence of circulation, and for a strictly adiabatic domain, (4.23) predicts the Taylor–Proudman law $\Omega = \Omega(\varpi)$. This will also be a good approximation if Ta is sufficiently large; but consistency with Ω as in Fig. 4.2(b) – constructed from the toroidal equation (4.17) – puts both a constraint on Ta and on the thermodynamic term

$$\frac{1}{\rho^2}(\nabla \rho \times \nabla p) = -\frac{1}{\rho T}\nabla \Delta T \times \nabla p, \tag{4.25}$$

where again $\nabla \Delta T$ is the superadiabatic temperature gradient. Since the deviation of the stratification from the adiabatic is so slight, the superadiabatic part of ∇p in (4.25) may be neglected; and with $\nabla p \approx \rho \mathbf{g}$, (4.23) reduces to

$$\mathcal{D}(\mathbf{v}_\mathrm{p}) = r \sin\theta \frac{\partial \Omega^2}{\partial z} - \frac{g}{T}\frac{\partial \Delta T}{\partial \theta}. \tag{4.26}$$

To keep $\mathbf{v}_\mathrm{p}$ small, the two terms on the right of (4.26) should be nearly in balance (the adiabaticity assumption would lead to the same problems as before). The equatorial acceleration implies that $\partial\Omega^2/\partial z < 0$, so we need $\cos\theta\, \partial(\Delta T)/\partial\theta < 0$ – 'hot poles' (Stix 1989*a*). The energy equation (with the viscous heating ignored) is

$$\nabla \cdot (\mathbf{F}^{\mathrm{conv}} + \mathbf{F}^{\mathrm{rad}}) + c_p \rho \mathbf{v}_p \cdot \nabla \Delta T = 0, \tag{4.27}$$

where $\mathbf{F}^{\mathrm{rad}}$ is the usual radiative heat flux, and

$$F_i^{\mathrm{conv}} = -c_p \rho \chi_{ij} \frac{\partial}{\partial x_j}(\Delta T) \tag{4.28}$$

with χ_{ij} the anisotropic thermal conductivity tensor. To generate a hot pole the anisotropy must not be based on the preferred radial direction. In the mixing-length representation, the influence of rotation on the turbulent convection yields

$$\begin{aligned} \chi_{ij} &= \chi_t \delta_{ij} + \chi_\| \frac{\Omega_i \Omega_j}{\Omega^2}, \\ \chi_t &= \chi_0 \Phi(\Omega^*), \qquad \chi_\| = \chi_0 \Phi_\|(\Omega^*), \end{aligned} \tag{4.29}$$

with $\chi_0 = \tau_c \overline{v'^2}/3$ the isotropic thermal conductivity for a non-rotating fluid. The functions Φ and $\Phi_\|$ are positive, with $\Phi(0) = 1$, $\Phi_\|(0) = 0$, and with both falling off slowly to zero for large Ω^* (Kitchatinov *et al.* 1994). The $\chi_\|$-term in (4.29) yields a larger radial thermal conductivity at the poles than at the equator, resulting in a poleward horizontal component of the convective heat flux, so that a hot pole can indeed be expected.

The simultaneous solution of (4.16), (4.26) and (4.27) shows that for small Ω^*, the circulation-free model remains a good approximation, because the Taylor number Ta is relatively small ($< 10^7$). As Ta is increased, the Taylor–Proudman result $\Omega = \Omega(\varpi)$ is reached, with $\Omega'(\varpi) < 0$ – equatorial *de*celeration. General agreement with the solar data is shown when $\Omega^* \approx 2.3$. The actual solar value is estimated as 6, suggesting that a deeper study of the relation between superadiabaticity and the turbulence is required; but the results to date are encouraging.

The distribution of rotation through the convection zone is crucial for our understanding of the solar dynamo (Chapters 6 and 8). The Lorentz forces will clearly exert some back-reaction on the rotation field. The predicted change in $\Omega(\varpi, z)$ at higher rotation already makes one cautious about too cavalier an extrapolation of results for the solar dynamo to younger and so more rapidly rotating late-type stars (Section 8.3).

Any such parametrization procedure depends on a distinction being drawn between the laminar large-scale motions and the primarily small-scale turbulent motions. With the turbulent mixing-length of the order of a scale-height, at least deep down in the convective zone, the separation of scales is not as sharp as one would like. It may be that the approach of Rüdiger and co-workers to these problems will ultimately be overtaken by the use of supercomputers. Nevertheless, the analysis begun by Lebedinski and Biermann will always retain its value for its qualitative insights.

4.3 The effect of convective motions on an imposed magnetic field

We now introduce a magnetic field into the superadiabatic domain, but at first suppose the field to be too weak to have any significant effect on the onset of convection, studying rather the effect of the motions on the field. As the simplest example, consider the interaction of a prescribed steady flow – simulating a granule or a supergranule – on a weak magnetic field threading the cell. If the imposed magnetic energy density $B^2/8\pi$ is much less than the kinetic energy density, and if also the MRN $\mathrm{R_m} \gg 1$, then initially the field is dragged and wound up by the 'inexorable' flow: the forces that drive the flow against the weak Lorentz forces pump energy into the field (cf. Section 2.4). At some epoch, either the magnetic energy becomes comparable with the kinetic, and the Lorentz forces now slow up the motion, or the twisting of the field so reduces the local length-scales that the local MRN becomes of order unity, and the field is no longer effectively frozen in but slips through the plasma. An asymptotic state is reached with the flux compressed against the boundaries of the cell.

This picture was inferred by Spitzer as early as 1957 (cf. Section 5.8). The process is followed in detail in a number of studies, beginning with Parker (1963), who studied the one-way interaction of the incompressible flow

$$\mathbf{v} = v_0[\sin(kx)\hat{\mathbf{x}} - kz\cos(kx)\hat{\mathbf{z}}] \tag{4.30}$$

on the uniform vertical field $B_0\hat{\mathbf{z}}$. The z-component of the induction equation for perfect conductivity is

$$\frac{\partial B_z}{\partial t} = -\frac{\partial}{\partial x}(v_0 \sin(kx) B_z), \tag{4.31}$$

which may be solved by the method of characteristics to yield

$$B_z = \frac{B_0\,\mathrm{e}^{-kv_0t}}{\cos^2(kx/2) + \sin^2(kx/2)\,\mathrm{e}^{-2kv_0t}}. \tag{4.32}$$

For definiteness, take $v_0 < 0$ and $z > 0$; then at $x = 0$ where there is upwelling, the field is steadily dispersed, whereas at $x = \pi/k$ there is a downdraft and the field is concentrated like $B_z = B_0\,\mathrm{e}^{kv_0t}$. A similar analysis for an initially horizontal field $B_0\hat{\mathbf{x}}$ yields

$$B_x = B_0[\cos^2(kx/2) + \sin^2(kx/2)\,\mathrm{e}^{-2kv_0t}]\,\mathrm{e}^{kv_0t}, \tag{4.33}$$

so the field is concentrated at the upwelling ($x = 0$).

Weiss (1966) and later Galloway and Weiss (1981) followed this kinematic flux expulsion numerically, again neglecting back-reaction of the Lorentz forces but retaining finite resistivity. The two-dimensional incompressible flow velocity in an eddy is simulated by the stream function

$$\psi = \frac{UL}{\pi}\cos\frac{\pi x}{L}\cos\frac{\pi z}{L}, \qquad \mathbf{v} = \nabla\psi \times \hat{\mathbf{y}}, \tag{4.34}$$

where U and L are characteristic speed and length. The time-development of an initially uniform field $B_0\hat{\mathbf{z}}$ is then followed from the induction equation

$$\frac{\partial \mathbf{B}}{\partial t} = \nabla\times(\mathbf{v}\times\mathbf{B}) + \eta\nabla^2\mathbf{B}, \tag{4.35}$$

subject to the symmetry conditions $B_x = 0$ on all boundaries. As usual it is the value of the MRN $\mathrm{R_m} = UL/\eta$ which fixes the solution for all time (Fig. 4.3). For moderately large values (e.g. $\mathrm{R_m} = 250$), the numerical results show clearly how the clockwise motion winds up the field, concentrating the flux at the sides of the cell. Near the cell centre the field is at first amplified, but then reconnection takes place and the field begins to decay. In a symmetrical system, a steady state with nearly all the flux expelled from the centre of the eddy is established after about 6 × the turnover time L/U.

One can estimate as follows the maximum field B_1 and the steady-state field B_m in the boundary layers at the edges of the cell. The rate of increase of B_1 through the effect of the flow U on B_0 is $\approx UB_0/L$. From flux conservation, the transverse length-scale of the wound-up field in this two-dimensional geometry is $l \approx (B_0/B_1)L$, so that the diffusion term in (4.35) is $\eta B_1/(B_0/B_1)^2L^2$. The increase in B_1 will be halted when these two terms balance, yielding for the maximum steady-state field

$$B_1 \approx \mathrm{R_m^{1/3}}B_0 \tag{4.36}$$

(cf. Weiss 1966; Moffatt and Kamkar 1982).

The boundary layers have a typical thickness d fixed by a balance between the diffusion speed η/d out of the layer against the flow speed into it, given by continuity to be $\approx (dU/L)$:

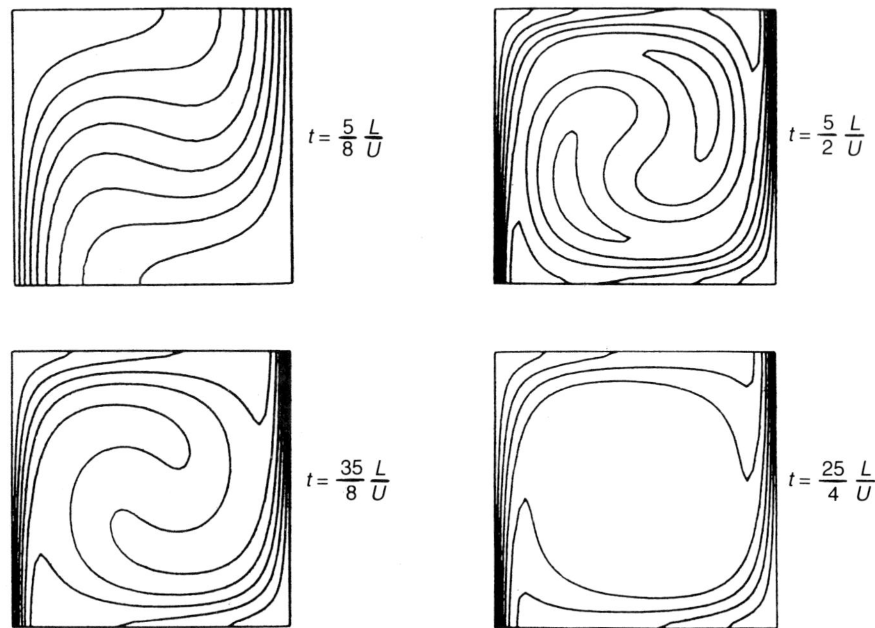

FIG. 4.3. Successive stages in the distortion, reconnection and expulsion of an initially uniform field by an eddy, for $\mathrm{R_m} = 250$. Lines of force are shown at multiples 5/8, 5/2, 35/8, 25/4 of the turnover time L/U. (From Galloway and Weiss 1981.)

$$d \approx \mathrm{R_m^{-1/2}} L, \tag{4.37}$$

whence again by flux conservation,

$$B_\mathrm{m} \approx \frac{L}{d} B_0 \approx \mathrm{R_m^{1/2}} B_0. \tag{4.38}$$

Care must be taken, however, not to extrapolate these two-dimensional results into three dimensions. For example, in a three-dimensional axisymmetric cell, the flux will be concentrated preferentially at the centre. Further, flux conservation in a tube of radius d replaces (4.38) by the much larger value

$$B_\mathrm{m} \approx \left(\frac{L}{d}\right)^2 B_0 \approx \mathrm{R_m} B_0. \tag{4.39}$$

Similar behaviour is found for kinematic flow when strict axisymmetry is replaced by motion in hexagonal cells, but with $\mathbf{B}_0$ still parallel to the axis (Galloway and Proctor 1983). However, when the imposed field is horizontal, but the flow has the same topology, the asymptotic field structure is much more complicated

(Arter *et al.* 1982; Arter 1983). Although strong fields are found to be confined to isolated regions at the axis and on the upper and lower boundaries, there is now no clear sign of complete flux expulsion.

In spite of these caveats, it is nevertheless customary to describe the redistribution of flux by inexorable motions as due to a macroscopic resistivity. A weak field can be tangled in a few turnover times L/U, and if this is written in the form L^2/η, then $\eta \simeq LU$. When dealing with a spontaneously arising turbulent field, with eddies of scale l and turnover time $\tau_c = l/v_t$, this is replaced by a *turbulent resistivity*

$$\eta_t = l v_t, \tag{4.40}$$

of the same order as the eddy viscosity introduced in Section 4.2. This turns out to be an essential feature of conventional dynamo theory (cf. Sections 6.3 and 6.4).

One clearly wants to extend the kinematic results for the different cases by coupling (4.35) with the equations of motion and of energy, so including the back-reaction of the Lorentz force on the motion. Weiss and collaborators have carried out a systematic programme of numerical simulations of magnetoconvection, initially in the *Boussinesq* approximation (Proctor and Weiss 1982). This assumes incompressibility except in the buoyancy term, where account is taken of the finite coefficient of expansion of the fluid. The approximation is made simply for convenience: as a stellar convective zone extends normally over more than a local scale-height, a realistic model should allow for compressibility. Nevertheless, the Boussinesq treatment is instructive. Galloway and Moore (1979) have explored further the three-dimensional axisymmetric cell problem, finding different asymptotic regimes for different initial field strengths. Thus if

$$v_{A0} \equiv \frac{B_0}{(4\pi\rho)^{1/2}} < \frac{U}{R_m}, \tag{4.41}$$

the concentrated field B_m given by (4.39) is weaker than the *equipartition* field B_{eq}, satisfying

$$\frac{B_{eq}^2}{8\pi} = \frac{1}{2}\rho U^2. \tag{4.42}$$

In this domain, the flow is hardly affected by the field, so the kinematic solutions are valid. At higher field strengths (measured by v_{A0}), the asymptotic regime again has the flux concentrated into the axis of the cell, but now two separate flow domains emerge: the flux rope domain without any motion, and the field-free external region in which normal (laminar) convection proceeds, with hardly any effects due to the presence of the rope. But again, this pleasingly simple picture does not persist when there is a strong departure from axisymmetry; with the imposed field horizontal, Arter (1985) finds an asymptotic structure similar to that found in the earlier kinematic studies already cited.

Compressibility enters the theory in two ways: first, stratification of the medium must be taken into account, and second, the fluid does not react passively to changes in its pressure, but adjusts its density. In an early study, Gough and Tayler (1966) applied the energy principle to estimate the effect of the field on the Schwarzschild criterion. (It is assumed implicitly that the relevant length-scales are so large that the inhibition of convective instability due to viscosity and heat conduction – as measured by the Rayleigh number (4.4) – is negligible.) They found stability against convection guaranteed provided

$$\frac{\mathrm{d}\ln T}{\mathrm{d}\ln p} < \frac{\gamma - 1}{\gamma} + \frac{B_{\mathrm{v}}^2}{B_{\mathrm{v}}^2 + 8\pi\gamma p}, \tag{4.43}$$

where B_{v} is the *vertical* component of $\mathbf{B}$. When B_{v} is small the Schwarzschild criterion is unaffected, as rolling motions can take place about the horizontal field. In general, a vertical field component able to interfere with incipient convection would need to have a magnetic pressure approaching the thermal pressure ('$\beta \approx 1$' in the plasma physicist's language).

Local interference with convection can be expected in sunspot zones, where strong fields are present in low-density surface regions (Section 4.6). The result (4.43) shows how difficult it would be to try and suppress convection through the bulk of an unstable zone. One might have expected a sufficient criterion to be that $B^2/8\pi \approx \rho v_{\mathrm{t}}^2/2$, where v_{t} is the Biermann–Cowling turbulent speed; and since their estimated v_{t} is very subsonic, this equipartition field has a pressure much below the thermal (the Alfvén speed much less than the sound speed). But the Biermann–Cowling estimate for v_{t} and the associated $\Delta(\nabla T)$ assumes no interference with free convection. If turbulence is supposed suppressed, the same heat flux must be carried by radiative transfer alone, requiring a much larger $\Delta(\nabla T)$. A self-consistent criterion requires that the vertical magnetic field have an energy density comparable with that of the much more energetic turbulent field that *would* develop if the large gradient $\Delta(\nabla T)$ were maintained in a field-free domain. As the buoyancy forces would then be comparable with gravity, the velocities would be nearly sonic, and the necessary magnetic field would indeed have a pressure comparable with the thermal pressure.

This argument (Mestel 1970), validated in a detailed study by Moss and Tayler (1969), is a *reductio ad absurdum*: there is no reason for believing that stars contain anywhere near this amount of magnetic flux (cf. Section 5.1 and Chapters 8 and 9). One rather awaits eagerly the results of the ongoing studies of compressible magnetoconvection, which should show in further detail how models of convecting domains containing a more realistic magnetic flux, either externally imposed or dynamo-generated within, prefer to distribute the flux throughout the zone, simultaneously giving structure to the density and kinetic fields. We accept that prima facie there will be a marked tendency for some local flux concentration, and now turn to the complementary problem: the effect of a local strong field on the otherwise spontaneous convective instability of a superadiabatic domain. The link with one of the most striking features of solar magnetic activity – the sunspot – is manifest and will be pursued in Section 4.6.

4.4 A strong imposed field and the onset of convection

We now survey the complementary problem: the effect of a magnetic field on the criteria for the onset of convection. A full treatment of the stability of a superadiabatic domain in the presence of a strong, externally imposed field must go beyond the hydromagnetic energy principle, which assumes no dissipative processes active. There is an extensive literature on the linear problem in the Boussinesq approximation (Thompson 1951; Chandrasekhar 1952, 1961); a concise summary of the essentials is given by Cowling (1976).

4.4.1 *Imposed field vertical*

Consider a liquid heated from below, confined within planes $z = 0$ and $z = L$, with gravity g acting in the negative z-direction. The undisturbed temperature distribution is $T_0 - \beta_T z$, with z the height above the reference level where $T = T_0$. After a small disturbance, the fluid has the velocity $\mathbf{v}$, and the temperature becomes

$$T = T_0 - \beta_T z + \theta. \tag{4.44}$$

The heat equation becomes

$$\frac{\partial T}{\partial t} + (\mathbf{v} \cdot \nabla) T = \kappa \nabla^2 T, \tag{4.45}$$

where κ is the (uniform) thermal diffusivity. To first order, (4.45) reduces to

$$\frac{\partial \theta}{\partial t} = \beta_T v_z + \kappa \nabla^2 \theta. \tag{4.46}$$

Similarly, the motions generate a perturbation field $\delta \mathbf{B}$ superposed on the uniform vertical field $\mathbf{B}_0$, from (2.85) satisfying to first order

$$\frac{\partial\, \delta \mathbf{B}}{\partial t} = (\mathbf{B}_0 \cdot \nabla) \mathbf{v} + \eta \nabla^2 \delta \mathbf{B}. \tag{4.47}$$

In the Boussinesq approximation, not only $\delta \mathbf{B}$ but also $\mathbf{v}$ is divergence-free. Because of the small but finite coefficient of expansion α, the density is related to the zero-order value ρ_0 by

$$\rho = \rho_0 (1 + \alpha(\beta_T z - \theta)). \tag{4.48}$$

The variation of ρ is taken into account only in the gravitational term in the equation of motion (2.35), yielding the buoyancy force; in linearized form

$$\frac{\partial \mathbf{v}}{\partial t} - \frac{1}{4\pi \rho_0} (\mathbf{B}_0 \cdot \nabla)\, \delta \mathbf{B} = \nabla \Pi + \nu \nabla^2 \mathbf{v} - \alpha \theta \mathbf{g} \tag{4.49}$$

where ν is the uniform kinematic viscosity, and

$$\Pi = \frac{p}{\rho_0} + \frac{B^2}{8\pi\rho_0} + g\left(z + \frac{1}{2}\alpha\beta_T z^2\right). \tag{4.50}$$

As always in the Boussinesq approximation, the fluid pressure is passive, merely responding to the other forces (including inertial forces).

The set of linear equations (4.46), (4.47), (4.49) plus $\nabla \cdot \mathbf{v} = 0$ combine into a single high-order partial differential equation for the vertical velocity component v_z. A rigorous treatment involves a detailed discussion of boundary conditions (Chandrasekhar 1961). However, a 'local' analysis that assumes solutions behaving like

$$\exp[\omega t + \mathrm{i}(lx + my)] \sin\left(\frac{\pi z}{L}\right) \tag{4.51}$$

is similar to the Chandrasekhar treatment with the idealized boundary conditions appropriate to a shear-free system, yielding a dispersion relation that retains the principal qualitative features and is reasonably close to the exact results. If all the diffusive coefficients are put equal to zero, exponential growth occurs for large l, m (i.e. horizontal wavelengths small compared with L) if

$$\pi B_0^2 < 4\rho_0 g\alpha\beta_T L^2. \tag{4.52}$$

(In a paper written in 1949 and not now easily accessible, but noted in Cowling 1953, Alfvén's collaborator Walén derived essentially the same result for an unstably stratified medium permeated by a horizontal magnetic field, by estimating the ratio of the buoyancy felt by a disturbed element to the restoring Lorentz curvature force.) With the inequality reversed, the system oscillates. With diffusion retained, the behaviour depends primarily on the ratio $\zeta \equiv \eta/\kappa$. If $\zeta > 1$, and viscosity is ignored, then again modes with small horizontal wavelength grow monotonically if (4.52) is replaced by

$$\pi B_0^2 < \zeta(4\rho_0 g\alpha\beta_T L^2) : \tag{4.53}$$

in this parameter range, diffusion facilitates normal turnover convection because the loss in buoyancy due to heat transfer is proportionately less than the reduction in the Lorentz restoring force due to departures from strict flux freezing. When $\zeta < 1$ the system can become *overstable*: if

$$\pi B_0^2 < \zeta^{-1}(4\rho_0 g\alpha\beta_T L^2), \tag{4.54}$$

then because of the excess of thermal over magnetic diffusion, the restoring magnetic force dominates over the buoyancy force and can generate oscillations with an exponentially growing amplitude (Thompson 1951). Chandrasekhar's full treatment (1961) includes the effects of viscosity and of varying boundary conditions; the domains respectively of steady and oscillatory motions are determined in terms of the familiar Rayleigh number and the Chandrasekhar number $Q = B^2L^2/4\pi\rho_0\nu\eta$. In a medium of low Prandtl number ν/κ it is more convenient to follow Weiss (1964, 1968) and use, instead of these, a modified

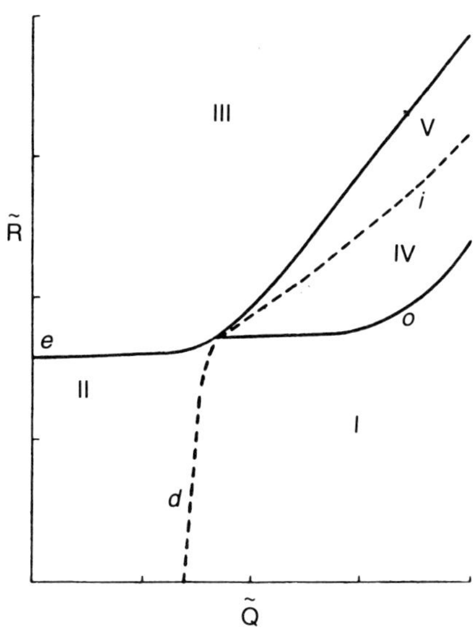

FIG. 4.4. A strong imposed vertical field and the onset of convection: a typical stability diagram. Beneath curve *di* the eigenmodes are oscillatory; above they increase or decrease exponentially with time. The curve *o* corresponds to the onset of overstability (oscillatory modes with exponentially increasing amplitude). Everything above *eo* is unstable. In region IV the unstable modes oscillate; in regions III and V they grow steadily. Relaxing the upper boundary condition tends to move curve *i* down and curve *o* up, reducing or even suppressing completely the overstable region. (From Weiss 1964, 1968.)

Rayleigh number $\tilde{R} = g\alpha\,\Delta(\nabla T)L^4/\pi^4\kappa^2$ and the square of the Hartmann number $\tilde{Q} = B^2L^2/4\pi^3\rho\kappa^2$. Figure 4.4 is a typical stability diagram for a Boussinesq fluid confined between rigid boundaries.

Compressible convection is more complicated even in the linear domain, (where one might have hoped that one needed just to replace the temperature gradient β_T by the adiabatic excess $\Delta(\nabla T)$ used in (4.1) and (4.2), and the thermal diffusivity κ by that appropriate to radiative transfer). Again, when $\zeta < 1$, the dominant thermal diffusion leads to oscillatory convective modes, but the modes that are destabilized depend on the plasma $\beta = 8\pi p/B^2$. The case $\beta \gg 1$ is the Boussinesq limit, with the overstable oscillations being slow magnetosonic modes, corresponding to transverse hydromagnetic waves. At the other extreme with $\beta \ll 1$, the destabilized modes are again slow magnetosonic modes but now correspond to sound waves travelling along the field lines (cf. Section 3.1; Syrovatsky and Zhugzhda 1967; Zhugzhda 1970). When $\beta \approx 1$, the stratification couples the slow and fast modes, and even a subadiabatic layer may become unstable (Antia and Chitre 1979; Cattaneo 1984). As in the Boussinesq problem, overstability appears only through double-diffusive effects, which introduce phase differences between the thermal and magnetic perturbations.

The analogue for a compressible medium of the Boussinesq approximation for a convecting liquid is the *anelastic* approximation (Gough 1964; Nordlund 1982, 1986), in which the zero-order density stratification is retained everywhere except in the gravity term, so again leading to a buoyancy force. This effectively suppresses sound waves; like the Boussinesq approximation, it assumes that the sound speed is much greater than all other propagation speeds, a good approximation so long as $\beta \gg 1$ (Hughes and Proctor 1988). In fact, in a sunspot the Alfvén and sound speeds are comparable, so neither the Boussinesq nor the anelastic approximations are strictly applicable, but much can be learnt by using the results to hand (Section 4.6).

4.4.2 *Imposed field horizontal*

Of importance for some recent developments in dynamo theory (Sections 6.7, 6.8) is the effect in a superadiabatic domain of a strong uniform *horizontal* field $\mathbf{B}_0$ – simulating a toroidal field – on the Schwarzschild criterion for convective instability in the dissipation-free limit. If the wavelengths are small compared with the scale-height – so that the treatment is Boussinesq – then for a disturbance $\propto \exp(\omega t + \mathrm{i}\mathbf{k}\cdot\mathbf{x})$, the linearized kinematic equation (2.88) yields

$$\omega\delta\mathbf{B} = \mathrm{i}(4\pi\rho)^{1/2}[\mathbf{v}(\mathbf{k}\cdot\mathbf{v}_\mathrm{A}) - \mathbf{v}_\mathrm{A}(\mathbf{k}\cdot\mathbf{v})], \tag{4.55}$$

and the inviscid momentum equation (2.35)

$$\rho\omega\mathbf{v} = -\mathrm{i}\mathbf{k}[\delta p + (\mathbf{v}_\mathrm{A}\cdot\delta\mathbf{B})(\rho/4\pi)^{1/2}] + \delta\rho\,\mathbf{g} + \mathrm{i}(\rho/4\pi)^{1/2}(\mathbf{k}\cdot\mathbf{v}_\mathrm{A})\,\delta\mathbf{B}, \tag{4.56}$$

where $\delta\mathbf{B}$, δp, $\delta\rho$ are perturbed quantities and again $\mathbf{v}_\mathrm{A} = \mathbf{B}_0/(4\pi\rho)^{1/2}$. Substitution of $\delta\mathbf{B}$ from (4.55) yields for the combined Lorentz force terms $\mathbf{L}$ in (4.56) the expression

$$\begin{aligned}\mathbf{L} &= \frac{\rho}{\omega}[\{(\mathbf{k}\cdot\mathbf{v}_\mathrm{A})(\mathbf{v}\cdot\mathbf{v}_\mathrm{A}) - v_\mathrm{A}^2(\mathbf{k}\cdot\mathbf{v})\}\mathbf{k} - (\mathbf{k}\cdot\mathbf{v}_\mathrm{A})\{(\mathbf{k}\cdot\mathbf{v}_\mathrm{A})\mathbf{v} - \mathbf{v}_\mathrm{A}(\mathbf{k}\cdot\mathbf{v})\}] \\ &= \frac{\rho}{\omega}[(\mathbf{k}\cdot\mathbf{v}_\mathrm{A})(\mathbf{v}_\mathrm{A}\times(\mathbf{k}\times\mathbf{v})) - (\mathbf{k}\cdot\mathbf{v})(\mathbf{v}_\mathrm{A}\times(\mathbf{k}\times\mathbf{v}_\mathrm{A}))].\end{aligned} \tag{4.57}$$

In Cartesians, with $\mathbf{g} = -g\hat{\mathbf{z}}$, $\mathbf{v}_\mathrm{A} = v_\mathrm{A}\hat{\mathbf{x}}$, and $\mathbf{k} = (l, m, n)$,

$$\mathbf{L} = \frac{\rho v_\mathrm{A}^2}{\omega}[0,\ -m(mv_y + nv_z) - l^2 v_y,\ -n(mv_y + nv_z) - l^2 v_z], \tag{4.58}$$

$$\nabla\cdot\mathbf{L} = -\mathrm{i}\frac{\rho v_\mathrm{A}^2}{\omega}k^2(mv_y + nv_z). \tag{4.59}$$

The Boussinesq condition $\nabla\cdot(\rho\mathbf{v}) = 0$ then yields

$$k^2\,\delta p - \mathrm{i}gn\,\delta\rho + \nabla\cdot\mathbf{L} = 0. \tag{4.60}$$

Elimination of δp between (4.60) and each of (4.56) yields relations between $\delta\rho$ and each component of $\mathbf{v}$.

By essentially the same argument as in (3.122) et seq., for adiabatic changes with pressure equilibrium with the ambient medium maintained,

$$\frac{\mathrm{d}p}{\mathrm{d}t} = (\mathbf{v}\cdot\nabla p) = c^2\left(\frac{\partial\rho}{\partial t} + \mathbf{v}.\nabla\rho\right), \tag{4.61}$$

or

$$\omega c^2\,\delta\rho = v_z\left(\frac{\mathrm{d}p}{\mathrm{d}z} - c^2\frac{\mathrm{d}\rho}{\mathrm{d}z}\right) = -(c^2/g)N^2\rho v_z, \tag{4.62}$$

where $c = (\gamma p/\rho)^{1/2}$ is the local sound speed, and (cf. 3.126)

$$N^2 = -g\left(\frac{\rho'}{\rho} + \frac{g}{c^2}\right). \tag{4.63}$$

Since the unperturbed field $\mathbf{B}_0$ is assumed uniform, it does not contribute to the zero-order equilibrium, and we again arrive at the 'Brunt–Väisälä' frequency N, but it is now imaginary, as we are concerned with a superadiabatic medium, and internal gravity waves are replaced by exponentially growing convective modes. If B_0 is zero, then the above equations yield the dispersion relation

$$\omega^2 = \omega_0^2 = -N^2\left[1 - \frac{(\mathbf{k}\cdot\mathbf{g})^2}{k^2g^2}\right] \tag{4.64}$$

With B_0 retained, one finds instead

$$\omega = [\omega_0^2 - (\mathbf{k}\cdot\mathbf{v}_\mathrm{A})^2]^{1/2}. \tag{4.65}$$

It is seen that within the linear domain, the modes predicted by the Schwarzschild criterion to be unstable, and with wave-vector $\mathbf{k}$ perpendicular to both $\mathbf{g}$ and $\mathbf{B}_0$, suffer no interference from the magnetic field, consistent with the result (4.43).

4.5 Non-linear theory: recent developments

Over the last two decades, the escalation in computing power has made feasible the study of fully three-dimensional, compressible convection, in both the non-magnetic and magnetic cases. Conceivably, we may be in sight of the ultimate goal – realistic simulations of stellar convective zones. However, because of the varying assumptions and approximations made, inevitably the results emanating from the active groups will sometimes differ on non-trivial issues. We present summaries of what seem to be among the most important studies to date, bringing out as far as possible the qualitative reasons for differences, and anticipating that the steady extension of our knowledge of what the equations say will probably accelerate over the next few years.

4.5.1 *The non-magnetic problem*

The studies by Nordlund, Stein and colleagues are noted particularly for their careful simulation of physical conditions in the outer solar regions. From the computed degree of atomic ionization and of molecular dissociation, they can construct an accurate equation of state, allow for the latent heat of ionization in the energy balance, and take explicit account of the details of radiative transfer, important at the edges of convective zones, where convective transport is giving way to radiative. They also adopt the realistic condition of an open boundary.

Stein and Nordlund (1989) were able to tackle fully three-dimensional, turbulent, Rayleigh–Bénard convection in a highly non-uniform medium, simulating a superadiabatic, subphotospheric region as in the Sun. Their results demonstrate how the topology of the convection is dominated by the density stratification. The motions are in fact well described by the anelastic approximation (e.g. Nordlund *et al.* 1996), with the continuity equation yielding small temporal density variations in the Eulerian frame:

$$\frac{\partial}{\partial t}\log\rho = -\frac{1}{\rho}\nabla\cdot(\rho\mathbf{v}) \simeq 0; \tag{4.66}$$

but equivalently, in the Lagrangian frame the changes are large:

$$\frac{\mathrm{d}}{\mathrm{d}t}\log\rho = -\nabla\cdot\mathbf{v} \simeq \mathbf{v}\cdot\nabla\log\rho \tag{4.67}$$

from (4.66). Since the horizontal variation of ρ is much smaller than the vertical, (4.67) can be written

$$\frac{\mathrm{d}}{\mathrm{d}t}\log\rho \simeq v_z\frac{\mathrm{d}}{\mathrm{d}z}\log <\rho> \tag{4.68}$$

where $<\rho>$ is the horizontal and temporal average. Thus the vertically moving fluid must follow effectively the dominantly vertical density variations – all ascending fluid expands and descending fluid contracts. If one performs a Fourier transformation, ignoring the vertical variation of the velocity compared with that of the density, then the horizontal wavelength λ_{h} is related to the vertical density scale-height H_ρ by

$$\lambda_{\mathrm{h}} \simeq 4\pi H_\rho(|v_{\mathrm{h}}|/|v_z|). \tag{4.69}$$

The pressure variations available to accelerate the horizontal flow are set essentially by the temperature fluctuations (through near hydrostatic balance in the vertical direction), which are largely independent of the horizontal scale. However, the product of the vertical velocity fluctuations and the temperature fluctuations are also constrained by the requirement that the convective flux be nearly equal to the total flux. Together, these require that $|v_{\mathrm{h}}|$ and $|v_z|$ be comparable, whence from (4.69), λ_{h} exceeds H_ρ by roughly an order of magnitude, as is indeed confirmed by observations of the solar granulation. Deeper down, the increasing scale-height supports correspondingly larger horizontal scales, a basic kinematic result well appreciated over the decades. In particular, Simon and Weiss (1968) suggested that the observed supergranular motions are a kinematic manifestation at the surface of large-scale motions occurring deep down, and driven by a local mixing-length convective model. A similar argument can be invoked to account for the mesogranulation. (The authors argued that one should also observe 'giant cells' with length-scales of the order of the depth of the whole zone, but they subsequently changed their views (Simon and Weiss 1991).)

Nordlund and Stein's work is carried further in a series of papers and conference reports (Nordlund and Stein 1990, 1996; Nordlund and Dravins 1990; Nordlund *et al.* 1994, 1996; Stein and Nordlund 1998*a*), leading to a new global picture of the solar convection zone. They find gently expanding, structureless, warm upflows, and strong, converging, filamentary, cool downdrafts. It is the radiative loss at the surface which drives the circulation, through its continual provision of entropy-deficient gas. The horizontal velocity field has a hierarchical cellular appearance, with small cells at the surface and successively larger cells at greater depth. The observed solar granulation is predicted to be a surprisingly shallow surface phenomenon. The pattern is recognizable in temperature over a range less than 400 km, much less than the horizontal scale of the phenomenon (cf. Fig. 3 in Nordlund and Stein 1991). Because of the downward increase in the local density scale-height, the deeper layers support successively larger cells. In their picture, the downflows of small cells near the surface merge into filamentary downdrafts of larger cells at greater depths, which they surmise continues through the whole convective zone. The small cells are visible directly as granulation; the larger cells are detectable through their horizontal velocity fields which advect the smaller cells. To quote Nordlund and Stein (1990), p. 197: 'the pressure fluctuations which drive the horizontal components of large scale velocity fields extend over a height range comparable to or larger than the horizontal size of the fluctuations. Since the aspect ratio (ratio of horizontal size to distance from the surface) of these flows is larger than unity, this implies that the horizontal velocity fields of larger-scale flows extend up to the surface.' Thus these essentially non-local computations vindicate the argument of Simon and Weiss (1968, 1991): one is indeed forced to relate the meso- and supergranulation to motions at certain depths, with the distribution of surface horizontal velocity amplitudes with horizontal scale reflecting the dependence of vertical velocity amplitudes on depth.

The picture as a whole seems to bear little resemblance to the Kolmogoroff picture of turbulence in a homogeneous, incompressible liquid, with the hierarchy of eddies being due to a downward cascade (e.g. Chandrasekhar 1949). It is true that like all other workers, Nordlund and co-workers are forced to introduce a macroviscosity, for it would be quite hopeless to attempt to perform computations down to the length-scales on which the microviscosity becomes important. But since it is on these scales that the dissipation built into the models must occur, there is again implicit in the formulation a cascade of energy through the 'inertial range', extending from the smallest macroscales appearing explicitly down to the viscous microscales. However, the remarkable advances in their

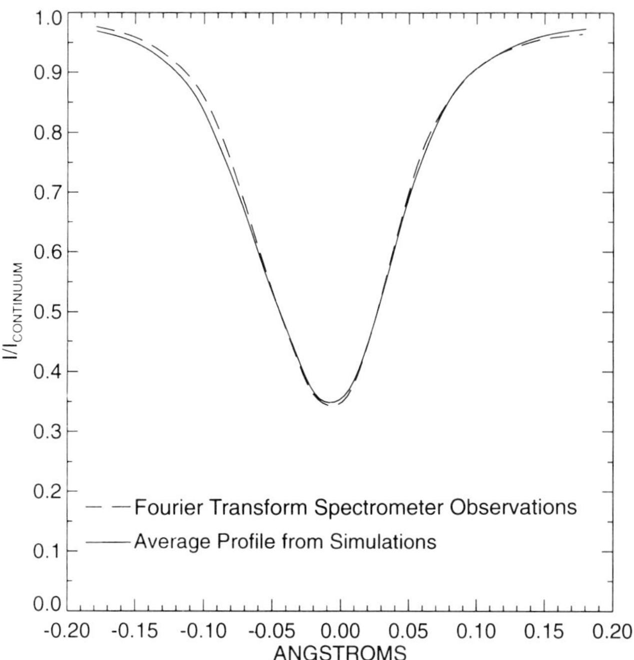

FIG. 4.5. Comparison of computed spectral lines with observation. (From Lites *et al.* 1989.)

numerical techniques enable the authors to resolve up to an order of magnitude of the former inertial range: i.e. most of the power is in the *resolved* motion, and it is now only a fraction that remains to be described phenomenologically by the turbulent viscous term.

The computations of Nordlund and co-workers – which we recall include details of the physics that are of special importance for the surface layers – are able to predict to within a few percent the strength and shape of the spectral lines. In particular, the distortion of the shape by the mass motions – put into standard theory as part of the phenomenology under the names 'macro-' and 'microturbulence' – are satisfactorily explained in terms of the resolved motions. The review by Spruit *et al.* (1990) emphasizes the impressively close resemblance of these computations to images of the solar convection zone (cf. Fig. 4.5). The mean vertical structure derived from these models is in close agreement with the deductions from helioseismological evidence; the remaining apparent discrepancy (Rosenthal 1998) now appears to have been eliminated (Rosenthal *et al.* 1998). Other submitted papers (Nordlund and Stein 1998*b*; Stein and Nordlund 1998) study in further detail the interaction of solar oscillations with convection.

Nigel Weiss sums up as follows (personal communication): 'Nordlund and colleagues are remarkably successful at describing photospheric convection (dominated by the almost laminar expansion of the rising plumes) and so of modelling the atmosphere above the convecting layer. However, *no-one* has yet supplied a correspondingly precise description of what happens in the deep convection zone, which is what matters for dynamos.'

The parallel computations by Cattaneo *et al.* (1991) are also fully three-dimensional, but are based on a more traditionally fluid dynamical approach. The authors consider convection in a polytropic ideal gas within a layer, with vanishing vertical velocity and tangential stress at top and bottom. As noted by Nordlund *et al.* (1994), the conclusions of the two groups are in many ways similar, but there are also important differences. Both stress that it is the non-adiabatic surface effects that drive the flow. A worker in stellar structure would not dissent, saying that it is the surface boundary conditions, together with the equation of state of the gas, that enforce the superadiabatic temperature gradient which maintains the convection. Again, both find large-scale up- and downdrafts, of widely differing structure because of the stratification. However, the most important differences appear in the description of the flow in the bulk of the convective zone away from the boundaries. Cattaneo *et al.* describe the convective motions as showing a complex mixture of long-range order and small-scale disorder: strong, coherent long-lived downflows surrounded by disorganized weaker motions ('turbulence'). In the authors' words: 'The disorganized ingredient is characterized by a *low* degree of spatial correlation of vertical velocity with fluctuations in the thermodynamic quantities, and so to a correspondingly low level of energy transport. In contrast, the stronger downflows are highly correlated with thermodynamic fluctuations and so contribute significantly to the net vertical enthalpy flux.' But in the large Reynolds number limit, the strong downdrafts appear to transport enthalpy upwards at about the same rate that they transport kinetic energy downwards; 'they reorganize the various energetic ingredients within the convection, without contributing to the net energy flux'. Thus an unexpected conclusion of the Cattaneo *et al.* computations is that most of the energy transport comes again from the disorganized component.

Some of the divergences between the claims of the different groups can be traced to their particular physical assumptions. Characteristic of the Nordlund *et al.* work is the explicit calculation of the degree of ionization. In the surface regions, the latent heat associated with the ionization contributes to the convective flux, changing markedly the balance between convective and kinetic energy flux, so that, in contrast to the Cattaneo *et al.* result, a significant net energy transport by the large-scale motions does in fact survive (Rast and Toomre 1993; Rast *et al.* 1993). The effects of differing boundary conditions on the bulk of the zone will be the less marked, the greater the depth of the zone. A criticism of some of the Cattaneo *et al.* models is of their adoption of a dynamical turbulent viscosity μ that is *constant*, implying a viscous diffusivity $\nu = \mu/\rho \propto 1/\rho$. If ν is just large enough to handle the turbulence at the lower boundary, then the number of scale-heights that can in practice be incorporated into the computations is limited, as otherwise the viscous time-scales near the low density upper boundary become unmanageably short, and the behaviour in those layers becomes very viscous. Thus the effects of the lower boundary tend to be felt throughout the whole model, and it is not immediately clear which of the detailed conclusions would survive an increase in the depth of the domain to a more realistic value.

In neither set of computations does one find behaviour reminiscent of the mixing-length picture as outlined in Section 4.1, with fluid parcels moving through a distance l before mixing with the surroundings. In fact, near the solar photosphere, the macroscopic scale of variation in the Nordlund *et al.* models is found to be smaller than the dimensions of the granulation cells, contrary to an implicit assumption of the mixing-length model. However, once such a model has been constructed, one can compute the *average* local temperature gradient, which will necessarily exceed in modulus the local adiabatic gradient, and by an amount greater than the temperature gradient in the very nearly adiabatic upflows. One can then write the energy flux in terms of the mean superadiabaticity, analogous to the canonical form (4.2). But this is of course no guarantee that mixing-length theory is other than a convenient *ex post facto* calibration of the numerical results. Some would question whether in the light of the picture emerging from the new computations, the classical fluid dynamical division into a mean flow plus a fluctuation field is any longer helpful. However, one feels that the mixing-length formalism will remain part of the language, simply because within its explicit limitations, it has been remarkably successful. Aake Nordlund and colleagues are world leaders in the search for a non-phenomenological, deductive model of the solar convective zone; yet in their words (personal communication): 'If one uses the classical mixing-length formulae, one gets velocities and fluxes that are of the right order of magnitude. Also, the mixing-length parameter that one needs to use to fit various observational constraints (all the way from helioseismology to the giant branches in the Hertzsprung–Russell diagram) does not vary much (perhaps by a factor 2). So the local mixing-length model is a good scaling relation; but equally it is meaningless to compare a local relationship in detail with the non-local dependence of the real thing.'

As already noted, the larger scale motions observable at the solar surface contain information on the flow deeper down. The outstanding difficulty is the absence of an obvious reason why certain distinct scales, associated with motions at a few particular depths, should be preferentially selected (Simon and Weiss 1968, 1991). It is therefore vital for the observational situation to be clarified. To an outsider, the evidence for the mesogranulation appears less strong than that for the supergranulation. Several workers have reported that they do not find a local peak in the power spectrum at the mesogranular scales, but rather that the horizontal velocity power spectrum appears to vary smoothly from a peak at granulation scales all the way past supergranulation scales. Nordlund argues that the 'observed' mesogranular scale is an artefact of the data-processing; whereas Weiss argues that there is certainly some structure on a scale intermediate between granules and supergranules. The evidence for supergranular motions and the associated concentration of magnetic flux in the supergranular network has always been considerably stronger. The ongoing observational programme, beginning with Hagenaar *et al.* (1997), will yield dopplergrams and magnetograms that describe flow and magnetic fields on scales from 100 to 200 000 km, and should be able to confirm the existence of a peak at a typical supergranular scale, and hopefully resolve the disagreement about the mesogranulation.

Clearly, a correct interpretation of the observed surface motions requires a reliable detailed global model. Nordlund (personal communication) is emphatic that the oft-cited ionization of helium has little effect on what appears at the surface, and in particular cannot be the reason for any preferential selection of length-scales. However, it is conjectured by Nordlund *et al.* that the next series of numerical simulations, with a 30 000 km horizontal scale and 10 000 km scale in depth will be able to reproduce 'supergranulation properties' – i.e. observable manifestations, superposed on the locally generated granulation, of flows well below the surface.

We await eagerly the results of more numerical simulations before we can be certain of the precise wording of Spruit's 'change in paradigm', as quoted in Section 4.1. Other work that pays particular attention to classical concepts is that by Chan and Sofia (1989, 1996), who interpret the results of their numerical simulations in terms of the moments of the fluid dynamical equations. Such an approach requires some closure hypothesis so as to yield a finite set of reduced equations. The authors use their integrations to test various closure prescriptions in the literature.

4.5.2 *Magnetoconvection*

The Boussinesq studies of Section 4.4 have been extended into the non-linear domain by Galloway and Moore (1979) and by Weiss (1981) (see the review by Proctor and Weiss 1982) for various values of the parameter $\zeta \equiv \eta/\kappa$. They found that for $\zeta < 1$ and a sufficiently strong field, there exists the extension of the thermally destabilized hydromagnetic oscillations found in the linear domain. If the superadiabatic gradient is allowed to increase, the non-linear oscillations give way to overturning convection; the magnetic flux is then segregated from the motion, being confined to stagnant sheets or tubes. For $\zeta > 1$, however, this segregation was no longer found.

More recently, several groups have tackled the formidable problem of fully non-linear compressible magnetoconvection. An important pioneering paper was by Hurlburt and Toomre (1988), who studied non-linear two-dimensional compressible convection in a polytropic layer (Fig. 4.6). They assumed constant resistivity, thermal conductivity and dynamic viscosity in a layer with stress-free upper and lower boundaries. The imposed magnetic field is vertical, with a constant flux. Comparison with the Boussinesq results quoted above shows how the stratification destroys the up–down symmetry, and displaces the centre of an eddy towards the bottom of the layer. The vertical motion is dominated by narrow, rapidly sinking sheets of cool dense fluid, balanced by broad, diffuse, slow upwellings. This in turn affects the magnetic field, which is concentrated primarily in the sinking plumes. The consequent increase in the magnetic pres-

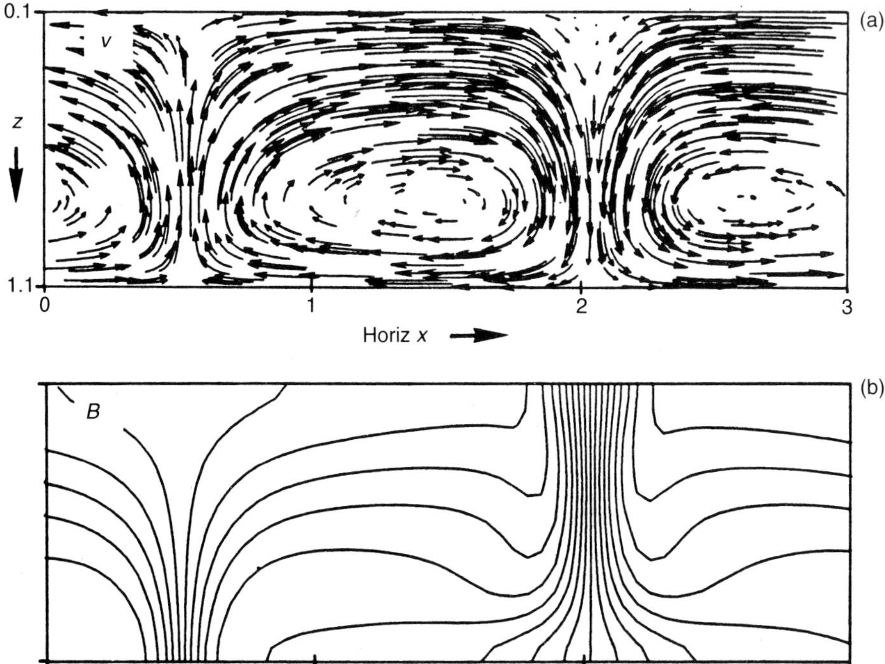

FIG. 4.6. Two-dimensional compressible magnetoconvection. (a) Streamlines and (b) field lines for a steady solution with $\zeta = 0.25$, $Q = 400$ and $R = 10^5$ (after Hurlburt and Toomre 1988). Comparison with Boussinesq results shows the marked asymmetry between rising and falling fluid that is introduced by compressibility.

sure within the plumes must be accompanied by a decrease in the gas pressure to maintain approximate horizontal magnetohydrostatic balance: the flux tube within a plume is partially evacuated (cf. Section 4.6).

Other numerical experiments on two-dimensional compressible magnetoconvection (Hurlburt *et al.* 1989; Weiss *et al.* 1990; Proctor *et al.* 1994; Brownjohn *et al.* 1995) have revealed a variety of both steady and time-dependent patterns of behaviour. For what is now seen to be an interim survey of the wealth of new results on fully *non-linear*, compressible convection in the presence of a strong magnetic field, the reader is referred to Proctor, in Thomas and Weiss (1992).

As emphasized by some of the authors, the restriction to two-dimensional perturbations can be very severe, and it is not a priori clear which features will survive once three-dimensional perturbations are admitted, as in some of the most recent ongoing numerical studies. Of the papers from the Cambridge group, that by Matthews *et al.* (1995*a*) stresses the pure fluid dynamical aspects. We begin by summarizing that by Weiss *et al.* (1996) on photospheric convection in strong magnetic fields, which is aimed at comparison with high-resolution solar observations. The paper explores the effect of an imposed magnetic field on the pattern of convection in a deep stratified layer of a perfect monatomic gas of depth d, with fixed temperatures T_0 and $T_0 + \Delta T$ at its upper and lower boundaries. The z-axis points downwards in the direction of the uniform gravitational acceleration $\mathbf{g}$, with the origin chosen so that the upper boundary is at $z_0 = T_0 d/\Delta T$. Attention is restricted to the region $(0 \leq x \leq \lambda d;\ 0 \leq y \leq \lambda d;\ z_0 \leq z \leq z_0 + d)$ – a square box with aspect ratio λ. In the absence of any motion, the gas is in a uniformly stratified equilibrium structure with

$$T = \Delta T\, z/d, \qquad \rho = \rho_0 (z/z_0)^m, \tag{4.70}$$

where $\rho_0 = \rho(z_0)$, the polytropic index $m = (gd/\mathcal{R}\,\Delta T) - 1$, and the superadiabatic gradient is

$$(\nabla - \nabla_{\mathrm{ad}}) = \frac{1}{(m+1)} - \frac{\gamma - 1}{\gamma}, \tag{4.71}$$

with $\gamma = 5/3$ for the assumed monatomic gas.

Radiative heat transfer is simulated by assuming a constant thermal conductivity K. Also assumed constant are the resistivity η, the shear viscosity μ, the magnetic permeability μ_0 and the heat capacity c_p. However, the viscous diffusivity $\nu = \mu/\rho$ and the thermal diffusivity $\kappa = K/c_p\rho$ both vary like $1/\rho$ and so decrease with depth: thus while the Prandtl number ν/κ remains constant, the crucial diffusivity ratio (cf. Section 4.4)

$$\zeta(z) \equiv \eta/\kappa = \zeta_0(\rho/\rho_0) \tag{4.72}$$

increases with depth. (The drawbacks of the constant μ assumption in a strongly stratified medium have already been noted.) Thus although some of the important physics – such as the varying degree of ionization and its effect on γ – is not explicitly included, the rapid inward increase in opacity is effectively simulated. The strength of the imposed magnetic field is measured by the Chandrasekhar number Q, introduced in Section 4.4.1.

In the authors' words, the changing patterns computed are summarized by saying that as the imposed field strength is reduced, the spatio-temporal structure becomes more complex, while the horizontal scale increases (Weiss *et al.* 1996). If Q is large enough, convection is completely suppressed. As Q is decreased, there is a bifurcation leading to steady, three-dimensional convection on a deformed hexagonal lattice, with a marked asymmetry between upward and downward motion, and a magnetic network at the upper boundary. A further reduction in Q produces an oscillatory bifurcation, followed by spatially modulated oscillations in which alternate plumes wax and wane periodically. As Q is decreased further, the oscillations become more violent and irregular, and their horizontal scale increases. For high Q the Lorentz force is dominant, and magnetic flux at the top of the layer is confined to a network enclosing the rising plumes; at the bottom, flux is swept into the rising plumes to produce fields that are locally intense. At low Q, strong fields form at junctions in the network but the magnetic structure changes in response to the evolving pattern of convection, with magnetic flux moving rapidly along the network that encloses the ephemeral plumes to form new field concentrations, at both the top and the bottom of the layer. When the imposed field is relatively weak, intense fields appear at junctions in the network, with the magnetic pressure comparable with the gas pressure and an order of magnitude greater than the dynamic pressure.

In this 1996 paper the value of the aspect ratio λ was constrained by the limitations of the computer not to exceed 2. In the most recent offering from the Cambridge school (Tao *et al.* 1998), a modern supercomputer allows study of convection in much wider boxes: the problem is identical with that of the 1996 paper, but now solutions are obtained with $\lambda = 4$ and 8 respectively. It is found that the results are remarkably sensitive to the value of λ. As the Chandrasekhar number Q is reduced at fixed Rayleigh number, in all cases convection sets in as steady overturning motion. For $Q = 2000$, Weiss *et al.* had found that the rising plumes form a close-packed, roughly hexagonal pattern. As Q is steadily reduced, this magnetically dominated regime will ultimately go over into one in which convection sweeps the field aside. Before that, at $Q \approx 1500$, there appear spatially modulated oscillations with adjacent plumes that alternate in strength (Weiss 1997). When Q is reduced to 1000, the pattern changes completely; however, whereas for $\lambda = 4/3$ there appear spatially modulated, two-dimensional, strictly periodic oscillations, when $\lambda = 2$ or $8/3$, the motion is chaotic, with broad vigorous plumes surrounded by a constantly changing magnetic network. As Q is further reduced, the plumes grow wider and more active.

In the newest computations, with λ increased to 4, it is found that some plumes amalgamate to form an almost field-free clump, surrounded by strong fields with smaller scale convection. When λ is put up to 8, this new pattern shows up far more clearly. The model is set up by slight perturbations to the static solution, with convection developing initially as an array of small-scale cells (corresponding to the most rapidly growing linear mode). The cells then combine to form larger isolated plumes, similar to those found previously for $\lambda = 2$, which in turn gradually amalgamate into clusters. At the upper boundary, magnetic flux is expelled from these clusters and confined to regions between them, where the field is so strong that only small-scale, close-packed plumes arise: the flux separates from the motion. Plate 2 shows the thermal and magnetic structure in this regime. The upper surface is divided into regions with vigorous convection and no magnetic fields, and regions with strong fields and small convective plumes; while at the bottom, magnetic flux is concentrated by inflows at the centres of these plumes and also to some extent beneath the larger clusters. Further, the pattern evolves continually with time. Where the field is strong, the small plumes undergo chaotically modulated oscillations, while within a cluster the plumes divide, split and recombine, while apparently retaining their identities.

This gratifyingly clear picture of spontaneous self-organization into strong-field, weakly convecting domains, coexisting with almost field-free, efficiently convecting domains, had indeed been anticipated in earlier studies of Boussinesq magnetoconvection (cf. Proctor and Weiss 1982; Section 4.3 above). The new results demonstrate the importance of using wide boxes for numerical experiments. Similar results have been found in a systematic study of two-dimensional magnetoconvection (Blanchflower *et al.* 1997). To one not personally involved in these very sophisticated computations but keenly interested in their outcome, what is impressive is not only the prediction of departures from the traditional picture of convection, but also the confirmation of how much of the qualitative inferences from the earlier work survives in the three-dimensional, fully compressible computations.

Nordlund and collaborators have carried out a parallel series of studies, again laying stress on the details of radiative transport and the equation of state (Nordlund 1984*a*,*b*, 1985; Nordlund and Stein 1989, 1990; Nordlund *et al.* 1994). The later papers summarize some time-dependent computations, aimed largely at the detailed modelling of sunspot umbras (cf. Section 4.6). As already noted, Nordlund *et al.* (1992) have also found the strong downward pumping of flux into the overshoot domain. Subsequent work with Brandenburg and colleagues is part of the ongoing attempt to understand the solar dynamo, and is referred to in Chapters 6 and 8.

Another set of computations is by Steiner *et al.* (1994) and subsequent papers and are discussed by Weiss (1997). Their new code is aimed especially at the study of small-scale structures – magnetic flux concentrations, current-layers and shocks in three-dimensional, time-dependent simulations. These authors again stay close to the physics in describing ionization and radiative transport at both low and high optical depths. The first results are simulations of radiative, non-stationary convection in the solar atmosphere but in *two* dimensions, so that only flux sheets and convective rolls are represented. Turbulent viscous stresses, heat transport and magnetic diffusion are introduced. The novel feature of the work is the development of appropriate numerical techniques that can resolve narrow structures and follow highly dynamic interactions between a magnetic flux sheet and turbulent convection. The video representations of the computations illustrate strikingly the formation of intense magnetic fields associated with convective downflows. A spectacular new phenomenon demonstated is the strong bending of a flux sheet by asymmetric convective flows, followed by a rapid sweeping back due to buoyancy and magnetic tension. Such violent events may lead to the excitation of transverse MHD waves that contribute to the heating of the upper solar atmosphere.

4.6 Sunspots, pores and isolated flux tubes

The origin of the study of magnetoconvection can reasonably be traced to a remarkable correspondence between Ludwig Biermann and Tom Cowling during the 1930s, especially on the problems of sunspots. Implicitly or explicitly they put aside the question of the origin of the observed locally strong fields, focusing instead on the mechanical and thermal equilibrium of the spot in the presence of the field. In the simplest picture, beneath the photosphere the spot is taken to be a cylindrical flux tube containing the uniform field $\mathbf{B} = B\hat{\mathbf{z}}$, separated from the field-free surroundings by a current sheet, so that the internal and external pressures p_i, p_e are related by

$$p_e - p_i = B^2/8\pi. \tag{4.73}$$

The reduced p_i required by (4.73) is consistent with the observed coolness of the spot, but to explain the *cause* of the cooling one must consider the energy sources and sinks. The early suggestion by H.N. Russell of adiabatic cooling of elements of gas rising through a *stably* stratified layer is clearly inapplicable to spots located in a superadiabatic and so unstable layer. Cowling quotes Biermann as proposing that the cooling is caused by the reduced heat supply from below, due to local magnetic interference with the convection that would otherwise proceed. Biermann published in 1941 a brief note in which he ascribes to Cowling the necessary criterion that the magnetic energy should be comparable with the turbulent energy. He also states (translation by Thomas and Weiss 1992) that 'this interpretation considers the magnetic field as the primary cause of sunspots'.

This interchange occurred before the formal statement of flux freezing by Alfvén in 1942 (though studies such as Ferraro (1937) on magnetism and rotation implicitly involve the concept), or the discovery of the Alfvén wave. In 1946, Cowling made a detailed study of both the growth – under a hypothetical 'battery' – and the Ohmic decay of the magnetic field of a sunspot *in a medium at rest.* He confirmed that both processes would take the same, unacceptably long time (2.96), and inferred that a sunspot field 'does not grow *in situ* any more than it decays *in situ*', but is frozen into moving material. Accepting that, one can emulate Biermann and Cowling by postponing for later the construction of a plausible model (kinematical and dynamical) of the mass motions responsible for the sunspot cycle, and concentrate on the structure of an individual spot during its observable lifetime.

The various studies on magnetoconvection cited above point to a modified Biermann picture, with the strong umbral field interfering with normal convection but not suppressing it. In more realistic models of a spot (e.g. Schlüter and Temesváry 1958), it is accepted that the field lines will not be cylindrical but will splay out, so that the convective heat flow is not only reduced but is also spread over an area that increases upwards (Hoyle 1949). Numerical studies of the thermal structure of a spot (Chitre 1963; Deinzer 1965, Chitre and Shaviv 1967; Yun 1970; Jahn and Schmidt 1991; Jahn, in Thomas and Weiss 1992) do in fact show that the assumption of complete suppression of convection within a spot umbra, and so of purely radiative transfer, does not yield solutions which behave properly deep down; however, acceptable phenomenological models can be constructed with convective heat transport given by the standard formula (4.2), but with the mixing-length a decreasing function of $B^2/8\pi p$, and with the thermal and magnetic fluxes proportional.

For a while there was controversy over the fate of the heat flux diverted from its normal outlet by the magnetic interference with convection by the spot field. The non-appearance of a bright ring surrounding the spot led Parker (1974) to argue that the cooling of a spot is not due to the reduction in heat supply from below through the Biermann effect, but rather to excess energy loss carried by Alfvén waves propagating away from the solar surface. Both observational and theoretical work has cast doubt on whether this could in fact be a strong enough cooling process (Thomas and Weiss 1992). Further, study of heat transport in the convection zone has shown that the 'missing heat' problem is itself spurious. Spruit (1977) showed that in a hypothetical steady state, the lateral convective heat transport would be so efficient that the diverted heat flux is spread over a very large area of the Sun, so no detectable bright ring would be expected. Later, Spruit pointed out that the observed lifetime of a spot (a few months at most) is far shorter than the time-scale for thermal adjustment in the solar convective zone, so that a steady state is never reached: the energy flux blocked by the spot is stored in the convective zone, not reappearing at the surface in any time-scales of observational relevance (Spruit 1982 and in Thomas and Weiss 1992; Foukal *et al.* 1983; Chiang and Foukal 1984).

More ambitious studies (e.g. Meyer *et al.* 1974; Weiss *et al.* 1990) attempt to apply the cited results on magnetoconvection. It is agreed that in the umbra, the ratio ζ is small down to about 2000 km ($\approx 10^{-3}$ at the photosphere), and we expect to find oscillatory convection in narrow elongated cells with motions that are predominantly vertical (Cowling 1976). This domain with its highly superadiabatic stratification and dynamically strong field is the site of the 'magnetic throttle' that limits heat transport. The temperature difference between the spot umbra and the mean undisturbed atmosphere at the same level is about 9000 K at the level of the spot photosphere, but is estimated to fall to 500 K at a depth of 2000 km, to be compared with a typical umbral radius of up to 10,000 km. At depths between 2000 and 20,000 km, the same increase in opacity (through ionization of H and He) that is responsible for the initial onset of convective instability increases ζ to above unity, so that double-diffusive instabilities should lead to overturning convection deep down, coupled with the oscillatory convection higher up (Weiss *et al.* 1990). Other workers (Choudhuri, Parker, in Thomas and Weiss 1992 argue for a separation between the strong flux tube regions, heated by radiative transport only, and field-free regions in which oscillatory convection transports energy to the surface. Again, one may have to wait for modelling by the supercomputers to resolve the issues.

The sunspot is not the only site of strong solar magnetic fields. High resolution observations over the last thirty years have shown that nearly all the magnetic flux threading the photosphere outside sunspots and pores is in the form of localized magnetic elements with field strengths in the range of 1000–1500 G and diameters less than 200 km. The simple picture of a thin cylindrical flux tube, satisfying the pressure balance condition (4.73), yields a maximum field strength at the solar surface of $B \approx 1700$ G for a completely evacuated tube with $p_i = 0$. Since the observed fields approach this limiting value, the flux tube must in fact be largely evacuated. Its photosphere – defined by optical depth $\tau \approx 1$ – will be depressed, analogously to the Wilson depression of a sunspot umbra.

Note that this field is again markedly larger than the 'equipartition' field

$B_{\rm eq} = (4\pi\rho v_{\rm t}^2)^{1/2}$, given by the balance of magnetic and turbulent energy densities, which yields 400 – 600 G for photospheric values $\rho \simeq 3 \times 10^{-7}\,{\rm g\,cm^{-3}}$, $v_t \simeq 2 - 3{\rm km\,s^{-1}}$. By reducing the heat supply, the modified Biermann process will lower the temperature and so also the pressure, leading both to further lateral compression of the field and also to a downflow of gas – 'convective collapse of the flux tube' (Parker 1978; Zwaan 1978). A linear stability analysis (Spruit and Zweibel 1979) shows that this process is halted when the surface field strength is about 1350 G, in good agreement with observation.

This treatment of an isolated flux tube as essentially a cylinder, with lateral equilibrium given essentially by (4.73), can be applied also to a pore – a local region of about 1500–2000 km in diameter and a field strength of about 1500 G, often found within a sunspot umbra. It is, however, clearly too simple a model to be applied to a sunspot as a whole, as it implicitly assumes there is no penumbra with its nearly horizontal field lines and the associated Evershed motion. Also, as the umbral photosphere is depressed by 500–700 km below the level outside the spot, (4.73) would yield $B \approx 5000$ G, far greater than the 3000 G typically observed in a sunspot umbra. In the more realistic models, starting with that of Schlüter and Temesváry (1958), there is a horizontal thermal pressure gradient, balanced largely by the curvature part of the Lorentz force which must be included along with the magnetic pressure gradient. For an up-to-date account of the present state of our understanding of the whole sunspot phenomenon, including the complications of the penumbral region, the reader is referred to Thomas and Weiss (1992).

4.7 Magnetic buoyancy

The basic observations of the sunspot cycle – Hale's laws of bipolar pairs, aligned approximately but not exactly with the equator, the polarity reversal for the second half-cycle – suggest strongly a kinematic picture with alternating sub-photospheric toroidal flux tubes which buckle so as to form an upper-case Greek Ω, so that a segment of a tube emerges above the surface, its intersections with the photosphere defining a sunspot pair with their opposite polarities. Many questions immediately arise. Does the toroidal part of the solar field consist primarily of tubes, or is much of the field – poloidal and toroidal – in diffuse form? Are the flux tubes generated in the convection zone, with its zero-order non-uniform rotation field (cf. Section 4.2)? If so, how do they interact with the turbulence; in particular, do newly created toroidal flux tubes spontaneously rise through the convection zone, or are they pumped down into the radiative core? Answers to these questions are clearly a necessary preliminary for a full understanding of the general solar field, presumed to be generated by a quasi-periodic dynamo (see Chapters 6 and 8). In this section, we begin by summarizing work on magnetic buoyancy.

4.7.1 *Flux tubes*

The simplest problem concentrates first on the equilibrium and then the stability of a toroidal tube in pressure equilibrium with its non-magnetic surroundings.

The tube is supposed thin enough to be approximated by a cylinder, so that

$$p_i + \frac{B_t^2}{8\pi} = p_e, \tag{4.74}$$

where p_i, p_e are respectively the gas pressures internal and external to the tube and B_t is a suitably defined mean field within the tube. In prima facie the simplest case, the tube and its surroundings are in thermal equilibrium and so at the same temperature; hence if the mean molecular weight μ is uniform, $\rho_i < \rho_e$, and by Archimedes' principle the tube will feel a net upward force, as the weight of the 'fluid displaced' exceeds that of the fluid in the tube. For a cylindrical tube of radius R_f, at the local temperature T, the upward force per unit length is

$$\pi R_f^2 g \frac{(B_t^2/8\pi)}{\mathcal{R}T/\mu} = \frac{B_t^2 R_f^2}{8H_p}, \tag{4.75}$$

where H_p is the local scale-height. This is the original idea of 'magnetic buoyancy', introduced by Parker (1955) and Jensen (1955). A more rigorous treatment allows for the departure of a toroidal tube from the cylindrical approximation, and the consequent magnetic curvature force (Section 6.8.1).

Once such motion starts, then the first question is whether it continues indefinitely, perhaps accelerating, or whether it approaches a state in mechanical equilibrium, with the same density ρ within as without, but with the pressure balance condition (4.74) maintained by the temperature defect

$$\Delta T = (T_e - T_i) = B_t^2 \mu / 8\pi \mathcal{R} \rho \equiv T_i/\beta. \tag{4.76}$$

where β is the ratio within the tube of gas pressure to magnetic pressure. Equivalently, one can ask whether the equilibrium state defined by (4.74) and (4.76) is stable against adiabatic disturbances that move the tube bodily without any bending ('interchange'). If the medium is approximated as plane-parallel rather than spherical, Spruit and van Ballegooijen (1982) found the criterion for instability to be

$$\delta \equiv \frac{\mathrm{d}\ln T_e}{\mathrm{d}\ln p_e} - 1 + \frac{1}{\gamma} > \frac{1-\gamma/2}{\gamma(1+\beta\gamma/2)}. \tag{4.77}$$

From the Schwarzschild criterion in the form (3.129), in a subadiabatic, convectively stable domain δ is negative, and so the criterion (4.77) is never satisfied (since $\gamma < 5/3$); and equally, in a subadiabatic domain, an initially buoyant, adiabatically moving flux tube will approach the equilibrium state (4.76). Thus the interchange instability can set in only in domains which are already convectively unstable.

Consider such a tube with the necessary temperature defect, in mechanical equilibrium in a radiative domain. It will clearly gain heat by radiative transfer from the surroundings, and so will slowly rise, but at a rate determined by the heat flow. For the problem as formulated so far, the appropriate ordering for a subadiabatic, radiative domain is: the modification to hydrostatic equilibrium due to the Lorentz forces yields a non-spherically symmetric correction to the temperature gradient; the consequent motion of the gas plus the inductively coupled magnetic field is fixed by the energy equation (2.54), with both radiative transfer and compressional heating (positive or negative) playing crucial roles. We call this the 'Eddington–Sweet' picture for radiative domains; it will be discussed in some detail for rotating magnetic stars in Sections 5.4 and 9.4.

The constraint of axial symmetry – which assumes that the tube remains horizontal – is very restrictive. Again in the plane-parallel case, Spruit and van Ballegooijen (1982) found that a mode with wave number k_y along a horizontal tube is unstable if

$$k_y^2 < \frac{(1/\gamma + \beta\delta)}{2H_p^2(1 + 1/\beta)}, \tag{4.78}$$

where H_p is the pressure scale-height of the external medium. In a radiative domain, with $\delta < 0$, whereas the interchange modes are stable, sufficiently long wavelengths are now unstable if $\beta(-\delta) < 1/\gamma$. However, unless β is very small, (4.78) requires wavelengths to be at least $5H_p$, and the straight cylindrical approximation becomes questionable.

In a convectively unstable domain, with $\delta > 0$, by (4.77) a tube with the same temperature defect (4.76) can be unstable to interchange as well as undulatory modes. Deep in the zone, with δ having the very small Biermann–Cowling estimate, β would need to be very large for the instability to occur, and a reduction in β by an increase in the field tends to stabilize. However, heat flow into the tube will yield a density defect and so strong buoyancy effects arise and *persist*, tending to drive the tube upwards. Near the surface, the stratification is markedly superadiabatic, so δ is not small, and also $H_p/r \ll 1$; hence from (4.78), a semi-wavelength $\simeq \pi/k_y \approx \pi H_p$ will be unstable. Prima facie, the theory allows a toroidal flux tube that has managed to reach the surface to buckle, as required.

Originally it was thought that any flux tube generated deep in the convective zone would indeed spontaneously rise, and in fact so rapidly that magnetic buoyancy was considered a serious embarrassment to the dynamo theory (cf. Chapters 6 and 8). However, a new consensus has now arisen: the belief is that at least some of the diffuse flux generated in at least the lower part of the convective zone is pumped downwards into a layer between the convective zone and the fully stable radiative domain below. Direct hydrodynamical support comes from recent three-dimensional simulations (Nordlund *et al.* 1992), which find that magnetic flux generated in the solar convection zone does indeed tend to be transported downwards towards the stably stratified region below, confirming earlier results (Jennings *et al.* 1991); and in fact in one subsequent study of the solar dynamo (Brandenburg *et al.* 1992*a*), a phenomenological turbulent downward-pumping term is introduced. These results vindicate the similar tentative conclusions of Spiegel and Weiss (1980) and of Galloway and Weiss (1981), drawn from pioneering two-dimensional Boussinesq studies.

4.7.2 *Instability in magnetically supported domains*

We accept provisionally the consensus view on the formation below the convective zone of a domain with a stratified, large-scale, essentially horizontal magnetic field, contributing through the Lorentz force to hydrostatic equilibrium against gravity. The term 'magnetic buoyancy', introduced originally to describe the Parker–Jensen phenomenon, is also used to describe the different though related onset of instability in such a domain with partial magnetic support. We have already considered this problem in Section 3.6 in order to illustrate the use of the MHD energy principle. In the linear domain, any such instabilities are of the Rayleigh–Taylor type, resulting from the release of gravitational energy. Again, one is particularly interested in instabilities that occur in domains which are subadiabatic, so that it is the magnetic field that is directly or indirectly responsible for any instability. It is convenient to summarize what appear to be the most relevant landmarks in the theoretical studies over the last half-century, and especially the exciting ongoing non-linear work (cf. Hughes and Proctor 1988; Hughes 1991, and in Thomas and Weiss 1992).

In a classical paper, Kruskal and Schwarzschild (1954) considered an isothermal plasma supported above a vacuum against the gravitational field $g\hat{\mathbf{z}}$ by a uniform magnetic field $\mathbf{B} = B_0\hat{\mathbf{y}}$. With no diffusive terms included, their local analysis found instability when the wave numbers k_x, k_y satisfy

$$2k_y^2 < |k_x|/H_p \tag{4.79}$$

where H_p is the pressure scale-height of the plasma. Note that for this system with a discontinuity in the field, the most easily destabilized modes are interchanges, with $k_y = 0$, for which there is no buckling of the field lines, so no work has to be done against the magnetic tension. Parker (1979) studied the generalized problem of a magnetized gas of uniform density ρ and uniform field $B_0\hat{\mathbf{y}}$ supporting a non-magnetized gas of uniform density $\rho + \Delta\rho$, with $\Delta\rho/\rho = v_{\mathrm{A}}^2/2v_{th}^2$, v_{A} being an Alfvén speed and v_{th} a notionally defined thermal speed. For incompressible, non-diffusive perturbations, (4.79) is replaced by

$$k_y^2/(k_x^2 + k_y^2)^{1/2} < (g/v_{\mathrm{A}}^2)(\Delta\rho/\rho). \tag{4.80}$$

Again, interchanges ($k_y = 0$) are the most readily destabilized. For these modes, the growth rate is identical with that for a fluid of density $\rho + \Delta\rho$ superposed on one of density ρ: the magnetic field enters only in the unperturbed state through the definition of $\Delta\rho$. Other linear work includes in particular that by Acheson (1978) which takes account of the curved geometry; neglect of such effects is reasonable in the solar convection zone and the outer parts of the radiative core.

4.7.3 *Non-linear developments*

Just as for the traditional convection problem (cf. Section 4.5), a new richness is emerging from numerical extension of the linear theory into the non-linear domain. Cattaneo and Hughes (1988) followed the evolution of interchange modes, including both diffusion and compressibility. Their equilibrium state had again a unidirectional horizontal magnetic field embedded in a convectively stable atmosphere, but with a destabilizing density jump at the upper magnetic interface. The initial Rayleigh–Taylor-type instability has the usual rise of buoyant magnetized gas and the fall of the denser, field-free gas above. A strong shear therefore develops, causing the onset of *Kelvin–Helmholtz* instabilities (cf. Section 3.6.4) that wrap the gas into regions of strong vorticity. Once most of the available gravitational energy has been released, it is the vorticity distribution that dominates the motion. It is remarkable that the interactions between pairs of vortices can be strong enough to pull down pockets of strong field, even though they are less dense than the neighbouring gas, contrary to the simple picture of the magnetically induced buoyancy in an initially non-turbulent medium. However, the net effect of the vortex-driven motion is to disperse the field in a time short compared with the magnetic diffusion time – describable as the effect of 'turbulent resistivity'.

Cattaneo *et al.* (1990) introduced a weak poloidal field, so that the initial field though still horizontal is no longer unidirectional:

$$\mathbf{B}_0 = [B_x(z),\, B_y(z),\, 0]. \tag{4.81}$$

They studied instabilities that are the Cartesian analogue of axisymmetric modes: all three components of the velocity are functions of t and of $(x,\, z)$ but not of y. They found that the instability is greatly affected by the distribution and strength of the poloidal field. Both the horizontal and vertical scales of the motions are controlled by the location of the 'resonant surface' on which the poloidal field vanishes. In the non-linear regime, a resonant surface close to the interface between the magnetized and field-free fluid localizes the instability, so that only a fraction of the magnetic region is disrupted by the motions, whereas a deeply seated resonant surface leads to a complete disruption of the layer and to the formation of large, helical magnetic fragments which maintain their identity during the whole simulation. It is not yet clear how severe is the 'axisymmetric' constraint.

Matthews *et al.* (1995*b*) return to the problem with a purely toroidal field, but now study the three-dimensional problem. The *initial* evolution remains essentially two-dimensional, as predicted by the linear analysis. The non-linear development is again controlled by the interactions between the strong vortices that form, but now there is a strong three-dimensional instability that causes arching of the vortices and their associated magnetic fields, and is driven by the interactions between anti-parallel vortices on neighbouring magnetic flux elements. The authors conjecture that it is this vortex-driven instability that is the crucial factor in the formation of isolated flux tubes from large-scale ordered fields, central to the sunspot problem (cf. Section 4.8).

4.8 Solar activity

Though there remain some sceptics, who rightly point to difficulties and unanswered questions, most workers accept that the basic magnetic process witnessed in the Sun and other late-type stars is a dynamo operating in the convective envelope and adjacent domains. The general dynamo problem in both its kinematical and dynamical aspects is discussed in Chapter 6, and its application to late-type stars in Chapter 8. The various studies on magnetoconvection and magnetic buoyancy summarized in the present chapter show how the whole area is being rapidly transformed, with new qualitative features being introduced which are sure to affect our picture of the solar dynamo process and of the intimately related sunspot cycle.

As is clear from Section 4.6, the sunspot is the principal photospheric phenomenon that arises when the interaction between the turbulence and the magnetic field yields domains near the surface with the magnetic and thermal energy densities comparable. For a more rapidly-rotating solar-type star, anticipating that the total dynamo-built flux will then tend to be higher, we may expect a larger fraction of the photosphere to be covered with spots (cf. Section 8.2).

There is now so much observational material on sunspots that the demands on the theorist have become much more exacting. The most recent comprehensive source is that edited by Thomas and Weiss (1992), but much can be learnt also from earlier books, such as the classical text by Bray and Loughhead (1964), from Stix (1989*b*) and Priest (1982). The last two books in particular discuss in detail how the presence of a significant magnetic flux over the solar photosphere, whether in isolated flux tubes or in pores and sunspots, is linked to a wealth of other phenomena – prominences, flares, generation of high-energy particles, emission in the radio, optical, UV and X-ray domains – in which the magnetic presence is crucial, as noted in the Introduction.

In this section we first discuss recent work on the dynamics of flux tubes that generalizes the linear stability work summarized in Section 4.7.1, and attempts to relate the properties of solar active regions and sunspots to the eruption of flux that has been stored in the overshoot domain below the convective envelope. In Section 4.8.2 the scene is set for magnetohydrodynamic activity in the low-density but strongly magnetic chromospheric and coronal regions above a sunspot region.

4.8.1 *Flux tube dynamics*

It is very likely that in future work on stellar MHD, not least in the dynamo problem, the flux tube will play an increasing role. It is therefore convenient to summarize here the essentials of a systematic study of erupting flux tubes, following closely the work of Caligari *et al.* (1995). This in turn is based on the dynamics of thin flux tubes as treated by Spruit (1981), van Ballegooijen (1983) and van Ballegooijen and Choudhuri (1988), and developed later by Ferriz-Mas and Schüssler (1993, 1994, 1995) to include the effects of (differential) rotation, stratification and arbitrary latitude.

For the thin tube approximation to be valid, the tube radius must be much less than the other relevant lengths – the radius of curvature of the tube, the local scale-height, the rotational shear scale and the wavelengths of the dominant perturbations. The tube is parametrized by a Lagrangian coordinate a (the integrated mass across the tube) that labels the individual mass elements. The vector $\mathbf{r}(a,t)$ defines the space curve describing the path of the flux tube at time t. At each given time the element of arc length along the curve is given by

$$\mathrm{d}s = \left|\frac{\partial \mathbf{r}}{\partial a}\right| \mathrm{d}a. \tag{4.82}$$

At each point of the curve there is set up the orthonormal Frénet triad of unit vectors, consisting of the tangent $\mathbf{l}$, the normal $\mathbf{n}$ and the binormal $\mathbf{b}$, defined by

$$\mathbf{l} = \frac{\partial \mathbf{r}}{\partial s}, \qquad \mathbf{n} = R\frac{\partial \mathbf{l}}{\partial s}, \qquad \mathbf{b} = \mathbf{l} \times \mathbf{n}, \tag{4.83}$$

with the radius of curvature R defined by

$$R = \left|\frac{\partial \mathbf{l}(t)}{\partial s}\right|^{-1}. \tag{4.84}$$

In the perfect conductivity limit, the equations of continuity and induction combine into (cf. 2.93)

$$\frac{\mathrm{d}}{\mathrm{d}t}\left(\frac{\mathbf{B}}{\rho}\right) = \left(\frac{\mathbf{B}}{\rho} \cdot \nabla\right) \mathbf{v}, \tag{4.85}$$

where the Lagrangian derivative $\mathrm{d}/\mathrm{d}t \equiv (\partial/\partial t)_a$, and $\mathbf{B}$, ρ, $\mathbf{v}$ are quantities within the flux tube. On writing $\mathbf{B}(a,t) = B(a,t)\mathbf{l}$, and defining $v_l = \mathbf{v}\cdot\mathbf{l}$, $v_n = \mathbf{v}\cdot\mathbf{n}$, we reduce the tangential component of (4.85) to

$$\frac{\mathrm{d}}{\mathrm{d}t}\left(\frac{B}{\rho}\right) = \frac{B}{\rho}\left(\frac{\partial v_l}{\partial s} - \frac{v_n}{R}\right). \tag{4.86}$$

The crucial pressure balance assumption (4.74) is here written as

$$p_e = p + \frac{B^2}{8\pi}. \tag{4.87}$$

It is convenient to write the equations of motion as seen in a frame rotating with a particular value Ω. The external pressure distribution p_e is fixed by assuming a steady state with no motions $\mathbf{v}_e$ other than the rotational velocity $\varpi\Omega_e\mathbf{t}$ in the inertial frame (equivalent to $\varpi(\Omega_e - \Omega)\mathbf{t}$ as measured in the chosen rotating frame):

$$\frac{\nabla p_e}{\rho_e} = \mathbf{g} + \Omega_e^2 \varpi. \tag{4.88}$$

In the thin tube approximation, the Frénet components of the equations of motion for mass elements inside the flux tube are given by

$$\rho \frac{\mathrm{d}\mathbf{v}}{\mathrm{d}t} \cdot \mathbf{l} = \frac{\partial}{\partial s}\left(\frac{B^2}{8\pi}\right) + (\Delta\rho\, \mathbf{g}_{\text{eff}} + \mathbf{F}_{\text{rot}}) \cdot \mathbf{l} \tag{4.89}$$

$$\rho \frac{\mathrm{d}\mathbf{v}}{\mathrm{d}t} \cdot \mathbf{n} = \frac{B^2}{4\pi R} + (\Delta\rho\, \mathbf{g}_{\text{eff}} + \mathbf{F}_{\text{rot}} + \mathbf{F}_{\text{D}}) \cdot \mathbf{n}, \tag{4.90}$$

$$\rho \frac{\mathrm{d}\mathbf{v}}{\mathrm{d}t} \cdot \mathbf{b} = (\Delta\rho\, \mathbf{g}_{\text{eff}} + \mathbf{F}_{\text{rot}} + \mathbf{F}_{\text{D}}) \cdot \mathbf{b}, \tag{4.91}$$

where

$$\Delta\rho = \rho - \rho_{\text{e}}, \tag{4.92}$$

$$\mathbf{g}_{\text{eff}} = \mathbf{g} + \Omega_{\text{e}}^2 \boldsymbol{\varpi}, \tag{4.93}$$

$$\mathbf{F}_{\text{rot}} = 2\rho(\mathbf{v} \times \boldsymbol{\Omega}) + \rho(\Omega^2 - \Omega_{\text{e}}^2)\boldsymbol{\varpi}. \tag{4.94}$$

(The corresponding equations for a general $\mathbf{v}_e$-field are the set (2.10) in Ferriz-Mas and Schüssler (1993)).

The motion is seen to be governed by buoyancy, magnetic pressure and curvature force, the joint Coriolis and differential centrifugal force, and the new term $\mathbf{F}_{\text{D}}$ = the aerodynamic drag force due to the relative motion, which becomes important for the thinnest tubes. The form adopted by Caligari *et al.* is that for a straight rigid cylinder in the limit of high fluid-dynamical Reynolds number (Batchelor 1967):

$$\mathbf{F}_{\text{D}} = -C_{\text{D}} \frac{\rho_{\text{e}} v_{\perp} \mathbf{v}_{\perp}}{\pi R_T}, \tag{4.95}$$

where $\mathbf{v}_{\perp}$ is the perpendicular part of of the velocity of the tube with respect to the external fluid, R_T is the tube radius, and the drag coefficient C_{D} is ≈ 1. The equations are completed by the ideal gas law and by the condition of adiabatic evolution (entropy conservation by the fluid elements), justified by the large radiative exchange time deep in the convective zone.

Before the Caligari *et al.* paper, several aspects of the general problem had been discussed by a number of authors (see Schüssler 1996 and references therein). Consider again a subadiabatic region, such as the convective overshoot domain. Figure 4.7 illustrates an example of the undulatory instability of a flux tube in spherical rather than plane geometry, studied initially by Spruit and van Ballegooijen (1982*a*,*b*). The zero-order flux-ring in equilibrium is perturbed by a displacement with azimuthal wave number $m = 4$. A downflow of plasma along the field lines within the flux tube yields an upward buoyancy force on the outward displaced parts and a downward force on the troughs, so that the perturbation grows.

Attention is focused on parameter values which yield linear growth times of less than a year, so that the predicted flux tube eruption is of interest for the solar cycle and the solar dynamo (cf. Sections 8.5 and 8.6). For the Sun, it is found that such growth rates are reached first at low latitudes and for field strengths of the order of 10^5 G, results insensitive to detailed modelling of the overshoot layer.

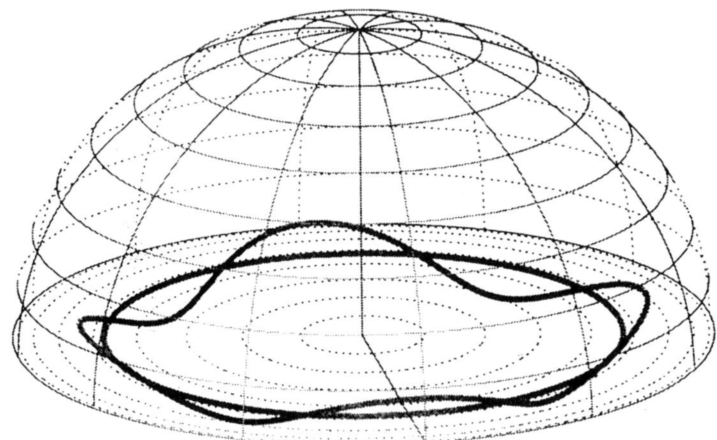

FIG. 4.7. Undulatory instability of a toroidal flux tube, with the initial flux ring perturbed by a displacement of azimuthal wave number $m = 4$. A downflow of plasma from the crests into the troughs lets the summits rise while the valleys sink. (From Schüssler 1996.)

This result is at first somewhat disconcerting, as it is about an order of magnitude above the equipartition field strength, implying a magnetic energy density about one hundred times the turbulent energy density. Equipartition had seemed a reasonable assumption to make, at least for the initial state of flux tubes rising from the bottom of the convective zone. However, as pointed out by Moreno-Insertis (1987, 1992), the rising flux tubes have to adjust to the steadily decreasing gas pressure while still satisfying (4.74). Thus a flux tube beginning its rise with just the equipartition field strength would find that its field strength would rapidly decline to well below the *local* equipartition value. Because of flux conservation, the associated cross-section of the tube increases accordingly – in fact the process is described as an *explosion* of the tube (e.g. Moreno-Insertis *et al.* 1995). The tubes would then be subject to distortion by the surrounding turbulence, and so could not be expected to retain the regularity required to explain the observations of the active regions; rather, they would be a possible source of flux to be acted upon by the turbulent convection.

A further complication arises when the solar rotation is taken into account (Choudhuri and Gilman 1987). Consider the simplest example of a buoyant axisymmetric flux ring, located in the equatorial plane and rising with velocity v_ϖ through the convective zone. Viewed from a frame rotating with the local solar angular velocity Ω, the Coriolis force causes a velocity v_ϕ opposite to the velocity of rotation (as seen in the inertial frame), reflecting angular momentum conservation; and this in turn leads to an inward directed Coriolis force (Fig. 4.8). Thus the total Coriolis acceleration is

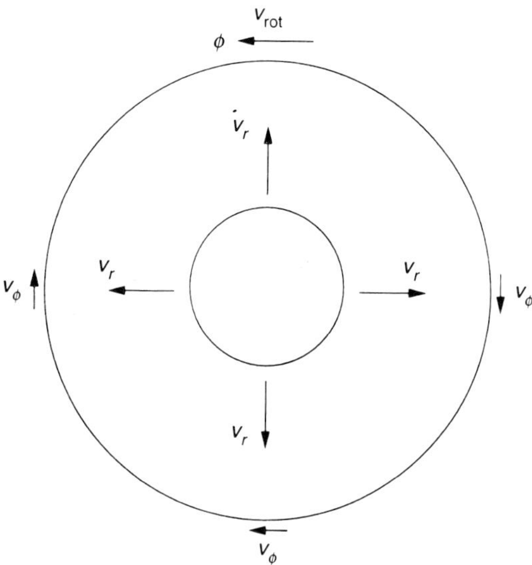

FIG. 4.8. Effect of rotation on an expanding flux ring, viewed from above along the axis of rotation. The outward motion v_r causes a flow v_ϕ against the direction of rotation due to angular momentum conservation, leading to an *inward*-directed Coriolis force. (From Schüssler 1996.)

$$-2\Omega\mathbf{k}\times\mathbf{v} = 2\Omega(v_\phi,\ -v_\varpi,\ 0). \tag{4.96}$$

If no other forces are acting, the linearized equation of motion $\dot{\mathbf{v}} = -2\Omega\mathbf{k}\times\mathbf{v}$ yields

$$v_\varpi,\ v_\phi \propto \sin(2\Omega t). \tag{4.97}$$

The tube feels a buoyancy force per unit mass, given roughly from (4.75) by $B_\mathrm{t}^2/8\pi\rho H_\mathrm{p}$, so that the ratio of buoyancy to Coriolis force is

$$\frac{B_\mathrm{t}^2/8\pi\rho H_\mathrm{p}}{2\Omega v_\varpi} = \left(\frac{B_\mathrm{t}}{B_\mathrm{eq}}\right)^2 \frac{v_\mathrm{t}}{v_\varpi}\frac{\mathrm{Ro}}{2}, \tag{4.98}$$

where v_t is again the turbulent speed, the equipartition field $B_\mathrm{eq} = (4\pi\rho)^{1/2}v_\mathrm{t}$, and $\mathrm{Ro} = v_\mathrm{t}/2\Omega H_\mathrm{p}$ is the local value of the Rossby number, ≈ 0.2 in the lower solar convective zone. Thus once the rise velocity v_ϖ exceeds v_t, then in order that buoyancy should dominate over the Coriolis force, we require that $B_\mathrm{t} > 3B_\mathrm{eq} \approx 3\times 10^4$ G; and if v_ϖ reaches the Alfvén speed, the ratio (4.98) becomes $(B_\mathrm{t}/B_\mathrm{eq})(\mathrm{Ro}/2)$, and we require that $B_\mathrm{t} > 10^5$ G. If B_t is significantly less, then a tube strictly in the equatorial plane will oscillate according to (4.97), with the buoyancy force fixing the amplitude. A tube located outside the equatorial plane feels a component of buoyancy parallel to $\mathbf{\Omega}$ and so rises in this direction uninhibited by Coriolis force, whatever the value of the field strength. However, unless $B_t \gg 3 \times 10^4$ G, the argument predicts the emergence of the flux and hence the appearance of the active regions preferentially located near the polar regions rather than in the lower latitudes as observed. The same result was found in numerical work by Choudhuri (1989) and by Fan *et al.* (1993) for buoyant, non-axisymmetric flux loops.

These effects all emerge together in the numerical solution of (4.86–4.91) by Caligari *et al.* (1995). The authors apply the equations to a systematic study of a model that may account both for the overall features and some of the details of solar activity. As described in Section 4.7.1, flux is supposed to have been pumped down to be stored in the subadiabatic, overshoot region at the base of the convective envelope, from which it can emerge through the undulatory magnetic Rayleigh–Taylor instability. It is confirmed that a field strength of $\approx 10^5$ G is needed for the instability to develop. Significant growth rates are first reached for azimuthal wave numbers $m = 1, 2$ at latitudes below 40°. For such strong fields, the poleward Coriolis deflection is negligible for the Sun: the rise through the convective envelope is fast enough – taking about 30 d – for the loop to appear at essentially the same latitude as that of emergence. The emerging loops are inclined with respect to the east–west direction; as shown earlier by D'Silva and Choudhuri (1993), the strong fields are necessary to ensure that the magnitude of the corresponding tilt angles as a function of emergence latitude are in reasonable agreement with observation. In addition, as is shown clearly in Fig. 4.9, there is a marked asymmetry between the two legs of the erupting loop: the Coriolis force causes the preceding parts of a loop to have a smaller angle of inclination to the projection of the tube on to the spherical surface. This asymmetry may account for some properties of young active regions, such as the asymmetric position of the magnetically neutral line, and the differing proper motions of preceding and following spots.

The authors argue that their results lend further support to the requirement of super-equipartition fields of order 10^5 G in the solar overshoot region, to which models of the solar dynamo must adjust. We return to this question in Sections 6.8.1 and 8.6.

A new research programme (Schüssler *et al.* 1996) applies the above theory to rapidly rotating solar-mass stars both on and to the right of the main sequence. The main conclusion is that the internal structure of the star is almost as important as the rotation in determining the latitude of emergence of flux at the surface. Only the youngest objects with their very deep Hayashi convective zones should show spots emerging at the stellar poles. A pre-main sequence star at age 10^7 yr – already on the Henyey branch – with the convective zone reaching to half the radius, should show high-latitude but not polar spots; while a main sequence star of one solar mass even at high rotation rates shows only intermediate latitude spots. The theoretical predictions appear to be in good agreement with Doppler images of young rapid rotators (cf. the references in Schüssler *et al.* 1996).

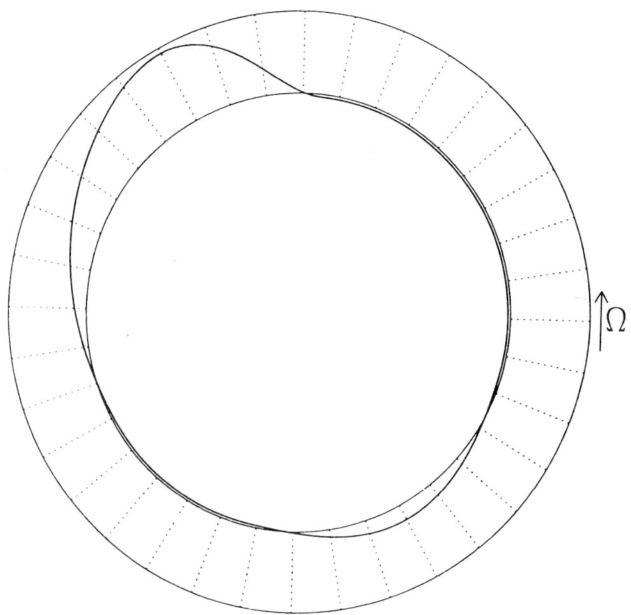

FIG. 4.9. Projection of an $m = 2$ mode on the equatorial plane, showing a marked asymmetry between the two legs of the emerging loop. (From Schüssler 1996.)

4.8.2 *Chromospheric and coronal MHD*

The basic reason for the dominance above the photosphere of even a low-flux magnetic field is just the divergence condition, which demands that the field decrease only algebraically with height, whereas the density will begin by exponentiating down, so ensuring that the mean plasma β decreases from moderate photospheric values to much lower values in the chromosphere. Well below the surface the mean field is passive, easily concentrated into thin loops. At the solar surface, at sunspot maximum about 0.1 per cent of the area is covered by regions with β of order unity; while in the chromosphere and over much of the corona, the field is dominant. (In the most active stars, up to 60 per cent of the photosphere is estimated to have $\beta \simeq 1$.)

The large-scale physics of the corona – especially the spontaneous development of a stellar wind – is discussed in Chapter 7. We conclude this chapter by looking at the equilibrium of the chromospheric/coronal gas near strong photospheric sources of magnetic flux, and in particular the response of the equilibrium structure to changes in the source distribution due to motions in the convective zone. The classical paper in this area is by Sweet (1958), who first drew attention to the formation of current sheets as a consequence of the flux-freezing constraint. We follow the later discussion by Low (1990, 1996). In a low

β-domain, the equation of hydrostatic equilibrium reduces to the force-free condition (3.65). Following earlier workers, Low considers the simplest problem in Cartesian geometry, with $z > 0$ representing the domain above the stellar photosphere $z = 0$, and the field structure independent of the coordinate x. From the condition $\nabla \cdot \mathbf{B} = 0$,

$$\mathbf{B} = \left(B_x, \frac{\partial A}{\partial z}, -\frac{\partial A}{\partial y} \right) \tag{4.99}$$

with both B_x and A independent of x. The current density maintaining (4.99) is then given by

$$\frac{4\pi \mathbf{j}}{c} = \left(-\nabla^2 A, \frac{\partial B_x}{\partial z}, -\frac{\partial B_x}{\partial y} \right). \tag{4.100}$$

The force-free condition $\mathbf{j} \times \mathbf{B} = 0$ requires

$$B_x = B_x(A), \tag{4.101}$$

and

$$\nabla^2 A + B_x \frac{\mathrm{d}B_x}{\mathrm{d}A} = 0. \tag{4.102}$$

If both $B_x(A)$ and also $A(y,\, z = 0)$ – corresponding to a definite component $B_z = -\partial A/\partial y$ on $z = 0$ – are prescribed, then solutions of (4.102) can be sought subject to the vanishing of A at infinity. Suppose that initially the field is described by one of these solutions, with a quadrupolar distribution on $z = 0$, and without any discontinuities with their maintaining current sheets. Then as in Sweet's original argument, suppose the field line footprints are moved on $z = 0$, with quadrupolar symmetry being maintained, but without the introduction of any new flux, so that a new function $A(y,\, z = 0)$ is imposed on the system. The displacement is continued into the domain $z > 0$, which is initially supposed perfectly conducting, so that the field is deformed according to (2.88). The deformed field will not in general satisfy hydrostatic equilibrium, and in particular will not be force-free. Now let the field adjust to equilibrium, with excess energy being dissipated, e.g. by viscosity, but subject still to the perfect conductivity constraint, so that there is no change in the field topology. The interesting question is whether the new equilibrium state is again force-free *everywhere*, or whether there are discontinuities, implying current sheets and local pinching.

The simplest examples have $B_x = 0$, so that the field lines lie in the (y, z)-planes, and the force-free condition (4.102) reduces to the curl-free condition

$$\nabla^2 A = 0. \tag{4.103}$$

Consider the initially continuous, quadrupolar solutions either as in Fig. 4.10(a), with the line $y = 0$, $z = z_n$ of neutral points, or as in Fig. 4.11(a), without any neutral points. We refer first to Fig. 4.10. Footpoints labelled α, β, γ, δ, displaced by photospheric motions to new positions as in Fig. 4.10(b), are supposed frozen into $z = 0$. Construct again the curl-free solution, without any discontinuities, subject to the new boundary function $A(y, z = 0)$ resulting from the footpoint motions. The solution is as in Fig. 4.10(c); and as Sweet pointed out, for the new field to conform to this structure, the system must in general have violated the flux-freezing constraint, for now footpoints such as α and δ are connected, and likewise β and γ. Thus as long as flux-freezing is imposed, the solution in Fig. 4.10(c) is not accessible from the initial state as in Fig. 4.10(a). Approach to a steady state will in general require excess energy dissipation, but if this occurs through viscous dissipation alone, without any resistive dissipation, then the structure actually taken up will be as in Fig. 4.10(b). The field is curl-free nearly everywhere, but there will be a line along the z-axis where B_z changes discontinuously to $-B_z$, and with equilibrium maintained by the thermal pressure $p = B_z^2/8\pi$ in the sheet carrying the current and separating the oppositely directed field lines (cf. Section 3.7). Specially chosen motions of the footprints could yield a continuous curl-free field with the same topology as in the initial state, in which case evolution with again non-resistive dissipation of energy will allow this state to be reached without the formation of a current sheet; but in general, an initial state as in Fig. 4.10(a), will evolve under strict flux-freezing into Fig. 4.10(b).

Low emphasizes that the formation of the current sheet does not depend on the presence of the X-type neutral point. This is shown by the example in Fig. 4.11, where the initial global field is again a quadrupolar solution of (4.103), but with the photosphere $z = 0$ chosen so that the predicted X-type point is in $z < 0$. This is a high-β domain where the initial field is *not* constrained to satisfy (4.103); the subsequent arguments do not apply in $z < 0$, nor indeed need there be a neutral point there. Motion of the photospheric footpoints again deforms the field. With relaxation to a steady state again subject to perfect conductivity, the Lorentz forces exerted by the deformed field push the two shaded bipolar fields under the separatrix line Γ towards each other, so expelling the flux located above Γ holding the two bipolar fields apart. The current sheet in Fig. 4.11(b) forms when the two shaded bipolar fields come into line contact, with the tangential field jumping discontinuously. In the final state – still assuming no resistive dissipation and diffusion – the field is curl-free in $z > 0$ except at the current sheet; again, equilibrium is maintained through a plasma pressure in the sheet balancing the same magnetic pressure $B_z^2/8\pi$ on $y = \pm 0$. The process is in essence no different from the case of Fig. 4.10 – flux expulsion leading to the formation of a current sheet, with the presence or absence of a neutral point in the initial field of no fundamental significance. The actual finite resistivity will allow diffusion near the current sheets: the field is then permitted to change topology as it relaxes towards the curl-free state everywhere above $z = 0$ (Fig. 4.10(c)).

Low and Wolfson (1988) generalized the problem by allowing the footprint motions to shear the initially two-dimensional quadrupolar field, so generating a component $B_x \neq 0$, with associated currents in the (y, z)-planes. To relate the footpoint displacement Δx to the generated component B_x, they integrate along

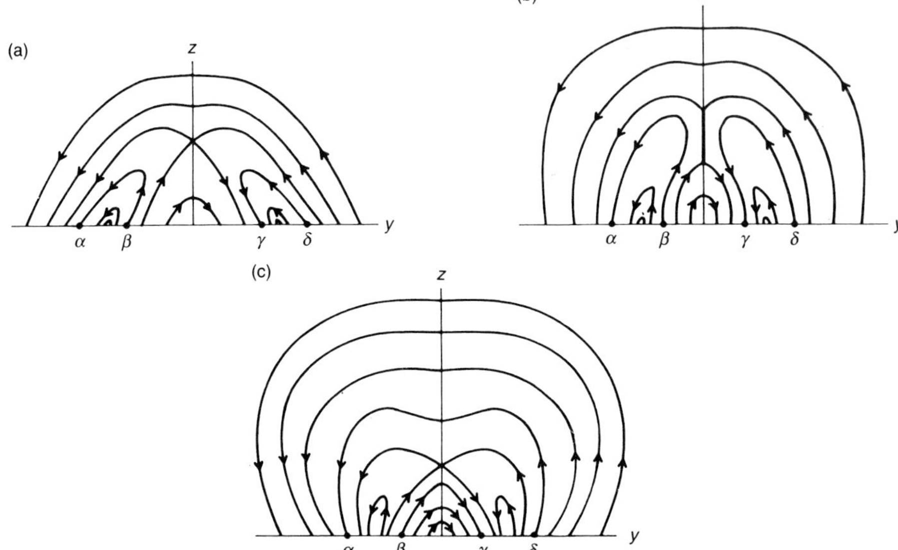

FIG. 4.10. Formation of an electric current sheet in a field with an X-type magnetic null-point (Low and Wolfson 1988, following Sweet 1958). (a) Quadrupolar potential field with an X-type null-point. (b) The field resulting from smooth footpoint displacement in y-direction under strict flux-freezing is curl-free nearly everywhere, but has the current-sheet represented by the vertical thick line. There has been no change in the field line connectivities. (c) The field that is curl-free everywhere and with the same normal flux distribution at $z = 0$ as in (b) has different field line connectivities.

a given bipolar field line, defined by a prescribed value of A:

$$\Delta x(A) = \int_A \frac{B_x}{|\nabla A|}\,\mathrm{d}s, \tag{4.104}$$

where $\mathrm{d}s = [(\mathrm{d}y)^2 + (\mathrm{d}z)^2]^{1/2}$ along the line $A =$ constant. If we now demand that the field be force-free, so that by (4.101) $B_x = B_x(A)$, then (4.104) becomes

$$\Delta x(A) = B_x(A) \int_A \frac{\mathrm{d}s}{|\nabla A|}. \tag{4.105}$$

Substitution into (4.102) yields

$$\nabla^2 A + \frac{1}{2}\frac{\mathrm{d}}{\mathrm{d}A}\left[\Delta x(A) \Big/ \int \frac{\mathrm{d}s}{|\nabla A|}\right]^2 = 0. \tag{4.106}$$

A strictly force-free field – with no disc ontinuities and associated current sheets – is given by a continuous solution of (4.106) which vanishes at infinity and is

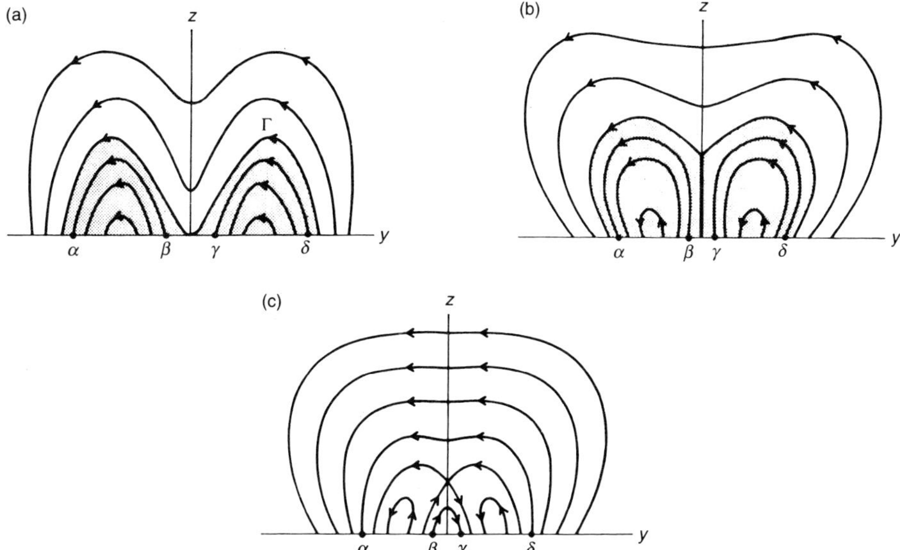

FIG. 4.11. Formation of an electric current sheet in a quadrupolar magnetic field *without* a null-point, by a converging footpoint displacement in the y-direction (Low and Wolfson 1988). (a) The separatrix line Γ separates the flux into two shaded bipolar fields and an overlying third bipolar field. Under strict flux-freezing, the deformed field again settles into a state (b) that is curl-free nearly everywhere and with unchanged field line connectivities, but with a vertical current sheet. (c) The field that is curl-free everywhere and with the same normal flux distribution at $z = 0$ as in (b) again has different field line connectivities.

subject as before to prescribed $A(y, z = 0)$ and also prescribed $\Delta x(A)$.

The integro-partial differential equation (4.106) is formidable. Low considers a special case in which the field in Fig. 4.11 is subject to a footpoint displacement directed strictly in the x-direction, zero in $y < 0$, increasing from zero at $y = 0$ to non-zero values in $y > 0$. As before, the footpoint displacements are supposed frozen, and the deformed field is allowed to relax through non-resistive dissipation into a new equilibrium state, preserving its sheared topology. Low now shows that again the field cannot be force-free everywhere – there must again be surfaces of discontinuity. Recall that the force-free condition requires $B_x(A)$ to be constant along a line of A. The separatrix line Γ divides the field into three zones: (1) where $\Delta x(A) = 0$ and so by (4.105) $B_x(A) = 0$; and (2) and (3) where $\Delta x(A)$, $B_x(A) \neq 0$; hence B_x must be discontinuous across Γ. The relaxed field takes on the shape with field line projections on the (y, z)-plane as in Fig. 4.12, where the points P and R are the original footpoints of the separatrix. As the point Q is approached, $B_y \to \infty$, with B_z staying finite and the separatrix becoming horizontal. Both the magnetic energy and Lorentz force densities associated with the singularity

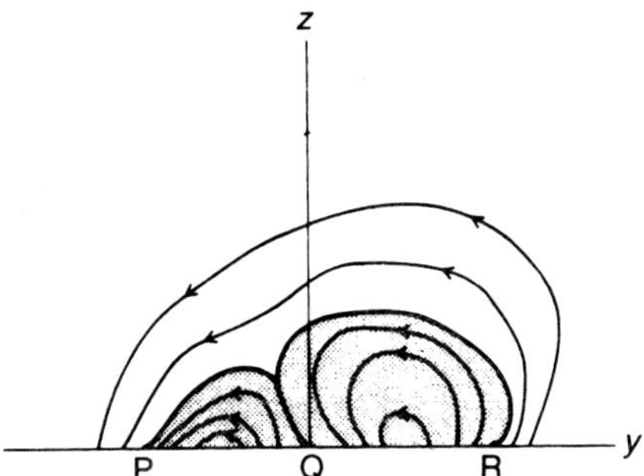

FIG. 4.12. The end-state field (taken from Low 1990), resulting from subjecting the quadrupolar initial field in Fig. 4.11(a) to a continuous footpoint displacement in the x-direction, of zero magnitude in $y < 0$ and non-zero in $y > 0$. The lines of force are shown projected onto the y-z plane, with the separatrix line Γ separating the magnetic flux into the two shaded bipolar fields and the third overlying bipolar field. The field has a non-vanishing x-component everywhere except in the left shaded bipolar field region. The separatrix line now has a triple point connected by a line extension to the origin Q; P and R are the original two footpoints of the separatrix line.

in B are integrable.

The motivation behind Sweet's 1958 paper was the search for a means by which energy pumped into the magnetic field by motions in the high-β, sub-photospheric convective domains would be made available ultimately for explosive release as a flare. The mutual destruction of oppositely-directed field lines through finite resistivity is assisted by the squeezing-out of pinched plasma (the magnetic pressure will vary along a separatrix in the above examples, and so also will the pressure in the domain between the field lines – cf. Fig. 3.18). The Sweet model was developed by Parker (1957) and later by Petschek (cf. Section 3.7.2). Later, in a series of papers beginning in 1972 and culminating in his monograph (1994), Parker argued that the spontaneous formation of current sheets in a corona is a much more general phenomenon than for the simple example just summarized. The motions of magnetic footpoints in the photosphere will normally be much more elaborate, and in the low-β corona, the force-free condition $\nabla\times\mathbf{B} = \alpha\mathbf{B}$ with α constant along a field line cannot hold everywhere, without discontinuities and current sheets, if evolution from one steady state to another is through non-resistive dissipation. Again, the crucial feature is topological: distinct magnetic flux bundles are brought into contact, in general by the expulsion of a third which initially separates them, each bundle being dragged by the bulk motion of the gas. Infinite conductivity is of course an idealization, but because of the high conductivity, intense volume current densities – idealized as current sheets – do form: the local magnetic Reynolds number does not fall to unity, allowing significant trans-field flow, until scales of variation of the field are correspondingly small. This is one way in which high conductivity domains such as the solar corona can be both hot and magnetized. Gold (1965) described the process of heating up the corona as 'dissipation at a distance': energy fed into the field below the photosphere is propagated outwards, to be released as heat higher up.

Parker's 'topological dissipation' argument was not immediately accepted. For references to the literature and a survey of the controversy, the reader is referred to Galsgaard and Nordlund (1996). These authors report on their numerical experiments with the resistive terms in the MHD equations retained. Their model consists of an initially straight **B**-field, subject to ongoing random, large-scale shearing motions on two boundaries. They confirm the rapid formation of effective tangential discontinuities, in a time that scales logarithmically with the resistivity, and is of the order of a few times the inverse shearing rate for any realistic resistivity. Thus although in the limit of strictly zero resistivity, no mathematical discontinuity would form in a finite time, in practice Parker's philosophy is vindicated. The subsequent reconnection drives supersonic and super-Alfvénic jet flows, which in turn cause the formation of smaller scale current sheets. The continuing motion on the boundaries yields a mean Poynting flux, which is balanced by the mean Joule dissipation; but both the boundary work and the Joule dissipation fluctuate strongly in space and time, being only weakly correlated. For a related study, see Schnack and Mikić (1994).

For a slightly earlier but comprehensive survey of mechanisms of chromospheric and coronal heating, see Ulmschneider *et al.* (1991). Priest (1998) gives an up-to-date review of coronal heating by magnetic reconnection, emphasizing the much greater richness of three-dimensional as compared with two-dimensional models.

References

Acheson, D.J. (1978). *Philosophical Transactions Royal Society A*, **289**, 459.

Alfvén, H. (1942). *Arkiv Matematik, Astronomi och Fysik*, **29B**, 2.

Antia, H.M. and Chitre, S.M. (1979). *Solar Physics*, **63**, 67.

Arter, W. (1983). *Journal Fluid Mechanics*, **132**, 25.

Arter, W. (1985). *Geophysical Astrophysical Fluid Dynamics*, **31**, 311.

Arter, W., Proctor, M.R.E. and Galloway, D.J. (1982). *Monthly Notices Royal Astronomical Society*, **201**, 57P.

Batchelor, G.K. (1967). *Fluid dynamics*, p. 339. Cambridge University Press.

Biermann, L. (1932). *Zeitschrift Astrophysik*, **5**, 117.

Biermann, L. (1951). *Zeitschrift Astrophysik*, **28**, 304.

Blanchflower, S.M., Rucklidge, A.M. and Weiss, N.O. (1998). *Monthly Notices Royal Astronomical Society*, in press.

Böhm-Vitense, E. (1958). *Zeitschrift Astrophysik*, **46**, 108.

Brandenburg, A., Moss, D., Rüdiger, G. and Tuominen, I. (1990). *Solar Physics*, **128**, 243.

Brandenburg, A., Moss, D. and Tuominen, I. (1992*a*). In *The solar cycle*, NSO/ Sacramento Peak 12th Summer Workshop (ed. K.L. Harvey), *ASP Conf. Proceedings*, **27**, 536.

Brandenburg, A., Moss, D. and Tuominen, I. (1992*b*). *Astronomy and Astrophysics*, **265**, 328.

Bray, J.R. and Loughhead, R.E. (1964). *Sunspots.* Chapman and Hall, London.

Brownjohn, D.P., Hurlburt, N.E., Proctor, M.R.E. and Weiss, N.O. (1995). *Journal Fluid Mechanics*, **300**, 287.

Caligari, P., Moreno-Insertis, F. and Schüssler, M. (1995). *Astrophysical Journal*, **441**, 886.

Canuto, V.M. (1991). In *The Sun and cool stars* (eds I. Tuominen, D. Moss and G. Rüdiger), p. 27. Springer, Berlin.

Canuto, V.M. (1993). *Astrophysical Journal*, **416**, 331.

Canuto, V.M. and Mazitelli, I. (1991). *Astrophysical Journal*, **370**, 295.

Canuto, V.M. and Mazitelli, I. (1992). *Astrophysical Journal*, **389**, 724.

Cattaneo, F. (1984). In *The hydromagnetics of the Sun*, p. 47. ESA SP-220, Noordwijk.

Cattaneo, F. and Hughes, D.W. (1988). *Journal Fluid Mechanics*, **196**, 323.

Cattaneo, F., Chiueh, T. and Hughes, D.W. (1990). *Journal Fluid Mechanics*, **219**, 1.

Cattaneo, F., Brummell, N.H., Toomre, J., Malagoli, A. and Hurlburt, N.E. (1991). *Astrophysical Journal*, **370**, 282.

Chan, K.L. and Sofia, S.S. (1989). *Astrophysical Journal*, **336**, 1022.

Chan, K.L. and Sofia, S.S. (1996). *Astrophysical Journal*, **466**, 372.

Chandrasekhar, S. (1949). *Astrophysical Journal*, **110**, 329.

Chandrasekhar, S. (1952). *Philosophical Magazine (7th Series)*, **43**, 501.

Chandrasekhar, S. (1961). *Hydrodynamic and hydromagnetic stability.* Oxford University Press.

Chiang, W.H. and Foukal, P.V. (1984). *Solar Physics*, **97**, 9.

Chitre, S.M. (1963). *Monthly Notices Royal Astronomical Society*, **126**, 431.

Chitre, S.M. and Shaviv, G. (1967). *Solar Physics*, **2**, 150.

Choudhuri, A.R. (1989). *Solar Physics*, **123**, 217.

Choudhuri, A.R. and Gilman, P.A. (1987). *Astrophysical Journal*, **316**, 788.

Cowling, T.G. (1935). *Monthly Notices Royal Astronomical Society*, **96**, 42.

Cowling, T.G. (1946). *Monthly Notices Royal Astronomical Society*, **106**, 218.

Cowling, T.G. (1953). *Solar electrodynamics.* In *The Sun* (ed. G.P. Kuiper), p. 532. Chicago University Press.

Cowling, T.G. (1976). *Magnetohydrodynamics.* Adam Hilger, London. (Interscience, New York edition 1957.)

Deinzer, W. (1965). *Astrophysical Journal*, **141**, 548.

D'Silva, S. and Choudhuri, A.R. (1993). *Astronomy and Astrophysics*, **272**, 621.

Durney, B. and Roxburgh, I.W. (1971). *Solar Physics*, **16**, 3.

Fan, Y., Fisher, G.H. and DeLuca, E.E. (1993). *Astrophysical Journal*, **405**, 390.

Ferraro, V.C.A. (1937). *Monthly Notices Royal Astronomical Society*, **97**, 458.

Ferriz-Mas, A. and Schüssler, M. (1993). *Geophysical Astrophysical Fluid Dynamics*, **72**, 209.

Ferriz-Mas, A. and Schüssler, M. (1994). *Astrophysical Journal*, **433**, 852.

Ferriz-Mas, A. and Schüssler, M. (1995). *Geophysical Astrophysical Fluid Dynamics*, **81**, 233.

Foukal, P.V., Fowler, P. and Livshits, M. (1983). *Astrophysical Journal*, **267**, 863.

Galloway, D.J. and Moore, D.R. (1979) *Geophysical Astrophysical Fluid Dynamics*, **12**, 73.

Galloway, D.J. and Proctor, M.R.E. (1983). *Geophysical Astrophysical Fluid Dynamics*, **24**, 109.

Galloway, D.J. and Weiss, N.O. (1981). *Astrophysical Journal*, **243**, 945.

Galsgaard, K. and Nordlund, A. (1996). *Journal Geophysical Research* (A6), **101**, 13445.

Gold, T. (1965). In *Solar and stellar magnetic fields* (ed. R. Lüst), p. 390. North-Holland, Amsterdam.

Gough, D.O. (1964). *Journal Atmospheric Science*, **26**, 448.

Gough, D.O. (1977). In *Problems of stellar convection* (eds E.A. Spiegel and J.-P. Zahn), p. 15. Lecture Notes in Physics **71**. Springer, Berlin.

Gough, D.O. and Tayler, R.J. (1966). *Monthly Notices Royal Astronomical Society*, **133**, 85.

Gough, D.O. and Weiss, N.O. (1976). *Monthly Notices Royal Astronomical Society*, **176**, 589.

Grossman, S.A. (1996). *Monthly Notices Royal Astronomical Society*, **279**, 305.

Hagenaar, H.J., Schrijver, C.J. and Title, A.M. (1997). *Astrophysical Journal*, **475**, 328.

Hoyle, F. (1949). *Some recent researches in solar physics.* Cambridge University Press.

Hughes, D.W. (1991). In *Advances in solar system MHD* (eds E.R. Priest and A.W. Hood), p. 77. Cambridge University Press.

Hughes, D.W. and Proctor, M.R.E. (1988). *Annual Review Fluid Mechanics*, **20**, 187.

Hurlburt, N.E. and Toomre, J. (1988). *Astrophysical Journal*, **327**, 920.

Hurlburt, N.E., Proctor, M.R.E., Weiss, N.O. and Brownjohn, D.P. (1989). *Journal Fluid Mechanics*, **209**, 587.

Iroshnikov, R.S. (1969). *Astronomical Journal SSSR*, **46**, 97.

Jahn, K. and Schmidt, H.U. (1991). *Astronomy and Astrophysics*, **290**, 295.

Jennings, R.L., Brandenburg, A., Nordlund, A., Stein, R.F. and Tuominen, I. (1991). In *The Sun and cool stars* (eds I. Tuominen, D. Moss and G. Rüdiger), p. 92. Springer, Berlin.

Jensen, E. (1955) *Annales d'Astrophysique*, **18**, 127.

Kippenhahn, R. (1963). *Astrophysical Journal*, **137**, 664.

Kitchatinov, L.L. (1987). *Geophysical Astrophysical Fluid Dynamics*, **38**, 273.

Kitchatinov, L.L. and Rüdiger, G. (1993). *Astronomy and Astrophysics*, **276**, 96.

Kitchatinov, L.L., Pipin, V.V. and Rüdiger, G. (1994). *Astronomische Nachrichten*, **315**, 157.

Köhler, H. (1970). *Solar Physics*, **13**, 2.

Kruskal, M.D. and Schwarzschild, M. (1954). *Proceedings Royal Society A*, **223**, 348.

Kuhn, T.S. (1970). *The structure of scientific revolutions* (2nd edn). University of Chicago Press.

Küker, M., Rüdiger, G. and Kitchatinov, L.L. (1993). *Astronomy and Astrophysics*, **279**, L1.

Lebedinski, A.I. (1941). *Astronomical Journal SSSR*, **18**, 10.

Lites, B.W., Nordlund, A. and Scharmer, G.S. (1989). In *Solar and stellar granulation* (eds R.J. Rutten and G. Severino), p. 349. NATO **263**. Kluwer, Dordrecht.

Low, B.C. (1990). *Annual Review Astronomy Astrophysics*, **28**, 491.

Low, B.C. (1996). In *Solar and astrophysical magnetohydrodynamic flows* (ed. K.C. Tsinganos), p. 109. Kluwer, Dordrecht.

Low, B.C. and Wolfson, R. (1988). *Astrophysical Journal*, **324**, 574.

Matthews, P.C., Proctor, M.R.E. and Weiss, N.O. (1995*a*). *Journal Fluid Mechanics*, **305**, 281.

Matthews, P.C., Hughes, D.W. and Proctor, M.R.E. (1995*b*). *Astrophysical Journal*, **448**, 938.

Mestel, L. (1970). *Mémoires Société Royale Liège*, **19**, 167.

Meyer, F., Schmidt, H.U., Weiss, N.O. and Wilson, P.R. (1974). *Monthly Notices Royal Astronomical Society*, **169**, 35.

Moffatt, H.K. and Kamkar, H. (1982). In *Stellar and planetary magnetism* (ed. A.M. Soward), p. 91. Gordon and Breach, New York.

Moreno-Insertis, F. (1987). In *The role of fine-scale magnetic fields on the structure of the solar atmosphere* (eds E. Schröter, M. Vázquez and A. Wyller), p. 167. Cambridge University Press.

Moreno-Insertis, F. (1992). In *Sunspots – theory and observations* (eds J.H. Thomas and N.O. Weiss), p. 385. Kluwer, Dordrecht.

Moreno-Insertis, F., Caligari, P. and Schüssler, M. (1995). *Astrophysical Journal*, **452**, 894.

Moss, D.L. and Tayler, R.J. (1969). *Monthly Notices Royal Astronomical Society*, **145**, 217.

Moss, D. and Vilhu, O. (1983). *Astronomy and Astrophysics*, **119**, 47.

Nordlund, A. (1982). *Astronomy and Astrophysics*, **107**, 1.

Nordlund, A. (1984*a*). In *Variations of the solar irradiance in active region timescales* (ed. G. Chapman), p. 121. NASA Conference Publications.

Nordlund, A. (1984*b*). Magnetoconvection. In *4th European meeting on solar physics. ESA-SP*. **220**, p. 37. ESA.

Nordlund, A. (1985). *Max-Planck-Institut Astrophysik Publications*, **212**, 101.

Nordlund, A. (1986). *Solar Physics*, **100**, 209.

Nordlund, A. and Dravins, D. (1990). *Astronomy and Astrophysics*, **228**, 155.

Nordlund, A. and Stein, R.F. (1989). In *Solar and stellar granulation* (eds R.J. Rutten and G. Severino), p. 453. NATO **263**. Kluwer, Dordrecht.

Nordlund, A. and Stein, R.F. (1990). In *Solar photosphere: structure, convection and magnetic fields* (ed. J.O. Stenflo), p. 191. Kluwer, Dordrecht.

Nordlund, A. and Stein, R.F. (1991). In *Stellar atmospheres: beyond classical models* (eds L. Crivellari, I. Hubeney and D.G. Hummer), p. 263. Kluwer, Dordrecht.

Nordlund, A. and Stein, R.F. (1996). In *Proceedings 32 Liège Colloquium* (eds A. Noels *et al.*) p. 75. Institut d'Astrophysique, Université de Liège.

Nordlund, A. and Stein, R.F. (1998). Submitted to *ApJ.*

Nordlund, A., Brandenburg, A., Jennings, R.L., Rieutord, M., Ruokolainen, J., Stein, R.F. and Tuominen, I. (1992). *Astrophysical Journal*, **392**, 647.

Nordlund, A., Galsgaard, K. and Stein, R.F. (1994). In *Solar surface magnetic fields* (eds R.J. Rutten and C.J. Schrijver), p. 471. NATO ASI Series **433**. Kluwer, Dordrecht.

Nordlund, A., Stein, R.F. and Brandenburg, A. (1996). In *Windows on the Sun's interior* (eds H.M. Antia and S.M. Chitre), p. 261. Bulletin Astronomical Society, India.

Parker, E.N. (1955). *Astrophysical Journal*, **122**, 293.

Parker, E.N. (1957). *Journal Geophysical Research*, **62**, 509.

Parker, E.N. (1963). *Astrophysical Journal*, **138**, 552.

Parker, E.N. (1972). *Astrophysical Journal*, **174**, 499.

Parker, E.N. (1974). *Solar Physics*, **36**, 249.

Parker, E.N. (1978). *Astrophysical Journal*, **221**, 368.

Parker, E.N. (1979). *Cosmical magnetic fields.* Oxford University Press.

Parker, E.N. (1994). *Spontaneous current sheets in magnetic fields.* Oxford University Press.

Priest, E.R. (1982). *Solar magnetohydrodynamics.* Reidel, Dordrecht.

Priest, E.R. (1998). *Astrophysics Space Science*, in press.

Proctor, M.R.E. and Weiss, N.O. (1982). *Reports Progress Physics*, **45**, 1317.

Proctor, M.R.E., Weiss, N.O., Brownjohn, D.P. and Hurlburt, N.E. (1994). *Journal Fluid Mechanics*, **280**, 227.

Pulkkinen, P., Tuominen, I., Nordlund, A. and Stein, R.F. (1993). *Astronomy and Astrophysics*, **267**, 265.

Rast, M.P. and Toomre, J. (1993). *Astrophysical Journal*, **408**, 224 and 240.

Rast, M.P., Nordlund, A., Stein, R.F. and Toomre, J. (1993). *Astrophysical Journal Letters*, **408**, L53.

Roberts, P.H. and Soward, A.M. (1975). *Astronomische Nachrichten*, **296**, 49.

Rosenthal, C.S. (1998). In *Sounding solar and stellar interiors*, IAU Symposium 181 (eds F.X. Schmider and J. Provost), p. 121. Nice Observatory.

Rosenthal, C.S., Christensen-Dalsgaard, J., Nordlund, A., Stein, R.F. and Trampedach, R. (1998). Submitted to *Astronomy and Astrophysics.*

Rüdiger, G. (1989). *Differential rotation and stellar convection: Sun and solar-type stars.* Gordon and Breach, New York.

Rüdiger, G. and Kitchatinov, L.L. (1994). In *The solar engine and its influence on terrestrial atmosphere and climate* (ed. E. Ribes), p. 27. NATO Workshop, Paris. Springer, Berlin.

Schlüter, A. and Temesváry, St. (1958). In *Electromagnetic phenomena in cosmical physics* (ed. B. Lehnert), p. 263. Cambridge University Press.

Schnack, D.D. and Mikić, Z. (1994). In *Solar active region evolution: comparing models with observation* (eds K.S. Balasubramanian and G.W. Simon), p. 180. Astronomical Society Pacific, San Francisco.

Schüssler, M. (1996). In *Solar and astrophysical magnetohydrodynamical flows* (ed. K.C. Tsinganos), p. 17. Kluwer, Dordrecht.

Schüssler, M., Caligari, P., Ferriz-Mas, A., Solanki, S.K. and Stix, M. (1996). *Astronomy and Astrophysics*, **314**, 503.

Schwarzschild, M. (1958). *Structure and evolution of the stars*, Princeton University Press, Princeton.

Simon, G.W. and Weiss, N.O. (1968). *Zeitschrift Astrophysik*, **69**, 435.

Simon, G.W. and Weiss, N.O. (1991). *Monthly Notices Royal Astronomical Society*, **252**, 1P.

Spiegel, E.A. and Weiss, N.O. (1980). *Nature*, **B287**, 616.

Spruit, H.C. (1974). *Solar Physics*, **34**, 277.

Spruit, H.C. (1977). *Solar Physics*, **55**, 3.

Spruit, H.C. (1981). *Astronomy and Astrophysics*, **98**, 155.

Spruit, H.C. (1982). *Astronomy and Astrophysics*, **108**, 348 and 356.

Spruit, H.C. (1998). *Memorie Societa Astronomica Italiana*, **68**, 397.

Spruit, H.C. and van Ballegooijen, A.A. (1982*a*). *Astronomy and Astrophysics*, **106**, 58.

Spruit, H.C. and van Ballegooijen. A.A. (1982*b*). *Astronomy and Astrophysics*, **113**, 350.

Spruit, H.C. and Zweibel, E.G. (1979). *Solar Physics*, **62**, 15.

Spruit, H.C., Nordlund, A. and Title, A.M. (1990). *Annual Review Astronomy Astrophysics*, **28**, 263.

Stein, R.F. and Nordlund, A. (1989). *Astrophysical Journal*, **342**, L95.

Stein, R.F. and Nordlund, A. (1998*a*). *Astrophysical Journal*, **499**, 914.

Stein, R.F. and Nordlund, A. (1998*b*). Submitted to *Astrophysical Journal.*

Steiner, O., Knölker, M. and Schüssler, M. (1994). In *Solar surface magnetism* (eds R.J. Rutten and C.J. Schrijver), p. 441. Kluwer, Dordrecht.

Stix, M. (1989*a*). *Reviews Modern Astronomy*, **2**, 248.

Stix, M. (1989*b*). *The Sun.* Springer, Berlin.

Sweet, P.A. (1958). In *Electromagnetic phenomena in cosmical physics* (ed. B. Lehnert), p. 123. Cambridge University Press.

Syrovatsky, S.I. and Zhugzhda, Y.D. (1967). *Soviet Astronomy*, **11**, 945.

Tao, L., Weiss, N.O., Brownjohn, D.P. and Proctor, M.R.E. (1998). *Astrophysical Journal Letters*, **496**, L39.

Thomas, J.H. and Weiss, N.O., eds (1992). *Sunspots: theory and observation.* Kluwer, Dordrecht.

Thompson, W.B. (1951). *Philosophical Magazine (7th Series)*, **42**, 1417.

Ulmschneider, P., Priest, E.R. and Rosner, R., eds (1991). *Mechanisms of chromospheric and coronal heating.* Springer, Berlin.

van Ballegooijen, A.A. (1983). *Astronomy and Astrophysics*, **118**, 275.

van Ballegooijen, A.A. and Choudhuri, A.R. (1988). *Astrophysical Journal*, **404**, 773.

Wasiutynski, J. (1946). *Astrophysica Norvegica*, **4**.

Weiss, N.O. (1964). *Philosophical Transactions Royal Society A*, **256**, 99..

Weiss, N.O. (1965). *Observatory*, **85**, 37.

Weiss, N.O. (1966). *Proceedings Royal Society A*, **293**, 310.

Weiss, N.O. (1968). In *Plasma instabilities in astrophysics* (eds. D.G. Wentzel and D.A. Tidman), p. 153. Gordon and Breach, New York.

Weiss, N.O. (1981). *Journal Fluid Mechanics*, **108**, 247 and 273.

Weiss, N.O. (1991). *Geophysical Astrophysical Fluid Dynamics*, **62**, 229.

Weiss, N.O. (1997). In *Advances in physics of sunspots* (eds B. Schmieder, J.C. del Toro Iniesta and M. Schüssler). P.A.S.P. Conference Series.

Weiss, N.O., Brownjohn, D.P., Hurlburt, N.E. and Proctor, M.R.E. (1990). *Monthly Notices Royal Astronomical Society*, **245**, 434.

Weiss, N.O., Brownjohn, D.P., Matthews, P.C. and Proctor, M.R.E. (1996). *Monthly Notices Royal Astronomical Society*, **283**, 1153.

Yun, H.S. (1970). *Astrophysical Journal*, **162**, 975.

Zhugzhda, Y.D. (1970). *Soviet Astronomy*, **14**, 274.

Zwaan, C. (1978). *Solar Physics*, **60**, 213.

5

MAGNETIC FIELDS IN STELLAR INTERIORS

5.1 General considerations

The scalar virial theorem (3.64) suggests an immediate first estimate for the likely effect of a magnetic field on the gross equilibrium of a star. A significant deviation from the normal balance between pressure and gravity requires a magnetic energy approaching the thermal and gravitational energies, implying an extreme upper limit for the mean field $\overline{B}$ given by

$$(\overline{B}^2/8\pi)(4\pi R^3/3) \approx GM^2/R \tag{5.1}$$

so that

$$\overline{B} \approx 1.3 \times 10^8 (M/M_\odot)/(R/R_\odot)^2. \tag{5.2}$$

By contrast, Babcock's most strongly magnetic star HD215441, with $M/M_\odot \approx 2$ and $R/R_\odot \approx 2$, has a surface polar field $B_s \approx 34\,000$ G, while a more typical early-type magnetic star has $B_s \approx 10^3$ G. The magnetic white dwarfs ($R/R_\odot \approx 10^{-2}$) have $B_s \approx 10^6$–10^8 G, and pulsars with $R/R_\odot \approx 10^{-4} - 10^{-5}$ have $B_s \approx 10^{12}$ G. This raises immediately the question as to what determines the fraction of a star's total flux that appears above the surface; but we can anticipate that models with a plausible degree of central concentration of flux fail to yield anywhere near the virial limit (5.2). We provisionally accept that over the bulk of a star, whereas in a rapid rotator the centrifugal force density $\rho\Omega^2\varpi$ may be a significant perturbation, the Lorentz force density $|(\nabla\times\mathbf{B}) \times \mathbf{B}/4\pi|$ is likely to be small compared with the gravitational force density ρg. However, in the low density surface regions of an early-type magnetic star the magnetic energy density $\approx 4 \times 10^4 (B_s/10^3)^2$, as compared with a photospheric thermal energy density $\approx 2 \times 10^5 (\rho_s/10^{-7})(T_s/10^4)$, and in the very low density circumstellar regions the magnetic energy may dominate out to many stellar radii (cf. Chapter 7).

As already discussed in Chapter 4, one expects to find strong contrasts between subadiabatic ('radiative') zones and superadiabatic and therefore turbulent convective zones. A primeval field may find itself tangled and even expelled from a turbulent zone, which however may generate and maintain its own dynamo field. We shall see that a field of too simple a structure in a radiative domain may itself be spontaneously unstable, converting magnetic energy into kinetic; and a field of topology sufficiently complex to be free of adiabatic instabilities may still evolve through secular instabilities, dependent on heat exchange and perhaps also Ohmic diffusion.

In the rest of this chapter, the emphasis will be primarily on stellar radiative zones. For simplicity, most detailed discussion is initially of systems that are symmetric about an axis that coincides with the rotation axis, discussion in detail of the more realistic oblique rotator model being postponed to Chapter 9. A field is then conveniently resolved into the mutually orthogonal meridional and azimuthal components (cf. Fig. 5.1), the axisymmetric forms of the more general poloidal and toroidal vectors (cf. the Appendix to Chapter 3):

$$\mathbf{B} = \mathbf{B}_{\mathrm{p}} + \mathbf{B}_{\mathrm{t}}, \tag{5.3}$$

$$\mathbf{B}_{\mathrm{p}} \cdot \mathbf{t} = 0, \qquad \mathbf{B}_{\mathrm{t}} = B_\phi \mathbf{t}, \tag{5.4}$$

where $\mathbf{t}$ is the unit toroidal vector. (The analysis is performed sometimes in vector or tensor notation and sometimes in terms of spherical polar (r, θ, ϕ) or cylindrical polar (ϖ, ϕ, z) components.) To satisfy the divergence-free condition, $\mathbf{B}_{\mathrm{p}}$ is written in terms of the flux function P, the analogue of the Stokes stream function in fluid dynamics:

$$\mathbf{B}_{\mathrm{p}} = -\nabla \times (P\mathbf{t}/\varpi) = -\nabla P \times \mathbf{t}/\varpi \tag{5.5}$$

since $\nabla \times (\mathbf{t}/\varpi) = 0$. The vector potential of the field $\mathbf{B}_{\mathrm{p}}$ is then

$$\mathbf{A} = A\mathbf{t} = -(P/\varpi)\mathbf{t} \tag{5.6}$$

and the lines of $\mathbf{B}_{\mathrm{p}}$ are given by $P =$ constant. Only degenerate fields have lines of $\mathbf{B}_{\mathrm{p}} + \mathbf{B}_{\mathrm{t}}$ that close in space; in general, a line of an axisymmetric field covers a toroidal surface defined by $P =$ constant and $0 \le \phi \le 2\pi$.

By Ampère's law, the currents maintaining the fields are respectively

$$\mathbf{j}_{\mathrm{t}} = (c/4\pi)\nabla \times \mathbf{B}_{\mathrm{p}} \tag{5.7}$$

and

$$\mathbf{j}_{\mathrm{p}} = (c/4\pi)\nabla \times \mathbf{B}_{\mathrm{t}} = (c/4\pi)\nabla(\varpi B_\phi) \times \mathbf{t}/\varpi. \tag{5.8}$$

The Lorentz force density is conveniently written as the sum of poloidal and toroidal components:

$$\begin{aligned} \mathbf{j} \times \mathbf{B}/c = (\mathbf{j}_{\mathrm{p}} \times \mathbf{B}_{\mathrm{t}}/c + \mathbf{j}_{\mathrm{t}} \times \mathbf{B}_{\mathrm{p}}/c) \qquad & \text{poloidal} \qquad (5.9) \\ + (\mathbf{j}_{\mathrm{p}} \times \mathbf{B}_{\mathrm{p}}/c) \qquad & \text{toroidal} \qquad (5.10) \end{aligned}$$

The equation of hydrostatic support is then

$$-\nabla p + \rho\nabla V + \rho\Omega^2\varpi + (\nabla \times \mathbf{B}_{\mathrm{p}}) \times \mathbf{B}_{\mathrm{p}}/4\pi + (\nabla \times \mathbf{B}_{\mathrm{t}}) \times \mathbf{B}_{\mathrm{t}}/4\pi = 0 \tag{5.11}$$

where $\Omega(\varpi, z)$ is the local angular velocity and V the gravitational potential. As already noted, the magnetic terms are likely to be small compared with pressure

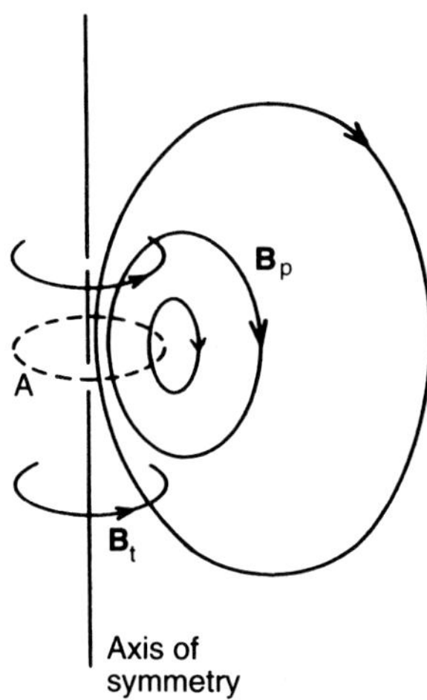

FIG. 5.1. Axisymmetric magnetic fields. The meridional (poloidal) components $\mathbf{B}_p$ are maintained by azimuthal (toroidal) currents $\mathbf{j}_t$; the toroidal component $\mathbf{B}_t$ by poloidal currents $\mathbf{j}_p$. $\mathbf{B}_p$ has at least one null line A; $\mathbf{B}_t$ vanishes on the axis. The Lorentz force has a toroidal component $\mathbf{j}_p \times \mathbf{B}_p/c$ that is in general non-zero.

and gravity, except in the surface regions and above. A non-vanishing toroidal component (5.10) will, however, react on the angular momentum distribution. Even though this component will also in general be small compared with gravity, that means just that the time in which changes in Ω occur is much longer than the free-fall time, but it can still be much shorter than a stellar evolution time. Much attention is for this reason focused on the mutual interaction of stellar magnetism and rotation. Moreover, we shall see in Chapter 9 that in an oblique rotator, even very small poloidal perturbing forces can have significant effects over a stellar lifetime.

5.2 Magnetic fields and stellar rotation

5.2.1 *Axisymmetric states*

The mutual interaction of a magnetic field and a rotation field in a perfectly conducting fluid provides one of the simplest applications of classical magnetohydrodynamics. Consider again a rotating star with a magnetic field symmetric about the axis of rotation, and written as in (5.3) and (5.4) as the sum of poloidal and toroidal components. Choose cylindrical polar coordinates (ϖ, ϕ, z) with the z-axis along the rotation axis. The rotational velocity is written as $\mathbf{v}_t = \varpi\Omega\mathbf{t}$ where $\Omega(\varpi, z)$ is the local angular velocity. The ϕ-component of the induction equation (2.88) then yields

$$\partial B_\phi/\partial t = \varpi(\mathbf{B}_\mathrm{p} \cdot \nabla)\Omega. \tag{5.12}$$

In the simplest case, at time t = 0 the star has a purely poloidal large-scale field. A prescribed rotation field that varies locally along the direction of the poloidal component generates a toroidal component that increases linearly with time. The poloidal current density (5.8) in general has a component flowing perpendicular to $\mathbf{B}_\mathrm{p}$, so exerting a torque which changes the Ω-field according to

$$\rho\varpi^2\,\partial\Omega/\partial t = \varpi(\mathbf{j}_\mathrm{p} \times \mathbf{B}_\mathrm{p}/c)\cdot\mathbf{t} = \mathbf{B}_\mathrm{p}\cdot\nabla(\varpi B_\phi/4\pi). \tag{5.13}$$

Because we are for the moment considering non-turbulent domains, any viscosity will be microscopic and so has negligible effect on the shear. Both the centrifugal force and the contribution $(\nabla \times \mathbf{B}_\mathrm{t}) \times \mathbf{B}_\mathrm{t}/4\pi$ to the poloidal magnetic force will vary with time, so causing associated small poloidal motions as the star adjusts to maintain hydrostatic equilibrium, but over the bulk of the star the consequent changes in ρ and $\mathbf{B}_\mathrm{p}$ will be modest even in a rapid rotator and will therefore be ignored. Equations (5.12) and (5.13) then yield linear wave equations for Ω and B_ϕ :

$$\rho\varpi^2\,\partial^2\Omega/\partial t^2 = \mathbf{B}_\mathrm{p}\cdot\nabla[\varpi^2(\mathbf{B}_\mathrm{p}\cdot\nabla)\Omega]/4\pi \tag{5.14}$$

and

$$\partial^2(\varpi B_\phi)/\partial t^2 = \varpi^2\mathbf{B}_\mathrm{p}\cdot\nabla[(4\pi\rho\varpi^2)^{-1}\mathbf{B}_\mathrm{p}\cdot\nabla(\varpi B_\phi)]. \tag{5.15}$$

Detailed solution would be analytically exacting (cf. analogous problems, with idealized ρ and $\mathbf{B}_\mathrm{p}$ fields, arising in the theory of star formation, discussed by Gillis *et al.* (1974, 1979) and summarized in Appendix B of Chapter 11). However, if the scale of variation of Ω along $\mathbf{B}_\mathrm{p}$ is small compared with ϖ and the scale of variation of $\mathbf{B}_\mathrm{p}$, then (5.14) approximates to

$$\partial^2\Omega/\partial t^2 = (B_\mathrm{p}^2/4\pi\rho)\,\partial^2\Omega/\partial s^2 = v_\mathrm{A}^2\,\partial^2\Omega/\partial s^2 \tag{5.16}$$

where ∂s is the element of length along $\mathbf{B}_\mathrm{p}$, and v_A is the local Alfvén speed, determined by the poloidal field. Although (5.16) is the standard one-dimensional wave equation, the variations in Ω and B_ϕ do not propagate without change of form, because of the variation of v_A with s. Nevertheless, we may infer that the magnetohydrodynamic equations (5.12) and (5.13) predict that a shear of given scale D will be reversed in a braking time τ_D which is of the same order as the Alfvénic time

$$\tau_b = D/v_\mathrm{A}. \tag{5.17}$$

This time is very short compared with a stellar evolution time, unless B_p is itself very small. For example, with $B = 1$ G, $D = 0.2R_\odot = 1.4\times10^{10}$ cm and $\rho = 30$ g cm^{-3} (about 20 times the mean solar density and one-third of the estimated central value), $\tau_b \approx 10^4$ yr; and to increase τ_b to 3×10^9 yr, B_p would need to be as low as 3×10^{-6} G. The problem described by (5.12) and (5.13) is strictly non-dissipative: τ_b is the time in which the shear is reversed rather than destroyed, and if no dissipation were to occur, the Ω-field would oscillate indefinitely. In fact, waves along field lines that are coupled with a convective zone via a boundary layer will be rapidly damped by turbulent viscosity. Waves along field lines that lie completely in a radiative zone will also be damped, but more slowly, by finite resistivity, and also through non-adiabaticity in the associated poloidal motions (cf. Sections 5.5 and 9.6). In an asymptotic steady state, and in the absence of other, non-rotatory motions, from (5.12) Ω is subject to the constraint

$$\mathbf{B}_{\mathrm{p}} \cdot \nabla\Omega = 0. \tag{5.18}$$

With $\mathbf{B}_{\mathrm{p}}$ written as in (5.5) in terms of the flux function P, (5.18) becomes $\nabla\Omega \times \nabla P = 0$, or

$$\Omega = \alpha(P) \tag{5.19}$$

– constancy of Ω along poloidal field lines. This law of isorotation (Ferraro 1937) was a landmark in the history of cosmical electrodynamics, and it remains a simple, intuitively clear consequence of field-freezing. Accompanying this kinematical constraint is a dynamical constraint: in a steady state (5.13) yields

$$\mathbf{B}_{\mathrm{p}} \cdot \nabla(\varpi B_\phi) = 0, \tag{5.20}$$

whence, again on use of (5.5),

$$\varpi B_\phi = \beta(P). \tag{5.21}$$

From (5.13), the meaning of (5.20) and (5.21) is that the currents $\mathbf{j}_{\mathrm{p}}$ must flow parallel to $\mathbf{B}_{\mathrm{p}}$ in order that the total field be torque-free (Lüst and Schlüter 1954; Chandrasekhar 1956*a*):

$$\mathbf{j}_{\mathrm{p}} = k\mathbf{B}_{\mathrm{p}}, \qquad k = -c\beta'(P)/4\pi \tag{5.22}$$

from (5.8), (5.5) and (5.21). Equivalently, a vanishing magnetic torque about the z-axis implies zero net magnetic transport of the z-component of angular momentum across a closed surface. In an axisymmetric system, the magnetic pressure term in the Lüst–Schlüter tensor (2.46) and (2.42) automatically makes zero contribution to the flow of z-angular momentum, while the condition on the magnetic tension term is that $-\varpi B_\phi/4\pi$ is constant along a poloidal flux tube, as is required by steady-state conditions in the absence of any material transport (cf. Section 5.5). The area A, defined by the segments $\mathrm{P_1Q_1}$, $\mathrm{P_2Q_2}$ of two neighbouring poloidal field lines, (Fig. 5.2(a)), will be threaded by the toroidal flux

$$\begin{aligned} \mathrm{d}F_{\mathrm{t}} \equiv \int\!\!\int_A B_\phi \,\mathrm{d}l\,\mathrm{d}s &= \oint \varpi B_\phi[(\varpi B_{\mathrm{p}}\,\mathrm{d}l)/(\varpi^2 B_{\mathrm{p}})]\,\mathrm{d}s \\ &= (\mathrm{d}F_{\mathrm{p}}/2\pi)\oint[(\varpi B_\phi)/(\varpi^2 B_{\mathrm{p}})]\,\mathrm{d}s. \end{aligned} \tag{5.23}$$

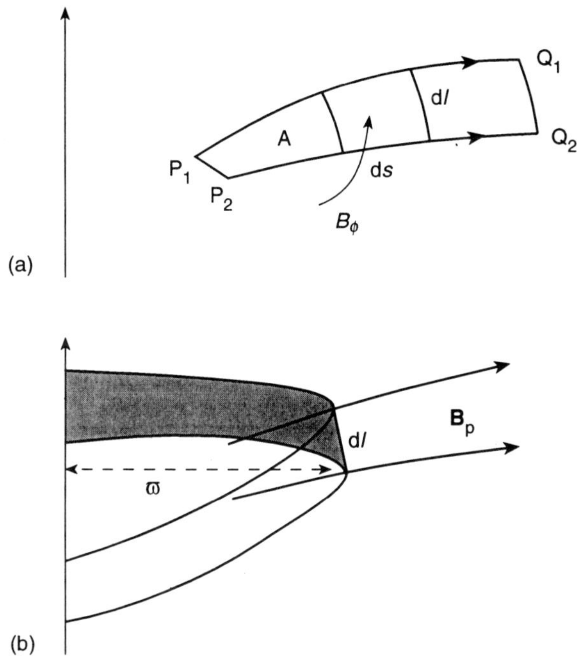

FIG. 5.2. (a) The toroidal flux (5.23) threading the area A. (b) The poloidal flux (5.24) crossing the ribbon $2\pi\varpi\,\mathrm{d}l$.

Here $\nabla \cdot \mathbf{B}_\mathrm{p} = 0$ is used in the form (cf. Fig. 5.2(b))

$$2\pi\varpi B_\mathrm{p}\,\mathrm{d}l = \mathrm{d}F_\mathrm{p} \tag{5.24}$$

where $\mathrm{d}F_\mathrm{p}$ is the poloidal flux crossing the ribbon $2\pi\varpi\,\mathrm{d}l$, generated by rotation about the axis of the two poloidal field-lines.

In a torque-free field, (ϖB_ϕ) is by (5.21) also constant along a field line, and

$$\mathrm{d}F_\mathrm{t} = [(\varpi B_\phi)\,\mathrm{d}F_\mathrm{p}/2\pi] \oint \mathrm{d}s/(\varpi^2 B_\mathrm{p}) \tag{5.25}$$

which is manifestly non-zero as long as $\varpi B_\phi = \beta(P) \neq 0$. In a non-steady state, but with the changes in B_ϕ due to motion under strict flux-freezing, then $\mathrm{d}F_\mathrm{t}$ must be an invariant. The illustrative problem discussed at the beginning of this section can be generalized to fields that at time $t = 0$ have both $\mathbf{B}_\mathrm{p}$ and $\mathbf{B}_\mathrm{t}$ components, but which are torque-free, and so with $\varpi B_\phi = \beta(P)$ and $\mathrm{d}F_\mathrm{t} \neq 0$. The effect of the non-uniform Ω is then to generate an extra component B_ϕ which is not torque-free but which again reacts back on the rotation field, redistributing angular momentum within the flux-tube, but leaving $\mathrm{d}F_\mathrm{t}$ invariant.

The anticipated strong interaction between magnetism and rotation has thus already imposed the severe constraint of isorotation on steady, high conductivity systems with purely rotatory motions. The generalization to cases with poloidal motions is given in Section 5.5. The effect of finite resistivity on the allowed steady states will be small when the MRN $\gg 1$, as is the case when $\mathbf{v}_\mathrm{t}$ is a typical rotation velocity. (In fact, if the field were purely poloidal, so that $\mathbf{j}_\mathrm{p} = 0$, then in a steady state isorotation would continue to hold, since the poloidal component of Ohm's law (2.84) would still have the form $\mathbf{E} + \Omega\varpi\mathbf{t} \times \mathbf{B}_\mathrm{p}/c = 0$, with its curl again yielding the isorotation condition (5.19).)

5.2.2 *'Quasi-steady', non-axisymmetric states*

It must be remembered that even the limited freedom of isorotation in a steady state is allowed only because of the assumption of strict axisymmetry. The relative shearing of field lines then makes no difference to the density field and the magnetic field; the only consequence of isorotation rather than uniform rotation is a change in the (curl-free) electric field $\mathbf{E} = -\Omega(P)\varpi\mathbf{t} \times \mathbf{B}_\mathrm{p}$ and hence in the polarization charge density $\rho_e = \nabla \cdot \mathbf{E}/4\pi$, which in any case is smaller than the ion or electron charge density by a very small factor of the order of $(B^2/4\pi\rho c^2)(\Omega/(eB/m_H c))$. In a non-axisymmetric system, however, relative shearing of field lines will normally alter both ρ and $\mathbf{B}$, and only very special field structures can allow kinematically steady states. A non-axisymmetric rotating magnetic star can be said to be 'in a steady state' if there exists a frame rotating with uniform angular velocity α in which all quantities are time-independent. It is sometimes more convenient to use the equivalent definition of a 'quasi-steady' system, meaning one in which quantities such as scalars or cylindrical and spherical polar components of vectors, when measured in the inertial frame, depend on azimuthal angle ϕ and time t in the combination $(\phi - \alpha t)$, so that the operators $\partial/\partial t$, $\partial/\partial\phi$ are related by $\partial/\partial t = -\alpha\,\partial/\partial\phi$. Let the star have a rotation field $\Omega(\mathbf{r})$. Then the induction equation (2.88) yields

$$(\Omega - \alpha)\,\partial(B_\varpi, B_z)\,/\partial\phi = -(B_\varpi, B_z)\partial(\Omega - \alpha)/\partial\phi \tag{5.26}$$

and

$$(\Omega - \alpha)\,\partial B_\phi/\partial\phi = \varpi(B_\varpi\,\partial/\partial\varpi + B_z\,\partial/\partial z)(\Omega - \alpha), \tag{5.27}$$

with the condition $\nabla\cdot\mathbf{B} = 0$ used. In the axisymmetric case, the two components (5.26) are satisfied identically and (5.27) is just the isorotation condition. With $\partial/\partial\phi \neq 0$, (5.26) requires

$$(\Omega - \alpha)B_\varpi = f(\varpi, z), \qquad (\Omega - \alpha)B_z = g(\varpi, z); \tag{5.28}$$

hence if $(\Omega - \alpha) \neq 0$,

$$B_\varpi/B_z = f(\varpi, z)/g(\varpi, z) \tag{5.29}$$

and so must be independent of ϕ. The constraint (5.29) is very severe: it is certainly not satisfied by a simple oblique rotator field, such as a dipolar field inclined to the rotation axis (cf. Chapter 9). Further, the non-axisymmetric Ω-field implied by (5.27) must satisfy the quasi-steady continuity conditions:

$$\alpha\, \partial\rho/\partial\phi = -\partial\rho/\partial t = \nabla\cdot(\rho\mathbf{v}) = \partial(\rho\Omega)/\partial\phi, \tag{5.30}$$

whence it follows that $(\Omega-\alpha)\rho$ is independent of ϕ. The departures of the density field from symmetry about the rotation axis are of the order of the small quantity $\bar{B}^2R^4/6GM^2$ (cf. Section 5.1). Thus if $\Omega \neq \alpha$, the conditions (5.28) allow only very small deviations of B_ϖ and B_z (and so also of B_ϕ) from axisymmetry; but if $\Omega = \alpha$, no constraint is imposed on the field, which is just swung round as a whole.

We have seen that in an axisymmetric system, departures from isorotation generate magnetic tensions with moments about the rotation axis, which drive the torsional Alfvén waves that redistribute angular momentum along a given flux-tube. In a non-axisymmetric system, significant departures from uniform rotation will generate, in addition, azimuthal magnetic pressure gradients which will interchange angular momentum between different flux-tubes. Order-of-magnitude arguments yield again the estimate (5.17) for the characteristic time. We return to these questions again in the study of the oblique rotator model in Chapter 9. Meanwhile, we recall that (5.12) and (5.14), which yield the estimate (5.17) for the time in which a shear is reversed, hold strictly only in the limit of infinite MRN. With $\eta = c^2/4\pi\sigma$ the microresistivity, this is an excellent approximation; however, study of the stability of magnetic fields will lead us to ask whether sometimes η should be replaced by an effective macroresistivity (cf. Sections 5.3 and 5.6).

5.3 Stability

In Section 3.6.2, the energy principle without the magnetic terms was used to show that a superadiabatic stellar zone is spontaneously unstable, with radial motions releasing gravitational energy to be converted into turbulent kinetic energy. The relevant term is then the volume integral of

$$\rho(\xi\cdot\mathbf{g})(\xi\cdot\nabla\rho/\rho)[(\gamma-1)-1/n]/\gamma \tag{5.31}$$

where ξ is the displacement vector, n the local polytropic index, and γ the usual ratio of specific heats. In a superadiabatic zone, $n < 1/(\gamma-1)$, and since $\mathbf{g}$ and $\nabla\rho/\rho$ are parallel or nearly parallel vectors, the term (5.31) is strongly negative for all displacements other than those nearly perpendicular to $\mathbf{g}$, implying instability. Equally, in a subadiabatic zone, the term (5.31) is strongly positive and so stabilizing, except again for displacements satisfying $\xi\cdot\mathbf{g} = 0$. As shown in the classical work of Biermann and Cowling, summarized in Section 4.1, in a superadiabatic zone the consequent turbulent heat transport is normally so efficient that the temperature gradient is reduced to a value only marginally greater than the adiabatic value – typically 1 part in 10^6 – except in low density regions near a stellar photosphere. Prima facie, one might expect an external magnetic field to be able to penetrate a convective zone and to interfere with or suppress the turbulence if the field energy density exceeds the turbulent energy density, as calculated from the Biermann–Cowling theory. However, the studies by Gough and Tayler (1966) and Moss and Tayler (1969), cited in Section 4.3, showed that much stronger fields would in fact be required, with energy density of the same order as the thermal. The reason is that the energy to be transported through the convection zone is essentially unaltered. Interference by magnetic stresses with the normal turbulent heat transport has the effect of forcing up the temperature gradient to values that are correspondingly more superadiabatic than in the standard theory, leading to a more energetic turbulence. An upper limit is given by supposing the turbulence to be sonic rather than highly subsonic, so requiring a magnetic energy density close to the thermal in order to interfere seriously with the turbulence.

Now consider a subadiabatic zone, permeated by a large-scale magnetic field, with an energy density well below the thermal, and again for the the moment ignore rotation. Then the stability of the magnetic field may be tested through the energy principle. It is immediately clear that with $B^2/8\pi\rho a^2 \ll 1$, any unstable modes must be such that $\xi \cdot \mathbf{g} \approx 0$, so that the strongly stabilizing term (5.31) is nullified. We therefore restrict the rest of the discussion to displacement fields satisfying this constraint.

An axisymmmetric poloidal magnetic field $\mathbf{B}_\mathrm{p}$, maintained by the toroidal current field $\mathbf{j}_\mathrm{t} = (c/4\pi)(\nabla \times \mathbf{B}_\mathrm{p})$, will have at least one O-type neutral point O (Figs 5.1 and 5.7). Near O the field structure is similar to that in a laboratory discharge, which is known to be unstable to 'sausage' and 'kink' instabilities (cf. Figs 3.10–3.13). It was conjectured that there would be a stellar magnetic instability, analogous to the kink mode, but again with ξ restricted by $\xi \cdot \mathbf{g} = 0$. It was indeed shown by Wright (1973) and by Markey and Tayler (1973, 1974) that modes that are incompressible – satisfying in addition $\nabla \cdot \xi = 0$ – yield a negative value of

$$\delta W = (1/8\pi)\mathrm{Re} \int [\delta\mathbf{B} \cdot \nabla \times (\mathbf{B} \times \xi^*) + \delta\mathbf{B}^2]\, \mathrm{d}\tau \tag{5.32}$$

where

$$\delta\mathbf{B} = \nabla \times (\xi \times \mathbf{B}) \tag{5.33}$$

and the asterisk denotes a complex conjugate. The variation of the gravitational field produced by the perturbation has been neglected, leading to an overestimate of the stability. The general discussion is due to to Van Assche *et al.* (1982). They Fourier-analyse the displacements according to

$$(\xi_r, \xi_\theta, \xi_\phi) = (R, S, \mathrm{i}T/m)\, \mathrm{e}^{\mathrm{i}m\phi} \tag{5.34}$$

where (r, θ, ϕ) are spherical polar coordinates and R, S, T are functions of (r, θ). The higher m, the more likely is the system to be unstable, and in the limit $m \to \infty$, δW depends on R, S and T but not on their derivatives, so that a separate stability criterion can be obtained on each magnetic surface and in particular on the neutral line. Hence instability is proved if the integral in (5.32) is negative on the neutral line for $m \to \infty$. In their proof of instability, Van Assche *et al.* did not in fact need to use the conditions that $\xi \cdot \mathbf{g}$ and $\nabla \cdot \xi$ should vanish, other than dropping the terms in the energy principle in which these expressions occur.

It is again known from laboratory studies that the instabilities of the pinched discharge may sometimes be removed if there is a field component parallel to the current, of strength comparable to the maximum value of the discharge field. This immediately suggests that stable stellar fields must have a similar complex topology, with toroidal flux linking the poloidal loops. The spontaneous wriggle distortion of the poloidal field, which we have seen to release energy, then implies an increase in the energy of the toroidal field, which may be sufficient to restore stability. Studies by Wright (1973), by Markey and Tayler (1973, 1974) and by Tayler (1980) go some way towards establishing this result in general. The problem is simplified by replacing the axisymmetric geometry by a cylindrical system, with coordinates (r, θ, z) centred on the neutral point. A uniform current density out to a radius r_0, parallel to z, yields a poloidal field with circular field lines:

$$B_\theta = B_0 r / r_0. \tag{5.35}$$

The toroidal field is simulated by a superposed uniform axial field

$$B_z = B_0 b. \tag{5.36}$$

In the perturbations, the factor $\mathrm{e}^{\mathrm{i}m\phi}$ is replaced by $\mathrm{e}^{\mathrm{i}kz}$ where $z = R_0\phi$ and R_0 is the distance of the neutral point from the axis of the star. A typical incompressible disturbance (satisfying $\nabla \cdot \xi = 0$) and normal to $\mathbf{g}$ can be taken of the form

$$\xi_r = X \sin\theta \sin kz, \qquad \xi_\theta = X \cos\theta \sin kz, \qquad \xi_z = (\mathrm{d}X/k\,\mathrm{d}r) \sin\theta \cos kz. \tag{5.37}$$

When δW is evaluated for such a disturbance and is then minimized, X is found to have the form

$$X \propto \mathrm{J}_0 \left\{ [2(1 - b^2k^2r_0^2)/(1 + b^2k^2r_0^2)]^{1/2} kr \right\}. \tag{5.38}$$

A local disturbance is discussed, with X vanishing at $r = r_0$. The most unstable case corresponds to the first zero of the Bessel function J_0, so that marginal stability is given by the vanishing of the argument in (5.38). When b is small there is a range of values of kr_0 for which instability occurs, but when $b > 0.244$ all wave numbers are stable. Tayler (1980) extends the analysis to toroidal rather than cylindrical geometry by the adoption of plausible trial fields; the results suggest strongly that stability holds when $b > 0.3$.

Tayler (1973) and Goossens and Tayler (1980) discuss also the stability of purely toroidal fields $B_\phi \mathbf{t}$. A general conclusion is that there is an unstable compressible mode close to the symmetry axis, if the electric current density is non-zero on the axis, i.e. if

$$B_\phi = B(\varpi/\varpi_0)^n \tag{5.39}$$

with $n = 1$. Again the instability is topological, depending only on the shape of the field but not on its strength. An important later result (Tayler 1980) shows that even mixed poloidal–toroidal fields are subject to an analogous instability near the symmetry axis, again provided B_ϕ is as in (5.39) with $n = 1$. This instability again depends on compressibility: the unstable mode is of the form $\exp[i(m\phi + kz)]$ with $(m + bk\varpi_0) = 0$. However, if $n > 1$, so that $j_r = 0$ on the axis, then it seems likely that if $8\pi\gamma p/B_\phi^2 \gg 0$ near the axis, then the mixed poloidal–toroidal field is stable.

The energy principle enables one to make some general statements about stable and unstable fields, as summarized above, but its use depends on the system's having negligible zero-order motions, in particular that the system is non-rotating. Frieman and Rotenberg (1960) showed that the typical rotational velocity would need to be at least comparable with the hydromagnetic wave speeds if it is to affect stability theory, and we have seen that in a star this is normally the case. The above-quoted results for non-rotating systems are highly suggestive, but we must now see what conclusions can be drawn from the direct study of the differential equations for perturbed, uniformly rotating systems. (The justification for adopting near uniform rotation as the most plausible mean state derives from the work of Section 5.2.) Pitts and Tayler (1985) have studied a limited number of model problems. They find that although an acceptable rotation rate does exert some stabilizing influence, it does not appear likely that it can remove most of the instabilities. In some cases for which the energy principle predicts instability when $\Omega = 0$, a finite Ω may slow up the growth rate, but the time of growth is still small compared with any evolutionary time-scale. But equally, as yet there is no case known in which a stable equilibrium is rendered unstable by the introduction of a reasonable uniform rotation. Their results are a tentative vindication of the use of the energy principle for a non-rotating magnetic star as a reasonable guide to the stability of rotating magnetic stars.

It is tempting to extrapolate these results into the non-linear domain and argue that any linear instability with a small wavelength will develop so that accelerated flux destruction occurs locally. This may often be plausible, though there are problems in which linear stability analysis is a misleading guide to the non-linear behaviour (cf. Section 11.3). In some unpublished numerical work, Moss (personal communication) has studied the effects of local instabilities on global decay rates, replacing the microdiffusivity by an analogous macrocoefficient based on the local instability growth rates. He concludes that as long as the instabilities are truly local, the global decay rates are little affected. We shall proceed by taking cognizance of the important but necessarily limited results to date, and argue tentatively that there do exist large-scale stellar magnetic fields which are either strictly dynamically stable, or for which recurrent local instabilities have little effect on the global decay (cf. Section 5.6).

Hydromagnetic instabilities may reduce somewhat the effectiveness of a magnetic field in reversing a rotational shear, sometimes replacing the estimate (5.17) by rather longer times. This will come up in Chapter 8, when we consider the suggestion that the Sun's radiative core may have a long-lived strong local shear.

5.4 Laminar meridian flow in radiative domains

A subadiabatic stellar domain may develop slow, large-scale laminar motions if subject to non-spherical perturbations such as centrifugal force or magnetic force. The seminal discussion in this area is that by H. von Zeipel (see Eddington 1926, 1959), who considered a uniformly rotating, chemically homogeneous, non-magnetic star in strict radiative equilibrium. The equation (5.11) of hydrostatic support can be written in terms of the joint gravitational–centrifugal potential:

$$\psi \equiv V + \Omega^2 \varpi^2/2 \tag{5.40}$$

and

$$\nabla p/\rho = \nabla\psi. \tag{5.41}$$

Poisson's equation can then be written

$$\nabla^2\psi = -4\pi G\rho + 2\Omega^2. \tag{5.42}$$

From (5.41), it follows that p is constant on 'level surfaces' $\psi =$ constant: i.e.

$$p = p(\psi), \qquad \rho(\psi) = \mathrm{d}p/\mathrm{d}\psi, \qquad T = (\mu/R)(p/\rho) = T(\psi). \tag{5.43}$$

The flux of radiation transported through a medium of opacity $\kappa(p, \rho)$ is

$$\mathbf{F} = -(4acT^3/3\kappa\rho)\,\nabla T = -f(\psi)\,\nabla\psi, \tag{5.44}$$

where

$$f(\psi) = (4acT^3/3\kappa\rho)\,\mathrm{d}T/\mathrm{d}\psi. \tag{5.45}$$

This result – that $\mathbf{F}$ varies over a level surface like the effective gravity $\nabla\psi$ – is often referred to as 'von Zeipel's theorem'. The net efflux of radiation per unit volume is

$$\nabla\cdot\mathbf{F} = -f(\psi)\,\nabla^2\psi - f'(\psi)(\nabla\psi)^2. \tag{5.46}$$

Now impose the condition of strict radiative equilibrium – meaning that the local nuclear energy generation $\rho\epsilon$ per cm^3 is balanced by the net efflux of radiation:

$$\rho\epsilon = \nabla\cdot\mathbf{F}. \tag{5.47}$$

With the help of (5.42), this leads to

$$[\rho\epsilon - (4\pi G\rho - 2\Omega^2)f(\psi)] = -f'(\psi)(\nabla\psi)^2. \tag{5.48}$$

The left-hand side of (5.48) is a pure function of ψ and so is constant over level surfaces, whereas the inevitable oblate structure of a uniformly rotating star ensures that $|\nabla\psi|$ decreases monotonically over a level surface from pole to equator. Von Zeipel argued that the left and right sides of (5.48) must therefore vanish separately: $f(\psi)$ is constant through the stellar radiative zone, and

$$\epsilon \propto (1 - \Omega^2/2\pi G\rho). \tag{5.49}$$

This conclusion –'von Zeipel's paradox' – is clearly a *reductio ad absurdum*, for the actual dependence of ϵ on ρ and T is part of the microphysical data, to which the macrophysics has to adjust. The assumption of strict local radiative equilibrium is seen to be too strong a constraint to impose on the system. Similar conclusions hold if Ω is supposed non-uniform but restricted to be of the form $\Omega(\varpi)$, so that again there exists a centrifugal potential $\int \Omega^2\varpi\,\mathrm{d}\varpi$. Relaxation of this restriction to allow Ω to have an arbitrary form $\Omega(r,\theta)$ is in general of no help. The basic reason is that the imposition of a non-radial perturbing force requires the equation of support to satisfy a non-trivial horizontal component in addition to the radial component. If we write the state variables as

$$p = p_0 + p_1(r,\theta), \qquad \rho = \rho_0 + \rho_1(r,\theta), \tag{5.50}$$

$$T = T_0 + T_1(r,\theta), \qquad V = V_0 + V_1(r,\theta), \tag{5.51}$$

(where for the moment p_1 etc. are not necessarily small perturbations), then whereas Poisson's equation, the equation of state and the equation of radiative equilibrium all continue to impose just one condition on p_1 etc., the hydrostatic equation being a vector equation imposes two, so overdetermining the problem. If strict local radiative equilibrium is imposed, then Ω is no longer arbitrarily prescribable but must be allowed to adjust its structure. Typically, Ω is found to be a function of r and θ that varies by a factor ≈ 2 over a radiative envelope (Schwarzschild 1947; Roxburgh 1964).

In fact, one would expect the rotation field to be fixed by dynamical conditions; it is rather the requirement of strict local radiative equilibrium that must be challenged, both in a non-magnetic and in a magnetic star. As pointed out by Vogt (1925) and Eddington (1929), the local excess or defect of the radiative heat supply found by von Zeipel should generate buoyancy forces which cause the gas to move respectively up and down the gradient of the entropy S. In an effectively inviscid radiative zone, the energy equation is

$$c_v\rho\,\mathrm{d}T/\mathrm{d}t = (p/\rho)\,\mathrm{d}\rho/\mathrm{d}t + (\rho\epsilon - \nabla\cdot\mathbf{F}) \tag{5.52}$$

where $\mathrm{d}/\mathrm{d}t \equiv \partial/\partial t + \mathbf{v}\cdot\nabla$ is as usual the derivative following the motion of an element of fluid, and c_v is the specific heat at constant volume. In the simplest example, which may in fact turn out to be the most relevant, the zone is again taken as uniformly rotating and in a steady state, so that (5.52) becomes with use of $\mathcal{R}/\mu = (\gamma - 1)c_v$,

$$\rho A(\psi)(\mathbf{v}\cdot\nabla\psi) = \rho\epsilon + f(\psi)(2\Omega^2 - 4\pi G\rho) + f'(\psi)(\nabla\psi)^2, \tag{5.53}$$

where $\mathbf{v}$ is the circulation velocity, and

$$\begin{aligned} A(\psi) = T\,\mathrm{d}S/\mathrm{d}\psi &= c_v[\mathrm{d}T/\mathrm{d}\psi - (\gamma-1)(T/\rho)\,\mathrm{d}\rho/\mathrm{d}\psi] \\ &= c_v T\,\mathrm{d}[\log(T/\rho^{\gamma-1})]/\mathrm{d}\psi. \end{aligned} \tag{5.54}$$

Division of (5.53) by $|\nabla\psi|$ and integration over a level surface ψ = constant yields

$$\begin{aligned} (\rho\epsilon + f(\psi)(2\Omega^2 - 4\pi G\rho))\int \mathrm{d}S/|\nabla\psi| + f'(\psi)\int|\nabla\psi|\,\mathrm{d}S \\ = \rho A\int \mathbf{v}\cdot\nabla\psi/|\nabla\psi|\,\mathrm{d}S = 0, \end{aligned} \tag{5.55}$$

since in a steady state there can be no net flux of gas across a closed surface. Equation (5.55) fixes the function $f(\psi)$. Substitution back into (5.53) yields the velocity component normal to the level surface:

$$\rho A(\psi)\mathbf{v}\cdot\nabla\psi = \Big[f(\psi)(4\pi G\rho - 2\Omega^2) - \rho\epsilon\Big] \times \left[-1 + |\nabla\psi|^2\frac{\int \mathrm{d}S/|\nabla\psi|}{\int|\nabla\psi|\,\mathrm{d}S}\right]. \tag{5.56}$$

In a convectively stable region $A \propto \mathrm{d}S/\mathrm{d}\psi < 0$, and so the velocity component normal to ψ is upwards if the right-hand side is positive. Since a level surface is shaped roughly like an oblate spheroid, $|\nabla\psi|$ on any particular level surface has its maximum at the poles and decreases monotonically towards the equator; hence the second bracket in (5.56) is positive at the poles and negative at the equator, and with one zero in between.

If the rotation is slow, the deviations p_1 etc. in (5.50) and (5.51) are everywhere small, and perturbation techniques may be applied (Chandrasekhar 1933; Sweet 1950*a*) to determine $\nabla\psi$ and so also the integrals in (5.56) — see Section 9.2. Even in a rapid rotator, the second bracket is clearly of the order of the local value of $\Omega^2 r/g$ which in a uniform rotator is well below unity through the centrally condensed bulk of the star, but may reach about 1/4 at the surface. As the surface regions contain little mass, the structure of a rapidly rotating star may be treated by a combination of small perturbation techniques for the bulk and a Roche approximation for the outer regions (Jackson *et al.* 1971, and references therein).

For definiteness, consider an early-type star with ϵ non-zero only in the convective core, and with the radiative envelope extending to the photosphere. Over the bulk of the star v_r has a simple $\mathrm{P}_2(\cos\theta)$ form, up at the poles and down at the equator, and the speed is of order $(L/Mg)(\Omega^2 r/g) \approx 7.5\ \times$

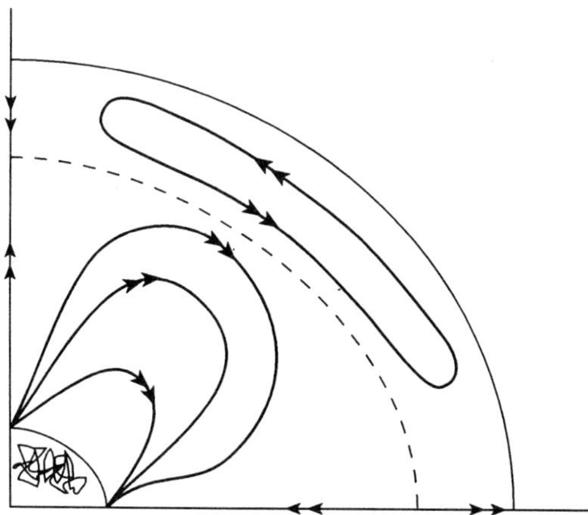

FIG. 5.3. The Eddington–Vogt–Sweet circulation in a uniformly rotating, early-type star.

$10^{-5}(\Omega^2 r/g)(L/L_\odot)(R/R_\odot)^2/(M/M_\odot)^2$ cm s^{-1} – the Kelvin–Helmholtz contraction velocity multiplied by the local centrifugal factor $\Omega^2 r/g$. However, at the level where $\rho = \Omega^2/2\pi G = 4.3 \times 10^{-2}/\mathrm{P}_d^2$ (P_d being the period in days), (5.56) shows that the radial velocity changes sign. Thus the simple theory as developed so far predicts a two-zone structure for the circulation, as in Fig. 5.3, with velocities in the outer zone quadratic in the perturbing parameter but becoming large like $1/\rho$ (Öpik 1951; Mestel 1966).

The case of uniform rotation is rather special. The perturbation treatment developed by Sweet can be applied to determine to first order the instantaneous non-spherical ρ, p, T, V distributions, and so also the local value of $\nabla.\mathbf{F}$, given an arbitrary perturbing field (centrifugal and/or magnetic). If again the system is assumed steady, then (5.52) can be written to first order, in an obvious notation, as

$$c_v \rho_0 T_0 v_r \frac{\mathrm{d}}{\mathrm{d}r}\left(\log \frac{T_0}{\rho_0^{\gamma-1}}\right) = -\nabla \cdot \mathbf{F}_1, \tag{5.57}$$

so fixing the component v_r, whence v_θ follows by continuity. (Note that if Ω is changing due to the circulation, then (5.57) should strictly include a term $c_v\, \rho_0\, \partial T_1/\partial t$, but this would be of second order in perturbing parameters.) In (5.56) we have retained the term $f(\psi)2\Omega^2$ even though this yields a term $\propto \Omega^4$ in $\mathbf{v}$, because the term $4\pi G\rho\, f(\psi)$ will become small with the density near the surface. It turns out that in general there are terms in $-\nabla \cdot \mathbf{F}_1$, *linear* in Ω^2, which are independent of ρ, so justifying for a general perturbing field the dropping of terms $\propto \Omega^4$ in (5.57). Thus except in the very special case of uniform rotation and negligible magnetic forces, the theory predicts for the surface layers velocities larger than those deep down by the factor $\overline{\rho}/\rho$ where $\overline{\rho}$ is the mean stellar density. Again, in a region where Ω is supposed to have a large gradient, the contributions of the derivatives of Ω can make $-\nabla \cdot \mathbf{F}_1$ much larger locally than its average over the domain. It is therefore not obvious that these thermally-driven motions are very slow over the whole star. Further, it is found that the structure of the circulation field as given by (5.57) and the accompanying continuity equation can be quite sensitive to the form of the perturbing field. Thus if Ω is generalized to $\Omega(r)$, the circulation remains of P_2 form but v_r may change sign many times, yielding a set of cellules. This clearly underlines the need to ask how a rotation field can be kept approximately steady, in particular against the advection of angular momentum by the very Eddington–Vogt–Sweet (E–V–S)-type circulation which a general rotation field generates.

The discussion naturally bifurcates at this point. In a problem in which magnetic effects are assumed negligible not only for hydrostatic equilibrium, but also for the distribution of angular momentum, then attention is focused on the viscous forces that will arise if the 'stable' radiative zone is nevertheless weakly 'turbulent'. Zahn (1983) has argued that shear instabilities in a non-uniformly rotating star would quite rapidly force the rotation field to adjust to a nearly radial dependence $\Omega(r)$. One could then picture an initially uniform rotation field evolving via the (E–V–S)-type circulation and Zahn's horizontal instability into the unique form which yields $\nabla \cdot \mathbf{F} = 0$. However, the centrifugal field is non-conservative if $\Omega(r)$ is non-uniform, and it is then subject to other instabilities studied by Goldreich and Schubert (1967) and Fricke (1968, 1969), which depend on heat exchange. Estimates for the effectiveness of this weak turbulence on the rotation field suggest that it will transfer significant angular momentum again in an E–V–S time (Kippenhahn 1969). If the stellar surface is subject to a braking torque (Chapters 7–9), then this will react strongly on the internal rotation field if the braking time is comparable with or shorter than the circulation time (Sakurai 1991; Zahn 1992).

A somewhat different philosophy underlies the treatment of analogous problems in geophysics (e.g. Pedlosky 1979, 1982). Within the rotating frame, it is again assumed that any velocities are slow compared with the rotatory velocities measured in the inertial frame (small Rossby number – cf. Chapters 6 and 8). Away from boundary layers the ratio of frictional to Coriolis force (measured by the Ekman number) is certainly small if micro- rather than turbulent viscosity is adopted. The basic *geostrophic* approximation is taken to be a balance between the Coriolis force acting on the slow velocities, the effective gravity and the pressure gradient. If the depth of the ocean or atmosphere is small, then the vertical component reduces to normal hydrostatic balance, and in an axisymmetric system, the θ-component is made to fix the toroidal velocity by the balance of the Coriolis component against the the pressure gradient per unit mass. Equivalently, the equation of motion *enforces* a non-uniform rotation as seen in the inertial frame by an amount proportional to $\partial p/\partial \theta$, which in turn depends on the horizontal temperature gradient, whence the name 'thermal wind' given to the shear fixed in this way. There is a partial analogy with the Schwarzschild–Roxburgh solution for a star in radiative equilibrium, in that the *toroidal* shear is determined for a state with negligible poloidal velocities. The point to be stressed is that in the absence of even a weak magnetic field, and also with friction assumed weak, again the system can develop shears without experiencing a strong dynamical back reaction.

The problem is radically altered when even a modest magnetic field is introduced. We have seen in Section 5.2 how a field that has negligible effects on the poloidal equilibrium of a star can nevertheless react strongly on the rotation field in well under a stellar lifetime. In the simplest problem, thermal effects enter through the non-vanishing of $\nabla \cdot \mathbf{F}$, which yields the E–V–S meridional circulation, and this advects angular momentum (equivalently, yields a toroidal Coriolis force component in the rotating frame). The crucial difference is that now the toroidal shear generates magnetic torques which react back strongly on the rotation field and so cannot be dropped from the toroidal component of the equation of motion. Prima facie, for the field to be able to offset the transfer of angular momentum by the E–V–S circulation, one requires that the local velocity of a transverse Alfvén wave should exceed the circulation speed. Over the bulk of the star, this requires B_{p} to be no greater than $\approx 3 \times 10^{-4}(\bar{L})(\bar{R}/\bar{M}^3)^{1/2}$ G even in a rapid rotator. A formal treatment of this problem is given below.

A stronger field can contribute significantly to hydrostatic equilibrium, especially in the low-density surface regions, and so modify the thermally-driven circulation. This is discussed further in Section 9.4.

5.5 The interaction between rotation, magnetism and circulation

5.5.1 *Steady-state integrals*

We now study the modifications to the steady-state integrals (5.19) and (5.21) due to flow $\mathbf{v}_{\mathrm{p}}$ in meridian planes (Chandrasekhar 1956*b*; Mestel 1961). In general, $\mathbf{v}_{\mathrm{p}}$ may be a complicated functional of both the centrifugal and magnetic forces. As we are concerned to elucidate the hydromagnetic consequences, we keep the discussion simple by adopting for $\mathbf{v}_{\mathrm{p}}$ an equatorially and axially symmetric one-zone structure. The total velocity field is now written as the sum of poloidal and toroidal components:

$$\mathbf{v} = \mathbf{v}_{\mathrm{p}} + \mathbf{v}_{\mathrm{t}} = \mathbf{v}_{\mathrm{p}} + \Omega \varpi \mathbf{t} \tag{5.58}$$

where $\Omega(\varpi, z)$ is the local angular velocity. If the perfect conductivity approximation is adequate, then in a steady state

$$\mathbf{v} \times \mathbf{B} \equiv (\mathbf{v}_{\mathrm{p}} \times \mathbf{B}_{\mathrm{p}}) + (\mathbf{v}_{\mathrm{p}} \times \mathbf{B}_{\mathrm{t}} + \mathbf{v}_{\mathrm{t}} \times \mathbf{B}_{\mathrm{p}}) = c\nabla\phi. \tag{5.59}$$

In an axisymmetric state, the gradient of the single-valued electric potential ϕ can have no toroidal component, so $\mathbf{v}_{\mathrm{p}} \times \mathbf{B}_{\mathrm{p}}$ must vanish, yielding

$$\mathbf{v}_{\mathrm{p}} = \kappa \mathbf{B}_{\mathrm{p}} \tag{5.60}$$

where κ is a scalar function of position. With the circulation field $\mathbf{v}_\mathrm{p}$ supposed prescribed, then (5.60) also fixes the shape of the poloidal field lines. In particular, an equatorially symmetric $\mathbf{v}_\mathrm{p}$ constrains $\mathbf{B}_\mathrm{p}$ to have the topology of a quadrupolar rather than a dipolar field; and equally, a circulation field flowing horizontally just beneath the photosphere would not be consistent with flux emanating from the surface. The obvious conclusion is that near the equator and photosphere, one must in general include explicitly the finite resistivity which will allow deviation from (5.60). With this caveat (cf. the discussion below and in Chapter 9), we nevertheless adopt (5.60) as applicable to the bulk of a radiative zone.

The poloidal component of (5.59) is

$$c\nabla\phi = (\Omega\varpi - \kappa B_\phi)\mathbf{t} \times \mathbf{B}_\mathrm{p} = -(\Omega - \kappa B_\phi/\varpi)\,\nabla P \tag{5.61}$$

on use of (5.60) and (5.5). Thus $\phi = \phi(P)$ – the poloidal field lines are equipotentials – and

$$\Omega - \kappa B_\phi/\varpi = -c\,\mathrm{d}\phi/\mathrm{d}P \equiv \alpha(P). \tag{5.62}$$

In this generalization of Ferraro's law (5.19), Ω is now no longer uniform along each field line, but the consequent change in B_ϕ due to twisting of the poloidal field is offset by the advection of B_ϕ by the meridian circulation. Equivalently, one can write (5.60) and (5.62) jointly as

$$\mathbf{v} = \kappa\mathbf{B} + \varpi\alpha(P)\mathbf{t}\,: \tag{5.63}$$

in the presence of meridian circulation, the general solution of the perfect conductivity condition consists merely of the addition to isorotation of a flow parallel to the total field $\mathbf{B} = \mathbf{B}_\mathrm{p} + \mathbf{B}_\mathrm{t}$. Each field line continues to be swung round at its proper angular velocity $\alpha(P)$, but the gas velocity has in addition the parallel flow $\kappa\mathbf{B}$.

The continuity equation relates the density ρ to the scalar κ:

$$0 = \nabla\cdot(\rho\mathbf{v}) = \nabla\cdot(\rho\mathbf{v}_\mathrm{p}) = \nabla\cdot(\rho\kappa\mathbf{B}_\mathrm{p}) = \mathbf{B}_\mathrm{p}\cdot\nabla(\rho\kappa) \tag{5.64}$$

or

$$\rho v_\mathrm{p}/B_\mathrm{p} = \rho\kappa \equiv \eta(P) = \text{constant on each field–streamline.} \tag{5.65}$$

The advection of angular momentum by the circulation is balanced by the magnetic torque (cf. (5.20)):

$$\rho\mathbf{v}_\mathrm{p}\cdot\nabla(\Omega\varpi^2) = \varpi\mathbf{t}\cdot(\nabla\times\mathbf{B}_\mathrm{t})\times\mathbf{B}_\mathrm{p}/4\pi = \mathbf{B}_\mathrm{p}\cdot\nabla(\varpi B_\phi/4\pi), \tag{5.66}$$

whence from (5.60) and (5.65)

$$-\varpi B_\phi/4\pi + \rho\kappa\Omega\varpi^2 = -\beta(P)/4\pi. \tag{5.67}$$

This is the generalization of (5.21): there is a constant flow of angular momentum along a unit flux tube, carried in part by the circulation and in part by the moment of the Maxwell stresses.

Equations (5.62), (5.65) and (5.67) combine into

$$\Omega = (\alpha + \eta\beta/\rho\varpi^2)/(1 - 4\pi\eta^2/\rho), \tag{5.68}$$

$$\varpi B_\phi = (\beta + 4\pi\eta\alpha\varpi^2)/(1 - 4\pi\eta^2/\rho), \tag{5.69}$$

where

$$4\pi\eta^2/\rho = 4\pi\rho v_{\mathrm{p}}^2/B_{\mathrm{p}}^2 = v_{\mathrm{p}}^2/v_A^2. \tag{5.70}$$

As expected, the local ratio of the circulation speed to the Alfvén speed v_A (defined by the poloidal field B_{p}) appears automatically. In the special case with $\mathbf{B}_{\mathrm{p}}$ of quadrupolar topology, and with field lines closing in the radiative zone, the toroidal flux $\mathrm{d}F_{\mathrm{t}}$ linked by segments of two neighbouring poloidal field lines is again given by (5.23); likewise, the angular momentum within the solid defined by rotating these loops about the axis is

$$\mathrm{d}J = 2\pi \int\int \rho(\Omega\varpi^2)\varpi \,\mathrm{d}l\,\mathrm{d}s = \mathrm{d}F_{\mathrm{p}} \oint (\rho\Omega\varpi^2/B_{\mathrm{p}})\,\mathrm{d}s. \tag{5.71}$$

Substitution of (5.62) and (5.67) into (5.23) and (5.71) shows how the parameters α and β, which enter as constants of integration for the field line considered, are related simply to $\mathrm{d}F_{\mathrm{t}}/\mathrm{d}F_{\mathrm{p}}$ and $\mathrm{d}J/\mathrm{d}F_{\mathrm{p}}$.

We have already seen that even a field as low as 10^{-2} G would yield $4\pi\eta^2/\rho \approx 10^{-4}$ over the bulk of a stellar radiative envelope, so the denominator in (5.68) and (5.69) is effectively unity. If $|\beta| < 4\pi\eta\alpha\varpi^2$, then $|\eta\beta|/\alpha\rho\varpi^2 \leq 4\pi\eta^2/\rho$ is also much less than unity, and (5.68) predicts only a small deviation from the isorotational value $\Omega = \alpha$. If instead $|\varpi B_\phi| \approx |\beta| \gg 4\pi\eta\alpha\varpi^2$, then (5.68) still yields Ω/α significantly different from unity only if $|B_\phi/B_{\mathrm{p}}| \approx |\beta|/\varpi B_{\mathrm{p}} \geq \alpha\rho\varpi^2/\eta\varpi B_{\mathrm{p}} \approx \alpha\varpi/v_{\mathrm{p}}$, which is typically 10^{11} or so. Such an enormous ratio of toroidal to poloidal field strength is almost certainly ruled out on stability grounds (cf. Section 5.3), and in any case, if B_{p} were greater than 10^{-2} G, the B_ϕ component would have too much energy to be gravitationally contained by the star. Thus the analysis confirms the conclusion of the last section: as long as $v_{\mathrm{p}}^2/v_A^2 \ll 1$, then the advection of angular momentum by the flow $\mathbf{v}_{\mathrm{p}}$ along the $\mathbf{B}_{\mathrm{p}}$-lines is easily offset by the torque exerted by a small toroidal twist in the field $\mathbf{B}_{\mathrm{p}}$. If also the parameter α is the same along different field lines (cf. below), then the magnetic torques keep the whole radiative zone rotating nearly uniformly. The E–S circulation field constructed in Section 5.4 assumes Ω uniform, and also neglects the effect of the poloidal magnetic forces on the $\rho - T$ field. This is prima facie an excellent approximation provided $B_{\mathrm{p}}^2 \ll 4\pi\rho\Omega^2\varpi^2$, a condition satisfied over the bulk of the star, and also manifestly consistent with $4\pi\rho v_{\mathrm{p}}^2 \ll B_{\mathrm{p}}^2$, since the estimated $v_{\mathrm{p}} \ll \Omega\varpi$.

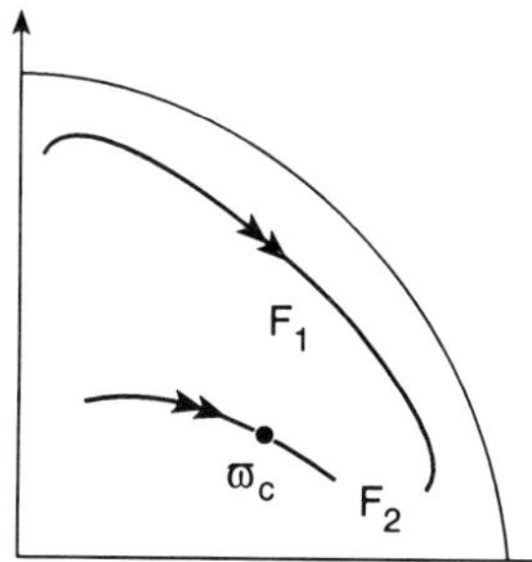

FIG. 5.4. 'Equatorial acceleration' in a rotating magnetic star with meridian circulation.

Again, the discussion will need modification in the low-density surface regions, where the magnetic force density may contribute significantly to hydrostatic equilibrium – cf. Section 9.4. The crucial point we wish to emphasize is that the introduction of a weak magnetic field into the classical problem of thermal balance in a uniformly rotating star allows the construction of a model satisfying all dynamical constraints.

5.5.2 *Equatorial acceleration*

We now digress for a moment to note some more general conclusions from the solutions (5.68) and (5.69). If the local circulation and Alfvén speeds were to become equal, at a point of density $\rho' = 4\pi\eta^2$, then non-singular Ω and B_ϕ would require that simultaneously the numerators vanish: $\beta = -4\pi\eta\alpha\varpi'^2$, where $\varpi' = r'\sin\theta'$ is the axial distance of such an 'Alfvénic' point (r', θ'). However, a quadrupolar-type field line would then in general have $\rho' = 4\pi\eta^2$ at two points, since ρ is primarily a function of r even in a rotationally distorted star; but these points will have different values of ϖ', so that the numerators cannot vanish simultaneously with the denominators at both such points. Thus for a dissipation-free steady state as discussed above to be possible, v_p must be either sub-Alfvénic or super-Alfvénic all the way round a poloidal loop.

Now suppose that the square of the 'Alfvénic Mach number' $= v_\mathrm{p}^2/v_\mathrm{A}^2$ is less than unity everywhere, and consider a loop which approaches the photosphere near the pole, and so has a nearly horizontal stretch F_1 from pole to equator (Fig. 5.4). Along the horizontal part ρ is nearly constant, and so the variation of Ω in (5.68) is due just to the increase in ϖ; and if $\eta\beta/\alpha$ is negative, Ω/α increases fron pole to equator – 'equatorial acceleration'.

This result can be generalized to apply to other field lines, such as F_2 in Fig. 5.4, in which again the axial distance ϖ increases monotonically, provided again that β/α is negative and can be written as $-4\pi\eta\alpha\varpi_c^2$, where ϖ_c is a distance to a point within the star. With $4\pi\eta^2/\rho \ll 1$, (5.68) can be approximated by $\Omega/\alpha = 1 + (4\pi\eta^2/\rho)(1 - \varpi_c^2/\varpi^2)$. Beyond $\varpi = \varpi_c$ there is again an outward increase in Ω, assisted by the increase in $4\pi\eta^2/\rho$; and in fact Ω/α may continue to decrease inwards within ϖ_c, depending on the precise rate of variation of ρ along the field line considered.

The prediction of equatorial acceleration is surprising, because one could have argued qualitatively as follows. If there is no magnetic constraint $(v_{\rm p}^2/v_{\rm A}^2 \to \infty)$, then in a steady state of circulation, each element of gas conserves its angular momentum: $\Omega\varpi^2$ = constant, predicting strong equatorial *de*celeration. When on the other hand the star has such a strong magnetic field that $v_{\rm p}^2/v_{\rm A}^2 \to 0$, then the magnetic stresses keep Ω effectively uniform. With $v_{\rm p}^2/v_{\rm A}^2 \leq 1$, one expects something in between: a modest equatorial deceleration in a steady state. The fallacy in this intuitive argument is the implicit assumption of a continuum of dissipation-free steady states, whereas we have seen that there is a gap, comprising the parameter range in which $(v_{\rm p}^2/v_{\rm A}^2 - 1)$ changes sign around a loop, for which no solutions exist. Thus the result, though unexpected, does not contradict any basic principles.

As noted above, the magnitude and sign of the parameter $\eta\beta/\alpha$ is related to the total toroidal flux $\mathrm{d}F_{\rm t}$ threading neighbouring poloidal loops and to the angular momentum $\mathrm{d}J$ in the associated volume. One very special case which yields $\eta\beta/\alpha < 1$ has $\mathrm{d}F_{\rm t} = 0$: the toroidal field is due purely to the presence of non-uniform rotation engendered by the circulation. Equivalently, the lines of the total field $\mathbf{B}_{\rm p}+\mathbf{B}_{\rm t}$ close in space, with zero net change in the azimuthal angle ϕ as one travels all the way round the loop. However, this is only one example of the parameter range that yields equatorial acceleration. Systems with non-ergodic fields lines form a set of measure zero; and as discussed in Section 5.3, a $\mathbf{B}_{\rm p}$-field without a comparable amount of linking toroidal flux would be unstable.

The interesting unanswered question is whether the surprising result found for the simple axisymmetric system – in which there is no angular momentum interchange between neighbouring field lines – contains the hint of an explanation of the observed equatorial acceleration in the Sun and other cosmical bodies. One's instinctive feeling is that, for the bulk of the solar convective zone, it is the Reynolds stresses of the inhomogeneous and anisotropic turbulence that dominate in determining the distribution of angular momentum, as discussed in Section 4.2, with the Lorentz force exerted by a dynamo-generated field just a perturbation, showing up in the modulation of the Ω-field over a solar cycle (cf. Chapter 8). Nevertheless, the tendency of the turbulence sometimes to concentrate the flux into quasi-laminar tubes, with nearly field-free flow outside (Sections 4.3 and 4.5), may suggest that the conclusions drawn from a hydromagnetic model with purely laminar flow need not be irrelevant to the dynamics of such tubes, even if the negligible microresistivity is replaced by the much larger turbulent resistivity. And we recall from Chapter 4 that our picture of convection is undergoing significant changes, so that arguments based on the earlier models may need scrutiny.

5.5.3 *The approach to a quasi-steady state*

Return now to the problem with $4\pi\eta^2/\rho \ll 1$, so that $\Omega/\alpha \approx 1$ is indeed the appropriate steady state solution for each field line. As in the work of Section 5.2, the constraint of strict axisymmetry is necessary for the existence of steady state solutions in which the parameters for the different field lines are independent. For the moment we retain axisymmetry but consider the effect of dissipative processes. We have noted already that if the stellar magnetic field has a dipolar component, then in a steady state there is necessarily flow across $\mathbf{B}_\mathrm{p}$ – allowed by finite resistivity – in the sub-photospheric and equatorial regions. Near the equator the slow $\mathbf{v}_\mathrm{p}$ requires only a modest local $\nabla \times \mathbf{B}_\mathrm{p}$. The analogous problem in the low-density surface regions, discussed briefly in Chapter 9, is more complex, but again there do not seem to be any essential difficulties.

The slow trans-field flow necessarily causes exchange of angular momentum between different field lines. A much stronger coupling occurs between field lines which penetrate at least the surface regions of a neighbouring convective domain, e.g. the convective core of an early-type magnetic star, or the envelope of a solar-type star. In a strictly axisymmetric system, the field lines such as QPQ′ in Fig. 5.5 would be expected to rotate at their bases with the local angular velocity of the convective zone – i.e. any latitude-dependence would be propagated into the neighbouring radiative zone. The simplest example takes the convective zone to rotate rigidly. One can then formulate the appropriate time-dependent problem to study the approach to an asymptotic steady state from an arbitrary initial rotation field in the envelope (Moss *et al.* 1990; Moss 1992, following earlier work by Mestel *et al.* 1988 and Tassoul and Tassoul 1989). Both finite resistivity and viscosity are included in the equations for the radiative zone. For completeness, a meridian circulation field of prescribed (P_2) structure and magnitude is included. As already noted, this is clearly an approximation, as the actual circulation field is a complicated functional of the perturbing forces; however, in a realistic case the evolution of the rotation field is determined far more by magnetic and dissipative effects than by the slow circulation. The results depend to some extent on the parameter values, but the general conclusions are as follows. An initial non-isorotational Ω-field sets up torsional Alfvén waves in the radiative zone along the poloidal field, as discussed in Section 5.2. The waves do not interchange angular momentum between different field lines but redistribute it along the field lines. Damping by Ohmic dissipation – especially in the cooler outer regions – makes the rotation along each field line approach a form quite close to (5.68), with α and β again determined by the initial angular momentum and toroidal flux distributions (though the finite resistivity will not only damp the waves but cause relative diffusion of toroidal and poloidal flux). Any viscous interaction within the radiative zone will cause slow interchange of angular momentum, but a much more powerful interchange occurs through coupling with the convective core with its strong eddy viscosity. The evolution of the rotation field towards its asymptotic state of near uniformity occurs through states described approximately by (5.68) and (5.69) with steadily changing α

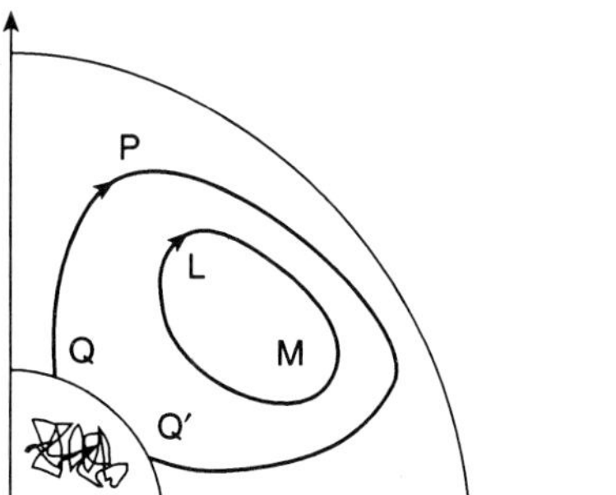

FIG. 5.5. Axially symmetric radiative zones: field lines like QPQ′ are coupled with the convective core; those like LM are uncoupled.

and β. Ambiguities persist, especially in the correct description of the boundary layer between convective and radiative domains, comprising the normal overshoot layer and the depth of penetration of the field lines, but the numerical results seem fairly insensitive to this.

In a strictly axisymmetric system, one could envisage a steady state as in Fig. 5.5, in which individual field lines such as LM retain their individual rotation rates, while those like QPQ′, coupled to the convection zone, are forced to corotate with it. However, such a state would probably not persist because of instabilities. Recall (from Section 5.4) Zahn's argument (1983) that in a non-magnetic star, shear instabilities would cause exchange of angular momentum over an equipotential surface, so tending to set up an near radially-dependent rotation law $\Omega(r)$. Balbus and Hawley (1994) argue that likewise in a star with a radial magnetic field component, essentially the same instability that occurs in magnetic accretion discs (Section 10.5) will act to wipe out horizontal shears. And as seen in Section 5.2.2, with a more realistic, non-axisymmetric, structure as in the oblique rotator, one expects all field lines to be dynamically coupled, so that in a steady state, differential rotation of different field lines is forbidden.

The numerical work has confirmed the expectation that a slow circulation, with $v_{\mathrm{p}}/v_{\mathrm{A}} \ll 1$ – whether wholly or only partly along the poloidal field – is of far less importance for the long-term evolution of the rotation field than either the magnetic torques, or the turbulent viscosity in a convective region. This encourages the belief that in considering the long-term effects of circulation in a rotating early-type star with a dynamically significant (though not necessarily observable) **B**-field, the case of nearly uniform rotation is astronomically the most reasonable as well as the simplest. It is worth noting that to account for the element abundances in rapidly rotating, early-type stars, one may need a steady laminar circulation, but without any associated turbulence (Charbonneau and Michaud 1988). Weak magnetic fields can easily be strong enough to keep the rotation nearly uniform, so that not only does the E–V–S P_2-type circulation persist, but also there is no tendency for shear turbulence to develop. For further work in this area, see Charbonneau and MacGregor 1992 and MacGregor and Brenner 1991.

5.6 Ohmic decay of primeval magnetic fields

5.6.1 *Decay of a purely poloidal field*

The decay of a poloidal stellar magnetic field without energy sources was first discussed by Cowling (1945). The rough estimate (2.96) yields 5×10^9 yr for the characteristic decay time of a large-scale stellar field in a radiative domain, with the maintaining currents flowing in regions of high temperature and so high microresistivity. This suggested that a 'fossil' theory of, for example, the solar magnetic field could not be ruled out from the start on energetic grounds. The first observation by Babcock (1959) of the reversal of the general solar field, and the subsequent confirmation that the poloidal field and the sunspots are both components of the solar cycle, forces one to think rather of an oscillatory dynamo model for solar-type stars (Chapters 6 and 8). However, a form of the fossil theory remains a plausible hypothesis for the fields of the early-type strongly magnetic stars, interpreted in terms of the oblique rotator model (Chapter 9). The necessary modifications to the theory, enforced by the stability considerations of Section 5.3, justify reappraisal of the decay problem.

From Faraday's law, an axisymmetric decaying poloidal field $\mathbf{B}_\mathrm{p}$, with the vector potential $\mathbf{A} = A\mathbf{t}$, induces a toroidal electric field

$$\mathbf{E} = -\partial \mathbf{A}/c\,\partial t \tag{5.72}$$

(there being no $\nabla\phi$ contribution because of the axisymmetry). Coupling with Ohm's law $\mathbf{E} = \mathbf{j}/\sigma$ for a stationary medium yields the decay equation

$$\partial \mathbf{A}/\partial t = -(c^2/4\pi\sigma)\nabla \times (\nabla \times \mathbf{A}) \tag{5.73}$$

or from (5.6),

$$\partial P/\partial t = -\eta\varpi[\nabla \times (\nabla P \times \mathbf{t}/\varpi)] \cdot \mathbf{t} = \eta D^2 P, \tag{5.74}$$

where the resistivity $\eta = (c^2/4\pi\sigma)$, and D^2 is the Stokes operator, defined by

$$D^2 P = \varpi(\nabla^2 - 1/\varpi^2)(P/\varpi) = (\nabla^2 - (2/\varpi)\partial/\partial\varpi)P = \nabla \cdot \mathbf{Q}, \tag{5.75}$$

where

$$\mathbf{Q} = \nabla P - (2/\varpi)P\hat{\varpi}, \tag{5.76}$$

and $\hat{\varpi}$ is the ϖ-coordinate unit vector.

In spherical polar coordinates (r, θ, ϕ), with $\mu = \cos\theta$, (5.74) becomes

$$\frac{\partial P}{\partial t} = \eta\left(\frac{\partial^2 P}{\partial r^2} + \frac{1 - \mu^2}{r^2}\frac{\partial^2 P}{\partial \mu^2}\right) \tag{5.77}$$

The eigensolutions of this equation are

$$P_{nl} = -\exp(-t/\tau_{nl})f_{nl}(r)(1-\mu^2)\,\mathrm{dP}_l/\mathrm{d}\mu \tag{5.78}$$

with

$$f''_{nl} - \frac{l(l+1)}{r^2}f_{nl} + \frac{f_{nl}}{\eta\tau_{nl}} = 0. \tag{5.79}$$

The conductivity can be written from (2.73) as $\sigma = \sigma_c(T/T_c)^{3/2}$, where T_c is the central temperature. If one uses the 'mathematical' boundary condition $T \to 0$ at the stellar surface, then Ohm's law enforces $(\nabla\times\mathbf{B}) \propto \mathbf{j} = 0$ outside the star. The P_{nl}-eigenmode then behaves externally like $r^{-l}(1-\mu^2)\,\mathrm{dP}_l/\mathrm{d}\mu$, yielding the boundary condition $(rf'_{nl} + lf_{nl})_R = 0$ at the surface R for the solution of (5.79). The warm or hot corona surrounding a star will not in fact have a low σ, and there are ample electrons available to carry the currents demanded by the macroscopic theory. However, because of the sharp outward decline in density above the photosphere, the ratio to the magnetic energy density of the thermal, gravitational and kinetic energy densities may become very small above the surface, so that the magnetic field adjusts to satisfy the force-free condition (cf. Section 3.4)

$$\nabla\times\mathbf{B} = k\mathbf{B}, \tag{5.80}$$

a stronger constraint than the torque-free condition (5.21); and if the field is purely poloidal, (5.80) reduces to the curl-free condition.

If (5.79) is applicable throughout the star, the requirement of finiteness at the centre enforces $f \propto r^{l+1}$. For an early-type star, a more plausible condition (cf. Section 5.8.2) is that the field is expelled from the convective core, so that $B_r = 0$ at the core boundary r_c, i.e. $f(r_c) = 0$; while for a late-type star, with a convective envelope bounded below at a radius r_c, $f(r_c) = 0$ and $f \propto r^{l+1}$ as $r \to 0$. In all cases, corresponding to a prescribed angular dependence $(1-\mu^2)\,dP_l/d\mu$, (5.79) and the appropriate boundary conditions fix the normalized eigenfunctions f_{nl} and the associated decay times τ_{nl}. As usual for a Sturm–Liouville problem, the nth eigenfunction f_{nl} has $(n-1)$ nodes within the star.

In the problem with σ uniform through a sphere R, studied by Lamb (1883) as an example in classical electromagnetism – 'the decay of eddy currents' – the functions f reduce to spherical Bessel functions. The Lamb–Cowling eigenfunctions are sometimes used as the base for the expansion of fields constructed to satisfy the dynamo equation (cf. Chapter 6).

As the simplest example, consider an early-type star with a dipolar field ($l = 1$). The most slowly decaying eigenmode is for $n = 1$, with the radial eigenfunction having no zeros. By (5.5) and (5.78),

$$\mathbf{B} = B_s\exp(-t/\tau_{11})\left(2\frac{f_{11}}{x^2}\cos\theta,\ -2\frac{f'_{11}(x)}{2x}\sin\theta,\ 0\right), \tag{5.81}$$

where $x = r/R$ and f_{11} is normalized so that B_s is the polar field strength at the surface. Figure 5.6 shows $2f_{11}/x^2$ plotted against x for the case when the

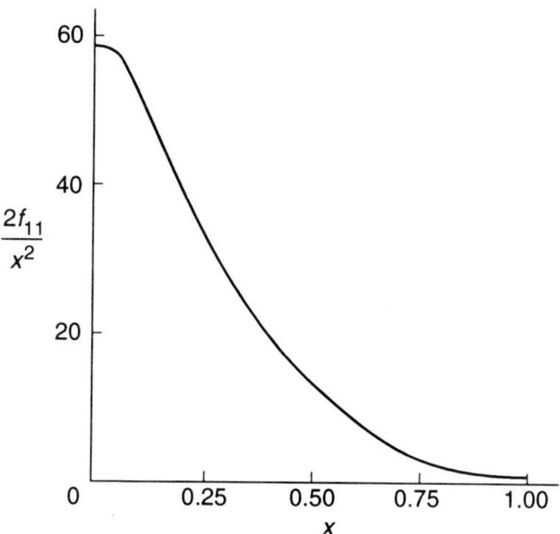

FIG. 5.6. Values of $2f_{11}/x^2$ against x for the slowest decaying Cowling eigenmode, assumed to penetrate the core.

field is assumed to penetrate the core, as in Cowling's original treatment. The precise e-folding time depends to some extent on the boundary conditions and the details of the stellar model; but a lifetime of $\approx 5 \times 10^9$ yr is typical. The total field strength decreases over the star by a factor of between 60 and 80.

An arbitrary initial poloidal field can be expanded in terms of the eigensolutions (5.78) with $t = 0$, and if its evolution were determined purely by Ohmic decay, its subsequent structure would be given by the same series with the appropriate $e^{-t/\tau_{nl}}$ factors introduced. Because of the strong decrease in σ towards the cool surface regions, in an early-type star (without a confining convective envelope) the currents maintaining the principal modes are still strongly concentrated to the stellar interior, so that $\mathbf{B}_{\mathrm{p}}$ is still nearly curl-free well below the surface.

5.6.2 *Decay of a mixed poloidal–toroidal field*

The identical problem for a purely toroidal field $\mathbf{B}_{\mathrm{t}}$ yields the decay equation

$$\begin{aligned}\frac{\partial B_\phi}{\partial t} &= -\mathbf{t}\cdot\nabla\times[\eta\nabla\times(B_\phi\mathbf{t})]\\ &= \frac{\partial}{\partial\varpi}\left(\eta\frac{\partial(\varpi B_\phi)}{\varpi\,\partial\varpi}\right)+\frac{\partial}{\partial z}\left(\eta\frac{\partial B_\phi}{\partial z}\right)\end{aligned} \tag{5.82}$$

For an early-type star, the surface boundary condition is again that of continuity with an external curl-free field, but now the vanishing of $\nabla\times\mathbf{B}_{\mathrm{t}} = \nabla(\varpi B_\phi)\times\mathbf{t}/\varpi$ requires that B_ϕ itself vanish to avoid a singularity on the axis. Again, in the principal eigenmodes the currents are concentrated deep in the star, but – in striking contrast to the poloidal problem – a weak $\mathbf{j}_\mathrm{p}$ implies a weak $B_\phi\mathbf{t}$, so that the toroidal field eigenmodes are much more strongly concentrated into the high temperature interior. As already emphasized, a field that is either purely poloidal or purely toroidal is dynamically unstable, so that an acceptable version of the fossil theory must therefore consider the simultaneous decay of both components. However, one cannot merely superpose the appropriate poloidal and toroidal solutions and allow them to decay independently of each other. Even if the composite field is initially torque-free, the very different distributions through the star of the $\mathbf{B}_\mathrm{p}$ and $\mathbf{B}_\mathrm{t}$ decay eigenmodes will ensure that evolution under Ohmic decay *alone* would quickly lead to a poloidal current field $\mathbf{j}_\mathrm{p} = (c/4\pi)\nabla \times \mathbf{B}_\mathrm{t}$ having in general a component across $\mathbf{B}_\mathrm{p}$. Such a field would immediately send out torsional Alfvén waves along each poloidal loop. Again, waves along field lines which enter a convective envelope will be rapidly damped, and even along loops that are fully in a radiative zone, the same resistivity that is causing the decay will also damp out the waves. For any reasonable field strength, both the travel time and the damping time are far shorter than the decay times of the largest scale modes. With the advection of angular momentum by the slow circulation negligibly slow over the bulk of a radiative zone, one is led to a picture in which the poloidal field decays essentially as in the Cowling calculation, but with the decaying toroidal loops instantaneously adjusting themselves so that the torque-free condition (5.21) is satisfied in the mean, possibly with some as yet undamped, weak torsional oscillations superposed. The elimination in the mean of the component of $\mathbf{j}_\mathrm{p}$ perpendicular to $\mathbf{B}_\mathrm{p}$ is easily achieved by a slight deviation $\tilde{\Omega}$ from isorotation, so that the resulting motional induction term cancels the component $\mathbf{E}_\perp$ of the electric field associated with the decay of $B_\phi\mathbf{t}$: $\tilde{\Omega} = (\mathbf{E}_\mathrm{p} \times \mathbf{B}_\mathrm{p})\cdot\mathbf{t}/B_\mathrm{p}^2\varpi$, the familiar $\mathbf{E}\times\mathbf{B}$ drift. (Only very near a neutral point of $\mathbf{B}_\mathrm{p}$ does the argument fail.)

The dynamical constraint (5.21) is paramount, forcing the poloidal currents to follow closely the lines of $\mathbf{B}_\mathrm{p}$. In a late-type star, where any such fossil field is in any case confined beneath the deep outer convection zone, this is of little consequence; but in an early-type star, the currents along field lines that reach and penetrate the surface are forced to flow through the cooler outer envelope, there to suffer higher Ohmic decay, and then into the coronal regions. For these stars the surface boundary condition is also modified, as the constraint on the field in the low-density exterior is that it satisfies the force-free condition (5.80) as well as the torque-free condition (5.21).

The simplest example studied (Mestel and Moss 1983*a*) adopts a prescribed structure for the poloidal field, e.g. that of Cowling's slowest decaying mode, described by its flux function P with $\exp(-t/\tau)$ removed. The toroidal decay problem now has as its dependent variable not the local value of B_ϕ, but the function $\mathrm{d}F_\mathrm{t}$ of (5.23), measuring the total toroidal flux between P and $P+\mathrm{d}P$, and with the torque-free constraint imposed in the form

$$\varpi B_\phi = \beta(P,t), \qquad \beta(0,t) = 0, \tag{5.83}$$

where the time-dependence due to Ohmic diffusion is made explicit. Equation (5.82) is integrated over the area between $P, P + \mathrm{d}P$ to yield

$$\partial(\mathrm{d}F_{\mathrm{t}})/\partial t = \partial\left(\int B_\phi \,\mathrm{d}s\,\mathrm{d}l\right)/\partial t = \oint^{P+\mathrm{d}P} [\eta\nabla(\varpi B_\phi) \times \mathbf{t}/\varpi]\cdot \mathrm{d}\mathbf{s}$$
$$= \mathrm{d}P\frac{\partial\left(\oint \eta\nabla(\varpi B_\phi)\cdot \mathbf{t}\times \mathrm{d}\mathbf{s}/\varpi\right)}{\partial P}, \tag{5.84}$$

whence from (5.83) and (5.5)

$$\frac{\partial}{\partial t}\left(\frac{\partial F_{\mathrm{t}}}{\partial P}\right) = -\frac{\partial}{\partial P}\left[\beta'(P)\oint \eta\mathbf{B}_{\mathrm{p}}\cdot \mathrm{d}\mathbf{s}\right]. \tag{5.85}$$

It is convenient to define

$$L(P) = \oint \eta\mathbf{B}_{\mathrm{p}}\cdot \mathrm{d}\mathbf{s}, \qquad M(P) = \oint \mathrm{d}s/\varpi^2 B_{\mathrm{p}}, \tag{5.86}$$

both calculable in terms of a prescribed poloidal field and a stellar model with a definite temperature field fixing the conductivity. Then by (5.24) and (5.83),

$$\partial F_{\mathrm{t}}/\partial P = -M(P)\beta(P,t) \tag{5.87}$$

and (5.85) becomes

$$M\,\partial\beta/\partial t = \partial[L\,\partial\beta/\partial P]/\partial P = L\,\partial^2\beta/\partial P^2 + L'\,\partial\beta/\partial P. \tag{5.88}$$

The decay equation (5.82), valid in the absence of a constraining field $\mathbf{B}_{\mathrm{p}}$, has B_ϕ a function of time and *two* spatial coordinates. In the present problem, the introduction of the torque-free constraint (5.21) replaces (5.82) by the equation (5.88) for β, which is a function of time and of *one* other variable, the flux-function P of $\mathbf{B}_{\mathrm{p}}$; once this is solved, (5.21) yields B_ϕ at different points of the same $\mathbf{B}_{\mathrm{p}}$-line. The procedure, which effectively decouples the decay of the poloidal field from the toroidal (but not vice versa), is only approximate as it does not take account of the changed surface boundary condition.

Equation (5.88) is valid over $0 \geq P \geq P_m$, where P_m defines the neutral point of the principal Cowling mode that we are adopting for $\mathbf{B}_{\mathrm{p}}$. The equation is separable and we can look for solutions of the form $\beta = \beta_0(P)\,\mathrm{e}^{-\gamma t}$, analogues of the pure toroidal modes, with the boundary condition $\beta = 0$ at $P = 0$. However, by the definitions (5.86), near $P = P_m$, $L \propto (P_m - P)$ and $M(P_m)$ is finite; hence (5.88) enforces its own boundary condition $[(\mathrm{d}\beta_0/\mathrm{d}P)(\mathrm{d}L/\mathrm{d}P) + \gamma M f_0] = 0$ at $P = P_m$. Since the eigenvalue γ appears in a boundary condition, the problem is not of the Sturm–Liouville type: in particular, there is no orthogonality property. For this reason the problem is best treated by direct numerical integration of (5.88), starting from assumed simple structures for $\beta(P)$. It was found (Mestel and Moss 1983*a*) that the decay times of toroidal fields deep in the star are not much affected by the dynamical condition (5.21). For the observable regions of the star, the results depend critically on the value of the coronal temperature. Recall that with the force-free condition replacing the curl-free, there is no longer a requirement that the poloidal currents and therefore the toroidal field be zero at the surface. It is then not surprising that if the stellar corona is at 10^6 K, the decay of the observable part of B_ϕ remains slow; whereas for a star with a cool corona (10^4 K), an observable B_ϕ field would require a local energy supply.

One can provisionally conclude that there exist field models as in Section 5.3, with linked poloidal and toroidal flux of comparable strengths, at least in the domains near the neutral point of $\mathbf{B}_\mathrm{p}$, which are both dynamically stable (prima facie), and have lifetimes against Ohmic decay long enough to leave the fossil theory still available as the explanation of the fields of some magnetic stars. The field could be a relic of the galactic field, pervading the gas from which stars formed (cf. Chapter 11), or of a field built up by dynamo action, e.g. during the pre-main sequence phases (Chapter 10).

We recall at this point Tayler's result (1980) that a mixed poloidal–toroidal field with non-vanishing $\mathbf{j}_\mathrm{p}$ on the axis is subject to an unstable mode. Tayler (1982) has also shown that the standard treatment of Ohmic diffusion transforms a stable toroidal field into an unstable structure within a stellar lifetime – i.e. the form (5.39) with $n > 1$ diffuses into a state with $n = 1$. Moss (1984) repeated Tayler's calculation but with the torque-free constraint (5.83) again supposed maintained, finding that Tayler's conclusion on the diffusion time is hardly affected. The important question is then again, what effect do such local instabilities have on the global decay times? As noted in Section 5.3, in some unpublished work, Moss has extrapolated the linear instability results into the non-linear domain, in particular studying the global diffusion problem but with a large macroscopic resistivity in the local unstable domains. His conclusion is that the global decay times are not much affected by genuinely local instabilities, so leaving the fossil theory a candidate available for the early-type magnetic stars (cf. Chapter 9).

5.7 The Biermann 'battery' process

As shown by (2.98), the simple two-fluid form of Ohm's law (2.56) contains the $\nabla p_e/n_e e$ term that allows for the generation of flux. This was exploited by Biermann (1950) as a possible source of a toroidal magnetic field in non-spherical stars. In a non-magnetic star, 'Ohm's law' – the equation of motion of the electron component – must still be satisfied, requiring

$$\mathbf{E} + \nabla p_e/n_e e = 0. \tag{5.89}$$

This is consistent with the continued absence of $\mathbf{B}$ only if $\nabla p_e/n_e e$ is curl-free. In a chemically homogeneous stellar region,

$$\frac{\nabla p_e}{n_e e} = \frac{Z}{Z+1}\frac{Am_H}{Ze}\frac{\nabla p}{\rho}, \tag{5.90}$$

which in turn is fixed by the equation of hydrostatic support (with magnetic forces assumed negligible)

$$\nabla p/\rho = \nabla V + \Omega^2(\varpi, z)\varpi. \tag{5.91}$$

In a non-rotating star, (5.89) and (5.91) yield the radially-directed Pannekoek–Rosseland electric field, with the scalar potential $\phi = (Z/Z+\ 1)(Am_H/Ze)V$, and so maintained by the very small fractional charge separation

$$\frac{\rho_e}{n_i Ze} = \left(-\frac{\nabla^2\phi}{4\pi}\right)\left(\frac{Am_H}{Ze\rho}\right) = \left(\frac{Z}{Z+1}\right)G\left(\frac{Am_H}{Ze}\right)^2 \approx 10^{-36}. \tag{5.92}$$

(Eddington 1959). The irrotational electric field serves to transfer the partial pressure of the electrons to act on the ions, which feel virtually all the gravitational force. However, in a rotating star, if Ω depends on z, $\Omega^2\varpi$ and so also $\nabla p_e/n_e e$ and hence $\mathbf{E}$ have a curl, and a toroidal magnetic field $\mathbf{B}_\mathrm{t}$ is built up, maintained by poloidal currents $\mathbf{j}_\mathrm{p}$. Cowling (1953) noted that the same enormous self-induction that yields a long decay time yields a similar time-scale for the currents to reach their asymptotic state, in which input of magnetic energy is balanced by Ohmic dissipation. Even so, the Biermann process promised to yield very large amounts of flux over a stellar lifetime. If the variation of Ω with z is of the same order as z itself, then the asymptotic field even in a slow rotator like the Sun is $\approx 10^3$ G; while in a very rapid rotator, the field can be large enough to require inclusion of terms $\propto \mathbf{j}_\mathrm{p} \times \mathbf{B}_\mathrm{t}$ in (5.91) and in Ohm's law (Mestel and Roxburgh 1962; Roxburgh and Strittmatter 1966).

5.7.1 *Coupling with a poloidal field*

Even though rotation laws dependent on z are subject to the secular Goldreich–Schubert instabilities, they could still last long enough not to invalidate the study of the initial build-up of the Biermann field. However, it had been shown earlier by Mestel and Roxburgh (1962) that if the star had even a very weak poloidal field, a dynamical interaction – similar to that arising in the decay problem of the last section – would effectively kill off the Biermann process before it got under way. We recall the argument as it again combines ideas and methods familiar both from magnetohydrodynamics and from classical electrodynamics, as developed originally for rigid conductors. For definiteness, we again adopt a poloidal field $\mathbf{B}_\mathrm{p}$ that is symmetric about the Ω-axis. The poloidal currents that will maintain a toroidal field $\mathbf{B}_\mathrm{t}$ satisfy

$$\mathbf{j}_\mathrm{p}/\sigma = \mathbf{E}_\mathrm{p} + \Omega\varpi\mathbf{t} \times \mathbf{B}_\mathrm{p}/c + \nabla p_e/n_e e - (\mathbf{j}_\mathrm{p} \times \mathbf{B}_\mathrm{t}/cn_e e + \mathbf{j}_\mathrm{t} \times \mathbf{B}_\mathrm{p}/cn_e e). \tag{5.93}$$

As we are interested in the maximum efficiency of the Biermann process, we ignore the slow decay of $\mathbf{B}_\mathrm{p}$ and study a steady state in which $\mathbf{B}_\mathrm{t}$ has reached its asymptotic value. The battery term $\nabla p_e/n_e$ in (5.93) will in general have a component perpendicular to $\mathbf{B}_\mathrm{p}$, yielding a current that exerts a torque. As in the decay problem of the last section, the dynamical torque-free condition

$$\mathbf{j}_\mathrm{p} = k\mathbf{B}_\mathrm{p}, \qquad \mathbf{B}_\mathrm{p} \cdot \nabla k = 0. \tag{5.94}$$

must be satisfied (again the advection of angular momentum by circulation being slow over the bulk of the star). The cancellation of the component of $\mathbf{j}_\mathrm{p}$ across $\mathbf{B}_\mathrm{p}$ is again achieved by a slight departure from isorotation, yielding a small non-irrotational term $\Omega\varpi\mathbf{t} \times \mathbf{B}_\mathrm{p}/c$ which must be added to the components perpendicular to $\mathbf{B}_\mathrm{p}$ of the battery and Hall terms. The required fractional variation $\Delta\Omega/\Omega$ is very small: on comparison of the motional induction and battery terms in (5.93), $\Delta\Omega/\Omega$ is estimated as at most of the order of the ratio Ω/ω_g, where $\omega_\mathrm{g} = (ZeB_\mathrm{p}/Am_\mathrm{H}c)$. It is thus an excellent approximation to retain the law of isorotation when considering the equilibrium of the star, while accepting that there are present the very slight deviations that ensure that Ohm's law in the form (5.93) is consistent with the torque-free condition (5.94).

The surviving battery currents are given by integrating (5.93) around a loop of $\mathbf{B}_\mathrm{p}$:

$$\oint \mathbf{j}_\mathrm{p} \cdot \mathrm{d}\mathbf{s}/\sigma = k \oint \mathbf{B}_\mathrm{p} \cdot \mathrm{d}\mathbf{s}/\sigma = \oint \frac{\nabla p_e}{n_e e} \cdot \mathrm{d}\mathbf{s}, \tag{5.95}$$

use being made of the constancy of k along each $\mathbf{B}_\mathrm{p}$-loop. Thus since $\oint \mathbf{B}_\mathrm{p} \cdot \mathrm{d}\mathbf{s}/\sigma > 0$, we require that the right-hand side of (5.95) – the 'emf' of the battery – must not vanish if the process is to survive. The value of k for each $\mathbf{B}_\mathrm{p}$-line, and hence the whole $\mathbf{B}_\mathrm{t}$-field, is then fixed by the ratio of the two integrals in (5.95). As discussed above, Biermann's original treatment assumed a chemically homogeneous, non-magnetic star with a non-conservative centrifugal field, so that

$$\nabla p_e/n_e e = \{Am_\mathrm{H}/(Z+1)e\}\nabla p/\rho = \{Am_\mathrm{H}/(Z+1)e\}[\nabla V + \Omega^2(\varpi, z)\varpi]. \tag{5.96}$$

In the absence of a constraining $\mathbf{B}_\mathrm{p}$-field, the currents $\mathbf{j}_\mathrm{p}$ find their own circuits; in particular, if the Ω-field is symmetric about the equator, then the currents have a quadrupolar-type parity. But in the presence of $\mathbf{B}_\mathrm{p}$, the currents are forced to flow round the $\mathbf{B}_\mathrm{p}$-loops, which are also to a very high approximation lines of constant Ω. If $\mathbf{B}_\mathrm{p}$ were strictly of dipolar parity in the equator, this would be enough to ensure that the emf integrals vanish. But even with $\mathbf{B}_\mathrm{p}$ of lower symmetry, the isorotation condition yields

$$k \propto \oint (\nabla p_e/n_e e) \cdot \mathrm{d}\mathbf{s} \propto \Omega^2 \oint \varpi \cdot \mathrm{d}\mathbf{s} = 0. \tag{5.97}$$

A small effect survives due to the very small departure from isorotation enforced by the necessity to kill off $(\mathbf{j}_\mathrm{p})_\perp$, but the resulting toroidal fields are negligible. Inclusion of a meridian circulation field is shown to leave the conclusions unaltered.

Kato and Nakagawa (1969) and Stix (1970) pointed out that a larger effect could survive in a star in which radiation pressure contributes significantly to the total pressure, for then there is no simple proportionality between $\nabla p_e/n_e e$ and

$\nabla p/\rho$ as in (5.96). This again requires the $\mathbf{B}_\mathrm{p}$-field to lack dipolar parity in the equator, and for Ω to vary from one line of $\mathbf{B}_\mathrm{p}$ to another. Recall, however, that we regard the axisymmetric theory as illustrative, being a simplified version of the realistic rigidly rotating non-symmetric model such as the oblique rotator. From now on we therefore impose uniform rotation in the mean, so that the battery field (5.96), acting in a chemically homogeneous star, would lead just to an extra polarization of the medium, superposed on the Pannekoek–Rosseland field, and in fact far less (again by the factor Ω/ω_g) than the polarization associated with the rotation of the star in the field $\mathbf{B}_\mathrm{p}$.

5.7.2 *The effect of chemical inhomogeneities*

Dolginov (1977) noted that even in a uniformly rotating star, chemical inhomogeneities would yield a modified Biermann battery, with

$$\nabla p_e/n_e e = (m_\mathrm{H}/e)\nabla[(\mu+2)p]/[(\mu+2)/\mu]\rho \tag{5.98}$$

where μ is the local mean molecular weight. In the present problem, substitution of (5.98) and (5.93) into (5.95) yields

$$k\oint \mathbf{B}_\mathrm{p}\cdot \mathrm{d}\mathbf{s} = (m_\mathrm{H}/e)\oint \nabla[(\mu+2)p]\cdot \mathrm{d}\mathbf{s}/[(\mu+2)/\mu]\rho. \tag{5.99}$$

Again, if $\mathbf{B}_\mathrm{p}$ has odd parity in the equator and also (as we shall assume from now on) the μ-distribution is symmetric in the equator, then the line integrals on the right vanish identically, so that the battery effect is again killed. But if $\mathbf{B}_\mathrm{p}$ is, for example, a quadrupole, or, more realistically, a dipole displaced along the axis, as in the model originally proposed by Landstreet (1970) for the Ap star fields, then in general one should find $k \neq 0$. The currents flowing along field lines passing near or through the photosphere suffer proportionately more dissipation, so there is some tendency for the constructed $\mathbf{B}_\mathrm{t}$-field to be concentrated in the regions where the poloidal field lines close well beneath the surface.

In a uniformly rotating star, with the magnetic forces over the bulk negligible compared with the centrifugal, the equation of hydrostatic support (5.11) reduces to (5.41), and so the battery integral on the right-hand side of (5.99) can be written

$$\oint \left(\mu\frac{\nabla p}{\rho} + \frac{\mu}{\mu+2}\frac{p}{\rho}\nabla\mu\right)\cdot \mathrm{d}\mathbf{s}. \tag{5.100}$$

Provided $\mathbf{B}_\mathrm{p}$ does not have dipolar parity in the equator, this will not in general vanish (unless μ is, like p and ρ, a function of ψ.) For definiteness, we adopt a standard Cowling model star with a convective core and a simple Kramers-type opacity for the radiative envelope. The star has mass M and radius R; radial distance is measured in units of $x = r/R$. The field $\mathbf{B}_\mathrm{p}$ is taken to be the slowest decaying dipolar mode solution of the decay equation (5.74), but displaced from the star's centre by a distance $d\times R$ along the magnetic axis. From the torque-free condition (5.83) and from (5.99),

$$-(c/4\pi)R\beta'(P)\oint \mathbf{B}_{\mathrm{p}}\cdot \mathrm{d}\mathbf{s}/\sigma = (m_{\mathrm{H}}/e)\oint \frac{\nabla[(\mu+2)p]\cdot \mathrm{d}\mathbf{s}}{(\mu+2)\rho/\mu}. \tag{5.101}$$

Integration across poloidal loops determines $\beta(P)$, subject to $\beta(0)=0$.

One example, studied in detail in Mestel and Moss (1983*b*), fixes the μ-distribution by postulating strict local radiative equilibrium in a uniformly rotating star. A non-spherical μ-distribution, written in the form

$$\mu(r,\theta) = \mu_0[1 + a_0(r) + a_2(r)\mathrm{P}_2(\cos\theta) + \ldots] \tag{5.102}$$

requires an associated perturbation T_μ to the temperature distribution, over and above that associated with the centrifugal distortion to the star, and this in turn yields in general a non-zero contribution $(\nabla\cdot\mathbf{F})_\mu$ to the divergence of the radiation flux and so to a 'μ-current' velocity field (Mestel 1953). If local radiative equilibrium does hold, and if any magnetic disturbance to the $(p,\ \rho,\ T)$ fields is negligible, then this term must cancel the term (5.46) which if unimpeded yields the Eddington–Vogt–Sweet circulation: i.e. the μ-current velocity must be equal and opposite to the E–V–S velocity. The resulting equation for a_2 is (Moss, personal communication; McDonald 1972)

$$a_2'' + \left(\frac{8.5T_0'}{T_0} - \frac{T_0''}{T_0}\right) - \frac{6a_2}{x^2} = \lambda_\Omega \frac{(n-1.5)}{(n+1)}\frac{\rho_0|V_0'|}{L}\frac{4\pi r^2 T_0'}{T_0}p_\Omega(r). \tag{5.103}$$

where primes represent differentiation with respect to x, n is the local polytropic index, increasing outwards from 1.5 near the convective core to 3.25 at the surface, and the radial component of the E–V–S velocity is written

$$(v_r)_\Omega = p_\Omega(r)\lambda_\Omega \mathrm{P}_2(\cos\theta), \qquad \lambda_\Omega = \Omega^2 R^3/GM. \tag{5.104}$$

The function $p_\Omega(r)$ is non-zero throughout the whole envelope. At the convective core it is reasonable to take $a_2 = 0$, as the convection should be an efficient mixer in both horizontal and vertical directions. Near the star's surface, the conditions of finiteness require that a_2 – satisfying (5.103) – has a part $\propto (1-x)^{9/4}$, forced by the E–V–S term and a finite contribution from the complementary function. Smooth link-up of the outward and inward solutions of (5.103) yield a unique function $a_2(x)$. A typical value for $|a_2|$ is $10^{-2}\lambda_\Omega$. With a_2 constructed, the integrations in (5.99) along the lines of the displaced dipolar field can be performed, whence k, $\mathbf{j}_{\mathrm{p}}$ and so $\mathbf{B}_{\mathrm{t}}$ can be constructed.

Details are given in Mestel and Moss (1983*b*). It is found that the asymptotic field is non-trivial, though somewhat weaker than that given by the original Biermann model, based on a non-conservative centrifugal field. In a rapid rotator, B_ϕ deep down would be comparable with the estimated *internal* B_{p}-field within an Ap star with a surface field of the observed value. An approximate treatment for a perpendicular rotator yields similar results. It is, however, doubtful whether any observed surface toroidal fields can be explained plausibly as the 'tip of an iceberg'.

The usual explanation of the periodically varying abundance anomalies in an Ap star surface appeals to abundance patches, perhaps built up by a combination of radiation driving and gravitational settling in a magnetically stabilized atmosphere (cf. Section 9.7). Dolginov (1977) argued that the associated non-spherical μ-distribution would again yield a Biermann battery. Some examples constructed by Mestel and Moss show that the process could indeed be important for producing surface toroidal fields of the inferred strength.

5.8 An introduction to the stellar dynamo problem

5.8.1 *Cowling's anti-dynamo theorem*

As noted by Sweet (1950*a*), the decay of an axisymmetric poloidal field in a stationary medium may be pictured as the diffusion of field lines with the local velocity

$$\mathbf{v}_{\mathrm{d}} = \frac{c(\mathbf{j}_{\mathrm{t}} \times \mathbf{B}_{\mathrm{p}})}{\sigma B_{\mathrm{p}}^2} = \frac{(\nabla \times \mathbf{B}_{\mathrm{p}}) \times \mathbf{B}_{\mathrm{p}}}{(4\pi\sigma/c^2) B_{\mathrm{p}}^2}; \tag{5.105}$$

for if the field is supposed frozen into a fluid moving with this velocity, one recovers Ohm's law for a stationary medium:

$$0 = \mathbf{E} + \mathbf{v}_{\mathrm{d}} \times \mathbf{B}_{\mathrm{p}}/c = \mathbf{E} - \mathbf{j}_{\mathrm{t}}/\sigma, \tag{5.106}$$

since $\mathbf{j}_{\mathrm{p}}.\mathbf{B}_{\mathrm{t}} = 0$. Likewise, a toroidal field decays as if the field lines have the velocity $c(\mathbf{j}_{\mathrm{p}} \times \mathbf{B}_{\mathrm{t}})/\sigma B_{\mathrm{t}}^2$. An axisymmetric poloidal field has lines which are closed curves in meridian planes and so contains at least one O-type neutral point O. By Ampère's law, near O the field lines circle in the right-handed sense with respect to $\mathbf{j}_{\mathrm{t}}$, so that $\mathbf{j}_{\mathrm{t}} \times \mathbf{B}_{\mathrm{p}}$ is directed locally towards O, yielding a diffusion velocity into the neutral point from all angles (Fig. 5.7). To offset this and keep $\mathbf{B}_{\mathrm{p}}$ steady, the velocity vectors of the gas would have to emerge from O, implying creation of matter out of O, and would be infinite in magnitude at O. Likewise, to offset the disappearance of toroidal loops into the axis, matter would need to flow out of the axis.

These statements provide a simple intuitive 'proof' of Cowling's anti-dynamo theorem: an axisymmetric field cannot be maintained against Ohmic decay by the inductive effects of mass motions. Cowling's formal proof (1934) for a poloidal field notes that the toroidal component of Ohm's law

$$\mathbf{j}_{\mathrm{t}}/\sigma = \mathbf{E}_{\mathrm{t}} + \mathbf{v}_{\mathrm{p}} \times \mathbf{B}_{\mathrm{p}}/c \tag{5.107}$$

reduces with use of (5.5), (5.74) and (5.75) to

$$\mathbf{v}_{\mathrm{p}} \cdot \nabla P = \eta D^2 P = \eta[\nabla^2 P - (2/\varpi)\, \partial P/\partial \varpi], \tag{5.108}$$

since in a steady state that is axisymmetric, the curl-free electric field has no $\mathbf{E}_{\mathrm{t}}$-component. At a neutral point that is a normal extremum of the flux function P, $\nabla P = 0$ but $\nabla^2 P \neq 0$, and so (5.108) cannot be satisfied unless σ is infinite,

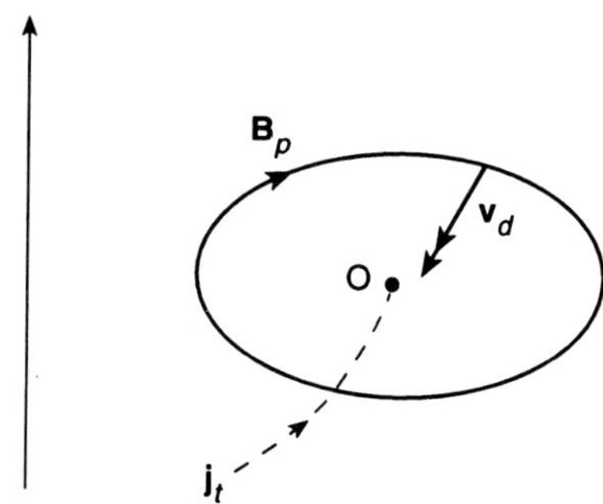

FIG. 5.7. Cowling's anti-dynamo theorem.

or unless outflow of gas from the neutral point yields v_{p} infinite at O. If P is an analytic function, then the proof is extensible to cases in which $\nabla^2 P$ also vanishes at O, for the left-hand side of (5.108) will always vanish to a higher order than the right. A more general formal proof was given by Backus and Chandrasekhar (1956).

An axisymmetric toroidal field has its neutral points on the axis, but now there is a curl-free electric field in Ohm's law:

$$\mathbf{j}_{\mathrm{p}}/\sigma = -\nabla\phi + \mathbf{v}_{\mathrm{p}} \times \mathbf{B}_{\mathrm{t}}/c. \tag{5.109}$$

A similar anti-dynamo argument follows by constructing the 'Kirchhoff integral' along the closed curve $XYZX$ in Fig. 5.8. The loops of $\mathbf{B}_{\mathrm{t}}$ circle around the axis in the same sense between the two points X, Y; also, $B_{\mathrm{t}} = 0$ on the arc YZX. (The simplest case will have YZX as the circle at infinity, so that $\mathbf{B}_{\mathrm{t}}$ is everywhere either right- or left-handed with respect to the axis.) Along the whole loop, by Ampère's law, $\mathbf{j}_{\mathrm{p}}$ points in the same direction, so that $\oint \mathbf{j}_{\mathrm{p}} \cdot \mathrm{d}\mathbf{s}/\sigma \neq 0$; but since $\mathbf{v} \times \mathbf{B} \cdot \mathrm{d}\mathbf{s}$ vanishes along the loop, so does the induction integral. A curl-free electric field clearly cannot drive a current all the way round a closed loop, so in a steady state $\mathbf{j}_{\mathrm{p}}$ and $\mathbf{B}_{\mathrm{t}}$ must vanish.

As discussed at length above, a rotation field leads in general to the generation of a toroidal component from a poloidal through the term $\mathbf{v}_{\mathrm{t}} \times \mathbf{B}_{\mathrm{p}}/c$. However, in a strictly axisymmetric system there is no corresponding term that generates a poloidal component from a toroidal: the term $\mathbf{v}_{\mathrm{p}} \times \mathbf{B}_{\mathrm{t}}/c$ merely advects toroidal loops. It is this 'topological asymmetry' (Elsasser 1955) that prevents the operation of an axisymmetric dynamo, in which toroidal flux could be used to replenish the poloidal flux that is diffusing into the O-type neutral points.

The essentially topological nature of these first anti-dynamo theorems is brought out by the generalization due to Bullard (1955, and personal communication): a field cannot be maintained against Ohmic decay by motional induction, due to flow of gas across the field, if there exists a closed curve C around which

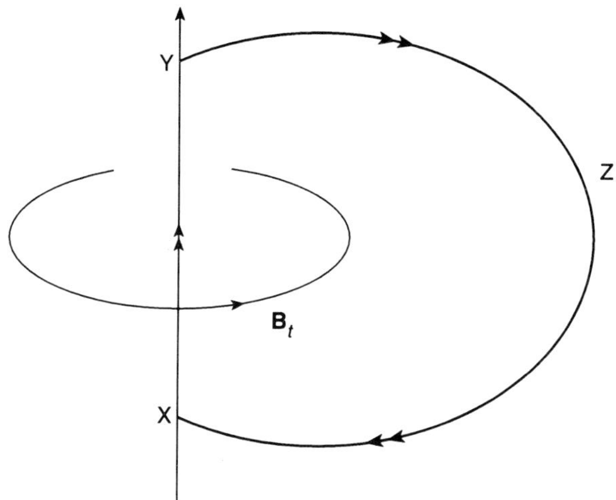

FIG. 5.8. The anti-dynamo theorem for purely toroidal fields.

the fields lines circle in the same sense, so that $\oint(\mathbf{j}_\mathrm{P}/\sigma)\cdot d\mathbf{s} \neq 0$. The curve C is either a null curve of the field, or is itself a limiting field line; in either case, $\oint(\mathbf{v}\times\mathbf{B}/c)\cdot d\mathbf{s} = 0$, so the same contradiction arises. With fields of this simple topology, there is no way to off-set the inexorable Ohmic diffusion of field lines towards the curve C, which is a sink of flux.

Cowling's anti-dynamo theorem has been generalized by Hide and Palmer (1982) to apply to non-steady axisymmetric fields, and to media in which the (scalar) permeability and resistivity depend on position and time. An arbitrary bulk velocity field $\mathbf{v}$ will drag field lines about, so that the neutral points of the field need not remain at fixed points in space. The new proof focuses on the overall maximum of $|P|$ in the (ϖ, z)-plane, showing that $|P|$ must decay to zero.

An alternative proof, applicable to homogeneous and incompressible media, which exploits the global rather than the local properties of the field, has been given by Cowling (1957) and Braginsky (1965). In a time-dependent, axisymmetric state, (5.107) becomes

$$\frac{\partial P}{\partial t} + \mathbf{v}\cdot\nabla P = \eta D^2 P. \tag{5.110}$$

(The decay problem studied in Section 5.6 has $\mathbf{v} = 0$, while the above proofs of Cowling's dynamo theorem show that a contradiction arises near a neutral point when one tries to put $\partial P/\partial t = 0$.) Moffatt (1978) relaxes the constraints slightly by requiring η to be constant not over the whole domain but just along the prescribed streamlines, so that $\mathbf{v}\cdot\nabla\eta = 0$. Since for divergence-free flow, $(P\mathbf{v}\cdot\nabla P)/\eta = \nabla\cdot(P^2/2\eta)\mathbf{v}$, multiplication of (5.110) by P/η and integration over all space yields

$$-\frac{\mathrm{d}}{\mathrm{d}t}\int \frac{P^2}{2\eta}\,\mathrm{d}\tau = \int PD^2P\,\mathrm{d}\tau \tag{5.111}$$

if the fluid motions are confined to a finite volume, so that $\int_\infty (P^2/2\eta)\mathbf{v}\cdot\mathbf{n}\,\mathrm{d}S = 0$. Similarly, by (5.75),

$$\begin{aligned}\int PD^2P\,\mathrm{d}\tau &= \int P\nabla\cdot\mathbf{Q}\,\mathrm{d}\tau = \int (\nabla\cdot(P\mathbf{Q}) - \mathbf{Q}\cdot\nabla P)\,\mathrm{d}\tau \\ &= -\int [(\nabla P)^2 - P\nabla P\cdot\hat{\varpi}]\,\mathrm{d}\tau,\end{aligned} \tag{5.112}$$

since $\nabla.(\hat{\varpi}/\varpi) = 0$. Thus

$$\frac{\mathrm{d}}{\mathrm{d}t}\int \frac{P^2}{2\eta}\,\mathrm{d}\tau = -\int (\nabla P)^2\,\mathrm{d}\tau = -\int \varpi^2 B^2\,\mathrm{d}\tau. \tag{5.113}$$

Hence P^2 must ultimately vanish everywhere.

5.8.2 *Mass motions and the rate of decay*

The requirement that the field remains axisymmetric or topologically similar puts an implicit constraint on the motions. Accepting this, one can go beyond Cowling's theorem and investigate how the rate of Ohmic decay – described in a medium at rest by (5.105) – is affected by mass motions. Cowling (1945) and Sweet (1950*b*) suggested that turbulent motions would tangle up the field, reducing the length scales and so accelerating the rate of decay. In general dynamo theory, 'turbulent resistivity' is indeed estimated, simultaneously with the efficacy of more general mass motions in bypassing the constraints set by the various anti-dynamo theorems (Chapter 6). It is at least prima facie plausible that even if the motions are so restricted that dynamo action is ruled out, the turbulent resistivity will still lead to expulsion of the externally generated flux from the turbulent domain.

The contrary suggestion is that if the motions are as far as possible directed opposite to the local velocity (5.105), then perhaps the decay could be significantly slowed up. As seen above, near the O-type neutral points the inexorable diffusion continues, so the question can be answered only by a global analysis. Backus (1957) showed formally that the best that could be expected is an increase in the decay time by a factor of about 4. A simpler parallel treatment by Spitzer (1957) runs essentially as follows. In spherical polars (5.110) can be written

$$\partial P/\partial t + \mathbf{v}\cdot\nabla P = r\sin\theta(cj_\phi/\sigma). \tag{5.114}$$

Multiplication of (5.114) by ρ and integration over the volume of the star yields

$$\frac{\partial}{\partial t}\int \rho P\,\mathrm{d}\tau = 2\pi\int\!\!\int \rho(c/\sigma)j_\phi r^3\sin^2\theta\,\mathrm{d}r\,\mathrm{d}\theta = 2c^2\int(\rho/\sigma)\,\mathrm{d}p_m. \tag{5.115}$$

Here $\mathrm{d}p_m = (\pi r^2\sin^2\theta)(j_\phi r\,\mathrm{d}\theta\,\mathrm{d}r)/c$ is the dipole moment due to the toroidal current $j_\phi r\,\mathrm{d}\theta\,\mathrm{d}r$ flowing in the circuit enclosing the area $\pi r^2\sin^2\theta$. Thus

$$\frac{\partial}{\partial t}\int \rho P \, \mathrm{d}\tau = 2\langle c^2\rho/\sigma\rangle p_m \tag{5.116}$$

where $\langle c^2\rho/\sigma\rangle$ is an average over the stellar volume. Note that (c^2/σ) is here the *microscopic* resistivity $\propto T^{-3/2}$, from (2.73). In a convection zone satisfying the adiabatic law $\rho \propto T^{1/(\gamma-1)}$, the combination $c^2\rho/\sigma \propto T^{(5-3\gamma)/2(\gamma-1)}$ which is nearly constant when $\gamma \simeq 5/3$, as in a star of moderate mass.

Outside the star, the field is again taken as curl-free, with P satisfying

$$P(r,\theta) = \sum \frac{c_n}{r^n}(1-\mu^2)\frac{\mathrm{dP}_n}{\mathrm{d}\mu} \tag{5.117}$$

with c_1 just the stellar dipole moment p_m. Now define

$$P_c(r) = \int_0^{\pi} P(r,\theta)\sin\theta \, \mathrm{d}\theta, \tag{5.118}$$

so that (5.116) becomes

$$\frac{\partial}{\partial t}\int_0^R \rho r^2 P_c(r,\theta)\,\mathrm{d}r = (p_m/\pi)\langle c^2\rho/\sigma\rangle \tag{5.119}$$

provided the deviations of ρ from sphericity are modest. Outside the star, (5.118) yields

$$P_c = (4/3)p_m/r, \tag{5.120}$$

all the non-dipolar terms vanishing identically. As P and therefore P_c are continuous at $r = R$ (to ensure continuity of B_r), condition (5.120) relates the star's dipole moment to the surface value of P_c. If P_c within the star were independent of r, then (5.119) would become

$$\frac{\partial p_m}{\partial t} = -3\frac{\langle c^2\rho/\sigma\rangle}{M}Rp_m \tag{5.121}$$

– decay with an e-folding time $\tau_d = M/(3R\langle c^2\rho/\sigma\rangle)$. If we introduce the quantity

$$\xi = \frac{\int_0^R \rho r^2 P_c\,\mathrm{d}r}{P_c(R)\int_0^R \rho r^2\,\mathrm{d}r}, \tag{5.122}$$

then this e-folding time becomes

$$\tau_d = \frac{\xi M}{3R\langle c^2\rho/\sigma\rangle}. \tag{5.123}$$

Note how the velocity has disappeared through use of the continuity equation: the influence of the motions on the decay rate manifests itself purely through their effect on the parameter ξ.

In the absence of motions, insertion into (5.122) of P_c as given, for example, by the slowest decaying Cowling mode, yields ξ near unity. A slow, laminar flow such as the E–V–S circulation will modify the field structure somewhat by forcing field lines into near parallelism, except near the equator and the surface, where the local distortion will be large enough to allow quasi-steady trans-field flow through finite resistivity. No large change in the value of ξ is to be expected.

The numerical work on the effect of imposed rapid motions on a given field, discussed in Chapter 4, suggests that a strong, essentially isotropic turbulence will compress the flux into regions where it is strong enough to resist further tangling and compression: e.g. in a fully convective star, towards the axis and the surface. But Spitzer's analysis shows that compression of the axisymmetric field does not in itself increase significantly the rate of decay of the dipole moment. This is seen from the form (5.122) for ξ. For example, suppose all the flux of a dipolar field is compressed into a thin cylindrical region near the axis, of width ϖ_c. Then $\nabla P = 0$ outside this domain, and so P is constant over the bulk of the star, $\int \rho r^2 P_c \, \mathrm{d}r \approx P_c(r) \int \rho r^2 \, \mathrm{d}r$, and (5.122) yields $\xi \approx 1$; the *axisymmetric* fluid motions have no significant effect on the decay rate of the external dipole moment. As noted, a slowly decaying, dynamically stable 'fossil' field remains a plausible model for the early-type magnetic stars (cf. Chapter 9), but for the Sun and other solar-type stars, and other bodies showing field reversals over decades or centuries, one is forced to turn to dynamo action. Dynamo theory in general is discussed in Chapter 6, and applied to late-type stars in Chapter 8.

References

Babcock, H.D. (1959). *Astrophysical Journal*, **130**, 364.

Backus, G.E. (1957). *Astrophysical Journal*, **125**, 500.

Backus, G.E. and Chandrasekhar, S. (1956). *Proceedings National Academy Sciences*, **42**, 105.

Balbus, S.A. and Hawley, J.F. (1994). *Monthly Notices Royal Astronomical Society*, **266**, 769.

Biermann, L. (1950). *Zeitschrift Naturforschung*, **5a**, 65.

Braginsky, S.I. (1965). *Soviet Physics JETP*, **20**, 726.

Bullard, E.C. (1955). *Proceedings Royal Society A*, **233**, 289.

Chandrasekhar, S. (1933). *Monthly Notices Royal Astronomical Society*, **93**, 390.

Chandrasekhar, S. (1956*a*). *Proceedings National Academy Sciences*, **42**, 1.

Chandrasekhar, S. (1956*b*). *Astrophysical Journal*, **124**, 232.

Charbonneau, P. and MacGregor, K.B. (1992). *Astrophysical Journal*, **387**, 639.

Charbonneau, P. and Michaud, G. (1988). *Astrophysical Journal*, **327**, 809.

Cowling, T.G. (1934). *Monthly Notices Royal Astronomical Society*, **94**, 39.

Cowling, T.G. (1945). *Monthly Notices Royal Astronomical Society*, **105**, 166.

Cowling, T.G. (1953). *Solar electrodynamics*, in *The Sun* (ed. G.P. Kuiper), p. 532. University of Chicago Press.

Cowling, T.G. (1957). *Quarterly Journal Mechanics Applied Mathematics*, **10**, 129.

Dolginov, A.Z. (1977). *Astronomy and Astrophysics*, **54**, 17.

Eddington, A.S. (1929). *Monthly Notices Royal Astronomical Society*, **90**, 54.
Eddington, A.S. (1959). *The internal constitution of the stars.* Dover, New York. (Cambridge University Press edition 1926).
Elsasser, W.M. (1995). *American Journal Physics*, **23**, 590.
Ferraro, V.C.A. (1937). *Monthly Notices Royal Astronomical Society*, **97**, 458.
Fricke, K. (1968). *Zeitschrift Astrophysik*, **68**, 317.
Fricke, K. (1969). *Astronomy and Astrophysics*, **1**, 388.
Frieman, E.A. and Rotenberg, M. (1960). *Reviews Modern Physics*, **32**, 898.
Gillis, J., Mestel, L. and Paris, R.B. (1974). *Astrophysics Space Science*, **27**, 167.
Gillis, J., Mestel, L. and Paris, R.B. (1979). *Monthly Notices Royal Astronomical Society*, **187**, 311.
Goldreich, P. and Schubert, G. (1967). *Astrophysical Journal*, **150**, 571.
Goossens, M. and Tayler, R.J. (1980). *Monthly Notices Royal Astronomical Society*, **193**, 833.
Gough, D.O. and Tayler, R.J. (1966). *Monthly Notices Royal Astronomical Society*, **133**, 85.
Hide, R. and Palmer, T.N. (1982). *Geophysical Astrophysical Fluid Dynamics*, **19**, 301.
Jackson, S., Sanderson, A.D., Smith, R.C. and Hazlehurst, J. (1971). *Astrophysical Journal*, **165**, 223.
Kato, S. and Nakagawa, Y. (1969). *Astrophysics Space Science*, **5**, 171.
Kippenhahn, R. (1969). *Astronomy and Astrophysics*, **2**, 309.
Lamb, H. (1883). *Philosophical Transactions Royal Society A*, **174**, 519.
Landstreet, J.D. (1970). *Astrophysical Journal*, **159**, 1001.
Lüst, R. and Schlüter, A. (1954). *Zeitschrift Astrophysik*, **34**, 263.
MacGregor, K.B. and Brenner, M. (1991). *Astrophysical Journal*, **376**, 204.
Markey, P. and Tayler, R.J. (1973). *Monthly Notices Royal Astronomical Society*, **163**, 77.
Markey, P. and Tayler, R.J. (1974). *Monthly Notices Royal Astronomical Society*, **168**, 505.
McDonald, B.E. (1972). *Astrophysics Space Science*, **19**, 309.
Mestel, L. (1953). *Monthly Notices Royal Astronomical Society*, **113**, 716.
Mestel, L. (1961). *Monthly Notices Royal Astronomical Society*, **122**, 473.
Mestel, L. (1966). *Zeitschrift Astrophysik*, **63**, 196.
Mestel, L. and Moss, D.L. (1983*a*). *Monthly Notices Royal Astronomical Society*, **204**, 575.
Mestel, L. and Moss, D.L. (1983*b*). *Monthly Notices Royal Astronomical Society*, **204**, 557.
Mestel, L. and Roxburgh, I.W. (1962). *Astrophysical Journal*, **136**, 615.
Mestel, L., Moss, D.L. and Tayler, R.J. (1988). *Monthly Notices Royal Astronomical Society*, **231**, 873.
Moffatt, H.K. (1978). *Magnetic field generation in electrically conducting fluids.* Cambridge University Press.
Moss, D. (1984). *Astrophysics Space Science*, **104**, 253.
Moss, D. (1992). *Monthly Notices Royal Astronomical Society*, **257**, 593.

Moss, D.L. and Tayler, R.J. (1969). *Monthly Notices Royal Astronomical Society*, **145**, 217.

Moss, D.L., Mestel, L. and Tayler, R.J. (1990). *Monthly Notices Royal Astronomical Society*, **245**, 550.

Öpik, E.J. (1951). *Monthly Notices Royal Astronomical Society*, **111**, 278.

Pedlosky, J. (1979, 1982). *Geophysical fluid dynamics*. Springer, New York.

Pitts, E. and Tayler, R.J. (1985). *Monthly Notices Royal Astronomical Society*, **216**, 139.

Roxburgh, I.W. (1964). *Monthly Notices Royal Astronomical Society*, **128**, 157 and 237.

Roxburgh, I.W. and Strittmatter, P.A. (1966). *Monthly Notices Royal Astronomical Society*, **133**, 1.

Sakurai, T. (1991). *Monthly Notices Royal Astronomical Society*, **248**, 457.

Schwarzschild, M. (1947). *Astrophysical Journal*, **106**, 427.

Spitzer Jr., L. (1957). *Astrophysical Journal*, **125**, 525.

Stix, M. (1970). *Astronomy and Astrophysics*, **4**, 161.

Sweet, P.A. (1950*a*). *Monthly Notices Royal Astronomical Society*, **110**, 548.

Sweet, P.A. (1950*b*). *Monthly Notices Royal Astronomical Society*, **110**, 69.

Tassoul, J.-L. and Tassoul, M. (1989). *Astrophysical Journal*, **345**, 472.

Tayler, R.J. (1973). *Monthly Notices Royal Astronomical Society*, **161**, 365.

Tayler, R.J. (1980). *Monthly Notices Royal Astronomical Society*, **191**, 151.

Tayler, R.J. (1982). *Monthly Notices Royal Astronomical Society*, **198**, 811.

Van Assche, W., Goossens, M. and Tayler, R.J. (1982). *Astronomy and Astrophysics*, **109**, 166.

Vogt, H. (1925). *Astronomische Nachrichten*, **223**, 229.

Wright, G.A.E. (1973). *Monthly Notices Royal Astronomical Society*, **162**, 329.

Zahn, J.P. (1983). In *Astrophysical processes in upper main-sequence stars* (eds A.N. Cox, S. Vauclair and J.-P. Zahn), p. 253. Swiss Society Astronomy Astrophysics, Geneva.

Zahn, J.-P. (1992). *Astronomy and Astrophysics*, **265**, 115.

6

STELLAR DYNAMOS

6.1 Introduction

The term 'dynamo action' appears in the literature describing two related but distinct processes. As noted in Section 2.4, when the velocity $\mathbf{v}$ of a conducting fluid is driven by the non-magnetic forces in the direction opposite to that of the Lorentz force, the rate of working $(\mathbf{j} \times \mathbf{B}/c) \cdot \mathbf{v}$ is negative, implying input of energy into the magnetic field; and in the high magnetic Reynolds number domain, the input on balance exceeds the Ohmic dissipation, so that an already existing field is *amplified.* A clear and important example (discussed in Section 4.3) is the local compression of flux by an inexorable circulation, which plausibly describes the formation of the granular and supergranular magnetic networks. Likewise, studies of the early phases of star formation (Chapters 11 and 12) involve the compression of the global flux in a gas cloud or fragment, with much of the gravitational energy released being converted into magnetic energy. But in neither example is there an increase in the total flux $\int_C \mathbf{B} \cdot \mathbf{n}\, \mathrm{d}S$, with C a circuit frozen into the gas.

In a technological dynamo, a small fraction of the current generated is diverted to flow in circuits wound so as to maintain the dynamo magnetic field according to Ampère's law. It is this aspect – the 'self-exciting dynamo' – that is the central issue of cosmical dynamo theory. We prefer to reserve the term 'dynamo action' for cases in which there are no sources of the magnetic field $\mathbf{B}$ other than the currents generated by fluid motions in the presence of $\mathbf{B}$; in contrast to 'amplification processes', which assume explicitly or implicitly the presence of externally maintained flux. As already indicated, amplifiers are important, but they are not identical with self-exciting dynamos, the topic of this chapter and of much of Chapters 8–10.

Turbulent motion of a highly conducting fluid, acting on an initially weak 'seed' field, will again amplify the field. If the fluid motions are nearly divergence-free, then by (2.90), increase of the local field strength is associated with *stretching* of field lines. In turbulent motion it is overwhelmingly probable that two initially contiguous points will rapidly acquire a large mutual separation, so yielding strong field amplification, but accompanied also by a tangling of the field, with energy being fed in at smaller and smaller wavelengths. On some scale the Lorentz forces can become strong enough to react back on the turbulence and limit the tangling, just as the compressed field in a sunspot resists further compression by the supergranulation. It is, however, not obvious that equipartition of energy between the turbulent and magnetic fields will subsequently be built up on all scales. An associated question is again whether the system can settle into a steady state, in which no external source of flux is required. If so, we have a genuine turbulent dynamo, with the asymptotic state virtually independent of the seed field. But if not, we have a turbulent amplifier: if the 'battery' generating the initial flux were switched off, the whole field would ultimately decay.

The answers may very well depend on the detailed symmetry properties of the turbulence. It is true that 'order does not arise spontaneously out of chaos' (Cowling 1965), but the question then becomes: 'What degree of departure from complete chaos may suffice to yield a large-scale, ordered field?'

We return to these questions in Section 6.4. Meanwhile, we note that in a dynamo (laminar or turbulent) that succeeds in generating and maintaining a large-scale field, destruction of unwanted small-scale fields is essential, if a large-scale galactic field is to be detectable through Faraday rotation. Further, as pointed out by Bondi and Gold (1950), if the fluid motions are restricted to a finite volume of *perfectly* conducting fluid, for example to the interior of an idealized star, then diffusion of newly generated flux out of the volume is strictly forbidden. In particular, the externally observed dipole moment hardly changes: it can be maximized only by concentrating the points of exit and entry of field lines at the North and South poles.

This draconian limitation could be bypassed if the field is strong enough to be able to tear the fluid. Equally, it can be regarded as a further *reductio ad absurdum* of taking field freezing too literally (cf. Sections 3.8, 4.8.2, 5.8.2 and 6.3 et seq.).

A picturesque heuristic dynamo model, discussed in Zel'dovich *et al.* (1983), is often referred to as the Zel'dovich 'stretch–twist–fold' (STF) process. In Fig. 6.1, the transformation from (a) to (b) stretches the loop, for example doubling its circumference, and – because of flux-freezing at constant density – decreasing by half the cross-section of the resulting torus. Simultaneously, the loop is twisted into the configuration shown. A subsequent fold transforms (b) into (c): roughly speaking, two loops are now present, with the field pointing in the same direction along each. However, to produce a field that is topologically identical with the initial field, but with the field strength scaled up, one must appeal to Ohmic dissipation, which will destroy the unwanted small-scale components as the two loops merge into one. Without diffusion, the amplification due to the STF process of the flux through a *geometric* circuit could be reversed. With subsequent diffusion, the STF process in general becomes irreversible, while the flux through a *material* circuit is no longer invariant but may be amplified. This has been emphasized in personal communication particularly by Kandu Subramanian and also by Aake Nordlund, who argues for renaming the Zel'dovich scheme the 'STFM-process', where M stands for 'merge'.

6.2 Laminar kinematic dynamos

Cowling's theorem, discussed in Section 5.8, was the first and most celebrated of a class of anti-dynamo theorems which eliminate either the most easily-pictured field structures or the very simplest velocity fields (cf. Roberts 1994). For a while,

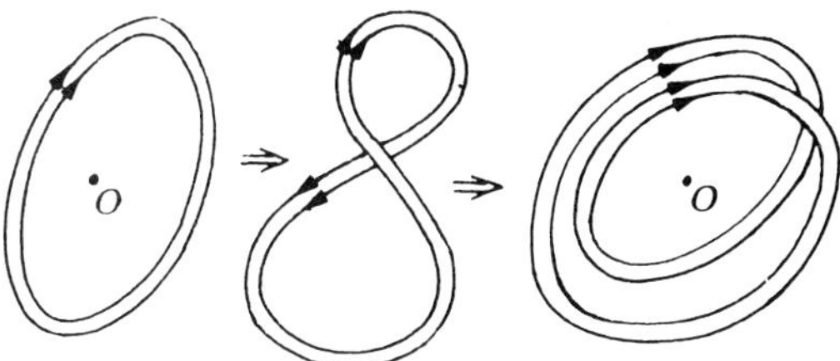

FIG. 6.1. The Zel'dovich stretch–twist–fold process.

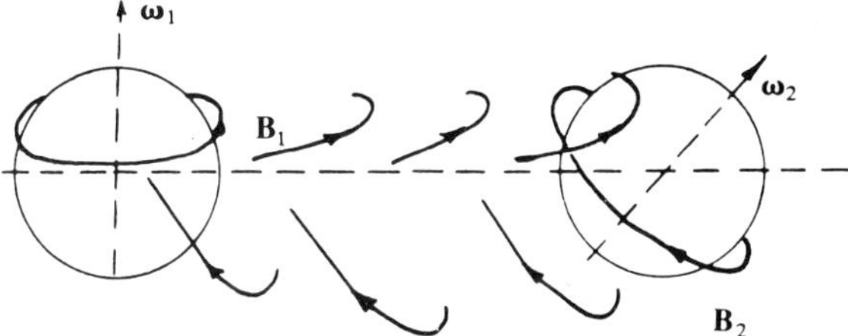

FIG. 6.2. The Herzenberg dynamo.

they were thought by some to be precursors of a more general theorem forbidding all genuine dynamo action in a more or less homogeneous medium. This pessimism was dispelled when rigorous existence proofs were produced for specific models (Backus 1958; Herzenberg 1958). The Herzenberg model (Fig. 6.2) involves two small spheres rotating about inclined axes within a bounded conducting medium. The toroidal field from one sphere produces a poloidal component at the other, which can be amplified by the shear in angular velocity. Lowes and Wilkinson (1968) succeeded in producing an experimental realization of this model, using two cylindrical rotors, which acts as a self-exciting dynamo when the rotors are spun fast enough: if one rotor is operated manually, then as the critical speed is reached one can feel the increased torque needed to overcome the back reaction of the growing Lorentz force. Brandenburg *et al.* (1998) have simulated numerically the Herzenberg dynamo, extending the parameters to a range not accessible to the original asymptotic analysis. For some inclination angles they find a new, unpredicted class of oscillatory solutions.

There is now a plethora of 'kinematic dynamo models', i.e. models with velocity fields that are not constructed solutions of the equations of motion (2.35) and continuity (2.38), but are just kinematically prescribed. Once rigorous existence proofs had been produced, there was an incentive to experiment with a variety of velocity fields, not all physically plausible, which often turn out able to maintain magnetic fields that are inevitably sufficiently complicated in structure to bypass the anti-dynamo theorems. Of particular interest are those which involve two characteristic length-scales. Childress (1969, 1970) and G.O. Roberts (1969, 1970, 1972) have shown that spatially periodic velocities, with a small length-scale, can maintain large-scale magnetic fields. The simplest examples include the Cartesian velocity field

$$\mathbf{v} = (\sin y - \cos z,\ \sin z - \cos x,\ \sin x - \cos y)\,, \tag{6.1}$$

(Childress 1969), and the still simpler flow – independent of x –

$$\mathbf{v} = (\cos y - \cos z,\ \sin z,\ \sin y) \tag{6.2}$$

(G.O. Roberts 1972). Significantly, the flows (6.1) and (6.2) are both solenoidal Beltrami flows, satisfying $\nabla\times\mathbf{v} = \mathbf{v}$, with the *helicity* $\mathbf{v}\cdot(\nabla\times\mathbf{v}) > 0$ at each point and hence with a non-zero average over the domain (cf. Section 6.4).

A realistic astrophysical dynamo cannot be infinite in extent but must be confined to a finite body, with the field falling off appropriately at infinity. Childress (1970) was in fact able to pack his periodic motions into a sphere. G.O. Roberts used an analogue of (6.2) in spherical polar coordinates (r, θ, ϕ) to show that a velocity field, *independent of* ϕ but again confined within a sphere, can generate an exponentially growing magnetic field with an $\exp(im\phi)$-dependence. Other such models, in which an *axisymmetric* velocity field maintains a *non-axisymmetric* field include one by Gailitis (1970) (described in detail by P.H. Roberts (1971) and Moffatt (1978)), in which the flow consists of a pair of ring vortices (cf. Section 9.6).

In general, the kinematic dynamo problem studies the basic equation

$$\frac{\partial\mathbf{B}}{\partial t} = \nabla\times(\mathbf{v}\times\mathbf{B}) - \nabla\times(\lambda\nabla\times\mathbf{B}), \tag{6.3}$$

where the resistivity $\lambda = c^2/4\pi\sigma$ is a known function of the state variables (primarily the temperature). (As we are concerned mainly with gaseous plasmas, the permeability μ is put equal to unity – there is no difficulty in restoring it, e.g. for study of planetary dynamos.) For the moment the fluid velocity field $\mathbf{v}$ is prescribed (though preferably with at least qualitative account being taken of the constraints imposed by the equation of motion). We are interested initially in steady-state solutions in which the induction and diffusion terms balance.

Cowling's anti-dynamo theorem and its extensions are limited to topologically simple field structures. An early attempt to produce a formalism that could in principle be applied to more general stellar fields was by Elsasser (1946, 1947). He studied possible dynamo action due to prescribed motions in a homogeneous incompressible sphere, satisfying $\nabla\cdot\mathbf{v} = 0$. Within the sphere solutions are sought of the equation

$$\lambda\nabla^2\mathbf{B} = \frac{\partial\mathbf{B}}{\partial t} - \nabla\times(\mathbf{v}\times\mathbf{B}), \tag{6.4}$$

where $\nabla^2\mathbf{B} \equiv -\nabla\times(\nabla\times\mathbf{B})$, since $\nabla\cdot\mathbf{B} = 0$. Outside the sphere the currents are supposed to vanish, so $\nabla\times\mathbf{B} = 0$. Elsasser expanded in terms of the Lamb–Cowling decay eigenmodes $\mathbf{B}_r \exp(-t/\tau_r)$, satisfying (6.4) with $\mathbf{v} = 0$ (cf. Section 5.6), which form a complete set: i.e. he wrote

$$\mathbf{B} = \sum_r b_r(t)\mathbf{B}_r \tag{6.5}$$

where

$$\nabla^2\mathbf{B}_r = -\frac{1}{\tau_r\lambda}\mathbf{B}_r. \tag{6.6}$$

The fields $\mathbf{B}_r$ satisfy an orthogonality relation

$$\int \mathbf{B}_r\cdot\mathbf{B}_s\,\mathrm{d}\tau = 0 \quad (r \neq s), \tag{6.7}$$

the integration being over all space. Since $\mathbf{j}_r = \sigma\mathbf{E}_r$,

$$\begin{aligned}\frac{4\pi}{c\sigma}\int \mathbf{j}_r\cdot\mathbf{j}_s\,\mathrm{d}\tau &= \frac{4\pi}{c}\int \mathbf{E}_r\cdot\mathbf{j}_s\,\mathrm{d}\tau = \int \mathbf{E}_r\cdot\nabla\times\mathbf{B}_s\,\mathrm{d}\tau\\ &= \int \nabla\cdot(\mathbf{B}_s\times\mathbf{E}_r)\,\mathrm{d}\tau + \int \nabla\times\mathbf{E}_r\cdot\mathbf{B}_s\,\mathrm{d}\tau.\end{aligned} \tag{6.8}$$

The tangential component of $\mathbf{B}$ will be continuous on the spherical surface, for in a medium of finite conductivity there can be no surface currents; hence the divergence integral can be replaced by $-(4\pi/c)\int_\infty \mathbf{s}\cdot\mathbf{n}\,\mathrm{d}S$ with $\mathbf{s}$ the Poynting vector, and this integral vanishes in the 'quasi-static' approximation which drops the displacement current. Substitution of $\nabla\times\mathbf{E}_r = -\partial\mathbf{B}_r/c\,\partial t = \mathbf{B}_r/c\tau_r$ into (6.8) yields

$$\begin{aligned}\frac{4\pi}{\sigma}\int \mathbf{j}_r\cdot\mathbf{j}_s\,\mathrm{d}\tau &= \frac{1}{\tau_r}\int \mathbf{B}_r\cdot\mathbf{B}_s\,\mathrm{d}\tau,\\ &= \frac{1}{\tau_s}\int \mathbf{B}_r\cdot\mathbf{B}_s\,\mathrm{d}\tau,\end{aligned} \tag{6.9}$$

the last step following by symmetry. As the Lamb–Cowling eigenvalues are all different, (6.7) follows. With the normalization,

$$\int \mathbf{B}_r^2\,\mathrm{d}\tau = 1, \tag{6.10}$$

we may write

$$\int \mathbf{j}_r\cdot\mathbf{j}_s\,\mathrm{d}\tau = \frac{\sigma}{4\pi\tau_r}\delta_{rs}. \tag{6.11}$$

Now restore the velocity field $\mathbf{v}$ to (6.4). Then by a similar argument,

$$\frac{4\pi}{\sigma}\int \mathbf{j}\cdot\mathbf{j}_r\,\mathrm{d}\tau = -\int \frac{\partial \mathbf{B}}{\partial t}\cdot\mathbf{B}_r\,\mathrm{d}\tau + \frac{4\pi}{c}\int (\mathbf{v}\times\mathbf{B})\cdot\mathbf{j}_r\,\mathrm{d}\tau. \tag{6.12}$$

Substitution of the expansion (6.5) and the associated expansion for $\mathbf{j}$ and use of (6.7), (6.10) and (6.11) yields finally

$$\frac{\mathrm{d}b_r}{\mathrm{d}t} = -\frac{b_r}{\tau_r} + \sum_s a_{rs}b_s, \tag{6.13}$$

where

$$a_{rs} = (4\pi/c)\int (\mathbf{v}\times\mathbf{B}_s)\cdot\mathbf{j}_r\,\mathrm{d}\tau. \tag{6.14}$$

In kinematic dynamo theory one may write $\mathbf{v} = k\mathbf{v}_0$, where $\mathbf{v}_0$ is a plausible flow pattern, conveniently normalized, and k is an adjustable coefficient. Redefine the matrix a_{rs} in terms of $\mathbf{v}_0$. If this flow pattern is to maintain a field in a steady state then there must exist non-zero values of the coefficients b_r, requiring that k be an eigenvalue of

$$\frac{b_r}{\tau_r} = k\sum_s a_{rs}b_s, \tag{6.15}$$

satisfying the infinite determinantal equation

$$|ka_{rs} - \delta_{rs}/\tau_r| = 0. \tag{6.16}$$

The procedure is acceptable if there exist *real* eigenvalues k in the limit as $r \to \infty$. The matrix a_{rs} is not necessarily symmetric, so there is no guarantee that k will be real. Cowling (1955) illustrated this by applying the formalism to a two-dimensional field in Cartesian geometry, in which the field lines and flow lines are both confined to rectangular cells. An anti-dynamo theorem applies to this field, so it is no surprise that the matrix a_{rs} is found to be *anti*-symmetric, yielding pure imaginary values for k.

Any practical calculation is inevitably by successive approximation, and an apparent convergence towards real eigenvalues has sometimes turned out to be illusory. Dudley and James (1989) give a comprehensive survey of the problem, again for steady, incompressible flow. By studying the time-dependent hydromagnetic equation, they are able to distinguish between cases when all magnetic field modes decay, and those for which dynamo maintenance is possible. They generalize the expansion procedure introduced by Bullard and Gellman (1954), writing $\mathbf{v}$ and $\mathbf{B}$ in the forms

$$\mathbf{v} = \sum_\alpha(\mathbf{t}_\alpha + \mathbf{s}_\alpha), \qquad \mathbf{B} = \sum_\alpha(\mathbf{T}_\alpha + \mathbf{S}_\alpha), \tag{6.17}$$

with

$$\mathbf{t}_\alpha = \nabla\times[t_\alpha Y_\alpha(\theta,\phi)\hat{\mathbf{r}}],$$

$$\mathbf{s}_\alpha = \nabla \times \nabla \times [s_\alpha Y_\alpha(\theta, \phi)\hat{\mathbf{r}}], \tag{6.18}$$

and similarly for $\mathbf{T}_\alpha$, $\mathbf{S}_\alpha$. The functions Y_α are surface harmonics; the scalar functions t_α, s_α depend just on r, while S_α, T_α depend on both r and t. Again, solutions are sought for a variety of prescribed velocity fields. The interactions between different magnetic and velocity modes are non-zero only when certain selection rules are satisfied. Dudley and James's results incorporate those of Bullard and Gellman (1954), Braginsky (1965), Gibson *et al.* (1969), Lilley (1970), G.O. Roberts (1970, 1972), Gubbins (1973), Pekeris *et al.* (1973) and of Kumar and Roberts (1975). They find dynamo action by an axisymmetric single roll flow, even simpler than that discussed by Gailitis, in their words 'dispelling the belief that dynamo maintenance relies on the supporting flow being complex, and having length scale significantly less than that of the conducting fluid volume'.

6.3 The Parker model

As seen in Section 5.8, the difficulty with strictly axisymmetric fields is that there is no way to replenish the poloidal flux that is steadily being lost through diffusion into an O-type neutral point. As before, we write $\mathbf{B} = \mathbf{B}_\mathrm{p} + \mathbf{B}_\mathrm{t}$, with the poloidal and toroidal components given by

$$\mathbf{B}_\mathrm{p} = \nabla \times \mathbf{A} = \nabla \times (A\mathbf{t}), \qquad \mathbf{B}_\mathrm{t} = B_\phi \mathbf{t}, \tag{6.19}$$

and the velocity $\mathbf{v} = \mathbf{v}_\mathrm{p} + \Omega\varpi\mathbf{t}$. The induction equation (6.4) for a homogeneous medium then breaks into poloidal and toroidal parts:

$$\frac{\partial \varpi A}{\partial t} + \mathbf{v}_\mathrm{p} \cdot \nabla(\varpi A) = \lambda\varpi(\nabla^2 - \varpi^{-2})A \tag{6.20}$$

and

$$\frac{\partial B_\phi}{\partial t} + \varpi\nabla \cdot \left(\frac{B_\phi}{\varpi}\mathbf{v}_\mathrm{p}\right) = \varpi\mathbf{B}_\mathrm{p} \cdot \nabla\Omega + \lambda(\nabla^2 - \varpi^{-2})B_\phi. \tag{6.21}$$

The term $\mathbf{B}_\mathrm{p} \cdot \nabla\Omega$ in (6.21) represents the familiar generation of $\mathbf{B}_\mathrm{t}$ by the shearing of $\mathbf{B}_\mathrm{p}$. The 'topological asymmetry' between the poloidal and toroidal field shows up through the absence of an analogous term in (6.20). In a seminal paper (1955), Parker argued that in a rotating star, convective motions would be able to complete the cycle by generating a poloidal component from a toroidal. The process is illustrated in Fig. 6.3 with the toroidal field modelled by a horizontal field $\mathbf{B}_0$ (Fig. 6.3(a)). Seen from a frame that rotates with the local angular velocity, as a rising blob of gas expands it feels a Coriolis force, which gives it an anti-cyclonic motion – just conservation of angular momentum (Fig. 6.3(b)). The motion is left-handed in the northern hemisphere and right-handed in the southern, as defined by the rotation axis. The nearly frozen-in toroidal field is thus twisted so as to yield a poloidal component (Fig. 6.3(c)). The sinking blobs have a reverse effect, but because of the negative radial density gradient, the rising blobs dominate. The small-scale poloidal loops so formed are pictured as coalescing through reconnection, yielding a large-scale field.

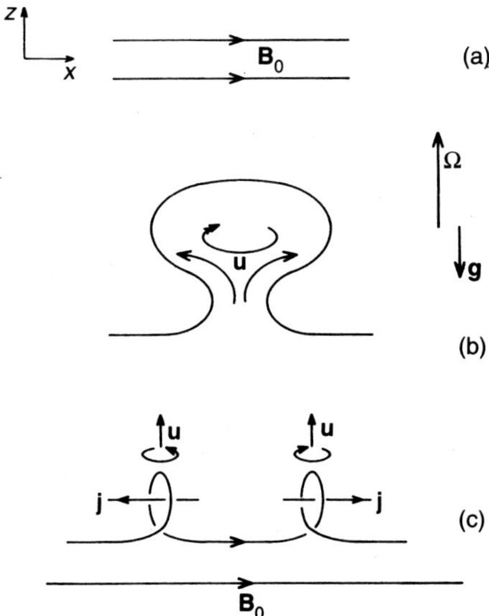

FIG. 6.3. The basic Parker mechanism. (a) The toroidal field modelled by a horizontal component $\mathbf{B}_0$. (b) The velocity $\mathbf{u}$, measured in the local rotating frame, in an upwelling in a strongly compressible stratified rotating medium. Initially vertical motions quickly become largely horizontal (single-headed arrows), so that Coriolis force yields a cyclonic rotation (double-headed arrow). (c) The initial horizontal field $\mathbf{B}_0$ is twisted by the velocity with a cyclonic component. On the left, $\mathbf{u}$ is parallel to $\nabla\times\mathbf{u}$, and $\mathbf{j}\cdot\mathbf{B}_0 < 0$; on the right, $\mathbf{u}$ is antiparallel to $\nabla\times\mathbf{u}$, and $\mathbf{j}\cdot\mathbf{B} > 0$. (After Moffatt 1978 and Moss 1986.)

Since the rate of generation of poloidal flux is proportional to B_ϕ, Parker modelled the net effect by inserting an extra term into (6.20); in the notation (slightly different from Parker's) which has now become standard,

$$\frac{\partial A}{\partial t} + \frac{1}{\varpi}\mathbf{v}_{\mathrm{p}} \cdot \nabla(\varpi A) = \alpha B_\phi + \eta(\nabla^2 - \varpi^{-2})A. \tag{6.22}$$

The coefficient α has the dimensions of velocity. From (6.19) and (6.21), to maintain a basically dipolar field, the form $\alpha = \alpha_0 \cos\theta$ is the simplest with the correct parity properties. In addition (cf. Sections 3.8, 4.3, 5.8 and 6.4), the microresistivity λ must be replaced by a much larger turbulent resistivity η.

Because of the 'α-effect' term, the kinematic dynamo equations can predict exponentially growing, axisymmetric fields with both poloidal and toroidal components. Cowling's theorem is not violated: the necessary non-axisymmetric motions occur but on a small scale, and their effects are parametrized by the

α-term. The simplest illustrative example (Parker 1955) adopts Cartesian geometry, with the z-axis normal to the local solar surface and the y-axis in the easterly (toroidal) direction. In the unperturbed state, $\partial/\partial y = 0$ in the mean is the analogue of axisymmetry, and $\mathbf{v} = (0, V(z), 0)$ is the analogue of non-uniform rotation. We satisfy $\nabla \cdot \mathbf{B} = 0$ by writing $\mathbf{B} = (-\partial A/\partial z, B_y, \partial A/\partial x)$; η is taken as constant. Equations (6.21) and (6.22) then become

$$\frac{\partial B_y}{\partial t} = V' \frac{\partial A}{\partial x} + \eta \nabla^2 B_y, \tag{6.23}$$

$$\frac{\partial A}{\partial t} = \alpha B_y + \eta \nabla^2 A, \tag{6.24}$$

where $V' = \mathrm{d}V/\mathrm{d}z$. Solutions proportional to $\exp(\mathrm{i}kx + pt)$ exist with

$$(p + \eta k^2)^2 = \mathrm{i}\alpha V' k = (1 \pm \mathrm{i})^2 |D| \eta^2 k^4 \tag{6.25}$$

where D is the *dynamo number*

$$D = \frac{\alpha V'}{2\eta^2 k^3}. \tag{6.26}$$

The $\pm$ signs refer to $D > 0$, $D < 0$ respectively. Thus if $|D| > 1$, then there exist exponentially growing dynamo wave solutions of the form

$$\exp[\eta k^2 (|D|^{1/2} - 1)t] \exp[\mathrm{i}(kx \pm \eta k^2 |D|^{1/2} t)]. \tag{6.27}$$

If $D > 0$, the growing waves travel in the negative x-direction (polewards); if $D < 0$, they travel equatorwards. Further, because of the enormous turbulent resistivity, there is no difficulty in finding a period

$$2\pi / \eta k^2 |D|^{1/2} \tag{6.28}$$

of the order of the solar cycle period rather than the Cowling decay time (cf. Section 8.6).

Parker's picture became clearly relevant to the Sun when H.D. Babcock (1959) discovered that the solar poloidal field was reversing along with the sunspot cycle. H.W. Babcock (1961) and Leighton (1964, 1969) developed phenomenological models that implicitly incorporate and illustrate Parker's ideas. More recently, Leighton's model has been revived by Wang and Sheeley (1991) and by Wang *et al.* (1989, 1991), and by Parker (1994). We return to this '$\alpha\Omega$ dynamo' picture of the solar cycle in Section 6.5 and in Chapter 8.

An analogous model can be applied also in disc-like geometry, whether to an accretion disc surrounding a star (Section 10.6) or to a disc-like galaxy (Parker 1971; Zel'dovich *et al.* 1983; Ruzmaikin *et al.* 1988; Beck *et al.* 1996, and many other papers). In the galactic case, the problem is that the e-folding time predicted may be as long as 2×10^8 yr, and attention has to be focused on how to generate a sufficiently large cosmological 'seed field' (e.g. Rees 1994).

6.4 Mean-field electrodynamics

Implicit in the Parker picture is the division of the motions into the large-scale non-uniform rotation which generates the toroidal field from the poloidal, and the somewhat smaller-scale convective eddies in which the toroidal→poloidal conversion occurs. The axisymmetric equation (6.22) depends on an implicit averaging over the small scales. This suggests a formal attack on the general problem of the turbulent dynamo by a two-scale analysis (Steenbeck *et al.* 1966; papers by Steenbeck *et al.*, translated by Roberts and Stix 1971; Krause and Rädler 1980; Moffatt 1978; Parker 1970, 1979; Zel'dovich *et al.* 1983; and many papers since). In particular, the systematic study of 'mean-field electrodynamics' by the group originally at Jena and later at Potsdam (Steenbeck, Krause, Rädler and colleagues) has been a strong influence on the world-wide development of dynamo theory.

In the simplest problem, there is a small-scale turbulent velocity field $\mathbf{v}$, with a characteristic length l, statistically steady and spatially homogeneous, *but not necessarily isotropic.* There is an associated fluctuating small-scale magnetic field $\mathbf{b}$, due to the interaction of the turbulence with a field $\mathbf{B}$ that varies over a characteristic length $L \gg l$; but the essence of the dynamo is that $\mathbf{B}$ is itself maintained by the averaged electromotive force due to the interaction of the $\mathbf{v}$ and $\mathbf{b}$ fields. The induction equation becomes

$$\frac{\partial}{\partial t}(\mathbf{B}+\mathbf{b}) = \nabla\times[\mathbf{v}\times(\mathbf{B}+\mathbf{b})] + \lambda\nabla^2(\mathbf{B}+\mathbf{b}). \tag{6.29}$$

Over a scale intermediate between l and L, the averages $\overline{\mathbf{v}}$, $\overline{\mathbf{b}}$ vanish, so from (6.29),

$$\frac{\partial \mathbf{B}}{\partial t} = \nabla\times\mathcal{E} + \lambda\nabla^2\mathbf{B}, \tag{6.30}$$

where the mean electromotive force $\mathcal{E}$ is defined by

$$\mathcal{E} \equiv \overline{\mathbf{v}\times\mathbf{b}}. \tag{6.31}$$

Subtraction of (6.30) from (6.29) yields

$$\frac{\partial \mathbf{b}}{\partial t} = \nabla\times(\mathbf{v}\times\mathbf{B}) + [\nabla\times(\mathbf{v}\times\mathbf{b}) - \nabla\times(\overline{\mathbf{v}\times\mathbf{b}})] + \lambda\nabla^2\mathbf{b}. \tag{6.32}$$

Equation (6.32) relating $\mathbf{b}$ to $\mathbf{B}$ is linear: apart from any transients, the small-scale field exists ultimately through the interaction of the turbulence with the large-scale field. Thus the dynamo term $\overline{\mathbf{v}\times\mathbf{b}}$ in (6.30) must be linearly dependent on $\mathbf{B}$. If as assumed $L \gg l$, one can write

$$\mathcal{E}_i = \alpha_{ij}B_j + \beta_{ijk}\frac{\partial B_k}{\partial x_j} + \ldots \tag{6.33}$$

where α_{ij}, $\beta_{ijk}, \ldots$ depend on the statistical properties of the turbulence. If the turbulence is pseudo-isotropic (i.e. invariant under rotation of axes, but not necessarily under reflection in the origin) then we may write

$$\alpha_{ij} = \alpha\delta_{ij}, \quad \beta_{ijk} = -\beta\epsilon_{ijk}, \tag{6.34}$$

since the Kronecker delta and the alternating tensor are the only isotropic tensors respectively of the second and third rank; whence from (6.33)

$$\mathcal{E} \simeq \alpha\mathbf{B} - \beta\nabla \times \mathbf{B}. \tag{6.35}$$

Note that $\mathcal{E}$ like $\mathbf{E}$ is a polar vector, with components that change sign on reflection of the axes in the origin, whereas $\mathbf{B}$ is an axial vector; hence whereas β is a pure scalar, α is a pseudo-scalar, changing sign on reflection in the origin. If in addition the turbulence were strictly isotropic and so also mirror-symmetric – invariant for reflections of the axes in the origin – then α, being itself a property of the turbulence, would have to be zero; but if the turbulence is pseudo-isotropic, α will not vanish and genuine dynamo action can occur.

The calculation of $\mathcal{E} \equiv \overline{\mathbf{v} \times \mathbf{b}}$ is greatly simplified if one can neglect the two non-linear terms in (6.32) – the 'first-order smoothing approximation' – even though this amounts to suppressing the transfer of energy by cascade between different wave numbers, characteristic of hydrodynamic turbulence. One case for which this is a good approximation is the low magnetic Reynolds number limit, unlikely to be a good model for stellar applications, but useful as an illustration of principles. The following treatment follows closely that given by Moffatt (1970, 1978). When $\mathrm{R_m} \ll 1$, then $v_0 l_0/\lambda \simeq v_0^2 t_0/\lambda \simeq l_0^2/t_0\lambda \ll 1$, where l_0, t_0 are characteristic length- and time-scales and $v_0 \simeq (\overline{v^2})^{1/2}$. The time-derivative in (6.32) is then similarly small, so there is an effectively instantaneous balance between inductive and diffusive effects:

$$\begin{aligned}\lambda\nabla^2\mathbf{b} = -\nabla \times (\mathbf{v} \times \mathbf{B}) &= -(\mathbf{B}\cdot\nabla)\mathbf{v} + (\nabla\cdot\mathbf{v})\mathbf{B} - (\mathbf{v}\cdot\nabla)\mathbf{B} \\ &\approx -(\mathbf{B}\cdot\nabla)\mathbf{v},\end{aligned} \tag{6.36}$$

where in the last line the turbulence is assumed to be very subsonic, so that $\nabla \cdot \mathbf{v} \approx 0$, and again $L \gg l$.

The velocity $\mathbf{v}$ and its Fourier transform $\tilde{\mathbf{v}}$ are related by

$$\tilde{\mathbf{v}}(\mathbf{k}) = \frac{1}{(2\pi)^3}\int \mathbf{v}(\mathbf{x}) \exp(-\mathrm{i}\mathbf{k}\cdot\mathbf{x})\, \mathrm{d}\mathbf{x}, \tag{6.37}$$

and the inverse relation

$$\mathbf{v}(\mathbf{x}) = \int \tilde{\mathbf{v}}(\mathbf{k}) \exp(\mathrm{i}\mathbf{k}\cdot\mathbf{x})\, \mathrm{d}\mathbf{k}. \tag{6.38}$$

Since $\mathbf{v}$ is real,

$$\tilde{\mathbf{v}}(-\mathbf{k}) = \tilde{\mathbf{v}}^*(\mathbf{k}), \tag{6.39}$$

where the asterisk denotes the complex conjugate. From $\nabla \cdot \mathbf{v} = 0$,

$$\mathbf{k} \cdot \tilde{\mathbf{v}}(\mathbf{k}) = 0. \tag{6.40}$$

By (6.36), $\tilde{\mathbf{v}}$ and the Fourier transform $\tilde{\mathbf{b}}$ of $\mathbf{b}$ are related by

$$\lambda k^2 \tilde{\mathbf{b}} = \mathrm{i}(\mathbf{B} \cdot \mathbf{k})\tilde{\mathbf{v}}, \tag{6.41}$$

so that the turbulent emf may be written

$$\mathcal{E}_k(\mathbf{x}) = (\overline{\mathbf{v} \times \mathbf{b}})_k = \epsilon_{kij} \int\int (\mathrm{i}\mathbf{B} \cdot \mathbf{k}/\lambda k^2)\overline{\tilde{v}_i(\mathbf{k})\tilde{v}_j^*(\mathbf{k}')} \exp \mathrm{i}(\mathbf{k} - \mathbf{k}') \cdot \mathbf{x} \, \mathrm{d}\mathbf{k} \, \mathrm{d}\mathbf{k}', \tag{6.42}$$

where

$$\overline{\tilde{v}_i(\mathbf{k})\tilde{v}_j^*(\mathbf{k}')} = \frac{1}{(2\pi)^6} \int\int \overline{v_i(\mathbf{x})v_j(\mathbf{x}')} \exp -\mathrm{i}(\mathbf{k} \cdot \mathbf{x} - \mathbf{k}' \cdot \mathbf{x}') \, \mathrm{d}\mathbf{x} \, \mathrm{d}\mathbf{x}'. \tag{6.43}$$

If the turbulence is homogeneous, then the correlation tensor of the field $\mathbf{v}$, defined by

$$R_{ij} \equiv \overline{v_i(\mathbf{x})v_j(\mathbf{x}')}, \tag{6.44}$$

must depend just on $\mathbf{r} \equiv (\mathbf{x} - \mathbf{x}')$, so that (6.43) becomes

$$\overline{\tilde{v}_i(\mathbf{k})\tilde{v}_j^*(\mathbf{k}')} = \frac{1}{(2\pi)^3} \int R_{ij}(\mathbf{r}) \exp -\mathrm{i}(\mathbf{k} \cdot \mathbf{r}) \, \mathrm{d}\mathbf{r} \equiv \Phi_{ij}(\mathbf{k}), \tag{6.45}$$

where $\Phi_{ij}(\mathbf{k})$ is the spectrum tensor, satisfying from (6.39) the Hermitian relations

$$\Phi_{ij}(\mathbf{k}) = \Phi_{ji}(-\mathbf{k}) = \Phi_{ji}^*(\mathbf{k}), \tag{6.46}$$

and from (6.40)

$$k_i \Phi_{ij}(\mathbf{k}) = 0, \quad k_j \Phi_{ij}(\mathbf{k}) = 0. \tag{6.47}$$

Hence (6.42) can be written $\mathcal{E}_i = \alpha_{ij} B_j$, with

$$\alpha_{ij} = (\mathrm{i}/\lambda)\epsilon_{ilm} \int \Phi_{lm} k^{-2} k_j \, \mathrm{d}\mathbf{k}, \tag{6.48}$$

(it follows from (6.46) that α_{ij} is real, as required). Imposing the extra condition of pseudo-isotropy yields $\alpha_{ij} = \alpha\delta_{ij}$ with $\alpha = \alpha_{ii}/3$.

The energy spectrum $E(k)$ of the turbulence is defined by

$$E(k) = \frac{1}{2} \int_{S_k} \Phi_{ii}(\mathbf{k}) \, \mathrm{d}S, \tag{6.49}$$

where S_k is the sphere of radius k in $\mathbf{k}$-space. The mean kinetic energy per unit mass is

$$\frac{1}{2}\overline{\mathbf{v}^2} = \frac{1}{2}R_{ii}(0) = \frac{1}{2}\int \Phi_{ii}(\mathbf{k})\,\mathrm{d}\mathbf{k} = \int E(k)\,\mathrm{d}k. \tag{6.50}$$

The vorticity field $\omega = \nabla \times \mathbf{v}$ has the Fourier transform $\tilde{\omega} = i\mathbf{k} \times \tilde{\mathbf{v}}$ and so its spectrum tensor has the form

$$\Omega_{ij}(\mathbf{k}) = \epsilon_{imn}\epsilon_{jpq}k_m k_p \Phi_{nq}. \tag{6.51}$$

By (6.47),

$$\Omega_{ii}(\mathbf{k}) = k^2 \Phi_{ii}, \tag{6.52}$$

whence

$$\frac{1}{2}\overline{\omega^2} = \int k^2 E(k)\,\mathrm{d}k. \tag{6.53}$$

By analogy with (6.49), the *helicity spectrum function* $F(k)$ is defined by

$$F(k) = \mathrm{i}\int_{S_k} \epsilon_{ijl}k_j \Phi_{il}(\mathbf{k})\,\mathrm{d}S, \tag{6.54}$$

whence

$$\overline{\mathbf{v}\cdot\omega} = \mathrm{i}\epsilon_{ijl}\int k_j \Phi_{il}\,\mathrm{d}\mathbf{k} = \int F(k)\,\mathrm{d}\mathbf{k}. \tag{6.55}$$

By (6.46), $F(k)$ is a real pseudo-scalar, and so is non-zero if the turbulence is non-mirror-symmetric. Comparison of (6.54) and (6.48) shows that under pseudo-isotropy,

$$\alpha = \alpha_{ii}/3 = -(\mathrm{i}/3\lambda)\epsilon_{lim}\int\int_{S_k} k_i \Phi_{lm} k^{-2}\,\mathrm{d}S\,\mathrm{d}k = -(1/3\lambda)\int k^{-2}F(k)\,\mathrm{d}k. \tag{6.56}$$

This formula is qualitatively significant in that it shows explicitly the relation between the α-effect and helicity, which is the simplest manifestation of the lack of mirror symmetry. If the turbulence were strictly isotropic, the dynamo would vanish. However, the magnitude and the sign of α depend on further details of the turbulent field.

Similar analysis (Moffatt 1978) yields

$$\beta = (2/3\lambda)\int k^{-2}E(k)\,\mathrm{d}k \tag{6.57}$$

for the turbulent resistivity. This shows formally (for the special case of low $\mathrm{R_m}$ turbulence) that there is an effective resistivity roughly proportional to the energy in both isotropic and pseudo-isotropic turbulence (cf. the discussions in Sections 4.3 and 5.8), but it is the pseudo-isotropy as measured by the helicity that yields the more subtle α-effect that in principle can lead to the generation and maintenance of large-scale magnetic fields.

Note that from (6.41), $\tilde{\mathbf{b}}$ and $\tilde{\mathbf{v}}$ are out of phase (by $\pi/2$), which ensures that $\mathcal{E}$ given by (6.42) can be non-zero (cf. Krause 1965 and Moffatt 1978). With first-order smoothing still assumed valid, retention in (6.36) of the $\partial\mathbf{b}/\partial t$ term from (6.32) replaces (6.48) by

$$\alpha_{ij} = \mathrm{i}\lambda\epsilon_{ilm}\int\int(\omega^2+\lambda^2k^4)^{-1}k^2k_j\Phi_{lm}(\mathbf{k},\omega)\,\mathrm{d}\mathbf{k}\,\mathrm{d}\omega, \tag{6.58}$$

and (6.56) by

$$\alpha = \alpha_{ii}/3 = -\frac{\lambda}{3}\int\int\frac{k^2F(k,\omega)}{\omega^2+\lambda^2k^4}\,\mathrm{d}k\,\mathrm{d}\omega, \tag{6.59}$$

again on use of (6.54) (Moffatt 1978). The low $\mathrm{R_m}$ case assumes that most of the turbulent energy is in the wave number band for which $\omega \simeq vk \ll \lambda k^2$. If $\Phi(\mathbf{k},\omega) = O(\omega^2)$ as $\omega \to 0$ – i.e. if there is no energy in zero-frequency modes – then the phase difference between $\mathbf{b}$ and $\mathbf{v}$ and so also the α-effect vanish in the limit $\lambda \to 0$. But if $\Phi_{lm}(\mathbf{k},0) \neq 0$, then as $\lambda \to 0$, (6.58) yields

$$\alpha_{ij} \simeq \pi\mathrm{i}\epsilon_{ilm}\int k_j\Phi_{lm}(\mathbf{k},0)\,\mathrm{d}\mathbf{k}, \tag{6.60}$$

implying that dynamo action continues at a finite rate even in the limit of vanishing micro-resistivity. However, as emphasized by Moffatt, as the asymptotic expression (6.60) is determined entirely by the spectral density at $\omega = 0$, the limiting procedure is not compatible with the first-order smoothing approximation, so the result (6.60) is no more than a hint.

It is intuitively satisfactory that in order to build a *large-scale* field, there needs to be some overall lack of isotropy in the velocity field, as measured by the non-vanishing mean helicity. Earlier, Batchelor (1950) had suggested that strictly isotropic turbulence would also act as a dynamo, with the energy cascading down from large to small scales being partly converted into magnetic energy, yielding a turbulent magnetic field that is in approximate equipartition with kinetic energy on small scales. The case was based on the similarity between the kinematic equation (6.4) for $\mathbf{B}$ and Helmholtz's equation for $\omega = \nabla\times\mathbf{v}$ in non-magnetofluid mechanics. The argument is questionable on several grounds. If it were valid, it would apply equally to two-dimensional systems, for which we know that genuine dynamo action is impossible (Zel'dovich *et al.* 1983). The equations for $\mathbf{B}$ and ω are not identical, because of the non-linearity of the Helmholtz equation. Further, a steady turbulent cascade implies a continuous input of vorticity at large scales. Thus if the $\omega - \mathbf{B}$ analogy is taken literally, it would imply a necessary continuous input of large-scale magnetic flux from a 'battery': the system would be acting as an amplifier rather than a fully self-excited dynamo (Section 6.1; Saffman 1963; Mestel 1965). More recently, Meneguzzi *et al.* (1981) carried out direct numerical simulations of three-dimensional incompressible magnetohydrodynamic turbulence, at kinetic and magnetic Reynolds numbers up to 100, considering both helical and non-helical flows. They confirmed that small-scale helical flow produces strong, large-scale magnetic fields, that have $|\mathbf{j}\cdot\mathbf{B}|/jB \approx 1$ and so are nearly force-free. In the non-helical cases, with no α-effect, they found evidence for spatially intermittent('fibril') fields, with a slight excess of magnetic energy at high wave numbers, but with integrated magnetic energy only 10% of the kinetic. (See also Molchanov *et al.* 1985; Zel'dovich *et al.* 1988; and references in Belyanin *et al.* 1993.)

6.5 Kinematic models of the turbulent dynamo

6.5.1 *General discussion*

The formal mean-field electrodynamic expansion, applied to turbulence lacking mirror-symmetry, together with the first-order smoothing approximation, applied in the low R_m limit, yields the two new terms: the α-effect – the emf $\mathcal{E} = \alpha\mathbf{B}$ along $\mathbf{B}$ – and the resistive term $-\eta\nabla\times\mathbf{B}$ (where the symbol η rather than β is used from now on). The 'standard dynamo' equations for axisymmetric systems with a laminar circulation $\mathbf{v}_p$ are now Parker's equation (6.22) – a modification of (6.20) – together with a similar modification to (6.21):

$$\frac{\partial B_\phi}{\partial t} + \varpi\nabla\left(\frac{B_\phi}{\varpi}\mathbf{v}_p\right) = \varpi\mathbf{B}_p\cdot\nabla\Omega + \eta(\nabla^2 - \varpi^{-2})B_\phi + \frac{1}{\varpi}\nabla\eta\cdot\nabla(\varpi B_\phi)$$
$$- \alpha(\nabla^2 - \varpi^{-2})A - \frac{1}{\varpi}\nabla\alpha\cdot\nabla(\varpi A). \tag{6.61}$$

From (6.56), (6.57), (6.55) and (6.50), the order-of-magnitude relation

$$\alpha \simeq -\eta\frac{\overline{\mathbf{v}\cdot\omega}}{\overline{\mathbf{v}^2}} \tag{6.62}$$

holds. Although a rigorous justification is available only in the low MRN limit, standard dynamo theory adopts the same equations but with the turbulent resistivity η of the order of lv, where $l = \tau_c v$ is the typical length-scale of a turbulent eddy of turnover time τ_c. Note that if the microresistivity λ is replaced by lv, then by (6.50), and with $k^{-1} \simeq l$, (6.57) yields $\beta \simeq \eta$, as expected.

The idea of a greatly enhanced rate of dissipation of weak fields – described by a 'turbulent resistivity' – goes back to Cowling (1945), Sweet (1950) and Spitzer (1957) (cf. Section 5.8). The new feature is the α-effect. The term αB_ϕ in (6.22) is essential in order to complete the dynamo cycle by generating a mean poloidal field from a toroidal. The terms in α in (6.61) are a bonus, contributing along with the familiar $\nabla\Omega$ term to the generation of a toroidal field from a poloidal: one can study not only '$\alpha\Omega$' but also 'α^2' and '$\alpha^2\Omega$' dynamos.

The kinematic dynamo equations (6.22) and (6.61) have an intuitive appeal primarily through their completing the poloidal→toroidal→poloidal cycle, but also because an enormously increased effective resistivity can yield periods of the order of the solar cycle (cf. Sections 8.5 and 8.6). However, the attempted rigorous justification of the equations in Section 6.4 is valid only under conditions which do not in fact hold inside stars (Cowling 1981; Weiss 1983). First, there is no clear-cut large separation of scales between that of the turbulent motions and that of the non-uniform rotation or the generated field. And second, the first-order smoothing approximation is justified when either the magnetic Reynolds number of the turbulence is small or the lifetime of convective eddies is long compared with the turnover time τ_c: neither assumption is valid inside stars.

These 'standard kinematic dynamo equations', with the pseudo-scalar α proportional to the local helicity, are best looked on as the simplest plausible parametrizations that incorporate the essence of the physics. They assume implicitly that 'fast dynamos' exist, as suggested by the non-rigorous result (6.60), i.e. that in the limit of vanishingly small micro-resistivity, an effective macro-resistivity still describes the necessary dissipation of unwanted flux, occurring on correspondingly small scales. (For a formal definition of a 'fast dynamo', and a detailed discussion of the associated mathematical problems, see Childress and Gilbert 1995.) Even with the limitations of purely kinematic theory, with no cognizance of the back-reaction of the Lorentz forces on the flow on any scale, there is ample scope for generalization. More sophisticated treatments, both analytic (Rüdiger and Kitchatinov 1993) and numerical (Brandenburg *et al.* 1990), which take account of the gradients of both density and turbulent intensity, yield more complicated equations, with a non-isotropic α-tensor, sometimes with differing signs for the components in the vertical and horizontal directions (cf. Section 8.6). The growing Lorentz forces will in fact modify both the macroscopic flow, and – more significantly – react on the microphysics, crucial to dynamo action, encapsulated in the α-effect and the turbulent resistivity. In most of the rest of this chapter we shall present some of the more interesting consequences of both the unmodified, formally kinematic, standard dynamo equations, and of 'quasi-kinematic' equations which include phenomenological modifications to the dynamo terms. We conclude by looking at the criticisms of standard dynamo theory which are enforcing a reappraisal of the basics of the theory.

Most effort has been devoted to the $\alpha\Omega$ dynamo, of which Parker's original paper (1955) and the Babcock–Leighton papers discussed in Section 8.5 are the paradigms. The effect of the differential rotation is measured by a magnetic Reynolds number $\mathrm{R}_\Omega = |\nabla\Omega| d^3/\eta$ with d an appropriate length scale. Likewise, the non-dimensional number measuring the α-effect is $\mathrm{R}_\alpha = \alpha d/\eta$; and the parameter determining dynamo action is again the dynamo number D, given in modulus by

$$|D| = \mathrm{R}_\alpha \mathrm{R}_\Omega = \alpha|\nabla\Omega| d^4/\eta^2. \tag{6.63}$$

Again, for turbulent eddies with a typical velocity v, turnover time τ_c, and length-scale $l \approx \overline{1/k} \approx v\tau_c$, then $\eta \approx \tau_c v^2 = lv = l^2/\tau_c$, and by (6.62),

$$\alpha \approx -\overline{\mathbf{v}\cdot(\nabla\times\mathbf{v})}\eta/\mathbf{v}^2 = -\tau_c\overline{\mathbf{v}\cdot(\nabla\times\mathbf{v})}. \tag{6.64}$$

The Coriolis force per unit mass $-\boldsymbol{\Omega}\times\mathbf{v}$ acting on a turbulent element rising with velocity $\mathbf{v}$ for a time $\simeq \tau_c$ generates a velocity $\approx \Omega v\tau_c$ with a curl $\simeq \Omega v\tau_c/l \simeq \Omega$, a helicity $\simeq v\Omega$ and so a value for $|\alpha| \simeq \Omega l$. Hence

$$|D| \simeq \Omega^2(|\nabla\Omega|/\Omega)ld^4/\eta^2 \simeq (\Omega\tau_c)^2(d/l)^4(|\nabla\Omega|l/\Omega). \tag{6.65}$$

The simplest provisional assumption is that there is just one length-scale, so that $d \simeq l$ and $|\nabla\Omega|l/\Omega \simeq 1$, yielding

$$|D| \simeq \sigma^2 = (\Omega\tau_c)^2, \tag{6.66}$$

where $\sigma = \Omega\tau_c$ is an inverse Rossby number (Durney and Latour 1978). The predicted growth rate, analogous to (6.27), is proportional to

$$|D|^{1/2} \propto (\Omega\tau_c) \tag{6.67}$$

– a result that is prima facie highly significant for the 'solar–stellar connection' (cf. Section 8.3). Further, anticipating that global models will also predict a cycle period $P_{\rm cycl}$ similar to (6.28), we find that for stars of the same structure this yields

$$P_{\rm cycl} \propto \Omega_{\rm cycl}^{-1} \propto |D|^{-1/2} \propto \Omega^{-1} \propto P. \tag{6.68}$$

It is not surprising that this linear model predicts a cycle frequency that decreases with the rotation frequency; however, for comparison with observation one awaits the predictions of the non-linear theories (cf. Sections 6.6 and 8.2).

Even at this preliminary, purely kinematic stage, there is clearly scope for a wide variety of dynamo models, described by these simplest dynamo equations. One is impelled to explore the sensitivity of the kinematic $\alpha\Omega$-dynamo solutions to prescribed variations in the spatial distribution of the α, Ω and η fields. One anticipates some dependence on the sign of the appropriately defined dynamo number. The simplest global model (Steenbeck and Krause 1966; P.H. Roberts 1972) adopts

$$\alpha = \alpha_0 \cos\theta, \qquad \Omega = \Omega' r, \tag{6.69}$$

with α_0, Ω' constants, holding over the interior of a sphere of radius R. This model yields only oscillatory dynamos (global analogues of the Parker dynamo waves of Section 6.3) with fields either symmetric (quadrupole type) or anti-symmetric (dipole type) about the equatorial plane. Other models, prima facie more plausible, restrict both the α-effect and the Ω-effect to the solar convection zone $0.7R_\odot < r < R_\odot$. Some of the models (Roberts and Stix 1972) include meridian circulation. All models can produce convincing looking simulations of the solar cycle with its butterfly diagram (Fig. 1.1). Quite generally, it emerges from these studies that to predict equatorwards migration of the dynamo waves, as suggested by the motion of the sunspot zones, the dynamo number D must be negative in the northern hemisphere. Detailed calculations for convective turbulence show that over the bulk of the solar convective zone the helicity is negative, yielding α positive and so requiring negative $\mathrm{d}\Omega/\mathrm{d}r$. This sets a problem for the kinematic dynamo, to which we return in Chapter 8.

6.5.2 *A model with separate shear and α-effect zones*

Most models, even when they allow α, η and $\nabla\Omega$ to vary with position, still have the α-effect and the shearing occurring in domains which overlap. For reasons, both observational and theoretical, to be discussed in Section 8.6, for the solar dynamo there has developed a near consensus which locates the generation by shear of the toroidal field in the overshoot region at the base of the convective envelope. It is then likely that this field will be so strong that it can suppress locally both the eddy diffusivity and the α-effect which, however, may remain effective in the convective domain (cf. Section 6.6). Several workers (Steenbeck and Krause 1969; Levy 1972*a,b*; Deinzer and Stix 1971; Stix 1973) have constructed dynamo models with separated regions of shear and cyclonic convection, but with a uniform diffusivity throughout. Brandenburg *et al.* (1992) have constructed a model with large variations in η but with the shear broadly distributed. We now summarize a recent illustrative model in Cartesian geometry (Parker 1993), with a dynamo surface wave on the interface separating two domains: one with strong shear, zero α-effect and weak diffusivity, simulating the overshoot domain bounding the solar radiative core, and another with zero shear, strong diffusivity and a normal α-effect, simulating the convective envelope. As with the simple dynamo wave model (6.23)–(6.26), the y-axis represents the azimuthal (toroidal) direction and the z-axis the vertical. The field is written

$$B_x = -\partial A/\partial z,\ B_y = B(x,z,t),\ B_z = \partial A/\partial x, \quad \text{for } z > 0, \tag{6.70}$$

$$b_x = -\partial a/\partial z,\ b_y = b(x,z,t),\ b_z = \partial a/\partial x, \quad \text{for } z < 0, \tag{6.71}$$

where $A(x,z,t)$ and $a(x,z,t)$ are the azimuthal vector potentials describing the poloidal fields in $z > 0$, $z < 0$ respectively. The dynamo equations are

$$\left[\frac{\partial}{\partial t} - \eta\nabla^2\right] B = 0, \tag{6.72}$$

$$\left[\frac{\partial}{\partial t} - \eta\nabla^2\right] A = \alpha B, \tag{6.73}$$

in $z > 0$, and

$$\left[\frac{\partial}{\partial t} - n\nabla^2\right] b = G\frac{\partial a}{\partial x}, \tag{6.74}$$

$$\left[\frac{\partial}{\partial t} - n\nabla^2\right] a = 0 \tag{6.75}$$

in $z < 0$. Here $G = \mathrm{d}v_y/\mathrm{d}z$ is a uniform shear and n the reduced eddy diffusivity. Because of the analogue of axisymmetry, $\nabla^2 \equiv \partial^2/\partial x^2 + \partial^2/\partial z^2$. The boundary conditions are the continuity of thc normal and tangential field components across $z = 0$, so that $b = B$, $a = A$ and $\partial a/\partial z = \partial A/\partial z$. Conservation of flux diffusing across $z = 0$ requires that

$$n\,\partial b/\partial z = \eta\,\partial B/\partial z. \tag{6.76}$$

The fields all go to zero at $z = \pm\infty$, since neither G nor α acting alone yields dynamo action.

Parker constructs plane wave solutions of the form in $z > 0$

$$B = C\exp(\sigma t - Sz)\exp\mathrm{i}(\omega t + kx - Qz), \tag{6.77}$$

$$A = (F + Ez)\exp(\sigma t - Sz)\exp\mathrm{i}(\omega t + kx - Qz), \tag{6.78}$$

with C an arbitrary real amplitude, σ, S, ω, k and Q all real, and with $S > 0$ to ensure that the field vanishes at $z = \infty$. (The coefficient F rather than Parker's D is used in (6.78) to avoid confusion with our standard use of D for the dynamo number.) In $z < 0$, the solutions are of similar form, but with Q replaced by $-q$ and S by $-s < 0$. Substitution into the above equations and the link-up conditions at $z = 0$ determine the solutions, which are conveniently written in terms of the ratio of diffusivities n/η, and a hybrid dynamo number $D = \alpha G/\eta^2 k^3$, depending on the parametrized turbulence in $z > 0$ and the shear in $z < 0$.

Two limiting cases are both simpler and of particular interest. The 'standard case', with uniform diffusivity $n = \eta$, yields $s = S$, $q = Q$, satisfying for exponentially growing modes ($\sigma > 0$) the relation $S + \mathrm{i}Q = (kD^{1/4}/2)\exp(\mathrm{i}\pi/8)$. In $z > 0$ the solutions are

$$\begin{aligned} B &= C\,\exp(\sigma t - Sz)\cos(\omega t + kx - Qz), \\ A &= C\,\frac{\alpha}{\eta k^2 D^{1/2}}\exp(\sigma t - Sz) \\ &\quad\times\left[\cos\left(\omega t + kx - Qz - \frac{\pi}{4}\right) + D^{1/4}kz\cos\left(\omega t + kx - Qz - \frac{\pi}{8}\right)\right]. \end{aligned} \tag{6.79}$$

In $z < 0$,

$$\begin{aligned} b &= C\exp(\sigma t + Sz)\left[\cos(\omega t + kx + Qz) - D^{1/4}kz\cos\left(\omega t + kx + Qz + \frac{\pi}{8}\right)\right], \\ a &= C\frac{\alpha}{\eta k^2 D^{1/2}}\exp(\sigma t + Sz)\cos\left(\omega t + kx + Qz - \frac{\pi}{4}\right). \end{aligned} \tag{6.80}$$

Since $Q/S = 0.4142$, within a quarter of a wavelength $Qz = \pi/2$ in both the $\pm z$-directions, the amplitude falls by the factor 44.37, showing that the wave is effectively confined to near the surface $z = 0$.

The total magnetic flux in the y-direction, per unit length in the x-direction, is

$$\Phi_B = \int_0^\infty B\,\mathrm{d}z = \frac{2C\exp\sigma t}{kD^{1/4}}\cos\left(\omega t + kx - \frac{\pi}{8}\right) \tag{6.81}$$

in $z > 0$, and

$$\Phi_b = \int_0^\infty b\,\mathrm{d}z = \frac{6C\exp\sigma t}{kD^{1/4}}\cos\left(\omega t + kx + \frac{\pi}{8}\right) \tag{6.82}$$

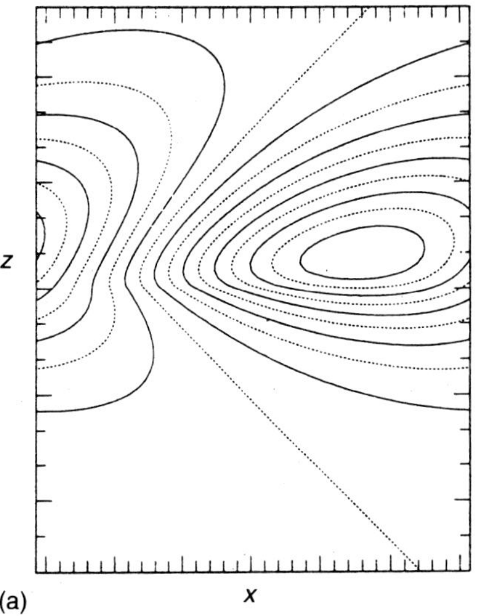

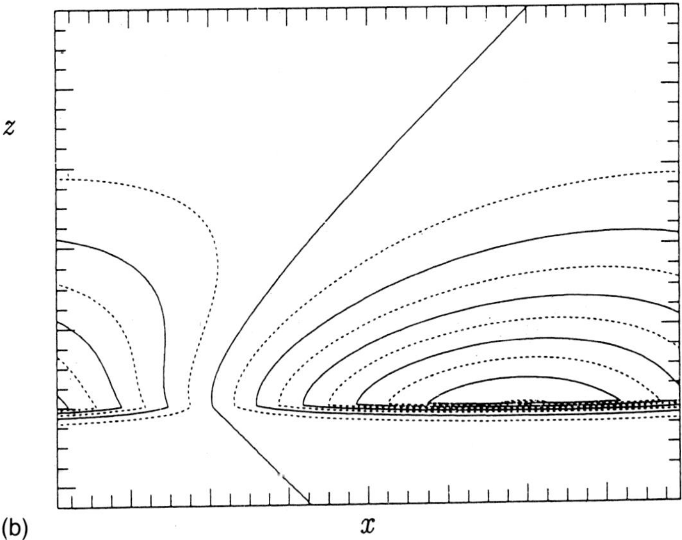

FIG. 6.4. The Parker (1993) dynamo surface wave. Contour map of the poloidal field for (a) the standard case $n = \eta$ and (b) with $n \ll \eta$.

in $z < 0$. Thus on average there is three times as much toroidal flux in the shear layer than in the convective layer: not surprisingly, as the toroidal flux is generated by the shear, and has to diffuse in order to appear at $z > 0$.

The growth-rate and frequency are given by

$$\sigma = \eta k^2[(D/32)^{1/2} - 1], \quad \omega = \eta k^2(D/32)^{1/2}. \tag{6.83}$$

Thus growing modes require $D > 32$, whereas the simple solution (6.26), corresponding to $Q = 0$, for the model with uniform α-effect and Ω-effect fully overlapping, requires $D > 2$. Thus even in this case of uniform diffusivity, the separation of the regions of cyclonic convection and shear reduces markedly the efficiency of the dynamo, which now depends for its operation on the diffusion into $z > 0$ of toroidal flux on which the convection can act, and into $z < 0$ of poloidal flux on which the shear can act. Poloidal field lines are plotted in Fig. 6.4(a).

The other limiting case, of more likely relevance to the Sun, is with $\mu^2 \equiv n/\eta \ll 1$, implying very weak flux diffusion in the shear region. If one writes σ, ω in terms of the real non-dimensional parameters β, ν,

$$\sigma = \eta k^2(\beta - 1), \quad \omega = \eta k^2\nu, \tag{6.84}$$

then for $\mu^2 \ll \beta - 1$,

$$\nu = \pm\beta^{1/2}(\beta - 1)^{1/2}, \quad \mathcal{D} \equiv \mu^2 D = \pm 8(\beta - 1/2)\beta^{1/2}(\beta - 1)^{1/2}. \tag{6.85}$$

Thus the parameter β (> 1) fixes the growth rate, the frequency and the associated dynamo number. (The case $\beta = 1$ is singular, requiring separate treatment; we quote just for the cases sufficiently above the bifurcation point $\beta = 1$ for $\mu^2 \ll (\beta - 1)$ to hold.) Then in $z > 0$, the toroidal field is again

$$B = C\exp(\sigma t - Sz)\cos(\omega t + kx - Qz), \tag{6.86}$$

but in $z < 0$,

$$b \approx -C\exp(\sigma t + sz)\{2\beta(\beta - 1/2)\}^{1/4}(kz/\mu^2)\cos(\omega t + kx + qz + \theta). \tag{6.87}$$

Further, whereas in the case with $\mu^2 = 1$, $S = s$ and $Q = q$, in the present case S/s, Q/q are both of order $1/\mu$, with $S > Q$, $s > q$, S, $Q \simeq k$, and s, $q \simeq k/\mu$. From 6.87, $|b|$ has a large maximum $O(1/\mu)$ at $z_m \approx -1/s$. Analogously to (6.81) and (6.82), one can construct at given x the fluxes of B and b per unit length in the x-direction. Apart from two cosines with their arguments differing by a constant phase, the ratio $\Phi_b/\Phi_B = [\beta/(\beta-1)]^{1/2}$ which is of order unity (except near the bifurcation point, when it becomes large as β approaches unity). Thus for small ratio of diffusivities μ^2, the toroidal field in $z < 0$ is confined to a thin layer of thickness $O(\mu/k)$ near $z = 0$, while that in $z > 0$ is in a layer of thickness $1/S$; but the ratio of toroidal field strengths is such that the toroidal flux in $z > 0$ is usually not much less than that in $z < 0$.

As $\mu \equiv (n/\eta)^{1/2} \to 0$, $\mu b_{\max} = f(\beta)B_{\max}$, or

$$nb_{\max}^2 = f^2(\beta)\eta B_{\max}^2 \tag{6.88}$$

with

$$f(\beta) = \frac{[8\beta(\beta - 1/2)]^{1/4}}{\{[2(\beta - 1/2)(\beta - 1)]^{1/2} + \beta - 1\}^{1/2}}. \tag{6.89}$$

It is remarkable that this linear kinematic model fixes the maximum energy densities in the two zones as inversely proportional to the diffusivities. Poloidal field lines are shown in Fig. 6.4(b).

The possible relevance of this model and of its non-linear extension by Tobias (1996) to the Sun and other solar-type stars will be noted in Section 8.6.

6.6 Non-linear feedback: quasi-kinematic models

As already stressed, the simplest mean-field equations are best looked on as convenient parametrizations that capture the essence of the physics in the *kinematic* domain, when the motions on all scales may be considered as prescribed, and in particular as independent of the Lorentz forces exerted by the very field which is being generated. However, the exponential growth in the field amplitude, predicted when the dynamo number D is supercritical, must clearly be halted at some stage by a back-reaction of the growing Lorentz forces on the motions driving the dynamo. An adequate theory should suggest at least an estimate for $F_p(\Omega)$, the mean poloidal flux emerging as a function of the angular velocity Ω, which plays a crucial role in the rotational history of the star (cf. Section 8.3).

6.6.1 *Buoyancy-limited growth*

A simple example due to Durney and Robinson (1982) is a good illustration of what is required. They considered a dynamo operating essentially within one scale-height $H_p \equiv L$ at the bottom of the convective zone, located at radius R_c. If the dynamo number is strongly supercritical, then (6.22), (6.61) with the η-terms neglected yield a growth time

$$t_A \simeq (L/\alpha\,\Delta_L\Omega)^{1/2}, \tag{6.90}$$

where $\Delta_L\Omega$ is the shear over the length L. They adopt

$$\Delta_L(\Omega) = c_2(L/R_c)^2\Omega, \tag{6.91}$$

fixing the constant c_2 by the assumption $\Delta_L(\Omega)/L \simeq \Delta\Omega_s/R_s$, where $\Delta\Omega_s$ is the observed surface variation of the solar rotation. This amounts to assuming that angular velocity gradients are of the same order throughout the convection zone. This is in fact contradicted by the more recent helioseismological results (cf. Section 8.4), but nevertheless, the model retains its pedagogic interest. The authors follow the turbulent model of an earlier paper (Durney and Spruit 1979), which yields $|\nabla\times\mathbf{v}| \simeq \Omega L/R_c$, rather than $\simeq \Omega$ as assumed in deriving (6.65); hence from $\alpha = -\tau_c(\nabla\times\mathbf{v})\cdot\mathbf{v}$ they find

$$\alpha = c_1 L^2\Omega/R_c, \tag{6.92}$$

fixing c_1 from the solar cycle period and from (6.28) and (6.26).

In this model, the asymptotic level of dynamo activity is fixed by equating the above growth time to an estimate for the rise time t_b of a buoyant flux tube through L. The motion of a cylindrical tube of radius R_f, moving with speed v_b through the turbulent zone with eddy viscosity $\nu_t = v_t l/3$, depends on the value of the hydrodynamic Reynolds number $\mathrm{Re} = 3v_b R_f/v_t l$. If R_f is of order l, then if $v_b \ll v_t$, $\mathrm{Re} \ll 1$, and one can take the estimate for the drag force F_d per unit length of the flux tube from standard hydrodynamics for low Re, confirmed by wind-tunnel data (Parker 1979):

$$F_d = 4\pi\rho v_b \nu_t/Q \tag{6.93}$$

where Q depends only weakly on Re. From (4.75), the buoyancy force per unit length on a tube in thermal equilibrium with its surroundings is $R_f^2 B_\mathrm{t}^2/8H_p$, and equating this to (6.93) yields

$$v_b = (3Q/8)v_\mathrm{A}^2(R_f^2/H_p l)/v_t \approx (R_f/H_p)^2 v_\mathrm{A}^2/v_t \tag{6.94}$$

since $3Q/8 \approx 1$, $l \simeq H_p$ in standard mixing-length theory, and $v_\mathrm{A} = B_\mathrm{t}/(4\pi\rho)^{1/2}$ is the Alfvén speed defined by the toroidal field generated by the shear. The flux tube radius is taken as $L/2$. The small Re assumption $v_b \ll v_t$ is seen to be valid as long as $v_\mathrm{A} \ll v_t$, i.e. as long as the field in the tube is well below the equipartition value. From (6.90), (6.91), (6.92) and (6.94), the condition $t_A = t_b = L/v_b$ yields for B_t the estimate

$$B_t \propto \Omega\rho_c^{1/2}(L/R_c)^{3/2}. \tag{6.95}$$

As the emergent poloidal flux is a surface manifestation of the toroidal flux generated deep down by the shear, the observed field will be proportional to B_t: the emerging flux increases linearly with Ω, and also increases with spectral type, since $\rho_c^{1/2}(L/R_c)^{3/2}$ increases as the convective zone deepens. This admittedly rather crude argument is of interest because a linear $B_\mathrm{p}(\Omega)$ relation appears to be tentatively favoured by the observed rotations of oldish late-type stars (Section 8.3).

6.6.2 *Magnetic back-reaction: modulated cycles*

There have been many studies in the literature which accept kinematical mean-field theory as applied to stars as an illuminating parametrization, incorporating much of the essential physics, and in the same spirit proceed to modify the equations to take some account of the back-reaction of the growing Lorentz forces on both large and small scales. As noted, the Parker (1993) model summarized in Section 6.5, though formally kinematic, does anticipate reduction of both the α-effect and the diffusivity in a region of strong magnetic field (cf. Section 6.7), but the treatment remains linear, so there is no cut-off in the exponential growth of the field. Earlier models (e.g. Stix 1972; Jepps 1975; Ivanova and Ruzmaikin 1977; Yoshimura 1975; Kleeorin and Ruzmaikin 1981; Moss *et al.* 1990*a,b*) introduced either or both of explicit, B-dependent 'α-quenching', and flux loss through buoyancy. Yoshimura (1978*a*) and Robinson and Durney (1982) simulated quenching of the shear, implicit also in the models of Brandenburg *et al.* (1989, 1992) which incorporate the 'Λ-effect' as a generator of shear (cf. Section 4.2). These modifications yield *steady*, non-linear, periodic $\alpha\Omega$ dynamo models, replacing the exponentially growing modes predicted in the linear domain for supercritical dynamo numbers.

The actual magnetic behaviour of the Sun and solar-type stars is much more complex. The solar cycle is both aperiodic and modulated to give intervals of reduced activity, such as the famous seventeenth century Maunder minimum (cf. Section 8.4). Without hoping at this stage to make a detailed comparison with observation, there is thus a further incentive to explore the potentialities of the modified mean-field equations, in particular to look for physical processes that will yield systems of higher-order equations with correspondingly richer behaviour. Yoshimura (1978*b*) was able to generate multiply periodic and aperiodic behaviour by introducing a time-lag into the magnetic buoyancy effect. We shall now summarize a toy model due to Weiss *et al.* (1984) (see also Jones *et al.* 1985), which includes a modelling of the dynamical interaction of the magnetic field with the differential rotation.

The problem is again treated in Cartesian geometry, with the dynamo equations essentially (6.23, 6.24), but modified to incorporate α-quenching, described by writing

$$\alpha = \alpha_0/(1 + \kappa|\mathbf{B}|^2), \tag{6.96}$$

and toroidal flux loss through magnetic buoyancy, described by replacing η in (6.24) by

$$\eta(1 + \lambda|\mathbf{B}|^2). \tag{6.97}$$

However, the expected reduction in η in regions of strong B – which would affect both (6.23) and (6.24) – is not included.

The model is restricted further by making B_y vanish at the top and bottom ($z = \pm d$) of the magnetic layer. The dynamo equations (6.23, 6.24) are again adopted, with the modifications (6.96), (6.97) temporarily dropped, but $\alpha(z)$ is taken to be antisymmetric in $z = 0$, and $V(z)$ is replaced by a space- and time-dependent function $v(x, z, t)$, which now includes the back-reaction of the macroscopic Lorentz force on the *mean flow*, sometimes called the 'Malkus–Proctor effect' (1975). The fluid is taken as Boussinesq, of density ρ_0, and the poloidal field is taken as vertical ($A = A(x, t)$). The y-component of the equation of motion is then

$$\frac{\partial v}{\partial t} = \frac{1}{4\pi\rho_0}\frac{\partial A}{\partial x}\frac{\partial B}{\partial z} + \nu\nabla^2 v + F(z), \tag{6.98}$$

where ν is the turbulent kinematic viscosity, and $F(z)$ is a function parametrizing the effect of the convection in driving a toroidal flow (cf. Section 4.2). The zero-order, non-magnetic state has the velocity $V_0(z)$ satisfying

$$\nu\nabla^2 V_0(z) + F(z) = 0. \tag{6.99}$$

The boundary conditions have B and $\partial v/\partial z$ vanishing at $z = \pm d$.

Solutions $A(x,t)$, $B(z,x,t)$ are sought, periodic in x with wavelength $2\pi/k$. Weiss *et al.* construct solutions in the form of truncated Fourier series, with the number of terms retained being the minimum needed to satisfy the boundary conditions:

$$A = \frac{1}{2}\{a(t)\,\mathrm{e}^{\mathrm{i}kx} + a^*(t)\,\mathrm{e}^{-\mathrm{i}kx}\}, \tag{6.100}$$

$$B = \frac{1}{2}\{b(t)\,\mathrm{e}^{\mathrm{i}kx} + b^*(t)\,\mathrm{e}^{-\mathrm{i}kx}\}[\mathrm{P}_1(z/d) - \mathrm{P}_3(z/d)], \tag{6.101}$$

$$= [V_0 + v_0(t) + \frac{1}{2}\{v_2(t)\,\mathrm{e}^{\mathrm{i}2kx} + v_2^*(t)\,\mathrm{e}^{-\mathrm{i}2kx}\}] \times[10\mathrm{P}_2(z/d) - 3\mathrm{P}_4(z/d)], \tag{6.102}$$

where the asterisk represents the complex conjugate, and P_k etc. are Legendre polynomials. Thus $V_0(z) = V_0[10\mathrm{P}_2(z/d) - 3\mathrm{P}_4(z/d)]$ with V_0 a constant. Equation (6.99) is then satisfied exactly if the forcing term $F(z)$ in this illustrative model is chosen to have the special form $(105\nu V_0/d^2)[\mathrm{P}_2(z/d)]$. Equations (6.100–6.102) are substituted into the dynamo equations (6.23, 6.24) and the equation of motion (6.98), and appropriate averages taken over z; e.g. (6.24) becomes

$$\mathrm{d}a/\mathrm{d}t = \alpha_0 b - \eta k^2 a, \quad \alpha_0 = \int_0^1 \alpha(z/d)[\mathrm{P}_1(z/d) - \mathrm{P}_3(z/d)]\,\mathrm{d}(z/d). \tag{6.103}$$

The resulting equations are non-dimensionalized, and the α-quenching and buoyancy terms restored:

$$\begin{aligned} \mathrm{d}A/\mathrm{d}t &= 2D(1+\kappa|B|^2)^{-1}B - A, \\ \mathrm{d}B/\mathrm{d}t &= \mathrm{i}(1+\omega_0)A - (\mathrm{i}A^*\omega/2) - (1+\lambda|B|^2)B, \\ \mathrm{d}\omega_0/\mathrm{d}t &= (\mathrm{i}/2)(A^*B - AB^*) - \nu_0\omega_0, \;\; \mathrm{d}\omega/\mathrm{d}t = -\mathrm{i}AB - \nu\omega, \end{aligned} \tag{6.104}$$

where

$$\omega_0 = v_0/V_0, \;\; \omega = v_2/V_0, \tag{6.105}$$

D is a suitably defined dynamo number, and ν, ν_0, related by

$$\nu = \nu_0(1 + 0.4k^2d^2), \tag{6.106}$$

are parameters proportional to the turbulent viscosity.

This system – a complex generalization of the Lorenz system (Lorenz 1963) – is able to yield a variety of solutions. The simplest class has A, B and ω varying harmonically while ω_0 is constant. This is most easily demonstrated by going to the high-viscosity limit with ν_0, ν infinite, so that ω, ω_0 are both zero, because the magnetic back-reaction is then negligible compared with the viscous force on any extra shear. (This limit implicitly assumes a large convective forcing term F in order to balance the viscous force on the zero-order shear.) Writing

$$B = b\,e^{\mathrm{i}pt}, \quad A = 2Dab\,e^{\mathrm{i}(pt+\phi)} \tag{6.107}$$

with a, b and p all real constants then yields

$$p^2 = 1 + \lambda b^2 \geq 1 \tag{6.108}$$

and

$$D = p(1+\kappa b^2)(1+0.5\lambda b^2) \geq 1. \tag{6.109}$$

If $\lambda = 0$ (so the only non-linearity is that due to α-quenching), then p stays constant at unity for all $D > 1$, and $b = [\kappa^{-1}(D-1)]^{1/2} \to (D/\kappa)^{1/2}$ for large D, and the phase-lag $\phi = -\pi/4$. On the other hand, if $\kappa = 0$, so only buoyancy losses are active, then

$$p^3 + p - 2D = 0, \tag{6.110}$$

and for large D, $p \simeq (2D)^{1/3}$ and $B \simeq \lambda^{-1/2}(2D)^{1/3}$.

Qualitatively different solutions exist when ν and ν_0 are finite. For simplicity, κ and λ are initially put equal to zero. Although the relevant solutions have $\nu > \nu_0$, the equations are in fact solved for unrestricted ν, ν_0. When $\nu_0 = \infty$ but ν is finite, there exists again a non-linear periodic solution for all $D > 1$, but when $\nu < 1$ this solution loses stability through a series of bifurcations as D is increased. For example, with $\nu = 0.5$, at $D = 3.0$ the system is doubly periodic; at $D \approx 3.806$ there begins a period-doubling cascade, and beyond $D \approx 3.84$ the solutions are apparently chaotic: the oscillatory solutions are aperiodic and their amplitude is irregularly modulated, with recurrent episodes of reduced activity (Fig. 6.5). A similar pattern of activity seems to persist for all larger values of D.

Finally, solutions are sought with ν fixed at 0.5 but with ν_0 being steadily reduced from infinity. The first bifurcation at $D = 1$ persists, but subsequent bifurcations are displaced to higher values of D as ν_0 is decreased; for example, at $\nu_0 = 1.0$, when $D = 24$ an aperiodic chaotic solution appears, but there is now no apparent modulation, and no period of low activity. The same behaviour persists as ν_0 decreases to below ν, as is strictly required. As ν_0 is reduced to zero – corresponding by (6.106) to the small wavelength limit $k \to \infty$ – the system reverts to the stable non-linear oscillator mode for all $D > 1$.

More recently, Schmalz and Stix (1991) have studied a similar system, except that the extra equation treats the evolution of α rather than of Ω. They again predict a transition from a limit cycle to a chaotic mean field, occurring at relatively modest dynamo numbers. It is more than tempting to see in these studies at least the genesis of an explanation of the complex long-term behaviour of the Sun and other solar-type stars (cf. Section 8.4).

Truncated models are open to the criticism that the most striking effects may disappear as the level of truncation is increased. However, Tobias *et al.* (1995),

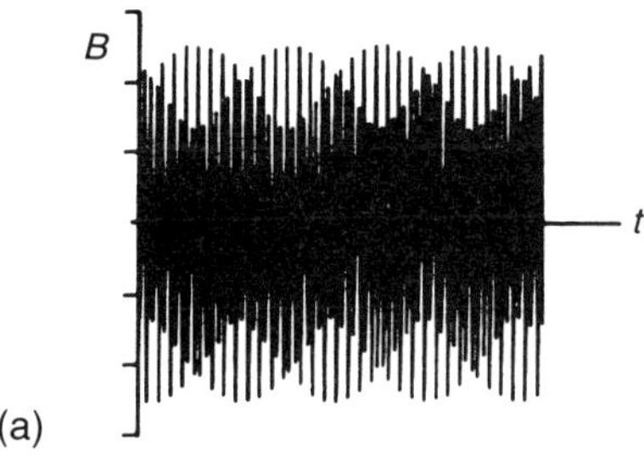

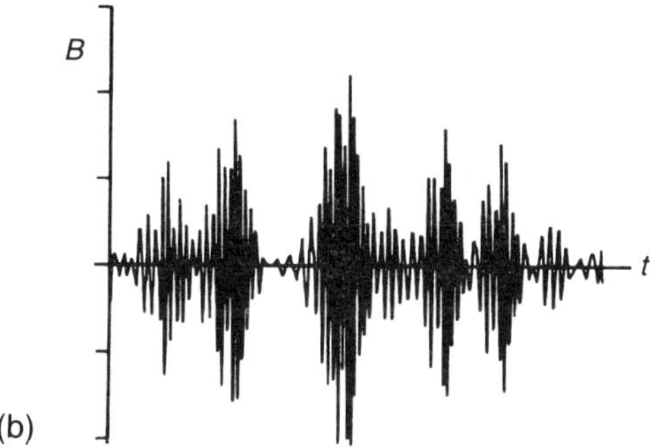

FIG. 6.5. Examples of solutions found by Weiss *et al.* (1984). (a) Doubly periodic solution for $\nu = 0.5$, $D = 3.0$. (b) Chaotic solution when $D = 16$, showing episodes of reduced activity.

who have constructed a third-order model of an oscillatory dynamo, then rely on bifurcation theory to argue that the properties found in their model are indeed generic and therefore robust.

For the quasi-periodic solutions, the dependence on Ω of the predicted cycle frequency Ω_{cycl} is of particular interest. Tobias (1995) finds that for waves of infinite lateral extent, as in the Weiss *et al.* model just summarized, Ω_{cycl} increases with increasing dynamo number $|D|$ – and so with Ω – if the amplification is limited by the non-linear variation of Ω due to the large-scale Lorentz forces (the 'Malkus–Proctor effect'); but the introduction of just α-quenching or η-quenching yields the opposite effect – Ω_{cycl} decreasing with increasing Ω. However, Tobias argues that the increase in magnetic buoyancy with growth of at least the toroidal component – exploited by Durney and Robinson (1982) in the model described in Section 6.6 – should be parametrized in the present exercise as η-enhancement, and this can reproduce again a positive correlation between Ω_{cycl} and Ω. Tobias's computations show also that this correlation appears to be sensitive to the assumption of infinite lateral extent: imposition of boundaries can yield a positive correlation, regardless of which particular quenching mechanism is invoked to convert the exponential growth of a linear dynamo wave into a finite-amplitude wave.

Other recent studies include that by Brooke and Moss (1994), who found chaotic solutions in a two-dimensional axisymmetric $\alpha\Omega$-dynamo in toroidal geometry with α-quenching the only non-linearity. Tavakol *et al.* (1995) find a remarkable difference between models computed for a sphere, simulating a fully convective star, and those for a convective envelope bounded below by a stable domain. The fully spherical models are much more 'stable', not liable to chaotic behaviour, and insensitive to changes in the algebraic form modelling α-quenching, whereas the shell models show chaotic behaviour over a wide parameter range, and are sensitive to modest changes in the lower boundary condition. Apparently, stellar models with similar parameters, including their Ω-values, can nevertheless sometimes evolve rapidly from one dynamo mode to others that are very different in structure and strength.

6.7 Fundamental problems of dynamo theory

We focus now on the more fundamental question: the effect of the growing magnetic field on the α-effect and the turbulent resistivity η. Over the decades there have been critics of the basics of dynamo theory, especially of the presumed turbulent resistivity (Piddington 1972, 1973; Layzer *et al.* 1979). More recently, their arguments have been revived and extended (Cattaneo and Vainshtein 1991; Vainshtein and Rosner 1991; Kulsrud and Anderson 1992; Vainshtein and Cattaneo 1992; Vainshtein *et al.* 1993). The claim is that both α and η will be reduced to values much below those appropriate for kinematic dynamo theory, when the *large-scale field* that is being maintained by dynamo action is orders of magnitude below that of the observed fields. As remarked by Moss (personal communication), it is not impossible that a dynamo number (6.26) could still remain supercritical, but $\eta|D|^{1/2} \propto \alpha^{1/2}$ would still be greatly reduced, so there would be great difficulty in predicting periodicities of the order of a solar cycle period.

The implicit picture of turbulent resistivity is borrowed from studies of turbulence in non-magnetic fluids (e.g. Heisenberg 1948; Chandrasekhar 1949; Batchelor 1953). The Kolmogoroff cascade of energy from large to small scales occurs in a time fixed by the large-scale eddies in the 'inertial range'. The micro-dissipative process, viscosity, fixes the scale on which conservation of energy gives way to thermalization: essentially, the scale for which the viscous Reynolds number is of order unity. Likewise, one can picture the tangling of a frozen-in large-scale field $\mathbf{B}_0$ continuing until scales are reached at which Ohmic dissipation and the associated field line reconnection can occur. The crucial difference is of course that tangling of the field increases its strength. Let us nevertheless demand that the consequent Lorentz forces remain small all the way down to the scale at which flux freezing ceases. With v_t and l again a typical turbulent velocity and eddy scale, the scale δ of the smallest magnetic structures is then given by $v_t/l \simeq \eta/\delta^2$, or

$$\delta \simeq l/(v_t l/\eta)^{1/2} = l/\mathrm{R}_m^{1/2}, \tag{6.111}$$

where $R_m = v_t l/\eta$ is the magnetic Reynolds number defined by the turbulent parameters. (We recall from Section 6.4 that the formal justification for the simple treatment of the α-effect through first-order smoothing requires $R_m \ll 1$, whereas in a realistic star $R_m \gg 1$.) A large-scale field B_0 subject to inexorable two-dimensional tangling will be increased because of flux freezing to the value $\bar{B} = B_0(l/\delta) = B_0 R_m^{1/2}$ from (6.111). This will be *dynamically* allowed provided $\bar{B} < B_{\rm eq}$, where $B_{\rm eq} = (4\pi\rho v_t^2)^{1/2}$, the equipartition field, so imposing on B_0 the condition

$$B_0 < R_m^{-1/2} B_{\rm eq}, \tag{6.112}$$

for the system to remain kinematic (Vainshtein and Rosner 1991). Since typically $R_m \approx 10^{12}$ in the Sun, and $B_{\rm eq} \approx 10^2$ G at the solar surface, B_0 from (6.112) is orders of magnitude below the observed large-scale fields. Numerical simulations by Cattaneo and Vainshtein (1991) and Vainshtein and Cattaneo (1992) for two-dimensional flows have confirmed that a realistic value for B_0 will lead to a disruption of the cascade.

The α-effect requires efficient field line reconnection. Vainshtein and Cattaneo argue that if the magnetic stresses are able to hold up the cascade before dissipative scales are reached, so that the simple picture of 'turbulent resistivity' is at the least questionable, simultaneously the α-effect will be suppressed. If (6.112) is the criterion, then this result 'strikes at the heart of standard mean field theory' (Rosner and Weiss 1992).

As noted in Section 6.6, a phenomenological 'α-quenching' has been employed by many workers. For consistency, a similar η-quenching should also be introduced. A popular *ansatz* has been

$$\alpha = \alpha_0/(1 + \overline{\mathbf{B}}^2/B_{\rm eq}^2) \tag{6.113}$$

where α_0 is the kinematic value (e.g. Brandenburg *et al.* 1989). However, the above discussion suggests that α-quenching will already be severe when

$$(\overline{\mathbf{B}}^2 + \overline{\mathbf{B}'^2})/B_{\rm eq}^2 = O(1) \tag{6.114}$$

– i.e. that not only the mean field but also the root-mean-square of the fluctuating field $\mathbf{B}'$ must be included (Vainshtein and Cattaneo 1992).

Two-dimensional simulations by Jones and Galloway (1993) do indeed show marked α-quenching for $B/B_{\rm eq}$ as much as 100 times smaller than that predicted by (6.113). The crucial question then becomes whether results from two-dimensional studies are a reliable guide to more realistic three-dimensional systems.

Three-dimensional computations have been performed by Tao *et al.* (1993) and by Brandenburg *et al.* (reported by Brandenburg 1993, 1994). Figure 6.6 shows the results of the different authors with a particular choice 0.2 for the turbulent magnetic Prandtl number $P_m = \nu/\eta$. The results of Rüdiger and Kitchatinov (1993) appear to be closest to the predictions of (6.113). The other

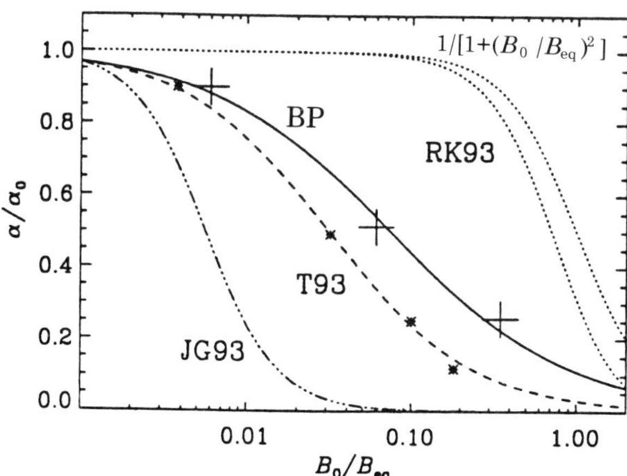

FIG. 6.6. 'α-quenching', as found by different authors: Jones and Galloway 1993; Tao *et al.* 1993; Brandenburg *et al.* unpublished; Rüdiger and Kitchatinov 1993; and the simple formula (6.113). (After Brandenburg *et al.*, personal communication; see also Brandenburg 1993.)

three-dimensional calculations show reduction in α setting in when $B_0 < B_{\text{eq}}$, suggesting that a formula somewhere in between (6.113) and (6.114) is appropriate.

Brandenburg *et al.* claim their results as evidence that 'the level at which the mean magnetic field saturates is not so small that it contradicts solar dynamo theory', as the α-effect and turbulent diffusivity are not quenched as much as had been previously suggested. However, some authors (Golub *et al.* 1981 and references therein; Moreno-Insertis 1992) argue for toroidal fields in the solar convection zone of 10^4–10^5 G, well above the equipartition strength, so enforcing a radical rethink of standard dynamo theory; for it is difficult to see how the turbulent convection could remain locally unaffected by such strong fields, continuing to yield the same regenerative α-effect and the equally important turbulent diffusion as if the back-reaction of the Lorentz forces were negligible, as assumed in the treatment of Section 6.4.

6.8 Different responses to the dilemma

The 1993 Parker model (summarized in Section 6.5), remains within standard dynamo theory, but hypothesizes that the generation of a strong toroidal field by rotational shear takes place in one domain, while the α-effect and the turbulent dissipation are restricted to a contiguous domain with a weak field. It will be seen in Chapter 8 that other workers suggest for the solar dynamo something akin to the classical Babcock–Leighton models, in which there is again a separation of the respective Ω- and α-zones, but with the α-effect being visible as part of the magnetic activity occurring above the photosphere. For the moment we summarize two alternative responses to the dilemma, both of which accept that the strong toroidal fields generated by shear will indeed react back on the motions, but not in a way so as to halt dynamo action but rather to guide it.

It is remarkable that computations into the non-linear domain that use the standard dynamo equations, for which the provisional justification as in Section 6.4 derives from mean-field theory, tend to produce fields that have an intermittent, 'fibril' structure for which mean-field theory may not be the most appropriate formalism (but cf. Parker 1982). This is indeed an encouragement towards describing developed dynamos in terms of discrete flux tubes, as in the two models to be discussed.

6.8.1 *A dynamo driven by the instability of strong flux tubes*

Following the earlier work of Schmitt (1984, 1987) and Schüssler (1993), Ferriz-Mas *et al.* (1994) argue that the spontaneous non-axisymmetric instability of a strong toroidal field in a rotating system will itself lead to dynamo action that again completes the poloidal–toroidal–poloidal loop: the non-axisymmetric modes – essentially the Rayleigh–Taylor–Parker modes discussed in Sections 3.6.2 and 4.7.2, as modified by $\mathbf{\Omega}$ – have perturbation fields $\mathbf{v}$ and $\mathbf{b}$ that yield a non-vanishing α-effect. The starting point is again the mildly subadiabatic overshoot domain at the base of a convective envelope, with a strong axisymmetric toroidal field, generated perhaps by a rotational shear acting on a poloidal field pumped downwards by the turbulence (but cf. Section 8.6.3). The analysis is considerably simplified if the field may be represented by an ensemble of isolated flux tubes, each describable by the *thin-tube approximation* (Spruit 1981 and references therein; Section 4.8.1). Recall that this assumes the radius of the tube to be small compared with all the other relevant lengths – the radius of curvature of the tube, the local scale-height, the rotational shear scale and the wavelengths of the dominant perturbations. Such a model is able to describe convincingly the properties of sunspot groups as the consequence of eruption through the photosphere of flux tubes (e.g. Moreno-Insertis 1992; Moreno-Insertis *et al.* 1992; Caligari *et al.* 1995); however, a similar description of the unobservable field in the overshoot later is necessarily conjectural.

Detailed linear analysis in general of non-axisymmetric instabilities of toroidal flux tubes is presented in Ferriz-Mas and Schüssler (1993, 1994, 1995) (following earlier work by Spruit and van Ballegooijen (1982), van Ballegooijen (1983) and van Ballegooijen and Choudhuri (1988)). We follow Ferriz-Mas *et al.* (1994) in simply quoting the most important results.

Consider first an axisymmetric toroidal flux tube in a plane parallel to the stellar equator (i.e. at constant latitude), and distant ϖ_0 from the symmetry axis. Let the gas within the tube rotate with angular velocity Ω about the axis, while the gas outside rotates with Ω_e. Take cylindrical polar coordinates (ϖ, ϕ, z) centred on the intersection of the flux tube plane with the axis. The suffices i and e refer respectively to the interior of the tube and to the local exterior. In equilibrium, the gas within the tube satisfies

$$-\rho_i\Omega^2\varpi = -\nabla\left(p_i + \frac{B_\phi^2}{8\pi}\right) - \frac{B_\phi^2}{4\pi\varpi}\hat{\varpi} + \rho_i\mathbf{g}, \tag{6.115}$$

where $\mathbf{g}$ is gravity, and both the magnetic pressure and the magnetic curvature term $B\mathbf{t}\cdot\nabla(B\mathbf{t})/4\pi = -(B_\phi^2/4\pi\varpi)\hat{\varpi}$ are included. Just outside the tube, the equation is

$$-\rho_e\Omega_e^2\varpi = -\nabla p_e + \rho_e\mathbf{g}. \tag{6.116}$$

From integration of (6.115) over the volume $\mathrm{d}\tau$ defined by the tube and the angle $\mathrm{d}\phi$, together with use of the boundary condition $p_i + B^2/8\pi = p_e$,

$$\left[\rho_i(\Omega^2\varpi_0 + \mathbf{g}_0) - \frac{B^2}{4\pi\varpi_0}\hat{\varpi}\right]\mathrm{d}\tau = \int p_e\mathbf{n}\,\mathrm{d}S. \tag{6.117}$$

Integration of (6.116) over the same volume, and subtraction from (6.117) then yields

$$(\rho_i\Omega^2 - \rho_e\Omega_e^2)\varpi_0 - \frac{B^2}{4\pi\varpi_0}\hat{\varpi} = (\rho_i - \rho_e)\mathbf{g}_0. \tag{6.118}$$

Since only the buoyancy force on the right-hand side of (6.118) has a component in the z-direction, then except for flux tubes lying strictly in the equatorial plane, in equilibrium both sides must vanish separately: $\rho_e = \rho_i$, and (6.118) reduces to the balance of magnetic curvature force per unit mass and the difference between the internal and external centrifugal forces (Moreno-Insertis *et al.* 1992):

$$\varpi_0^2(\Omega^2 - \Omega_e^2) = v_A^2, \tag{6.119}$$

where $v_A = B_0/(4\pi\rho_0)^{1/2}$ is the Alfvén speed in the tube. Note that the treatment differs from the simple ‘magnetic buoyancy’ argument of Section 4.7 in its taking account of the small but non-vanishing magnetic curvature force. The result (6.119) could clearly be derived as a very special case of the set (4.89)–(4.94).

Since the joint thermal-plus-magnetic pressure is continuous at the tube surface, continuity of density requires that the inner and outer temperatures must differ as well as the angular velocities. Thus if the tubes were dynamically stable, evolution would be at a rate determined by heat flow – the ‘Eddington–Sweet’-type effect discussed in Sections 4.7.1 and 5.4 and later in Section 9.4.

Ferriz-Mas *et al.* study the stability using a similar cylindrical coordinate system, fixed in the frame rotating with the angular velocity Ω of the gas in the tube. The unperturbed arc-length $s_0 = \varpi_0\phi_0$ is used as a Lagrangian coordinate, defining the initial position of a mass particle. All of the (isentropic) perturbations are expressed in terms of the Lagrangian displacement vector $\xi(s_0, t)$, which in turn is Fourier analysed into components

$$\hat{\xi}\exp\mathrm{i}(m\phi_0 - \omega t), \tag{6.120}$$

where m is an integer, the azimuthal wave number, and ω is the complex frequency. The resulting equations are cast into non-dimensional form by writing all lengths in terms of the pressure scale-height within the tube, $H = p_{i0}/\rho_0 g_0$, with the associated unit of time $\tau = \sqrt{2}H/v_A$, and dimensionless frequency $\tilde{\omega} = \omega\tau$. The equilibrium state then depends on a few dimensionless parameters: the ratio of specific heats γ; the dimensionless curvature $f = H/\varpi_0$; the plasma $\beta = 8\pi p_{i0}/B_0^2$; the degree of superadiabaticity $\delta = \nabla - \nabla_{\mathrm{ad}}$; and $\tilde{\Omega}_e = \Omega_e\tau$ and its local gradient. The resulting set of six homogeneous equations has a non-trivial solution provided the associated determinant vanishes, yielding for the dispersion relation a sixth-order algebraic equation with real coefficients. Instability is implied by the occurrence of a conjugate complex pair of $\tilde{\omega}$.

The problem is of interest for dynamo theory because some unstable modes yield an α-effect: i.e. the average over ϕ of $\mathbf{v} \times \mathbf{b}$ is not zero but proportional to the unperturbed field $\mathbf{B}_\phi = B_0\mathbf{t}$:

$$\overline{\mathbf{v} \times \mathbf{b}} = \alpha \mathbf{B}_\phi. \tag{6.121}$$

Corresponding to an eigenvalue $\omega = \omega_R + \mathrm{i}\omega_I$, α has the value

$$\alpha = \frac{m\omega_I}{2\varpi_0} \frac{\mathrm{i}(a^*b - b^*a)}{aa^* + bb^*} (\hat{\xi}_\varpi \hat{\xi}_\varpi^* + \hat{\xi}_z \hat{\xi}_z^*) \exp 2\omega_I t, \tag{6.122}$$

where the complex coefficients a, b are defined by the relation $a\hat{\xi}_z + b\hat{\xi}_\varpi = 0$ holding between the (ϖ, z)-components of the corresponding eigenfunction. Since $\mathrm{i}(a^*b - b^*a)$ is a real quantity, it follows that a real, exponentially growing α exists if and only if the mode is non-axisymmetric ($m \neq 0$) and unstable ($\omega_I > 0$), for it is only then that there is the required phase-difference between $\mathbf{v}$ and $\mathbf{b}$. The coefficient α is anti-symmetric with respect to the equatorial plane, vanishing at the equator. A change from m to $-m$ leaves α unaltered, since $m\omega_R > 0$ (unstable modes always propagate in the direction of rotation), and $\mathrm{i}(a^*b - b^*a)$ also changes its sign. Of course, the magnitude of α cannot be predicted from a linear theory.

The authors emphasize that the analogy with the conventional α-effect, due to non-mirror-symmetric turbulence, is quite close. The dominant toroidal flux is unstable to the formation of loops that travel as growing waves in the ϕ-direction. Because of the non-vanishing Ω, the loops are both twisted and rotated by the Coriolis force, so that the waves become helical, with a poloidal field component. The helical loops are then pictured as coming into contact with each other and reconnecting, so as to form a closed poloidal loop (Schüssler 1993), ready to generate a new toroidal field through the action of the angular velocity gradient. Both the Ω and the α-effects occur in the same domain, but it is now the Lorentz forces driving the instability that are thereby responsible for the α-effect: excess energy in the toroidal field is in part used to build a new poloidal field.

The significant difference from the standard picture is that it is only for B_ϕ sufficiently strong (greater than a few times 10^4 G) that the flux tube instability sets in: the process must be linked with a truly 'self-excited' dynamo that generates the strong flux tube from a weak seed field, operating below the instability threshold, e.g. a conventional $\alpha\Omega$-dynamo operating in the convection zone.

6.8.2 *A two-dimensional flux tube model*

A model by Vainshtein *et al.* (1993) which again exploits the anisotropy introduced into the dynamics and electrodynamics by a strong toroidal field is perhaps appropriate to a system for which the toroidal flux tubes are not strong enough for the above-cited instability to occur. From Section 4.4, the discussion in the Boussinesq approximation of gravity waves/convective instability in the presence of a homogeneous horizontal field yields the dispersion relation (4.65)

$$\omega = [\omega_0^2 - (\mathbf{k} \cdot \mathbf{v}_A)^2]^{1/2}, \tag{6.123}$$

where $\mathbf{v}_A$ is the Alfvén speed and ω_0 is the growth rate (4.64) in the absence of $\mathbf{B}$. When $v_A \approx \omega_0/k$ the field will interfere with all motions except those for which $\mathbf{k}$ and $\mathbf{v}_A$ are perpendicular: the strong field forces the turbulence to consist of two-dimensional motions in planes normal to $\mathbf{B}$, which interchange straight field lines, while the normal three-dimensional turbulence that distorts field lines is either suppressed (or, more reasonably, replaced by a field of Alfvén waves).

These two-dimensional motions will transport toroidal loops bodily. In a typical $\alpha\Omega$ dynamo the mean toroidal fields are of opposite sign in the two hemispheres, and so random two-dimensional motions will sometimes bring oppositely directed toroidal field lines into juxtaposition. This can yield accelerated field diffusion and decay: the rate is given in terms of the *microresistivity*, but the time-scales can be greatly reduced because of the locally small length-scales. Thus an enhanced dissipation rate can exist even if the field is strong. Further, it was noted in Section 3.7 that a structure with oppositely-directed, pinching field lines is subject to resistive instabilities that yield loops detached from the straight field lines. Once such loops are formed, then the familiar Parker Coriolis effect will be able to rotate them into meridian planes: the reconnection process frees the flux that will form the new poloidal flux from the strong constraining effect of the stresses that would otherwise be generated (cf. Fig. 6.7). The proposal reverses the ordering of the classical picture. There, the field is weak and therefore passive: the combination of convective motion and vortical motion due to Coriolis force is pictured as producing a poloidal loop which *subsequently* detaches from the parent toroidal field. A toroidal field that is too strong will prevent this. The new picture has the loops forming and *detaching* during the reconnection of oppositely-directed toroidal loops; the Coriolis-driven α-effect can then act on loops no longer constrained by the strong toroidal field. Instead of being subject to strong 'α-quenching', the system finds a way for the process to persist.

With this picture in mind, Vainshtein *et al.* produce a suitably modified version of the phenomenological dynamo equations (6.21) and (6.22). In this

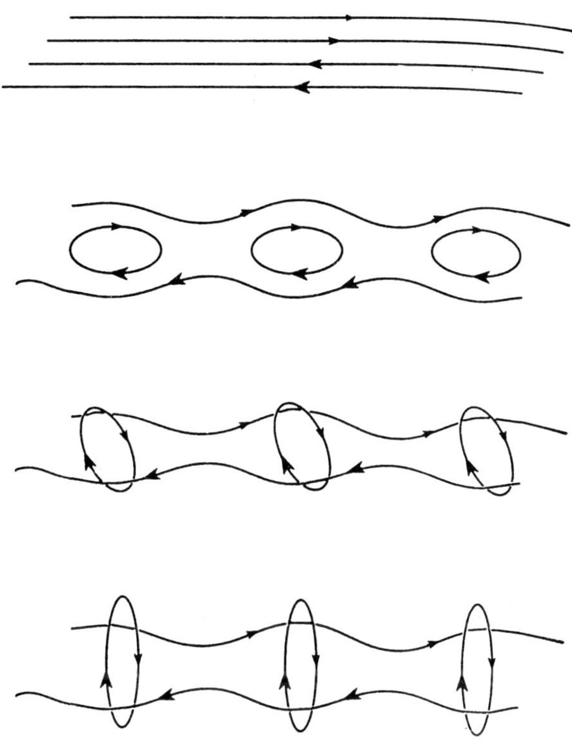

FIG. 6.7. Reconnection preceding the Coriolis effect. (After Vainshtein *et al.* 1993.)

exploratory study, the geometry used is Cartesian, but the description is in language appropriate to a spherical star. Thus the segments of the x-axis $0 < x < L/2$, $L/2 < x < L$ represent notionally the northern and southern hemispheres of a stellar convection zone, with $x = 0$, $x = L$ the poles and $x = L/2$ the equator. The planes $z = 0$, $z = z_0$ represent the bottom and top of the convection zone. Each hemisphere has a 'poloidal' field $B_z = \partial A/\partial x$ and a 'toroidal' field B_y. Again, it is non-uniform rotation that is the generator of the toroidal field, so in the absence of turbulent diffusion, B_y would be of one sign in the northern hemisphere and of the opposite sign in the southern. However, an essential part of the model is the diffusion across the equator of toroidal flux, leading to occasional juxtaposition of oppositely signed field lines, with consequent efficient micro-diffusion and loop formation. This is allowed for by writing $B^+\hat{\mathbf{y}}$ and $-B^-\hat{\mathbf{y}}$ for the positive and negative parts of the toroidal field in each hemisphere, with B^+, B^- defined to be positive. In the northern hemisphere, both B_z and $B^+ - B^-$ are positive, in the southern both are negative. The dynamo is then described by the equations

$$(\partial/\partial t - \eta\,\partial^2/\partial x^2)B^+ = G\,\partial A/\partial x - B^-/\tau, \tag{6.124}$$
$$(\partial/\partial t - \eta\,\partial^2/\partial x^2)B^- = -B^-/\tau, \tag{6.125}$$
$$(\partial/\partial t - \eta\,\partial^2/\partial x^2)A = \alpha B^- \tag{6.126}$$

in the northern hemisphere, and a similar set with B^+, B^- exchanged and G, α replaced by $-G$, $-\alpha$ in the southern hemisphere. As before, G is the vertical shear coefficient; τ is the reconnection time-scale for B^+, B^- fields, and αB^- is just the α-effect.

In these equations, the terms in η represent the modified turbulent diffusion of the field. In the northern hemisphere, the toroidal field B^+ is generated by shear, is subject to this diffusion, and is reduced by reconnection with the B^--field that has been advected in from the southern hemisphere; whereas the B^--field in the north feels only the diffusion and reconnection. The poloidal field potential A is again subject to diffusion, but is regenerated through the α-effect, which occurs at the rate fixed by the intrusion of B^- from the south, for by hypothesis it is the reconnection process which generates the detached loops which are then rotated by the Coriolis force. A similar description applies to the southern hemisphere. An attractive feature of the model is the explicit inclusion not only of the familiar turbulent diffusivity η, but also the reconnection rate $1/\tau$ which must involve the micro-resistivity.

Solutions are sought of the form $\propto \exp ft$, subject to boundary conditions at the poles and the equator. In the relevant parameter range, the approximate northern hemisphere forms are

$$\begin{aligned}
B_z &= K\left[\frac{16}{GL}\left(\frac{\eta}{\tau}\right)^{1/2}\right]\exp(9ct/\tau),\\
B^- &= K\frac{x}{(\eta\tau)^{1/2}}\exp(9ct/\tau),\\
B^+ &= K\frac{x}{(\eta\tau)^{1/2}}\left[1+\frac{10}{3}\left(1-\frac{x}{L/2}\right)\right]\exp(9ct/\tau)
\end{aligned} \tag{6.127}$$

with $c \approx 1$ and K a constant. For the reconnection time, the authors adopt the formula $\tau = (L/v_{\mathrm{A}})(v_{\mathrm{A}}L/\eta_m)^{1/2}$, where η_m is the microresistivity $c^2/4\pi\sigma$ (cf. (3.179)). It is remarkable that insertion of typical solar values yields an estimate for the parameter α close to that for the classical $\alpha\Omega$ dynamo (Parker 1979).

6.9 Numerical simulations

We conclude with a brief summary of recent work, done in the spirit of the various studies on convection and magnetoconvection cited in Section 4.5. Brandenburg *et al.* (1996), following the earlier paper by Nordlund *et al.* (1992), simulate three-dimensional compressible convection in a rotating system, with overshoot into surrounding stable layers, and with an initially weak magnetic field. A perfect gas law is assumed with $\gamma = 5/3$, and with a constant kinematic viscosity ν and magnetic diffusivity η – both necessarily macroscopic – throughout the convective zone.

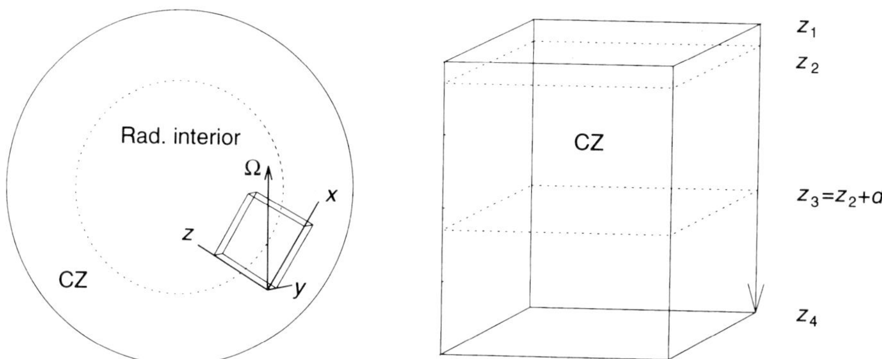

FIG. 6.8. A Cartesian box in the convection zone at a southern latitude, with the z-axis parallel to $\mathbf{g}$ and the rotation axis in the (x, z)-plane.

Whether the dynamo action occurring is associated with the local helicity, or rather with some other measure of anisotropy, is not clear: dynamo action is inferred from the results of the computations, without any explicit computation of an α-effect. Likewise, the questions raised above of the possible reduction of the turbulent resistivity and of the mean emf along the field lines are not directly addressed. Instead, the effects of a particular physical process, on all scales down to the limit set by the grid, are automatically included in the computations. Thus the growing Lorentz forces are calculated along with the velocity field at any instant; subsequent Fourier expansion could in principle separate the back reaction of the large-scale force on the mean flow (the Malkus–Proctor effect) from the interference of the small-scale force components on the turbulent tangling of the field lines. Likewise, once $\mathbf{v}$ and $\mathbf{B}$ have been computed, the motional induction term in Ohm's law would yield the large scale $\mathbf{v} \times \mathbf{B}$ term but also extra terms, represented in the standard theory discussed above by the α-effect, and the turbulent resistivity.

The computations are 'local', the convection zone being confined to a Cartesian box of height d (Fig. 6.8). The z-axis is chosen to point downwards in the direction of gravity $\mathbf{g} = g\hat{\mathbf{g}}$. The frame rotation vector $\boldsymbol{\Omega}$ is inclined to $\mathbf{g}$ at an angle θ, i.e. $\hat{\mathbf{g}} \cdot \hat{\boldsymbol{\Omega}} = \cos\theta$. The particular value $\theta = \pi/3$ is chosen, so that the domain studied is 30° south of the equator. The computational domain has an unstable layer between $z = z_2$, $z = z_2 + d = z_3$, and with the stable overshoot domains $z_1 < z < z_2$, $z_3 < z < z_4$ respectively above and below. The x-axis points from south to north along the meridian, the y-axis in the azimuthal direction of rotation. The solutions constructed are periodic in x and y; in most cases, of horizontal scale $2d$ in each direction.

Radiative heat conduction and surface losses are included, along with viscous and Ohmic heating. The coefficient of heat conduction is chosen to ensure that the three domains in z are indeed convectively stable or unstable as stated. The boundaries $z = z_1$, z_4 are taken to be impenetrable, stress-free perfect conductors, with a constant temperature at the top and a constant energy flux at the bottom.

The aim is to show that an initial seed field can sometimes be both amplified and permanently maintained by the dynamo action of the self-consistently constructed motions, and to discover its asymptotic spatial structure. Although the authors have the solar dynamo in mind, they accept that they are still far from realistic solar conditions: spatial resolution again limits the maximum stratification attainable, and forces η and ν to be taken far above their real microscopic values.

The principal conclusions that are relevant to the dynamo problem are as follows.

(1) The fluid flow includes swirling downdrafts which are more turbulent than the more laminar upwellings, and are reminiscent of the turbulent plumes believed to exist in the solar convective zone (Rieutord and Zahn 1995). The evolution of the magnetic field appears to be controlled primarily by these downdrafts. It is convenient to define the normalized dimensionless helicity density

$$H = \frac{\omega \cdot \mathbf{v}}{\omega_{\mathrm{rms}} v_{\mathrm{rms}}}, \tag{6.128}$$

where the suffix rms refers to the root-mean-square value. Within the downdrafts H is large but of either sign. Taken over the whole box, cancellations ensure that the relative helicity is quite small ($\simeq -0.03$).

(2) For large enough magnetic Reynolds number, the non-magnetic state is unstable: an initial seed field grows until saturated by non-linear effects. In the computations, the magnetic energy increases by 3–5 orders of magnitude from its seed value until it saturates. An important parameter is the magnetic Taylor microscale

$$\lambda_{\mathrm{M}} = \left(\frac{5 < \mathbf{B}^2 >}{< \nabla \times \mathbf{B} >^2} \right)^{1/2} . \tag{6.129}$$

When $\lambda_{\mathrm{M}} \ll d$, the field is locally concentrated into flux tubes. The magnetic diffusion time-scale based on λ_{M} is $\tau_{\mathrm{M}} = \lambda_{\mathrm{M}}^2/10\eta$, as compared with the global diffusion time d^2/η. As the dynamo saturates, it is found that $\lambda_{\mathrm{M}} \to 0.07d$ approximately. It is fairly constant throughout the convection zone, but is larger near the surface and in the lower overshoot layer where the convective motions are weak. The results are insensitive to the value of the magnetic Prandtl number ν/η.

There is always a query to be raised about numerical dynamo simulations: is the machine producing a genuine dynamo, or is it a transient amplifier, with the field energy ultimately decaying to the order of the seed field energy? Cattaneo *et al.* (1991) argue that genuine dynamo action is virtually proven provided the simulations have run for longer than $\tau_{\mathrm{M}}(L/l)^2$, where L is the box scale and l an appropriate small scale, such as the kinetic Taylor microscale

$\lambda_K = (5\langle \mathbf{v}\rangle^2/\langle \mathbf{\Omega}\rangle^2)^{1/2}$. In the Brandenburg *et al.* computations this criterion is satisfied. Further, the simulations show that after an e-folding of the magnetic energy, the magnetic structures look statistically the same – they are just stronger. This supports the claim that the model is a genuine dynamo and not a simple amplification that could be reversed (cf. Section 6.1)

(3) The detailed results show the formation of large-scale, coherent flux tube structures. They become longer and more clearly defined near saturation than during growth (Fig. 6.9). As already noted in Section 4.7, in these computations, magnetic buoyancy is found to be less important than the advection of flux in the downdrafts: the expected buoyancy appears to be counterbalanced by drag forces. However, the authors emphasize that it is an open question how far these conclusions would carry over to a more realistic simulation of the solar case, for which the Mach numbers at the bottom of the convection zone are likely to be two orders smaller, and where a maintained differential rotation in the transition layer will generate strong toroidal fields.

(4) In spite of the stratification of the medium, the root-mean-square value of the Alfvén speed varies little through the convection zone. There is a high probability of the vorticity ω and the magnetic field $\mathbf{B}$ being parallel or antiparallel, especially when $\mathbf{B}$ is strong, whereas strong vortex tubes and strong magnetic fields avoid each other. The effect appears to be dynamical, with the strong magnetic fields 'enslaving' the weaker vorticity. The cosines of the angles between all other fields appear to be more or less random. When $\mathbf{B}$ is weak, $\mathbf{j}$ and $\mathbf{B}$ are predominantly perpendicular, but as the dynamo saturates this is no longer the case: it is found that $-\overline{\mathbf{v}\cdot(\mathbf{j}\times\mathbf{B}/c)}$ – the mean rate of working of the fluid against the Lorentz force – is reduced, because near saturation the conversion of kinetic energy into magnetic is reduced, as the field moves somewhat closer to a force-free structure.

(5) The solutions are models of *local* dynamo action. In fact, the boundary conditions on the z-surfaces force the field to conform to the Bondi–Gold theorem, modified to apply to a fluid of finite resistivity but bounded in z by perfectly conducting walls (cf. Section 6.1). Within the fluid the magnetic energy can increase indefinitely (until limited by dynamical back-reaction). Also, if one prescribes a surface, for example a section of the plane z with $z_1 < z < z_4$ and with x, y dimensions less than the periodicity $2d$, then the initially weak flux across this can simultaneously grow: the process is a genuine dynamo, with the increase in field energy due not only to tangling but to an increase in the largest scale components. But with perfectly conducting z-boundaries, the external dipole moment and the associated field hardly changes.

We look forward eagerly to the next-stage computations, hopefully of a realistic global simulation of the solar convection zone, with relaxation of the unnecessarily restrictive perfectly conducting boundary condition. We hazard a guess that many of the interesting features of the present solutions will reappear – the powerful downdrafts, the concentration of flux into tubes, the details on helicity and vorticity, but also with external flux, growing, declining and changing sign as witnessed in the Sun.

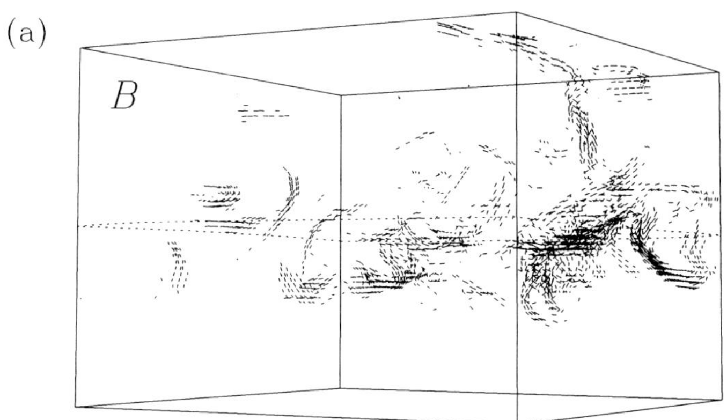

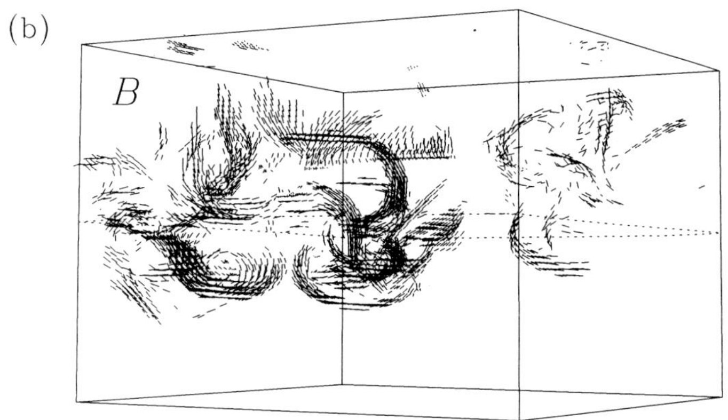

FIG. 6.9. Magnetic field (a) during build-up and (b) at saturation. Only vectors above a certain threshold are plotted. The interface between stably and unstably stratified layers is marked by a dotted line.

6.10 Conclusions

Enough has been presented to make clear that the standard – kinematic or quasi-kinematic – dynamo equations have a lot to say that is relevant to stellar magnetism, but that also the basics of the theory will be subjected to an ongoing critical scrutiny over the next decade or more. Recent surveys of the area up to 1993 include the articles by Brandenburg, Roberts and Weiss in the volume *Lectures on solar and planetary magnetism* (1994), eds Proctor and Gilbert, and by Brandenburg and by Schmitt in *The cosmic dynamo* (1993), eds Krause, Rädler and Rüdiger.

Parallel numerical simulations, with their prediction of flux tube structure, show up the inevitable limitations of analysis based on mean-field assumptions. More insight may be gained by comparison with the models explicitly based on a highly non-uniform distribution of flux.

In Sections 8.4–8.6 we return to problems of the solar dynamo with special regard to the most recent observations.

References

Babcock, H.D. (1959). *Astrophysical Journal*, **130**, 364.

Babcock, H.W. (1961). *Astrophysical Journal*, **133**, 572.

Backus, G.E. (1958). *Annals Physics*, **4**, 372.

Batchelor, G.K. (1950). *Proceedings Royal Society A*, **201**, 405.

Batchelor, G.K. (1953). *The theory of homogeneous turbulence.* Cambridge University Press.

Beck, R., Brandenburg A., Moss, D., Shukurov, A. and Sokoloff, D. (1996). *Annual Review of Astronomy and Astrophysics*, **34**, 155.

Belyanin, M., Sokoloff, D. and Shukurov, A. (1993). *Geophysical Astrophysical Fluid Dynamics*, **68**, 237.

Bondi, H. and Gold, T. (1950). *Monthly Notices Royal Astronomical Society*, **110**, 607.

Braginsky, S.I. (1965). *Soviet Physics JETP*, **20**, 726.

Brandenburg, A. (1993). In *The cosmic dynamo* (eds F. Krause, K.-H. Rädler and G. Rüdiger), p. 111. Kluwer, Dordrecht.

Brandenburg, A. (1994). In *Lectures on solar and planetary dynamos* (eds M.R.E. Proctor and A.D. Gilbert), p. 117. Cambridge University Press.

Brandenburg, A., Krause, F., Meinel, R., Moss, D. and Tuominen, I. (1989). *Astronomy and Astrophysics*, **213**, 411.

Brandenburg, A., Nordlund, A, Pulkkinen, P., Stein, R.F. and Tuominen, I. (1990). *Astronomy and Astrophysics.* **232**, 277.

Brandenburg, A., Moss, D. and Tuominen, I. (1992). In *Proc. NSO/Sacramento Peak 12th Summer Workshop* (ed. K.L.Harvey), p. 256. ASP, San Francisco.

Brandenburg, A., Jennings, R.L., Nordlund, A., Rieutord, M., Stein, R.F. and Tuominen, I. (1996). *Journal Fluid Mechanics*, **306**, 325.

Brandenburg, A., Moss, D. and Soward, A.M. (1998). *Proceedings Royal Society A*, **454**, 1283.

Brooke, J. and Moss, D. (1994). *Monthly Notices Royal Astronomical Society*, **266**, 733.

Bullard, E.C. and Gellman, H. (1954). *Philosophical Transactions Royal Society A*, **247**, 213.

Caligari, P., Moreno-Insertis, F. and Schüssler, M. (1995). *Astrophysical Journal*, **441**, 886.

Cattaneo, F. and Vainshtein, S.I. (1991). *Astrophysical Journal*, **376**, L21.

Cattaneo, F., Hughes, D.W. and Weiss, N.O. (1991). *Monthly Notices Royal Astronomical Society*, **253**, 479.

Chandrasekhar, S. (1949). *Astrophysical Journal*, **110**, 329.

Childress, S. (1969). *The application of modern physics to the earth and planetary interiors* (ed. S.K. Runcorn), p. 629. Wiley, London.

Childress, S. (1970). *Journal Mathematical Physics*, **11**, 3063.

Childress, S. and Gilbert, A.D. (1995). *Stretch, twist and fold: the fast dynamo.* Springer, Berlin.

Cowling, T.G. (1945). *Monthly Notices Royal Astronomical Society*, **105**, 166.

Cowling, T.G. (1955). *Vistas in astronomy*, Vol. 1, p. 313. Pergamon, London.

Cowling. T.G. (1965). In *Stellar and solar magnetic fields* (ed. R. Lüst), p. 405. North-Holland, Amsterdam.

Cowling, T.G. (1981). *Annual Review Astronomy Astrophysics*, **19**, 115.

Deinzer, W. and Stix, M. (1971). *Astronomy and Astrophysics*, **12**, 111.

Dudley, M.L. and James, R.W. (1989). *Proceedings Royal Society A*, **425**, 407.

Durney, B.R. and Latour, J. (1978). *Geophysical Astrophysical Fluid Dynamics*, **9**, 241.

Durney, B.R. and Robinson, R.D. (1982). *Astrophysical Journal*, **253**, 290.

Durney, B.R. and Spruit, H.C. (1979). *Astrophysical Journal*, **234**, 1067.

Elsasser, W.M. (1946). *Physical Review*, **69**, 106; **70**, 202.

Elsasser, W.M. (1947). *Physical Review*, **72**, 821.

Ferriz-Mas, A. and Schüssler, M. (1993). *Geophysical Astrophysical Fluid Dynamics*, **72**, 209.

Ferriz-Mas, A. and Schüssler, M. (1994). *Astrophysical Journal*, **433**, 852.

Ferriz-Mas, A. and Schüssler, M. (1995). *Geophysical Astrophysical Fluid Dynamics*, **81**, 233.

Ferriz-Mas, A., Schmitt, D. and Schüssler, M. (1994). *Astronomy and Astrophysics*, **289**, 949.

Gailitis, A. (1970). *Magnetohydrodynamics*, **6**, 19.

Gibson, R.D., Roberts, P.H. and Scott, S. (1969). In *The application of modern physics to the Earth and planetary interiors* (ed. S.K. Runcorn), p. 577. Wiley, New York.

Golub, L., Rosner, R., Vaiana, G.S. and Weiss, N.O. (1981). *Astrophysical Journal*, **243**, 309.

Gubbins, D. (1973). *Philosophical Transactions Royal Society A*, **274**, 493.

Heisenberg, W. (1948). *Proceedings Royal Society A*, **195**, 402.

Herzenberg, A. (1958). *Philosophical Transactions Royal Society A*, **250**, 543.

Ivanova, T.S. and Ruzmaikin, A.A. (1977). *Soviet Astronomy*, **21**, 479.

Jepps, S.A. (1975). *Journal Fluid Mechanics*, **67**, 625.

Jones, C.A. and Galloway, D.J. (1993). In *Solar and planetary dynamos* (eds M.R.E. Proctor, P.C. Matthews and A.M. Rucklidge), p. 161. Cambridge University Press.

Jones, C.A., Weiss, N.O. and Cattaneo, F. (1985). *Physica*, **14D**, 161.

Kleeorin, N.I. and Ruzmaikin, A.A. (1981). *Geophysical and Astrophysical Fluid Dynamics*, **17**, 281.

Krause, F. (1965). In *Stellar and solar magnetic fields* (ed. R. Lüst), p. 426. North-Holland, Amsterdam.

Krause, F. and Rädler, K.-H. (1980). *Mean-field magnetohydrodynamics and dynamo theory.* Pergamon, Oxford.

Kulsrud, R.M. and Anderson, S.W. (1992). *Astrophysical Journal*, **396**, 606.

Kumar, S. and Roberts, P.H. (1975). *Proceedings Royal Society A*, **344**, 235.

Layzer, D., Rosner, R. and Doyle, H.T. (1979). *Astrophysical Journal*, **229**, 1126.

Leighton, R.B. (1964). *Astrophysical Journal*, **140**, 1559.

Leighton, R.B. (1969). *Astrophysical Journal*, **156**, 1.

Levy, E.H. (1972*a*). *Astrophysical Journal*, **171**, 635.

Levy, E.H. (1972*b*). *Astrophysical Journal*, **175**, 573.

Lilley, F.E.M. (1970). *Proceedings Royal Society A*, **316**, 153.

Lowes, F.J. and Wilkinson, I. (1968). *Nature*, **219**, 717.

Malkus, W.V.R. and Proctor, M.R.E. (1975). *Journal Fluid Mechanics*, **67**, 417.

Meneguzzi, M., Frisch, U. and Pouquet, A. (1981). *Physical Review Letters*, **47**, 1060.

Mestel, L. (1965). In *Stellar and solar magnetic fields* (ed. R. Lüst), p. 424. North-Holland, Amsterdam.

Moffatt, H.K. (1970). *Journal Fluid Mechanics*, **41**, 435.

Moffatt, H.K. (1978). *Magnetic field generation in electrically conducting fluids.* Cambridge University Press.

Molchanov, S.A., Ruzmaikin, A.A. and Sokoloff, D.D. (1985). *Soviet Physics Uspenski*, **28**, 307.

Moreno-Insertis, F. (1992). In *Sunspots: theory and observation* (eds J.H. Thomas and N.O. Weiss), p. 385. Kluwer, Dordrecht.

Moreno-Insertis, F., Schüssler, M. and Ferriz-Mas, A. (1992). *Astronomy and Astrophysics*, **264**, 686.

Moss, D. (1986). *Physics Reports*, **140**, 1.

Moss, D., Tuominen, I. and Brandenburg, A. (1990*a*). *Astronomy and Astrophysics*, **228**, 284.

Moss, D., Tuominen, I. and Brandenburg, A. (1990*b*). *Astronomy and Astrophysics*, **240**, 142.

Nordlund, A., Brandenburg, A., Jennings, R.L., Rieutord, M., Ruokolainen, J., Stein, R.F. *et al.* (1992). *Astrophysical Journal*, **392**, 647.

Parker, E.N. (1955). *Astrophysical Journal*, **122**, 293.

Parker, E.N. (1970). *Astrophysical Journal*, **162**, 665.

Parker, E.N. (1971). *Astrophysical Journal*, **163**, 255.

Parker, E.N. (1979). *Cosmical magnetic fields.* Oxford University Press.

Parker, E.N. (1982). *Astrophysical Journal*, **256**, 292, 302, 736, 746.

Parker, E.N. (1993). *Astrophysical Journal*, **408**, 707.

Parker, E.N. (1994). In *Cosmical magnetism* (ed. D. Lynden-Bell), p. 123. Kluwer, Dordrecht.

Pekeris, C.L., Accad, Y. and Shkoller, B, (1973). *Philosophical Transactions Royal Society A*, **275**, 425.

Piddington, J.H. (1972). *Solar Physics*, **22**, 3.

Piddington, J.H. (1973). *Astrophysics Space Science*, **24**, 259.

Rees, M.J. (1994). In *Cosmical magnetism* (ed. D. Lynden-Bell), p. 155. Kluwer, Dordrecht.

Rieutord, M. and Zahn, J.-P. (1995). *Astronomy and Astrophysics*, **296**, 127.

Roberts, G.O. (1969). *The application of modern physics to the earth and planetary interiors* (ed. S.K. Runcorn), p. 603. Wiley, London.

Roberts, G.O. (1970). *Philosophical Transactions Royal Society A*, **266**, 535.

Roberts, G.O. (1972). *Philosophical Transactions Royal Society A*, **271**, 411.

Roberts, P.H. (1971). *Mathematical problems in the geophysical sciences* (ed. W.H. Reid), p. 129. American Mathematical Society, Providence.

Roberts, P.H. (1972). *Philosophical Transactions Royal Society A*, **271**, 411.

Roberts, P.H. (1994). In *Lectures on solar and planetary dynamos* (eds M.R.E. Proctor and A.D. Gilbert), p. 1. Cambridge University Press.

Roberts, P.H. and Stix, M. (1971). *The turbulent dynamo*, translations of papers by Steenbeck *et al.* NCAR-TN/IA-60.

Roberts, P.H. and Stix, M. (1972). *Astronomy and Astrophysics*, **18**, 453.

Robinson, R.D. and Durney, B.R. (1982). *Astronomy and Astrophysics*, **108**, 322.

Rosner, R. and Weiss, N.O. (1992). In *The solar cycle* (ed. K. Harvey), p. 511. Astronomical Society of the Pacific.

Rüdiger, G. and Kitchatinov, L.L. (1993). *Astronomy and Astrophysics*, **269**, 581.

Ruzmaikin, A.A., Shukurov, A.A. and Sokoloff, D.D. (1988). *Magnetic fields of galaxies.* Kluwer, Dordrecht.

Saffman, P.G. (1963). *Journal Fluid Mechanics*, **16**, 545.

Schmalz, S. and Stix, M. (1991). *Astronomy and Astrophysics*, **245**, 654.

Schmitt, D. (1984). In *The hydromagnetics of the Sun* (eds T.D. Guyenne and J.J. Hunt), p. 223. ESA SP-220.

Schmitt, D. (1987). *Astronomy and Astrophysics*, **174**, 281.

Schmitt, D. (1993). In *The cosmic dynamo* (eds F. Krause, K.-H. Rädler and G. Rüdiger), p. 1. Kluwer, Dordrecht.

Schüssler, M. (1993). In *The cosmic dynamo* (eds F. Krause, K.-H. Rädler and G. Rüdiger), p. 27. Kluwer, Dordrecht.

Spitzer Jr., L. (1957). *Astrophysical Journal*, **125**, 525.

Spruit, H.C. (1981). *Astronomy and Astrophysics*, **102**, 109.

Spruit, H.C. and van Ballegooijen, A.A. (1982). *Astronomy and Astrophysics*, **106**, 58.

Steenbeck, M. and Krause, F. (1966). *Zeitschrift Naturforschung*, **21a**, 1285.

Steenbeck, M. and Krause, F. (1969). *Astronomische Nachrichten*, **291**, 49.

Steenbeck, M., Krause, F. and Rädler, K.-H. (1966). *Zeitschrift Naturforschung*, **21a**, 369.

Stix, M. (1972). *Astronomy and Astrophysics*, **20**, 9.

Stix, M. (1973). *Astronomy and Astrophysics*, **24**, 275.

Sweet, P.A. (1950). *Monthly Notices Royal Astronomical Society*, **110**, 69.

Tao, L., Cattaneo, F. and Vainshtein, S.I. (1993). In *Solar and planetary dynamos* (eds M.R.E. Proctor, P.C. Matthews and M. Rucklidge), p. 303. Cambridge

University Press.

Tavakol, R., Tworkowski, A.S., Brandenburg, A., Moss, D. and Tuominen, I. (1995). *Astronomy and Astrophysics*, **296**, 269.

Tobias, S.M. (1996). *Astronomy and Astrophysics.* **307**, L21.

Tobias, S.M., Weiss, N.O. and Kirk, V. (1995). *Monthly Notices Royal Astronomical Society*, **273**, 1150.

Vainshtein, S.I. and Cattaneo, F. (1992). *Astrophysical Journal*, **393**, 163.

Vainshtein, S.I. and Rosner, R. (1991). *Astrophysical Journal*, **376**, 199.

Vainshtein, S.I., Parker, E.N. and Rosner, R. (1993). *Astrophysical Journal*, **404**, 773.

van Ballegooijen, A.A. (1983). *Astronomy and Astrophysics*, **118**, 275.

van Ballegooijen, A.A. and Choudhuri, A.R. (1988). *Astrophysical Journal*, **333**, 965.

Wang, Y.-M. and Sheeley Jr., N.R. (1991). *Astrophysical Journal*, **375**, 761.

Wang, Y.-M., Nash, A.G. and Sheeley, N.R. (1989). *Astrophysical Journal*, **347**, 529.

Wang, Y.-M., Sheeley Jr., N.R. and Nash, A.G. (1991). *Astrophysical Journal*, **383**, 431.

Weiss, N.O. (1983). *Solar magnetism.* In *Stellar and planetary magnetism* (ed. A.M. Soward), p. 115. Gordon and Breach, New York.

Weiss, N.O. (1994). In *Lectures on solar and planetary dynamos* (eds M.R.E. Proctor and A.D. Gilbert), p. 59. Cambridge University Press.

Weiss, N.O., Cattaneo, F. and Jones, C.A. (1984). *Geophysical Astrophysical Fluid Dynamics*, **30**, 305.

Yoshimura, H. (1975). *Astrophysical Journal*, **201**, 740.

Yoshimura, H. (1978*a*). *Astrophysical Journal*, **220**, 692.

Yoshimura, H. (1978*b*). *Astrophysical Journal*, **226**, 706.

Zel'dovich, Ya.B., Ruzmaikin, A.A. and Sokoloff, D.D. (1983). *Magnetic fields in astrophysics.* Gordon and Breach, New York.

Zel'dovich, Ya.B. (1988). *Soviet Science Reviews*, **C7**, 2.

7

MAGNETIC BRAKING OF STARS BY WINDS

7.1 Introduction

The chromosphere and corona of the Sun and other late-type stars are heated by the dissipation of energy originating below the surface. Biermann (1946) and Schwarzschild (1948) pointed out that ordinary sound waves generated by the turbulence in the subphotospheric convective zone will steepen into shocks as they propagate upwards into a domain with declining density. This purely gas-dynamical heating remains the probable explanation for the lower chromosphere, but for the upper chromosphere and corona, the heating process is now believed to be hydromagnetic (e.g. Osterbrock 1961). In regions with closed field lines near solar active regions, the dissipation is plausibly located in the spontaneously forming current sheets and filaments (cf. Section 4.8). For the open field line regions (see below), dissipation of magnetohydrodynamic waves appears to be viable. A complete theory of the corona will include the heat sources, radiation losses, and heat transfer by thermal conduction. For detailed accounts, with references to previous literature, see the cited monographs by Priest (1982) and Stix (1989). In this chapter, the heating of the low-density corona to a temperature above 10^6 K which declines slowly outwards is taken as an observational datum and its dynamical consequences are explored.

It is known that in the absence of a magnetic field, a sufficiently hot stellar corona must expand as a more or less spherically symmetric wind. This is because the condition of hydrostatic equilibrium yields a pressure far from the star that is orders of magnitude greater than the combined thermal, turbulent, cosmic ray and magnetic pressures of the interstellar medium: there is no 'lid' capable of holding in a hot static corona (Parker 1963). The picture is modified by the presence of a magnetic field, satisfying the modest requirement that at the coronal base, its energy density should be at least comparable with the thermal energy density. Field loops that do not extend far from the star will keep gas trapped. Field lines that in the absence of the hot gas would close far from the star will be unable to restrain the hot gas, which will instead flow out, dragging the field with it. One is led to a picture of a multi-component corona, with 'dead zones' of hot trapped gas, and 'wind zones' of cooler gas that flows out along 'open' field lines (Mestel 1967*a,b*, 1968*a*; Okamoto 1974; Pneuman and Kopp 1972; Rowse and Roxburgh 1981; Mestel and Spruit 1987; Tsinganos and Low 1989). Observations in X-rays of the Sun and other late-type stars do confirm that the wind is associated with the darker 'coronal holes' (see e.g. Priest 1982; Stix 1989).

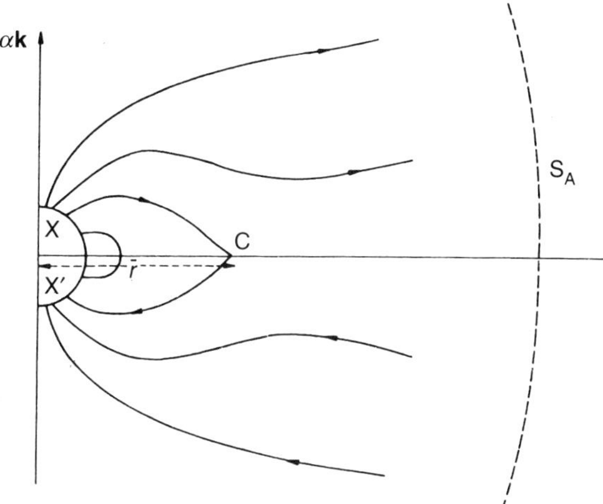

FIG. 7.1. Schematic coronal model with a dipole-like flux distribution over the stellar surface: the wind zone and the dead zone.

The magnetic field lines emanating from a star try to bring the surrounding highly conducting gas into corotation with the star. If there is a continuous outflow of gas, the consequent drain of angular momentum from the star can be far greater than that carried away if the gas were to feel no torque after leaving the stellar surface (Schatzman 1962). The continuous angular momentum loss will be primarily from the wind zone. (The 'dead' zone may not in fact be completely dead: there is evidence for at least sporadic emission of gas, carrying high angular momentum (Section 8.2), but we concentrate here on the wind zone.) We again begin with the simplest case, with $\mathbf{B}_{\rm p}$ symmetric about the rotation axis, and usually with a more or less dipolar flux distribution over the stellar surface, so that the coronal structure is schematically as in Fig. 7.1. If the field were to remain strictly poloidal it would exert zero torque, and each element of the outflowing gas would carry just the angular momentum with which it left the star. However, the consequent rotational shear would generate a toroidal component $\mathbf{B}_{\rm t}$ from $\mathbf{B}_{\rm p}$, and the twisted field now exerts a torque which tries to establish corotation with the star. The process is thus a competition between the speed $v_{\rm p}$ of the outflowing wind which generates the shear and the Alfvén speed $v_{\rm A} = B_{\rm p}/(4\pi\rho)^{1/2}$ at which the field lines try to straighten. Intuitively one expects that the gas will be kept in near corotation as long as $v_{\rm p} \ll v_{\rm A}$, so that a moderate rate of mass loss can yield a disproportionately large rate of loss of angular momentum.

In the following sections these qualitative ideas are made more precise by the detailed mathematical theory.

7.2 The braking of axisymmetric systems

Much of the necessary formalism with the appropriate notation has already been developed in Section 5.5 for the study of internal stellar rotation in the presence of meridian circulation. The poloidal magnetic field is again described by the flux function P:

$$\mathbf{B}_{\mathrm{p}} = -\nabla \times (P\mathbf{t}/\varpi) = -\nabla P \times \mathbf{t}/\varpi. \tag{7.1}$$

The kinematic perfect conductivity equation

$$\nabla \times (\mathbf{v} \times \mathbf{B}) = 0 \tag{7.2}$$

yields

$$\mathbf{v}_{\mathrm{p}} = \kappa \mathbf{B}_{\mathrm{p}}, \tag{7.3}$$

$$\Omega - \kappa B_{\phi}/\varpi = \alpha(P), \tag{7.4}$$

equivalent to

$$\mathbf{v} = \kappa \mathbf{B} + \alpha(P)\varpi \mathbf{t}. \tag{7.5}$$

This is the generalization of Ferraro's law of isorotation: for the moment we are allowing the possibility that α varies from field line to field line. Note that $\mathbf{v}_{\mathrm{p}} \cdot \nabla P = \kappa \mathbf{B}_{\mathrm{p}} \cdot \nabla P = 0$: under perfect conductivity the flux function P defines both the poloidal field lines and the poloidal wind streamlines. The continuity equation

$$\nabla \cdot (\rho \mathbf{v}) = 0 \tag{7.6}$$

yields

$$\rho\kappa = \rho v_{\mathrm{p}}/B_{\mathrm{p}} = \eta(P), \tag{7.7}$$

where $\eta(P)$ is the flow of gas per flux-tube of unit strength, defined by P (cf. (5.65)). From the torque equation (5.66)

$$\rho \mathbf{v} \cdot \nabla(\Omega\varpi^2) = \varpi(\mathbf{j}_{\mathrm{p}} \times \mathbf{B}_{\mathrm{p}}/c) \cdot \mathbf{t} = \nabla \cdot (\varpi B_{\phi} \mathbf{B}_{\mathrm{p}}/4\pi) \tag{7.8}$$

we derive with the help of (7.3) the torque integral, which expresses the rate of transport $-\beta(P)/4\pi$ per unit poloidal flux-tube of angular momentum, carried jointly by the magnetic stresses and the gas:

$$-\varpi B_{\phi}/4\pi + \eta\Omega\varpi^2 = -\beta(P)/4\pi \tag{7.9}$$

(cf. (2.48)). Equations (7.4), (7.7) and (7.9) combine into

$$\Omega(1 - 4\pi\eta^2/\rho) = \alpha + \eta\beta/\rho\varpi^2, \tag{7.10}$$

where

$$4\pi\eta^2/\rho = v_{\mathrm{p}}^2/v_{\mathrm{A}}^2 = 4\pi\rho v_{\mathrm{p}}^2/B_{\mathrm{p}}^2. \tag{7.11}$$

Near the star $v_p \ll v_A$ for even weak surface fields, but with ρ decreasing outwards, $4\pi\eta^2(P)/\rho$ increases along a field-streamline, becoming unity at the Alfvénic point P_A where

$$v_p = v_A = B_p/(4\pi\rho)^{1/2}, \tag{7.12}$$

or

$$\rho = \rho_A = 4\pi\eta^2. \tag{7.13}$$

By (7.10), Ω (and likewise B_ϕ) will become singular at P_A unless

$$-\beta/4\pi = \alpha\rho_A\varpi_A^2/4\pi\eta = \eta(\alpha\varpi_A^2) = (\rho v_p/B_p)(\alpha\varpi_A^2). \tag{7.14}$$

Recall that $-\beta/4\pi$ is the joint flow of angular momentum per second along a unit poloidal flux-tube. The 'eigenvalue' (7.14) shows that this flow is equivalent to that carried by the steady matter flux η per unit flux tube, if the gas were kept rotating with the angular velocity $\alpha(P)$ out to P_A – 'effective corotation'.

At any point, the actual values of Ω and B_ϕ are given by

$$\frac{\Omega}{\alpha} = \frac{(1-\rho_A\varpi_A^2/\rho\varpi^2)}{(1-\rho_A/\rho)} = \frac{[1-(v_p/v_A)(B_A\varpi_A^2/B_p\varpi^2)]}{(1-\rho_A/\rho)} \tag{7.15}$$

on use of (7.7) (with $(B_p)_A, (v_p)_A$ shortened to B_A, v_A), and

$$\varpi B_\phi = -4\pi\eta\alpha\varpi_A^2\frac{(1-\varpi^2/\varpi_A^2)}{(1-\rho_A/\rho)}. \tag{7.16}$$

The quantity $B_p\varpi^2$ is approximately constant along a field-streamline, while $\rho_A/\rho, v_p/v_A$ both decrease strongly towards the coronal base; hence close to the star $\Omega \simeq \alpha(P)$, and by (7.16) the magnetic transport term in (7.9) dominates by the factor $\simeq (\varpi_A/\varpi)^2$. Sufficiently far beyond P_A, we anticipate that $\rho\varpi^2/\rho_A\varpi_A^2$ is small, so that (7.15) becomes $\Omega\varpi^2 \simeq \alpha\varpi_A^2$, and the material transport term in (7.9) ultimately dominates. Over a wide domain encompassing P_A the two terms are comparable, and even well within P_A, Ω already lags markedly behind α. Note that the frequently heard statement 'the gas is kept corotating with the star out to the Alfvénic surface S_A (defined by the points P_A)' is inaccurate and misleading. The correct statement is as above: 'effective corotation', as given by (7.14), is an exact prescription for computing the angular momentum transport along a field-streamline, carried *jointly* by the magnetic and material stresses.

Observations by Pizzo *et al.* (1983) of the solar wind confirm the basic predictions of the theory and infer a value $\simeq 12R_\odot$ for the Alfvénic radius r_A.

7.3 The wind theory

The full vector equation of motion of the inviscid, conducting outstreaming gas

$$-\frac{\nabla p}{\rho} + \nabla V + \frac{\mathbf{j}\times\mathbf{B}}{c\rho} = \nabla\left(\frac{\mathbf{v}^2}{2}\right) - \mathbf{v}\times(\nabla\times\mathbf{v}) \tag{7.17}$$

has the toroidal component that yields the integral (7.9). The scalar product of (7.17) with $\rho\mathbf{v}$ yields

$$\rho\mathbf{v}\cdot\nabla\left(\frac{\mathbf{v}^2}{2}-V\right)+\rho\mathbf{v}\cdot\frac{\nabla p}{\rho}=\mathbf{v}\cdot\left(\frac{\mathbf{j}\times\mathbf{B}}{c}\right)=\alpha(P)\varpi\mathbf{t}\cdot\left(\frac{\mathbf{j}\times\mathbf{B}}{c}\right)$$
$$=\rho\mathbf{v}\cdot\nabla(\alpha\Omega\varpi^2) \tag{7.18}$$

with the help of (7.5) and (7.8). A complete theory must supplement the equation of state $p=(\mathcal{R}/\mu)\rho T$ by a formula, explicit or implicit, giving the net local heat supply to the corona – the algebraic sum of that supplied by heat conduction, radiated by the optically thin medium, and fed in by dissipation of energy supplied from the convective zone (waves, current sheets – cf. Section 4.8). This physical problem is formidable and unsolved. The simplest way of bypassing it is to adopt a polytropic law. The solar corona has a temperature that declines fairly slowly from its maximum above the coronal base, and at least for the domain determining the angular momentum loss, the approximation $p\propto\rho^{\Gamma}$, with Γ a constant well below the adiabatic value 5/3, is not unreasonable. With the polytropic assumption, (7.7) and (7.18) then yield the generalized Bernoulli integral:

$$\frac{1}{2}(v_{\mathrm{p}}^2+\Omega^2\varpi^2)+\frac{\Gamma}{\Gamma-1}\frac{p}{\rho}-\frac{GM}{r}-\alpha\Omega\varpi^2=E(P). \tag{7.19}$$

The new terms are the local rotational energy $\Omega^2\varpi^2/2$ per gram and the term $\alpha\Omega\varpi^2$ that comes from the rate of working of the same magnetic torques that give angular momentum to the outflowing gas.

Substitution from (7.9) and (7.14) yields equivalently

$$\eta\left[\frac{1}{2}(v_{\mathrm{p}}^2+\Omega^2\varpi^2)+\frac{\Gamma}{\Gamma-1}\frac{p}{\rho}-\frac{GM}{r}\right]-\frac{\alpha\varpi B_\phi}{4\pi}=\eta(E+\alpha^2\varpi_{\mathrm{A}}^2)$$
$$\equiv\eta\tilde{E}(P), \tag{7.20}$$

where the second term on the left is immediately recognized as the Poynting energy flux along a unit poloidal flux-tube; for on substitution of the perfect conductivity relation $\mathbf{E}=-\mathbf{v}\times\mathbf{B}/c$ and of (7.5), the Poynting vector $(c/4\pi)(\mathbf{E}\times\mathbf{B})$ yields the poloidal component $(-\alpha\varpi B_\phi/4\pi)\mathbf{B}_{\mathrm{p}}$. Equation (7.20) can be derived directly by writing the term $\mathbf{v}\cdot(\mathbf{j}\times\mathbf{B}/c)$ in (7.18) as $\mathbf{j}\cdot\mathbf{E}$ and then performing the Poynting transformation.

From now on, the sound speed a_{w} in the wind zone is taken as constant, so that the polytropic relation is replaced by the isothermal law $p=a_{\mathrm{w}}^2\rho$, and $a_{\mathrm{w}}^2\log\rho$ replaces the term $(\Gamma/\Gamma-1)p/\rho$ in (7.19). An alternative approach to the problem of the pressure and density distributions is presented in the Appendix.

If the structure of the poloidal field $\mathbf{B}_{\mathrm{p}}$ is supposed known, then substitution of (7.7) and (7.15) converts (7.19) into a relation

$$H(\varpi,\rho)=E(P) \tag{7.21}$$

between ρ and the convenient monotonic coordinate ϖ for the particular field-streamline considered. Once the choice (7.14) has ensured that Ω and B_{t} are non-singular, then all non-singular solutions for ρ pass automatically through the Alfvénic point $\mathrm{P_A}$; for when $\rho = \rho_{\mathrm{A}}, H(\varpi, \rho)$ diverges except at $\varpi = \varpi_{\mathrm{A}}$, where it is finite (Weber and Davis 1967; Goldreich and Julian 1970). In Sakurai's words (1985): '$\rho = \rho_{\mathrm{A}}$ is an infinitely high wall in (ϖ, ρ)-space with a hole at $(\varpi_{\mathrm{A}}, \rho_{\mathrm{A}})$, so all solutions which go from sub-Alfvénic ($\rho > \rho_{\mathrm{A}}$) to super-Alfvénic ($\rho < \rho_{\mathrm{A}}$) regions automatically pass through the Alfvénic point $(\varpi_{\mathrm{A}}, \rho_{\mathrm{A}})$ in the (ϖ, ρ)-plane.'

Along each field-streamline $H(\varpi, \rho) = E$, the density gradient is

$$\frac{\mathrm{d}\rho}{\mathrm{d}\varpi} = -\frac{\partial H/\partial \varpi}{\partial H/\partial \rho}. \tag{7.22}$$

The critical points occur when numerator and denominator in (7.22) vanish simultaneously. The curve $\partial H/\partial \varpi = 0$ is the locus of the Laval nozzles in the (ϖ, ρ)-plane. With the help of (7.4) and (7.7), it is convenient to write H – the left-hand side of (7.19) – as

$$H = \frac{\eta^2 B_{\mathrm{p}}^2}{2\rho^2} + a_{\mathrm{w}}^2 \log \rho - \frac{GM}{r} + \frac{1}{2}\left[\alpha\varpi + \frac{\eta B_\phi}{\rho}\right]^2 - \alpha\varpi\left[\alpha\varpi + \frac{\eta B_\phi}{\rho}\right]. \tag{7.23}$$

Then from (7.23) and (7.16),

$$\rho\frac{\partial H}{\partial \rho} = -v_{\mathrm{p}}^2 + a_{\mathrm{w}}^2 - \frac{\eta^2 B_\phi^2}{\rho^2(1-\rho_{\mathrm{A}}/\rho)}. \tag{7.24}$$

From (7.13),

$$\frac{\rho_{\mathrm{A}}}{\rho} = \frac{4\pi\eta^2}{\rho} = \left[\frac{v_{\mathrm{p}}}{(B_{\mathrm{p}}/4\pi\rho)^{1/2}}\right]^2 \equiv \left(\frac{v_{\mathrm{p}}}{v_{\mathrm{Ap}}}\right)^2, \tag{7.25}$$

where v_{Ap} is the Alfvén speed defined by the local value of B_{p}. The *Alfvénic point* introduced in (7.12) is where $v_{\mathrm{p}} = v_{\mathrm{Ap}}$. Then (7.24) is easily converted into

$$\rho\frac{\partial H}{\partial \rho} = -\frac{v_{\mathrm{p}}^4 - v_{\mathrm{p}}^2(a_{\mathrm{w}}^2 + v_{\mathrm{Ap}}^2 + v_{\mathrm{A}\phi}^2) + a_{\mathrm{w}}^2 v_{\mathrm{Ap}}^2}{v_{\mathrm{p}}^2 - v_{\mathrm{Ap}}^2} \tag{7.26}$$

where $v_{\mathrm{A}\phi}^2 = B_\phi^2/4\pi\rho$. Refer now to (3.9), and identify the field $\mathbf{B}_0$ with the total field $(\mathbf{B}_{\mathrm{p}} + \mathbf{B}_{\mathrm{t}})$ in the present problem, and let the wave be travelling in the direction $\mathbf{k}$ parallel to $\mathbf{B}_{\mathrm{p}}$, so that $V_{\mathrm{A}}^2 \cos^2\theta = v_{\mathrm{Ap}}^2$. It is seen that $\partial H/\partial \rho = 0$ when v_{p} equals either v_{f} or v_{s}, respectively the speeds of fast and slow MHD waves, determined by the total field (poloidal plus toroidal) and travelling along $\mathbf{B}_{\mathrm{p}}$. In general, v_{Ap}, v_{s} and v_{f} all differ, but on the axis, where B_ϕ and so also $v_{\mathrm{A}\phi}$ vanish, one of v_{s}, v_{f} coincides with v_{Ap}.

As in Bondi's spherical accretion flow (1952) and Parker's fundamental wind solution (1963), we can define a fundamental solution with smooth, monotonically accelerating flow when H_ρ and H_ϖ vanish together; in an obvious notation

$$H(\varpi_{\mathrm{f}}, \rho_{\mathrm{f}}) = E = H(\varpi_{\mathrm{s}}, \rho_{\mathrm{s}}). \qquad (7.27)$$

For each field-streamline, the physical parameters of the problem are the sound speed a_{w}, the angular velocity α, the specific mass flux η, the field strength at the coronal base, the Alfvénic distance ϖ_{A} and the energy E. Of these, the first four are extrinsic, defining the problem to be solved. The two intrinsic conditions (7.27) then suffice to determine the problem completely (Weber and Davis 1967). Replacement of isothermality by a polytropic simulation makes no essential difference, for in both cases the thermal term depends only on the density.

In most cases of interest, at the slow point $v_{\mathrm{Ap}} \gg v_{\mathrm{A}\phi}, a_{\mathrm{w}}$, so that $v_{\mathrm{s}}^2 \simeq a_{\mathrm{w}}^2(1 - v_{\mathrm{A}\phi}^2/v_{\mathrm{Ap}}^2) \simeq a_{\mathrm{w}}^2$: the wind speed at the slow point hardly differs from the sound speed. However, at the fast point the speed is equal approximately to the local Alfvén speed determined by the *total* magnetic field $(B_{\mathrm{p}}^2 + B_{\mathrm{t}}^2)^{1/2}$, and so depends explicitly on the rotation α.

In a slow rotator such as the Sun, the terms in Ω in (7.19) are small compared with the thermal pressure term, and so merely perturb slightly Parker's (1963) thermally-driven wind: the gas undergoes strong acceleration through the slow point (where as noted the speed is essentially the sound speed), followed by a much slower subsequent acceleration. By contrast, in a moderately rapid or very rapid rotator, the terms in Ω take over before the Alfvénic point is reached, yielding a net extra driving term $\alpha^2\varpi^2(\Omega/\alpha)(1 - \Omega/2\alpha)$. Well within $\mathrm{P_A}$, where the magnetic stresses keep $\Omega \simeq \alpha$, this term $\simeq \alpha^2\varpi^2/2$, and (see below) even at $\mathrm{P_A}$ it is typically of the same order. This 'centrifugal wind' effect must be included in estimates of the dependence of the rate of braking on the star's rotation (Sections 8.3 and 10.8).

7.4 The structure of the poloidal field

7.4.1 *General discussion*

From (7.7) and (7.13), at a general point

$$(\rho v_{\mathrm{p}}^2/2)/(B_{\mathrm{p}}^2/8\pi) = 4\pi(\rho v_{\mathrm{p}}/B_{\mathrm{p}})^2/\rho = \rho_{\mathrm{A}}/\rho = (v_{\mathrm{p}}/v_{\mathrm{pA}})(B_{\mathrm{pA}}/B_{\mathrm{p}}). \qquad (7.28)$$

Anticipating that at the Alfvénic surface $\mathrm{S_A}$ (defined by the points $\mathrm{P_A}$), the toroidal and poloidal components are of the same order, it is seen that $\mathrm{S_A}$ separates the domains nearer the star where the magnetic energy dominates from the more distant regions where the kinetic energy dominates. This suggests strongly that whereas well within $\mathrm{S_A}$ the flow in the wind zone will be channelled parallel to the field, well beyond $\mathrm{S_A}$ the field will be passive, being pulled out to follow the wind. In the simplest model, with $\mathbf{B}_{\mathrm{p}}$ having a prescribed dipolar structure over the star's surface, field lines leaving the star at high colatitude will reach the equator at points with $r \ll r_{\rm A}$ and so with the curl-free approximation still valid, whereas those leaving near the poles will extend to regions where the forces driving the wind will be able to distort the field to follow the wind. Thus we are led to the picture of the multi-component corona (Fig. 7.1), as outlined above, with a wind zone and a dead zone. Near enough to the star the magnetic energy dominates over the thermal energy and the wind kinetic energy, so that in the wind zone as well as in the dead zone the field structure will still approximate to that of a vacuum dipole.

In most studies in the literature, it is assumed that the integrals derived under the perfect conductivity approximation remain valid all the way to infinity, and also that a polytropic or isothermal $p-\rho$ relation holds. The detailed construction of $\mathbf{B}_{\rm p}$ then requires solution of the poloidal trans-$\mathbf{B}_{\rm p}$ component of the equation of motion (7.17) for the wind zone, subject to the above integral constraints (Okamoto 1975; Heinemann and Olbert 1978; Sakurai 1985, 1990), and of the analogous simpler equation for the dead zone, subject to the boundary condition of continuity of $(p + \mathbf{B}^2/8\pi)$ across the separatrix between the two zones. We consider first the wind zone, again assumed isothermal, and also restrict the discussion to the case of uniform rotation α of the field lines rather than the more general isorotation case $\alpha(P)$. The poloidal component of (7.17) can be written

$$\rho\left(\mathbf{v}_{\rm p} \times (\nabla \times \mathbf{v}_{\rm p})\right) + \frac{(\nabla \times \mathbf{B}_{\rm p})\times\mathbf{B}_{\rm p}}{4\pi} + \rho\left(\mathbf{v}_{\rm t} \times (\nabla \times \mathbf{v}_{\rm t})\right) + \frac{(\nabla \times \mathbf{B}_{\rm t})\times\mathbf{B}_{\rm t}}{4\pi}$$

$$= \rho\nabla\left(\frac{1}{2}(\mathbf{v}_{\rm p}^2 + \mathbf{v}_{\rm t}^2) + a_{\rm w}^2 \log\rho - \frac{GM}{r}\right) \qquad (7.29)$$

$$= \rho E'(P)\nabla P + \rho\alpha\nabla(\Omega\varpi^2) \qquad (7.30)$$

on use of the Bernoulli integral (7.19), (with $(\Gamma/\Gamma-1)K\rho^{\Gamma-1}$ replacing $a_{\rm w}^2\log\rho$ in the polytropic case). With the help of (7.4), (7.7), (7.9) and (7.14), the terms in $\mathbf{v}_{\rm t}$ and $\mathbf{B}_{\rm t}$ in (7.30) conveniently combine with the last term:

$$-\rho\Omega\varpi\mathbf{t} \times \left[\nabla \times \left\{\Omega\varpi^2\left(\frac{\mathbf{t}}{\varpi}\right)\right\}\right] - \nabla \times \left[(\varpi B_\phi) \times \left(\frac{\mathbf{t}}{\varpi}\right)\right] \times \left(\frac{B_\phi}{4\pi}\right)\mathbf{t}$$

$$+\rho\alpha\,\nabla(\Omega\varpi^2) = -\rho(\Omega-\alpha)\,\nabla(\Omega\varpi^2) + \frac{B_\phi}{4\pi\varpi}\,\nabla(\varpi B_\phi)$$

$$= \varpi B_\phi\left[\Omega\eta' - \alpha\frac{(\eta\varpi_{\rm A}^2)'}{\varpi^2}\right]\nabla P, \qquad (7.31)$$

where the prime denotes differentiation with respect to P. The terms in $\mathbf{v}_{\rm p}$, $\mathbf{B}_{\rm p}$ in (7.30), with the help also of (7.1) and (7.3), yield (on the left-hand side):

$$\left[\left(\frac{4\pi\eta^2}{\rho} - 1\right)\frac{(\nabla \times \mathbf{B}_p)\cdot\mathbf{t}}{4\pi\varpi} + \frac{\eta\eta' B_{\rm p}^2}{\rho} - \frac{\eta^2}{\rho^2\varpi^2}\nabla\rho\cdot\nabla P\right]\nabla P. \qquad (7.32)$$

Thus all the terms in (7.30) are force densities in meridian planes parallel to ∇P, i.e. normal to the poloidal field lines $\mathbf{B}_\mathrm{p}$; they combine to yield the transfield equation for P:

$$\left(\frac{4\pi\eta^2}{\rho}-1\right)\frac{1}{\varpi^2}\left(P_{\varpi\varpi}+P_{zz}-\frac{P_\varpi}{\varpi}\right) \\ +\frac{4\pi\eta\eta'}{\rho\varpi^2}(\nabla P)^2-\frac{4\pi\eta^2}{\rho^2\varpi^2}\nabla\rho\cdot\nabla P-4\pi\rho E'(P) \\ +16\pi^2\eta\alpha^2\varpi_\mathrm{A}^2\frac{(1-\varpi^2/\varpi_\mathrm{A}^2)}{(1-4\pi\eta^2/\rho)}\left[\eta'\frac{(1-4\pi\eta^2\varpi_\mathrm{A}^2/\rho\varpi^2)}{(1-4\pi\eta^2/\rho)}-\frac{(\eta\varpi_\mathrm{A}^2)'}{\varpi^2}\right]=0. \tag{7.33}$$

Equation (7.33) contains $\nabla\rho\cdot\nabla P$, proportional to the gradient of ρ across the field-streamline P, and ρ is an implicit function of r,θ,P and ∇P through the Bernoulli integral (7.19). On use of (7.1) and (7.7), (7.19) becomes

$$H=\frac{\eta^2}{2\rho^2\varpi^2}\left[\left(\frac{\partial P}{\partial\varpi}\right)^2+\left(\frac{\partial P}{\partial z}\right)^2\right]+a_\mathrm{w}^2\log\rho-\frac{GM}{r} \\ +\frac{\alpha^2r^2\sin^2\theta}{2}\left[\left(\frac{\Omega}{\alpha}\right)\left(\frac{\Omega}{\alpha}-2\right)\right]=E(P). \tag{7.34}$$

Since from (7.15) and (7.13), Ω/α depends on η and ϖ_A which are functions of P, the gradient of (7.34) becomes

$$\frac{\partial H}{\partial\rho}\nabla\rho+\frac{\eta^2}{\rho^2\varpi^2}\left[\left(\frac{\partial P}{\partial\varpi}\frac{\partial^2P}{\partial\varpi^2}+\frac{\partial P}{\partial z}\frac{\partial^2P}{\partial z\,\partial\varpi}\right),\frac{\partial P}{\partial\varpi}\frac{\partial^2P}{\partial\varpi\,\partial z}+\frac{\partial P}{\partial z}\frac{\partial^2P}{\partial z^2}\right] \\ +\left(\frac{\partial H}{\partial\varpi},\frac{\partial H}{\partial z}\right)+\left(\frac{\partial H}{\partial\eta}\frac{\partial\eta}{\partial P}+\frac{\partial H}{\partial\varpi_\mathrm{A}}\frac{\partial\varpi_\mathrm{A}}{\partial P}\right)\nabla P=\frac{\partial E}{\partial P}\nabla P, \tag{7.35}$$

and the scalar product with ∇P yields $\nabla\rho\cdot\nabla P$, to be eliminated from (7.33). One arrives finally at the the quasi-linear transfield equation, which may be written (Heinemann and Olbert 1978):

$$(v_\mathrm{p}^2-v_\mathrm{Ap}^2)\Bigg(\left[(v_\mathrm{p}^2-v_\mathrm{f}^2)(v_\mathrm{p}^2-v_\mathrm{s}^2)v_\mathrm{Ap}^2-v_\mathrm{p}^4v_{\mathrm{A}z}^2\right]\frac{\partial^2P}{\partial\varpi^2} \\ +\left[(v_\mathrm{p}^2-v_\mathrm{f}^2)(v_\mathrm{p}^2-v_\mathrm{s}^2)v_\mathrm{Ap}^2-v_\mathrm{p}^4v_{\mathrm{A}\varpi}^2\right]\frac{\partial^2P}{\partial z^2} \\ +2v_\mathrm{p}^4v_{\mathrm{A}z}v_{\mathrm{A}\varpi}\frac{\partial^2P}{\partial\varpi\,\partial z}\Bigg)=F(\varpi,z,P,\nabla P), \tag{7.36}$$

where $v_\mathrm{A}=(1/4\pi\rho)^{1/2}B_\mathrm{p}$, and F is a known function.

A partial differential equation with the highest order terms of the form $(aP_{\varpi\varpi}+cP_{zz}+2bP_{\varpi z})$ has real characteristics if $(b^2-ac)>0$. Applied to (7.36), this reduces to

$$(v_\mathrm{p}^2-v_\mathrm{f}^2)(v_\mathrm{p}^2-v_\mathrm{s}^2)\{v_\mathrm{p}^2(v_\mathrm{f}^2+v_\mathrm{s}^2)-v_\mathrm{f}^2v_\mathrm{s}^2\}>0. \tag{7.37}$$

Thus as the gas accelerates outwards, the equation is elliptic out to the cusp surface defined by $v_{\rm p}^2 = v_{\rm f}^2 v_{\rm s}^2/(v_{\rm f}^2 + v_{\rm s}^2)$; then hyperbolic as far as the slow surface defined by $v_{\rm p} = v_{\rm s}$; elliptic again through the Alfvénic surface out to the fast surface; and finally hyperbolic to infinity.

The corresponding equation for the dead zone (assumed isothermal, but with a higher sound speed $a_{\rm d}$) is much simpler:

$$P_{\varpi\varpi} + P_{zz} - \frac{P_\varpi}{\varpi} + 4\pi\varpi^2 \rho E'(P) = 0 \tag{7.38}$$

with

$$E(P) = -\frac{1}{2}\alpha^2\varpi^2 - \frac{GM}{r} + a_{\rm d}^2 \log \rho. \tag{7.39}$$

However, the task of constructing a solution that satisfies equations and boundary conditions is formidable, not least because the shape and location of the different surfaces depend on the initially unknown field structure. Intuitively-based simple models, such as that of Section 7.5 below, estimate that the dead zone ends at an equatorial radius $\bar{r}$ well within the Alfvénic surface (cf. Fig. 7.1), so that (7.36) is supposed valid at all latitudes from $r = \bar{r}$ through $\rm S_A$ to infinity. Equation (7.36) is in fact singular at the Alfvénic surface $\rm S_A$ where $v_{\rm p} = v_{\rm Ap}$ – $4\pi\eta^2/\rho_{\rm A} = 1$ – and the requirement that the second derivatives be finite on $\rm S_A$, with no kinking of field lines, reduces to the Heyvaerts–Norman Alfvén regularity condition, which is usually a cubic in the slope of the Alfvén number at the Alfvén point (Heyvaerts and Norman 1989, and in Tsinganos 1996; Sauty and Tsinganos 1994, eqn (4.4); Camenzind 1991):

$$\frac{(\nabla P)^2}{2\varpi^2}\frac{\mathbf{n}\cdot\nabla(\eta^2/\rho^2)}{|\nabla P|} + E' - \frac{2\eta}{\rho}\alpha B_\phi \varpi_{\rm A}' + \frac{\eta' B_\phi^2}{4\pi\eta\rho} = 0, \tag{7.40}$$

where all quantities are computed on $\rm S_A$, primes are derivatives with respect to P, and $\mathbf{n}$ is the unit normal across $\mathbf{B}_{\rm p}$ in the direction of increasing P (with our sign convention, towards the axis). Once this is satisfied, then in principle one can continue the outwards integration of (7.36).

At this point a query arises: is it obvious that this electrodynamically non-dissipative, non-linear equation, with the polytropic or isothermal condition imposed on the pressure, necessarily has non-singular solutions that extend to infinity? Prima facie, in general there may not be enough freedom available to ensure continuity of both P and its normal derivative across $\rm S_A$, subject to the constraint (7.40), and also to allow proper behaviour at infinity.

An analogous problem arises in the structure of the pulsar magnetosphere, with the light-cylinder replacing the Alfvénic surface (Section 14.3). It appears that, in general, the cold, dissipation-free equation to the flux function P has solutions which are smooth at the light-cylinder but blow up at a finite distance beyond. It is suggested that in reality there is a dissipation zone beyond the light-cylinder where the system spontaneously builds up an effective macroresistivity, allowing a locally strong component $E_{\|}$. The likely instability of the dominant toroidal field that would otherwise build up is a plausible physical process yielding such a resistivity. The situation has an echo of Riemann's classical work on gas dynamics, in which the postulated dissipation-free equations spontaneously build up a density field which enforces local breakdown in entropy conservation and the consequent formation of a shock. The suggested analogous feature here is the system's refusal to accept that the severe constraint $\mathbf{E} \cdot \mathbf{B} = 0$ can hold everywhere, but instead insisting on there being a local domain with a strong $E_{\|}$-component, called by Alfvén (1981) a 'double layer'.

For the stellar wind problem, Sakurai (1985, 1990) has published a global numerical solution of the polytropic, dissipation-free equation. His field is topologically similar to the Weber–Davis split monopole, without a dead zone even well within $\mathrm{S_A}$; instead, there is a strongly pinched equatorial zone all the way back to the star. By contrast, Bogovalov (1996), also assuming a basic split monopole field, does find numerical evidence for breakdown in continuous solutions of the perfectly conducting equations, occurring at the point where the fast magnetosonic surface crosses the equator. One wonders whether the cause of the differing conclusions is numerical: Bogovalov stresses the need for further numerical simulations with a much higher resolution. Other workers (e.g. Heyvaerts and Norman) who have studied the solutions of an equation equivalent to (7.36) (cf. Section 7.4.2) report in a personal communication that there are indeed difficulties in the way of a smooth link-up of an appropriate asymptotic solution with one valid near and within $\mathrm{S_A}$. We return briefly to this query in the Appendix.

7.4.2 *Asymptotic behaviour*

It could very well be that the global properties of the solution will not be much affected by such a localized dissipative zone. A major concern of this chapter – the estimate of the torques exerted on a star through coupling with a magnetized stellar wind – is dealt with adequately by the theory of the previous sections combined with a plausible approximate structure for the magnetic and flow fields out to and beyond $\mathrm{S_A}$, as discussed in Section 7.5. However, the observations of winds and in particular of collimated jets, especially from pre-main sequence stars, justifies the focusing of attention on the asymptotic behaviour; an accurate link-up between the near and distant domains, including the location of a possible intervening dissipative zone, can reasonably be left for future work.

The new feature emerging from Sakurai's numerical work is the prediction of the asymptotic poleward deflection of the flow. The reason is that well beyond the Alfvénic surface, where Ω has fallen well below α, from (7.3) and (7.4),

$$\frac{|B_\phi|}{B_\mathrm{p}} \approx \frac{\alpha\varpi}{v_\mathrm{p}}. \tag{7.41}$$

Thermal driving is already weak beyond the slow point, and centrifugal driving falls off beyond the Alfvénic point, so the gas will asymptotically coast, with the ratio (7.41) proportional to ϖ. Thus if the poloidal component B_{p} were to remain radial and continue falling off like $1/r^2$, (7.41) would increase like ϖ. Since B_ϕ must vanish on the axis, just off the axis the B_ϕ-contribution to the Lorentz force will pinch, so causing some degree of collimation parallel to the axis (an effect first observed in solar wind calculations by Suess and Nerney 1975).

The collimation problem is in fact a good deal more subtle. It is discussed further in Section 10.9, where it will emerge that there are serious arguments against appealing to the toroidal field as the prime cause of the observed collimated jets. Nevertheless, for completeness we shall summarize what the asymptotic properties of the complete polytropic wind equations have to say on the question. The discussion follows closely that by Heyvaerts and Norman in their 1989 paper and especially in their 1996 paper (with some minor changes to conform to our notation).

Collimation – cylindrical focusing about the z-axis – implies that the coordinate ϖ along a poloidal field line stays finite as $z \to \infty$. The procedure adopted is to assume the reverse, i.e. that $\varpi \to \infty$ so that collimation does *not* occur, and then to draw appropriate inferences. Then $\rho/\rho_{\mathrm{A}} \ll 1$, $\varpi/\varpi_{\mathrm{A}} \gg 1$ asymptotically, so that from (7.16) and (7.15),

$$\varpi B_\phi \simeq -\frac{\alpha\rho\varpi^2}{\eta} \tag{7.42}$$

and

$$\frac{\Omega}{\alpha} \simeq -\frac{\rho}{\rho_{\mathrm{A}}} + \frac{\varpi_{\mathrm{A}}^2}{\varpi^2}. \tag{7.43}$$

No assumption is made about the behaviour of $\rho\varpi^2$; instead, the quantity

$$R_\infty^2 = \lim\left(\frac{\rho\varpi^2}{4\pi\eta^2}\right) \tag{7.44}$$

is defined, where 'lim' refers to the asymptotic value along the field line considered.

Consider now the Bernoulli integral, and recall that for simplicity we have assumed the expanding gas to remain isothermal (or polytropic with a fixed index Γ close to unity). For studies of the wind out to and some way beyond $\mathrm{S_A}$, this can be justified by appealing to observations of the solar corona, where it is clear that the temperature decline outwards is much below the adiabatic rate, implying that thermal conduction and input of heat through magnetic dissipation of energy fed in from the convective zone remain important out to quite large distances. However, these energy sources must ultimately die out, and the equation of state $p = \rho a_{\mathrm{w}}^2$ with a_{w} a constant sound speed must be replaced by $p = K\rho^\gamma$ with $\gamma \simeq 5/3$. The term $a_{\mathrm{w}}^2 \log\rho$ in Bernoulli's equation will then be replaced by $(\gamma/\gamma-1)K\rho^{\gamma-1}$, which vanishes instead of diverging as $\rho \to 0$ at ∞. From now on, the thermal term (like the gravitational term) will be assumed asymptotically negligible, so that the Bernoulli integral in either form (7.19) or (7.20), combined respectively with (7.15) or (7.16), becomes

$$\frac{v_{\mathrm{p}}^2}{2} = \frac{\eta^2(\nabla P)^2}{2\rho^2\varpi^2} \simeq \tilde{E} - \frac{\rho\alpha^2\varpi^2}{\rho_{\mathrm{A}}}, \tag{7.45}$$

the surviving terms on the right being just the total energy flux and the Poynting flux. Thus (7.45) fixes an upper bound on $\rho\varpi^2$:

$$\lim \frac{\alpha^2}{\rho_{\mathrm{A}}}(\rho\varpi^2) < \tilde{E}. \tag{7.46}$$

Suppose first that $R_\infty^2 = 0$, i.e. that $\rho\varpi^2 \to 0$ as one moves to infinity along a particular field line P. Then by (7.42), $\varpi B_\phi \to 0$ along the line, implying a vanishing total poloidal current within the surface of revolution defined by P. From (7.45),

$$\varpi|\nabla P| \le (2\tilde{E})^{1/2}\rho\varpi^2/\eta, \tag{7.47}$$

and so if $\lim(\rho\varpi^2) = 0$, then there exists a function $\lambda(P, \varpi)$ which $\to 0$ as $\varpi \to \infty$, and such that

$$\varpi|\nabla P| < \lambda(P, \varpi). \tag{7.48}$$

Then since

$$\left(\frac{\partial P}{\partial z}\right) \le |\nabla P| < \frac{\lambda}{\varpi}, \tag{7.49}$$

it follows that

$$z > \varpi \int_P \frac{\mathrm{d}P'}{\lambda(P', \varpi)} = \varpi\Lambda(P, \varpi) \tag{7.50}$$

at fixed ϖ, with $\Lambda \to \infty$ as $\varpi \to \infty$. Hence

$$\lim\left(\frac{z}{\varpi}\right) \to \infty: \tag{7.51}$$

when $\rho\varpi^2 \to 0$, there is no net current to infinity, and the magnetic surfaces formed by rotating field lines about the z-axis are paraboloids.

Now turn to the transfield equation (7.29, 7.30). Far beyond $\mathrm{S_A}$, the term in $\mathbf{B}_{\mathrm{p}}$ is small compared with that in $\mathbf{v}_{\mathrm{p}}$ by the factor ρ/ρ_{A}. Write $\mathbf{v}_{\mathrm{p}} = v_{\mathrm{p}}\mathbf{b}$, where $\mathbf{b}$ is the unit vector along the local direction of a poloidal field-streamline, and the unit normal $\mathbf{n}$ is defined by $\mathbf{n} = \mathbf{b} \times \mathbf{t}$. Let ψ be the angle between the directions of $\mathbf{b}$ and $\hat{\varpi}$ (cf. Fig. 7.2). Then from (7.31), (7.42) and (7.43),

$$\mathbf{v}_{\mathrm{p}}\times(\nabla\times\mathbf{v}_{\mathrm{p}}) = \left(-\frac{\partial\psi}{\partial s}v_{\mathrm{p}}^2 + \mathbf{n}\cdot\nabla(v_{\mathrm{p}}^2/2)\right)\mathbf{n} = \left(\tilde{E}' + \frac{\alpha^2\rho\varpi^2\eta'}{4\pi\eta^3}\right)(-\nabla P), \tag{7.52}$$

where s is the arc-length along the field-streamline P. On use of $\varpi B_{\mathrm{p}} = \mathbf{n}\cdot\nabla P$ and of (7.45) for v_{p} and the definition (7.44) for R_∞^2, one arrives finally at

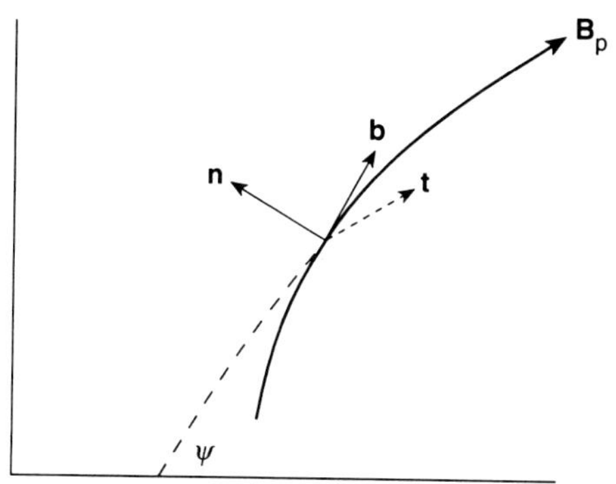

FIG. 7.2. Field line geometry.

$$\varpi \frac{\mathrm{d}\psi}{\mathrm{d}s} = -\frac{(\alpha^2 R_\infty^4)(4\pi\eta)}{v_\infty}\left(\frac{2R_\infty'}{R_\infty} + \frac{\eta'}{\eta}\right) \equiv k(P), \tag{7.53}$$

where k is constant along a field-streamline. (The minus sign in (7.53) comes from its presence in the definition (7.1) of the flux function P. If the uniform rotation α is replaced by isorotation $\alpha(P)$, then the extra term α'/α appears in the bracket in (7.53).)

Heyvaerts and Norman then show that not only the curvature $\mathrm{d}\psi/\mathrm{d}s$ goes to zero at infinity but also its product with ϖ. With the help of the two auxiliary equations

$$\mathrm{d}\varpi = \cos\psi\,\mathrm{d}s, \qquad \mathrm{d}z = \sin\psi\,\mathrm{d}s, \tag{7.54}$$

(7.53) is solved to yield

$$\varpi = \varpi_0 = \exp\left(\frac{\sin\psi}{k}\right), \tag{7.55}$$

$$z = z_0 + \frac{\varpi_0}{k}\int^{\psi} \exp\left(\frac{\sin\psi'}{k}\right)\sin\psi'\,\mathrm{d}\psi'. \tag{7.56}$$

Thus if $k \neq 0$, ϖ is a periodic function of the parameter ψ which cannot reach infinity, contradicting the assumption made. If however $k = 0$, then from (7.53), either (a)

$$R_\infty^2 = 0, \tag{7.57}$$

or (b)

$$\lim \frac{\mathrm{d}}{\mathrm{d}P}\left(\frac{\rho\varpi^2}{\eta}\right) \to 0. \tag{7.58}$$

By (7.44), (a) implies $\rho\varpi^2 = 0$, the case already discussed, with zero poloidal current at infinity and paraboloidal magnetic surfaces. If (b) holds, then, by

(7.42), ϖB_ϕ and so also the total poloidal current within the surface P is a constant independent of P. Thus in either case there is an asymptotically zero current enclosed between magnetic surfaces for which $\lim \varpi \to \infty$ at fixed P.

If case (b) were to hold for all P up to the value $P = 0$ pertaining to the z-axis, then a system with a non-zero Poynting flux to infinity but having paraboloidal magnetic surfaces would need to have a line current along the axis. More realistically, we conclude that a non-zero Poynting flux would be carried by a cylindrically collimating core, for which by definition ϖ stays finite as $z \to \infty$.

Both questions – field structure and collimation – and the crucial question of stability are taken up again in the Appendix and in Section 10.9.

7.5 A simple field model

Although, as seen, the problem of the precise structure of the global magnetic field raises an important question of principle – the possible spontaneous local breakdown in strictly dissipation-free flow – this may occur beyond the Alfvénic surface and so will not affect crucially estimates of the rate of braking, for which one can proceed more intuitively, as in Mestel and Spruit (1987). The approximations made are supported by the subsequent numerical work of Washimi and Shibata (1993). The overall field structure is supposed of the general form illustrated in Fig. 7.1, with the flux distribution on the stellar surface again assumed dipolar. The field $\mathbf{B}_\mathrm{p}$ is taken to be a vacuum dipole from the star's surface R out to a radius $\bar{r}$, and radial beyond $\bar{r}$, which is also the equatorial boundary of the dead zone. As the angular momentum flux is fixed by conditions at the Alfvénic surface, well within the distances at which collimation becomes important, the radial-field approximation for $\bar{r} < r < r_\mathrm{A}$ is acceptable. Thus for $R < r < \bar{r}$,

$$P = -\frac{1}{2} B_0 R^3 \sin^2\theta, \qquad \mathbf{B}_\mathrm{p} = B_0 \left(\frac{R}{r}\right)^3 \left(\cos\theta, \frac{1}{2}\sin\theta\right),$$
$$B_\mathrm{p}^2 = B_0^2 \left(\frac{R}{r}\right)^6 \frac{(3\cos^2\theta + 1)}{4}. \tag{7.59}$$

In an early treatment (Mestel 1968*a*), the most conservative assumption was made, with the field supposed curl-free all the way to the Alfvénic surface, i.e. with $\bar{r}$ put equal to the equatorial value of r_A. Detailed numerical work then showed that the angular momentum transfer process saturates with increasing surface field strength B_0 – the increase in the 'lever arm' of the Alfvénic surface is effectively balanced by the decrease in the fraction of the stellar magnetic flux entering the wind zone. In the more realistic Mestel–Spruit model, the dead zone ends at $\bar{r}$ well within SA. The field has a cusp on the equator at $\bar{r}$, where the field lines diverge to form the separatrix S between the wind and dead zones. Beyond $\bar{r}$, the radial field is expected to adjust itself to be nearly independent of θ, except in a pinched equatorial zone where B_r decreases rapidly to zero and where the thermal pressure on the equator balances the external magnetic pressure $B_r^2(r)/8\pi$; i.e. for $r > \bar{r}$

$$\mathbf{B}_{\mathrm{p}} = \bar{B}(\bar{r}/r)^2(1,0). \tag{7.60}$$

The dipolar field (7.59) at $\bar{r}$ has a strength that varies between pole and equator by a factor 2. The numerical results discussed below are not in fact sensitive to such factors, so for definiteness we link up the fields (7.59) and (7.60) by the choice $\bar{B} = B_0(R/\bar{r})^3$.

The boundary condition $(p + B^2/8\pi)$ continuous across S implies a discontinuity in B_{p}, which becomes especially marked in rapid rotators. Just outside S, in the wind zone, we have seen that the centrifugal force of near corotation assists in the outward acceleration of the gas, so reducing ρ and p below that in a thermally driven wind. By contrast, in the dead zone hydrostatic balance must hold along each field line, so the centrifugal force acting on the corotating gas must be balanced by a pressure gradient. Again we take the dead zone isothermal, with the sound speed a_{d}. At the point (r, θ) along the dipolar field line that leaves the star at (R, θ_0),

$$\frac{\rho}{(\rho_0)_{\mathrm{d}}} = \exp\left[-\frac{GM}{Ra_{\mathrm{d}}^2}\left(1 - \frac{R}{r}\right) + \frac{\alpha^2 R^2}{2a_{\mathrm{d}}^2}\left(\frac{r^2 \sin^2\theta}{R^2} - \sin^2\theta_0\right)\right], \tag{7.61}$$

where $(\rho_0)_{\mathrm{d}}$ is the coronal base density at the point (R, θ_0). (Variations of $(\rho_0)_{\mathrm{d}}$ with θ_0 will be ignored.) Beyond the point where the components along $\mathbf{B}$ of gravity and centrifugal force balance, ρ and p exponentiate outwards, requiring by the boundary condition on S a corresponding increase in the relative jump in B_{p}. At the cusp, the boundary condition can be approximated by

$$(B^2/8\pi)_{\mathrm{w}} = p_{\mathrm{d}} \tag{7.62}$$

in an obvious notation. This suffices to fix $\bar{r}$.

We now apply the general theory for the steady outflow of gas and angular momentum through the wind zone, with the field having the above simple structure. The fast magnetosonic point is beyond $\mathrm{S_A}$ and so in the radial field domain, with $B_r r^2$ = constant on a given field-streamline; hence the Bernoulli integral (7.19) can be written

$$H(r,\rho) \equiv \frac{\eta^2 B_{\mathrm{A}}^2 r_{\mathrm{A}}^4}{2\rho^2 r^4} + \frac{\alpha^2 r^2 \sin^2\theta}{2}\left[\left(\frac{\Omega}{\alpha}\right)\left(\frac{\Omega}{\alpha} - 2\right)\right] - \frac{GM}{r} + a_{\mathrm{w}}^2 \log\frac{\rho}{\rho_{\mathrm{A}}} = E, \tag{7.63}$$

with

$$\frac{\Omega}{\alpha} = \frac{1 - \rho_{\mathrm{A}} r_{\mathrm{A}}^2/\rho r^2}{1 - \rho_{\mathrm{A}}/\rho} \tag{7.64}$$

from (7.15). Define

$$X = r_{\mathrm{f}}/r_{\mathrm{A}}, \qquad D = \rho_{\mathrm{f}}/\rho_{\mathrm{A}}, \qquad V = 1/DX^2 = v_{\mathrm{f}}/v_{\mathrm{A}},$$
$$\mu = \alpha^2 r_{\mathrm{A}}^2 \sin^2\theta/v_{\mathrm{A}}^2, \qquad \nu = a_{\mathrm{w}}^2/v_{\mathrm{A}}^2. \tag{7.65}$$

(Recall that v_A is defined to be $(v_p)_A$ – the value of the wind speed at the point P_A on the field-streamline considered where the wind speed catches up with the local Alfvén speed $B_p/(4\pi\rho)^{1/2}$.) Then the critical point conditions $\partial H/\partial r = 0, \partial H/\partial \rho = 0$ at r_f, ρ_f yield

$$2V^2 = \mu X^2 \frac{-V^2 + 2VX^2 - 1}{(VX^2 - 1)^2} \tag{7.66}$$

and

$$V^2 - \nu = \mu V^2 X^2 \frac{(X^2 - 1)^2}{(VX^2 - 1)^3}. \tag{7.67}$$

The gravitational term has been dropped from (7.66), as in any case of interest it is small compared with v_f^2.

Division of (7.66) by (7.67) yields a quadratic equation for $X^2 \equiv (r_f/r_A)^2$ in terms of $V \equiv (v_f/v_A)$ and just the parameter ν. Consider the case of rapid rotation, for which centrifugal driving forces up v_A to well above a_w, so that $\nu \ll 1$. With V anticipated to be close neither to 1 or 3 (cf. (7.73) below), the relevant root of the quadratic is $X \simeq [V(V-1)(3- \ V)/2\nu]^{1/2}$ (showing that V is limited by $1 < V < 3$). Substitution back into (7.66) yields $V \simeq \mu^{1/3}[1+\mathrm{O}(\nu)]$, equivalent to

$$v_f^3 \simeq \alpha^2 \varpi_A^2 v_A. \tag{7.68}$$

From (7.15), (7.16) and (7.63),

$$\Omega_f \varpi_f^2 \simeq \alpha \varpi_A^2 (1 - v_A/v_f), \tag{7.69}$$

$$\varpi_f (B_t)_f \simeq -4\pi\eta\alpha\varpi_A^2 (v_A/v_f), \tag{7.70}$$

and

$$H(\varpi_f, \rho_f) = v_f^2/2 - (\alpha\varpi_A)^2(1 - v_A/v_f) \tag{7.71}$$

$$= v_A^2[(3/2)(v_f/v_A)^2 - (v_f/v_A)^3] \tag{7.72}$$

on use of (7.68).

To complete this solution for rapid rotation (parameter ν small), the flow must be made to pass through the slow magnetosonic point, where the Laval nozzle effect is due to the gravitational field. In all cases of interest the slow point is in the domain where the dipolar approximation is valid, with $B \propto 1/r^3 \propto 1/\varpi^2$. Also, except for very rapid rotators, the Ω-terms in (7.19) can be dropped. On substitution of (7.7), the critical point conditions $\partial H/\partial r = 0$, $\partial H/\partial \rho = 0$ yield $v_s = a_w$ (as anticipated) and $r_s = GM/3a_w^2$. With standard solar values inserted, r_s is above the coronal base. The condition $H = E =$ constant yields $v_0 \simeq 0.15 a_w$ as a typical coronal base velocity. (When $\alpha/\alpha_\odot > 40$ – for which $\alpha^2 R^3/GM > 3 \times 10^{-2}$ – the Ω-terms begin to be significant for streamlines not close to the axis, forcing the slow point in closer to the star and increasing v_0 somewhat, but the effect on the transport of angular momentum is not large.) With the same normalization as used near the fast point, $H \simeq a_w^2[-5/2 + \log(\rho_s/\rho_A)] = O(v_A^2\nu)$. Thus when $\nu \ll 1$, the final condition (7.27) together with (7.68) yields

$$\begin{aligned} v_f/v_A &= V \simeq 3/2, & v_A &\simeq (2\sqrt{6}/9)\alpha\varpi_A \simeq 0.54\alpha\varpi_A, \\ r_f/r_A &= X \simeq 3/4\nu^{1/2}, & \Omega_f\varpi_f^2 &\simeq \alpha\varpi_A^2/3. \end{aligned} \tag{7.73}$$

It is seen that for a 'rapid' rotator, defined by $\nu \ll 1$, centrifugal driving forces the wind speed at P_A up to about one-half the corotation speed, far greater than the sound speed; also, the fast point is well beyond the Alfvénic point P_A. Substitution of $v_A/\alpha\varpi_A$ from (7.73) into (7.63) then yields $(\Omega/\alpha)_A = (1 - \sqrt{19/27}) \simeq 0.16$ plus a term of order ν.

In the other limiting case of a slow rotator, the parameter $\mu \ll 1$ and centrifugal driving is weak: the Ω terms in (7.19) are small even at the Alfvénic point P_A, and the fast magnetosonic point is close to the Alfvénic point. Equations (7.66) and (7.67) now yield $V = v/v_A = 1+\mu\nu(1-\nu)$ and $X = r_f/r_A = 1+\mu(1-\nu)^2/2$. Equating H-values at r_A and r_f yields $\Omega/\alpha \simeq \nu$ at P_A. The wind speed $v_{th}(r)$ is given essentially by the simple Parker theory with no centrifugal driving. Once gas is well past the sonic point, the subsequent acceleration by thermal pressure steadily declines. The wind speed at the Alfvénic point depends slightly on the value of $\bar{r}$ where the field changes from dipolar to radial. Typical estimates show that $v_A/a_w \equiv \nu^{-1/2}$ varies from 2.1 at $r_A/R = 12$ to no more than 3.6 at $r_A/R = 90$. At the Alfvénic point the angular velocity $\alpha\nu$ of the outstreaming gas will again be markedly below that of the star, but the 'effective corotation' result (7.14) fixes the total flow of angular momentum. For all parameter values, application of l'Hôpital's rule at r_A, ρ_A to (7.64) yields $(\rho'/\rho)_A = -(2/r_A)/[1 - (\Omega/\alpha)_A]$. The derivative of (7.63) then yields $(\Omega/\alpha)'_A$.

Even in a rapid rotator, however large the ratio $(\alpha r_A/a_w)^2$, there is always a cone about the axis within which $\alpha^2\varpi_A^2/a_w^2 \ll 1$, so that again the flow reduces to Parker's thermally driven wind, with speed $v_{th}(r)$. For numerical estimates of the angular momentum transport, the approximate form

$$\frac{v_A}{a_w} = \left[\left(\frac{v_{th}(r_A/R)}{a_w}\right)^2 + \frac{\alpha^2 r_A^2}{3a_w^2}\sin^2\theta\right]^{1/2} \tag{7.74}$$

turns out to be reasonable for the whole parameter range.

7.6 The rate of braking

In the wind zone, the appropriate non-dimensional parameters that fix the flow are

$$l_w = \frac{GM}{Ra_w^2}, \quad \kappa = \frac{\alpha^2 R^3}{GM}, \quad \kappa l_w = \frac{\alpha^2 R^2}{a_w^2}, \quad \zeta_w = \frac{B_0^2}{8\pi(\rho_0)_w a_w^2}, \tag{7.75}$$

where $(\rho_0)_\mathrm{w}$ is assumed constant over the wind zone base, but with the suffix w allowing for the anticipated differences in both a and ρ_0 between wind and dead zones. From now on, the suffix w will be dropped for brevity. The radial field-streamline $\theta =$ constant cuts the Alfvénic surface $\mathrm{S_A}$ at r_A given by combining (7.12), (7.13) and (7.74):

$$\begin{aligned} 4\pi\eta &= \left(\frac{B_\mathrm{p}}{v_\mathrm{p}}\right)_\mathrm{A} \\ &= \frac{B_0(R/\bar{r})^3(\bar{r}/r_\mathrm{A})^2}{a[\{v_\mathrm{th}(r_\mathrm{A}/R)/a\}^2 + (\alpha^2 r_\mathrm{A}^2/3a^2)\sin^2\theta]^{1/2}} \end{aligned} \tag{7.76}$$

or

$$\begin{aligned} \left(\frac{r_\mathrm{A}}{R}\right)^2 \left[\left(\frac{v_\mathrm{th}(r_\mathrm{A}/R)}{a}\right)^2 + \frac{\kappa l}{3}\left(\frac{r_\mathrm{A}}{R}\right)^2 \sin^2\theta\right]^{1/2} &= \left(\frac{R}{\bar{r}}\right)\frac{B_0}{4\pi\eta a} \\ &= \left(\frac{R}{\bar{r}}\right)\frac{2\zeta}{(v_0/a)}, \end{aligned} \tag{7.77}$$

where v_0 is the wind speed at the coronal base, and v_0/a is typically about 0.15. Equation (7.77) fixes r_A/R for each field line in the wind zone in terms of the parameters (7.75) and $\bar{r}/R$.

In general, from the effective corotation result (7.14), the total rate of transport of angular momentum, carried jointly by the flow and the magnetic stresses, is the integral

$$\begin{aligned} -\dot{J} &= 2(2\pi)\alpha \int_0^{\pi/2} \rho_\mathrm{A}(\mathbf{v}\cdot\mathbf{n})_\mathrm{A}(r_\mathrm{A}^2\sin^2\theta) r_\mathrm{A}^2 \sin\theta\, \mathrm{d}\theta \\ &= 4\pi\alpha \int_0^{\pi/2} \eta(\mathbf{B}\cdot\mathbf{n})_\mathrm{A} r_\mathrm{A}^4 \sin^3\theta\, \mathrm{d}\theta, \end{aligned} \tag{7.78}$$

where $\mathbf{n}$ is the unit vector normal to the Alfvén surface $r_\mathrm{A}(\theta)$. With ρ_0 taken as constant over the wind zone base, the quantity $\eta = \rho_0 v_0/B_0 = \rho_\mathrm{A} v_\mathrm{A}/B_\mathrm{A}$ will be approximately constant over the whole wind zone. The distance r_A will vary from field line to field line, especially when centrifugal driving is strong. A lower limit to the rate of outflow of angular momentum results from taking the flow beyond $\bar{r}$ to be spherically symmetric, with v the speed along the field-streamline that follows the equator, so that $\mathrm{S_A}$ is approximated as a sphere of radius $r_\mathrm{A} = \varpi_\mathrm{A}(\pi/2)$. For θ smaller, (7.77) yields r_A correspondingly larger, but as the lever-arm $\varpi_\mathrm{A} = r_\mathrm{A}\sin\theta$, the radial field lines near to the axis contribute relatively little to the braking rate (7.78), which can now be written in a variety of ways:

$$-\dot{J} = \frac{8\pi}{3}(\rho_\mathrm{A} v_\mathrm{A} r_\mathrm{A}^2)\alpha r_\mathrm{A}^2 = \frac{8\pi}{3}\eta B_\mathrm{A}\alpha r_\mathrm{A}^4$$

$$= \frac{4}{3}\eta\Phi\alpha r_{\mathrm{A}}^2 = \frac{\Phi^2\alpha}{6\pi^2 v_{\mathrm{A}}}, \tag{7.79}$$

where use is made of the Alfvénic surface condition in the form (7.13)

$$\rho_{\mathrm{A}} = 4\pi\eta^2, \tag{7.80}$$

and

$$\Phi = 2\pi B_{\mathrm{A}} r_{\mathrm{A}}^2 = 2\pi\bar{B}\bar{r}^2 \tag{7.81}$$

is the flux of the 'open field lines' that form the wind zone. The associated mass loss rate can be written

$$-\dot{M} = 4\pi\rho_{\mathrm{A}} v_{\mathrm{A}} r_{\mathrm{A}}^2 = 4\pi\eta(B_{\mathrm{A}} r_{\mathrm{A}}^2) = 2\eta\Phi. \tag{7.82}$$

Since $\rho_{\mathrm{A}} v_{\mathrm{A}} r_{\mathrm{A}}^2 = \bar{\rho}\bar{v}\bar{r}^2$, and also, by continuity with the dipolar region $R < r < \bar{r}$, $\rho(\bar{r})v(\bar{r})/(\rho_0)_{\mathrm{w}} v_0 = \bar{B}/B_0 \simeq (R/\bar{r})^3$, (7.79) can be written (normalized in terms of solar quantities) as

$$-\dot{J} = \left[\frac{8\pi}{3}\alpha_{\odot}(\rho_0)_{\odot} R^4 a \left(\frac{v_0}{a}\right)\right] K\left(\frac{\alpha}{\alpha_{\odot}}\right), \tag{7.83}$$

where

$$K(\alpha/\alpha_{\odot}) = (R/\bar{r})(r_{\mathrm{A}}/R)^2(\alpha/\alpha_{\odot})[\rho_0/(\rho_0)_{\odot}]. \tag{7.84}$$

Equations (7.83) and (7.79) predict for a late-type star of given spectral type a variety of braking laws, depending in particular on how $\Phi, a, \rho_0, v_{\mathrm{A}}$ and $\bar{r}$ vary with the rotation α. The simplest case has ρ_0 and $\bar{r}$ constant, and α low enough for the wind to be well approximated as purely thermally-driven, with the terms in Ω in the Bernoulli equation (7.19) negligible. The gas then flows along the field essentially as in Parker's simplest model (1963): after being driven through the slow magnetosonic point into the supersonic domain, it is subsequently accelerated very slowly, reaching the Alfvénic surface with the speed v_{A} that is always $\approx (2-3)a$. Reference to (7.79) then shows that $-\dot{J}$ depends essentially only on Φ and α, but not on $-\dot{M}$; for with the centrifugal coefficient κl negligible, (7.77) and (7.82) yield $(r_{\mathrm{A}}/R)^2 \propto \eta^{-1} \propto (-\dot{M})^{-1}$, showing that in this case – purely thermal driving, with a consequent nearly constant wind speed, a fixed dead zone radius $\bar{r}$, and a radial magnetic field beyond – an increase in ρ_0, and so also in η and $-\dot{M}$, is exactly compensated by a reduction in r_{A}, leaving $-\dot{J}$ unchanged. With strong coupling assumed between radiative core and convective envelope, so that the star is spun down as a whole, a linear dynamo law $\Phi \propto \alpha$ then predicts

$$\dot{\alpha} \propto -\alpha^3 \tag{7.85}$$

with the coefficient dependent just on the spectral type of the star.

Things are different for rotation rates high enough for centrifugal driving to be dominant at the Alfvénic surface. From (7.76), $r_A \propto [\Phi^2/(-\dot{M})\alpha]^{1/3}$, and

$$\begin{aligned} \dot{J} &\propto -\dot{M}\alpha r_A^2 \propto \Phi^{4/3}\alpha^{1/3}(-\dot{M})^{1/3} \\ &\propto \alpha^{5/3}(-\dot{M})^{1/3} \end{aligned} \tag{7.86}$$

for a linear dynamo law. The dependence on $-\dot{M}$ derives from the centrifugal wind velocity $v_A \propto r_A$: the outflowing gas reaches the Alfvénic surface earlier than it would if $v_A \simeq v_{th} \simeq$ constant, so a change in $-\dot{M}$ is now no longer exactly compensated by the change in r_A. In Section 8.3 we discuss in some detail the comparison of the theory with the accumulating wealth of observations of late-type stars in young stellar clusters.

7.7 A digression on the microphysics

Whatever the details of the process by which angular momentum is transported from the star, the actual braking is achieved by the flow through part or all of the star of a poloidal current with a component perpendicular to the poloidal field. It is instructive to consider in detail how the magnetic torque density $\varpi(\mathbf{j}_p \times \mathbf{B}_p)/c$ – felt by the electrons that carry virtually all the poloidal current $\mathbf{j}_p$ – is transferred to act on the ions, which carry the bulk of the mass.

In general, it is the electric field that keeps ions and electrons in a plasma closely coupled together, severely limiting the degree of charge-separation (cf. Chapter 2). However, in a steady, axisymmetric system, there is no macroscopic electric field in the toroidal direction, so that coupling between the two gases must be via their mutual friction, requiring the electrons to have an extra toroidal drift $(v_d)_t$ relative to the ions. For simplicity, suppose that the ions are rotating uniformly with angular velocity Ω, so that

$$(\mathbf{v}_d)_t = (\mathbf{v}_e)_t - \Omega\varpi\mathbf{t} \tag{7.87}$$

where $\mathbf{v}_e$ is the electron velocity in space. In the notation of Chapter 2, the toroidal frictional force per unit volume felt by the electrons is

$$-\frac{n_e m_e (\mathbf{v}_d)_t}{\tau_{ei}} = \frac{n_e e \mathbf{j}_t}{\sigma} \tag{7.88}$$

since $\mathbf{j} = -n_e e \mathbf{v}_d$. As the electron inertia is negligible, this frictional force is balanced by the toroidal Lorentz force component:

$$\frac{\mathbf{j}_t}{\sigma} = -\frac{\mathbf{j}_p \times \mathbf{B}_p}{c n_e e} \tag{7.89}$$

or

$$(v_e)_t - \Omega\varpi = \frac{\sigma}{n_e e^2}\mathbf{t}\cdot\left(\frac{\mathbf{j}_p \times \mathbf{B}_p}{c}\right). \tag{7.90}$$

Equation (7.89) is seen to be just the balance in the $\mathbf{t}$-direction of the Ohmic and Hall terms in the generalized Ohm's law (2.55), when $\mathbf{E}_t = 0$ and $\mathbf{v}_i = \Omega\varpi\mathbf{t}$.

When the star is being steadily braked,

$$\left(\frac{\mathbf{j}_{\mathrm{p}} \times \mathbf{B}_{\mathrm{p}}}{c}\right) \cdot \mathbf{t} = \rho\varpi\frac{\partial\Omega}{\partial t} \equiv -\rho\varpi\frac{\Omega}{\tau_b} \tag{7.91}$$

where the last equality defines the 'instantaneous' spin-down time τ_b. Hence from (7.90) and (7.91),

$$(v_e)_{\mathrm{t}} = \Omega\varpi\left(1 - \frac{\sigma\rho}{n_e^2 e^2 \tau_b}\right) : \tag{7.92}$$

because the electrons have a poloidal motion they feel the toroidal component of the Lorentz force directly and so they lag behind the ions. However, since

$$\frac{\sigma\rho}{n_e^2 e^2 \tau_b} = \left(\frac{n_i}{n_e}\right)\left(\frac{m_i}{m_e}\right)\left(\frac{\tau_{ei}}{\tau_b}\right) \ll 1 \tag{7.93}$$

the lag is very small – $|(v_e)_d|/\Omega\varpi \ll 1$.

The poloidal component of the electron equation of motion is

$$\mathbf{E}_{\mathrm{p}} + \frac{(\mathbf{v}_e)_{\mathrm{t}} \times \mathbf{B}_{\mathrm{p}}}{c} + \frac{(\mathbf{v}_e)_{\mathrm{p}} \times \mathbf{B}_{\mathrm{t}}}{c} = \frac{\mathbf{j}_{\mathrm{p}}}{\sigma}. \tag{7.94}$$

For the moment, neglect the third term; then

$$(\mathbf{v}_e)_{\mathrm{t}} = \frac{c}{B_{\mathrm{p}}^2}[\mathbf{E}_\perp - (\mathbf{j}_{\mathrm{p}})_\perp/\sigma] \times \mathbf{B}_{\mathrm{p}} \tag{7.95}$$

where $\perp$ denotes the component perpendicular to $\mathbf{B}_{\mathrm{p}}$. Note that the second term on the right is in the direction of $-\mathbf{j}_{\mathrm{p}} \times \mathbf{B}_{\mathrm{p}}/c$, whereas from (7.90), $[(v_e)_{\mathrm{t}} - \Omega\varpi]\mathbf{t}$ is in the direction of $\mathbf{j}_{\mathrm{p}} \times \mathbf{B}_{\mathrm{p}}/c$. This is energetically consistent. The scalar product of (7.94) with $\mathbf{j}_{\mathrm{p}}$ yields

$$\frac{\mathbf{j}_{\mathrm{p}}^2}{\sigma} = \mathbf{j}_{\mathrm{p}} \cdot \mathbf{E} - \left(\frac{\mathbf{j}_{\mathrm{p}} \times \mathbf{B}_{\mathrm{p}}}{c}\right) \cdot (\mathbf{v}_e)_{\mathrm{t}}. \tag{7.96}$$

In a quasi-steady state $\int \mathbf{j}_{\mathrm{p}} \cdot \mathbf{E}\, \mathrm{d}\tau$ over the whole domain reduces to the Poynting integral over the sphere at infinity which vanishes (cf. Section 2.4). Thus $\mathbf{j}_{\mathrm{p}} \times \mathbf{B}_{\mathrm{p}}/c$ must be primarily opposed to $(\mathbf{v}_e)_{\mathrm{t}}$: it acts energetically as a dynamo rather than an electric motor, with the electrons gaining energy to balance the dissipated energy j_{p}^2/σ.

The curl of (7.94) (with the third term still neglected) yields with the help of (7.90)

$$\nabla \times \left(\frac{\mathbf{j}_{\mathrm{p}}}{\sigma}\right) = \nabla \times \left[\frac{\sigma}{c^2 n_e^2 e^2}(\mathbf{j}_{\mathrm{p}} \times \mathbf{B}_{\mathrm{p}}) \times \mathbf{B}_{\mathrm{p}}\right], \tag{7.97}$$

since in a steady state $\nabla \times \mathbf{E} = 0$, and $\nabla \times (\Omega\varpi\mathbf{t} \times \mathbf{B}) = 0$ when Ω is uniform. The effect of the right-hand side of (7.97) is small, for the ratio of the terms within the curl operators is $\simeq (\sigma B_{\rm p}/cn_e e)^2 = (\omega_e \tau_{ei})^2 \ll 1$ within the star. Thus by Ampère's law the equation (7.97) to B_ϕ is effectively

$$\nabla \times [\nabla \times (B_\phi \mathbf{t}/\sigma)] = 0 \tag{7.98}$$

with the surface boundary condition given by braking theory.

In a steady state one can always define an 'angular velocity $\Omega_{\rm f}$ of the magnetic field', but because of finite resistivity the rotation of the ions will vary slightly along individual field lines. Thus from (7.94), again with neglect of the third term,

$$\mathbf{E}_{\rm p} = -\Omega_{\rm f}\varpi \mathbf{t} \times \mathbf{B}_{\rm p}/c = \mathbf{j}_{\rm p}/\sigma - (\mathbf{v}_e)_{\rm t} \times \mathbf{B}_{\rm p}/c, \tag{7.99}$$

whence with the help of (7.90)

$$\Omega_{\rm f} - \Omega = \frac{1}{\varpi}(\mathbf{j}_{\rm p} \times \mathbf{B}_{\rm p}/c) \cdot \mathbf{t} \left[\frac{1}{(\sigma/c^2)B_{\rm p}^2} + \frac{\sigma}{n_e^2 e^2}\right]. \tag{7.100}$$

Again, the second term on the right (which comes from (7.89)) is smaller than the first by the ratio $(\omega_e \tau_{ei})^2$.

As long as σ is a microresistivity, $|(\Omega_{\rm f} - \Omega)/\Omega| \ll 1$. It can also easily be shown that the energy dissipation occurring within the star is small compared with that being converted into the kinetic energy of a centrifugally driven wind, by the ratio

$$\frac{[R(4\pi\rho)^{1/2}/B_{\rm p}]^2}{(4\pi\sigma R^2/c^2)(\tau_b)} = \frac{(\text{Alfvén time})^2}{(\text{Cowling decay time})(\text{braking time})} \ll 1. \tag{7.101}$$

Likewise the neglect of the term $(\mathbf{v}_e)_{\rm p} \times \mathbf{B}_{\rm t}/c$ in (7.94) is fully justified; and the current $\mathbf{j}_{\rm t}$, given by (7.89), yields a trivial modification to $\mathbf{B}_{\rm p}$. An axisymmetric field $\mathbf{B}_{\rm p}$ decaying as in Cowling's discussion (Section 5.6) will have an associated current density $\mathbf{j}_{\rm t} = c\nabla \times \mathbf{B}_{\rm p}/4\pi$ that is much greater than that given by the momentum transfer requirement (7.89); it is maintained by the decay-induced electric field $\mathbf{E}_{\rm t}$ which acts on both electrons and ions, balancing the equal and opposite Ohmic frictional forces associated with the current.

To sum up: collisions between electrons and ions are necessary so that the braking torque on the electron current is transferred so as to act on the ions, but the effect on the basic MHD approximation is trivial.

7.8 Magnetic braking of the oblique rotator

Consider now a non-axisymmetric system such as the *oblique rotator*, in which even if the magnetic field has an axis of symmetry $\mathbf{p}$, the axis does not coincide with the angular momentum vector. The appropriate generalization of the steady state discussed above is the '*quasi-steady*' state, for which there exists a frame which rotates with angular velocity $\alpha\mathbf{k}$ with respect to the inertial frame and in which all quantities are steady. In Chapter 9 we study the internal dynamics of such a star when subject to zero external torque, so that $\mathbf{k}$ coincides with the angular momentum vector. A magnetically-controlled wind emitted by the star will again exert a braking torque about $\mathbf{k}$, but because of the reduced symmetry, the Lüst–Schlüter angular momentum transport tensor (2.46) will yield also a *precessional torque* which acts to alter the obliquity angle χ between $\mathbf{k}$ and $\mathbf{p}$ (cf. Fig. 7.3). The precessional time-scale (like the braking time-scale) will certainly be long enough for $\mathbf{k}$ again to be taken as identical with the instantaneous angular momentum vector. The analogue of the Eulerian nutation or the Chandler wobble, the consequent internal motions and their effect on the basic magnetic field are studied in Chapter 9. Here, we take the magnetic flux distribution over the surface of the uniformly rotating star to be time-independent, and see what effects the reduced symmetry has on the dynamics of the wind (Mestel 1968*b*; Mestel and Selley 1970).

We seek solutions steady in the rotating frame, with the velocity in this frame (written upper case) parallel to the field: in Cartesian tensor notation

$$V_i = \kappa B_i. \tag{7.102}$$

(Recall from Section 5.2 that in a non-axisymmetric system, the steady state of isorotation – allowed in axial symmetry – reduces normally to uniform rotation.) The equation of motion is

$$\frac{\partial}{\partial x_k}(\rho V_i V_k) + 2\alpha\epsilon_{ijk}k_j\rho V_k = -\frac{\partial p}{\partial x_i} + \rho\frac{\partial \Phi}{\partial x_i} - \frac{\partial T_{ij}}{\partial x_j} + \rho\epsilon_{ijk}\alpha k_j\epsilon_{klm}x_l\alpha k_m \tag{7.103}$$

where as before the Maxwell stress-tensor $T_{ij} = (\mathbf{B}^2/8\pi)\delta_{ij} - B_iB_j/4\pi$, and ϵ_{ijk} is the Levi-Civita alternating tensor. (To avoid confusion with the velocity V, we use here Φ for the gravitational potential.) The first term in (7.103) incorporates the continuity equation

$$\frac{\partial(\rho V_k)}{\partial x_k} = 0. \tag{7.104}$$

The second term is the Coriolis force $2\rho\alpha\mathbf{k} \times \mathbf{V}$, and the terms on the right are the pressure gradient, the gravitational force density, the Lorentz force density written in terms of the Maxwell stress tensor, and the centrifugal force density ($\rho\alpha^2[\mathbf{k} \times (\mathbf{r} \times \mathbf{k})]$ in vector notation).

It is convenient first to derive an integral result. Consider a volume τ, fixed in the rotating frame. Let S be the surface bounding τ and n_l the outward drawn normal. Construct the quantity

$$\int_S F_{il} n_l \,\mathrm{d}S \equiv \int_S \epsilon_{ijk}x_j[T_{kl} + p\delta_{kl} + \rho V_l(V_k + \alpha(\mathbf{k} \times \mathbf{r})_k)]n_l \,\mathrm{d}S. \tag{7.105}$$

This is the outflow of the i-component of angular momentum across S, due to magnetic stresses, thermal pressure and gas flow. Since the first three terms in the square brackets are symmetric, those parts of the integral can be transformed by Gauss's theorem into

$$\int \epsilon_{ijk} x_j \frac{\partial}{\partial x_l}(T_{kl} + p\delta_{kl} + \rho V_k V_l)\, \mathrm{d}\tau, \tag{7.106}$$

which on use of (7.103) becomes

$$\int \epsilon_{ijk} x_j \left[\rho \frac{\partial \Phi}{\partial x_k} + \rho\alpha^2 [\mathbf{k} \times (\mathbf{r} \times \mathbf{k})]_k - 2\alpha\epsilon_{kpl} k_{\mathrm{p}} \rho V_l\right] \mathrm{d}\tau. \tag{7.107}$$

The centrifugal term in (7.107) reduces at once to

$$-\alpha \left[\mathbf{k} \times \int [\mathbf{r} \times \rho\alpha(\mathbf{k} \times \mathbf{r})]\, \mathrm{d}\tau\right]_i . \tag{7.108}$$

The Coriolis term in (7.107) combines with the last term in the outflow integral (7.105) to yield

$$-\alpha \left[\mathbf{k} \times \int (\mathbf{r} \times \rho\mathbf{V})\, \mathrm{d}\tau\right]_i . \tag{7.109}$$

Thus the total outflow of the i-component of angular momentum is

$$\int_S F_{il} n_l\, \mathrm{d}S \equiv \int_\tau (\mathbf{r} \times \rho\nabla\Phi)_i\, \mathrm{d}\tau - \left[\alpha\mathbf{k} \times \int \mathbf{r} \times \rho[\mathbf{V} + \alpha(\mathbf{k} \times \mathbf{r})]\, \mathrm{d}\tau\right]_i . \tag{7.110}$$

The first term on the right of (7.110) is the moment of the gravitational force density, which vanishes if $\mathbf{g}$ is radial; it will in any case be small in the problems of interest and so will be dropped.

Now apply the transform to the volume bounded by the double surface consisting of $\mathrm{S_1}$ at the coronal base and a general surface further from the star (later to be identified with the Alfvénic surface). Then

$$\int_{\mathrm{S_1}} F_{il} n_l\, \mathrm{d}S = \int_{\mathrm{S_A}} F_{il} n_l\, \mathrm{d}S + \left[\alpha\mathbf{k} \times \int_\tau \mathbf{r} \times \rho[\mathbf{V} + \alpha(\mathbf{k} \times \mathbf{r})]\, \mathrm{d}\tau\right]_i . \tag{7.111}$$

The term on the left has its sign changed because the normal over $\mathrm{S_1}$ is now taken away from the star. The second term on the right is a 'flywheel' term. The gas within τ has moment of momentum given by the volume integral, and an inertial observer sees this swung round at the rate $\alpha\mathbf{k}$, so accounting for the difference between the angular momentum flow integrals over $\mathrm{S_1}$ and $\mathrm{S_A}$ respectively.

We now take $\mathrm{S_A}$ to be the Alfvénic surface, as defined by $\mathbf{V}$ in the rotating frame and the total magnetic field $\mathbf{B}$. Thus on $\mathrm{S_A}$,

$$4\pi\rho\kappa^2 = 1 \tag{7.112}$$

(κ being the scalar in (7.102)), and the torque due to the Reynolds stresses is balanced by that due to the magnetic tensions:

$$\int_{S_A} \epsilon_{ijk} x_j (\rho\kappa^2 - 1/4\pi) B_k B_l n_l \, dS = 0. \tag{7.113}$$

Hence the total torque $\mathbf{L}$ acting on the star is given by

$$- L_i = \int_{S_1} F_{il} n_l \, dS = \left[\alpha\mathbf{k} \times \int \mathbf{r} \times \rho[\mathbf{V} + \alpha(\mathbf{k} \times \mathbf{r})] \, d\tau \right]_i$$
$$+ \int_{S_A} \left(p + \frac{B^2}{8\pi} \right) (\mathbf{r} \times \mathbf{n})_i \, dS + \int_{S_A} \rho \mathbf{V} \cdot \mathbf{n} [\mathbf{r} \times (\alpha\mathbf{k} \times \mathbf{r})]_i \, dS. \tag{7.114}$$

The first term is the flywheel effect of the rotating gas between the coronal base and the Alfvénic surface. The second term is the moment about the centre of the star of the thermal and magnetic pressures acting on the Alfvénic surface; it would disappear if S_A were spherical. The third term represents the angular momentum loss there would be if the gas at S_A had retained the angular velocity with which it left the star. In an axisymmetric state only this last term survives – the 'effective corotation' of (7.14) – and the total torque has only a z-component. In a system of lower symmetry such as the oblique rotator, all three terms survive in general, and also the net torque has components in all three directions. As noted, the Z-component again brakes the stellar rotation, while the X- and Y-components cause the instantaneous axis to precess through the star: in particular, the Y-component causes a precession about OX, so altering the obliquity angle χ, with a corresponding rotation in space of the frozen-in magnetic axis. In the realistic case (cf. Section 5.1) in which the centrifugal perturbations dominate, so that the star is nearly axisymmetric about $\mathbf{k}$, then (cf. Mestel 1968*b*)

$$\dot{\chi} = - \left(\frac{L_Y}{L_Z} \right) \frac{d}{dt} \log(C\alpha), \tag{7.115}$$

where C is the zero-order moment of inertia of the star.

The braking torque L_Z will depend only weakly on χ, but there are special values of χ for which $L_Y = 0$, e.g. when the system is axisymmetric – $\chi = 0$ – or equatorially symmetric – $\chi = \pi/2$.

To estimate L_Y/L_Z we need to solve the equations to some order of approximation. We again have a Bernoulli integral in the rotating frame:

$$\frac{1}{2}V^2 + a^2 \log\rho - \Phi - \frac{1}{2}\alpha^2\varpi^2 = K_B, \tag{7.116}$$

where the symbol K_B is shorthand for 'constant on a field-streamline'. Again, from (7.102) and (7.104) we have the continuity integral

$$\rho \left(\frac{V}{B} \right) = K_B. \tag{7.117}$$

There follows an equation deceptively similar to the Parker equation in an axisymmetric system:

$$\frac{1}{2}V^2 - a^2 \log V = \frac{GM}{r} - a^2 \log B + \frac{1}{2}\alpha^2 \varpi^2 + K_B. \qquad (7.118)$$

Again we expect a similar flow, with the constant for each field-streamline fixed by conditions at the slow magnetosonic (effectively the sonic) point, and the gas speed slow near the star but accelerating outwards under thermal pressure and centrifugal force. However, it must be noted that $\mathbf{V}$ is the total velocity as seen in the rotating frame. In the axisymmetric problem, $\mathbf{V} = \mathbf{V}_\mathrm{p} + \mathbf{V}_\mathrm{t}$, where the poloidal and toroidal parts are mutually orthogonal, so that $V^2 = V_\mathrm{p}^2 + V_\mathrm{t}^2$. In the oblique problem there is a break-up of $\mathbf{B}$ and $\mathbf{V}$ into analogues $\overline{\mathbf{B}} + \mathbf{B}', \overline{\mathbf{V}} + \mathbf{V}'$, with opposite parities on reflection in the plane OYZ, so that the contributions from opposite hemispheres are of the same sign, yielding a non-zero net torque, but the unprimed and primed vectors are not mutually orthogonal, and so even the definition (7.112) of $\mathrm{S_A}$ depends on $\mathbf{B}'$ as well as $\overline{\mathbf{B}}$. Thus one cannot strictly write V^2 in (7.118) in terms just of $\overline{V}$ and $\overline{B}$ in advance of the solution of the whole problem. This is of minor importance near the sonic point, but becomes serious as $\mathrm{S_A}$ is approached. Again, in the axisymmetric problem there exists the torque integral (7.9) measuring flow of Z-angular momentum along $\mathbf{B}_\mathrm{p}$; for the symmetry ensures that there is no transfer of Z-angular momentum between different field-streamlines. In the oblique problem, however, there is in general an interchange of all three components of angular momentum from one field line to another by the action of the magnetic stresses (and also the thermal pressure), so that the axisymmetric torque integral is replaced by a partial differential equation.

Only one case has been worked out rigorously, by a perturbation procedure (Mestel and Selley 1970). As a zero-order approximation the basic magnetic field is taken as a split monopole — radial and independent of angle, except for a change of sign at the magnetic equator. The rotation α is supposed slow enough for terms of order α^2 to be negligible, so that in particular the wind is purely thermal, and to zero order the Alfvénic surface $\mathrm{S_A}$ is a sphere of radius r_A. The perturbation $\mathbf{B}'$ – the analogue of the toroidal field in the axisymmetric problem – can be found fairly easily to order $\alpha r_\mathrm{A}/a$. One can then verify directly from integrating the Maxwell stresses over the stellar surface, or indirectly from the transform (7.114), that to the same order the precessional torques vanish: hardly surprisingly, as the magnetic stresses are quadratic in field strength and so take no cognizance of the sign changes at the equator. The solution is then perturbed by allowing the field at the surface of the star to have an angular-dependent part of order ϵ, expanded in terms of surface harmonics. A long and tedious calculation yields the contribution to the field of order $(\alpha r_\mathrm{A}/c)\epsilon$, with a consequent non-vanishing torque component $L_Y \propto \sin 2\chi$. However, it is much simpler to use again the transformation (7.114), for the pressure and 'flywheel' terms are of order ϵ^2, and both L_Y and L_Z are given to order $(\alpha r_\mathrm{A}/a)\epsilon$ by just the 'effective corotation' term

$$-\int_{\mathrm{S_A}} \rho(\mathbf{V}\cdot\mathbf{n})[\mathbf{r}\times(\alpha\mathbf{k}\times\mathbf{r})]\,\mathrm{d}S, \qquad (7.119)$$

as in the axisymmetric case. The ρ and $\mathbf{V}$ fields are required to zero order in $\alpha r_{\mathrm{A}}/a$, i.e. they are given by the stellar wind theory for a *non-rotating* star, with the flux distribution over the stellar surface as prescribed.

An interesting qualitative result emerges from this calculation: the sign of the precessional torque is such as to make the instantaneous axis of rotation seek out on the star's surface the region where the magnetic field strength is the strongest. One can understand this qualitatively by considering the extreme cases illustrated in Fig. 7.3. Outflowing gas leads to a field line's being twisted about $\mathbf{k}$ in the negative sense. In Fig. 7.3(a) the undistorted field lines form a magnetic tube about the axis $\mathbf{p}$; and when distorted the magnetic tensions exert a positive torque about OY in addition to the familiar negative (braking) torque about OZ. From (7.115), $\dot{\chi}$ then has the same negative sign as $\mathrm{d}(C\alpha)/\mathrm{d}t$, and the torque acts to *align* $\mathbf{p}$ and $\mathbf{k}$. In Fig. 7.3(b) the basic field is still symmetric about $\mathbf{p}$, but the field lines are compressed into a disc at the magnetic equator. The twisted field now yields L_Y as well as L_Z negative, whence from (7.115) $\dot{\chi}$ is positive: the magnetic axis rotates in space, and the instantaneous axis through the star, in the sense so as to make $\mathbf{p}$ perpendicular to $\mathbf{k}$. Thus the model can predict a tendency towards the highly aligned or the highly oblique rotator, depending on the surface distribution of flux. However, the effect will only be significant provided $\mid L_Y/L_Z \mid$ is not too small. We return to the general problem of the oblique rotator in Chapter 9, on the early-type magnetic stars.

7.9 Relativistic centrifugal winds

The magnetohydrodynamic problems of collapsed bodies, such as radio pulsars, and of active galactic nuclei require generalization of the plasma equations to allow for the effects of special relativity, while problems involving black holes necessarily involve general relativity (Blandford and Znajek 1977; Macdonald and Thorne 1982; Phinney 1982; Okamoto 1992; Horiuchi *et al.* 1995; and many other papers). The special problems of radio pulsars are discussed at length in Chapters 13 and 14. Nevertheless, it is instructive to extend the above non-relativistic theory in the simplest possible way, to cover perfectly conducting plasmas forced to move at speeds approaching c (Michel 1969; Goldreich and Julian 1970; Li and Melrose 1994). As the gravitational potential is often enormously greater than the thermal energy per gram, it is convenient to drop the pressure and study the pure centrifugal wind problem. The kinematic equations (7.3), (7.4), (7.5) and (7.7) remain unaltered, but the equation of motion is now

$$\nabla(\gamma c^2) - \mathbf{v} \times [\nabla \times (\gamma \mathbf{v})] = \nabla V + \frac{(\nabla \times \mathbf{B}) \times \mathbf{B}}{4\pi\rho}, \tag{7.120}$$

where ρ is related to the co-moving (rest frame) density ρ_0 by $\rho = \gamma\rho_0$. Equation (7.120) yields the torque integral

$$-\varpi B_\phi/4\pi + \eta\gamma\Omega\varpi^2 = -\beta/4\pi \tag{7.121}$$

and the energy integral

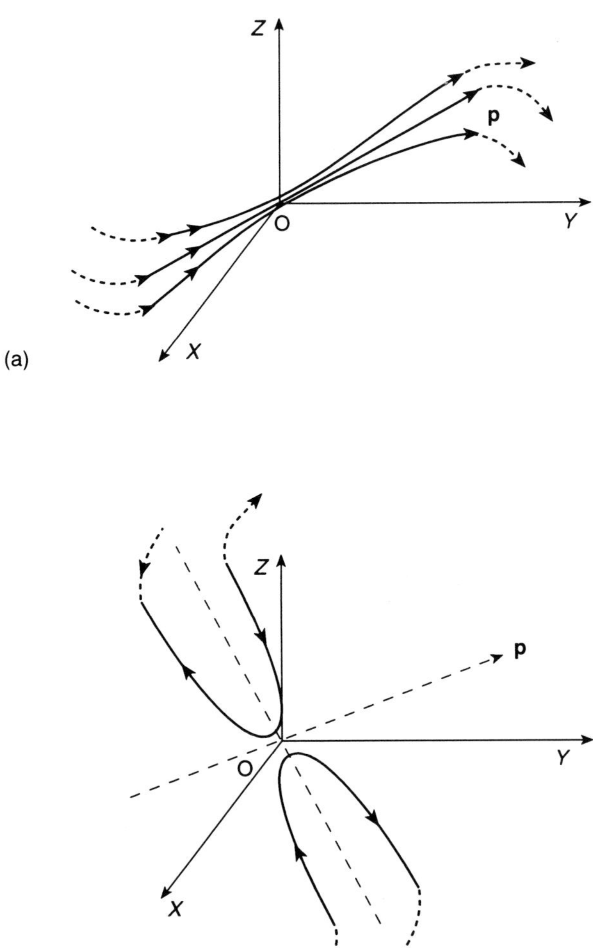

FIG. 7.3. Wind emitted by an oblique rotator: the braking and precessional torques. (a) Limiting case with the basic field forming a narrow tube symmetric about the magnetic axis **p**. The dotted extensions show the field line distortions due to the non-uniform rotation generated by the outflowing gas. The tensions along the field lines have both a negative moment about OZ (the familiar braking torque) and also a positive moment about OY (the precessional torque). (b) Limiting case with the basic field symmetric about **p** (dashed line), but compressed into a disc near the magnetic equator defined by **p** (dashed). The dotted extensions again show the rotational distortion of the field lines, with the tensions now exerting negative torques about OY as well as OZ. Both cases predict that the instantaneous axis of rotation seeks out the region on the surface where the field strength is largest.

$$\gamma c^2(1 - \alpha\Omega\varpi^2/c^2) - GM/r \equiv H(\varpi, z) = E(P) \tag{7.122}$$

in place respectively of (7.9) and (7.19). The analogue of (7.20) is

$$\eta\left[\gamma c^2 - GM/r\right] - \alpha\varpi B_\phi/4\pi = \eta E - \alpha\beta/4\pi \equiv \epsilon. \tag{7.123}$$

Elimination in turn of B_ϕ and Ω yields

$$\Omega = \left(\alpha + \eta\beta/\rho\varpi^2\right)/(1 - 4\pi\gamma\eta^2/\rho), \tag{7.124}$$

and

$$\varpi B_\phi = (\beta + 4\pi\eta\gamma\alpha\varpi^2)/(1 - 4\pi\gamma\eta^2/\rho). \tag{7.125}$$

In the domain of interest the centrifugal term dominates over the gravitational, which is therefore dropped from now on. Substitution from (7.124) into (7.123) then yields

$$\gamma = \frac{E}{c^2}\frac{\left[1 - (1 - \alpha\beta c^2/4\pi\eta E)(4\pi\gamma v\eta^2/\rho)\right]}{\left[1 - (4\pi\gamma\eta^2/\rho) - (\alpha^2\varpi^2/c^2)\right]} \tag{7.126}$$

It is easy to check that the non-relativistic formulae are recovered by writing $\gamma = 1 + (v_{\mathrm{p}}^2 + v_{\mathrm{t}}^2)/2c^2$ and taking the limit $c \to \infty$ with $(E - c^2)$ staying finite.

The 'effective corotation' result (7.14) is generalized: (7.124) and (7.125) are kept non-singular at the 'pure Alfvénic point' C at axial distance ϖ_c, where

$$4\pi\gamma_c\eta^2/\rho_c = [v_{\mathrm{p}}/(B_{\mathrm{p}}/(4\pi\gamma\rho)^{1/2}]_c^2 = 1 \tag{7.127}$$

by the requirement that β and ϖ_c be related by

$$-\beta/4\pi = \eta\gamma_c\alpha\varpi_c^2 = (\rho v_{\mathrm{p}}/B_{\mathrm{p}})\gamma_c\alpha\varpi_c^2. \tag{7.128}$$

However, no extra constraint is then imposed by the energy integral (7.126), which is identically satisfied when (7.127) and (7.128) are substituted (Li and Melrose 1994). Finiteness at the point (suffix A) defined by the vanishing of the denominator in (7.126) requires that

$$\frac{4\pi\gamma_{\mathrm{A}}\eta^2}{\rho_{\mathrm{A}}} = 1 - \frac{\alpha^2\varpi_{\mathrm{A}}^2}{c^2} = \frac{1}{1 - \alpha\beta c^2/4\pi E}. \tag{7.129}$$

In the non-relativistic limit, the points C and A coincide. In terms of

$$x = (\alpha\varpi/c)^2, \qquad y = 4\pi\eta^2\gamma/\rho, \qquad \mu = E/c^2, \tag{7.130}$$

(7.126) can be written

$$\gamma = \frac{\mu(1 - y/y_{\mathrm{A}})}{1 - x - y}, \tag{7.131}$$

and (7.125), (7.124) become

$$\varpi B_\phi = \frac{\beta(1 - x - y_{\mathrm{A}})}{(1 - y_{\mathrm{A}})(1 - x - y)}, \tag{7.132}$$

$$\Omega = \frac{\alpha\,[y_{\mathrm{A}}(x + y) - y]}{x(y_{\mathrm{A}} - y)}. \tag{7.133}$$

Note that each numerator has the factor $(1 - y)$ which cancels the denominator, so that Ω and B_ϕ automatically remain non-singular at the pure Alfvénic point (7.127), with their values determined by (7.133) and (7.132) with $y = 1$, $x = x_c$; and likewise γ_c is fixed by (7.131). The pure Alfvénic point retains significance as defining the extent of effective corotation, and in fact once β/α is fixed by (7.129) and (7.142) below, then (7.128) can be used to determine the position of ϖ_c: but it is the point A that is the generalization of the Alfvénic point of the non-relativistic theory.

As in Section 7.5, the theory is illustrated by the adoption of a radial poloidal field (Michel 1969; Goldreich and Julian 1970); in the simplest case, a split monopole structure, with $|B_r|$ independent of θ. If also ρ, v_r are assumed spherically symmetric near the star, then $\eta = \rho v_r/B_r$ will be constant over each hemisphere, but the latitude-dependent centrifugal driving will ensure that ρ, v_r, etc. are latitude-dependent in the outer magnetosphere. In fact, as θ is constant on each field line, one can allow both B_r and η to be functions of θ and develop essentially the same theory for each individual field-streamline. Again, the steady flow must pass smoothly through the slow and fast magnetosonic points. Anticipating that in a strongly centrifugal wind, the fast point will be far beyond the points A and C, expression (7.124) can be approximated by

$$\Omega = -\alpha\rho/4\pi\gamma\eta^2 - \beta/4\pi\gamma\eta\varpi^2 + \mathrm{O}(1/\varpi^4). \tag{7.134}$$

Expression (7.20) then becomes

$$H = \gamma c^2 + \alpha\beta/4\pi\eta + \alpha^2\rho r^2 \sin^2\theta/4\pi\eta^2 + \mathrm{O}(1/r^2), \tag{7.135}$$

where only terms finite at infinity are retained. From (7.134), the rotatory velocity $\Omega\varpi$ is small at infinity, so

$$\frac{1}{\gamma^2} = 1 - \frac{v^2}{c^2} = 1 - \frac{\eta^2\Phi^2}{c^2\rho^2 r^4} + \mathrm{O}\left(\frac{1}{r^2}\right), \tag{7.136}$$

where $\Phi = B_r r^2$ and $\rho v = \eta B = \eta\Phi/r^2$. The conditions defining the fast critical point become

$$0 = \frac{\partial H}{\partial r} = -\frac{2\eta^2\Phi^2\gamma^3}{\rho^2 r^5} + \frac{2r\alpha^2\rho\sin^2\theta}{4\pi\eta^2} + \mathrm{O}\left(\frac{1}{r^3}\right), \tag{7.137}$$

$$0 = \frac{\partial H}{\partial \rho} = -\frac{\eta^2 \Phi^2 \gamma^3}{\rho^3 r^4} + \frac{\alpha^2 r^2 \sin^2\theta}{4\pi\eta^2} + \mathrm{O}(1), \tag{7.138}$$

showing that in the zero sound speed limit the fast point is at infinity. The asymptotic value of γ is thus

$$\gamma_{as} \simeq \left(\frac{\alpha^2 \Phi}{4\pi\eta c^3}\right)^{1/3} \left(\frac{c}{v}\right) \simeq \sigma^{1/3} + \frac{1}{2\sigma^{1/3}}, \tag{7.139}$$

where (in the Michel/Goldreich–Julian notation)

$$\sigma = \frac{\alpha^2 \Phi}{4\pi\eta c^3} = \left(\frac{\alpha r_0}{c}\right)^2, \qquad r_0^2 = \frac{\Phi}{4\pi\eta c} = \left(\frac{c^2}{\alpha^2}\right)\sigma. \tag{7.140}$$

In the strongly relativistic case, $\sigma \gg 1$. From (7.135),

$$\begin{aligned} H/c^2 &= \gamma + \alpha\beta/4\pi\eta c^2 + \sigma \sin^2\theta (c/v) \\ &= \sigma^{1/3} + 1/2\sigma^{1/3} + \alpha\beta/4\pi\eta c^2 + \sigma \sin^2\theta(1 + 1/2\sigma^{1/3}). \end{aligned} \tag{7.141}$$

At the slow point near the star, $H/c^2 = 1 +$ terms of order $(a/c)^2$ where a is the sound speed; thus by the constancy of H

$$\begin{aligned} -\beta/4\pi &= (\eta c^2/\alpha)[\sigma \sin^2\theta + \sigma^{1/3}(1 + \sin^2\theta/2) - 1 + 1/2\sigma^{2/3}] \\ &\simeq \eta\alpha r_0^2[\sin^2\theta + \sigma^{-2/3}(1 + \sin^2\theta/2)]. \end{aligned} \tag{7.142}$$

Equation (7.142) expresses $-\beta/4\pi$ – the angular momentum flow per unit flux tube – in terms of the extrinsic parameters α, Φ, η. Equations (7.129), (7.124) and the definitions of η and γ then yield $\varpi_\mathrm{A}, \rho_\mathrm{A}, \Omega_\mathrm{A}$, etc. From the definition (7.140), the result (7.142) shows that the axial distance $r_0 \sin\theta$ would be the extent of effective corotation if the gas were non-relativistic (with $\gamma = 1$), rather than the distance ϖ_c of (7.128).

The solution predicts for the asymptotic ratio of angular momentum carried by the gas to the total outflow

$$\begin{aligned} \frac{\eta\gamma\Omega\varpi^2}{-\beta/4\pi} &= 1 + \frac{\alpha\rho\varpi^2}{\eta\beta} \\ &\simeq \frac{\sigma^{-2/3}}{\sin^2\theta} \end{aligned} \tag{7.143}$$

on use of (7.142) and of $c/v \simeq (1 + 1/2\gamma^2) = (1 + 1/2\sigma^{2/3})$. A principal issue in relativistic wind theory is to see how the outflow of angular momentum and energy is ultimately via the gas flow rather than the Poynting flow. We return to this question in Chapters 13 and 14 on pulsar electrodynamics.

APPENDIX

The axisymmetric magnetic rotator

(a) *The energetics*

In the discussion of Sections 7.3 and 7.4, some simplification is obtained by the imposition of the familiar uniform polytropic $p-\rho$ relation, leading to generalized Bernoulli integrals (7.19) and (7.20), but the solution of the consequent partial differential equation for the structure of the poloidal field remains formidable. Another approach, summarized below, starts rather with an *ansatz* that reduces the field structure problem to the solution of ordinary differential equations, allowing the pressure and density fields to emerge from the solution. It should be emphasized that both approximations are a substitute for a complete understanding of the energetics of the problem. It is instructive to return to the steady state momentum equation (7.17) and write down the general energy equation derivable:

$$\rho\mathbf{v}\cdot\nabla\left(\frac{1}{2}(v_{\rm p}^2+\Omega^2\varpi^2)-\frac{GM}{r}-\frac{\alpha(P)\varpi B_\phi}{4\pi\eta}\right)=-\rho\mathbf{v}\cdot\left(\frac{\nabla p}{\rho}\right), \tag{7.144}$$

where now the 'angular velocity of the field-lines' $\alpha(P)$ is not assumed a priori to be a constant. From the first law of thermodynamics, the local volume rate of heat supply q is given by the departure from adiabaticity: in standard notation

$$q=\rho\mathbf{v}\cdot\nabla(c_vT)-\frac{p}{\rho}\mathbf{v}\cdot\nabla\rho=\rho\mathbf{v}\cdot\nabla h-\rho\mathbf{v}\cdot\frac{\nabla p}{\rho}, \tag{7.145}$$

where, since $c_v=\mathcal{R}/(\gamma-1)\mu$, the enthalpy h per unit mass is seen to be

$$h=\frac{\gamma}{\gamma-1}\frac{\mathcal{R}T}{\mu}, \tag{7.146}$$

γ is the usual ratio of specific heats $c_p/c_v\approx 5/3$, and the mean molecular weight $\mu\approx(4-3X)/(3-X)$ for a fully ionized gas of H and He with the fraction X by number of H. Thus a solution of the equations with a prescribed distribution of p and ρ, necessarily implies the local heat supply per cm^3 s

$$q=\rho\mathbf{v}\cdot\nabla\bar{E} \tag{7.147}$$

with

$$\bar{E}\equiv\frac{1}{2}(v_{\rm p}^2+\Omega^2\varpi^2)+h-\frac{GM}{r}-\frac{\alpha\varpi B_\phi}{4\pi\eta}. \tag{7.148}$$

For example, in the case with a uniform polytropic law $p=K\rho^\Gamma$, satisfying the generalized Bernoulli equation (7.20), it is seen that

$$q=\frac{\Gamma-\gamma}{\gamma-1}K\rho^{\Gamma-1}\mathbf{v}\cdot\nabla\rho. \tag{7.149}$$

(Parker 1963). In general, $\bar{E}$ is not conserved along a field-streamline; instead, the conserved quantity is

$$\bar{H}(P) \equiv \bar{E} - \int^{s} \frac{q(s,P)}{\rho(s,P)v_{\mathrm{p}}(s,P)}\,\mathrm{d}s \tag{7.150}$$

where s is the path-length along a field line P (Sauty and Tsinganos 1994).

(b) *Latitudinally self-similar solutions*

As noted above, for treatment of the domain between the star and the Alfvén surface, the isothermal approximation $\Gamma = 1$ may not be too bad; and in fact one can hazard a guess that for estimation of the instantaneous rate of loss of angular momentum, as discussed in Section 8.3, the approximate treatments of the wind and the field structure in Sections 7.5 and 7.6 will turn out to have yielded tolerably accurate results. However, as pointed out by Parker (1963) and by others since, a polytropic equation of state with a constant value of Γ cannot be valid both near and far from the Sun: attempts to reproduce the observed conditions at 1 AU yield unacceptably high velocities and low densities near the solar surface. One can of course impose barotropic laws that simulate more closely the observations. Perhaps in time the escalation in computing power will enable direct iterative integration of the associated partial differential equations for the flux function P in both the wind and dead zones, with simultaneous determination of the critical surfaces and of the separatrix. Meanwhile, an alternative path for detailed exploration of the properties of the system is that developed over the years by Low, Tsinganos, Trussoni, Sauty, Priest, Lima and colleagues, in which a poloidal *self-similarity ansatz* reduces the transfield partial differential equations to a set of ordinary differential equations (see Chapters 17–19 in Tsinganos 1996; Lima *et al.* 1996; and extra references in both).

Inevitably one pays a price: the simple polytropic/isothermal assumption must be dropped, as self-similarity puts its own *mathematical* constraints on the p–ρ relation. (It can be argued that this is no worse than the situation in the standard treatment, which also bypasses satisfaction of a local physically-determined heat supply equation but instead makes appeal to observations of the solar corona.) More important is the difficulty in finding a self-similarity solution that is valid everywhere in the magnetosphere. For example, a *radially* self-similar class of solutions by Blandford and Payne (cf. Chapter 10), widely used to model winds from an accretion disc, shows pathological behaviour on the axis. We outline now a class of *latitudinally* self-similar solutions, discussed by Sauty, Trussoni and Tsinganos in a series of papers, and in general appropriate to a domain enclosing the axis but not too near the equator. (The notation is altered to conform to that of Sections 7.2 et seq.) The flux function $P(r,\,\theta)$ is assumed separable and of dipolar form

$$P = -f(r)\sin^2\theta. \tag{7.151}$$

The 'Alfvén number'

$$M = \frac{v_\mathrm{p}}{B_\mathrm{p}/(4\pi\rho)^{1/2}} = \left(\frac{4\pi\eta^2(P)}{\rho}\right)^{1/2} \tag{7.152}$$

is assumed to be a function of r only, so that the Alfvénic surface $M = 1$ is a sphere of radius r_*, i.e. $r_\mathrm{A} = r_*$ for all θ. The quantities v_*, B_* and

$$\rho_* = \frac{B_*^2}{4\pi v_*^2} = 4\pi\eta^2(0) \tag{7.153}$$

are defined to be values attained at the axial Alfvénic point $r = r_*$, $\theta = 0$, and the non-dimensional quantities

$$\tilde{r} = \frac{r}{r_*}, \qquad \tilde{P} = \frac{2P}{r_*^2 B_*} \equiv -\tilde{f}(\tilde{r})\sin^2\theta \tag{7.154}$$

are introduced. The density can then be written

$$\rho(\tilde{r},\theta) = \frac{4\pi\eta^2(\tilde{P})}{M^2(\tilde{r})} = \frac{\eta^2(\tilde{P})}{\eta^2(0)}\rho(\tilde{r},0) \equiv \rho(\tilde{r},\,\tilde{P}). \tag{7.155}$$

The density is now expanded up to the first power in P,

$$\rho(\tilde{r},\,\tilde{P}) = \frac{4\pi\eta^2(0)}{M^2(\tilde{r})}(1-\delta\tilde{P}) = \frac{\rho_*}{M^2(\tilde{r})}(1-\delta\tilde{P}), \tag{7.156}$$

with

$$\eta(\tilde{P}) = \eta(0)(1-\delta\tilde{P})^{1/2} = \frac{\rho_* v_*}{B_*}(1-\delta\tilde{P})^{1/2}. \tag{7.157}$$

The poloidal vectors are easily written down from the above equations:

$$B_r = B_*\frac{\tilde{f}(\tilde{r})}{\tilde{r}^2}\cos\theta, \qquad B_\theta = -\frac{B_*}{2}\frac{1}{\tilde{r}^2}\frac{\mathrm{d}\tilde{f}(\tilde{r})}{\mathrm{d}\tilde{r}}\sin\theta, \tag{7.158}$$

with $\tilde{f}(1) = 1$, and

$$v_r = v_* M^2(\tilde{r})\frac{\tilde{f}(\tilde{r})}{\tilde{r}^2}\frac{\cos\theta}{(1-\delta\tilde{P})^{1/2}}, \qquad v_\theta = -\frac{v_*}{2}\frac{M^2(\tilde{r})}{\tilde{r}^2}\frac{\mathrm{d}\tilde{f}(\tilde{r})}{\mathrm{d}\tilde{r}}\frac{\sin\theta}{(1-\delta\tilde{P})^{1/2}}. \tag{7.159}$$

(From now on the tilde on non-dimensional quantities will be dropped.)

The toroidal dynamics is again governed by (7.4), (7.9) and the effective corotation result (7.14) that ensures that Ω and B_ϕ are finite on the Alfvénic surface $M = 1$. The authors now impose the relation

$$\beta = -4\pi\eta(P)\alpha(P)\varpi_\mathrm{A}^2(P) = \lambda r_* B_*(-P). \tag{7.160}$$

Since the poloidal current flowing within the cylinder ϖ is $\propto \varpi B_\phi$, by (7.16), the assumption (7.160) implies that this current is $\simeq (-P)$, vanishing on the axis as required. Condition (7.157) then requires that the field line rotation function α is not uniform, but is constrained into the form

$$\alpha(P) = \frac{\lambda v_*}{r_*}\frac{1}{(1-\delta P)^{1/2}}, \tag{7.161}$$

rather than being prescribed from the dynamics of the convection zone. The toroidal functions become

$$v_\phi = \Omega\varpi = \lambda\frac{v_*}{r}\frac{r^2 - M^2(r)f(r)}{1-M^2(r)}\frac{\sin\theta}{(1-\delta P)^{1/2}}, \tag{7.162}$$

$$B_\phi = -\lambda B_*\frac{1}{r}\frac{f(r)-r^2}{1-M^2}\sin\theta. \tag{7.163}$$

The self-similarity *ansatz* with the imposed spherical Alfvénic surface must react on the pressure–density relationship; for the thermal and centrifugal fields are made to conspire to prevent the highly anisotropic centrifugal driving term from manifesting itself in the shape of the Alfvénic surface. As noted, the assumptions do not in general allow imposition of a polytropic or barotropic relation. Instead, the solution constrains both p and ρ as functions of position. In the present model, the pressure is written as a truncated expansion in P:

$$\frac{p(r,\,\theta)}{\rho_* v_*^2/2} = Q_0(r)(1-\kappa(r)P) = Q_0(r) + Q_1(r)\sin^2\theta, \tag{7.164}$$

where κ (quite distinct from that defined by (7.3)) is given by $Q_1 = Q_0\kappa(r)f(r)$. The θ-component of momentum balance yields an equation for $\mathrm{d}M^2/\mathrm{d}r$ in terms of Q_1 and of f and its first two derivatives; the r-component yields two equations, respectively for $\mathrm{d}Q_0/\mathrm{d}r$, $\mathrm{d}Q_1/\mathrm{d}r$ in terms of M^2 and its first derivative, and of f and its first two derivatives; and besides δ and λ there is the extra parameter $\nu^2 = 2GM/r_* v_*^2$. As explained above, the separate derivation of ρ and p fixes the required local heating function.

The model so far has three differential equations for the four functions M, f, Q_0 and Q_1. Two possible ways of closing the system are discussed.

(1) The two components of the pressure are left independent, but the function $f(r)$ fixing the slope of the field-streamlines is prescribed, leaving M, Q_0 and Q_1 to be determined from the three equations.

(2) The two components of the pressure are given a functional relation. The simplest example takes $\kappa(r) = \kappa =$ constant, now leaving the equations to determine Q_0, Q_1 and the function f for the field lines.

Case (1) allows one to choose $f(r) = 1$, so that the poloidal field lines are radial. (A more general class of model with radial field lines is discussed briefly below.) Since $v_\theta = 0$ everywhere, such a model can in principle extend all the way from the axis $\theta = 0$ to the equator $\theta = \pi/2$, with no dead zones, but with a discontinuity in B_ϕ (given by a change of sign in B_*), so there is an equatorial current sheet, requiring an equatorial plasma pressure balancing $B_\phi^2/8\pi$. It is implicitly assumed that the equatorial outflow of plasma is able to prevent the reconnection that would convert the field structure into one with dead zones. Other choices of $f(r)$ yield field-streamlines that do not become radial as they approach the equator and so are not acceptable models at low latitudes.

Case (2) yields first-order differential equations for Q_0, M^2 and the two conveniently defined functions

$$G \equiv \frac{r}{f^{1/2}}, \qquad F \equiv \frac{rf'}{f}. \tag{7.165}$$

$G(r)$ represents the radius of cross-section of a flux-tube in terms of the corresponding radius at the Alfvénic point, while $F(r)$ governs the opening of the field lines. (There is no danger of confusing this $G(r)$ with the gravitational constant.) For $F(r) = 0$ the poloidal field lines are radial, as noted. If $F(r) = 2$, $f(r) \propto r^2$, $G(r) =$ constant, corresponding to perfect collimation parallel to the axis. For $F(r) < 2$, the expanding tube does not attain full collimation, while if $F(r) > 2$, the field-streamlines curve towards the axis. Again, since in general the streamlines extrapolated to the equator will not become radial, the model is not applicable at low latitudes.

(c) *The critical points*

With $\kappa(r) = \kappa$, two of the equations have the form

$$\frac{\mathrm{d}F}{\mathrm{d}r} = \frac{N_F}{D}, \qquad \frac{\mathrm{d}M^2}{\mathrm{d}r} = \frac{N_M}{D}, \tag{7.166}$$

where N_F, N_M are lengthy expressions dependent on the variables r, G, F, M^2, Q_0 and on the parameters κ, δ, ν, λ, and D depends on r, G, F, M^2 and on κ, λ. The equations have a singularity at the Alfvén point $M = G = 1$, and another at the point where N_F, N_M, D simultaneously vanish. We have seen in Section 7.3 that if the transfield equation is neglected and the poloidal field structure just taken as given, then once the condition (7.14) is satisfied, the Alfvén point is no longer a singularity. In the present problem, however, the transfield equation – admittedly in a reduced form following the self-similarity assumption – is included in the analysis, so $M = 1$ is a real singularity. The slope $\mathrm{d}M^2/\mathrm{d}r$ must be determined from l'Hôpital's rule – the analogue of the Heyvaerts–Norman Alfvén regularity condition – in order that the solution can be continued outwards (Sauty and Tsinganos 1994).

When $\kappa \geq 0$, the second singularity – located at the point where D changes sign – is in the subAlfvénic domain. This is the analogue of the slow point in the standard (polytropic) treatment. It is a classical X-type neutral point, with position dependent on the strength of the toroidal component. A reduction in the parameter λ introduced in (7.160) moves the singularity closer to the Alfvénic point $M = 1$. It was noted in Section 7.3 that on the axis, where $B_\phi = 0$, one or other of the fast or slow points coincides with the poloidal Alfvénic point. As emphasized by Tsinganos (personal communication), it appears that the constraint of a spherical Alfvénic surface ensures that there are only two distinct critical points on *all* field lines.

Recall that in the present treatment, the field $\mathbf{B}_\mathrm{p}$ is not prescribed as in Section 7.3, but is constructed from the transfield equation (with the self-similarity *ansatz* imposed) simultaneously with the wind flow. Thus $\mathbf{B}$ is a function of ρ as well as of (r, θ), so that the analogues of (7.24), and (7.26) are changed. It is shown in Tsinganos *et al.* (1996) that with a suitable definition of the sound speed c_s for this non-polytropic problem, the condition holding at the upstream critical point is

$$v_r^4 - v_r^2(v_\mathrm{A}^2 + c_\mathrm{s}^2) + c_\mathrm{s}^2 v_{\mathrm{A},r}^2 = 0 : \tag{7.167}$$

although both the poloidal field and velocity vectors have θ-components, v_p is replaced by v_r and in the last term $v_{\mathrm{A,p}}$ by $v_{\mathrm{A},r}$. As the authors remark, since the system is symmetric in ϕ and self-similar in θ, an MHD wave that preserves these two symmetries should propagate perpendicular to the θ and ϕ directions, i.e. radially.

This last result is in fact a very special case of a general conclusion reached for axisymmetric MHD systems by Surlantzis *et al.* (1996) and Tsinganos *et al.* (1996) (following earlier work by their group) and by Bogovalov (1994). In a hyperbolic domain, the real characteristics include a special member that is crossed by one family of characteristics, but is a common tangent to the other family. This special characteristic is given the name *separatrix*, but it is not to be confused with the field line separating wind and dead zones. The critical surfaces (singularities) of the system coincide with the position of the separatrices. This is contrary to the statement often made that the singularities always occur where the flow velocity equals the fast or slow speed, which is seen to be an illicit extrapolation from the simple discussion of Section 7.3. Only in the special case when the critical surfaces are perpendicular to the streamlines does the speed there equal a classical MHD wave speed.

(d) *Asymptotic collimation*

The special assumption $\kappa =$ constant makes proportional to each other not only the two components of the pressure but also the corresponding components of the energy supply function q defined in (7.147), so that (7.150) yields a new integral $\epsilon =$ constant, valid on all field-streamlines (Sauty and Tsinganos 1994; Tsinganos and Sauty 1994):

$$\epsilon = \frac{M^4}{r^2G^2}\left(\frac{F^2}{4} - 1 - \kappa\frac{r^2}{G^2}\right) - (\delta - \kappa)\frac{\nu^2}{r} + \frac{\lambda^2}{1-M^2}\left[\frac{(M^2-G^2)^2}{G^2(1-M^2)} + 2(1-G^2)\right]. \tag{7.168}$$

This is equivalent to

$$\frac{\epsilon}{2\lambda^2} = \frac{\rho(r,P)\bar{H}(P) - \rho(r,P=0)\bar{H}(P=0)}{\rho(r,P)\alpha^2(P)\varpi_{\mathrm{A}}^2(P)}, \tag{7.169}$$

with $\bar{H}(P)$ given by (7.150). Thus ϵ measures the excess of energy per unit volume flowing along the field line P relative to that flowing along the axis.

Suppose now that these separable solutions remain relevant when extrapolated to infinity (e.g. suppose that there are no hidden instabilities). One is interested particularly in cases of self-collimation, satisfying the ϖ-component of the equation of motion:

$$\rho\Omega^2\varpi = \frac{\mathrm{d}}{\mathrm{d}\varpi}\left(\frac{B_\phi^2}{8\pi}\right) + \frac{B_\phi^2}{4\pi\varpi} + \frac{\mathrm{d}p}{\mathrm{d}\varpi}. \tag{7.170}$$

With vanishing asymptotic pressure, this reduces to a simple relation between the asymptotic values of M and G:

$$G_\infty^2 = M_\infty \frac{M_\infty + \sqrt{2}}{1 + \sqrt{2}M_\infty} \approx \frac{M_\infty}{\sqrt{2}} \tag{7.171}$$

for large values of M_∞. When $r \gg 1$, (7.168) becomes

$$\epsilon = \frac{\lambda^2}{(M_\infty^2 - 1)^2}\left[\frac{(M_\infty^2 - G_\infty^2)^2}{G_\infty^2} + 2(G_\infty^2 - 1)(M_\infty^2 - 1)\right] - \kappa\frac{M_\infty^4}{G_\infty^4}. \tag{7.172}$$

In the simplest case with $\kappa = 0$ (spherically symmetric pressure), it then follows that cylindrically collimated solutions with M_∞ and G_∞ finite exist only if ϵ is positive, and as ϵ goes through 0 to become negative, M_∞ and G_∞ become infinite, corresponding to radially opening field lines. The authors argue that this energy criterion distinguishes between strongly collimated winds – jets – emerging from rapid magnetic rotators, and non-collimated winds emerging from slow magnetic rotators such as the Sun.

When $\kappa > 0$, the higher exterior pressure helps to collimate the lower pressure jet. In fact, the computations show that even slowly rotating stars would have collimated winds at sufficiently large distances. Whether these toroidally pinched flows are the correct representation of observed jets depends on their stability (cf. Section 10.9).

(e) *Models with radial poloidal field-streamlines*

Another class of separable models has been treated by Lima *et al.* (1996 and 1997). The magnetic field structure is helicoidal – a radial poloidal field with the familiar shear-generated B_ϕ-component superposed. As again the Alfvénic surface is a sphere of radius r_*, the new set is a generalization of case (1) of the

Tsinganos–Trussoni (1991) models with $f = 1$. The various quantities have the forms

$$
\begin{aligned}
B_r &= B_0 \frac{(1+\mu \sin^{2\varepsilon}\theta)^{1/2}}{(r/r_0)^2}, & v_r &= v_0 Y(r/r_0)\left(\frac{1+\mu\sin^{2\varepsilon}\theta}{1+\delta\sin^{2\varepsilon}\theta}\right)^{1/2},\\
B_\phi &= \lambda B_0 \frac{\sin^\varepsilon\theta}{r/r_0}\left(\frac{r^2/r_*^2-1}{1-M_{\mathrm{A}}^2}\right), & v_\phi &= \lambda v_0 \frac{r/r_0 \sin^\varepsilon\theta}{(1+\delta\sin^{2\varepsilon}\theta)^{1/2}}\left(\frac{Y_*-Y}{1-M_{\mathrm{A}}^2}\right),\\
\rho &= \frac{\rho_0}{Y(r/r_0)^2}(1+\delta\sin^{2\varepsilon}\theta), & &\\
M^2 &= \left(\frac{4\pi\rho_0 v_0^2}{B_0^2}\right) Y(r/r_0)(r/r_0)^2, & &\\
p &= \frac{\rho_0 v_0^2}{2}\left[Q_0(r/r_0) + Q_1(r/r_0)\sin^{2\varepsilon}\theta\right]. & & \qquad (7.173)
\end{aligned}
$$

Here the suffix 0 refers to the coronal base. Again, one of the parameters of the problem is $\nu = (2GM/r_0 v_0^2)^{1/2}$, the ratio of the escape velocity to v_0. The balance of forces yields equations for the derivatives of Q_0, Q_1 and Y but not of P, which is fixed by the prescription of a radial poloidal field. This time, the Alfvén point is as in the standard (polytropic) problem: once B_ϕ and Ω have been made non-singular, no slope Y' is selected at the Alfvénic point – all solutions automatically pass through. The unique solution is filtered out by another, genuine X-type singular point that is now *down*stream of the Alfvénic point. The work of Tsinganos *et al.* (1996) again shows that (7.167) holds – the r-component of the flow speed equals the fast MHD wave speed in that direction; but as in this model $v_{\mathrm{p}} = v_r$, the condition is formally identical with that of Section 7.3. Again, the fixing of the ρ, p and hence the temperature field by the dynamics imposes the requirement (7.145) on the energy source q.

(f) *Resumé*

The latitudinally self-similar models with *radial* poloidal field-streamlines can be extended to the equator. They show that global solutions for P – extending from the star to infinity – can be constructed when the transfield equation is subjected to the self-similarity constraint, provided the attempt to impose a priori a p–ρ relation is abandoned. The non-radial models are acceptable in a domain encompassing the axis and extending to infinity, again provided the system is allowed to determine its own (p, ρ)-fields. This seems to leave open the query, raised in Section 7.4 and again in Chapter 14, whether, in general, the system of equations, with the perfect conductivity condition $\mathbf{E}\cdot\mathbf{B} = 0$ imposed but with *arbitrary* heat sources, has solutions satisfying the conditions at the star, at infinity and on the Alfvénic surface, or whether a local domain with non-vanishing $\mathbf{E}\cdot\mathbf{B}$ is required.

References

Alfvén, H. (1981). *Cosmic plasma.* Reidel, Dordrecht.

Biermann, L. (1946). *Naturwissenschaften*, **33**, 118.

Blandford, R.D. and Znajek, R. (1977). *Monthly Notices Royal Astronomical Society*, **179**, 433.

Bogovalov, S.V. (1994). *Monthly Notices Royal Astronomical Society*, **270**, 721.

Bogovalov, S.V. (1996). *Monthly Notices Royal Astronomical Society*, **280**, 39.

Bondi, H. (1952). *Monthly Notices Royal Astronomical Society*, **112**, 195.

Camenzind, M. (1991). *Reviews Modern Astronomy*, **3**, 234.

Goldreich, P and Julian, W.H. (1970). *Astrophysical Journal*, **160**, 971.

Heinemann, M. and Olbert, S. (1978). *Journal Geophysical Research*, **83**, 2457.

Heyvaerts, J. and Norman, C. (1989). *Astrophysical Journal*, **347**, 1055.

Heyvaerts, J. and Norman, C. (1996). In *Solar and astrophysical magnetohydrodynamic flows* (ed. K. Tsinganos), p. 459. Kluwer, Dordrecht.

Horiuchi, S., Mestel, L. and Okamoto, I. (1995). *Monthly Notices Royal Astronomical Society*, **275**, 1160.

Li, J. and Melrose, D.B. (1994). *Monthly Notices Royal Astronomical Society*, **270**, 687.

Lima, J.J.G., Tsinganos, K. and Priest, E.R. (1996). *Astrophysical Letters and Communications*, **34**, 281.

Lima, J.J.G., Priest, E.R. and Tsinganos, K. (1997). *The corona and solar wind near minimum of activity.* Proceedings fifth SOHO conference, p. 521. ESA SP-404.

Macdonald, D. and Thorne, K.S. (1982). *Monthly Notices Royal Astronomical Society*, **198**, 345.

Mestel, L. (1967*a*). *Mémoires Société Royale Liège*, **5**, 15, 351.

Mestel, L. (1967*b*). In *Plasma astrophysics* (ed. P.A. Sturrock), p. 185. Academic Press, London.

Mestel, L. (1968*a*). *Monthly Notices Royal Astronomical Society*, **138**, 359.

Mestel, L. (1968*b*). *Monthly Notices Royal Astronomical Society*, **140**, 177.

Mestel, L. and Selley, C.S. (1970). *Monthly Notices Royal Astronomical Society*, **149**, 197.

Mestel, L. and Spruit, H.C. (1987). *Monthly Notices Royal Astronomical Society*, **226**, 57.

Michel, F.C. (1969). *Astrophysical Journal*, **158**, 727.

Okamoto, I. (1974). *Monthly Notices Royal Astronomical Society*, **166**, 683.

Okamoto, I. (1975). *Monthly Notices Royal Astronomical Society*, **173**, 357.

Okamoto, I. (1992). *Monthly Notices Royal Astronomical Society*, **254**, 192.

Osterbrock, D. (1961). *Astrophysical Journal*, **134**, 347.

Parker, E.N. (1963). *Interplanetary dynamical processes.* Interscience, New York.

Phinney, S. (1982). Ph.D. dissertation, Cambridge University.

Pizzo, V., Schwenn, R., Marsch, E., Rosenbauer, H., Mühlhäuer, K.-H. and Neubauer, F.M. (1983). *Astrophysical Journal*, **271**, 335.

Pneuman, G. and Kopp, R.A. (1972). *Solar Physics*, **18**, 258.

Priest, E.R. (1982). *Solar magnetohydrodynamics.* Reidel, Dordrecht.
Rowse, D. and Roxburgh, I.W. (1981). *Solar Physics*, **74**, 165.
Sakurai, T. (1985). *Astronomy and Astrophysics*, **152**, 121.
Sakurai, T. (1990). *Computer Physics Reports*, **12**, 247.
Sauty, C. and Tsinganos, K. (1994). *Astronomy and Astrophysics*, **287**, 893.
Schatzman, E. (1962). *Astronomy and Astrophysics*, **25**, 18.
Schwarzschild, M. (1948). *Astrophysical Journal*, **107**, 1.
Stix, M. (1989). *The Sun.* Springer, Berlin.
Suess, S.T. and Nerney, S.F. (1975). *Solar Physics*, **40**, 487.
Surlantzis, K., Tsinganos, K. and Priest, E.R. (1996). *Astronomical Letters and Communications*, **34**, 251.
Tsinganos, K. (ed.) (1996). *Solar and astrophysical magnetohydrodynamic flows.* Kluwer, Dordrecht.
Tsinganos, K. and Low, B.C. (1989). *Astrophysical Journal*, **342**, 1028.
Tsinganos, K. and Sauty, C. (1994). In *Cosmical magnetism* (ed. D. Lynden-Bell), p. 45. Kluwer, Dordrecht.
Tsinganos, K. and Trussoni, E. (1991). *Astronomy and Astrophysics*, **249**, 156.
Tsinganos, K., Sauty, C., Surlantzis, E., Trussoni, E. and Contopoulos, J. (1996). *Monthly Notices Royal Astronomical Society*, **283**, 811.
Washimi, H. and Shibata, S. (1993). *Monthly Notices Royal Astronomical Society*, **262**, 936.
Weber, E.J. and Davis Jr., L. (1967). *Astrophysical Journal*, **148**, 217.

8

LATE-TYPE STARS

8.1 Introduction

For most main-sequence stars, and also for pre- or post-main-sequence stars that are not too far to the right in the Hertzsprung–Russell (H–R) diagram, the theory of stellar structure conforms closely to Eddington's (1926) prescription, as modified subsequently by Biermann and Cowling to allow for convective zones (e.g. Schwarzschild 1958; Kippenhahn and Weigert 1990). Hydrostatic support of the bulk of the star against its self-gravitation requires a temperature gradient, and the consequent outward flow of radiation is fixed essentially by the mean stellar opacity. The surface regions – linking the opaque, small photon mean-free-path interior with the nearly transparent interstellar medium – are passive: they simply adjust their temperature–density fields so as to transport the Eddington luminosity.

A low-mass main-sequence star has a radiative rather than a convective core. This is because the central temperature is low enough for the nuclear energy generation ϵ per gram to be via the proton–proton chain, for which $\epsilon \propto T^{3.5}$ approximately, rather than through the much more temperature-sensitive CN-cycle that dominates in an upper main-sequence star. The transition from the CN-cycle to the p–p chain occurs near type A, for which the surface temperature T_s is $\simeq 10^4$ K, near to the ionization temperature of hydrogen. A 'late-type' star is conveniently defined as one with $T_s < 10^4$ K, so that the photospheric hydrogen is neutral, but below the photosphere the temperature climbs rapidly. When T reaches $\simeq 10^4$ K, collisional ionization of H forces the ratio of specific heats γ down to near unity: the large latent heat of ionization depresses the adiabatic temperature gradient, so by the Schwarzschild criterion (3.129) the medium becomes convectively unstable (Unsöld 1930). Deeper down, the continuing superadiabaticity is aided by the first and then the second ionization of the substantial fraction (≈ 23 per cent) of helium. Even after virtually all the H and He is ionized, the high opacity ensures that the radiative temperature gradient continues to exceed the adiabatic gradient (Biermann 1935). In the case of the Sun, the radiative gradient becomes subadiabatic at $T \approx 2 \times 10^6$ K, with the consequent transition to a radiative core estimated to occur at a radius $r_c \approx 0.7 R_\odot$, a value confirmed by helioseismological observations (Christensen-Dalsgaard *et al.* 1991; Goode *et al.* 1991; Gough *et al.* 1996). In F-stars the convective envelope is shallower, in K-dwarfs considerably deeper, while late M-dwarfs are thought to be fully convective below the photosphere.

A pre-main sequence star, of mass equal to that of a low-main-sequence star, has a correspondingly lower value for T_s, and so will again have a sub-photospheric convective zone, but as long as it is not too far to the right in the H–R diagram its luminosity is again given essentially by Eddington's theory. Because of the form of the Kramers-type opacity law, the Eddington luminosity is a strong function of the mass but a weak function of radius, and a star completing its Kelvin–Helmholtz approach to the main sequence will move along the nearly horizontal 'Henyey line' (e.g. Kippenhahn and Weigert 1990). However, a crucial new feature enters at lower surface temperatures, because of the behaviour of the opacity κ. From the theory of stellar atmospheres (Eddington 1959), the condition that the temperature of the photospheric gas should agree with the surface temperature, defined by

$$L = 4\pi\sigma R^2 T_s^4 \tag{8.1}$$

requires that the optical depth $\tau = \int_R^\infty \kappa\rho\,\mathrm{d}r \simeq 2/3$. Combined with the local equation of hydrostatic support $\mathrm{d}p/\mathrm{d}r = -g\rho$, this yields as a good approximation for the photospheric pressure

$$p(R) = \frac{GM}{R}\frac{2}{3}\frac{1}{\kappa(R)}. \tag{8.2}$$

This radiative boundary condition (8.2) must be satisfied in all stars; however, in stars with radiative envelopes out to the photosphere, the effect on the interior structure and so on the luminosity is small. The importance of this radiative 'bottleneck' in a star that is wholly or largely convective seems to have been first appreciated by Cowling (1938). Its effect is felt in models of main-sequence dwarfs (Osterbrock 1952), but its most dramatic consequences occur in stars well to the right of the main sequence (Hayashi 1961). At low temperatures, the opacity $\kappa(R)$ is determined largely by the properties of the negative hydrogen ion (Wildt 1939; Chandrasekhar and Breen 1946), the extra electrons coming from metals with low ionization potentials. If the temperature becomes too low, collisional ionization is weak, the opacity becomes too low, and condition (8.2) cannot be satisfied. There is therefore a lower limit to the surface temperature: a 'photosphere' with a lower temperature would in fact be transparent, with the emerging radiation coming from a deeper, hotter layer. This lower limit to T_s implies that an extended star of radius R may have a luminosity given by (8.1) in excess of the Eddington luminosity. To supply this super-luminosity to the surface, the star acquires a still deeper convective envelope. One possible scenario for the pre-main-sequence phase is for a star to contract down the 'Hayashi line' at nearly constant T_s and so with $L \propto R^2$ until it reaches the Henyey line, contracting then at approximately constant L towards the main sequence. A star of mass $M \leq M_\odot$ will be fully convective on the Hayashi line. A star that is of 'early-type' on the main sequence will have a similar pre-main-sequence track, except that on the Hayashi line the star is not fully convective but retains a radiative core.

The normal evolution of an initially homogeneous star on the main sequence is towards an inhomogeneous structure, due to nuclear processing in the hotter inner regions. The contraction of a burnt-out core is accompanied by an expansion of the envelope. At first, the representative points in the H–R diagram move roughly horizontally, back into the domain to the right of the main sequence; but the same argument that leads to the Hayashi line yields now the nearly vertical giant and asymptotic giant branches (Hoyle and Schwarzschild 1955). It is seen that stars with deep sub-photospheric convective envelopes – characteristic of 'late-type' stars – are found on the lower main sequence and in both the pre- and post-main-sequence phases of normal stellar evolution. Prima facie, one anticipates magnetic activity as observed on the Sun to be present. In this chapter we concentrate on magnetic effects in lower main-sequence stars, touching finally on giant stars. The special problems of pre-main-sequence stars, such as the T Tauri stars, are postponed till Chapter 10.

For the summary of the relevant observational material I have drawn on several sources, in particular the review *Solar and stellar dynamos* by Weiss (1994) and the book *Sunspots: theory and observations* (Thomas and Weiss 1992).

8.2 The 'solar–stellar connection'

Following the first Zeeman measurements by Hale of strong magnetic fields in sunspots, and the early hints of a general solar magnetic field, a vast amount of observational material has accumulated on 'solar activity', with magnetic fields playing an essential role (cf. Sections 4.6, 4.8 and 8.4). More recently, observations of other late-type main-sequence stars at optical, X-ray and radio frequencies have confirmed the presence of similar activity, so defining the area of research aptly named the 'solar–stellar connection'. Stars that are much more active magnetically than the Sun are inferred from Zeeman broadening to have field strengths of about 2000 G over half of the unspotted fraction of their surfaces (Robinson *et al.* 1980; Marcy 1984; Saar and Linsky 1985; Mathys and Solanki 1988; Saar 1991). They also show photometric variations, plausibly interpreted as due to the rotation of the essentially non-axisymmetric surface structure, to be expected if there is extensive star-spot coverage. The luminosity of an active star may vary by up to 30 per cent, implying that spots cover up to 60 per cent of its surface area (Byrne 1992), whereas the solar luminosity varies by no more than ≈ 0.1 per cent. The solar corona is heated (almost certainly magnetically) to $\approx 10^6$ K and so emits thermal X-rays; similar emission is detected by satellites from stellar coronae (Vaiana *et al.* 1981; Pallavicini *et al.* 1981). Likewise, stellar flaring is detected at both radio and X-ray frequencies.

It has been known for many years that Ca^+ H and K emission is closely correlated with solar magnetic fields. In 1966 Wilson initiated a systematic programme at Mt. Wilson of monitoring Ca^+ emission from nearby stars (cf. Wilson 1978). When one plots for nearby field stars a suitable measure R_{HK} against the colour index $(B-V)$, defining the spectral type, it is found that for stars of given spectral type there is a wide range of Ca-activity. However, if one selects single stars that belong to a fairly old cluster such as the Hyades, and so are of essentially the same age, then there are found only modest variations between stars of the same type, or even between stars of different type; whereas within younger clusters such as the Pleiades there is again a wide scatter. Further, stars with high velocities normal to the galactic plane, normally thought to be old, are only weakly active. Thus for single stars the scatter in R_{HK} appears to be a function of age.

It has long been recognized that the stellar rotation is the crucial parameter fixing the strength of activity in a star of given spectral type (Kraft 1967; Durney 1972). The rotation period of an active star can be estimated from Doppler broadening of spectral lines, or can be determined more precisely from the rotational modulation of the Ca-activity. Noyes *et al.* (1984*a*) introduce as a parameter the inverse Rossby number $\sigma = \Omega\tau_c$, where τ_c is the computed convective turnover time at the base of the convective envelope. We recall from Section 6.5 that the simplest $\alpha\Omega$ dynamo yields the dynamo number $D \propto \sigma^2$. The results of Noyes *et al.* do show a striking correlation between the time-averaged Ca-activity and σ (Fig. 8.1). There has been some subsequent debate over the claimed τ_c-dependence. Rodono (1988, personal communication) has questioned whether the observations require the variation of τ_c between different spectral types to be taken into account. Stepien (1993) argues that the dependence is genuine for single main-sequence dwarfs with $0.5 \leq B - V \leq 0.8$, whereas for cooler dwarfs and for giants, activity is almost colour-independent. For hotter dwarfs, Stepien queries even the otherwise striking correlation with Ω.

If the Sun were observed as a nearby star, the solar cycle would be detected from variations in its Ca^+ emission. Similar cyclical Ca-activity of moderate strength has been detected in a dozen slowly rotating main sequence dwarfs (Fig. 8.2).

Following Noyes *et al.* (1984*b*), Saar and Baliunas (1992) plot the cycle frequency Ω_{cycl} against the inverse Rossby number σ (Fig. 8.3). From the limited number of cases with well-determined Ω, one infers the relation $\Omega_{\mathrm{cycl}} \propto \sigma^n$ with n tentatively estimated as ≈ 3. At higher rotations, the activity shows a more chaotic behaviour. For stars which do not show solar-type behaviour, but for which an approximate Ω_{cycl} can be inferred, the observed variation with σ suggests an index closer to unity.

The combination of basic dynamo theory (Chapter 6) and braking theory (Chapter 7) leads to a provisional description of the rotational history and the correlated activity. An $\alpha\Omega$ dynamo generates a large-scale magnetic field with a total flux that is systematically higher at higher Ω for at least some of the relevant range of Ω. The magnetic field couples the sub-photospheric convection zone with the chromosphere and corona: turbulent kinetic energy is converted into excess magnetic energy which is dissipated, so maintaining a hot stellar corona that tends to expand as a stellar wind, with consequent braking of the star's rotation. The decrease in Ω and so of the dynamo-maintained flux leads to a decline in chromospheric and coronal 'magnetic activity' in both the optical, radio and X-ray bands.

Support for the basic premise that rotation is a crucial parameter comes

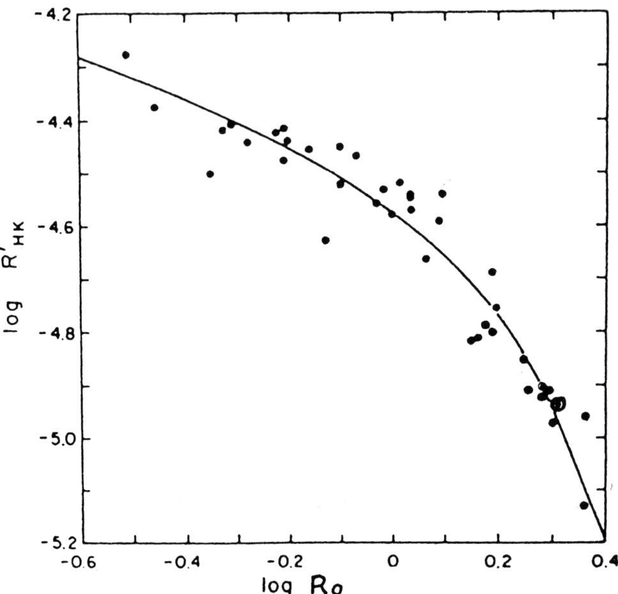

FIG. 8.1. Dependence of stellar magnetic activity on rotation. Logarithmic plot of the time averaged index R'_{HK} of calcium activity against the Rossby number $\mathrm{Ro} = 2\pi/\sigma$, with the Sun's position marked, and the solid curve an inferred functional relationship. (After Noyes *et al.* 1984*a*.)

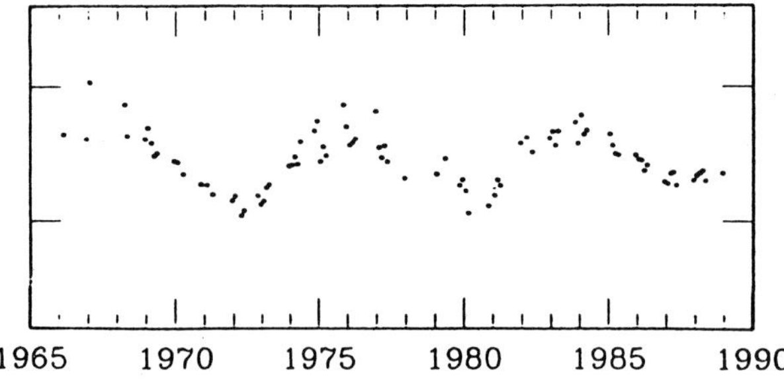

FIG. 8.2. Chromospheric Ca^{+} emission from 1966 to 1989 for the G2 star HD81809, showing cyclic activity with a period of 8.3 yr. (After Baliunas and Jastrow 1990.)

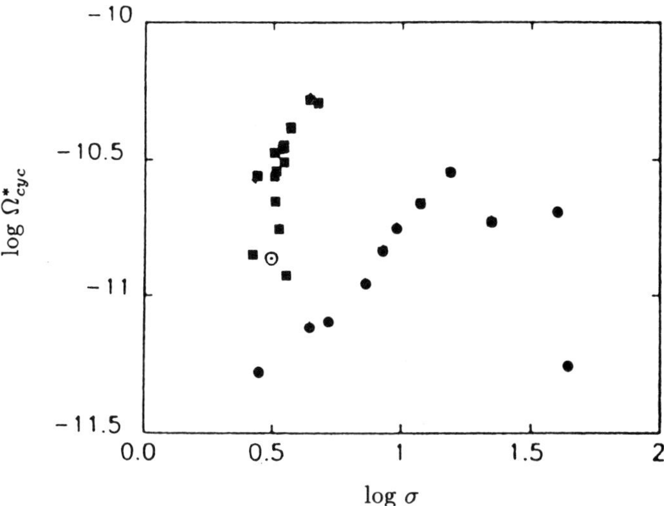

FIG. 8.3. Dependence of cycle frequency on angular velocity. Logarithmic plot of a scaled cycle frequency Ω_{cycl} as a function of the inverse Rossby number σ. The Sun's position is marked, squares indicate stars with solar-like cycles, and circles indicate other stars. There is a clear indication of two distinct families. (After Saar and Baliunas 1992.)

from observations of the RS CVn stars – evolved stars in close binary systems – which by contrast remain active as they age (e.g. Middelkoop and Zwaan 1981; Baliunas and Vaughan 1985). They are subject to the same braking process, but there is now available a source of angular momentum in the orbital motion. Coupling (tidal or magnetic) between spin and orbital motion ensures that near-synchronization is maintained. Spin angular momentum carried off by the wind is replenished from the orbital angular momentum, with a consequent modest mutual approach of the mass-centres of the two stars: loss of angular momentum now leads to a slight increase in Ω and so to a *maintenance* of magnetic activity.

In Chapter 7 it was argued on general grounds that a stellar corona should have a multicomponent structure, with one or more 'dead zones' of hot gas magnetically trapped, and 'wind zones' of cooler outflowing gas (the solar 'coronal holes'). Support for the presence of a dead zone comes from observations of non-thermal radio-emission from stellar coronae (Mutel and Morris 1988). Further support comes from combined optical and X-ray studies of the rapidly rotating ($P = 0.5$ d) G8-K0 dwarf star AB Doradus (Collier Cameron *et al.* 1988). The X-ray data from the satellite EXOSAT show that the part of the spectrum with photons of energy 1–10 keV is well fitted by thermal bremsstrahlung from gas at $T \approx 1.7 \times 10^7$ K, with an emission measure that yields an emission volume greater than 10 times the volume of the star. This emission does not vary significantly, whereas the part with photons of 0.1–1 keV does vary, and furthermore is *not* an extrapolation to lower frequencies of the harder component. The plausible inference is that the high energy emission comes from a dead zone extending out to $(2-3)R$, whereas the low energy emission is by a cooler coronal structure, relatively compact and near active regions, and like starspots, subject to rotational eclipse.

Observations in Hα (Collier Cameron and Robinson 1989) show transient absorption features, which are consistent with clouds of HI in *enforced corotation with the star*, transiting the stellar disc and scattering chromospheric Hα photons. The crucial observation is of the line-of-sight velocity component of the absorbing material, which always matches that of the obscured stellar limb as the cloud enters or emerges from transit, showing clearly that it is in corotation; for without the magnetic constraint, the gas would instead rotate differentially, in nearly Keplerian orbits. The cloud temperatures are estimated to be 5500–7500 K, and the masses $(10^{-16}$–$10^{-14})M_{\odot}$. There is some evidence of outward motion as well. From the rate at which they cross the profile, the clouds appear to form at (3–4)R, which is outside the Keplerian radius $r_{\mathrm{K}} = (GM/\Omega^2)^{1/3}$ at which corotation with the star yields a centrifugal acceleration balancing gravity. Refer now to the model of Section 7.5, in which hydrostatic support within a corotating dipolar dead zone yields the density (7.61) (with α replaced by Ω). Beyond $r \simeq r_{\mathrm{K}}$, the exponential outward *in*crease in ρ should lead to cooling and recombination, further compression and outflow from the star of more hot gas. With ρ high enough, the centrifugal force density $\rho\Omega^2\varpi$ should pull some gas out of the 'dead zone', which thus ceases to be completely 'dead'. In rapid rotators, the estimated angular momentum loss via the wind zone will be augmented by gas breaking out of the dead zone. Although the model of Section 7.5 is just a simple simulation of the real situation, nevertheless the semi-quantitative vindication of the picture encourages belief that its qualitative features will survive.

A bright, rapidly rotating star such as AB Doradus is an excellent candidate for a high-resolution spectroscopic mapping technique such as Doppler imaging (cf. Chapter 1). The observations to date (Unruh *et al.* 1995 and references therein) have already produced much more detailed information on the spot distribution on this star: a low-latitude band that produces most of the photometric modulation, and that appears to be stable over at least five stellar rotations; and a more variable high-latitude band centred on 60°. More recent observations (Donati and Collier Cameron 1997) have yielded a surface differential rotation that is of the same order as that in the Sun and so is a much smaller fraction of the mean surface rotation. Accumulating data such as these cannot fail ultimately to constrain stellar dynamo models appropriate to rapid rotators.

8.3 The rotational history of late-type main-sequence stars

We now study spin-down in more detail. We continue working with the simple model of Section 7.5, in which the field is taken as that of a vacuum dipole from the star's surface R out to a radius $\bar{r}$, and as radial beyond $\bar{r}$. The field line passing through the equatorial point $(\bar{r}, \pi/2)$ separates the wind zone – linked with the polar caps – from the 'dead zone'. From (7.79), the rate of loss of angular momentum J from a single star is reasonably well approximated by

$$-\frac{\mathrm{d}J}{\mathrm{d}t} = \frac{8\pi}{3}(\rho_{\mathrm{A}} v_{\mathrm{A}} r_{\mathrm{A}}^2)\Omega r_{\mathrm{A}}^2 = \frac{\Phi^2 \Omega}{6\pi^2 v_{\mathrm{A}}}, \tag{8.3}$$

where the symbol Ω is again used for the star's rotation, the suffix A refers to the Alfvénic surface (approximated as a sphere), and $\Phi \equiv 2\pi B_{\mathrm{A}} r_{\mathrm{A}}^2 = 2\pi \bar{B} \bar{r}^2$ is the flux of the 'open field lines' that form the wind zone.

Consider first the case with Ω low enough for the wind to be purely thermally-driven, as in Parker's original model (1963), except that within $\bar{r}$ the flow is along dipolar field lines. As noted in Section 7.6, if the coronal sound speed a is independent of Ω, then the nearly constant wind-speed in the supersonic domain yields $-\dot{J} \propto \Phi^2 \Omega$ from (7.79). For the moment it is assumed that coupling between the convective envelope and the radiative core is strong enough to ensure that the whole star spins down as an effectively rigid body: $J = Mk^2R^2\Omega$ with kR the radius of gyration of the whole star. To make further progress one needs a relation between Φ and Ω. A theoretical prediction requires a convincing dynamical model of the dynamo. As is clear from Chapter 6, even if there were extant such a model for the Sun, its generalization for stars with higher rotation rates is not obvious. It is safer to be guided by observation, recalling the striking correlation between the diagnostics of magnetic activity and the inverse Rossby number $\Omega\tau_c$. On the theoretical side, it is suggestive that in the simplest $\alpha\Omega$ linear dynamo wave model, for dynamo numbers well above the critical value the growth rate from (6.67) is proportional to $\Omega\tau_c$. Admittedly there is no clear link between a linear growth rate and the strength of the asymptotic field as fixed by non-linear feedback. Nevertheless, in a first attempt one can reasonably try the simplest case and take the surface flux $\Phi_0 \propto \Omega$ for the slower rotators of a given spectral type (as predicted, for example, by the simple Durney–Robinson model summarized in Section 6.6). For high Ω, such a law yields a starspot filling factor that approaches unity, consistent with the cited observations, and suggesting that the flux may ultimately saturate. Provisionally, the linear relation is therefore adopted up to $\Omega = \tilde{\Omega}$, with a constant value for Φ beyond. The saturation value $\tilde{\Omega}$ is a parameter for each spectral type, hopefully to be inferred after comparison with observation. Variation of the dead zone radius $\bar{r}$ with Ω is ignored, so that the wind zone flux $\Phi \propto \Phi_0 \propto \Omega$ also. For $\Omega < \tilde{\Omega}$ we then arrive at the braking equation (7.85)

$$-\dot{\Omega} = K\Omega^3 \tag{8.4}$$

with K depending just on the spectral type but not on the mass loss rate $-\dot{M}$.

Equation (8.4) integrates to yield

$$\Omega = (\Omega_0^{-2} + 2Kt)^{-1/2} \tag{8.5}$$

where Ω_0 is the zero-age main-sequence value (presumed less than the saturation value $\tilde{\Omega}$). Asymptotically, when the star has forgotten its initial rotation, (8.5) becomes

$$\Omega = (2K)^{-1/2}t^{-1/2}. \tag{8.6}$$

Thus this simple model yields a theoretical basis for Skumanich's law $\Omega \propto t^{-1/2}$, inferred (1972) from the then available observations of the rotations of late-type stars in the Hyades and Pleiades clusters, with their estimated ages respectively of 6×10^8 yr and 7×10^7 yr, and from the rotation of the Sun at its age of 4.5×10^9 yr.

There is now a much more extensive body of observational data. Whereas in the Hyades the spread in Ω for each spectral type has remained small, it is now clear that within the Pleiades and the still younger cluster α Persei, of estimated age 5×10^7 yr, the scatter is large, with many rapidly rotating dwarf stars (van Leeuwen and Alphenaar 1982; Soderblom *et al.* 1983; Stauffer *et al.* 1984; van Leeuwen *et al.* 1987). A plausible explanation is that there is a similar scatter in the zero-age main-sequence rotation Ω_0 which has not been forgotten at the Pleiades age: an asymptotic law such as (8.6) is appropriate for the Hyades, but for younger clusters, formulae such as (8.5) have to be used. In a first study, one can extend the above model by introducing dynamo saturation at $\tilde{\Omega}$, beyond which Φ stays fixed and (8.4) is replaced by

$$-\frac{\mathrm{d}\Omega}{\mathrm{d}t} = (K\tilde{\Omega}^2)\Omega. \tag{8.7}$$

If $\Omega_0 < \tilde{\Omega}$, the solution (8.5) holds for all time, but if $\Omega_0 > \tilde{\Omega}$, then

$$\Omega = \Omega_0 \exp(-K\tilde{\Omega}^2 t) \qquad 0 < t < \tilde{t}, \tag{8.8}$$

$$= [\tilde{\Omega}^{-2} + 2K(t - \tilde{t})]^{-1/2} \qquad \tilde{t} < t. \tag{8.9}$$

Observers have already argued for a limit in the increase of surface field strengths Ω. Vilhu (1984) estimated dynamo saturation to occur at $\approx$ 4–6 $\Omega_\odot$ from cut-off in the observed chromospheric UV and optical line flux. Likewise, Saar (1991) finds that the filling factor of surface flux saturates near Vilhu's chromospheric indicators. However, Saar's Zeeman technique is sensitive to the bright photospheric network, which is associated with chromospheric activity. *Photometric* variability shows up with bigger amplitudes in the rapid rotators in Pleiades and α Persei, implying that these stars have a bigger coverage of *cool spots* and so of kG-fields, which do not show up in the Zeeman measurements. For example, two stars, one with a 7-day and the other with a half-day period, show similar chromospheric activity, but the second has much greater spot coverage. A recent survey by O'Dell *et al.* (1995) finds a minimum saturation level of 6–10 times that inferred from chromospheric or transition zone indicators. These results indicate that there is a large-scale, dynamo-built field that does not saturate until a value $\tilde{\Omega}$ is reached much above the rather modest Vilhu/Saar value.

When extrapolating the braking theory to high rotation rates, it must be remembered that the braking formulae used so far are oversimplified. First of all, the likely variation of the extent of the dead zone needs exploration (Mestel and Spruit 1987). The boundary condition that should hold strictly on the separatrix between the dead and wind zones is the continuity of total pressure $(p+B^2/8\pi)$. For rapid rotators, the effect of centrifugal driving by the nearly corotating gas in the wind zone is to accelerate the gas and so to *reduce* both density and pressure (the Bernoulli effect). Within the contiguous dead zone, hydrostatic equilibrium must hold, with balance along the direction of $\mathbf{B}_{\rm p}$ of gravity, pressure gradient and centrifugal force. With the dipole field approximation again adopted, the density along a field line that leaves the star at (R, θ_0) is given by (7.61) (with α replaced by Ω). As already noted in Section 8.2, beyond the point where the centrifugal force on corotating gas balances gravity, the pressure and density increase exponentially *outwards*. An estimate for the limit $\bar{r}$ of the dead zone is given by the pressure balance condition (7.62) at the cusp. At low rotation, an increase in Ω at constant ρ_0 causes an increase in $\bar{r}$ because of the assumed increase $B_0 \propto \Omega$; but at higher Ω the centrifugal effect on the dead zone pressure takes over and the increase in $\bar{r}$ is halted and reversed. Any increase in ρ_0 with B_0 will likewise act to halt the growth in $\bar{r}$. In the parameter domain of greatest interest, when centrifugal driving is affecting the wind, it turns out that the variation of $\bar{r}$ with Ω is modest enough to be ignored in a preliminary treatment.

At high rotations, when centrifugal driving is dominant, from (7.73), $v_{\rm A} \simeq 0.54\Omega\varpi_{\rm A}$, leading to

$$r_{\rm A} \propto [\Phi^2/(-\dot{M})\Omega]^{1/3} \tag{8.10}$$

and

$$-\frac{\mathrm{d}J}{\mathrm{d}t} \propto \Phi^{4/3}(-\dot{M})^{1/3}\Omega^{1/3}, \tag{8.11}$$

showing the dependence (already noted in Section 7.6) of $-\dot{J}$ on $-\dot{M}$ and so on the coronal base density ρ_0. Note also that the dependence of $-\dot{J}$ on Ω is weakened: if Ω is high enough for the flux to saturate, and $-\dot{M}$ is constant, $-\dot{\Omega} \propto \Omega^{1/3}$ replaces the linear law (8.7). To keep the rate of braking near its maximum, one needs $-\dot{M}$ to increase markedly with Ω, reducing $r_{\rm A}$ sufficiently for the centrifugal driving term in Bernoulli's equation to remain modest, so that $v_{\rm A}$ stays close to $v_{\rm th}$.

An adequate theory must be able to explain the rotational history of other late-type stars as well as those of solar-type. The change in the density structure – due primarily to the increasing depth of the convective envelope in later spectral types – affects significantly the non-dimensional factor in the radius of gyration k. Recall also that at least in linear $\alpha\Omega$ dynamo models, Ω appears multiplied by τ_c, where τ_c is the convective turnover time at the base of the convective zone, varying by a factor 4–5 between the G and M stars. Thus one possible generalization of the earlier dynamo relation writes

$$B_0 = B_{0,\odot}\frac{\tau_c\Omega}{\tau_{c\odot}\Omega_\odot} \tag{8.12}$$

for $\Omega < \tilde{\Omega}$, and

$$B_0 = B_{0,\odot}\frac{\tau_c\tilde{\Omega}}{\tau_{c\odot}\Omega_\odot} \tag{8.13}$$

for $\Omega > \tilde{\Omega}$, with the saturation angular velocity $\tilde{\Omega}$ depending on spectral type.

The plethora of new observational results sets new challenges to the theory. Some authors (Soderblom *et al.* 1993) follow earlier work (Endal and Sofia 1981; Stauffer and Hartmann 1987) and urge that the standard magnetic braking theory without some degree of core–envelope decoupling is in conflict with observation. Li and Collier Cameron (1993) have argued strongly against this. In particular, to produce consistent results for both α Persei, Pleiades and Hyades, they find that the coupling would need to be so weak that angular momentum transfer from core to envelope would take place over several 10^9 yr, so setting upper limits on internal magnetic fields that are implausibly severe. In their later study, Collier Cameron and Li (1994) adopt near uniform rotation provisionally and study whether standard braking theory is able to cope with the observations. In addition to the generalized linear dynamo relation (8.12), (8.13), they allow for a possible variation in the temperature of the wind zone according to $T_{\mathrm{w}} \propto a_{\mathrm{w}}^2 \propto B_0^a$, with the index a in the range $0 < a < 1$. There is also some evidence from studies of lunar material (Geiss and Bochsler 1991) that the mass flux in the solar wind may have been up to an order of magnitude greater when the Sun was (1–2) $\times\, 10^9$ yr younger than today and so was rotating probably at about three times the present rate. A higher value for ρ_0 and hence for $-\dot{M}$ should be associated with a higher T_0 and a consequent larger scale-height. The mass-loss rate is therefore parametrized by

$$\dot{M} = \dot{M}'\left(\frac{B_0}{B_{0\odot}}\right)^b, \tag{8.14}$$

where $0 < b < 2$, and

$$\dot{M}' = \dot{M}_\odot\left(\frac{R}{R_\odot}\right)^2 \tag{8.15}$$

allows for the dependence of radius on spectral type.

With the relations (8.12)–(8.15) built in, we arrive at modified asymptotic braking laws. For purely thermal driving, the speed at the Alfvén surface is given by $v_{\mathrm{A}} = Ca_{\mathrm{w}}$, where C varies very slowly with changes in the other parameters. (In their computations, Collier Cameron and Li take $C = 2.5$, in the middle of the range found by Mestel and Spruit (1987).) Equations (8.4) and (8.7) for the unsaturated and saturated domains are replaced respectively by

$$-\dot{\Omega} = \kappa\Omega^{3-a/2} \qquad \Omega < \tilde{\Omega}, \tag{8.16}$$

$$-\dot{\Omega} = \kappa\tilde{\Omega}^{2-a/2}\Omega \qquad \Omega > \tilde{\Omega}, \tag{8.17}$$

with

$$\kappa = \frac{2R^2}{3CMk^2}\frac{(B_{0,\odot})^2}{a_\odot}\left(\frac{\tau_c}{\tau_{c,\odot}\Omega_\odot}\right)^{2-a/2}. \tag{8.18}$$

At the other limit of purely centrifugal driving, with v_A from (7.73) approximated by $\Omega r_A/\sqrt{3}$, for the unsaturated domain $\Omega < \tilde{\Omega}$, (8.11) is replaced by

$$-\dot{\Omega} = \kappa_c \Omega^{(5+b)/3} \tag{8.19}$$

with

$$\kappa_c = \frac{2R^{2/3}}{3Mk^2} B_\odot^{4/3} (-3\dot{M})^{1/3} \left(\frac{\tau_c}{\tau_{c,\odot}\Omega_\odot} \right)^{(4+b)/3} \tag{8.20}$$

In the saturated domain ($\Omega > \tilde{\Omega}$), (8.20) is replaced by

$$-\dot{\Omega} = \kappa_c \tilde{\Omega}^{(4+b)/3} \Omega^{1/3}. \tag{8.21}$$

Note that with $b > 0$, the Ω-dependence in (8.19) is steepened. However, (8.21) is likely to be the better approximation for a rapid rotator, for centrifugal driving may very well take over only when Ω is high enough for dynamo saturation to have set in.

For comparison with observation, the radii of gyration are taken from Rucinski (1988). Colours, masses and radii are interpolated from VandenBerg and Bridges (1984) and convective turnover times from Rucinski and VandenBerg (1986, 1990).

If there were available a complete theory for these stars – both for the dynamo generation and for the structure of the chromosphere and corona – the quantities κ, κ_c would be predictable; but as it is, the braking relations must be calibrated from observation. For details, see Collier Cameron and Li (1994). A typical $\Omega - t$ curve for a given $(B - V)$ is shown in Fig. 8.4.

The principal conclusions of their detailed survey are as follows:

(1) A wide spread in rotation rates is already present when the stars arrive on the zero-age main sequence.
(2) Standard braking theory when extended as above seems so far adequate with the stars kept in near solid-body rotation. Core–envelope decoupling is inherently implausible, and in fact does not help with interpretation of the observations.
(3) At low rotations, the surface magnetic field scales linearly with the inverse Rossby number.
(4) Dynamo saturation appears to occur at a value $\tilde{\Omega}$ that is a factor 4 or 5 greater than the value at which the chromospheric emission fluxes appear to saturate.
(5) The temperature in the inner wind (assumed isothermal) is independent of Ω at a given spectral type, but increases slightly towards later spectral types.
(6) The mass loss rate $-\dot{M} \propto \rho_0 R^2$ must scale either linearly or as the square of the surface-averaged field at the base of the wind zone.

The above discussion has assumed the existence of a saturation rotation $\tilde{\Omega}$ – necessarily below the upper limit set by the ratio of centrifugal to gravita-

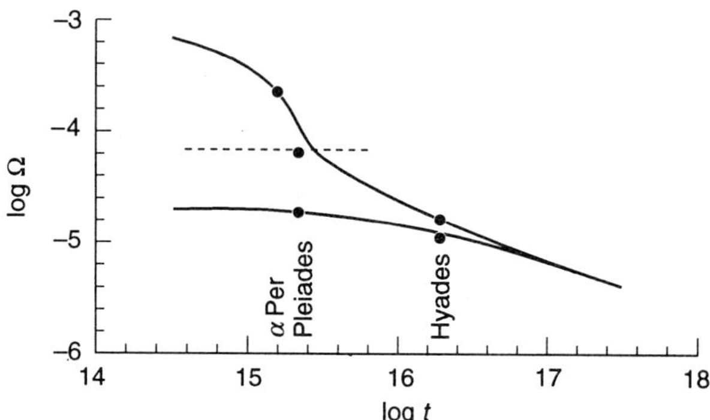

FIG. 8.4. Typical rotational evolution tracks for stars that are either very rapid or moderately rapid rotators on the zero-age Main Sequence. Braking is described by (8.16) and (8.18) – a radial poloidal magnetic field (no dead zones), purely thermal driving, and a linear B_0–Ω dynamo law up to the saturation value $\tilde{\Omega}$. Reasonable agreement with the solar parameters, and also with observations of stars with different values of Ω in clusters of different ages is achieved with the the values $\kappa = 2.2 \times 10^{-7}$ s and $\tilde{\Omega} = 6.8 \times 10^{-5}$ s^{-1}. (From Collier Cameron and Li 1994.)

tional acceleration – at which the postulated linear increase with Ω of dynamo-generated flux ceases. An alternative viewpoint (Solanki *et al.* 1997) argues that saturation is not necessary to explain the presence of rapid rotators in young clusters. As discussed in Section 4.8.1, in a rapid rotator, flux tubes rising from the overshoot layer at the base of the convective envelope will suffer a polewards Coriolis deflection. If the external dipole component of the solar field and the location of the sunspot zones are indeed closely linked, then the increase in the braking rate expected from a growth in the surface flux with Ω may be offset by a corresponding reduction in the torque lever-arm: saturation of the angular momentum loss may occur without saturation in the dynamo-generated flux. It remains to be seen if such an explanation can be made fully consistent with the detailed properties of the external field, such as the existence of a dead zone, and with our growing knowledge of starspot distributions on rapidly rotating cool stars (Schüssler *et al.* 1996).

As is clear from Chapter 6, the uncertainties in dynamo theory are such that one welcomes any tentative constraints on the parameters, emerging from cognate studies which link up with observation. The confirmation (1) of a wide ZAMS spread in Ω is also important as it focuses attention on the final 'pre-main-sequence' phase of star formation. Magnetic braking probably plays a crucial role in the early, diffuse molecular cloud phases by enabling condensing and fragmenting masses to lose several orders of magnitude of angular momentum;

but the most optimistic estimates still yield masses in the later opaque phases which rotate close to the centrifugal limit (cf. Chapters 11 and 12). The opaque Hayashi and Henyey phases have a much more relaxed time-scale in which further transfer of angular momentum can occur. The possible rotational history of pre-main sequence stars is discussed in Sections 10.2–10.4.

8.4 The solar cycle: recent observational material

We now give a resumé (based largely on Weiss 1994) of the observations of solar activity that appear relevant to the solar dynamo and its consequences. The most prominent features are the active regions, which include sunspots, and have been observed over the centuries. These active regions commonly occur in pairs, oriented at a small angle γ (about 10°) to parallels of latitude, with the preceding spot closer to the equator, and with $\sin\gamma \approx (\cos\theta)/2$, where θ is the colatitude. These bipolar regions have a consistent polarity over each hemisphere, changing sign at the equator. The polarity reverses after an interval of about 11 yr, so that the magnetic cycle has a 22-yr period.

The active regions occur within an equatorial zone extending to latitudes of $\pm 30^\circ$. At the rising phase of the cycle, sunspots appear at high latitudes; as the cycle proceeds towards maximum, the sunspot numbers increase and the zones of activity spread as they migrate towards the equator (Spörer's law), where they shrink away at sunspot minimum, while simultaneously new spots with reversed polarity appear at high latitudes. This pattern of activity yields the familiar 'butterfly diagram' (Fig. 1.1), which any viable solar dynamo model must reproduce.

In addition to the strong but irregular fields from active regions, the Sun has a weak general field near the poles, which extends far into the corona and whose effect on coronal structure has been monitored for a century. At sunspot minimum there is a strong polar field which disappears at sunspot maximum. H.D. Babcock (1959) reported that the field had reversed at the maximum in 1958. Since then the polar field has behaved less systematically – for a while there was the same polarity at both poles. The field around the North Pole reversed again in 1971, followed by that at the South Pole in 1972. Apparently the polar fields follow the solar cycle but are coupled rather loosely to fields in active regions near the equator.

It is now possible to resolve structures of a few hundred kilometres on the solar surface (cf. Fig. 1.4). Soft X-rays are a particularly useful probe of solar activity. They show that coronal emission is dominated by active regions and compact bright points, associated with reconnection events (cf. Section 3.7). Superposition of X-ray and optical images shows clearly the relation between coronal loops and sunspots. The magnetograms show that photospheric magnetic fields are highly intermittent: active regions contain sunspots and pores; within faculae the magnetic flux is confined to intergranular lanes and the weaker photospheric network is made up of isolated flux tubes with intense magnetic fields. These magnetic structures persist down to the smallest resolvable scales.

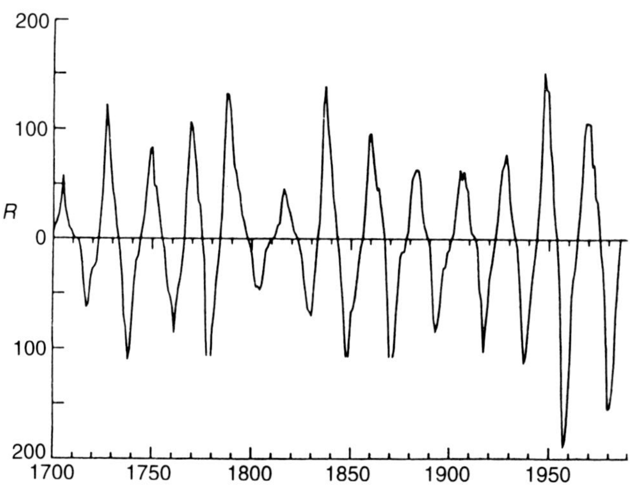

FIG. 8.5. The 22-yr solar magnetic cycle. A record of alternating magnetic activity, constructed from the records of sunspot numbers R from 1700 to 1985. (After Bracewell 1988.)

Solar activity is conveniently measured by the relative sunspot number R (Stix 1989). The records over the last 300 yr show that the cycle appears to be chaotic with an aperiodically modulated amplitude (Fig. 8.5). There is independent evidence that the celebrated 70–80 yr epoch in the seventeenth century of unusually low sunspot activity (the 'Maunder minimum') was not a one-off event, but rather that such minima are recurrent features of solar and stellar activity. Galactic cosmic rays impinging on the earth's atmosphere yield radioisotopes, including ^{10}Be and ^{14}C (Figs 8.6*a*,*b*). Since the cosmic ray flux is modulated by magnetic fields in the solar wind, there are fluctuations – anti-correlated with solar activity – in the rates at which these isotopes are produced. The abundance curve over the last 10^5 yr for ^{10}Be, preserved in polar ice-caps, shows not only the 11-yr activity cycle but also a long-term modulation, including both the Maunder minimum and several previous minima, spaced over 200–300 yr (Raisbeck *et al.* 1990). Whereas ^{10}Be is deposited within 2 yr, ^{14}C spends 30–40 yr in the atmosphere before being absorbed into trees. The curve for ^{14}C therefore lags behind that for ^{10}Be, and the 11-yr activity cycle cannot readily be picked up; but again, long-term modulation can be and is found. Since abundance anomalies in ^{14}C affect radiocarbon dating, they have been investigated in some detail (Stuiver and Braziunas 1989).

The records extend back 9000 yr: superposed on a long-term trend, plausibly associated with variations in the geomagnetic field, there are short-lived positive anomalies corresponding to grand minima in solar activity, which appear to occur

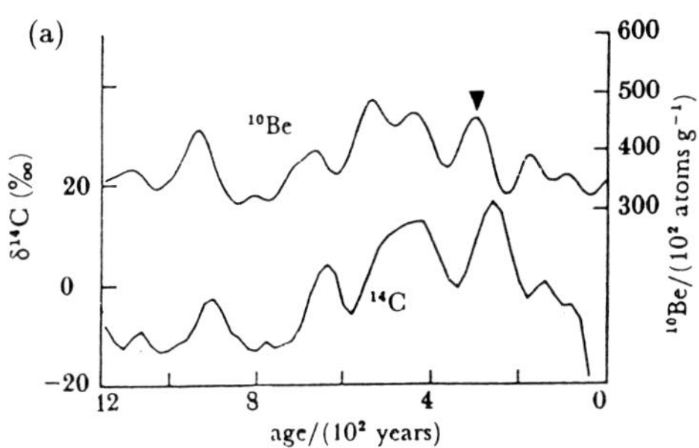

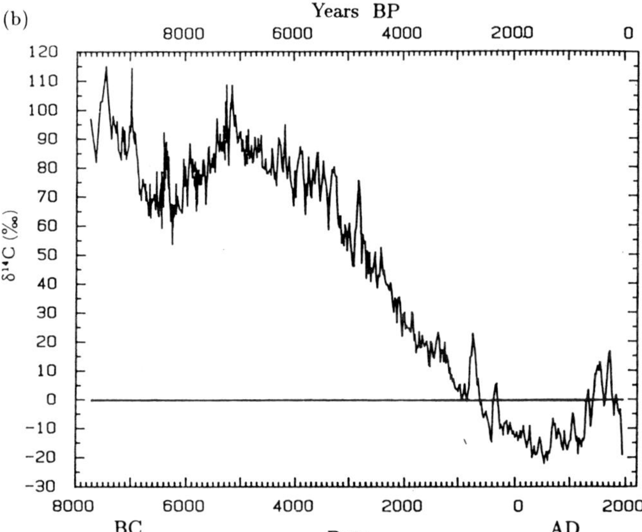

FIG. 8.6. (a) Variations in ^{10}Be and ^{14}C production over the past 1200 yr. The triangle marks the Maunder minimum. (After Raisbeck *et al.* 1990.) (b) Fluctuations in ^{14}C abundance over the last 9500 yr. (After Stuiver and Braziunas 1988, 1989.)

aperiodically, with a characteristic time-scale of about 200 yr (Damon and Sonett 1991).

Parallel to this, Baliunas and Jastrow (1990) have plotted the Ca$^+$-emission from a group of weakly active stars (Fig. 8.7). The distribution is bimodal, suggesting that the stars spend about 30 per cent of the time in a quiescent state, with Ca$^+$-emission at the level found in field-free regions of the Sun at sunspot minimum. Thus the evidence from both the solar record and from other solar-type stars demands that a realistic dynamo model account not only for the

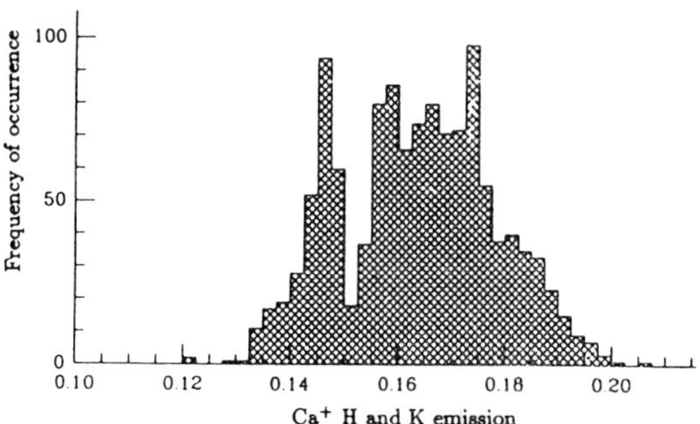

FIG. 8.7. The bimodal distribution of Ca^+ emission in a group of weakly active stars. The two humps may correspond to epochs of cyclic activity and grand minima. (After Baliunas and Jastrow 1990.)

generation of cyclic magnetic fields but also for their aperiodic modulation.

The details of the solar dynamo depend critically on the velocity fields, especially the rotation field. As already remarked, recent helioseismological measurements have led to a radical reappraisal (Goode *et al.* 1991). Five-minute oscillations at the solar surface, corresponding to high-degree acoustic modes, show rotational splitting which can be used to determine the variation of Ω with both latitude and depth. The standard interpretation of the results is that the surface differential rotation persists throughout the convection zone, with $\Omega \approx \Omega(\theta)$, but there is an apparent rapid transition to a uniform rotation in a layer (the 'tachocline') of thickness $\leq 5 \times 10^9$ cm $= 0.07\ R_\odot$ at the interface between the convective and radiative zones. Within the layer $|\partial\Omega/\partial r|$ is large, with $\partial\Omega/\partial r$ positive at the equator and negative at the poles.

It should be stated that some workers question whether these conclusions can be regarded as definitive. Sekii *et al.* (1995) suggest that the technique by which the spectral line measurements are inverted so as to yield $\Omega(r, \theta)$ is sensitive to the choice of base functions, and that there may be a built-in tendency to infer a weak r-dependence in the convective zone. The ambiguities will hopefully be resolved by analysis of more extensive new observations expected in the next few years. A more recent set of measurements by Tomczyk *et al.* (1995) confirm that near the base of the convective zone, the solar rotation profile does indeed undergo a rapid transition from a surface-like differential rotation to a rotation independent of latitude; and that below the convective zone base down to $0.2\ R_\odot$, the measurements are consistent with rigid rotation at a rate somewhat lower than the surface equatorial rate. Further confirmation comes from the GONG results (Fig. 8.8).

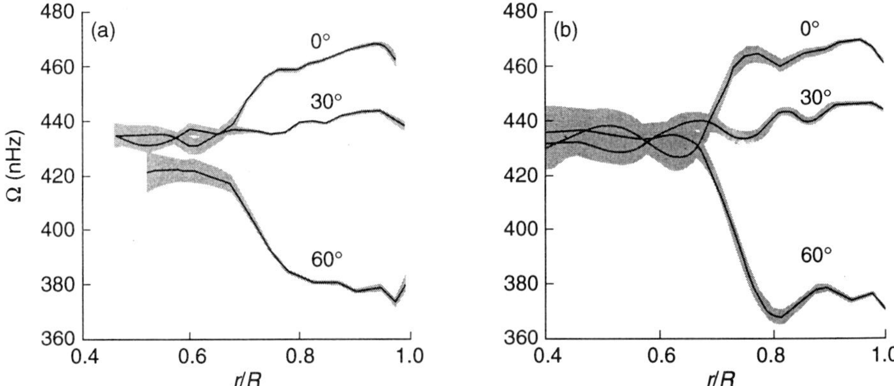

FIG. 8.8. The GONG results for the rotation rate at latitudes 0°, 30° and 60° as a function of solar radius, as given by two different inversion techniques. Formal error bars (±1 standard deviation) are indicated by the shaded regions. Both inversions indicate that the surface-like differential rotation persists through the bulk of the convection zone, with a transition near the base of the convection zone to a flow consistent with latitudinally independent rotation. Reprinted with permission from (Thompson *et al.* 1996.) Copyright 1996, American Association for the Advancement of Science

8.5 Phenomenological studies of the solar dynamo

A subphotospheric convection zone as in the Sun and other late-type stars appears prima facie as an obvious locale for the operation of an $\alpha\Omega$ turbulent dynamo (Sections 6.3, 6.5). The non-uniform rotation observed at the surface is consistent with there being a significant shear extending through the whole zone, so allowing for the generation of a toroidal field by the Ω-effect; while the spontaneous formation of convective cells is just what is required for the conversion of toroidal flux back into poloidal, as described in Parker's seminal (1955) paper and now thought of as a clear illustration of the α-effect.

Parker's ideas became strikingly relevant following H.D. Babcock's discovery (1959) that the general solar magnetic field appears to be reversing along with the sunspot cycle. H.W. Babcock (1961) produced an illuminating phenomenological description of the solar cycle, based on his magnetograph observations. The model was subsequently developed by Leighton (1964, 1969). The basic ideas link up with much of Parker's discussion. The non-uniform rotation of the solar convection zone, acting on a weak dipolar poloidal field, generates a strong subphotospheric toroidal field. The toroidal field at some epoch becomes unstable and so pops up through the surface to produce bipolar active regions, including sunspots. The toroidal field is twisted by Coriolis forces to yield a poloidal field opposite to that originally there; this reversed field cancels out the initial field, so starting a new half-cycle. In this picture, the α-effect manifests itself in the axial tilt of the bipolar active regions – the angle γ introduced above.

The Babcock–Leighton model has in recent years been revived in a modified form by observationally-oriented solar physicists (Wang and Sheeley 1991; Wang *et al.* 1991). The most important change is the introduction of a large-scale *laminar circulation*, superposed on the granulation, mesogranulation and supergranulation components of the turbulence. The authors (Wang *et al.* 1989) argue for a *poleward* motion at the surface of about 10^3 cm s^{-1}, in order to explain the newer observations, in particular the distribution of polar fields around sunspot minimum, and the poleward surges picked up in latitude–time plots of the photospheric field. The recent GONG helioseismological measurements (Hathaway *et al.* 1996) do in fact yield a surface motion of 2×10^3 cm s^{-1}. By mass conservation there must be a return flow, equatorwards, below the surface.

An important feature of the original Leighton model is the actual loss of toroidal flux from the Sun, an assumption criticized by Parker (1984), as it would require eruptions of bipolar magnetic regions to be strongly correlated over 360°. In the more recent work, there is no such loss term in the equations: the main effect of the eruption and transport of flux at the solar surface is the topological rearrangement of the subsurface field through reconnection.

We now summarize the model of Wang *et al.* Spherical polar coordinates (r, θ, ϕ) are used with the colatitude θ again defined with respect to the rotation axis and the longitude ϕ increasing westward (the direction of the solar rotation). The magnetic field vanishes below a radius R_h. The toroidal field component B_ϕ is confined to the subsurface layer $R_h < r < R_h + h$, within which it is a function just of θ and t. The thickness h is small compared with R_h (only a moderately good approximation for the solar convection zone). Observation indicates that at the solar surface $R_\odot$, $\mathbf{B}$ is nearly radial, so $B_\theta(R_\odot)$ is taken as zero. In line with recent helioseismological measurements (e.g. Goode *et al.* 1991), the angular velocity is taken as $\Omega(\theta)$. The component B_θ must then also be small outside the subsurface layer, otherwise the purely θ-dependent shear would again build up a large B_ϕ. In the layer, the horizontal meridional velocity component v_θ is taken as a function of θ alone and directed equatorwards; at the surface, it is directed polewards. The vertical component required by continuity will be small in a shallow zone and so is neglected.

The *longitudinally averaged* magnetic field $\mathbf{B}(r, \theta, t)$ obeys the induction equation in the form

$$\frac{\partial \mathbf{B}}{\partial t} = \nabla\times(\mathbf{v} \times \mathbf{B} - \eta\nabla\times\mathbf{B}) + \mathbf{S}. \tag{8.22}$$

The turbulent resistivity η (cf. Sections 3.8 and 6.4) is given a constant value in the layer, but vanishes below it. At the surface it is identified with the effective diffusivity due to the observed supergranular motion. The vector $\mathbf{S}$ is a divergence-free source term which describes the eruption of bipolar active regions at the surface. Because BMR eruptions do not reduce the net toroidal flux, S_ϕ

is taken to be zero for all r; at the surface, S_θ is also zero.

After suitable radial averaging there result the following equations. For the surface field $B_s(\theta,t) \equiv B_r(R_\odot,\theta,t)$

$$\begin{aligned}\frac{\partial B_s}{\partial t} &= -\frac{1}{R_\odot \sin\theta}\frac{\partial}{\partial\theta}[\sin\theta\, v_s(\theta) B_s] \\ &+ \frac{\kappa_s}{R_\odot^2 \sin\theta}\frac{\partial}{\partial\theta}\left(\sin\theta\frac{\partial B_s}{\partial\theta}\right) + S(R_\odot,\theta,t),\end{aligned} \tag{8.23}$$

where $v_s(\theta)$ and κ_s are respectively the meridional flow velocity and the supergranular diffusion rate at the solar surface. The divergence condition yields the relation between B_s and B_m, the radially averaged B_θ:

$$B_m(\theta,t) = -\frac{R_\odot^2}{hR_h \sin\theta}\int_0^\theta B_s(\theta',t)\sin\theta' \mathrm{d}\theta'. \tag{8.24}$$

The ϕ-component of (8.22) yields for the radially averaged B_ϕ:

$$\begin{aligned}\frac{\partial B_\phi}{\partial t} &= \sin\theta\frac{\mathrm{d}\Omega}{\mathrm{d}\theta}B_m(\theta,t) - \frac{1}{R_h}\frac{\partial}{\partial\theta}[v_h(\theta)B_\phi] \\ &+ \frac{\kappa_h}{R_h^2}\frac{\partial}{\partial\theta}\left[\frac{1}{\sin\theta}\frac{\partial}{\partial\theta}(\sin\theta B_\phi)\right],\end{aligned} \tag{8.25}$$

where $v_h(\theta)$ and κ_h represent the meridian flow velocity and the diffusion rate within the subsurface layer. The source term adopted is similar to that used by Leighton (1969):

$$S(R_\odot,\theta,t) = \frac{1}{\sigma}\left(\frac{hR_h}{R_\odot^2}\right)\frac{1}{\sin\theta}\frac{\partial}{\partial\theta}(B_\phi^3\cos\theta), \tag{8.26}$$

where the parameter σ specifies the rate at which flux erupts at the solar surface.

The authors adopt the empirical surface rotation profile of Newton and Nunn (1951): $\Omega(\theta) = 13.39 - 2.77\cos^2\theta$ deg day^{-1}. From their earlier modelling of the Sun's large-scale field, they take a supergranular diffusion rate of $\kappa_s = 6 \times 10^{12}$ cm^2 s^{-1} and a poleward surface flow given by $v_s = 10^3 \sin\theta|\cos\theta|^{0.01}$ cm s^{-1}. For the subsurface meridional flow they take $v_h(\theta) = v_0 \sin^p\theta|\cos\theta|^q$. The quantities σ, κ_h, v_0, p, q are free parameters of the model. The value $R_h = 0.7R_\odot$ is chosen, motivated by the helioseismological data. The choice $h = 0.1R_\odot$ keeps the toroidal field layer thin but leaves an uncomfortably large gap $R_h + h < r < R_\odot$ which makes no contribution to the the above space-averaged equations.

The model incorporates a number of processes that conspire to maintain the oscillations of the solar magnetic field. In an illustrative case, the values $v_0 \approx 10^2$ cm s^{-1} and $\kappa_h \approx 10^{11}$ cm^2 s^{-1} yield a 22-yr oscillation period. As in the earlier work, the toroidal field, generated below the surface by the rotational shear of the poloidal field, erupts in the form of BMRs; again the Coriolis force yields the

α-effect, shown by the axial tilt of the BMRs. The erupted flux acts to reverse the meridional field. The reversal of the polar field is due to the combined action of the surface transport processes: the diffusive annihilation of leading polarity flux around the equator leaves a surplus of trailing polarity flux, which is carried to the poles and concentrated there by the surface meridional flow. Implicit is the assumption that flux from the changing poloidal field is advected down by the part of the circulation linking the surface flow with that in the toroidal field layer.

The subsurface flow that steadily convects toroidal flux from midlatitudes towards the equator appears to be essential for the maintenance of globally periodic oscillations in the absence of a radial Ω-gradient. There are two other important differences from the Babcock–Leighton models: with the dropping of the assumption of actual expulsion of toroidal flux from the Sun, there is clearly no place for a critical toroidal field strength below which eruptions cannot occur.

Figure 8.9 displays strikingly the predictive power of the model. Clearly, no phenomenological model can be a substitute for a full solution of the magnetohydrodynamic equations. However, the Babcock–Leighton approach has been a great stimulus through its linkage of observation with qualitative concepts from MHD leading to the approximate quantification just summarized. We now return to a more formal, deductive approach to the solar-type dynamo. The difficulties that beset application of mean-field theory to real stars have been discussed in Section 6.7: they have led some theorists to work towards a synthesis with the phenomenological models (see the next section).

8.6 The solar dynamo revisited

It is recognized that the simplest treatment of the α-effect by mean-field electrodynamics (Section 6.4) makes approximations that are not applicable to the Sun. However, the demonstrated link-up (6.56) between non-zero helicity and the α-parameter illustrates convincingly how the alignment of small-scale current loops, consequent on the non-mirror symmetry of the turbulence in a rotating system, enables a large-scale poloidal field to be maintained. It is very plausible that a qualitatively similar effect should occur in general in rotating convective zones; but as is clear from Section 6.7, the dynamical basis of the process needs further elucidation.

8.6.1 *Historical resumé*

A complete theory should be able to infer from the basic equations both the dynamo-built field and the driving motions, in particular the non-uniform rotation field. There have in fact been studies of the dynamics of the convection zone which treat the dynamo problem from first principles, i.e. without introducing any explicit α-effect. Gilman (1979) used the Boussinesq approximation to study global convection by giant cells in a rapidly rotating annulus heated at its inner boundary. He found that Ω is approximately constant on cylinders (the Taylor–Proudman result – cf. Section 4.2), so that the observed surface rotation implies $\partial\Omega/\partial r > 0$ throughout the zone. A small seed magnetic field can

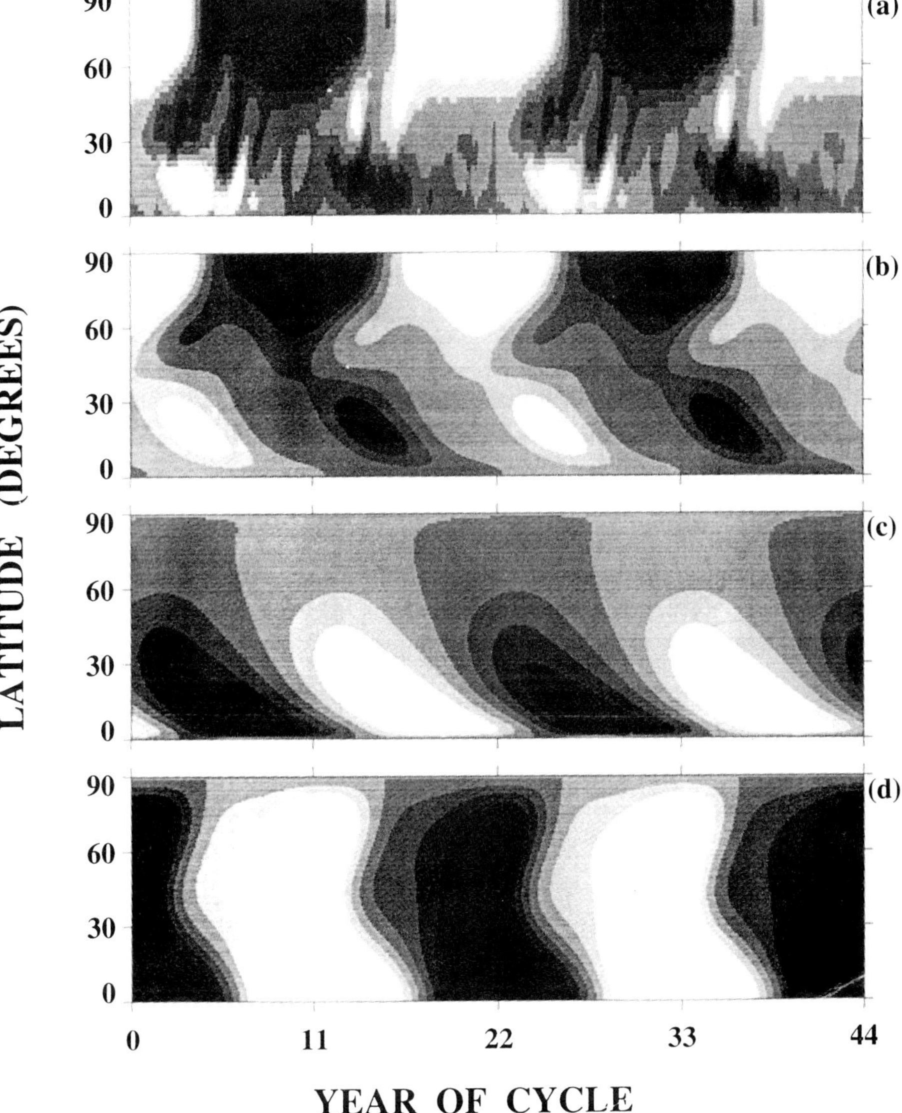

FIG. 8.9. Comparison of predictions of the Wang *et al.* solar dynamo model with observation. (a) A latitude–time plot of the line-of-sight photospheric field measured by the Wilcox Solar Observatory (Stanford) during 1976–97, plotted twice in succession. (b) In the same format, the calculated evolution of the radial (surface) field component. (c) The calculated evolution of the subsurface toroidal field component. (d) The calculated evolution of the subsurface meridional field component. Black indicates strong negative polarity, white strong positive polarity. (Courtesy of Y.-M. Wang.)

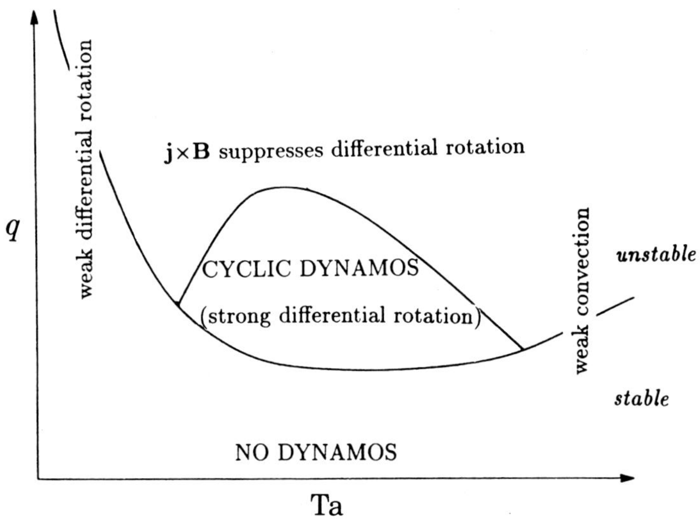

FIG. 8.10. Schematic representation of Gilman's non-linear dynamo calculations in the Ta/q plane.

grow, developing dynamo waves which, however, again propagate towards the poles rather than the equator (Gilman 1983). Calculations using the anelastic approximation yield essentially the same results (Glatzmaier 1985; Gilman and Miller 1986). This is in agreement with the results quoted in Section 6.5: in the language of mean-field theory, the computed values of α turn out to be positive in the northern hemisphere, and with Ω increasing outwards this yields poleward migration, contrary to the observed migration of the sunspot zones, a conclusion sometimes called the 'dynamo dilemma' (Parker 1987*a*).

Gilman's 1983 calculations, summarized in Fig. 8.10, followed the effects of varying the diffusivity and the rotation. He used as parameters the ratio q of thermal to magnetic diffusivity and the Taylor number $\mathrm{Ta} \equiv 4\Omega^2 R_{\odot}^2/\nu_t^4$ where ν_t is the viscosity (interpreted subsequently as a turbulent viscosity). The Rayleigh number, determining the heat flux, is kept fixed. As q is increased, dynamo action sets in as a magnetic instability at a critical value that depends on Ta. At low Ta, the differential rotation is too weak for dynamo action to occur; at higher Ta, convection is inhibited by rotation and the critical value of q rises. There is a restricted parameter range in which differential rotation is strong and so non-linear cyclic dynamos appear. As q is increased further the Lorentz forces suppress the differential rotation, so regular cyclic behaviour disappears.

With the limited computing power then available, the diffusivities adopted in the pioneering Gilman–Glatzmaier work were not only isotropic but had to be given large macrophysical values. The next steps generalize this macroscopic parametrization of the effects of the smaller-scale motions by taking account of the essential anisotropy and inhomogeneity of the Reynolds stresses, leading to the 'Λ-effect' discussed in Section 4.2. and the prediction (rather than the arbitrary assumption) of differential rotation. We recall that a major difficulty is then to achieve an Ω-field consistent with both the toroidal and poloidal dynamics and with the helioseismological observations (cf., for example, Brandenburg *et al.* 1992). The cited study by Rüdiger and Kitchatinov (1994) is very impressive in its ability to reproduce the currently accepted solar rotation law. It can, however, be argued that this should be regarded as using observation to learn something important about the turbulence: in the present case, that the modest deviations from strict adiabaticity are crucial. Much ongoing research on the solar dynamo is quite reasonably content to accept the inferred Ω-field as kinematic input. As noted in Chapter 4, something like a new consensus seems to have arisen, locating the Ω-effect part of the dynamo in the tachocline layer, in which the θ-dependent Ω-law in the convective zone goes over into a near uniform law in the radiative core.

There had in fact already been arguments in favour of locating the solar dynamo deep within the convective zone. Parker (1975) pointed to the danger that magnetic buoyancy (cf. Section 4.7) would cause the toroidal flux tubes generated by shear to rise through the convective zone at almost the local Alfvén speed. Only in the deepest part of the convective zone could the estimated time of rise be long enough for the shear to generate the necessary strong toroidal field strength before buoyancy took over. As noted in Section 6.6, Durney and Robinson (1982) used the rate of loss of toroidal flux, given by the balance of buoyancy against frictional drag (Parker 1979) to derive a simple law for the non-linear limitation of the growth of dynamo-built flux. Several other authors (Spiegel and Weiss 1980; Golub *et al.* 1981; Galloway and Weiss 1981; van Ballegooijen 1982) were led rather to postulate a shell dynamo in the domain of convective overshoot, at 2×10^{10} cm depth. It was also argued that this could help to resolve the dynamo dilemma; for both the global convective models of Gilman and Glatzmaier and an earlier model of Yoshimura (1972) predict that α in the northern hemisphere changes its sign to negative at the bottom of the convective zone, essentially because convergent flows dominate at the bottom of the convective cells.

As noted, the helioseismological observations gave some support to the location of at least one-half of the dynamo process near the core envelope interface. Somewhat ironically, quite early on a debate began on whether the magnetic buoyancy problem was as severe as suggested by Parker's original arguments. Unno and Ribes (1976) and Schüssler (1977) were able to increase the rise time of toroidal flux tubes by including a turbulent frictional drag. As against this, Moreno-Insertis (1983) argued that the actual superadiabatic stratification would act to reduce the rise time. Parker (1987*b*) modified the original simple picture of magnetic buoyancy by noting that a strong toroidal field in the convective zone would interfere with the flow of heat, causing a pile-up of heat below the layer of the field and a cool shadow above the layer; he argued that the weight of the cool shadow 'goes a long way towards cancelling the magnetic buoyancy'.

The contrary effect of downward concentration of flux through 'topological pumping' had been introduced by Drobyshevski and Yuferev (1974). They recalled that in Bénard cell convection, the hot fluid rises in a plume at the centre of a cell, while the cooler fluid descends along the sides of the cells. In the *three*-dimensional motion of a highly conducting fluid, magnetic flux is easily transported downwards, whereas only closed loops, with zero net flux, are carried up. In an Appendix to their paper, Moffatt demonstrates a similar though weaker downward concentration of flux even in turbulent fluids of low magnetic Reynolds number. The same effect should appear with any flow topologically similar to Bénard convection. The process can be regarded as another example of the flux expulsion picture treated by Parker and by Weiss – cf. Section 4.3. Petrovay (1991) applied the same argument to the convective overshoot domain, where it is the downward flow that is expected to be in isolated regions while the upward flow is in a three-dimensional linked domain.

Application of results derived from a picture of laminar convection to a domain of developed turbulence can only be suggestive. However, our picture of the motions in a convective zone – even without account being taken of the back-reaction of the Lorentz force exerted by a dynamo-built field – is becoming more complex (see the discussion in Section 4.5). In particular, numerical simulations by different groups of convection in strongly stratified media, though clearly differing on important questions, significantly agree that there are widely separated, concentrated downdrafts, with velocities several times those typical of the standard mixing-length convection models, and which overshoot the notional base of the convective envelope. Their effect on a weak and so passive magnetic flux will again be to pump it downwards. And in fact in Section 4.7, some recent magnetoconvective computations were cited which give support to the picture in which the turbulent downward pumping of flux generated in the convection zone dominates over spontaneous buoyancy. In some video animations of MHD turbulence, for example of a Cartesian box located at the bottom of the convective zone, the strong tendency for field to be sucked by the downdrafts is manifest, whereas the upward motion of buoyant flux tubes is hardly detectable (Brandenburg and Tuominen 1991). However, once sufficient flux has been concentrated into the overshoot domain, there will arise the counteracting instability of toroidal fields that contribute significantly to support against gravity (also, somewhat misleadingly, referred to as 'magnetic buoyancy' – cf. Section 4.7). One can thus picture a dynamic equilibrium, with downward turbulent pumping on average balanced by eruption of toroidal flux from the overshoot domain and its subsequent rise to the surface.

8.6.2 *Current ideas on the solar dynamo*

As discussed in Section 6.6, both the strong turbulent resistivity and the classical α-effect, expected when the field is weak, will surely be affected as the field approaches the equipartition strength and may be much reduced. This is the motivation behind the Parker (1993) model summarized in Section 6.5.2, in which the dynamo is an $\alpha\Omega$ surface wave on the interface between the convective zone and the shear (overshoot) layer. In the convection zone, the field is weak enough for the conventional α- and η-values to be adopted, but the shear is taken as zero, whereas in the shear layer a zero α and a much weaker resistivity n are adopted. An illustrative example that yields a 22-yr period has $\eta = 10^{11}$–10^{12} $\mathrm{cm^2\,s^{-1}}$, $\alpha = 2 \times 10^2$ $\mathrm{cm^2\,s^{-1}}$ in the convective zone, and $n = 0.01\eta$, $\mathrm{d}v/\mathrm{d}r = 10^{-5}$ $\mathrm{s^{-1}}$, i.e. 3×10^4 $\mathrm{cm\,s^{-1}}$ across the 3×10^9 cm depth of the shear layer.

Difficulties arise with this tentative model when one tries to relate it to the sunspots, the surface manifestations of the dynamo (Parker 1994). One again pictures the bipolar active regions arising when a toroidal field that has buckled into the form of a Greek Ω penetrates the photosphere. Recall the discussion of Section 4.8.1, which gives several reasons why the toroidal field loops that are able to reach the surface as coherent structures should have field strengths $\simeq 10^5$ G – in order to have a short enough eruption instability time-scale, to withstand tangling by the turbulence, and to avoid too large a poleward Coriolis deflection of the rising loops and too big a tilt of a sunspot pair with respect to the East–West direction. This immediately raises the question of how such a strongly super-equipartition field can be built up. To produce a field of this strength by the observed non-uniform rotation in say 5 yr would require B_{p} to be 10^2 G, and so would yield overwhelming Maxwell stresses $B_{\mathrm{p}}B_{\phi}/4\pi$ which would destroy the shear. The maintenance of the shear is a problem that has to be faced; prima facie, it seems that one can make the magnetic forces large enough in one domain, at the cost of making them much too large elsewhere, and vice versa!

Parker (1994) suggests instead that a toroidal field strength $\approx 10^3$ G in the overshoot layer is indeed a correct estimate, but he nevertheless expects some slow upward buoyant leakage of azimuthal flux, yielding a mean azimuthal field of $\simeq 10^2$ G in the convective zone, and described by the small diffusion coefficient $n \ll \eta$ adopted for the overshoot layer in his model. He then notes that deep in the convection zone, even a field of 0.5×10^5 G has a pressure smaller than the ambient thermal pressure by a factor $\simeq 1.7 \times 10^{-6}$; hence a reduction Δp of the thermal pressure within a flux loop by 1 part in 10^6 will lead to a lateral compression that builds up the field to 0.5×10^5 G. He argues that following the formation of an Ω-loop, the difference between the pressure of the adiabatically evolving gas within the vertical arms of the tube and that of the superadiabatic ambient gas will ensure that a toroidal flux tube with a mean field of say 2×10^3 G will spontaneously convert into a 'fibril field' of widely separated flux bundles of 10^5 G. Such a field structure is an essential element in Parker's sunspot model (e.g. Priest 1982).

Parker's 'Ω-amplification' argument has been criticized by Moreno-Insertis *et al.* (1995) as being not consistently formulated. They find that amplification is small unless one comes very close to the 'explosion height' (cf. Section 4.8.1), where the lateral pressure equilibrium is breaking down, with the subsequent evolution remaining conjectural. However, let us sidestep this for the moment and follow the argument further. Parker links up with the essentially kinematic

Babcock–Leighton models, as developed by Wang *et al.* (Section 8.5). If the primary processes of the dynamo – the shearing and the α-effect – were both deep-seated, then the observed solar phenomena would be valuable diagnostic tools, but secondary effects, epiphenomena in Cowling's words. The essence of the Babcock–Leighton–Wang *et al.* picture is rather that the α-effect is observed at the solar surface as the combination of diffusive annihilation at the equator of the leading flux of bipolar magnetic regions plus the poleward migration of the trailing flux. Parker (1994) suggests that although there may still be a mean-field-type α-effect operating on a weak toroidal field in the lower convective zone, a modified Babcock–Leighton picture is applicable. Following Spruit *et al.* (1987) and Wilson *et al.* (1990), he argues that the Ω-loops pinch off at their base through rapid reconnection, so forming free O-loops that can be easily rotated by Coriolis force into meridian planes, there to regenerate the poloidal field. A similar sequence is incorporated into the model of Vainshtein *et al.* (1993), summarized in Section 6.8.2. One suspects that a large-scale laminar meridian circulation, as postulated by Wang *et al.*, may be required to keep the cycle operating, e.g. to transport the newly generated poloidal field into the tachocline, where the dominant rotational shearing effect continues to operate.

Other workers (e.g. Nordlund *et al.* 1994) also argue for the revival of an updated Babcock–Leighton solar dynamo as in the Wang *et al.* picture, with transfer of at least an essential part of the α-effect to the visible regions. Equally, this is questioned by others, who continue to regard the surface phenomena just as diagnostics of a deep-seated dynamo process. In particular, Tobias (1995, 1996) develops further the same (1993) Parker kinematic model, described in Subsection 6.5.2, with its separation of the strong shear and strong α domains, but he now introduces the macrodynamic Lorentz force (the Malkus–Proctor effect) as a dominant non-linear process. The key parameters are as usual the appropriately defined dynamo number D – taken as negative so as to ensure the migration of the dynamo waves from pole to equator – and the magnetic Prandtl number $\mathrm{Pr_m} = \nu/\eta$, which is initially given the low value 0.02. Dipolar symmetry is again imposed. At a critical value of $-D$, dynamo action starts with the generation of a periodic field. A further increase in $-D$ leads to transition at a secondary bifurcation into doubly periodic solutions, with weak modulation of the average magnetic energy on a time-scale much longer than that of the basic oscillation. Further increase of $-D$ leads to solutions which spend long periods with much reduced mean magnetic energy – results similar to to those of Weiss *et al.* (1984) and Jones *et al.* (1985) (Section 6.6.2), and again reminiscent of the Maunder minimum (Fig. 8.11). In the low-$\mathrm{Pr_m}$ domain in which the computations are carried out, the modulation time-scale varies like $\mathrm{Pr_m}^{-1/2}$. At still larger values of $-D$, there is again a transition to chaotic behaviour.

There are now a growing number of parallel studies which return to first principles so as to reconstruct the fundamental vector $\mathcal{E} = \overline{\mathbf{v}' \times \mathbf{B}'/c}$ – which contains both the α-effect and the turbulent resistivity (cf. Section 6.4) – but for more realistic turbulent fields. They emphasize that the density stratification

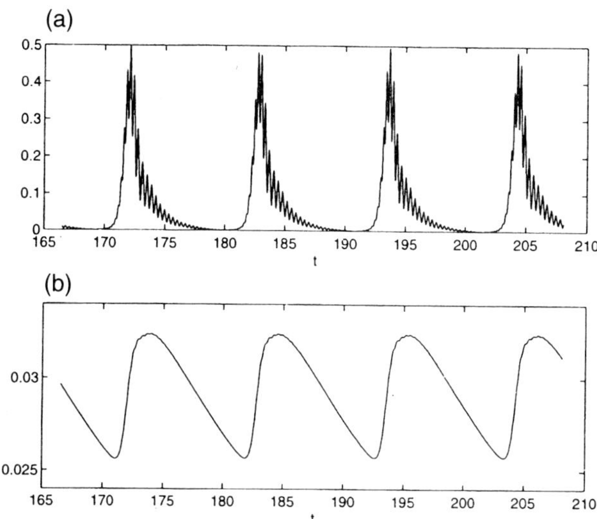

FIG. 8.11. The Parker kinematic model, modified by the macrodynamic Lorentz force. Doubly periodic solutions for $\tau = 0.01$ and $D = -1500$: (a) time series for average magnetic energy density, showing periods of reduced magnetic activity; (b) as for (a), but average kinetic energy. (From Tobias 1996.)

and the inhomogeneity in the intensity of the turbulence both contribute to the α-effect (e.g. Ferriere 1992, Rüdiger and Kitchatinov 1993). The isotropic α-tensor is replaced by a tensor with equal horizontal components but with the vertical component of opposite sign. Rüdiger and Brandenburg (1995) apply such a model to construct a dynamo operating in the overshoot layer, where the toroidal α-component – crucial for the $\alpha\Omega$ dynamo – is negative, as had indeed been suggested earlier. Their model is able to reproduce a cyclic dynamo with a realistic butterfly diagram. To obtain a 22-yr period, the turbulent energy density in the layer has to be diluted by the factor $\epsilon \simeq 0.2$-0.5 – a measure of the reduced interaction between the turbulent and magnetic fields, due to intermittency.

This work is in the spirit of classical dynamo theory, in which it is the non-magnetic forces that drive the motions which yield the α-effect. We recall also that of Ferriz-Mas *et al.* (1994), summarized in Section 6.8.1: the Coriolis force, acting on the spontaneous non-axisymmetric instability of a strong toroidal field, can lead to growing helical waves that yield a new α-effect. However, this strikingly new process does depend on there being an active standard dynamo, able to build up the required strong toroidal field.

One can also raise at least a query over the reality of the 'dynamo dilemma'. Is the prediction by some models of a poleward migration of the dynamo wave – contrary to the observed equatorwards migration of the sunspot zones – necessarily a disaster? Recall that the phenomenological models of Wang *et al.*

(1989, 1991) have a large-scale laminar meridional circulation built in. In a recent study, Choudhuri *et al.* (1995) construct a kinematic dynamo model, again with the shear restricted to the bottom of the convective zone and the α-effect to a layer near the surface, but with the two layers coupled by a circulation with a poleward-directed flow in the upper part and an equatorward flow in the deep layers of the convective zone. They find that even if the dynamo wave would, in the absence of circulation, propagate polewards, the toroidal field belts (which are responsible for the surface activity) nevertheless move equatorwards, yielding butterfly diagrams that are in qualitative agreement with observation. Premature attempts to reach even general agreement with observation can be misleading.

Whatever the details of the ultimate solar dynamo model, there will remain many crucial questions to be answered before the theory can be applied with confidence to other late-type stars. At what value of the mean rotation does the Ω-field go over into a form closer to the Taylor–Proudman constraint $\Omega(\varpi)$; and how does this affect the dynamo-built field? In particular, when does it go over from being quasi-a.c. to quasi-d.c.? In the a.c. domain, what is primarily responsible for the variation of Ω_{cycle} with Ω (cf. Section 6.5)? How does the total flux depend on the mean rotation, and when (if at all) does the dynamo saturate – questions crucial for a definitive rather than a tentative solution to the rotational history problem of Section 8.3? And is the long-term modulation (the Maunder minima, etc.) an example of 'deterministic chaos', resulting naturally from the non-linearity of the equations (e.g. Weiss and Tobias 1997), or is it a consequence of stochastic fluctuations in the convection zone, the preferred explanation of Schmitt *et al.* (1996)?

8.6.3 *The overshoot domain and the radiative interior*

The discussion has concentrated on the dynamics and electrodynamics of the convective envelope. One needs also an all-inclusive treatment of the radiative zone, especially of the tachocline with its inferred strong latitude-dependent $\mathrm{d}\Omega/\mathrm{d}r$. As noted in Section 5.4, even in a slow rotator, a locally large $\nabla\Omega$ can give rise to comparatively fast thermally-driven, Eddington–Sweet-type circulation currents, especially near the interface between convective and radiative zones where the modulus of the temperature gradient is only slightly less than the adiabatic value. In a quasi-steady state, there must again be a balance between the advection of angular momentum by the circulation, the torques exerted by the locally strong magnetic field, and the turbulent viscous torques associated with any overshoot from the convective zone.

An important constraint is supplied by the observed decline of lithium with age in the atmospheres of G stars. Li is destroyed at a temperature of about 2.5×10^6 K in stellar interiors, equal to the temperature at about half a scale-height below the base of the convective zone, and so implying some ongoing slight interchange of material from the convective zone with at least part of the radiative zone. Studies of this problem for both single and binary stars (e.g. Spiegel 1972; Spiegel and Zahn 1992; Zahn 1992, 1994) attempt to link the Li depletion with magnetically-controlled spin-down of the star (Chapter 7): the radiative zone has to adjust to the continual reduction in Ω at the base of the convective zone. As the shearing part of the solar dynamo is believed to be located just below the convective zone, a fully self-consistent treatment of the circulation in the radiative zone – which carries the Li – must include the reaction of the **B**-field on the Ω-field, and possibly also on the thermal field (cf. Sections 5.4 and 9.4).

An alternative possibility exploits again the predicted downdrafts in the convective envelope. Spruit (1997) emphasizes that one needs a mixing process which is mild compared with the mixing in a convective zone, because not all the solar Li observed in the convective zone has been burnt over the solar lifetime; but mild flows are not able to penetrate sufficiently into the stably stratified layers below the convective zone base. Standard models of convective overshoot do not satisfy the requirements. Spruit suggests that a small fraction of the downdrafts manage to conserve the low entropy they acquire at the surface, in spite of heating by thermal diffusion and mixing with the surroundings. They would then continue sinking through the stable layers until they reach the level with the same entropy. Spruit estimates that an entropy contrast corresponding to a density excess of $\delta\rho/\rho \approx 0.1$ would be sufficient for the gas to reach the Li burning depth. The new models of the turbulent convective envelope allow this as a possibility; but to find a stochastic model that definitely predicts the correct (small) fraction of downdrafts that reach the Li burning depth would clearly be a formidable task.

From time to time there are reports in the literature of a rapidly rotating inner core within the solar radiative zone. The uncertainties endemic in the inversion of helioseismological data have already been noted, and it is no surprise that the most recent appraisal concluded the existence of such a core to be at the best unproven. If such a core did exist, it could be interpreted as a relic of the high rotation of the Sun when it arrived on the Zero-Age-Main-Sequence, which has remained uncoupled from the rest of the radiative zone and the convective envelope during the subsequent magnetically-controlled spin-down of the Sun (Chapter 7). The difficulties that this would set for theorists (anticipated in Section 5.5) have been discussed by Mestel and Weiss (1987). If such a shear is to persist for 10^9 yr or more, the upper bounds that this will impose on any magnetic field penetrating this core are very severe. If strict flux freezing is assumed, the field cannot exceed 10^{-6} G. (Recall that the galactic field is estimated at 3×10^{-6} G.) It can be argued that this exaggerates the efficiency of magnetic redistribution of angular momentum; for once the shear has generated a toroidal field that is markedly larger than the poloidal, then the onset of instability (Section 5.3) will yield a large effective macroresistivity, so markedly lengthening the time of build-up of $\mathbf{B}_t$ and the associated torques. But even after making generous allowance for this, the upper limit on B_p is still $\simeq 10^{-4}$. (Conditions in a convective zone are radically different, for there the Reynolds stresses attempt to set up an essentially non-uniform Ω-field, as discussed in Section 4.2.) The extensive new observations expected over the next decade

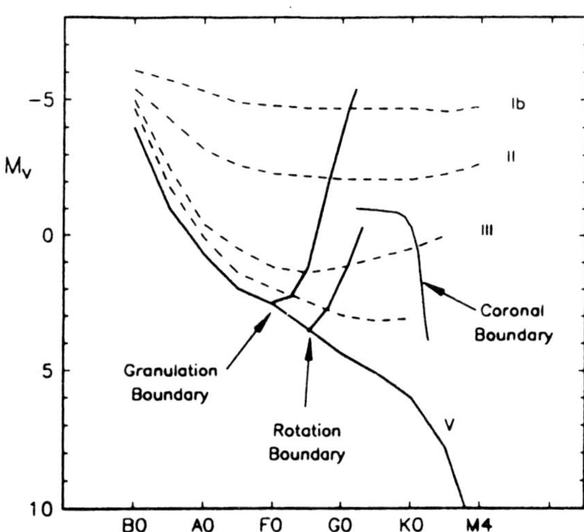

FIG. 8.12. The Hertzsprung–Russell diagram. The granulation boundary, the rotation boundary and the coronal boundary, discussed in the text.

should hopefully decide whether or not there is such a core.

8.7 Evolved stars

Dynamo action qualitatively similar to that in late-type dwarf stars is to be expected also in stars that have evolved into the giant domain. A succinct summary of some of the relevant observations is given by Gray (1991). Figure 8.12 (taken from Gray's paper) is a schematic Hertzsprung–Russell diagram, with three boundary curves delineated. The presence of solar-type, subphotospheric convection shows up in asymmetries in the spectral lines, due to the combination of upward-moving and so blue-shifted brighter material, and downward-moving and so red-shifted cooler material. The line bisector, i.e. the locus defined by the midpoints of horizontal line segments bounded by the sides of the spectral line profile, is used to specify the asymmetry present in a particular line. Solar-type lines have bisectors that sweep redward near the continuum. It is satisfactory that solar-type bisectors are found only in the cooler half of the H–R diagram. In the photospheres of hot stars, there is evidence for other (stronger) photospheric velocity fields, with the slope and curvature of their line-bisectors opposite to the solar-type cases.

The *granulation boundary* is defined by the vanishing of the asymmetry. It corresponds to the onset of deep subphotospheric convection, and significantly to the onset of magnetic activity indicators, as in dwarf stars. Even though the convection zones in stars very near the boundary are very thin, the activity starts up promptly at the granulation boundary, and also switches off promptly when a star evolves from the cooler to the hotter side, as the more luminous stars do during the blue-loop part of their evolution.

The sharp drop in rotation along the main sequence through the early F stars, corresponding to the increase in depth of the convective envelope and correlated with the increase in magnetic activity indicators, was pointed out in classical papers by Wilson (1966) and Kraft (1967). Dwarfs being essentially unevolved stars, the sequence in spectral type is a sequence in mass, whereas the giants define rather a constant-mass sequence in time. For giants of moderate luminosity, the spectral type increases monotonically with time. For single stars, there is a precipitous drop in rotation between types G0 and G3, defining the *rotation boundary* shown in Fig. 8.12. On the hot side of this boundary the mean rotation is large (≈ 60 km s^{-1}) and the distribution of rotations is Maxwellian, whereas on the cool side the rotation is low (typically 6 km s^{-1}) and is a single-valued function of spectral type. Gray argues that the observations point towards his *rotostat hypothesis*, a special form of dynamo-generated magnetic braking. Dynamo action is supposed to cease when the surface rotation falls below a limiting value. He postulates a source of angular momentum below the convective envelope which is tapped as the lower boundary of the convective zone deepens with time. As soon as the surface rotation increases sufficiently, the dynamo starts up and magnetic braking reduces the rotation below the limit. Detailed computations are needed to see if this picture can be made quantitatively convincing. The uncertainties of dynamo theory for stars of structure different from main sequence dwarfs are likely to be a stumbling block. The constraints set by internal magnetic fields on a large internal angular velocity gradient have been noted, e.g. in Section 5.5.

As always, measurements of line broadening yield directly not the rotation velocity v but $v \sin i$, where i is the inclination of the rotation axis to the line of sight. Since for the giants in the G3 to K2 III range the rotation has a unique value for each spectral type, the plausible assumption of a random orientation of rotation axes enables conversion of $v \sin i$ to v for each spectral type, just by multiplication by $4/\pi$; whence for each star, $(v \sin i)/v$ yields i. Magnetic activity indicators can then be plotted as a function of orientation. The clear conclusion is that stars seen pole-on ($i = 0$) are weak emitters, while those seen near equator-on ($i = \pi/2$) are the strongest. This is clearly reminiscent of the strong preference of solar activity for the equatorial regions, suggesting some similarities with the dynamo as we see it operating on the Sun.

The other curve in Fig. 8.12 is the *coronal boundary*. For stars below this curve, coronal indicators such as X-ray emission appear; but in stars which are either too luminous or have too low an effective temperature, none of the usual indicators appear. Gray again interprets this in terms of rotationally generated dynamos: stars outside this boundary just rotate too slowly. Stars of high luminosity rotate below the required limit across the domain to the right of the granulation boundary; this is suggested as due to evolutionary expansion that has greatly increased the stellar moment of inertia. Lower luminosity stars stay near the dynamo limit, until their normal evolution sends them up the asymptotic giant branch, and the corresponding increase in moment of inertia again reduces the rotation below the dynamo limit, so that if Gray's rotostat hypothesis is valid, magnetically maintained coronae cannot develop.

More recent observations – long-term photometry (Jetsu *et al.* 1993) and surface imaging (Piskunov *et al.* 1990, 1994) – suggest strongly that there are large-scale non-axisymmetric structures on active late-type stars, in particular on giant stars. Still in the spirit of exploiting the standard mean-field dynamo equations to their limit, one is led to ask when non-axisymmetric fields can be generated. The $\alpha\Omega$ modes predicted by the standard kinematic equations – with non-uniform rotation the only laminar flow – are preferentially axisymmetric: in linear theory the axisymmetric modes are excited at lower dynamo number than the non-axisymmetric, and also it is the axisymmetric modes that persist in the non-linear regime. For non-axisymmetric modes to be preferred, one has to choose rather special distributions of α and Ω, and with only modest differential rotation. Strong differential rotation inhibits non-axisymmetric field generation, for the persistent winding up leads to accelerated diffusive destruction of flux (cf. Section 9.3).

It is in any case desirable if possible to introduce velocity fields that are not arbitrarily chosen but which emerge from a plausible parametrization of the dynamics of the turbulence. Barker and Moss (1993) again exploited the Λ-effect of Section 4.2. They solved the axisymmetric, purely hydrodynamic problem for a rotating incompressible fluid in a spherical shell, and then used the resulting large-scale velocity fields (differential rotation, meridian circulation) as input into an essentially kinematic, non-axisymmetric dynamo calculation but with a simple α-quenching non-linear term (cf. Section 6.6). For a moderately supercritical dynamo number and a Taylor number $\mathrm{Ta} \approx 10^5$, they found a stable non-linear solution. For Ta smaller, only axisymmetric modes are stable, and with $\mathrm{Ta} = 10^6$, the increasing differential rotation leads again to the favouring of axisymmetric modes.

In the latest study, Moss *et al.* (1995) study a conventional distributed dynamo – i.e. with the α-effect active throughout the (incompressible) convective zone rather than just in the overshoot layer. The Λ-effect again operates to generate differential rotation, but now angular momentum is redistributed not only by the eddy-viscosity terms but also by the back-reaction of the macroscopic Lorentz forces, which act as the non-linear limit on field growth – no α-quenching non-linearity need be introduced. The authors again find that there are wide parameter ranges in which non-axisymmetric fields are the only stable solutions.

There is clearly enormous scope for future work, linking observation with dynamo theory, which we have made clear is still an infant, though a rapidly developing infant.

References

Babcock, H.D. (1959). *Astrophysical Journal*, **130**, 364.
Babcock, H.W. (1961). *Astrophysical Journal*, **133**, 572.

Baliunas, S.L. and Jastrow, R. (1990). *Nature*, **348**, 520.

Baliunas, S.L. and Vaughan, A.H. (1985). *Annual Review Astronomy Astrophysics*, **23**, 379.

Barker, D.M. and Moss, D. (1993). In *The cosmic dynamo* (eds F. Krause, K.-H. Radler and G. Rüdiger), p. 147. Kluwer, Dordrecht.

Biermann, L. (1935). *Astronomische Nachrichten*, **257**, 269.

Bracewell, R.N. (1988). *Quarterly Journal Royal Astronomical Society*, **29**, 119.

Brandenburg, A. and Tuominen, I. (1991). In *The Sun and cool stars* (eds I. Tuominen, D. Moss and G. Rüdiger), p. 223. Springer, Berlin.

Brandenburg, A., Moss, D. and Tuominen, I. (1992). *Astronomy and Astrophysics*, **265**, 328.

Byrne, P.B. (1992). In *Sunspots: theory and observations* (eds J.H. Thomas and N.O. Weiss), p. 63. Kluwer, Dordrecht.

Chandrasekhar, S. and Breen, F.H. (1946). *Astrophysical Journal*, **104**, 430.

Choudhuri, A.R., Schüssler, M. and Dikpati, M. (1995). *Astronomy and Astrophysics*, **303**, L29.

Christensen-Dalsgaard, J., Gough, D.O., and Thompson, M.J. (1991). *Astrophysical Journal*, **378**, 413.

Collier Cameron, A. and Li, J. (1994). *Monthly Notices Royal Astronomical Society*, **269**, 1099.

Collier Cameron, A. and Robinson, R.D. (1989). *Monthly Notices Royal Astronomical Society*, **236**, 57 and **238**, 657.

Collier Cameron, A., Bedford, D.K., Rucinski, S.M., Vilhu, O. and White, N.E. (1988). *Monthly Notices Royal Astronomical Society*, **231**, 131.

Collier Cameron, A., Li., J. and Mestel, L. (1991). In *Angular momentum evolution of young stars* (eds S. Catalano and J.R. Stauffer), p. 297. Kluwer, Dordrecht.

Cowling, T.G. (1938). *Monthly Notices Royal Astronomical Society*, **98**, 528.

Damon, P.E. and Sonett, C.P. (1991). In *The Sun in time* (eds C.P. Sonett, M.S. Giampapa and M.S. Matthews), p. 360. University of Arizona Press.

DeLuca, E.E., Werne, J., Rosner, R. and Cattaneo, F. (1990). *Physical Review Letters*, **64**, 2370.

Donati, J.-F. and Collier Cameron, A. (1997). *Monthly Notices Royal Astronomical Society*, **291**, 1.

Drobyshevski, E.M. and Yuferev, V.S. (1974). *Journal Fluid Mechanics*, **65**, 33.

Durney, B.R. (1972). *Asilomar conference on the solar wind* (eds C.P. Sonnett, P.J. Coleman Jr. and J.M. Wilcox), p. 282. NASA, Washington.

Durney, B.R. and Robinson, R.D. (1982). *Astrophysical Journal*, **253**, 290.

Eddington, A.S. (1959). *The internal constitution of the stars*. Dover, New York. (Cambridge University Press edition 1926.)

Endal, A.S. and Sofia, S. (1981). *Astrophysical Journal*, **243**, 625.

Ferriere, K. (1992). *Astrophysical Journal*, **380**, 286.

Ferriz-Mas, A., Schmitt, D. and Schüssler, M. (1994). *Astronomy and Astrophysics*, **289**, 949.

Galloway, D.J. and Weiss, N.O. (1981). *Astrophysical Journal*, **243**, 945.

Geiss, J. and Bochsler, P. (1991). In *The Sun in time* (eds C.P. Sonnett, M.S. Giampapa and M.S. Matthews), p. 98. University of Arizona Press.
Gilman, P.A. (1979). *Astrophysical Journal*, **231**, 284.
Gilman, P.A. (1983). *Astrophysical Journal Supplement*, **53**, 243.
Gilman, P.A. and Miller, J. (1986). *Astrophysical Journal Supplement*, **61**, 585.
Glatzmaier, G.A. (1985). *Astrophysical Journal*, **291**, 300.
Golub, L., Rosner, R., Vaiana, G.S. and Weiss, N.O. (1981). *Astrophysical Journal*, **243**, 309.
Goode, P.R., Dziembowski, W.A., Korzennik, S.G. and Rhodes Jr., E.J. (1991). *Astrophysical Journal*, **367**, 649.
Gough, D.O., Kosovichev, A.G., Toomre, J., Anderson, E., Antia, H.M., Basu, S. *et al.* (1996). *Science*, **272**, 1296.
Gray, D.F. (1991). In *The Sun and cool stars* (eds I. Tuominen, D. Moss and G. Rüdiger), p. 336. Springer, Berlin.
Hathaway, D.H., Gilman, P.A., Harvey, J.W., Hill, F.. Howard, R.F., Jones, H.P. *et al.* (1996). *Science*, **272**, 1306.
Hayashi, C. (1961). *Publications Astronomical Society Japan*, **13**, 450.
Hoyle, F. and Schwarzschild, M. (1955). *Astrophysical Journal Supplement*, **13**, 1955.
Jetsu, L., Pelt, J. and Tuominen, I. (1993). *Astronomy and Astrophysics*, **278**, 449.
Jones, C.A., Weiss, N.O. and Cattaneo, F. (1985). *Physica*, **14D**, 161.
Kippenhahn, R. and Weigert, A. (1990). *Stellar structure and evolution.* Springer, Berlin.
Kraft, R.P. (1967). *Astrophysical Journal*, **150**, 551.
Leighton, R.B. (1964). *Astrophysical Journal*, **140**, 1547.
Leighton, R.B. (1969). *Astrophysical Journal*, **156**, 1.
Li, J. and Collier Cameron, A. (1993). *Monthly Notices Royal Astronomical Society*, **261**, 766.
Marcy, G.W. (1984). *Astrophysical Journal*, **276**, 286.
Mathys, G. and Solanki, S.K. (1989). *Astronomy and Astrophysics*, **208**, 189.
Mestel, L. and Spruit, H.C. (1987). *Monthly Notices Royal Astronomical Society*, **226**, 57.
Mestel, L. and Weiss, N.O. (1987). *Monthly Notices Royal Astronomical Society*, **226**, 123.
Middelkoop, F. and Zwaan, C. (1981). *Astronomy and Astrophysics*, **101**, 26.
Moreno-Insertis, F. (1983). *Astronomy and Astrophysics*, **122**, 241.
Moreno-Insertis, F., Caligari, P. and Schüssler, M. (1995). *Astrophysical Journal*, **452**, 894.
Moss, D., Tuominen, I. and Brandenburg, A. (1991). *Astronomy and Astrophysics*, **245**, 129.
Moss, D., Barker, D.M., Brandenburg, A. and Tuominen, I. (1995). *Astronomy and Astrophysics*, **294**, 155.
Mutel, R.L. and Morris, D.H. (1988). *Properties of stellar magnetospheres deduced from radio observations of close binaries.* In *Activity in cool star*

envelopes (eds O. Havnes, B.R. Pettersen, J.H.M.M. Schmitt and J.E. Solheim), p. 283. Kluwer, Dordrecht.
Newton, H.W. and Nunn, M.L. (1951). *Monthly Notices Royal Astronomical Society*, **111**, 413.
Nordlund, A., Galsgaard, K. and Stein, R.F. (1994). In *Solar surface magnetism* (eds R.J. Rutten and C.J. Schrijver), p. 471. Kluwer, Dordrecht.
Noyes, R.W., Hartmann, L.W., Baliunas, S.L., Duncan, D.K. and Vaughan, A.H. (1984*a*). *Astrophysical Journal*, **279**, 763.
Noyes, R.W., Weiss, N.O. and Vaughan, A.H. (1984*b*). *Astrophysical Journal*, **287**, 769.
O'Dell, M.A., Panagi, P., Hendry, M.A. and Collier Cameron, A. (1995). *Astronomy and Astrophysics*, **294**, 715.
Osterbrock, D.E. (1952). *Astrophysical Journal*, **118**, 529.
Pallavicini, R., Golub, L., Rosner, R., Vaiana, G.S., Ayres, T. and Linsky, J.L. (1981). *Astrophysical Journal*, **248**, 279.
Parker, E.N. (1955). *Astrophysical Journal*, **122**, 293.
Parker, E.N. (1963). *Interplanetary dynamical process.* Interscience, New York.
Parker, E.N. (1975). *Astrophysical Journal*, **198**, 205.
Parker, E.N. (1979). *Cosmical magnetic fields.* Oxford University Press.
Parker, E.N. (1984). *Astrophysical Journal*, **281**, 839.
Parker, E.N. (1987*a*). *Solar Physics*, **110**, 11.
Parker, E.N. (1987*b*). *Astrophysical Journal*, **312**, 868.
Parker, E.N. (1993). *Astrophysical Journal*, **408**, 707.
Parker, E.N. (1994). In *Cosmical magnetism* (ed. D. Lynden-Bell), p. 123. Kluwer, Dordrecht.
Petrovay, K. (1991). In *The Sun and cool stars* (eds I.Tuominen, D. Moss and G. Rüdiger), p. 67. Springer, Berlin.
Piskunov, N.E., Tuominen, I. and Vilhu, O. (1990). *Astronomy and Astrophysics*, **230**, 363.
Piskunov, N.E., Huenemoerder, D.P. and Saar, S.H. (1994). In *Cool stars, stellar systems, and the Sun* (ed. J.-P. Caillault), p. 658. Astronomical Society Pacific Conference Series, Vol. 64.
Priest, E.R. (1982). *Solar magnetohydrodynamics.* Reidel, Dordrecht.
Raisbeck, G.M., Yiou, F., Jouzel, J. and Petit, J.R. (1990). *Philosophical Transactions Royal Society London A*, **330**, 463.
Robinson, R.D., Worden, P.W. and Harvey, J.W. (1980). *Astrophysical Journal*, **236**, L155.
Rucinski, S.M. (1988). *Astronomical Journal*, **95**, 1895.
Rucinski, S.M. and VandenBerg, D.A. (1986). *Publications Astronomical Society Pacific*, **98**, 669.
Rucinski, S.M. and VandenBerg, D.A. (1990). *Astronomical Journal*, **99**, 1279.
Rüdiger, G. and Brandenburg, A. (1995). *Astronomy and Astrophysics*, **296**, 557.
Rüdiger, G. and Kitchatinov, L.L. (1993). *Astronomy and Astrophysics*, **269**, 581.
Rüdiger, G. and Kitchatinov, L.L. (1994). In *The solar engine and its influence*

on terrestrial atmosphere and climate (ed. E. Ribes), p. 17. Nato Workshop, Paris. Springer, Berlin.

Saar, S.H. (1991). In *The Sun and cool stars* (eds I. Tuominen, D. Moss and G. Rüdiger), p. 389. Springer, Berlin.

Saar, S.H. and Baliunas, N.O. (1992). In *The solar cycle* (ed. K. Harvey) p. 150. Astronomical Society of the Pacific.

Saar, S.H. and Linsky, J.F. (1985). *Astrophysical Journal*, **299**, L47.

Schüssler, M. (1977). *Astronomy and Astrophysics*, **56**, 439.

Schüssler, M., Caligari, P., Ferriz-Mas, A., Solanki, S.K. and Stix, M. (1996). *Astronomy and Astrophysics*, **314**, 503.

Schmitt, D., Schüssler, M. and Ferriz-Mas, A. (1996). *Astronomy and Astrophysics*, **311**, L1.

Schwarzschild, M. (1958). *Structure and evolution of the stars.* Princeton University Press. (Dover edition 1965.)

Sekii, T., Gough, D.O. and Kosovichev, A.G. (1995). In *GONG'94: Helio- and astroseismology from Earth and space* (eds R.K. Ulrich, E.J. Rhodes Jr. and W. Däppen), *Astronomical Society Pacific Conference Series*, **76**, 59.

Skumanich, A. (1972). *Astrophysical Journal*, **171**, 565.

Soderblom, D.R., Jones, B.F. and Walker, M.F. (1983). *Astrophysical Journal*, **274**, L37.

Solanki, S.K., Motamen, S. and Keppens, R. (1997). *Astronomy and Astrophysics*, **325**, 1045.

Spiegel, E.A. (1972). In *Physics of the solar system* (ed. S.I. Rasool), p. 61. NASA, Washington.

Spiegel, E.A. and Weiss, N.O. (1980). *Nature*, **287**, 616.

Spiegel, E.A. and Zahn, J.-P. (1992). *Astronomy and Astrophysics*, **265**, 106.

Spruit, H.C. (1997). *Memorie Societa Astronomica Italiana*, **68**, 397.

Spruit, H.C., Title, A.M. and van Ballegooijen, A.A. (1987). *Solar Physics*, **110**, 115.

Stauffer, J.R. and Hartmann, L.W. (1987). *Astrophysical Journal*, **318**, 337.

Stauffer, J.R., Hartmann, L., Soderblom, D.R. and Burnham, N. (1984). *Astrophysical Journal*, **289**, 247.

Stepien, K. (1993). In *The cosmic dynamo* (eds F. Krause, K.-H. Rädler and G. Rüdiger), p. 141. Kluwer, Dordrecht.

Stix, M. (1989). *The Sun.* Springer, Berlin.

Stuiver, M. and Braziunas, T.F. (1988). In *Secular solar and geomagnetic variations in the last 10,000 years* (eds. F.R. Stephenson and A.W. Wolfendale), p. 245. Kluwer, Dordrecht.

Stuiver, M. and Braziunas, T.F. (1989). *Nature*, **338**, 405.

Thomas, J.H. and Weiss, N.O. (1992). *Sunspots: theory and observation.* Kluwer, Dordrecht.

Thompson, M.J., Toomre, J., Anderson, E.R., Antia, H.M., Berthomieu, G., Burtonclay, D. *et al.* (1996). *Science*, **272**, 1300.

Tobias, S.M. (1995). Ph.D. dissertation, Cambridge.

Tobias, S.M. (1996). *Astronomy and Astrophysics*, **307**, L21.

Tomczyk, S., Schou, J. and Thompson, M.J. (1995). *Astrophysical Journal*, **448**, L57.

Unno, W. and Ribes, E. (1976). *Astrophysical Journal*, **208**, 222.

Unruh, Y.C., Collier Cameron, A. and Cutispoto, G. (1995). *Monthly Notices Royal Astronomical Society*, **277**, 1145.

Unsöld, A. (1930). *Zeitschrift Astrophysik*, **1**, 138.

Vaiana, G.S., Cassinelli, J.P., Fabbiano, G., Giacconi, R., Golub, L., Gorenstein, P. *et al.* (1981). *Astrophysical Journal*, **245**, 163.

Vainshtein, S.I., Parker, E.N. and Rosner, R. (1993). *Astrophysical Journal*, **404**, 773.

van Ballegooijen, A.A. (1982). *Astronomy and Astrophysics*, **113**, 99.

van Leeuwen, F. and Alphenaar, P. (1982). *ESO Messenger*, **28**, 15.

van Leeuwen, F., Alphenaar, P. and Meys, J.J.M. (1987). *Astronomy Astrophysics Supplement*, **67**, 483.

Vilhu, O. (1984). *Astronomy and Astrophysics*, **133**, 117.

Wang, Y.-M. and Sheeley Jr., N.R. (1991). *Astrophysical Journal*, **375**, 761.

Wang, Y.-M., Nash, A.G. and Sheeley, N.R. (1989). *Astrophysical Journal*, **347**, 529.

Wang, Y.-M., Sheeley Jr., N. R. and Nash, A.G. (1991). *Astrophysical Journal*, **383**, 431.

Weiss, N.O. (1994). In *Lectures on solar and planetary dynamos* (eds M.R.E. Proctor and A.D. Gilbert), p. 59. Cambridge University Press.

Weiss, N.O. and Tobias, S.M. (1997). In *Solar and heliospheric plasma physics* (ed. G.M. Simnett). Springer Lecture Notes in Physics.

Weiss, N.O., Cattaneo, F. and Jones, C.A. (1984). *Geophysical Astrophysical Fluid Dynamics*, **30**, 205.

Wildt, R. (1939). *Astrophysical Journal*, **89**, 295.

Wilson, O.C. (1966). *Astrophysical Journal*, **144**, 695.

Wilson, O.C. (1978). *Astrophysical Journal*, **226**, 379.

Wilson, P.R., McIntosh, P.S. and Snodgrass, H.B. (1990). *Solar Physics*, **127**, 1.

Yoshimura, H. (1972). *Astrophysical Journal*, **178**, 863.

Zahn, J.-P. (1992). *Astronomy and Astrophysics*, **265**, 115.

Zahn, J.-P. (1994). *Astronomy and Astrophysics*, **288**, 829.

9

THE EARLY-TYPE MAGNETIC STARS

9.1 The basic observational data

In Chapters 4 and 5 we discussed various aspects of the interaction between the magnetic field and the thermal-gravitational-rotational fields in different stellar domains. In the present chapter we broaden the discussion with an eye to the understanding of those early-type stars which show surface magnetic fields, detectable through the Zeeman effect (cf. Section 1.2). The principal observational properties are as follows. (For summaries of the earlier work, see Babcock 1958, 1960; Ledoux and Renson 1966 and references therein; for more recent summaries, Borra *et al.* 1982; Wolff 1983; Mathys 1989; Landstreet 1992 and references therein.)

(1) The average line-of-sight field – the 'effective field' B_{eff}, determined by spectral line circular polarization measurements – lies between 2×10^4 G and the detectability limit (100–200 G), a typical value being a few hundred gauss.

(2) The effective field, the spectral line strengths and the stellar luminosity all vary with the same period P. Typical periods are 1–10 days; the median of the distribution is 3 days; some periods are as short as half a day, some as long as 100 days, and a few stars show periods of the order of a decade. About two-thirds of the well-studied cases show polarity reversal. Fields measured by photoelectric techniques (more straightforward to model than the older photographic data) show a fairly sinusoidal variation. In a minority of stars – those with very sharp lines and fields of at least 2.8 kG, and now numbering 42 (Mathys *et al.* 1997) – Zeeman splitting measurements yield the mean field B_{s}; this varies with the same period as B_{eff}, but by a factor of less than 2. The early suggestion of a marked correlation between surface field strength and period has not been confirmed by subsequent observations (cf. Fig. 50 in Mathys *et al.* 1997).

(3) All the magnetic stars appear to belong to the class of chemically peculiar (CP) stars of type A or B (Ap or Bp stars). Probably all Ap stars of peculiarity classes Si and Sr–Cr–Eu have observable fields, typically of a few hundred gauss. Until recently, none of the Hg–Mn class nor the related Am (metallic-line) stars showed B_{eff}. However, evidence is beginning to be found for incoherent fields in both classes (Mathys and Lanz 1990; Lanz and Mathys 1993; Mathys and Hubrig 1995). Fields have also been discovered in two classes of Bp stars, namely those with anomalously strong and anomalously weak He lines.

(4) The period is almost certainly a rotation period $P = 2\pi/\Omega$, since only short-period stars ever show rotational broadening, while long period stars invariably have sharp lines. Deutsch (1956) first pointed out that the line widths of Ap stars vary inversely with their periods. Preston (1971) sharpened the ar-

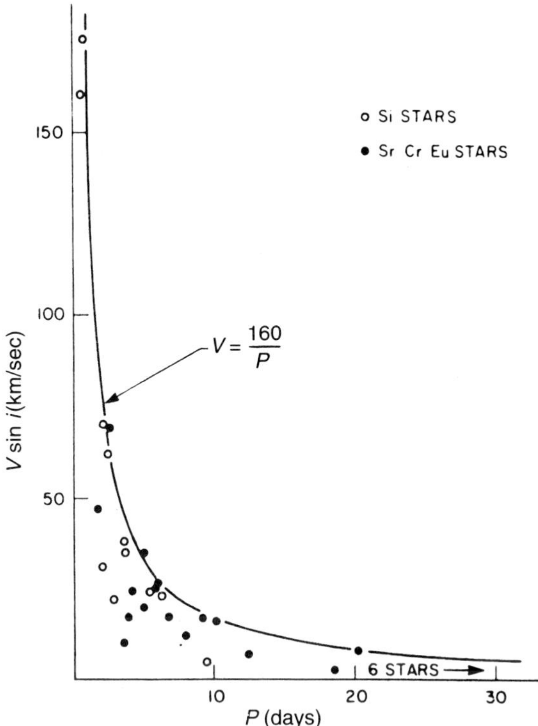

FIG. 9.1. Plot of $V \sin i$ against P. The curve represents the relation between equatorial rotational velocity and period for stars with a radius $R = 3.2R_{\odot}$, and is a reasonable upper envelope for the plotted points. (After Preston 1971.)

gument by plotting against P the spectroscopically-determined $V \sin i$ where V = equatorial speed and i the inclination of the rotation axis to the line-of-sight; the representative points all lie beneath a hyperbola, identified with the curve given by putting $i = \pi/2$ and assuming a constant typical radius R, so that $V = \Omega R = 2\pi R/P$ (Fig. 9.1).

These observational conclusions suggest strongly that the periodic variations are due to a basic magnetic structure, non-symmetric about the rotation axis, that is carried around with the rotating star. The simplest version has an identifiable magnetic axis $\mathbf{p}$ inclined at an obliquity angle χ to the rotation axis $\mathbf{k}$. This phenomenological *oblique rotator* model (Stibbs 1950; Deutsch 1958, 1970; Preston 1970) is adopted as a working hypothesis, often with the additional simplification that the basic magnetic field is symmetric about the axis $\mathbf{p}$ (Fig. 1.2). The observed departure from strict anti-symmetry at reversal can usually be covered by a simple generalization, for example by the displaced dipole model

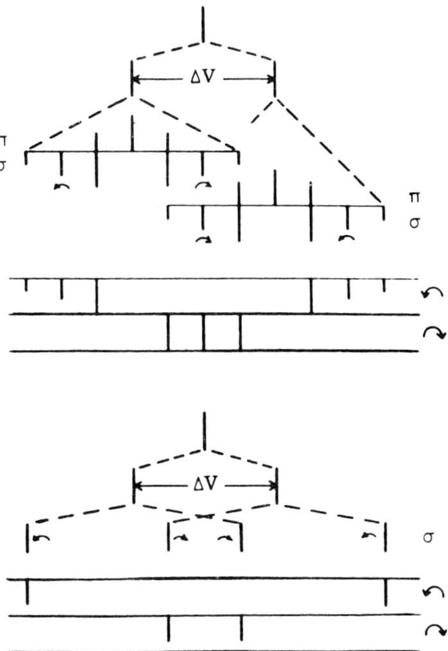

FIG. 9.2. The 'cross-over effect', illustrated for a nearly longitudinal field. The line is first split by the Doppler effect, corresponding to a differential rotational velocity ΔV between the two areas of the star with opposite polarities. If the two areas have longitudinal fields of opposite polarity, the lines are further split into their Zeeman patterns with the polarizations shown, yielding very different widths for the two oppositely polarized components. (After Babcock 1956, Gollnow 1965.)

(Landstreet 1970), though there are at least three maverick objects for which the quadrupole component is too large to be so explained (Thompson and Landstreet 1985; Landstreet 1990; Mathys and Hubrig 1997).

Further support for the oblique rotator model comes from its giving a simple explanation of the 'cross-over effect' (Babcock 1956). This is the difference in widths of the two circularly polarized Zeeman components of a spectral line, which varies with the magnetic period, becoming particularly large when the effective field is becoming small, changing from one polarity to the other (Fig. 9.2).

The angle χ is a parameter adjustable for the particular star under study. The earlier suggestion (Preston 1971) of a marked preference for either small or large χ has not been sustained, and most observers argue for at most a marginal non-randomness in χ (Hensberge *et al.* 1979; Borra and Landstreet 1980). A contrary view is urged by the Potsdam group (Krause and Oetken 1976; Oetken

1977, 1979; Krause 1983), who favour the perpendicular rotator ($\chi = \pi/2$) for theoretical reasons (cf. Section 9.6). The Potsdam group stress that mapping of the surface magnetic field does depend on knowledge of the distribution of the tracer element. Unambiguous observational evidence may emerge from development of the Borra–Landstreet hydrogen Balmer-line technique (1977, 1980), or from mapping of field structure carried out using more extensive data sets than observations just of $B_{\rm eff}$ (Landstreet 1996, personal communication).

The observations immediately set a series of challenges to theorists (see e.g. Mestel 1984 for a summary). What modifications to the theory of stellar structure follow from the departure from axisymmetry, and what are the likely observational consequences? What physical processes affect the obliquity? What braking process is responsible for the usually slow rotation, and during what epochs in the star's history? How does $\mathbf{B}$ vary through the star, and what determines how much flux appears above the surface, for small and large obliquity; and in particular, why do only a minority of early-type stars show strong or even just observable surface fields? And what is the basic cause of the abundance peculiarities, both in the two classes of magnetic CP stars and in the non-magnetic CP and the Am stars? And most fundamental of all, what is the origin of the magnetic field itself? Does the apparent quasi-stability of the fields allow them to be interpreted in terms of the 'fossil' theory, with the field a relic either of the galactic field threading the primeval gas from which the star formed (cf. Chapter 11), or of a dynamo-built field from an earlier epoch in the star's life (e.g. the Hayashi phase, cf. Chapter 10)? Or are the fields maintained by contemporary dynamo action, whether in the convective core or the radiative envelope; in which case, why are only a minority of early-type stars observably magnetic?

A further challenge to the theorist has emerged with the discovery of the 'roAp'-phenomenon – rapidly oscillating Ap stars (Kurtz 1978, 1982, 1990 (and references therein); see also Section 2.1 of Gautschy and Saio 1996). The oblique rotator again appears to offer a suitable basic phenomenological model (Section 9.8).

We shall proceed by further study of the problems in stellar structure that emerge from the oblique rotator model, including the special case of the aligned rotator, with $\chi = 0$; we then inquire whether understanding of the sometimes bewildering observational data is in sight.

9.2 The dynamics of the oblique rotator: the Eulerian nutation and the consequent internal motions

It was noted in Chapter 5 that even in an aligned, axisymmetric rotator, dissipative processes may effectively couple together different field lines, so that the allowed steady state of isorotation may in fact be restricted to near uniform rotation. In a non-axisymmetric star, the only plausible kinematically allowed steady state will have all field lines rotating together. The simplest generalization of the models of Chapter 5 then has the basic magnetic field a superposition of poloidal and toroidal fields (again so as to satisfy stability constraints), both assumed symmetric about the magnetic axis $\mathbf{p}$, which is inclined to the angular

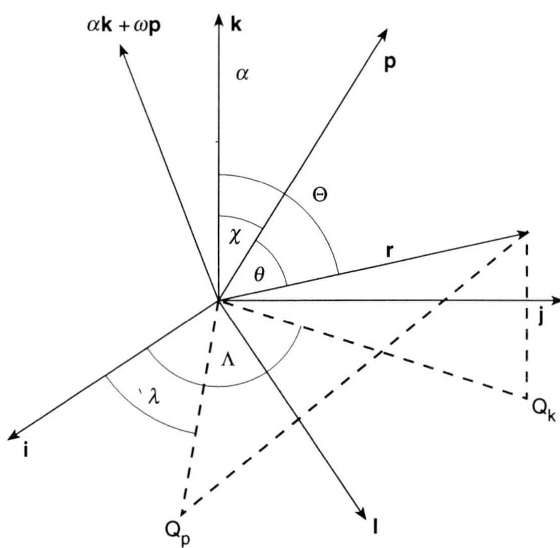

FIG. 9.3. The coordinate systems for the oblique rotator. The triad (**i**, **j**, **k**), with spherical polar coordinates (r, Θ, Λ), is rotating with angular velocity α about **k**, the direction of the angular momentum vector; Q_k is the projection of **r** on to the plane (**i**, **j**) perpendicular to **k**. The system (**i**, **l**, **p**), with spherical polar coordinates (r, θ, λ), defines the basic magnetic field with axis **p**; Q_p is the projection of **r** on to the plane (**i**, **l**), perpendicular to **p**.

momentum axis **k** by the angle χ. The angular momentum vector $\mathbf{h} \propto \alpha\mathbf{k}$ is nearly invariant over time-scales short compared with characteristic braking and precession times, e.g. as discussed in Chapter 7. One system of axes is completed by the unit vector **j** in the plane $(\mathbf{p}, \mathbf{k})$, orthogonal to **k**, and by the vector **i** in the right-handed triad $(\mathbf{i}, \mathbf{j}, \mathbf{k})$ which rotates with the angular velocity α. The spherical polar coordinate system (r, Θ, Λ) refers to **k**, the system (r, θ, λ) to **p** (Fig. 9.3).

It was first pointed out by Spitzer (1958) that the dynamics of the magnetic oblique rotator must have some features in common with the classical problem of a rigid body rotating about an axis other than a principal axis of inertia. This idea was developed later by application of perturbation theory (Mestel and Takhar 1972; Mestel *et al.* 1981; Nittmann and Wood 1981). If the magnetic forces were zero, then the density field in the star would be as in Section 5.4, symmetric about **k** and so appropriately written as $\rho_0(r) + \rho_\alpha(r, \Theta)$. Equally, if centrifugal forces could be ignored, the density would be symmetric about **p**, and so of the form $\rho_0(r) + \rho_B(r, \theta)$. Working to the first order in each of the perturbations, one therefore writes

$$\rho = \rho_0(r) + \rho_\alpha(r, \Theta) + \rho_B(r, \theta). \tag{9.1}$$

We anticipate that over the bulk of the star, $|\rho_B| \ll |\rho_\alpha|$; nevertheless, the presence of ρ_B introduces a qualitatively new feature. The density field (9.1), rotating with angular velocity $\alpha\mathbf{k}$, has an angular momentum vector $\int \rho\mathbf{r} \times (\alpha\mathbf{k} \times \mathbf{r})\,\mathrm{d}\tau$ with a non-vanishing component not only along $\mathbf{k}$ but also along $\mathbf{j}$, the latter given by

$$-\alpha \int (\mathbf{j}\cdot\mathbf{r})(\mathbf{k}\cdot\mathbf{r})\rho_B\,\mathrm{d}\tau \tag{9.2}$$

(the contributions from $\rho_0 + \rho_\alpha$ vanishing by symmetry). Since the vector $\alpha\mathbf{k}$ is defined to be that of the invariant angular momentum, the system must have extra internal motions with a $\mathbf{j}$-component equal and opposite to (9.2). If the star were axisymmetric, with $\rho_\alpha = 0$, then the problem would be the analogue of that of a rigid top, in which the angular velocity $\alpha\mathbf{k}$ is supplemented by the Eulerian nutation $\omega\mathbf{p}$. In our non-rigid star, we may still adopt this as the simplest first correction that yields an invariant angular momentum. To first order in perturbing quantities, the nutation $\omega\mathbf{p}$ has a $\mathbf{j}$-component of angular momentum that cancels (9.2) if

$$\omega = \frac{2\pi\alpha\cos\chi}{I_0}\int\int \rho_B(r,\mu)r^4\mathrm{P}_2(\mu)\,\mathrm{d}r\,\mathrm{d}\mu, \tag{9.3}$$

where $\mu = \cos\theta$ and I_0 is the moment of inertia of the spherically symmetric density field $\rho_0(r)$. If the field $\mathbf{B}$ were purely poloidal, of total flux F, then (9.3) would yield $\omega \simeq \alpha(F^2/\pi^2GM^2) \ll \alpha$. Near the axis, the toroidal linking flux exerts forces that pinch towards the axis, tending to make the star prolate rather than oblate, so that together the $\mathbf{B}_\mathrm{p} + \mathbf{B}_\mathrm{t}$ fields would be expected to yield a rather smaller deviation from sphericity than either acting alone; nevertheless, it would be only in exceptional cases that ω would be reduced to a small fraction of $\alpha(F^2/\pi^2GM^2)$.

The Eulerian frequency is thus seen to be always much below the basic rotation frequency, but it may still have significant consequences over a stellar lifetime. If the centrifugal perturbation ρ_α were negligible, then the Eulerian nutation could be a complete description of the internal motions, for the density field would be symmetric about $\mathbf{p}$, so that the nutation $\omega\mathbf{p}$ would not alter the density–pressure field. In fact, the oblique rotator is a body with three unequal axes of inertia, and $\rho_\alpha(r, \Theta)$ is normally the dominant perturbation. If the star were a rigid body, its motion as seen in the frame rotating with $\alpha\mathbf{k}$ could be described as a combination of the Eulerian nutation about $\mathbf{p}$ and a rocking motion of the obliquity angle χ, both with the frequency (9.3). However, the star must be able to remain in equilibrium without the support of non-hydrostatic stresses. The nutation acting alone changes the density–pressure field; hydrostatic equilibrium is maintained by an additional velocity field ('ξ-motions') which produce compensating changes in the density–pressure field. Referred to an inertial frame, instantaneously coinciding with the rotating frame defined by the unit vectors

k and **p**, an element has a velocity $(\alpha\mathbf{k}\times\mathbf{r})+(\omega\mathbf{p}\times\mathbf{r})+\dot{\xi}$. Provided the total magnetic energy is high enough to ensure that the nutation period is short compared with the Kelvin–Helmholtz time, then these ξ-motions may be taken as nearly adiabatic.

The analysis proceeds as follows. At time t the element at the point (r,θ,λ) has the density (9.1), which can be rewritten

$$\rho(r,\theta,\lambda)=\rho_0(r)+\rho_B(r,\theta)+\rho_\alpha(r,\theta,\lambda) \tag{9.4}$$

(cf. expression (9.22) below). At time δt later the Eulerian nutation has rotated the element through an angle $\delta\lambda=\omega\,\delta t$ about the **p**-axis, so that the density and pressure at the point $(r,\theta,\lambda+\delta\lambda)$ have changed by

$$\rho_\alpha(r,\theta,\lambda)-\rho_\alpha(r,\theta,\lambda+\delta\lambda)=-\delta\lambda\frac{\partial\rho_\alpha}{\partial\lambda}, \tag{9.5}$$

$$p_\alpha(r,\theta,\lambda)-p_\alpha(r,\theta,\lambda+\delta\lambda)=-\delta\lambda\frac{\partial p_\alpha}{\partial\lambda}. \tag{9.6}$$

The requirement that the star should stay in hydrostatic equilibrium enforces a field of ξ-motions with associated Eulerian density–pressure changes $\delta\rho$, δp. The total density–pressure changes

$$\rho^*=\delta\rho-\frac{\partial\rho_\alpha}{\partial\lambda}\delta\lambda,\qquad p^*=\delta p-\frac{\partial p_\alpha}{\partial\lambda}\delta\lambda. \tag{9.7}$$

The associated change V^* in the gravitational potential is non-zero only by virtue of perturbations of higher order than the basic centrifugal and magnetic perturbations, such as the extra magnetic force due to the distortion of **B** by the ξ-motions (cf. Subsection 9.2.2), and the inertial forces associated with the nutation and the ξ-motions. To the order of approximation which ignores such higher-order terms (along with non-linear effects of the two basic perturbations), the state defined by $(\rho_0+\rho^*)$, (p_0+p^*), (V_0+V^*) corresponds at most to an expansion or contraction of each spherical mass shell. As we are studying adiabatic motions consequent on the Eulerian nutation, and since ρ_0, p_0, V_0 define a dynamically stable state of hydrostatic equilibrium, we can be sure that ρ^*, p^*, V^* all vanish to our order of working, for they would fail to satisfy balance between pressure and gravity. Thus during time δt the density and pressure are kept to their hydrostatic values by adiabatic displacements which satisfy

$$-\frac{\partial\rho_\alpha}{\partial\lambda}\omega\,\delta t=-\delta\rho\simeq\nabla\cdot(\rho_0\,\delta\xi)=\delta\xi_r\,\rho_0'+\rho_0\nabla\cdot(\delta\xi), \tag{9.8}$$

$$-\frac{\partial p_\alpha}{\partial\lambda}\omega\,\delta t=-\delta p\simeq\delta\xi_r\,p_0'+\gamma p_0\nabla\cdot(\delta\xi), \tag{9.9}$$

where we have used the condition that the associated Lagrangian variations $\Delta\rho\equiv\delta\rho+\delta\xi\cdot\nabla\rho$, $\Delta p\equiv\delta p+\delta\xi\cdot\nabla p$ must be related adiabatically by $\Delta p=(\gamma p/\rho)\,\Delta\rho$. In a uniformly rotating star, $p_\alpha=\rho_\alpha(\mathrm{d}p_0/\mathrm{d}\rho_0)+\mathrm{O}(\rho_\alpha^2)$, and (9.9) becomes

$$(\delta\xi)_r\,\rho_0' + \frac{\gamma p_0}{(\mathrm{d}p_0/\mathrm{d}\rho_0)}\nabla\cdot(\delta\xi) = -(\partial\rho_\alpha/\partial\lambda)\omega\,\delta t. \tag{9.10}$$

Equations (9.8) and (9.10) yield

$$\left(\frac{\gamma p_0}{\rho_0(\mathrm{d}p_0/\mathrm{d}\rho_0)} - 1\right)\nabla\cdot\left(\frac{\delta\xi}{\delta t}\right) = 0. \tag{9.11}$$

In a sub-adiabatic, radiative zone the quantity inside the first bracket in (9.11) differs markedly from zero, so that (9.11) and (9.10) reduce to

$$\nabla\cdot\dot{\xi} = 0, \quad \dot{\xi}_r = -\omega(\partial\rho_\alpha/\partial\lambda)/\rho_0'. \tag{9.12}$$

The condition $\nabla\cdot\dot{\xi} = 0$ ensures that elements of gas move at constant density and suffer no compressional heating or cooling; to our order of working, no temperature gradients develop other than those associated with hydrostatic equilibrium under the basic centrifugal and magnetic perturbations, and so there is no extra dissipation by radiative conduction. The adiabaticity condition used to determine (9.12) is enforced if the nutation period $2\pi/\omega$ is less than the Kelvin–Helmholtz time τ_{KH}: any motions with $\nabla\cdot\mathbf{v}\neq 0$ would be part of a field of non-radial pressure oscillations superposed on the normally much slower nutation and divergence-free ξ-motions. If in fact $2\pi/\omega$ exceeds τ_{KH}, then any ξ-motions with $\nabla\cdot\dot{\xi}\neq 0$ would be subject to strong radiative damping within a few nutation periods. Thus even if the mean magnetic forces are so weak that $2\pi/\omega$ (given by (9.3)) is longer than τ_{KH}, the ξ-motions will spontaneously adjust to the divergence-free condition.

9.2.1 *The construction of a unique $\dot{\xi}$-field*

The second of the two equations (9.12) fixes $\dot{\xi}_r$, but to determine $\dot{\xi}_\theta$, $\dot{\xi}_\lambda$ uniquely one must expand the full equations to order $(\alpha^2 B^2)$. For our purposes it suffices to construct the unique $\dot{\xi}$-field which has the minimum kinetic energy, subject to the two constraints (9.12); the actual kinetic energy in the ξ-motions cannot be less than this minimum.

To construct the $\dot{\xi}$-field, and also to compute the nutation frequency (9.3), we need a technique for constructing the density field in a slightly non-spherical star. In the notation of Sweet (1950) (following earlier work by Chandrasekhar (1933)), the (r,θ) components of an axially symmetric perturbing force $\mathbf{f}$ per gram are expanded in Legendre functions:

$$f_r = \sum_0^\infty a_s \mathrm{P}_s(\cos\theta), \qquad f_\theta = -\sum_0^\infty b_s(r)\frac{\partial \mathrm{P}_s}{\partial\theta}. \tag{9.13}$$

The perturbed gravitational field is written as

$$\frac{\partial V}{\partial\theta} = -\sum_1^\infty c_s(r)\frac{\partial \mathrm{P}_s}{\partial\theta}. \tag{9.14}$$

The hydrostatic equation and Poisson's equation then yield

$$\frac{1}{r^2}\left(r^2 c_s'\right)' - \frac{s(s+1)}{r^2} c_s = \frac{4\pi G \rho_0' c_s}{(-V_0')} + \frac{4\pi G \rho_0}{(-V_0')}[(rb_s)' + a_s]. \tag{9.15}$$

The appropriate boundary conditions are that $c_s = 0$ at $r = 0$ and $c_s \to 0$ as $r \to \infty$, and c_s and c_s' are continuous at the unperturbed spherical surface of the star, where the density is supposed to go smoothly to zero. Thus provided the form of the perturbing forces – centrifugal or magnetic – may be considered known we can construct $\partial V/\partial\theta$ and so $\partial\rho/\partial\theta$ at each point of the star.

We first put $\theta = \Theta$ and apply the technique to a star rotating uniformly about the axis **k**, for which

$$f_r = \frac{2}{3}\alpha^2 r[1 - \mathrm{P}_2(\cos\Theta)], \qquad f_\Theta = -\frac{1}{3}\alpha^2 r \frac{\partial \mathrm{P}_2}{\partial\Theta}, \tag{9.16}$$

and (9.15) reduces to

$$h'' + \frac{6}{r}h' - \frac{4\pi G \rho_0' h}{(-V_0')} = 0, \tag{9.17}$$

where

$$h = 1 + \frac{3c_2}{r^2\alpha^2}. \tag{9.18}$$

The boundary conditions become:

$$\begin{aligned} &h \ \text{ finite at } r = 0,\\ &h' + \frac{5h}{r} - \frac{5}{r} = 0 \text{ at the stellar surface.} \end{aligned} \tag{9.19}$$

The non-spherical part of ρ_α is then given by

$$\rho_\alpha = \frac{\alpha^2 r^2 h}{3(-V_0')}\rho_0' \mathrm{P}_2(\cos\Theta). \tag{9.20}$$

Substitution of the relation

$$\cos\Theta = \cos\theta\cos\chi + \sin\theta\cos\lambda\sin\chi \tag{9.21}$$

yields ρ_α in terms of the polar angles (θ, λ) referred to the magnetic axis:

$$\begin{aligned} \rho_\alpha = \frac{\alpha^2 r^2 h}{6(-V_0')}\rho_0' \Big[&(1 + 2\mathrm{P}_2(\cos\theta))\cos^2\chi + (1 - \mathrm{P}_2(\cos\theta))(1 + \cos 2\lambda)\sin^2\chi \\ &+ (3/2)\sin 2\theta\cos\lambda\sin 2\chi - 1\Big], \end{aligned} \tag{9.22}$$

whence from (9.12)

$$\dot{\xi}_r = f(r)\,\omega\left(6\sin^2\theta\sin 2\lambda\,\sin^2\chi + 3\sin 2\theta\sin 2\chi\sin\lambda\right), \tag{9.23}$$

with

$$f(r) = \frac{\alpha^2 r^2 h}{12(-V_0')}. \tag{9.24}$$

Minimization of the energy of the $(\dot{\xi}_\theta, \dot{\xi}_\lambda)$ part of the motions

$$\frac{1}{2}\int \rho_0(r)[\dot{\xi}_\theta^2 + \dot{\xi}_\lambda^2]\,\mathrm{d}\tau \tag{9.25}$$

subject to the divergence-free constraint (9.12), yields the Euler–Lagrange equations

$$\partial(\mu\sin\theta)/\partial\theta - \rho_0\dot{\xi}_\theta = 0, \tag{9.26}$$

$$\partial(\mu\sin\theta)/\partial\lambda - \rho_0\sin\theta\dot{\xi}_\lambda = 0, \tag{9.27}$$

where μ is a Lagrange multiplier; whence

$$\frac{\partial}{\partial\lambda}(\dot{\xi}_\theta) - \frac{\partial}{\partial\theta}(\dot{\xi}_\lambda\sin\theta) = 0. \tag{9.28}$$

When combined with (9.12), this yields the unique, singularity-free solution (Mestel *et al.* 1981)

$$\dot{\xi}_\theta = [(r^2 f)'/r][\sin 2\theta\sin 2\lambda\sin^2\chi + \cos 2\theta\sin\lambda\sin 2\chi], \tag{9.29}$$

$$\dot{\xi}_\lambda = [(r^2 f)'/r][2\sin 2\theta\cos\lambda\sin 2\chi + \cos\theta\cos\lambda\sin 2\chi]. \tag{9.30}$$

To complete the construction of the ξ-motion field, we apply the same technique to construct $\rho_B(r,\theta)$. In the simplest illustrative example, the basic field $\mathbf{B}_0$ is written as the sum of a dipolar poloidal component with flux function $P = -\sin^2\theta B(r)$, and a toroidal component satisfying the torque-free condition: in spherical polars,

$$\mathbf{B}_0 = [2\cos\theta B(r)/r^2,\ -\sin\theta B'(r)/r,\ \beta(P)/r\sin\theta]. \tag{9.31}$$

This field exerts the poloidal force density $(\nabla\times\mathbf{B}_0)\times\mathbf{B}_0/4\pi$, which reduces to

$$-[B'' - 2B/r^2 - \beta(P)\beta'(P)/\sin^2\theta](B'\sin^2\theta, 2B\sin\theta\cos\theta/r, 0)/4\pi r^2. \tag{9.32}$$

This is further simplified by taking the linear form $\beta(P) = \beta P$ with β a constant, so that the total magnetic perturbation is just of P_2 form. (This requires a special structure for the external force-free field (cf. Section 3.4), but there is no essential difficulty in generalizing the present work to more realistic forms for $\beta(P)$.) Thus the perturbing functions a_2, b_2 in (9.13) become

$$a_2 = \frac{2}{3}B'\left[B'' - (2-\beta^2)B\right]/4\pi r^2, \tag{9.33}$$

$$b_2 = -\frac{2}{3}B\left[B'' - (2-\beta^2)B\right]/4\pi r^3. \tag{9.34}$$

If now $B(r)$ and the constant β are supposed known, (9.15) can again be solved for c_2, subject to the boundary conditions that c_2 and c_2' vanish at $r = 0$, and $3c_2 + rc_2' = 0$ at the stellar surface, to assure continuity with the vacuum potential outside. The non-spherical part of the density perturbation is then given from Poisson's equation by $\tilde{\rho}_B(r)\mathrm{P}_2(\cos\theta)$, where

$$\tilde{\rho}_B(r) = \frac{1}{(-V_0')}\left(c_2\rho_0' + \frac{4B(B'' - 2B/r^2)}{12\pi r^3} - \frac{2B(B''' - 2B''/r^2 + 4B/r^3)}{12\pi r^3}\right), \tag{9.35}$$

and the nutation frequency (9.3) becomes

$$\omega = \frac{4\pi\alpha}{5I_0}\cos\chi \int_0^R \tilde{\rho}_B(r)r^4\,\mathrm{d}r. \tag{9.36}$$

9.2.2 *Consequences of the ξ-motions*

From (9.12) and (9.22), the maximum displacement of an element of gas during one Eulerian nutation period is $\approx (\pi/\omega)\omega|\rho_\alpha/\rho_0'|$ which is $\mathrm{O}(\epsilon)$ with $\epsilon(r) \equiv \alpha^2 r^3/Gm(r)$. The above analysis is linear, depending on $\epsilon(r)$ being small, and so is strictly applicable only to slow rotators, or to regions deep within a rapid uniform rotator, for which as already noted the central mass condensation ensures that $\epsilon(r)$ is small over the bulk of the mass. However, it is clear that similar motions, but of larger amplitude, must arise in the outer regions of an oblique star in rapid rotation. One is then led to wonder whether these ξ-motions might be of importance for the evolution of rapid rotators, by mixing nuclear-processed material from deep in the star through the outer regions. The motions analysed above are strictly periodic, each element returning to its initial position after time $2\pi/\omega$; but in a less idealized model one would expect the same mechanical principles to produce a degree of randomness. Further, even early-type stars develop local convective regions during their evolution, and while on the main sequence have convective zones due to He-ionization. Mixing of nuclear-processed material brought up by ξ-motions through a convectively stable region into a local convection zone must affect stellar evolution. At one time it was thought that the thermally-driven Eddington–Vogt–Sweet circulation ('Ω-currents'), discussed in Section 5.4, would lead to continuous mixing between, for example, the convective core and much of the radiative envelope of a rapidly rotating early-type star. However, it was subsequently shown (Mestel 1953; Mestel and Moss 1986) that the non-spherical distribution of mean molecular weight μ set up by the circulation itself would yield analogous 'μ-currents' which would prevent continuous mixing by the Ω-currents. In Section 5.7, eqn (5.102) et seq., the μ-distribution is constructed that indeed yields $\nabla \cdot \mathbf{F} = 0$, so that no thermally-driven currents flow. However, if a star has an oblique magnetic field, the ξ-motions could kill the μ-effect by coupling radiative and weak convective zones. If the detailed comparison of stellar evolution computations with observation points to the need for some non-turbulent mixing in some phases, then this indirect effect of an oblique magnetic field should be kept in mind, remembering also that rapidly rotating stars may possess significant magnetic fields without necessarily being observably magnetic – cf. Section 9.6.

Since the frequency $\omega \simeq \mathrm{O}(F^2/\pi^2 GM^2)\alpha \ll \alpha$, the inertial effects of both the nutation and the associated ξ-motions are negligible corrections. Because the basic field $\mathbf{B}_0$ is assumed symmetric about the axis $\mathbf{p}$ it is undistorted by the nutation; however, the ξ-motions will necessarily distort $\mathbf{B}_0$, and the nutation will rotate the distorted field about $\mathbf{p}$. Write the field as

$$\mathbf{B} = \mathbf{B}_0 + \tilde{\mathbf{B}}; \tag{9.37}$$

then if Ohmic dissipation is small over a nutation period,

$$\frac{\partial \tilde{\mathbf{B}}}{\partial t} = \nabla \times (\mathbf{v} \times \mathbf{B}) = \nabla \times [(\omega \mathbf{p} \times \mathbf{r}) \times \tilde{\mathbf{B}} + \dot{\xi} \times \mathbf{B}_0], \tag{9.38}$$

where $\mathbf{v} = \omega \mathbf{p} \times \mathbf{r} + \dot{\xi}$ is the velocity as seen in the frame rotating with the angular velocity $\alpha \mathbf{k}$, and the higher-order term $\dot{\xi} \times \tilde{\mathbf{B}}$ has been dropped. Since from (9.12), $|\dot{\xi} \times \mathbf{B}_0| \simeq \mathrm{O}(\omega r \epsilon B_0)$, from (9.38), $\tilde{\mathbf{B}} \simeq \mathrm{O}(\epsilon B_0)$, with both terms on the right of (9.38) of the same order. Equation (9.38) can be solved for $\tilde{B}$, once $\mathbf{B}_0$ is given (Mestel *et al.* 1981; Nittman and Wood 1981). For example, the distortion of the poloidal part of (9.31) yields for $\tilde{B}_r$

$$\begin{aligned}&\left[8\frac{(r^2 f)'}{r^4}B(r) - 12\frac{f}{r^2}B'(r)\right]\sin^2\theta\cos\theta\big[\cos 2\lambda(\cos 2\omega t - 1) + \sin 2\lambda \sin 2\omega t\big]\sin^2\chi \\ &+ \left[-\frac{f}{r^2}B'(r)(18\sin\theta\cos^2\theta - 6\sin^3\theta) - \frac{(r^2 f)'}{r^4}B(r)(2\sin^3\theta - 14\sin\theta\cos^2\theta)\right] \\ &\times\big[\cos\lambda(\cos\omega t - 1) + \sin\lambda\sin\omega t\big]\sin 2\chi\end{aligned} \tag{9.39}$$

and similar unwieldy expressions for $\tilde{B}_\theta$, $\tilde{B}_\lambda$.

9.2.3 *Dissipation of the ξ-motions*

The neglect of all dissipative processes over a nutation period is almost certainly a good approximation on which to base the construction of the zero-order expressions for ξ and $\tilde{\mathbf{B}}$; however, over many nutation periods one expects a steady drain on the rotational energy of the star. Spitzer (1958) argued – again by analogy with the mechanics of almost rigid bodies – that dissipation should change the obliquity angle χ. The rotational energy is given by $\mathbf{h}^2/2I$, where $\mathbf{h}$ is the invariant angular momentum and

$$I \equiv \int \rho r^2 \sin^2 \Theta \, d\tau \tag{9.40}$$

is the moment of inertia about the instantaneous axis of rotation. The linear perturbation analysis allows I to be written as

$$I = I_0 + I_\alpha + I_B, \tag{9.41}$$

with I_0 the moment of the spherically symmetric density field of a non-rotating star, I_α the moment of the density distortion due to the rotation α about the instantaneous axis (identical with **k** to our order of working), and I_B the moment of the density distortion due to **B**. Both I_0 and I_α remain constant as the angle χ changes, for since the star is fluid, the centrifugal perturbations to the (p, ρ) fields steadily adjust themselves to the instantaneous axis of rotation as it precesses through the star. Thus even though the total magnetic energy is almost certainly much less than the centrifugal, it is the magnetic perturbation to the ρ-field that determines the asymptotic state. The contribution I_B to (9.41) has a χ-independent part – due to the spherically symmetric part of ρ_B – and a χ-dependent part conveniently written as

$$(I_B)_0 \mathrm{P}_2(\cos\chi) = -\frac{8\pi}{15}\mathrm{P}_2(\cos\chi) \int_0^R \tilde{\rho}_B(r) r^4 \, dr, \tag{9.42}$$

where $\tilde{\rho}_B(r)$ is defined by (9.35) and the relation (9.21) has been used in (9.40).

A star with a field that is dominantly poloidal will have essentially oblate isopycnic surfaces, yielding $\tilde{\rho}_B$ negative and $(I_B)_0 > 0$. Thus as energy is dissipated, $h^2/2(I_0 + I_\alpha + I_B) \simeq [(h^2/2(I_0 + I_\alpha)][1 - (I_B)/(I_0 + I_\alpha)]$ declines through I_B increasing to its maximum, i.e. by $\chi \to 0$. To our order of working, the energy available for dissipation is

$$\frac{h^2}{2(I_0 + I_\alpha)^2}[(I_B)_0 - (I_B)_\chi] \approx -(2\pi/5)\alpha^2 \sin^2\chi \int_0^R \tilde{\rho}_B r^4 \, dr. \tag{9.43}$$

(The energy in the ξ-motions themselves is smaller than this by the factor (F^2/GM^2) even for a rapid rotator.) Similarly, if the stellar field were dominantly toroidal, the star would be dynamically prolate with respect to the magnetic axis, so that $(I_B)_\chi < 0$, and dissipation of energy would cause precession towards the state with $\chi = \pi/2$. The energy available would be as in (9.43) but with the factor $\sin^2\chi$ replaced by $\cos^2\chi$.

The crucial question is the strength of the dissipative processes. As the ξ-motions are nearly divergence-free, radiative conduction in unlikely to be important. A more powerful process is the Ohmic dissipation of the currents that maintain the field $\tilde{\mathbf{B}}$, ((9.39) and its companions), arising from the action of the ξ-motions on the basic field (9.31). Over one nutation period, the total dissipation is

$$(c/4\pi)^2 \int_0^{2\pi/\omega} \mathrm{d}t \left(\int \frac{(\nabla \times \tilde{\mathbf{B}})^2}{\sigma} \mathrm{d}\tau \right). \tag{9.44}$$

An upper limit to the time in which χ reaches its asymptotic value is given by dividing (9.43) by (9.44). For brevity we shall call this time the 'time of alignment', but recall that the dissipation will indeed yield $\chi \to 0$ when the star is dynamically oblate, but $\chi \to \pi/2$ if it is dynamically prolate. And if the time of alignment is longer than the stellar evolution time, and no other process intervenes to alter χ, then the ξ-motions persist, with possibly significant effects on stellar evolution.

To make estimates of the nutation frequency (9.3) and the energy available for dissipation (9.44), one needs to know the density perturbation $\tilde{\rho}_B(r)$, determined by the structure of the Lorentz force density through the star. In Mestel *et al.* (1981) and Nittman and Wood (1981), several examples were studied with different surface fields and different degrees of central condensation. One set of calculations adopts for $\mathbf{B}_\mathrm{p}$ the principal Cowling decay mode (excluded from the convective core), with the field strength varying by about 13 from the photosphere to the core surface. Various studies over the years on the interaction between the magnetic, rotation and thermo-gravitational fields (cf. Section 9.7) had suggested that the internal fields might be much stronger, so some such models were also used, yielding larger $|\tilde{\rho}|$ and so larger values of ω, etc.

Estimates of both the nutation period and the 'alignment time' vary greatly. For a field of given radial structure, the nutation period will be longer the smaller the surface field; but as both the energy to be dissipated and the dissipation rate scale with the square of the basic field, the alignment time is independent of the surface field strength. A star as strongly magnetic as Babcock's star, with $B_\mathrm{s} = 35\,000$ G and with the poloidal field built on Cowling's decay model has $2\pi/\omega \simeq 4 \times 10^4 P_d$ where P_d is the rotation period in days, while the alignment time – calculated assuming just (9.44) for the dissipative process – varies typically from 4×10^7 yr to 10^{12} yr as P_d increases from 0.5 days to 100 days. Only for the shortest rotation periods is the estimated alignment time astronomically relevant; and for stars with weaker surface fields, even the nutation period will exceed the Kelvin–Helmholtz time. Markedly shorter times result if the fields are assumed to increase inwards much more strongly. Thus if B varies by a factor 35 from surface to convective core, $2\pi/\omega$ is typically $25 \times 10^3 P_d$ yr when $B_\mathrm{s} = 35\,000$ G and $7 \times 10^5 P_d$ yr when $B_\mathrm{s} = 3000$, and the alignment times vary typically from 6×10^4 yr when $P_d = 0.5$ days to 2.6×10^9 yr when $P_d = 100$ days. If $\bar{B}$ increases by 350 – a strongly centrally condensed field – then $2\pi/\omega$ would be $70 P_d$ yr for $B_\mathrm{s} = 3000$ and $6 \times 10^4 P_d$ yr for $B_\mathrm{s} = 100$, and the alignment times would vary from 3.7×10^3 yr for $P_d = 0.5$ days to 1.5×10^8 yr for $P_d = 100$ days (Nittmann and Wood 1981).

It should be re-emphasized that the estimated dissipation times are necessarily upper limits. In particular, viscous drag on the ξ-motions occurring inside a convective core – even with the **B**-field expelled from the core – could shorten the times. Nevertheless, the tentative conclusion is that if the field increases inwards only modestly, as in the principal Cowling decay mode, then only in the fastest rotators is dissipation likely to to alter an initial obliquity within a stellar lifetime, whereas the 'alignment time' would be quite short in a star with a strong central field concentration. We return to the implications in Section 9.9, when assessing the relative merits of the fossil and the contemporary dynamo expanations of the CP magnetic fields.

9.3 Non-uniform rotation and the oblique rotator model

It was pointed out in Sections 5.2 and 9.2 that isorotation – as opposed to uniform rotation – is a kinematically-allowed steady state only under strict axisymmetry. In particular, in an oblique rotator a general state of non-uniform rotation will immediately distort the magnetic field, generating azimuthal magnetic forces which will certainly interchange angular momentum between different flux tubes. As in the work reported in Sections 5.2 and 5.5, finite viscosity and resistivity will tend to damp out the motions, leading one might expect just to an asymptotic state of uniform rotation. There is, however, a new effect, pointed out by Rädler (1986). In the axisymmetric problem, the shearing of a poloidal field $\mathbf{B}_{\mathrm{p}}$ with dipolar parity generates a toroidal component $\mathbf{B}_{\mathrm{t}}$ that is antisymmetric in the equator. Finite resistivity slows up the generation of $\mathbf{B}_{\mathrm{t}}$ by allowing relative slippage of gas and $\mathbf{B}_{\mathrm{p}}$. It also causes some mutual annihilation of $\mathbf{B}_{\mathrm{t}}$-lines, described by the diffusion velocity $(\nabla \times \mathbf{B}_{\mathrm{t}}) \times \mathbf{B}_{\mathrm{t}}/4\pi\sigma B_{\mathrm{t}}^2$ (cf. (5.105)), but as the poloidal field is hardly affected by the shearing, there is no change in its rate of diffusion. By contrast, rotational distortion of a highly oblique field can lead to the juxtaposition of oppositely directed lines of the basic field, and so can cause accelerated Ohmic decay. Rädler studied just the kinematic problem with a prescribed non-uniform rotation, ignoring the back-reaction of the Lorentz forces. His results suggest that a weakish field, initially highly oblique to the rotation axis, could be converted into a more nearly aligned field, simply through the accelerated decay of the perpendicular component. A stronger field should, however, be able to reverse the initially imposed non-uniform rotation before the accelerated decay gets under way. The evolution of a field under an initial non-uniform rotation could then be sensitive to the value of the parameter measuring the ratio of magnetic to rotational energies.

A highly idealized illustrative model is discussed by Moss *et al.* (1990). In terms of cylindrical polar coordinates (ϖ, ϕ, z) based on the rotation axis $\mathbf{k}$, the 'star' is the cylindrical region $\varpi \leq R$, and the convective 'core' the region $\varpi < x_c R$. Again the turbulence is supposed to keep the core rotating rigidly with angular velocity $\Omega_0 \mathbf{k}$, and the fluid is assumed incompressible with uniform and constant kinematic viscosity ν and resistivity λ. Because of the essential non-axisymmetry of the system, it is convenient to use a coordinate frame rotating with the core. The divergence-free magnetic and velocity fields are written

$$\mathbf{B}(\varpi, \phi, t) = \nabla \times (a\mathbf{k}) = \nabla a \times \mathbf{k}, \tag{9.45}$$

$$\mathbf{u}(\varpi, \phi, t) = \nabla \times (\psi\mathbf{k}) = \nabla \psi \times \mathbf{k}, \tag{9.46}$$

both vectors being therefore perpendicular to $\mathbf{k}$, whereas the current-density $\mathbf{j}$, the electric field $\mathbf{E}$ and the vorticity

$$\boldsymbol{\omega} = \nabla \times \mathbf{u} = \omega \mathbf{k} \tag{9.47}$$

are all parallel to $\mathbf{k}$. The relations

$$\nabla a = (-B_\phi,\, B_\varpi,\, 0), \tag{9.48}$$
$$\nabla \psi = (-\Omega(\varpi)\varpi,\, 0,\, 0), \tag{9.49}$$

with $u_\phi = \Omega(\varpi)\varpi$, follow from (9.45) and (9.46).

The use of the rotating frame introduces an extra term $-\Omega_0 \partial \mathbf{B}/\partial\phi$ into the left-hand side of the induction equation, to be cancelled by the contribution $\nabla\times(\Omega_0 \varpi \mathbf{t} \times \mathbf{B})$ to the motional induction term on the right, as expected. The equation then integrates to

$$\partial a/\partial t = \mathbf{k}\cdot(\mathbf{u}\times\mathbf{B}) - \lambda\mathbf{k}\cdot\nabla\times\mathbf{B} = \mathbf{k}\cdot(\nabla\psi\times\nabla a) + \lambda\nabla^2 a. \tag{9.50}$$

(Although formally one could add a function $f(\varpi, \phi, t)$ to (9.50), this would correspond to an arbitrary z-independent potential electric field, requiring sources at $z = \pm\infty$ and thus of no relevance to our problem.) The curl of the equation of motion – to be solved simultaneously with (9.50) – reduces to the scalar equation

$$\partial\omega/\partial t = \mathbf{k}\cdot(\nabla\psi\times\nabla\omega) + \mathbf{k}\cdot[\nabla a\times\nabla(\nabla^2)/4\pi\rho]. \tag{9.51}$$

Note that the Coriolis term $2\Omega_0\nabla\psi$ has zero curl and so disappears from (9.51). From the curl of (9.46), the stream function ψ is related to ω by

$$\nabla^2\psi = -\omega. \tag{9.52}$$

Integration forward of (9.50) and (9.51) yields a and ω at a later time, and solution of (9.52) then enables the process to be continued.

The scalings appropriate for this problem differ somewhat from those made earlier:

$$\tau R = \Omega_0 t, \qquad x = \varpi/R, \tag{9.53}$$
$$p_1 = \frac{\lambda}{\Omega_0 R^2}, \qquad p_2 = \frac{\bar{B}^2}{4\pi\rho\Omega_0^2 R^2}, \qquad p_3 = \frac{\nu}{\Omega_0 R^2}, \qquad Pr = \frac{p_3}{p_1} = \frac{\nu}{\lambda}$$

Thus p_3 and p_1 are repectively inverse fluid and magnetic Reynolds numbers, and p_2 is a measure of $(V_A/V_{\text{rot}})^2$. The resulting dimensionless equations are solved in the shell $x_c \le x \le 1$, $0 \le \phi \le \pi$, with $x_c = 0.4$. An essential consequence of the non-axisymmetry is that even if the initial motions have only a u_ϕ (rotatory) component, the equations generate a component u_r in order to satisfy the continuity constraint (9.46). Boundary conditions are that $\mathbf{u}$ be stress-free at

$x = 1$ and that u_ϕ (measured in the rotating frame) vanish at $x = x_c$. At $x = 1$, $\mathbf{B}$ must fit smoothly on to a curl-free external field, and $a(x_c, \phi)$ is held fixed at its initial value. The initial magnetic field was chosen to be a two-dimensional equatorial dipole with

$$a = \sin\phi/x, \tag{9.54}$$

and the initial differential rotation is defined by

$$\psi = (1 - x_c)^2 - (x - x_c)^2. \tag{9.55}$$

For most of the calculations, p_1 was fixed at 0.02 and $Pr = 0.01$, except for the case $p_2 = 0.04$, $Pr = 0.1$; otherwise, p_2 varied over the wide range $0.001 \leq p_2 \leq 0.04$.

Some of the computational results are illustrated in Fig. 9.4 . They demonstrate that for a given p_1, there is indeed a critical value p_{2c} such that when $p_2 \gg p_{2c}$, the ϕ-Lorentz force suppresses the differential rotation, whereas when $p_2 \ll p_{2c}$ the field is tightly wound up and would suffer severe Ohmic diffusion before the Lorentz forces would have time to reverse the winding. A crude estimate for p_{2c} is given by first neglecting the term in λ in (9.50), which then has the solution $a = f[\Omega(\varpi)t - \phi]$, where the function f is fixed by the chosen initial form (9.54): in dimensional form,

$$a = -\frac{B_0 R_0^2}{\varpi} \sin[\Omega(\varpi)t - \phi]. \tag{9.56}$$

Let Ω have the shear $\Delta\Omega$ over the scale D. One then finds that the dissipation term $\lambda\nabla^2 a$ in (9.50) reaches the order of the induction term at time $t \approx (\Omega/\Delta\Omega)(D/\Omega^{1/2})(1/\lambda^{1/2})$, provided the MRN $\equiv (\Omega D)D/\lambda \gg 1$. If this time is longer than the Alfvén crossing time $D(4\pi\rho)^{1/2}/B_r$, then the dynamical effects dominate. When $\Delta\Omega \simeq \Omega$, this condition yields that the critical value p_{2c} should be close to p_1, in rough agreement with the numerical results. In a subsequent paper, Moss (1992) shows that similar results hold in a model with the fluid compressible.

The work gives tentative support to the argument that a weak perpendicular component acted upon by a strong shear would be destroyed, whereas a stronger field will persist and destroy the shear. The use of a microscopic value of λ would in fact yield such a small value for p_{2c} that virtually any significant field would be able to suppress the shear; however (cf. the discussion in Chapter 5), the hydromagnetic instability of dominant toroidal fields would again require use of a much larger macro-diffusivity. If a weak field were initially oblique rather than perpendicular, the Rädler process would act effectively to reduce the obliquity angle to zero, though without dynamo action, the poloidal field persisting would be weak. (It may be significant that only a small minority of the strongly magnetic CP stars seem to have fields with near zero obliquity.)

The discussion will clearly be modified if dynamo action occurs in the convective core. The destruction of a weak perpendicular field by shear could be

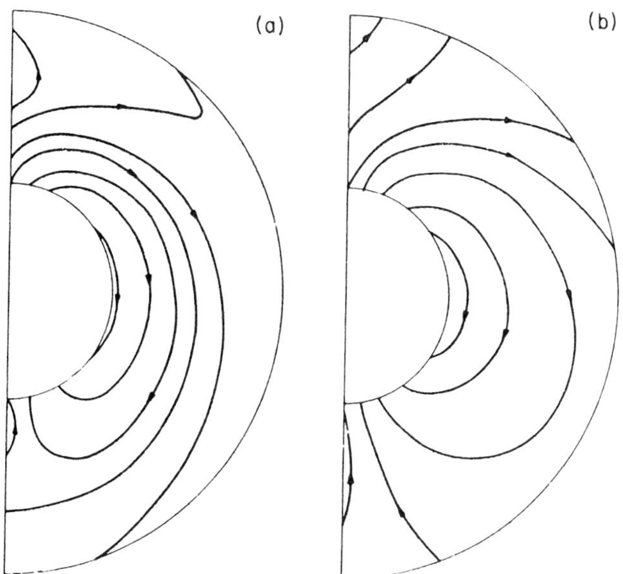

FIG. 9.4. The cylindrical perpendicular rotator model, with $p_1 \equiv \lambda/\Omega_0 R^2 = 0.02$, $Pr \equiv \nu/\lambda = 0.01$. (a) Rotation-dominated model: $p_2 \equiv \bar{B}^2/4\pi\rho\Omega_0^2 R^2 = 0.001$; field lines at $\tau = (\Omega_0/R)t = 20$. (b) Magnetic-dominated model: $p_2 = 0.02$; field lines at $\tau = 12$.

balanced by flux generation at the core surface (Chapter 6). Since the Lorentz torque exerted by a rotationally distorted field reacts on the shear – i.e. the magnetic energy continually destroyed comes in part from the kinetic energy of the shear – the field would again ultimately destroy the imposed non-uniform rotation. A crucial question is clearly that of the time-scale for dynamo-generated flux to penetrate the envelope (cf. Section 9.6).

9.4 Models of rotating magnetic stars

9.4.1 *Axisymmetric models in radiative equilibrium*

We suppose now that a magnetic star has been brought into a state with uniform angular velocity α, and turn to the problem of thermal equilibrium, in particular in the radiative envelope of an early-type star. We begin with the simplest case when the field is symmetric about the rotation axis $\mathbf{k}$, so that the density, pressure, temperature and potential may be written as

$$\rho = \rho_0(r) + \rho_\alpha(r,\theta) + \rho_B(r,\theta), \tag{9.57}$$

etc. as in (9.1), except that now $\Theta \equiv \theta$, and so there is no dependence on azimuth. For the moment the star is assumed to be chemically homogeneous, so that if first-order perturbation theory is adequate,

$$T = T_0 \left(1 + \frac{p_\alpha + p_B}{p_0} - \frac{\rho_\alpha + \rho_B}{\rho_0} \right). \tag{9.58}$$

As in Section 5.4, we study the effect of the non-spherical perturbations on radiation flux $\mathbf{F}$. Recall that with α uniform, $(-\nabla \cdot \mathbf{F})_\alpha$ is non-zero. Likewise, the force density $(\nabla \times \mathbf{B}) \times \mathbf{B}/4\pi$ due to an arbitrary $\mathbf{B}$-field will yield a non-vanishing magnetic contribution $(-\nabla \cdot \mathbf{F})_B$ to the net heat transport. In an obvious notation, this generalized first-order perturbation theory yields a superposition of meridian circulation velocity fields

$$\mathbf{v}_\mathrm{p} = \mathbf{v}_\alpha + \mathbf{v}_B, \tag{9.59}$$

determined by inserting the perturbations in T, ρ into the heat equation. The terms ρ_B etc. and so also $\mathbf{v}_B$ may easily be non-trivial in the low-density surface regions, but for the magnetic perturbations to be significant over the bulk of the star, the field must increase strongly inwards, as noted in Chapter 5.

The simplest problem assumes that the magnetic field is both strong enough and has the correct structure to enforce local radiative equilibrium: i.e. within the radiative envelope, $(-\nabla \cdot \mathbf{F})_B$ is everywhere equal and opposite to the von Zeipel term $(-\nabla \cdot \mathbf{F})_\alpha$. The first such models were constructed by Davies (1968) and Wright (1969), assuming a purely poloidal field of dipolar angular structure (see also Monaghan and Robson (1971) and other papers referred to in Moss (1986*a*)). The fields have an O-type neutral point on the equator at $(r_c, \pi/2)$. Of particular interest is the ratio of the flux $F_\mathrm{s} = 2\pi \int_0^{\pi/2} B_r(R,\theta) R^2 \sin\theta \, \mathrm{d}\theta$ that crosses the northern hemispherical surface (and so is observable), to the total flux $F_\mathrm{t} = 2\pi \int_0^{r_c} -B_\theta(r, \pi/2) r \, \mathrm{d}r$. Convenient parameters to describe rotating magnetic stars are

$$\lambda_\Omega = \Omega^2 R^3/GM, \qquad \lambda_B = \bar{B}^2 R^4/4GM^2, \tag{9.60}$$

where $\bar{B}$ is a measure of the surface field, e.g. the polar field strength. The Wright solutions are a function of the ratio λ_Ω/λ_B. The ratio of the central and surface polar fields, and so also the ratio $F_\mathrm{t}/F_\mathrm{s}$, increases with increasing λ_Ω/λ_B. The interpretation is clear. We are demanding that the perturbations to the thermal field due to the magnetic forces are large enough to cancel the corresponding centrifugal perturbations. If F_t is imagined reduced at constant Ω, then the flux must be used more 'efficiently' – i.e. the field has to be more centrally concentrated to yield a larger $\nabla \times \mathbf{B}$ in the dense regions deep in the star. The limit to such solutions comes when it is no longer possible to make $\nabla \cdot \mathbf{F}$ vanish everywhere, even with the surface flux becoming vanishingly small.

These models had some prima facie appeal, in that they illustrate an *anti-*correlation between the observable flux F_s and Ω for a given F_t. This is in very broad accord with observation of the magnetic CP stars, in that the normal

A stars, which do not show any Zeeman effect, are normally rapid rotators, whereas the rotations of the CP stars are distributed about a markedly lower mean value (cf. Section 9.1). The absence of a circulation velocity means that there is no constraint enforced by the hydromagnetic equation (2.88) other than that of isorotation (specialized to uniform rotation). Finite resistivity will cause the usual slow decay, but as long as Ω and B are large enough, the field can be pictured as steadily adjusting so as to satisfy $\nabla \cdot \mathbf{F} = 0$. To satisfy stability requirements, one needs to construct analogous models with mixed poloidal–toroidal fields (Section 5.3), but the results are qualitatively similar.

Field models such as these, constructed to satisfy radiative equilibrium in a rotating star, have a much larger ratio of internal to surface field strength than the principal Cowling decay eigenmode. One can expect similar results in a moderately oblique rotator. Stars with such fields would have nutation frequencies that are correspondingly larger and so also times of approach to alignment or perpendicularity that are very much shorter than for fields of moderate central concentration (cf. Subsection 9.2.3 and Section 9.6).

9.4.2 *Models with thermally-driven circulation*

More general quasi-steady solutions to the problem of thermal equilibrium can be sought with a non-vanishing $\mathbf{v}_{\mathrm{p}}$, satisfying the generalization of (5.57)

$$C_v \rho_0 T_0 \mathbf{v} \cdot \nabla \left(\log(T_0/\rho_0^{\gamma-1})\right) = -(\nabla \cdot \mathbf{F})_{\Omega} - (\nabla \cdot \mathbf{F})_B \tag{9.61}$$

and the steady state continuity condition $\nabla \cdot (\rho_0 \mathbf{v}) = 0$ in both differential and integral form. The advection of angular momentum by the circulation will again be effectively offset by magnetic stresses (cf. Section 5.5), so that Ω = uniform remains an excellent approximation. If perfect conductivity is assumed, then in a steady state $\rho v_{\mathrm{p}}/B_{\mathrm{p}}$ = constant (cf. (5.65)), and a circulation confined to below the photosphere would prevent the appearance of any flux above the surface. Even allowing for finite resistivity, an 'inexorable' circulation could limit severely the observable flux. Further, even if the field has dipolar parity, the Lorentz force will be of even parity with respect to the equator, and so the velocity $\mathbf{v}_B$ will have even parity (like $\mathbf{v}_{\Omega}$). As noted earlier, steady flow requires slippage across the dipolar field lines in the equatorial regions. Although by Cowling's anti-dynamo theorem, in this axisymmetric system there can be no strictly steady states, in a rapid rotator solutions can be constructed that are quasi-steady over times short compared with the decay time. They could be relevant provided the time of flow – either through the radiative envelope as a whole or through the low-density surface regions – is short compared with the main sequence lifetime of a CP star.

Some illustrative calculations have been performed for axisymmetric models with strictly poloidal fields with both parities (Mestel and Moss 1977). One class of solution is essentially the modification due to finite resistivity of the zero-circulation solutions described above. The magnetic field structure has adjusted so that $\nabla \cdot \mathbf{F} \simeq 0$, and so is strongly concentrated to the centre. The actual circulation speed is much below the Eddington–Sweet value. The other class of solution is rotationally dominated. Over the bulk of the radiative zone, $\mathbf{v}_\mathrm{p} \simeq \mathbf{v}_\Omega$, and the poloidal magnetic field is channelled to follow the Eddington–Sweet streamlines. As these models do not have $(\nabla \cdot \mathbf{F})_B \simeq -(\nabla \cdot \mathbf{F})_\alpha$ in the interior, the degree of central condensation is much less marked. However, in the surface regions $\mathbf{v}_\mathrm{p} \simeq 0$ – the magnetic and centrifugal perturbations again nearly cancel.

This result brings out the importance of considering together all the equations to the problem. As noted in Section 5.4, the perturbations to the thermal field due to an arbitrary magnetic field – or indeed an arbitrary rotation field – yield from (9.61) a radial velocity component, of *first* order in the perturbations, with $1/\rho$ in the denominator. This was first noted by Baker and Kippenhahn (1959), who pointed out that the case of uniform rotation was special, in that the first order term in $-\nabla \cdot \mathbf{F}$ is also proportional to ρ, so yielding a first order $(v_\Omega)_r$ that does not become large as $\rho \to 0$. (Equation (5.56) shows that there is a term *quadratic* in the centrifugal perturbation which also yields $v_r \propto 1/\rho$.) If the perturbing forces are supposed prescribed, then the prediction of thermally-driven velocities that become singular at low ρ clearly demands a modification of the Eddington–Sweet scheme of approximation. The neglect of inertial and viscous forces would cease to be valid at low density; the equation of hydrostatic support would need to be modified to include terms which involve the velocity, and a locally self-consistent solution would be a generalized boundary layer (Mestel 1953, Appendix; Roxburgh and Smith 1977; Tassoul and Tassoul 1982). In a magnetic star, however, it appears that there is enough freedom available in the magnetic field structure to deal with the potential singularity: finite, quasi-steady solutions can be found, provided explicit account is taken of finite resistivity, which allows trans-field flow in the surface regions. The current density $\mathbf{j} = (c/4\pi)\nabla \times \mathbf{B}$ can adjust itself so that $|\mathbf{j} \times \mathbf{B}|/c\rho g$ stays finite, but as noted this would in general yield a singular velocity as $\rho \to 0$. In fact, in the models studied so far, $\mathbf{j}$ adjusts to vanish more strongly than ρ, and not only does $-(\nabla \cdot \mathbf{F})_B$ vary like ρ – yielding a finite $(v_B)_r$ – but the total velocity $\mathbf{v}_B + \mathbf{v}_\Omega$ in the surface regions becomes small compared with $\mathbf{v}_\Omega$.

These models are qualitatively similar to those with near zero circulation throughout the radiative zone, in that they predict an increasing ratio $F_\mathrm{t}/F_\mathrm{s}$ as λ_Ω/λ_B increases. A typical ratio of interior to surface field is of the order of several hundred or a thousand. More realistic models which contain a stabilizing toroidal component (cf. Section 5.3) yield very similar results.

9.4.3 *Generalizations*

As noted above, the prediction from these various models of an anti-correlation between Ω and B_s could be seen as fitting in broadly with the observational data as of that time. However, the relevance of strictly steady-state models does depend on the time of approach to a steady state being short compared with a stellar lifetime. In a paper that now appears as a landmark in the development of this whole area, Moss (1984*a*; see also 1986*b*) developed a two-dimensional numerical treatment for the time-dependent equations. Moss's method has the great advantage of avoiding the errors that arise from the inevitable truncation of expansions, e.g. in Legendre polynomials. Solutions were started from a wide range of initial fluxes; the quasi-static solutions were recovered as special cases. Of particular interest are the calculations which start with a fairly uniform B-distribution through the interior of the star. Moss is able to define a critical rotation period P_c such that when $P < P_c$, a mean value $\overline{\mathrm{R}}_m$ of the magnetic Reynolds number for the Eddington–Sweet circulation is rather greater than unity, and when $P > P_c$, $\overline{\mathrm{R}}_m \leq 1$. Unless the field is strong, corresponding to mean strengths of more than about 10^5–10^6 G, the total magneto-centrifugal circulation is effectively just the centrifugally-driven Eddington–Sweet circulation, except in the outermost regions of the star. For typical CP star parameters, $P_c \sim 4$ days, increasing with increasing stellar mass. If $P > P_c$, then an initially fairly uniform flux distribution is found to persist more or less indefinitely, changing only slowly on approximately an Ohmic decay time for the envelope (typically a few $\times 10^9$ yr). By contrast, if $P < P_c$, the field is dragged below the surface and concentrated to the interior, the process occurring roughly on a circulation time scale. For relevant parameters this time is of order a few $\times 10^7$ to a few $\times 10^8$ yr, varying roughly as the square of the period and having a weak dependence on field strength, in that stronger fields are better able to resist burying. Main sequence lifetimes for Ap stars are typically a few $\times 10^8$ yr, so it is quite possible to observe even rapid rotators which have not yet had time to bury their field deep in their interiors. Calculations with initial fields corresponding to displaced dipole surface fields show very similar trends.

The tendency of an inexorable circulation that is symmetric about the axis of the field to bury the field deep in the star is easy to picture. However, Moss (1984*a*, 1986*a*) points out that quite different behaviour may be expected in a highly oblique rotator. Consider a star with an initial obliquity χ, and suppose that through the bulk of the radiative envelope the field is weak enough – $\bar{B} <$ 10^4–10^5 G – for the flow to be given approximately by the rotationally-driven Eddington–Sweet circulation, in meridian planes defined by the rotation axis. We work with the same coordinate system as in Section 9.2. The modified E–S circulation has radial component

$$v_r = V(r)\mathrm{P}_2(\cos\Theta) \tag{9.62}$$

with $V(r) \simeq V_\Omega(r)$ except near the convective core and the stellar surface, where the local magnetic force density complicates the flow structure. In terms of the (r, θ, λ)-system, based on the axis of the initial field, (9.62) with (9.21) substituted becomes

$$\begin{aligned} v_r(r,\theta,\lambda) = V(r)\mathrm{P}_2(\cos\theta)\mathrm{P}_2(\cos\chi) + (3/4)V(r)\sin^2\theta\cos 2\lambda\sin^2\chi \\ +(3/2)V(r)\sin\theta\cos\theta\cos\lambda\sin 2\chi. \end{aligned} \tag{9.63}$$

In the extreme case with $\chi = \pi/2$,

$$v_r = -(1/2)V(r)[\mathrm{P}_2(\cos\theta) - (3/2)\sin^2\theta\cos 2\lambda]. \tag{9.64}$$

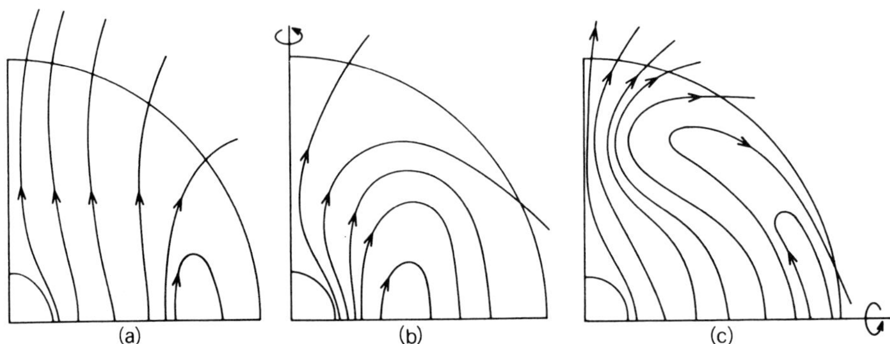

FIG. 9.5. The effect of an inexorable circulation on a large-scale stellar field. (a) Schematic field lines for initial 'quasi-uniform' field configuration. The field is assumed excluded from the convective core. (b) Aligned axes, field lines in quasi-steady configuration ($t \simeq$ few $\times\, 10^8$ yr), $P < P_c$, showing marked reduction of emerging flux. (c) Perpendicular axes, field lines in quasi-steady configuration, $P < P_c$, showing no reduction of emerging flux, but compression towards the equator. (After Moss 1984*a*.)

Averaged over λ, this becomes

$$\bar{v}_r = -(1/2)V(r)\mathrm{P}_2(\cos\theta) \tag{9.65}$$

– of the same form with respect to the field as the E–S circulation in the aligned ($\chi = 0$) case, but of opposite sense.

Figure 9.5 illustrates the interaction between a basically dipolar field and the modified E–S circulation in (a,b) the aligned case, and (c) the present case of a λ-averaged circulation in the perpendicular rotator. We see that if $\mathrm{R_m} \geq 1$ throughout the bulk of the radiative envelope, then in case (c) the field will not be buried but rather concentrated towards the magnetic equator. Thus the theory predicts radically different effects of an inexorable circulation for the aligned and perpendicular rotators respectively.

In general, the average over λ of $v_r(r, \theta, \lambda)$ given by (9.63) has the factor $\mathrm{P}_2(\cos\chi)$, so that for $\chi < \chi_c = \cos^{-1} 1/\sqrt{3} \simeq 55°$ the circulation has the same sense on average as for the aligned rotator, and for $\chi > \chi_c$ as for the perpendicular rotator. When $\chi < \chi_c$ the same argument predicts that for sufficiently rapid rotation the field will eventually be buried, but since the speed of the averaged rotation is reduced by the factor $\mathrm{P}_2(\cos\chi)$, the critical angular velocity is increased by the factor $(\mathrm{P}_2(\cos\chi))^{-1/2}$; whereas when $\chi > \chi_c$ the field is not buried. The detailed computations of Moss (1990*b*) again confirm the qualitative expectations.

The same, essentially kinematic, process affects the apparent distribution of obliquities. One anticipates that an E–V–S-type, rotationally-dominated circulation, acting on a large-scale field, would systematically increase a non-zero initial obliquity χ_0. This is again confirmed by Moss (1977, 1990*b*), whose computations use a self-consistently constructed, thermally-driven magneto-centrifugal circulation. Moss also calculates the predicted variation of the parameter $r = B_e(\min)/B_e(\max)$ for an arbitrary distribution of χ_0-values.

Another topic studied by Moss (1984*b*) is the influence of a non-uniform distribution of mean molecular weight μ on the distribution of flux. As noted above, a non-spherical μ-field in general yields a non-zero contribution $(\nabla \cdot \mathbf{F})_\mu$ to the divergence of the radiation flux, and the consequent 'μ-current' velocity $\mathbf{v}_\mu$ must be superposed on $\mathbf{v}_\alpha + \mathbf{v}_B$. The μ-currents act to choke the circulation near the convective core, and unless the ξ-motions are able to mix material (cf. Section 9.2.2), this ensures that in general even a rapid rotator will evolve into the giant domain to the right of the main sequence rather than up and slightly to the left (Mestel and Moss 1986 and references therein). Originally it was thought that the circulation would continue outside a 'μ-barrier' separating core and envelope, but difficulty in the attempted construction of a self-consistent model of the magnetic-plus-viscous boundary layer outside the barrier suggests that instead the circulation slowly kills itself by distributing helium-rich material through the radiative envelope. This 'creeping paralysis' leads ultimately to the state with

$$(\nabla \cdot \mathbf{F})_\alpha + (\nabla \cdot \mathbf{F})_B + (\nabla \cdot \mathbf{F})_\mu = 0. \tag{9.66}$$

In a star with a moderate poloidal flux, the magnetic contribution to this condition will be small except possibly in the surface regions. (The consequent slow generation of a *toroidal* field by the Biermann 'battery' process is outlined in Section 5.7.2.) The question of interest here is how this growing interference with the slow E–V–S circulation will affect the distribution of flux. Moss concludes from estimates of the relevant time-scales that the effects are likely to be small. The μ-effect will not seriously inhibit the central concentration of flux in a low-obliquity case nor the equatorial compression in a high-obliquity case. Equally, once such concentration has occurred, subsequent paralysis of the circulation by μ-gradients – whether arising from nuclear processing in the core or helium sinking in the surface – will indeed allow Ohmic diffusion of the field back to its undistorted structure, but again the time-scale is probably too long.

The implications of these various conclusions for interpretation of the observations are assessed in Section 9.6.

9.5 Magnetic braking of the oblique rotator; changes in obliquity

The evidence that the magnetic stars mostly rotate markedly more slowly than non-magnetic stars in the same part of the Hertzsprung–Russell diagram suggests that some form of magnetic braking has been active in some epoch. The process that has been studied in greatest detail is that through coupling with a stellar wind (Chapter 7, especially Section 7.8 on the oblique rotator). Because of the lower symmetry of the oblique as compared with the aligned rotator, the predicted torque on the star (7.114) has both a braking component along the rotation axis and also a precessional component, tending to change the obliquity angle.

In Section 8.3 the theory of braking by a magnetically-controlled wind was applied to late-type stars, for which there is strong evidence that the surface flux is primarily or wholly the consequence of contemporary '$\alpha\Omega$' dynamo action. For much of the star's rotational history, the field B_s – averaged over a 'stellar cycle' – decreases with decreasing Ω_s, yielding an algebraic law of variation of Ω_s with time. By contrast, in the early-type magnetic stars there is no clear correlation of B_s or F_s with Ω_s. The arguments for and against the dynamo or fossil interpretation of the CP-star magnetic fields are discussed in Section 9.6. For the moment we note that if the open flux $\Phi \propto F_s$ does not decrease with Ω_s, then (7.79) will predict much more efficient braking over the star's lifetime. If Ω_s is small enough for centrifugal driving to be unimportant, the wind is thermal with v_A nearly constant, and (7.79) yields an exponential braking law. When the star is rotating rapidly, centrifugal driving with Φ constant yields a somewhat reduced braking rate, given by $\dot{J} \propto -(-\dot{M})^{1/3}\Omega_s^{1/3}$, i.e. with a still weaker Ω_s-dependence (though with a smaller value of $-\dot{J}$ for a given Ω_s). Such laws are potentially embarrassing as they lack a built-in cut-off.

A few magnetic stars are known which show periods of the order of a decade. If these are genuine rotation periods, then perhaps these stars are mavericks which have suffered exponential braking via a strong primeval field coupled to a wind during the Hayashi phase. Schüssler (1980) has proposed that the less massive CP stars – in the Sr–Eu–Cr group – which have a longish Hayashi phase – do lose most of their angular momentum via a magnetically-controlled wind before reaching the main sequence. Those in the more massive Si-group have a shorter Hayashi phase and so reach the main sequence as rapid rotators. It may, however, be questioned whether subsequent braking on the main sequence is via a magnetically-controlled wind. It is significant that the X-ray satellite observations do not seem to find evidence of a hot corona around A-type stars. Hot coronas are observed around the more massive magnetic He-weak and He-strong stars, but their main-sequence lifetimes are short. We are impelled to look for a different main sequence braking process for the slowly rotating CP–stars.

The density in a cool corona will exponentiate rapidly to a very low value, suggesting that gravitational accretion from the local interstellar gas is a more appropriate model than expansion of a thermally-supported atmosphere. This in turn can yield 'accretion braking' of the stellar rotation (Mestel 1975; Arons and Lea 1976*a*,*b*; Wolff 1981). Consider first the highly idealized case of a non-magnetic star, at rest in a gaseous medium with effectively zero angular momentum about the star and negligible magnetic flux, and which has settled into a state of steady, spherically-symmetric accretion (Bondi 1952). The inflowing gas is supposed to contain sufficient coolants to keep it nearly isothermal, with sound speed c_s, in spite of the compressional heat generated. Apart from differences in the appropriate boundary conditions, the problem is the inverse of the simplest stellar wind problem and can be treated exactly, but the essence of the solution is as follows. The flow is divided into two domains by the Bondi sphere $r_B = 2GM/c_s^2$. Beyond r_B the gas is nearly in hydrostatic balance, with $\rho(r) \approx \rho_\infty$ and $v \propto 1/r^2$; within r_B the gas effectively falls freely, with $v^2 \simeq 2GM/r$ and $\rho/\rho_\infty \simeq (r_B/r)^{3/2}$, until the presence of the dense stellar surface causes inward motion to be terminated by formation of a shock.

Now suppose the star has a strong magnetic field. If the corona is cool, then the density – determined by hydrostatic equilibrium, holding along the closed field loops (cf. (7.61)) – will again show a rapid exponential decline, but now the magnetic stresses will interfere with the inflow. As long as field-freezing remains a good approximation, the inflowing gas will not cross the stellar field lines but will rather compress the field – a picture reminiscent of the classical Chapman–Ferraro work on terrestrial magnetic storms (e.g. Akasofu and Chapman 1972). The freely-falling gas within the Bondi sphere has a kinetic energy density $\rho GM/r \simeq GM\rho_\infty(r_B/r)^{3/2}(1/r)$, and inflow will be temporarily halted at a new Alfvénic distance $\tilde{r}$, where the kinetic and field energies are comparable. If the distortion of the field from the vacuum dipolar structure is modest, then $\tilde{r}$ is given approximately by

$$\frac{B^2}{8\pi} = \frac{B_s^2}{8\pi}\left(\frac{R}{\tilde{r}}\right)^6 \simeq \frac{GM\rho_\infty}{R}\left(\frac{r_B}{R}\right)^{3/2}\left(\frac{R}{\tilde{r}}\right)^{5/2}, \tag{9.67}$$

or

$$\left(\frac{\tilde{r}}{R}\right) \simeq \frac{10(B_s/10^3)^{4/7}(T_\infty/10^2)^{2/7}}{(\rho_\infty/10m_H)^{2/7}}. \tag{9.68}$$

Note that $\tilde{r}$ is not very sensitive to variations in the parameters. With typical values, the argument predicts that the field will prevent gas coming closer than about 10 stellar radii; whereas if the 'cool' corona has T as high as 10^5 the scale-height is still only $\approx 10^{-2}R$, suggesting that the magnetosphere within $\tilde{r}$ is nearly a vacuum. However, a state in which a magnetic field free of gas is holding up external gas against gravity is known to be Rayleigh–Taylor unstable (cf. Section 4.7.2), leading to a non-axisymmetric structure that allows further flow of gas towards the star down magnetic surfaces. In particular, fluting in the azimuthal direction defined by the rotation axis will allow inflowing gas of low angular momentum to be spun up by the Maxwell stresses into near corotation with the star. If $\Omega_s^2\tilde{r} > GM/\tilde{r}^2$, the gas will be shot back centrifugally into the interstellar medium, carrying off the angular momentum that has been transported from the star out to $\tilde{r}$ by the stresses of the azimuthally-distorted magnetic field. An upper limit to the rate of angular momentum loss by this process is of order

$$A\Omega_s\tilde{r}^2 \simeq 4\pi\rho_\infty c_s r_B^2\Omega_s\tilde{r}^2 \tag{9.69}$$

where A is the accretion rate, written in terms of ρ_∞, c_s and r_B. The angular velocity of a star with radius of gyration kR will decrease exponentially in the time-scale

$$t_b = \frac{kM}{A(\tilde{r}/R)^2} \propto \frac{c_s^{9/7}}{MB_s^{8/7}\rho_\infty^{3/7}}. \tag{9.70}$$

The braking of the star continues until

$$\tilde{\Omega}_s^2 \tilde{r}^3 \simeq GM, \tag{9.71}$$

after which accreted gas can flow through the radius $\tilde{r}$ towards the star, so braking ceases.

The accretion braking process is attractive, in that (9.71) yields a natural cut-off at a rotation period $\tilde{P} = 2\pi/\tilde{\Omega}_s$, typically of a few days; and the braking time estimated from (9.70) is $\approx 10^7$–10^8 yr in favourable circumstances, which is of satisfactory order. The most obvious weakness of the above very approximate treatment is the neglect of the angular momentum about the star of the inflowing gas (cf. Chapter 10).

As with braking by a wind, the reduced symmetry of the oblique rotator ensures that the same tensions along the field lines which exert a braking torque will also in general have non-vanishing moments about the two axes perpendicular to the rotation axis, yielding in particular a precessional torque which again alters the obliquity angle. One anticipates again the general result that the instantaneous axis of rotation seeks out the region on the star's surface where the field strength is greatest (cf. Section 7.8). The precessional torque will again be less than the braking torque, but is probably closer to it than in the wind problem, because the field distorted by inflow will not be forced into an approximately split monopole structure.

9.6 Recapitulation

The fossil-versus-contemporary-dynamo question remains a fundamental, unresolved issue for the early-type magnetic stars. Besides their intrinsic problems, both theories have to face the often bewildering observational data. Why are stars that are of similar spectral type often not magnetically similar? Of the A stars that are known from the period of their spectrum variations to be rotating slowly, why do only some show detectable fields? Although most of the A stars which show rapid rotation through line-broadening are spectroscopically normal and not observably magnetic, why does a minority of the magnetic Ap stars nevertheless rotate rapidly? And how can one understand the lack of correlation within the class of magnetic Ap stars: e.g. stars with similar Zeeman patterns differing in their periods by a factor 10 or more?

It is convenient to treat these questions historically, beginning with the axisymmetric 'aligned rotator'. As already noted, the steady-state models discussed in Section 9.4 – both those with their envelopes in strict local radiative equilibrium, and those with a persisting, magnetically-modified E–V–S circulation – do yield a broad anti-correlation between Ω and the observable flux F_s in a star of total flux F_t. If these models were to describe actual observed magnetic stars, then the argument has to be inverted: one infers (for a star of given spectral type) from the observed Ω and F_s the value of F_t consistent with the assumption of a thermally steady state. Typical values of several hundred to a thousand are found for the ratio of internal to surface field-strength – much greater than that given by the longest-living Cowling decay-mode. Analogous models but with linking toroidal flux, as required for dynamical stability, yield qualitatively similar results.

Simultaneously, there were observations that suggested a link-up with theory. We have already noted the early reports, subsequently discounted, of a correlation – within the class of observed magnetic stars – between rotation period and surface field. Further, it was reported tentatively that the obliquity angle χ had a bimodal distribution, with a marked preference for $\chi \simeq 0$ or $\chi \simeq \pi/2$. This suggested a link-up with the dynamical theory of the oblique rotator, in particular with the conclusion that dissipation of the ξ-motions would force the star to rotate about the axis of maximum moment of inertia. The idea was that stars with a dominant poloidal flux would be dynamically oblate about the magnetic axis and so would approach the state with $\chi = 0$, whereas those with a dominant toroidal flux could be dynamically prolate and so could approach $\chi = \pi/2$. The dissipation rate is sensitive to the nutation rate ω and hence to the total flux, as measured by the parameter ϵ. Stellar models with both a strong central concentration of flux and observable fields would have dissipation times short compared with the stellar evolution time, and so could reach the asymptotic χ-states.

However, the observational situation has changed; it is now agreed that there is at the most a marginal non-randomness in the χ-distribution, so that a theory that predicts a short time for either alignment or non-alignment to occur has become an embarrassment. The ξ-motion dissipation process could remain efficient, but be masked by other processes tending to randomize χ or acting in the opposite sense; but equally, the large F_t-values which yield the efficient dissipation could turn out to be spurious. It had always been recognized that the long time-scales of thermal adjustment within stellar radiative zones would restrict the relevance of steady-state solutions. This is shown convincingly in the cited computations by Moss. In particular, it is quite possible to observe even rapid rotators before they have reached a thermally steady state, for example before a rotator with small obliquity has finished concentrating its flux into the interior. Such a star, showing a strong surface field and the associated abundance peculiarities, but with a modest ratio $F_\mathrm{t}/F_\mathrm{s}$, would be caught in a non-steady state: if one were able to wait long enough, this ratio would increase, but then the surface field would be seen to become weak. Likewise, the results for a highly oblique rotator, which show that the 'inexorable', rotationally-driven circulation does not then bury the field, but rather compresses it towards the magnetic equator, have to be applied with account being taken of time-dependence.

These kinematical results do depend on the microresistivity being appropriate, as indeed one expects to be the case when the field is strong enough to suppress any incipient turbulence that could cause local tangling and so an effective large macroresistivity (Sections 3.8, 4.3 and 6.7). Moss, however, notes that it would not require an enormous increase in resistivity for the slow, thermally-driven motions to be able to flow across the field, so reducing markedly the changes that would be imposed on the field over a main sequence lifetime, even in a rapid rotator. Provisionally accepting the computed results as plausible, we see that a convincing link-up between theory and observation must take account of several processes, of which some are ill understood: the kinematic dragging process, with its effect on the obliquity and the surface flux, the ξ-motion dissipation effect, and the combined braking-plus-precessional torque.

The discussion has been implicitly in terms of the fossil theory, in the generalized sense that the internal flux and the initial obliquity are supposed prescribed extra parameters, fixed by the star's prehistory. An advantage of the fossil theory is its flexibility, demonstrated especially in the extensive studies by Moss (1984*a*, 1985, 1986*b*, 1987), which as discussed, follow the evolution of various initial flux distributions through the combined effects of Ohmic diffusion and advection by the self-consistently constructed, thermally-driven velocity field in the radiative envelope. For example, the Landstreet (1970) displaced dipole model is adequate for the interpretation of nearly all of the observed magnetic CP stars, but as noted in Section 9.1, there are at least three stars which appear to have too large a quadrupolar component. Moss (1987) shows that an initial flux distribution which is abnormal in having both points of entry and exit of field lines largely confined to one hemisphere will continue to give the appearance of having a strong quadrupolar component. His point is that flux distributions that are all topologically consistent with a fossil origin in the interstellar medium can yield a wide variety of observed fields. However, if further advances in stability theory were to shorten markedly the expected lifetimes of large-scale stellar fields, then one would be forced to look for an appropriate dynamo model.

As noted, Krause (1971) and Krause and Oetken (1976) have argued for a dynamo operating within the convective core, which generates a field with an axis that is either strictly perpendicular to the rotation axis, or at least with a non-trivial perpendicular component. Krause (1983) accounts for the existence of rapidly rotating, normal, *non*-magnetic stars by replacing the suggestion of a fossil magnetic field by a fossil, markedly non-uniform rotation field in the radiative envelope. A spontaneously-generated perpendicular magnetic field would then be rapidly sheared, and so would be subject to excess dissipation (cf. Section 9.3). Only in those stars with a weak shear would the dynamo-built perpendicular field in the core be able to rise to the surface. Borra and Landstreet (1978) have criticized the perpendicular rotator model model on observational grounds. Again, the crucial theoretical question concerns the time-scales. The energy being continuously dissipated comes from the postulated non-uniform rotation: the effect of the dynamo is to produce a mechanism for an efficient destruction of the shear energy. Will the initial state of rotation last very much longer than the Alfvén travel time, defined by the **B**-field that would be generated in the absence of the shear?

For all theories that locate the dynamo in the convective core, there is the difficulty in seeing how flux generated by a d.c. dynamo in the core can rise quickly enough through the subadiabatic, radiative envelope, ultimately to manifest itself at the star's surface as a large-scale ordered field with a well-defined obliquity angle. Moss (1989) considered individual flux tubes rising through magnetic buoyancy, essentially at the 'Eddington–Sweet' rate, typically in the rise time $\simeq 3 \times 10^{12}(R_B/H_t)^2/(B/10^6)^2$ yr, where H_t is the temperature scale height and R_B the radius of the flux-tube. The OB1 association contains magnetic CP stars which must have the age $\approx 5 \times 10^6$ yr, so putting severe upper limits on R_B/H_t, lower limits on $B_s/10^6$, or both. Moss estimates the value of B in the convective core first by assuming equipartition with the turbulence – $B_{eq} \approx (4\pi\rho)^{1/2}v_t \approx 10^5$ G. This requires that R_B/H_t be very small: the field escaping would need to be an extreme fibril field, consisting of flux-'threads' rather than flux-ropes, which have to maintain their identity through their ascent, but which nevertheless contrive to form ultimately a well-ordered large-scale surface field. An alternative estimate for B based on the balance of Coriolis and Lorentz forces gives similar results. Further, if a steady state has been reached, and if the core field is connected to the surface field, then the associated poloidal flux in the core is an upper limit to that at the surface, making it very difficult to account for those stars that have surface fields well above 10^4 G. Another difficulty is the likely inhibition of buoyancy by a developing negative μ-gradient near the core surface through nuclear processing.

These problems led Moss (1990*a*) to consider the prima facie more hopeful case of a Gailitis-type dynamo (Section 6.2), in which axisymmetric laminar motions – for example the simple E–V–S circulation, unmodified by magnetic forces – generate and maintain fields that are *non*-axisymmetric and so do not need to have the singular lines which by Cowling's theorem and its extensions are the Achilles heel of fields that are axisymmetric or topologically similar. Moss's conclusions are that such a model could operate in the more rapidly rotating CP stars, but that even with the shortest period of 0.5 days, typical generation times are only just short enough to be of potential interest. The process seems unable to account for the fields in the more slowly rotating magnetic stars, or for the fields in the younger objects.

The observed stability of the fields of individual CP stars and of the associated spectral and light variations implies that magnetic evolution occurs on much longer time-scales than for the Sun and solar-type stars, where we see manifestations of the interactions between turbulence and magnetism occurring in days or weeks, as well as the solar and stellar cycles that recur three or four times in a human lifetime. In spite of the frustrating ambiguity in the interpretation of the observations, and of the huge gaps in our theoretical understanding, one can draw some tentative if modest conclusions: (1) the oblique rotator remains a plausible phenomenological model for the early-type magnetic stars; (2) no overriding objections to the generalized fossil theory have as yet emerged; (3) the contemporary dynamo interpretation still seems to be faced with more severe difficulties than the fossil.

For an interesting alternative approach to the problems of the magnetic fields of the early-type stars, the reader is referred to Meyer (1994).

9.7 Abundance anomalies

As noted in the Section 9.1, the early-type magnetic stars are a sub-class of the chemically-peculiar stars. Details of the observational situation have been reviewed most recently in IAU Coll. **138** (Dworetsky *et al.* 1993). Two papers especially relevant to the magnetic Ap stars are by Landstreet (1988) and Landstreet *et al.* (1989).

Element segregation can arise spontaneously in stable stellar atmospheres through the relative gravitational settling of species of different mass (pressure diffusion), through thermal diffusion due to a temperature gradient, and through selective radiation pressure on individual particles. The basic ideas appear in Eddington (1959) and are treated rigorously in Chapman and Cowling (1970). It is reasonable to study whether diffusion alone is capable of accounting for the observations in detail, or whether some additional, hydrodynamic process is involved. Of the many theoretical papers and reviews, devoted specifically to the CP stars, we mention particularly Michaud (1970), Michaud *et al.* (1976, 1981), Vauclair and Vauclair (1982), Vauclair (1983), Mégessier (1984), Babel and Michaud (1991*a,b,c*), and Babel in Dworetsky *et al.* (1993).

Recent observational studies (Holweger *et al.* 1986; Hill and Landstreet 1993; Hill 1995) indicate that segregation is not entirely smothered in 'normal' A-stars, but it is certainly markedly reduced. One is led to ask: what is it about normal A-stars that inhibits the spontaneous development of 'abundance anomalies'? The observations suggest convincingly that slower rotation is a crucial factor that allows abundance anomalies to manifest themselves. One expects magnetic stars to suffer systematic braking (cf. Section 9.5), and the small minority of rapid rotators among the magnetic CP stars may be provisionally interpreted as those which have not yet had sufficient time to lose much angular momentum. Further, of the A stars in close binary systems, it is those of intermediate separation d that show the peculiarities of Am or Ap type (Abt 1965), whereas those in both close and wide binaries have normal atmospheres. The obvious suggested explanation is that both close and moderately close systems will have time to achieve synchronization of spin and orbital angular velocities Ω_s and Ω_d, due to dissipation of the tidal motions arising in a non-synchronized system (e.g. Zahn 1992*b*), even in the absence of magnetic coupling. A-type stars in synchronized close binaries will therefore rotate rapidly, since $\Omega_s = \Omega_d = G(M_1 + M_2)/d^3 \simeq G(M_1+M_2)/(R_1+R_2)^3$, and so should behave like normal, rapidly rotating single A stars. Synchronized moderately close binary members – e.g. with $d/(R_1+R_2) \simeq$ 2–3 – will rotate fairly slowly and so should show CP characteristics; while for wider binaries, the time for tidal synchronization will be too long, and A stars should retain the rapid rotation with which they are presumably born, and so should indeed have normal atmospheres. But there appear to remain subtleties requiring quantitative explanation; for example, from his studies of abnormal stars in open clusters, Abt (1979) concluded that abnormalities begin to develop in some moderate rotators, which subsequently suffer further braking (magnetic or tidal).

For stars that are not observably magnetic, a plausible explanation of the absence of the CP phenomenon in rapid rotators is that the low-density surface regions will then develop turbulent velocities that mix material and so counteract the effects of element diffusion. As noted in Section 5.4, for general perturbing forces, the Eddington–Sweet theory predicts circulation speeds that become fast like $\bar{\rho}/\rho$ in the stellar surface. An initial uniform rotation will then develop a violent shear and is likely itself to become unstable, probably returning to near uniform rotation, at least over horizontal surfaces (Zahn 1983, 1992*a*). By contrast, a moderately strong magnetic field will be able to keep not only the bulk of the star but also the surface regions in near uniform rotation, without any shear turbulence developing. The quasi-steady models of the atmospheres of rotating magnetic stars (Mestel and Moss 1977; Moss 1984*a*) show that even in a rapid rotator, any surviving magneto-centrifugo E–V–S circulation will be very slow and should not offset the effects of slow but persisting element diffusion.

Spontaneous thermal convection would also mix efficiently. In cooler A stars there will be a weak subphotospheric convection zone due to the completion of hydrogen ionization. A strong magnetic field with a vertical component can, however, stabilize both the convective overshoot and at least part of the superadiabatic domain (cf. Chapter 4). The many detailed studies of diffusion suggest that the 'simple diffusion model' – no mass loss, circulation or turbulence – can account for the observed abundance anomalies, *averaged* over the stellar surface, provided the convection zone is only partially suppressed, but not their distribution over the stellar surface (Babel and Michaud 1991*a*). In such a stabilized domain, the horizontal magnetic component will interfere to some extent with vertical diffusion. An instructive discussion of the simplest case is given by Babel and Michaud (1991*c*); for convenience we follow their notation. A pure hydrogen gas is assumed, with local densities n_e, n_1, n_0 of electrons, protons and H-atoms respectively, and associated partial pressures $p_e = n_e kT$, etc. There is a horizontal magnetic field $\mathbf{B}$. Only diffusion under the partial pressure gradients is considered. Quasi-steady equations are written down, reminiscent of those in Chapter 2:

$$-\nabla p_1 + n_1 m\mathbf{g} + n_1 e\left(\mathbf{E} + \frac{\mathbf{v}_1 \times \mathbf{B}}{c}\right) = mn_1\nu_{10}(\mathbf{v}_1 - \mathbf{v}_0) + mn_1\nu_{1e}(\mathbf{v}_1 - \mathbf{v}_e), \tag{9.72}$$

$$-\nabla p_0 + n_0 m\mathbf{g} = mn_0\nu_{01}(\mathbf{v}_0 - \mathbf{v}_1), \tag{9.73}$$

$$-\nabla p_e - n_e e\left(\mathbf{E} + \frac{\mathbf{v}_e \times \mathbf{B}}{c}\right) = m_e n_e \nu_{e1}(\mathbf{v}_e - \mathbf{v}_0), \tag{9.74}$$

where the gravitational force on the electrons and the small mutual electron-neutral gas friction have been ignored. As usual (cf. Sections 2.2 and 2.3), $\mathbf{E}$ given by (9.72) or (9.74) yields the net charge density $\rho_e = \nabla \cdot \mathbf{E}/4\pi$ satisfying $|\rho_e/n_e e| \ll 1$, implying $n_1 = n_e$ to a high approximation. We recover the bulk equation

$$-\nabla p + (n_1 + n_0)m\mathbf{g} + \frac{\mathbf{j} \times \mathbf{B}}{c} = 0. \tag{9.75}$$

since $n_0\nu_{01} = n_1\nu_{10}$ etc. by the law of action and reaction.

If local axes are chosen with z vertically outwards and $\mathbf{B}$ along the y-axis, then $\mathbf{v}_1 \times \mathbf{B}$ has a horizontal component along y. From the y-component of (9.73), $v_{0y} = v_{1y}$, whence from the y-component of (9.74)

$$j_y = n_e e(v_1 - v_e) = \frac{e^2 B_x}{m_e c}\frac{n_e}{\nu_{e1}} v_{1z} \tag{9.76}$$

– just Ohm's law (2.84) with $\sigma = n_e e^2/m_e\nu_{e1}$. Adding the z-components of (9.72) and (9.74), with substitution from (9.76), yields the vertical diffusion velocity v_{1z}

$$v_{1z} \approx \frac{v_1^0}{1 + (m_e/m\nu_{10})(\omega_e^2/\nu_{e1})(n_0/(n_0 + n_1))} \tag{9.77}$$

where $\omega_e = eB/m_e c$, and

$$v_1^0 = \frac{n_0 kT[-2\mathrm{d}\log p_1/\mathrm{d}z + ((2n_1 + n_0)/(n_1 + n_0))\,\mathrm{d}\log p/\mathrm{d}z]}{m\nu_{10}(n_0 + n_1)} \tag{9.78}$$

is the diffusion velocity, again driven by partial pressure, but with $B_x = 0$. The Lorentz force $\mathbf{f}_L \equiv \mathbf{j} \times \mathbf{B}/c$ has a z-component

$$-\frac{j_y B_x}{c} = -\frac{n_e\omega_e^2 m_e}{\nu_{e1}} v_{1z}. \tag{9.79}$$

(cf. 2.58). This local effective enhancement of gravity requires a remodelled stellar atmosphere.

In the limit when $\omega_e^2/\nu_{e1}\nu_{e0} \gg 1$ (in the notation of Chapter 2, when $(\omega_e\tau_{e1})(\omega_e\tau_{10}) \gg 1$), then $v_{1z}/v_1^0 \ll 1$: the magnetic field effectively blocks the diffusion of protons and electrons. In equilibrium, the mean velocity is zero, so in the same high-field limit, $v_{0z} = 0$. The electrons and protons then have a joint equation of equilibrium, given by the sum of (9.72) and (9.74)):

$$f_{Lz} = 2\frac{\mathrm{d}p_1}{\mathrm{d}z} + n_1 m g. \tag{9.80}$$

As there is now no frictional coupling between the charged and neutral components, the neutral component also must adjust to satisfy equilibrium independently. The overall effect is a large increase in the ionization region. In actual atmospheres, the limit of zero proton diffusion will not actually be reached, but the same qualitative changes will appear.

By contrast, in a domain in which the magnetic field is nearly vertical the upward drift of protons will persist and can impart momentum to ionized trace elements such as Ca (Babel and Michaud 1991*c*). Thus the presence or absence of this form of 'ambipolar diffusion' of hydrogen, related to local structure of the magnetic field, can be a major process affecting the abundances in different regions over a stellar surface, independently of the more familiar selective radiation pressure. However, Babel (1992, 1994) argues also for a small but inhomogeneous stellar wind outflow, with its structure for the star 53 Cam related to the observed magnetic structure.

A cognate question is that of the luminosity variations of the Ap stars, which could be due at least in part to the direct effect of the magnetic forces on the outer layers, affecting both the shape of the photosphere and the temperature variations. The same *displaced* dipole model – required to explain the lack of perfect anti-symmetry in the Zeeman measurements – should also yield asymmetry in the light curves, so that although the magnetic forces are quadratic in $\mathbf{B}$, luminosity variations should also have the period P rather than $P/2$. The earlier suggestion by Mestel (1967) that a magnetic star as a whole should be dynamically symmetric about the magnetic axis now seems implausible, but it is still possible that a realistic model will have the magnetic forces dominating over the centrifugal in the photospheric regions. Some phenomenological studies of the star α^2 CVn by Böhm-Vitense and Van Dyk (1987) are able to explain both the optical and ultraviolet variations by adopting a photosphere which is an oblate spheroid about the magnetic axis and has a lower temperature at the poles than at the equator. The authors suggest that there is an active cooling mechanism at work at the poles, e.g. through MHD waves, or a stellar wind. This is similar to the alternative view on sunspot cooling advanced by Parker (1979) and judged by most workers to be less plausible than the Biermann picture of magnetic interference with heat supply from below (cf. Section 4.6). It could, however, turn out to be relevant to early-type stars, which do not rely on a subphotospheric convective zone to transport heat: instead of inhibiting heat supply to the surface, the field may act as conduit for excess heat loss. However, an alternative view (Molnar 1973; Lanz, personal communication) argues that the photometric variations are primarily a consequence of rotational modulation of the far-ultraviolet absorption by the observed abundance inhomogeneities, leading to a periodic flux redistribution at longer wavelengths. It emerges that a proper understanding of the abundance anomalies and the luminosity variations requires better atmosphere models and further hydrodynamic studies.

9.8 The roAp phenomenon

A challenging new area of research was opened by the discovery of the rapidly oscillating Ap (roAp) stars (Kurtz 1978). A comprehensive review of their properties is given by Kurtz (1990); some updating will be found in Section 2.1 of Gautschy and Saio (1996). The stars belong to the Sr–Cr–Eu CP subclass. The pulsation periods are in the range 5–15 min. The amplitudes are modulated in time-scales identical with both the stellar rotation periods and the temporal variations of the strong stellar magnetic fields (from several hundred to several thousand G, typical of Ap stars in general).

The oblique pulsator model (Kurtz 1982) emerged early as a successful and popular phenomenological model, with the oscillations pictured as high-overtone, low-degree, non-radial p-modes, with the magnetic and pulsation axes aligned. An alternative proposal (Mathys 1985) is the 'spotted pulsator', which has the pulsation axis aligned rather with the rotation axis. The oscillation pattern is then always seen from the same aspect, but the amplitude modulation is due to an inhomogeneous flux distribution, a function of magnetic phase during a rotation period.

The necessary modifications to standard stellar pulsation theory turn out to be quite subtle (Biront *et al.* 1982; Roberts and Soward 1983; Campbell and Papaloizou 1986). Recall from the earlier sections that the gross magnetic energy of the star is almost certainly a small fraction of the gravitational, and equivalently, that over the dense bulk of the star the magnetic pressure is a small fraction of the thermal. Also, even in the low-density regions close to the photosphere where the magnetic and thermal pressures are comparable, in many of the quasi-steady models studied the current density adjusts itself so as to make $|\nabla \times \mathbf{B}|/\rho g$ much smaller than naively expected. Thus the effect of the zero-order field on the structure of the star is modest even in the surface regions. However, under the perturbations associated with a given oscillatory mode of a non-magnetic star, whereas over the bulk the Lorentz force exerted by the perturbed field will again be small compared with the perturbed gravity and pressure gradient, in the surface regions the distortion to the nearly curl-free field now exerts forces that are not negligible. In particular, a basically radial mode will acquire significant horizontal velocity components. The straightforward perturbation treatment of the oscillations becomes singular in the surface regions, and must be modified so as to match asymptotic approximations, respectively appropriate to the bulk of the star and to the surface boundary layer.

Roberts and Soward (1983) (see also Roberts 1984) pointed out that Alfvén-type waves are necessarily generated in the surface regions and will propagate inwards. Their rigorous analysis shows that in a strictly *non-dissipative* problem, the motions are sensitive to the strength and structure of the magnetic field deep in the star: in their words, 'a conclusion that seems as mathematically inevitable as it is astrophysically unrealistic'. They conclude that the modelling is at fault, and that in reality, the short-wavelength Alfvén waves will be damped by the action of radiative viscosity and conductivity. Further, the outer layers of the star will contain He and perhaps H ionization zones which will be turbulent and so will act as efficient dampers of the short-wavelength waves. They therefore reformulate the problem by incorporating this powerful dissipation, changing a pure standing wave problem into one with progressive waves. In the surface regions, the Alfvén waves make the dominant contribution to the horizontal motions, which, however, have typical amplitudes of about one-tenth of the vertical motions. The generation and subsequent dissipation of the Alfvén waves make an important contribution to the damping of the global pulsation. The work was extended by Campbell and Papaloizou (1986) to allow for angular dependence of the displacements over the stellar surface, and to cover a larger range of frequency and wavenumber.

As emphasized by Kurtz, a successful theoretical understanding of the observations should yield a valuable probe of the whole CP star phenomenon, analogous to helioseismology and its offshoot astroseismology. Many problems remain, especially the excitation process, and the related question as to why only one or two modes are excited. In some modes there is an observed frequency splitting with the components separated by a multiple of the rotation frequency Ω. Dolez and Gough (1982) pointed out that oscillation modes symmetric about the magnetic axis should precess about the rotation axis at a rate $\approx 0.99\Omega$, which is apparently not observed. They conjecture that the precession does occur, but excitation occurs only when the oscillation and rotation axes are approximately parallel, and that growth and decay rates are much greater than the precession rate. They suggest that the excitation of the oscillations is related to the spatial chemical variations, which have the symmetry of the magnetic field. Thus the non-radial modes grow rapidly when aligned with the magnetic field, but once the precession has destroyed the initial alignment the modes decay. Excitation of these oscillations through the 'κ-mechanism' can occur if, for example, there is an excess of He at the poles. Vauclair *et al.* (1991) argue that mass loss by a stellar wind flowing along the magnetic axis will indeed lead to the required helium excess.

The study of magnetic oscillations is motivated also by the *Blazhko effect*, observed in some of the RR Lyrae short-period variables stars (e.g. Gautschy and Saio 1996). This is an observed variation in form and amplitude of the fundamental mode in which RR Lyrae variables oscillate, in a period of between 20 and 100 days, as compared with a typical oscillation period of about 1 day. Cousens (1983) suggested that the Blazhko period is essentially the stellar rotation period, so that what we observe is the effect of the oblique surface magnetic field on the fundamental stellar mode. The motions are again radial deep down, but are significantly modified in the low-density surface regions. In the prototype star RR Lyrae there is indeed observed a magnetic field that varies with both the fundamental pulsation period and the Blazhko period. More recently, Takata and Shibahashi (1995) have rediscussed the oscillating dipole model, finding in contrast to Cousens a Blazhko amplitude that depends on the strength of the magnetic field.

References

Abt, H.A. (1965). *Astrophysical Journal Supplement*, **11**, 429.

Abt, H.A. (1979). *Astrophysical Journal*, **230**, 485.

Akasofu, S.-I. and Chapman, S. (1972). *Solar–terrestrial physics.* Clarendon Press, Oxford.

Arons, J. and Lea, S.M. (1976*a*). *Astrophysical Journal*, **207**, 914.

Arons, J. and Lea, S.M. (1976*b*). *Astrophysical Journal*, **210**, 792.

Babcock, H.W. (1956). *Astrophysical Journal*, **124**, 489.

Babcock, H.W. (1958). *Astrophysical Journal Supplement*, **3**, 141.

Babcock H.W. (1960). *Stellar atmospheres.* In *Stars and stellar systems* Vol. VI (ed. J.L. Greenstein), p. 282. University Chicago Press.

Babel, J. (1992). *Astronomy and Astrophysics*, **258**, 449.

Babel, J. (1994). In *Cosmical magnetism, contributed papers* (ed. D. Lynden-Bell), p. 63. Institute of Astronomy, Cambridge.

Babel, J. and Michaud, G. (1991*a*). *Astrophysical Journal*, **366**, 560.

Babel, J. and Michaud, G. (1991*b*). *Astronomy and Astrophysics*, **241**, 493.

Babel, J. and Michaud, G. (1991*c*). *Astronomy and Astrophysics*, **248**, 155.

Baker, N. and Kippenhahn, R. (1959). *Zeitschrift Astrophysik*, **48**, 140.

Biront, D., Cousens, A., Goossens, M. and Mestel, L. (1982). *Monthly Notices Royal Astronomical Society*, **201**, 619.

Böhm-Vitense, E. and Van Dyk, S.D. (1987). *Astronomical Journal*, **93**, 1527.

Bondi, H. (1952). *Monthly Notices Royal Astronomical Society*, **112**, 195.

Borra, E.F. and Landstreet, J.D. (1977). *Astrophysical Journal*, **212**, 141.

Borra, E.F. and Landstreet, J.D. (1978). *Astrophysical Journal*, **222**, 226.

Borra, E.F. and Landstreet, J.D. (1980). *Astrophysical Journal Supplement*, **42**, 421.

Borra, E.F., Landstreet, J.D. and Mestel, L. (1982). *Annual Review Astrononomy Astrophysics*, **20**, 191.

Campbell, C.G. and Papaloizou, J.C.B. (1986). *Monthly Notices Royal Astronomical Society*, **220**, 577.

Chandrasekhar, S. (1933). *Monthly Notices Royal Astronomical Society*, **93**, 390.

Chapman, S. and Cowling, T.G. (1970). *The mathematical theory of non-uniform gases*, (3rd edn). Cambridge University Press. (Paperback edition, Cambridge Mathematical Library Series 1990.)

Cousens, A. (1983). *Monthly Notices Royal Astronomical Society*, **203**, 1171.

Davies, G.F. (1968). *Australian Journal Physics*, **21**, 294.

Deutsch, A.J. (1956). *Publications Astronomical Society Pacific*, **68**, 92.

Deutsch, A.J. (1958). In *Electromagnetic phenomena in cosmical physics* (ed. B. Lehnert), p. 209. Cambridge University Press.

Deutsch, A.J. (1970). *Astrophysical Journal*, **159**, 985.

Dolez, N. and Gough, D.O. (1982). In *Pulsations in classical and cataclysmic variable stars* (eds J.P. Cox and C.J. Hansen), p. 248. JILA, Boulder.

Dworetsky, M.M., Castelli, F. and Farragiana, R. (eds). (1993). *Peculiar versus normal phenomena in A-type and related stars*, IAU Coll. **138**. San Francisco, Astronomical Society of the Pacific.

Eddington, A.S. (1959). *The internal constitution of the stars.* Dover, New York. (Cambridge University Press edition 1926.)

Gautschy, A. and Saio, H. (1996). *Annual Review Astronomy Astrophysics*, **34**, 551.

Gollnow, H. (1965). In *Stellar and solar magnetic fields* (ed. R. Lüst), p. 23. North-Holland, Amsterdam.

Hensberge, H., van Rensbergen, W., Goossens, M. and Deridder, G. (1979). *Astronomy Astrophysics*, **75**, 83.

Hill, G.M. (1995). *Astronomy and Astrophysics*, **294**, 536.

Hill, G.M. and Landstreet, J.D. (1993). *Astronomy and Astrophysics*, **276**, 142.

Holweger, H., Steffen, M. and Gigas, D. (1986). *Astronomy and Astrophysics*, **163**, 333.

Krause, F. (1971). *Astronomische Nachrichten*, **293**, 187.

Krause, F. (1983). In *Stellar and planetary magnetism* (ed. A. M. Soward), p. 205. New York, Gordon and Breach.

Krause, F. and Oetken, L. (1976). In *Physics of Ap stars* (eds W. Weiss, H. Jenkner and H.J. Wood), p. 29. Universitäts-Sternwarte, Vienna.

Kurtz, D.W. (1978). *Astrophysical Journal*, **221**, 869.

Kurtz, D.W. (1982). *Monthly Notices Royal Astronomical Society*, **200**, 807.

Kurtz, D.W. (1990). *Annual Review Astronomy Astrophysics*, **28**, 607.

Landstreet, J.D. (1970). *Astrophysical Journal*, **159**, 1001.

Landstreet, J.D. (1992). *Astronomy and Astrophysics Reviews*, **4**, 25.

Landstreet, J.D. (1988). *Astrophysical Journal*, **326**, 967.

Landstreet, J.D. (1990). *Astrophysical Journal*, **352**, L5.

Landstreet, J.D., Barker, P.K., Bohlender, D.A. and Jewison, M.S. (1989). *Astrophysical Journal*, **344**. 876.

Lanz, T. and Mathys, G. (1993). *Astronomy and Astrophysics*, **280**, 486.

Ledoux, P. and Renson, P. (1966). *Annual Review Astronomy Astrophysics*, **4**, 293.

Mathys, G. (1985). *Astronomy and Astrophysics*, **151**, 315.

Mathys, G. (1989). *Fundamentals Cosmical Physics*, **13**, 143.

Mathys, G. and Hubrig, S. (1995). *Astronomy and Astrophysics*, **293**, 810.

Mathys, G. and Hubrig, S. (1997). *Astronomy and Astrophysics Supplement Series*, **124**, 475.

Mathys, G. and Lanz, T. (1990). *Astronomy and Astrophysics*, **230**, L21.

Mathys, G., Hubrig, S., Landstreet, J.D., Lanz, T. and Manfroid, J. (1997). *Astronomy and Astrophysics Supplement Series*, **123**, 353.

Mégessier, C. (1984). *Astronomy and Astrophysics*, **138**, 267.

Mestel, L. (1953). *Monthly Notices Royal Astronomical Society*, **113**, 716.

Mestel, L. (1967). In *The magnetic and related stars* (ed. R.C. Cameron), p. 101. Mono Book Corporation, Baltimore.

Mestel, L. (1975). *Mémoires Société Royale de Science de Liège*, 6e Sér., **8**, 79.

Mestel, L. (1984). *Astronomische Nachrichten*, **305**, 301.

Mestel, L. and Moss, D.L. (1977). *Monthly Notices Royal Astronomical Society*, **178**, 27.

Mestel, L. and Moss, D.L. (1986). *Monthly Notices Royal Astronomical Society*, **221**, 22.

Mestel, L. and Takhar, H.S. (1972). *Monthly Notices Royal Astronomical Society*, **156**, 419.

Mestel, L., Nittman, J., Wood, W.P. and Wright, G.A.E. (1981). *Monthly Notices Royal Astronomical Society*, **195**, 979.

Meyer, F. (1994). In *Cosmical magnetism* (ed. D. Lynden-Bell), p. 67. Kluwer, Dordrecht.

Michaud, G. (1970). *Astrophysical Journal*, **160**, 641.

Michaud, G., Charland, Y., Vauclair, S. and Vauclair G. (1976). *Astrophysical Journal*, **210**, 447.

Michaud, G., Mégessier, C. and Charland, Y. (1981). *Astronomy and Astrophysics*, **103**, 244.

Molnar, M.R. (1973). *Astrophysical Journal*, **179**, 527.

Monaghan, J.J. and Robson, K.W. (1971). *Monthly Notices Royal Astronomical Society*, **155**, 231.

Moss, D. (1977). *Monthly Notices Royal Astronomical Society*, **178**, 27.

Moss, D. (1984*a*). *Monthly Notices Royal Astronomical Society*, **209**, 607.

Moss, D. (1984*b*). *Monthly Notices Royal Astronomical Society*, **210**, 489.

Moss, D. (1985). *Monthly Notices Royal Astronomical Society*, **213**, 575.

Moss, D. (1986*a*). *Magnetic fields in stars. Physics Reports*, **140**, 1.

Moss, D. (1986*b*). *Monthly Notices Royal Astronomical Society*, **223**, 727.

Moss, D. (1987). *Monthly Notices Royal Astronomical Society*, **226**, 297.

Moss, D. (1989). *Monthly Notices Royal Astronomical Society*, **236**, 629.

Moss, D. (1990*a*). *Monthly Notices Royal Astronomical Society*, **243**, 537.

Moss, D. (1990*b*). *Monthly Notices Royal Astronomical Society*, **244**, 272.

Moss, D. (1992). *Monthly Notices Royal Astronomical Society*, **257**, 593.

Moss, D., Mestel, L. and Tayler, R.J. (1990). *Monthly Notices Royal Astronomical Society*, **245**, 550.

Nittman, J. and Wood, W.P. (1981). *Monthly Notices Royal Astronomical Society*, **196**, 491.

Oetken, L. (1977). *Astronomische Nachrichten*, **298**, 197.

Oetken, L. (1979). *Astronomische Nachrichten*, **300**, 1.

Parker, E.N. (1979). *Cosmical magnetic fields.* Clarendon Press, Oxford.

Preston, G.W. (1970). In *Stellar rotation* (ed. A. Slettebak), p. 254. Dordrecht, Reidel.

Preston, G.W. (1971). *Publications Astronomical Society Pacific*, **83**, 571.

Rädler, K.-H. (1986). In *Plasma astrophysics* (ed. T.D. Guyenne), p. 569. ESA SP-251.

Roberts, P.H. (1984). *Astronomische Nachrichten*, **305**, 45.

Roberts, P.H. and Soward, A.M. (1983). *Monthly Notices Royal Astronomical Society*, **205**, 1171.

Roxburgh, I.W. and Smith, B. (1977). *Astronomy and Astrophysics*, **45**, 1.

Schüssler, M. (1980). *Nucleonika*, **25**, 1425.

Spitzer Jr., L. (1958). In *Electromagnetic phenomena in cosmical physics* (ed. B. Lehnert), p. 169. Dordrecht, Reidel.

Stibbs, D.W.N. (1950). *Monthly Notices Royal Astronomical Society*, **110**, 395.

Sweet, P.A. (1950). *Monthly Notices Royal Astronomical Society*, **110**, 548.

Takata, M. and Shibahashi, H. (1995). *Publications Astronomical Society Japan*, **47**, 219.

Tassoul, J.-L. and Tassoul, M. (1982). *Astrophysical Journal Supplement*, **49**, 317.

Thompson, I.B. and Landstreet, J.D. (1985). *Astrophysical Journal Letters*, **289**, L9.

Vauclair, S. (1983). In *Astrophysical processes in upper main sequence stars* (eds B. Hauck and A. Maeder), p. 167. Sauverny, Geneva Observatory.
Vauclair, S. and Vauclair, G. (1982). *Annual Review of Astronomy and Astrophysics*, **20**, 37.
Vauclair, S., Dolez, N. and Gough, D.O. (1991). *Astronomy and Astrophysics*, **252**, 618.
Wolff, S.C. (1981). *Astrophysical Journal*, **224**, 221.
Wolff, S.C. (1983). *The A-type stars*, NASA SP463, 93. NASA, Washington.
Wright, G.A.E. (1969). *Monthly Notices Royal Astronomical Society*, **146**,197.
Zahn, J.-P. (1983). In *Astrophysical processes in upper main-sequence stars* (eds A.N. Cox, S. Vauclair and J.-P. Zahn), p. 253. Swiss Society Astronomy Astrophysics.
Zahn, J.-P. (1992). *Astronomy and Astrophysics*, **265**, 115.
Zahn, J.-P. (1992). In *Binaries as tracers of stellar formation* (eds A. Duquennoy and M. Mayor), p. 253. Cambridge University Press.

10

PRE-MAIN SEQUENCE STARS

10.1 The later stages of star formation

The role of the galactic magnetic field during the early phases of star formation is discussed in detail in Chapters 11 and 12. A recurring theme is the 'angular momentum problem' – how condensation of gas by so many orders of magnitude in scale can occur without being impeded by spin-up. It is argued there that magnetic effects in general, and in particular magnetic transfer of angular momentum, play a crucial role in the contraction and fragmentation of cosmical gas clouds. Even so, in general it is more than likely that when fragments have reached the opaque phase – after which contraction occurs on the slower radiation loss time rather than in the free-fall time – further redistribution of angular momentum must still occur before the main sequence is reached.

The general problem has a long history, going back to Laplace's nebular hypothesis for the origin jointly of the Sun and of the planets (see e.g. Pringle 1981). In the first edition of *The Earth*, Jeffreys (1924) recognized that any effective viscosity would redistribute angular momentum within the early solar nebula, causing the inner parts to move in and the outer parts to be lost. Peek (1942) and von Weizsäcker (1943, 1948) stated explicitly that a turbulent viscosity would separate the nebula into a pressure-supported central core containing most of the mass and an extended disc containing most of the angular momentum. With the mixing length and so also the viscosity prescribed as a function of radius, a formal solution for the subsequent evolution can be constructed (Lüst 1952).

In this picture, the viscous redistribution of angular momentum is supposed to be crucial for the actual formation of the central star. We shall assume rather that a pressure-supported but rapidly rotating central body has formed, but that it is surrounded by a centrifugally supported disc.

The dynamical reasons forcing one to consider disc-like structures are so fundamental that it is no surprise that similar models have had fruitful application to X-ray stars (Prendergast and Burbidge 1968) and quasars (Lynden-Bell 1969) (just the pioneering papers being cited in both cases). 'Accretion disc' theory is now a well-developed area of study (e.g. Frank *et al.* 1992; Papaloizou and Lin 1995; Campbell 1997), with the central gravitating body sometimes an uncollapsed star, sometimes a white dwarf, neutron star or a black hole of stellar mass, and sometimes a massive galactic nucleus, with or without a black hole.

We concentrate on the problems of pre-main sequence stars, in particular the classical T Tauri stars (CTTS), with their strong, low-ionization emission-line spectra, powerful X-ray emission and UV and IR excesses. Their proximity to dark molecular cloud complexes is consistent with their being newly-formed stars that are contracting towards the main sequence through the Hayashi domain in the Hertzsprung–Russell diagram. Bertout (1989) and Bouvier (1991, 1994) have published reviews of their properties. The strong, variable optical and UV continuum and line emission spectra appear to form in a dense, partly-ionized stellar wind and/or extended chromosphere. The coronal temperatures appear to be too weak to drive the standard stellar wind. A plausible model is the Alfvén wave-driven wind, discussed in Section 10.4. In particular, the Hα emission lines have profiles with symmetrical wings, expected if they form in an unobscured wind within about a stellar radius from the surface. Many CTTS show also broad forbidden lines, formed in low-density regions at a few tens of AU from the star. These lines show the Doppler broadening and shifts consistent with the wind velocity field, but only the blue-shifted components are observed: the red-shifted components from the far side of the wind are obscured, plausibly by opaque, flattened disc-like structures of order 100 AU in radius (Appenzeller *et al.* 1984).

Further, infrared observations made with the *IRAS* satellite confirmed the presence of large amounts of dust, with the infrared luminosity correlated with the strength of the forbidden OI emission from the wind. Significantly, in many of the most active CTTs, the infrared luminosity is substantially greater than can be explained by the standard picture of simple reprocessing of the light from the central star by circumstellar dust. One seems driven to interpret the corpus of observations in terms of a star surrounded by a disc-like structure which is itself active, with some form of viscosity causing redistribution of angular momentum, and the consequent inflow of gas liberating the gravitational energy that appears as infrared radiation, as in the model of Lynden-Bell and Pringle (1974). It appears that both the infrared luminosities and the spectral distributions are well-fitted (Adams *et al.* 1987, 1988; Kenyon and Hartmann 1987). The inferred mass accretion rates range from $\simeq 10^{-5.5} M_\odot\ yr^{-1}$ for the rare but very active FU Ori objects, through $10^{-7} M_\odot\ yr^{-1}$ for the most active ten per cent of the CTTS, to $10^{-8} M_\odot\ yr^{-1}$ or lower for the bulk of the CTTS.

In the same region of the H–R diagram lie the weak-line T Tauri stars (WTTS), which show little or no UV/IR excess or emission lines, but in their X-ray emission are statistically indistinguishable from the CTTS – in fact they were discovered by observations from X-ray satellites. Both the CTTS and the WTTS show strong surface activity, qualitatively similar to solar magnetic activity, but orders of magnitude more intense. Although direct measurements of magnetic fields in T Tauri stars have so far been successful in only one or two cases (Basri *et al.* 1992), the presence of strong magnetic fields of order 1 kG is again shown by observations of star-spots, and (cf. Section 8.4) these are again a valuable direct diagnostic of stellar rotation as they produce periodic changes in brightness by up to 10 per cent. Other evidence for strong magnetic fields in both T Tauri classes comes from the X-ray emission, found in most of these stars. The soft X-ray spectrum can be explained as bremsstrahlung from a plasma at 10^7 K. The occasional strong X-ray flare points to the presence of magnetic loops extending to several stellar radii, with surface magnetic fields at the base of the loop of the order of 1 kG. More direct evidence for strong magnetic fields comes from centimetric radio emission, which is however found in a much smaller number of T Tauri stars. In marked contrast to the X-ray band, at centimetric wavelengths the radio emissions from CTTS and WTTS have quite different properties: the centimetric emission from CTTS is best explained as arising in an ionized wind, whereas the large radio flares observed in WTTS appear to be non-thermal gyrosynchrotron emission from ≈ 1 MeV electrons, trapped in the same extensive magnetic loops, with field strengths ranging from a few hundred to a few thousand gauss (e.g. André 1987).

The overall evidence suggests strongly that the essential difference is the absence of a massive accretion disc. The WTTS display enhanced solar-type magnetic activity similar to that in the CTTS, and the familiar *chromospheric* indicators (especially Ca H and K lines) remain good diagnostics for the WTTS, but not for the CTTS, where non-stellar contributions dominate, presumed to come from regions where material from the accretion disc impinges on the star. Much of this chapter is concerned with the magnetic effects of the accretion disc in the CTTS.

Study of the rotations of CTTS was pioneered by Herbig (1952, 1957, 1962). With the available spectroscopic technology, the projected equatorial velocity $V \sin i$ could be measured in only the brightest stars, and they were found to be rapid rotators. Herbig conjectured that these large values of $V \sin i$ would be confined to the more luminous cases, and this was confirmed subsequently by Vogel and Kuhi (1981). Bouvier *et al.* (1986) and Bouvier (1991) infer typical rotation speeds of 20 km s^{-1}, an order of magnitude lower than the centrifugal limit, again confirming the conclusions of Vogel and Kuhi (1981). Bouvier (1994) reports on recent measurements which suggest that the CTTS are usually slower rotators than the WTTS, on average by a factor 2.

It is reasonable to try and link the presence of strong fields with the now familiar $\alpha\Omega$ dynamo process. It is, however, essential that the evidence for correlation between, for example, X-ray luminosity and Ω be carefully scrutinized, both for lower main sequence and for pre-main sequence stars, given that there is clearly a strong motivation to find such a correlation. Bouvier (1994, and others cited therein) reports an overall X-ray/Ω correlation as evidence for dynamo maintenance of the magnetic fields. On the other hand, Montmerle *et al.* (1994) stress as somewhat contradictory evidence the recently confirmed difference by a mean factor 2 in the rotation rates of the CTTS and WTTS, in spite of there being no statistical difference in their X-ray properties. (Montmerle *et al.* also express some scepticism as to whether X-ray observations confirm convincingly the dynamo interpretation for late-type main sequence stars.) They suggest that at least part of the T Tauri magnetic flux is a 'fossil' (cf. Chapter 9). Earlier, Tayler (1987) had argued that the greater spread in activity found in pre-main sequence stars suggests that some pre-main sequence stars may have fields with both a dynamo-maintained part and also a fossil that has not yet completely decayed, even though the outer parts of the star have always been convective. In this chapter the simple dynamo picture – with magnetic flux a function of Ω – will be provisionally adopted, but with the possible enforced introduction of fossil flux as an extra parameter being borne in mind.

The two statements – CTTS but not WTTS show evidence for active accretion discs, but the CTTS on average rotate half as rapidly as the WTTS – are jointly indeed somewhat surprising. In the simple picture of disc accretion, the inward motion is driven by the outward transport of the high angular momentum of the disc material, but the rotation is always kept close to the local Keplerian rate; and in particular, gas impinging on the star at the equator and subsequently mixed with the convective stellar outer regions will steadily increase the star's specific angular momentum. Even if the star begins as a slow rotator at the top of the Hayashi track – a rather implausible assumption – accretion of angular momentum at the above rate over the Hayashi time would keep the star rotating near the centrifugal limit. The Alfvén wave-driven wind of Section 10.4 may carry off a significant angular momentum, but it is difficult to link this up convincingly with the presence or absence of an accretion disc. A more plausible alternative involves magnetic coupling of the star with the disc, especially with the outer parts beyond the 'corotation radius' where the local Keplerian rotation equals that of the star (Fig. 10.1). The net magnetic torque can then be opposite in sign to that of the accretion torque but possibly of the same order of magnitude, suggesting that interaction with a surrounding disc can make a T Tauri star rotate systematically more slowly (Bouvier 1990). Thus the combination of observation and tentative theorizing forces one to explore how the canonical models of accretion discs (Shakura and Sunyaev 1973; Lynden-Bell and Pringle 1974) are affected by coupling to the star via a large-scale magnetic field. Although the original motivation for study of magnetic accretion discs was for application to neutron star X-ray sources, the same analysis is appropriate for the study of the rotational history of the pre-main sequence T Tauri stars. Bouvier's suggestion has been explored by Königl (1991) and later by Collier Cameron, Campbell and Quaintrell (1993, 1995) and by Armitage and Clarke (1996) and colleagues (cf. Section 10.3).

10.2 Magnetic accretion discs

A full dynamical treatment is formidable, needing to incorporate several competing and complementary processes. The magnetic field may be due to dynamos active not only in the star but also in the disc (Sections 10.6 and 10.7). The arguments for turbulence in the disc were originally largely phenomenological, working back from the conclusion that only a macroscopic viscosity would be capable of yielding significant angular momentum evolution within cosmical time-scales; however, it is now recognized that a disc magnetic field can feed energy from the non-uniform rotation into a turbulent velocity/magnetic field (Section 10.5). As noted, angular momentum loss from the star can occur directly through an Alfvén wave-driven wind (Section 10.4), while a generalized centrifugally-driven wind may flow from the outer parts of the disc (Section 10.8), forcing re-examination of the collimation problem (Section 10.9). We are still some way from a comprehensive model which both incorporates all active processes and estimates their relative importance. However, the primary problem

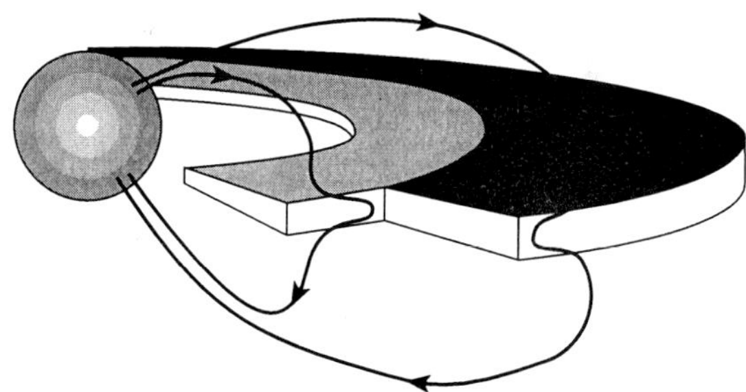

FIG. 10.1. Schematic model of a T Tauri star magnetically coupled with a surrounding disc. Within the corotation radius ϖ_c, the field lines are dragged forward and so assist the accretion torque in transferring angular momentum onto the star; beyond ϖ_c, the field lines are dragged back, so the fields acts to brake the star. (After Collier Cameron 1995; courtesy of A. Collier Cameron.)

is clearly that of the magnetic coupling between star and disc. The pioneering studies were by Davidson and Ostriker (1973), Pringle and Rees (1972), Lamb *et al.* (1973) and later by Ghosh and Lamb (1978, 1979), with subsequent reappraisal by Aly (1980), Anzer and Börner (1980), and by Wang (1987). The several more recent papers that pursue the topic divide fairly sharply into two classes. Those by Campbell (1992*a,b*), Livio and Pringle (1992), Clarke, Armitage and colleagues (1995, 1996), Li (1996) and Li *et al.* (1996), though differing on minor and sometimes on major points, follow Ghosh and Lamb in allowing the stellar magnetic field lines to penetrate the disc both within and without the corotation point. By contrast, those by Lovelace *et al.* (1995), Najita and Shu (1994), Shu *et al.* 1994*a,b* and Ostriker and Shu (1995) all severely limit the degree of penetration.

All authors accept that in general the differential rotation between star and disc will generate a toroidal field in the magnetospheric regions in between. The question at issue is how large the toroidal/poloidal ratio is allowed to become. A strong toroidal field will cause the same billowing-out of field lines as found by Aly (1991), Lynden-Bell and Boily (1994) and others (cf. Section 3.4), tending to convert a closed magnetosphere into an open structure, with much of the flux that would otherwise link star and disc now extending to infinity. Lovelace *et al.* and Shu *et al.* both adopt such a field structure: Ostriker and Shu (1995) argue that as a consequence, 'most of the interesting magnetohydrodynamics is initiated within a small neighbourhood of the corotation point', where the shear is small. By contrast, the papers by Livio and Pringle, by Campbell, and those derivative from them assume explicitly or implicitly that the generation of the toroidal component is limited through spontaneous development of instabilities that offset further shearing of the field.

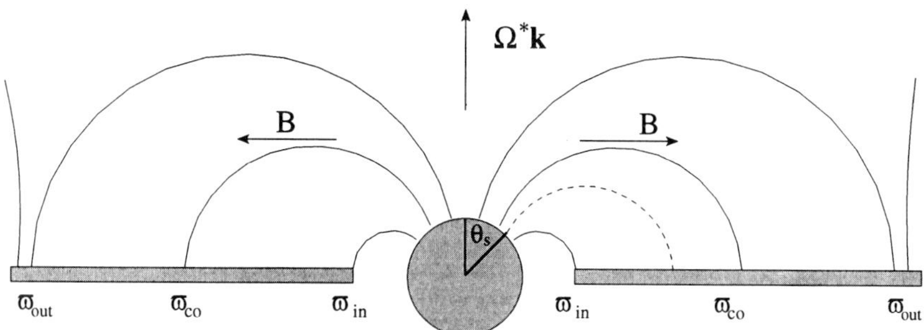

FIG. 10.2. An idealized steady state magnetic accretion disc model, with the dipolar magnetic axis coinciding with the rotation axis. (After Li 1996.)

Both types of model allow some degree of magnetic regulation of the star's spin. As a primary concern is the rotational history of T Tauri stars, we discuss in detail the case with star–disc linkage at a maximum so that interchange of angular momentum occurs with greatest efficiency.

10.2.1 *The magnetosphere*

The model studied – assumed symmetric about the rotation axis, and in a steady state – is illustrated in Fig. 10.2. For simplicity, the star is given the uniform angular velocity Ω^*. The stellar dynamo is taken as generating a field with a dipolar angular distribution on the star's surface. As in the canonical disc theory, outward transport of angular momentum – frictional and magnetic – leads to a steady inflow of gas, crossing field lines by virtue of the high macroresistivity in the disc. It will be found (cf. Section 10.2.4) that a plausible disc model has a fairly well-defined inner radius $\varpi_{\rm m}$ less than the corotation radius $\varpi_{\rm co}$, where the magnetic field forces breakdown in the disc approximation.

The 'magnetosphere' in this problem is defined to be the low-density, high-conductivity domain in between the star and the disc. It is assumed (see below) that the vacuum dipole approximation to the field is adequate. For each field line linking the star and the disc regions near $\varpi_{\rm m}$, the suffix 'd' refers to the point of transition on a field line, within which the gas has properties pertaining to the magnetosphere rather than to the dense disc. Near $\varpi_{\rm m}$ the inflowing gas in the disc acquires a significant component normal to the disc; at $\varpi_{\rm d}$ it will have left the dense disc and begun its flow into the star along the field line. In the magnetosphere the steady state kinematic, continuity and torque equations yield the now familiar integrals (Sections 5.5 and 7.2; Li 1996)

$$\mathbf{v}_{\rm p} = \kappa \mathbf{B}_{\rm p}, \tag{10.1}$$

$$v_\phi = \Omega \varpi = \varpi \alpha + \kappa B_\phi, \tag{10.2}$$

$$\frac{\rho v_{\rm p}}{B_{\rm p}} = \rho\kappa = \eta, \tag{10.3}$$

$$-\frac{\varpi B_\phi}{4\pi} + \eta\Omega\varpi^2 = -\frac{\beta}{4\pi}. \tag{10.4}$$

For convenience we take the dipole to be in a phase of alignment rather than counter-alignment with the rotation axis, so that in the northern hemisphere, inflowing gas corresponds to κ and so also η negative. A dipolar field has $B_{\rm p}\varpi^2 \propto B_{\rm p}r^3 \approx$ constant, so we may write (10.3) as

$$\rho v_{\rm p}\varpi^2 = \eta(B_{\rm p}\varpi^2) = \text{constant}, \tag{10.5}$$

provided the vacuum dipole model remains acceptable – see below.

Away from $\varpi_{\rm m}$, where the velocity of flow through the magnetosphere is small or zero, these equations would yield familiar simple results: $\Omega = \alpha = \Omega^*$ – corotation with the star; and $\varpi B_\phi = \beta$ – a torque-free field, transporting angular momentum via the Maxwell stresses at the rate $-\beta/4\pi$ along a unit flux tube. Within the corotation radius $\varpi_{\rm co}$, a section of a field line penetrating the disc is dragged ahead of the magnetosphere, generating a positive component B_ϕ that carries angular momentum inwards; and likewise, beyond corotation the generated B_ϕ is negative and the flow of angular momentum is outwards. Similar results are anticipated if there is inflow, provided the the density is low enough to keep the flow markedly sub-Alfvénic – i.e. provided $4\pi\rho v_{\rm p}^2/B_{\rm p}^2 = 4\pi\eta^2/\rho \ll 1$. Equations (10.2)–(10.4) then approximate to

$$\Omega \approx \alpha + \frac{\eta\beta}{\rho\varpi^2}, \tag{10.6}$$

$$\varpi B_\phi \approx \beta + 4\pi\eta\alpha\varpi^2. \tag{10.7}$$

The accreted gas may leave the disc at $\varpi_{\rm d}$ with a modest velocity, but it is gravitationally accelerated, so that asymptotically $v_{\rm p} \propto r^{-1/2} \propto \varpi^{-1/3}$, and by (10.5), $\rho \propto r^{-5/2} \propto \varpi^{-5/3}$; hence if the gas enters the magnetosphere with sub-Alfvénic speed, the inward increase in density ensures that the flow does indeed remain sub-Alfvénic until it reaches the star at ϖ_s. From (10.6) and (10.7), the constants α and β for each field line may be written

$$\alpha = \frac{(\Omega_{\rm d}|v_s| - \Omega_s|v_{\rm d}|)}{(|v_s| - |v_{\rm d}|)}, \tag{10.8}$$

$$\beta = \frac{(B_{\rm p}\varpi^2)(\Omega_{\rm d} - \Omega_s)}{(|v_s| - |v_{\rm d}|)}, \tag{10.9}$$

where allowance is made for $v_{\rm p}$ and η being negative. Thus from (10.9), with $|v_s| > |v_{\rm d}|$, for the total angular momentum flow $-\beta/4\pi$ along a particular unit flux tube to be inward or outward, the system must again adjust so that $(\Omega_{\rm d}-\Omega_s)$ is respectively positive or negative.

The gas impinging on the star at ϖ_s will be decelerated sharply at a shock front (e.g. Li *et al.* 1996). The integrals (10.1)–(10.4) are conserved through the shock front; hence in (10.8) and (10.9), Ω_s, v_s, etc. may be replaced by Ω^*, v^* etc. It is at least plausible to expect that behind the shock, $|v^*| \ll |v_{\rm d}|$, so that again $\alpha \approx \Omega^*$.

From (10.6) and (10.7), the respective deviations of Ω from α and of B_ϕ from β/ϖ are measured by the moduli of

$$\frac{\eta B_\phi}{\rho\varpi\alpha} = \left(\frac{v_{\rm p}}{\varpi\alpha}\right)\left(\frac{B_\phi}{B_{\rm p}}\right) \tag{10.10}$$

and

$$\frac{4\pi\eta\Omega\varpi^2}{\varpi B_\phi} \approx \left(\frac{B_{\rm p}}{B_\phi}\right)\left(\frac{4\pi\eta^2}{\rho}\right)\left(\frac{\alpha\varpi}{v_{\rm p}}\right). \tag{10.11}$$

We shall assume that $|B_\phi|/B_{\rm p}$ is at most of order unity; hence at $\varpi_{\rm d}$, where the gas enters the magnetosphere with a modest poloidal velocity, the modulus of (10.10) should be well below unity. It is thus a reasonable approximation to take $\Omega_{\rm d} \approx \alpha \approx \Omega^*$ as a boundary condition to be applied just outside the disc. We shall assume also that (10.11) is small, so that the torque-free approximation to (10.7) is valid.

Recall that the essence of the problem is that the (ϖ, z)-equilibrium conditions in the dense disc force the disc rotation $\Omega_{\rm d}(\varpi, z)$ in general to differ from Ω_s, so that a B_ϕ-component is generated. In a steady state, the growth in B_ϕ must be limited by one or more dissipative process. Campbell (1992*a*,*b*) appeals just to macroscopic processes in the disc – magnetic buoyancy and turbulent resistivity. The value of the consequent B_ϕ-component at $z_{\rm d}$ is a boundary condition to which the magnetic field in the low-density magnetosphere will try to adjust. In the low density magnetosphere, $\varpi B_\phi = \beta$ (the torque-free condition), and $\mathbf{B}_{\rm p}$ will be the appropriate solution of the poloidal component of the force-free equation

$$(\nabla \times \mathbf{B}_{\rm p}) \times \mathbf{B}_{\rm p} + (\nabla \times \mathbf{B}_{\rm t}) \times \mathbf{B}_{\rm t} = 0, \tag{10.12}$$

subject to the prescribed distribution of poloidal flux over the star's surface, and to the above constraints on B_ϕ arising from the essentially non-force-free behaviour in the accretion disc.

If it should turn out that $|B_\phi|/B_{\rm p} \ll 1$, then the force-free condition (10.12) for $\mathbf{B}_{\rm p}$ reduces to the curl-free condition, and the field pervading at least the inner magnetosphere and entering the $\pm z(\varpi)$-faces of the disc has indeed the form

$$B_\varpi = \frac{3B_0R^3}{2}\frac{\varpi z}{(\varpi^2+z^2)^{5/2}}, \qquad B_z = -\frac{B_0R^3}{2}\frac{(\varpi^2-2z^2)}{(\varpi^2+z^2)^{5/2}}, \tag{10.13}$$

with B_0 the surface polar field strength. The work by Aly, Lynden-Bell and Boily and others, discussed in Section 3.4 and cited above, demonstrates strikingly how an attempt by the system to increase arbitrarily the torques acting between star and disc, through a systematic increase in $|B_\phi|$, is likely to be limited by the response of the poloidal field in the intervening force-free domain, which quickly cuts down sharply the amount of stellar flux that succeeds in intersecting the disc rather than extending to infinity. However, Livio and Pringle (1992) argue that the increase in the magnetospheric $|B_\phi|/B_z$ is limited rather by the spontaneous development of instabilities, leading to enforced reconnection. They suggest that the poloidal component remains essentially the unperturbed dipolar field (13.12), but that the equilibrium value near the disc of B_ϕ is

$$B_\phi = B_z \left(1 - \frac{\Omega^*}{\Omega(\varpi)}\right), \tag{10.14}$$

where $\Omega(\varpi)$ is the value in the equatorial plane of the disc (in most cases of interest, the Keplerian angular velocity – see below). The component (10.14) is small near $\varpi_{\rm co}$, as required, and has the correct signs within and without the corotation radius $\varpi_{\rm co}$. Within corotation, the ratio $|B_\phi|/B_z \simeq 1$ as required, but beyond corotation it diverges for large ϖ. Wang (1995) accepts the argument of limitation by reconnection, and retains essentially the same form (10.14) within $\varpi_{\rm co}$, but replaces it by

$$B_\phi = B_z \left(\frac{\Omega(\varpi)}{\Omega^*} - 1\right) \tag{10.15}$$

for $\varpi > \varpi_{\rm co}$.

As we are initially interested in getting an upper limit to the star–disc coupling and the consequent angular momentum interchange, we shall follow Campbell and begin by estimating B_ϕ from an appropriate modification to the disc dynamics, with $\mathbf{B}_{\rm p}$ given by the unperturbed dipolar form (13.12). One might expect that the tentative model arrived at would be likely to give an overestimate of the magnetic interaction between disc and star, as for at least much of the relevant parameter range, the computed ratio $|B_\phi|/|B_z|$ will exceed unity in the magnetosphere. However, Wang (1995) finds that the magnetic torque is rather insensitive to the exact behaviour of this ratio, provided the star–disc magnetic coupling is not restricted just to a domain very close to the corotation radius (as is in the Shu *et al.* models). More work needs to be done to confirm that the system prefers to adopt a steady state solution, rather than performing irregular oscillations between states with differing amounts of closed and open field lines.

10.2.2 *Basic disc theory; angular momentum transport*

The stellar field that penetrates the disc will be modified by the motions present. Studies on disc dynamo generation are postponed to Sections 10.6 and 10.7. Early calculations (Aly 1980, Anzer and Börner 1980) assumed that the field was prevented from entering the disc by the turbulence, so that the dynamical interactions all occurred in surface layers. This would fit into a picture in which the discs are spontaneously turbulent, analogously to convective regions in stars, where it could plausibly be argued that the turbulence has a strong diamagnetic effect (Section 5.8). However, with opinion moving in favour of a magnetic source for disc turbulence (Section 10.5), it is more plausible to follow Ghosh and Lamb (1979) and the subsequent authors, and allow the field to penetrate the disc, there to be described as a superposition of a large-scale part that exerts macroscopic forces, feels the effect of the laminar flow, and acts as a source of the turbulence, plus a small-scale turbulent part.

The essence of the steady disc problem is the redistribution of angular momentum by the viscous and magnetic stresses, which must be balanced by the advection of angular momentum by the bulk motion with velocity (v_ϖ, v_z). In the simplest models, the angular velocity $\Omega(\varpi)$ depends only on ϖ except in a boundary layer near the disc surface. The kinematic turbulent viscosity ν is assumed to be isotropic, so the viscous force density in cylindrical polars can be constructed from the Cartesian form (2.51). Thus in a steady state, the ϕ-component of the equation of motion when multiplied by $\rho\varpi$ yields

$$\begin{aligned}\nabla \cdot (\rho\Omega\varpi^2\mathbf{v}) = \rho\mathbf{v}\cdot\nabla(\Omega\varpi^2) &= \mathbf{B}\cdot\nabla\left(\frac{\varpi B_\phi}{4\pi}\right) + \frac{1}{\varpi}\frac{\partial}{\partial\varpi}\left(\rho\nu\varpi^3\frac{\partial\Omega}{\partial\varpi}\right)\\ &= \nabla\cdot\left(\frac{\varpi B_\phi \mathbf{B}_\mathrm{p}}{4\pi}\right) + \nabla\cdot(\rho\nu\varpi^2\nabla\Omega), \qquad (10.16)\end{aligned}$$

where the steady-state continuity equation

$$\nabla\cdot(\rho\mathbf{v}) = \frac{1}{\varpi}\frac{\partial}{\partial\varpi}(\varpi\rho v_\varpi) + \frac{\partial}{\partial z}(\rho v_z) = 0 \qquad (10.17)$$

has been used.

Equation (10.16) describes a disc in which both the internal viscous torques and the torques exerted by the large-scale magnetic field contribute to the angular momentum transport and so to the evolution of the disc. Consider first the simplest classical disc models (Shakura and Sunyaev 1973; Lynden-Bell and Pringle 1974) which have no large-scale magnetic torque terms (although it was suggested early (Lynden-Bell 1969) that the physics underlying the turbulent viscosity may be hydromagnetic). In the thin disc approximation, with $|v_z| \ll |v_\varpi|$, (10.17) yields

$$\varpi\rho v_\varpi = \text{constant} = m, \qquad (10.18)$$

and (10.16) without the terms in $\mathbf{B}$ integrates to

$$\Omega\varpi^2 m = \rho\nu\varpi^3\frac{\partial\Omega}{\partial\varpi} + C, \qquad (10.19)$$

with $-2\pi m$ a constant measuring the rate of mass accretion per unit disc thickness and $-2\pi C$ the associated net inflow of angular momentum, carried jointly by the laminar motion and the viscous stresses. Note that $m < 0$ – accretion of gas – can be consistent with either $C < 0$ or $C > 0$ – respectively gain or loss of angular momentum. If Ω obeys something like the Keplerian law $\propto \varpi^{-3/2}$ (see below), then well beyond the inner edge of the disc, (10.19) implies

$$v_\varpi \simeq -\frac{3}{2}\frac{\nu}{\varpi}. \tag{10.20}$$

As the disc thickness varies with ϖ, there will be a v_z-component as well. The velocity (v_ϖ, v_z) will locally modify the field emanating from the star. In a steady state, the poloidal component of the kinematic equation integrates to

$$v_\varpi B_z - v_z B_\varpi = -\eta(\nabla \times \mathbf{B}_\mathrm{p})_\phi \approx -\eta \frac{\partial B_\varpi}{\partial z}, \tag{10.21}$$

where as before η is the turbulent resistivity. From the continuity condition (10.17), $|v_z| \simeq (z/\varpi)|v_\varpi|$, and from (13.12) for the unperturbed field in the disc, $|B_z| \simeq (\varpi/z)|B_\varpi|$; hence $|v_z B_\varpi| \simeq (z/\varpi)^2 |v_\varpi B_z|$ and so is negligible. However, from the divergence condition, the perturbation field $\mathbf{B}'$ superposed on (13.12) will in general satisfy $|B'_\varpi|/\varpi \simeq |B'_z|/z$, so that $|v_\varpi B'_z| \simeq |v_z B'_\varpi|$, and (10.21) yields $|v_\varpi B_z| \simeq \eta |B'_\varpi|/z \simeq \eta(\varpi/z^2)|B'_z|$, or

$$\frac{|B'_z|}{|B_z|} \simeq \frac{\varpi|v_\varpi|}{\eta}\left(\frac{z}{\varpi}\right)^2. \tag{10.22}$$

We are interested especially in cases where this ratio of the distortion to the z-component to the basic z-component is small. Suppose first that the flow is driven just by the turbulent viscous stresses, with v_ϖ typically given by (10.20), so that

$$\frac{|B'_z|}{|B_z|} \simeq \mathrm{P_m}\left(\frac{z}{\varpi}\right)^2, \tag{10.23}$$

where $\mathrm{P_m} = \nu/\eta$ is the turbulent magnetic Prandtl number. It is often argued (e.g. van Ballegooijen 1989) that since both ν and η are due to turbulent processes, $\mathrm{P_m}$ should be ≈ 1. This would be compelling if the turbulence were really more or less isotropic. However, as emphasized by Spruit (personal communication), in the present problem we are comparing the z-diffusion of the magnetic field, due to the z-component of the magnetic diffusivity, with the ϖ-diffusion of the angular momentum, due to the ϖ-component of the viscosity. If, for example, the small-scale flows that are parametrized by the viscous stresses are primarily in the (ϖ, ϕ) plane, the associated z-diffusion may be much weaker, corresponding to $\mathrm{P_m} \gg 1$. We shall provisionally take $\mathrm{P_m} \simeq 1$ as an example, but emphasize again that the results should be regarded as suggestive rather than definitive. Equation (10.23) does then confirm that $|B'_z|/|B_z| \ll 1$.

If instead the magnetic torques yield a faster inflow v_ϖ, then B'_z remains a small perturbation provided

$$|v_\varpi| \ll |\varpi/z|^2(\eta/\varpi). \tag{10.24}$$

It is assumed provisionally that the large inflow speeds that would violate (10.24) are not achieved, something that can be subsequently checked. Then from (13.12), to first order in (z/ϖ), the poloidal field in the disc is given by

$$B_\varpi = \frac{3B_0}{2}\left(\frac{R}{\varpi}\right)^3\frac{z}{\varpi}, \qquad B_z = -\frac{B_0}{2}\left(\frac{R}{\varpi}\right)^3 \tag{10.25}$$

Now consider the other extreme possibility, with the macroscopic magnetic torques essentially controlling the angular momentum transport, so that the dominant terms in (10.16) yield

$$\frac{v_\varpi}{\varpi}\frac{\partial}{\partial\varpi}(\Omega\varpi^2) + \frac{v_z}{\varpi}\frac{\partial}{\partial z}(\Omega\varpi^2) = \frac{B_z}{4\pi\rho}\frac{\partial B_\phi}{\partial z}. \tag{10.26}$$

(The v_z-term must be kept at least provisionally, because it is multiplied by a z-derivative.) The generation, diffusion and advection of the B_ϕ-component are described by the ϕ-component of the kinematic equation:

$$\eta\left(\nabla^2 B_\phi - \frac{B_\phi}{\varpi^2}\right) + \frac{1}{\varpi}\frac{\mathrm{d}\eta}{\mathrm{d}\varpi}\frac{\partial}{\partial\varpi}(\varpi B_\phi) = -\varpi\mathbf{B}_\mathrm{p}\cdot\nabla\Omega + \varpi\nabla\cdot\left(\frac{B_\phi}{\varpi}\mathbf{v}_\mathrm{p}\right). \tag{10.27}$$

Similar order-of-magnitude estimates show that the advection term is small compared with the dominant diffusion term by the same ratio $|B_z'|/|B_z|$, so that (10.27) reduces to

$$\eta\frac{\partial^2 B_\phi}{\partial z^2} = -\varpi B_z\frac{\partial\Omega}{\partial z}. \tag{10.28}$$

Equations (10.26) and (10.28) then yield an estimate for the magnetically-driven inflow speed

$$|v_\varpi| \simeq \frac{(B_z^2/4\pi\rho)}{\eta/\varpi}. \tag{10.29}$$

One can now formally justify for the mid-plane the customary approximation to the ϖ-equation of motion as a balance between centrifugal and gravitational force:

$$v_\phi = \Omega(\varpi)\varpi = v_K = (\varpi|\partial V/\partial\varpi|)^{1/2} = \left(\frac{GM}{\varpi}\right)^{1/2}, \tag{10.30}$$

where the suffix 'K' stands for 'Keplerian'. It is readily seen that in magnitude, the other inertial terms $\simeq (v_\varpi/v_K)^2|\partial V/\partial\varpi|$ and so are small as long as the inflow speed is modest. There remain the ϖ-components of the magnetic force and the thermal pressure. In the mid-plane $z=0$, $B_\phi=0$, and the distorted poloidal field exerts the force per unit mass $-B_z j_\phi'/c\rho_c \simeq -B_z(\partial B_\varpi'/\partial z)/4\pi\rho_c \simeq -v_\varpi(B_z^2/4\pi\rho_c)/\eta$ from (10.21), and this is also $\simeq v_\varpi^2/\varpi$ even if v_ϖ is given by (10.29) rather than by ν/ϖ. The pressure gradient term $-\partial p_c/\rho_c\partial\varpi$ is necessarily small compared with gravity, otherwise the disc-like structure would cease to be valid. Thus the Keplerian approximation (10.30) is vindicated for the mid-plane.

10.2.3 *Vertical equilibrium*

As the disc surface is approached, ρ will decrease and the effective resistivity becomes small as the disc fuses with the magnetosphere. Simultaneously $\Omega(\varpi, z)$ goes over from the near-Keplerian value to the corotation value Ω_s. A detailed solution will inevitably be numerical. Campbell (1992*a*) approximates by writing

$$\begin{aligned} \Omega(\varpi, z) = \Omega_K &= (GM/\varpi^3)^{1/2}, && |z| < z^+ - \Delta, \\ &= \Omega_s, && |z| > z^+ - \Delta, \end{aligned} \tag{10.31}$$

and

$$\begin{aligned} \eta &= \eta(\varpi), && |z| < z^+, \\ &\to 0, && |z| > z^+, \end{aligned} \tag{10.32}$$

with $\Delta/z^+ \ll 1$. In this idealization, the vertical shear is located just beneath the disc surface, but the toroidal field generated here diffuses into the whole disc because of the finite η.

A similar treatment of the z-component of the equation of motion

$$-\frac{\partial V}{\partial z} + v_\varpi \frac{\partial v_z}{\partial \varpi} + \frac{\partial}{\partial z}\left(\frac{v_z^2}{2}\right) + \frac{1}{\rho}\frac{\partial}{\partial z}\left(p + \frac{B_\phi^2}{8\pi}\right) + \frac{j_\phi B_\varpi}{c\rho} = 0 \tag{10.33}$$

shows that the dominant terms satisfy

$$\frac{\partial}{\partial z}\left(p + \frac{B_\phi^2}{8\pi}\right) = 0. \tag{10.34}$$

It is found that the simplification (10.31), (10.32) of the vertical problem yields physically acceptable solutions for the domain within the disc. Equation (10.28) now becomes

$$\frac{\partial^2 B_\phi}{\partial z^2} = -\frac{\varpi B_z}{\eta}[(\Omega_s - \Omega_K)\delta(z - z^+ + \Delta) + (\Omega_K - \Omega_s)\delta(z + z^+ - \Delta)]. \tag{10.35}$$

By symmetry, $B_\phi(\varpi, 0) = 0$. It is assumed that in the magnetosphere just outside the disc, (10.11) is well approximated by the torque-free condition $\varpi B_\phi = \beta$, or

$$\frac{\partial B_\phi}{\partial z} = 0, \qquad z^+ < |z| \ll \varpi. \tag{10.36}$$

The solution for B_ϕ, continuous at z^+, for the limit of an infinitesimal transition layer with $\Delta \to 0$, is

$$B_\phi(\varpi, z) = \frac{\varpi B_z}{\eta}(\Omega_s - \Omega_K)z, \qquad |z| < z^+, \tag{10.37}$$

$$B_\phi^+(\varpi) \equiv B_\phi(\varpi, z^+), \qquad z^+ \lesssim |z| \ll \varpi. \tag{10.38}$$

Equation (10.34) integrates to

$$p + \frac{B_\phi^2}{8\pi} = p(\varpi, 0) \equiv p_c(\varpi). \tag{10.39}$$

With a low thermal pressure in the magnetosphere, (10.39) yields

$$p_c(\varpi) = \frac{B_\phi^2(\varpi, z^+)}{8\pi}. \tag{10.40}$$

The continuity equation (10.17) and the unapproximated angular momentum transport equation (10.16) combine into

$$\frac{\partial}{\partial\varpi}(\varpi\rho v_\varpi \Omega\varpi^2) + \frac{\partial}{\partial z}(\varpi\rho v_z \Omega\varpi^2) = \frac{\partial}{\partial\varpi}\left(\rho\nu\varpi^3\frac{\partial\Omega}{\partial\varpi}\right) + \frac{\varpi^2}{4\pi}\frac{\partial}{\partial z}(B_z B_\phi). \tag{10.41}$$

It is convenient to define the vertical integrals through the disc of the viscous and magnetic force densities;

$$(\tilde{F}^m)_\phi = \int_{-z^+}^{z^+} (F^m)_\phi \,\mathrm{d}z \tag{10.42}$$

$$(\tilde{F}^\nu)_\phi = \int_{-z^+}^{z^+} (F^\nu)_\phi \,\mathrm{d}z \tag{10.43}$$

$$= \frac{1}{\varpi^2}\frac{\mathrm{d}}{\mathrm{d}\varpi}\left[\nu\varpi^3\Sigma\frac{\mathrm{d}\Omega}{\mathrm{d}\varpi}\right], \tag{10.44}$$

where

$$\Sigma(\varpi) = \int_{z^-}^{z^+} \rho \,\mathrm{d}z = 2\int_0^{z^+} \rho \,\mathrm{d}z, \tag{10.45}$$

is the area density. The associated steady mass transfer rate is

$$\dot{M} = -4\pi \int_0^{z^+} \varpi\rho v_\varpi \,\mathrm{d}z. \tag{10.46}$$

Integration of (10.41) vertically through the disc then yields

$$-\dot{M}\frac{\mathrm{d}}{\mathrm{d}\varpi}(\Omega_K\varpi^2) = \varpi^2 B_z B_\phi^+ + \frac{\mathrm{d}}{\mathrm{d}\varpi}\left(2\pi\varpi^3\nu\Sigma\frac{\mathrm{d}\Omega}{\mathrm{d}\varpi}\right)$$
$$\equiv 2\pi\varpi^2[(\tilde{F}^m)_\phi + (\tilde{F}^\nu)_\phi]. \tag{10.47}$$

Here $\nu(\varpi)$ is a notionally defined z-average. In a steady state in which there is zero net inflow of angular momentum into the star, (10.47) can be replaced by

$$-\dot{M}(\Omega_K \varpi^2) - 2\pi\varpi^3\nu\Sigma\frac{\mathrm{d}\Omega}{\mathrm{d}\varpi} - \int_{\varpi_\mathrm{m}}^{\varpi} \varpi^2 B_z B_\phi^+ \,\mathrm{d}\varpi = 0 \tag{10.48}$$

with the constant of integration zero.

If the inner edge of the disc ϖ_m lies within the corotation radius ϖ_co, defined by

$$\Omega_K(\varpi_\mathrm{co}) = (GM)^{1/2}/\varpi_\mathrm{co}^{3/2} = \Omega^*, \tag{10.49}$$

then in the domain $\varpi_\mathrm{m} < \varpi < \varpi_\mathrm{co}$, $B_\phi > 0$ and the magnetic torques transfer angular momentum from the disc to the star, so assisting the viscous stresses in causing inflow of gas and ultimate accretion by the star. Beyond ϖ_co, $B_\phi < 0$, and the magnetic torques transfer angular momentum to the disc. If a steady state of accretion is to be possible, then beyond the corotation point it is the viscous stresses that must dominate, their effect being reduced but not cancelled by the magnetic torques.

10.2.4 *The thermal structure of the disc, and its inner radius*

To elucidate further the structure of the disc, and to estimate its inner radius, one must study its thermal properties. The dominant internal heat source comes from the dissipation of the currents, given per unit volume per second by

$$Q = \frac{j_\varpi^2}{\sigma} = \frac{\eta}{4\pi}[(\nabla\times\mathbf{B})_\varpi]^2 \tag{10.50}$$

since $|j_\varpi|/|j_z| \simeq (\varpi/z) \gg 1$, and $|j_\varpi|/|j_\phi| \simeq v_\phi/|v_\varpi| \gg 1$; hence

$$Q = \frac{\eta}{4\pi}\left(\frac{\partial B_\phi}{\partial z}\right)^2 = \frac{\eta}{4\pi}\left(\frac{B_\phi^+}{z^+}\right)^2 = \frac{\varpi^2 B_z^2(\Omega_s - \Omega_K)^2}{4\pi\eta} \tag{10.51}$$

from (10.37). In an optically thick disc, the equation of radiative transfer holds in the familiar small photon mean-free-path form: with a Kramers opacity law $\kappa = K\rho/T^{7/2}$, the heat flux per unit area per second in the z-direction is

$$F = -\frac{16\sigma T^{13/2}}{3K\rho^2}\frac{\mathrm{d}T}{\mathrm{d}z} = Qz, \tag{10.52}$$

where σ is the Stefan–Boltzmann constant, and $\eta(\varpi)$ in (10.51) is approximated essentially as an average over z. From (10.37)–(10.40) and the equation of state,

$$p = \frac{\mathcal{R}}{\mu}\rho T = \frac{(B_\phi^+)^2}{8\pi}\left[1 - \left(\frac{z}{z^+}\right)^2\right]. \tag{10.53}$$

Equations (10.52) and (10.53) then integrate to yield

$$\rho_c = \frac{(\mu/\mathcal{R})^{15/19} p_c^{15/19}}{[(3K/4\sigma)(19/48)z^+F^+]^{2/19}}, \tag{10.54}$$

with the flux F^+ over each face given from (10.51), (10.52) and (10.53) by

$$z^+F^+ = \frac{\eta}{4\pi}(B_\phi^+)^2. \tag{10.55}$$

Consider now the inner part of the disc, within the corotation radius $\varpi_{\rm co}$. We first test the hypothesis that the magnetic torques are the dominant driver of the accretion flow through their removal of angular momentum from the disc. If η is taken as a prescribed function of the independent variable ϖ, and with a scale of variation of the order of ϖ, then the above equations combine to yield in particular v_ϖ, which is found to increase in magnitude strongly with decreasing ϖ. An extreme inner limit to the disc will be at the point where $|v_\varpi| = v_K$, by when the disc approximation has broken down. However, even with our very uncertain understanding of turbulent resistivity, it is more satisfactory to relate η to the dependent variables. We again provisionally take the Prandtl number ν/η to be near unity. Standard non-magnetic accretion disc theory (Shakura and Sunyaev 1973; Frank *et al.* 1992) takes for the mean turbulent speed a fraction ε of the sound speed, and the mixing length as the disc semi-thickness:

$$\eta = \nu = \varepsilon(p_c/\rho_c)^{1/2}z^+ = \frac{\varepsilon B_\phi^+}{(8\pi\rho_c)^{1/2}}z^+. \tag{10.56}$$

A similar form results when magnetic buoyancy is adopted for the dissipation mechanism (Wang 1987). When combined with the kinematic result (10.37), this yields

$$B_\phi^+ = (8\pi/\varepsilon^2)^{1/4}\varpi^{1/2}\rho_c^{1/4}|B_z|^{1/2}(\Omega_K - \Omega_s)^{1/2}. \tag{10.57}$$

If this twist in the field is to drive the accretion alone, then by (10.47) with $\nu = 0$,

$$B_\phi^+ = \frac{\dot{M}\mathrm{d}(\Omega_K\varpi^2)/\mathrm{d}\varpi}{\varpi^2|B_z|}, \tag{10.58}$$

whence (10.57) yields for the central density

$$\rho_c = \frac{\varepsilon^2 GM\dot{M}^4}{2\pi B_0^6 R^{18}}\frac{\varpi^9}{(1-\Omega_s/\Omega_K)^2}. \tag{10.59}$$

The equations (10.54), (10.55), (10.56) and (10.40) to thermal balance and vertical equilibrium then yield

$$z^+ \propto \frac{(B_\phi^+)^{12}}{\rho_c^9} \propto \frac{(1-\Omega_s/\Omega_K)^{18}}{\varpi^{75}}. \tag{10.60}$$

With electron scattering opacity, one finds $z^+ \propto \varpi^{-38}$. Similar results follow if η is due to buoyancy rather than turbulence.

These conclusions are a *reductio ad absurdum* of the assumption of pure magnetic driving. A first rough estimate for the inner radius $\varpi_{\rm m}$ is given rather by assuming purely viscous driving. Without the magnetic term, and with a Keplerian rotation law, (10.48) is equivalent to (10.18) and (10.19), integrated over the disc thickness and with $C = 0$, yielding the well-known result from viscous disc theory

$$\nu\Sigma = \dot{M}/3\pi. \tag{10.61}$$

One can then calculate the associated B_ϕ^+ and ask where the magnetic and viscous torques become equal. As predicted by Bath *et al.* (1974) and Wang (1987), the results make $\varpi_{\rm m}$ considerably larger than the spherical Alfvén radius based on free-fall (cf. Section 9.8) that has often been used as an estimate.

Thus if the magnetic torque were to become large compared with the viscous, then in a dynamically and thermally self-consistent treatment, z^+ increases rapidly to become comparable with ϖ, and there is a breakdown in the typical disc-like approximations, such as the one-dimensional treatment of hydrostatic and thermal equilibrium.

A recent further discussion of the problem is by Campbell and Heptinstall (1998). Within the corotation radius $\varpi_{\rm co}$ both the magnetic and the viscous torques remain active, with the magnetic torque transferring angular momentum to the star, but with the viscous torque now acting in the opposite sense and so moderating the magnetic. (This is the mirror image of the situation well beyond $\varpi_{\rm co}$, where as noted, the magnetic transfer of stellar angular momentum to the disc is less than the viscous transfer outwards, so that steady inflow persists.) However, the model is forced to have a much stronger inward increase of density, so that the disc is subject to the viscous instability of an accretion disc (cf. Frank *et al.* 1992).

A complete treatment must allow also for the effective merging of the disc into the magnetosphere as $\varpi \to \varpi_{\rm m}$, through the v_z component's becoming steadily more important. In a steady state, (10.48) would continue to hold but with $\dot{M}$ defined by (10.46) now a function of ϖ that vanishes at $\varpi_{\rm m}$, satisfying

$$\frac{\rm d}{{\rm d}\varpi}\dot{M}(\varpi) = 4\pi|\eta B_z|. \tag{10.62}$$

The component B_ϕ^+ will go zero as $\varpi_{\rm m}$ is approached. It is not clear by how much this will affect the ratio $|B_\phi^+/B_z|$ at larger values of ϖ.

10.2.5 *The estimated net torque*

The 'fastness parameter' ξ is defined by

$$\xi = \frac{\Omega_s}{\Omega_K(\varpi_{\rm m})}, \qquad \varpi_{\rm m} = \xi^{2/3}\varpi_{\rm co} \tag{10.63}$$

where $\varpi_{\rm co}$ is again the corotation radius (10.49). From the discussion in the last section, a reasonable approximation for the inner radius $\varpi_{\rm m}$ is given by writing

$$\tilde{F}_\phi^m = 2\tilde{F}_\phi^\nu \tag{10.64}$$

where the symbol $\tilde{F}$ refers to the vertical integral of the force densities, defined in (10.42) and (10.43). As before, the pure viscous approximation (10.20) is adopted and the associated magnetic torque calculated. Within $\varpi_{\rm co}$, the ratio

$$\frac{\tilde{F}^m_\phi}{\tilde{F}^\nu_\phi} = \frac{2\varpi B_z B_\phi^+}{\dot{M}\Omega_K} \tag{10.65}$$

is positive and monotonically decreasing to zero at $\varpi_{\rm co}$. For $\varpi > \varpi_{\rm co}$, the ratio is negative, tending to zero at infinity, and with its modulus reaching a maximum at $\Omega_K = (21/29)\Omega_s$ for buoyant diffusion and at $\Omega_K = (17/29)\Omega_s$ for turbulent diffusion. For high enough accretion rates, these maxima are below 2, so disruption of the disc occurs at the radius $\varpi_{\rm m} < \varpi_{\rm co}$ when the ratio (10.65) reaches 2, so that the parameter $\xi < 1$.

As indicated, we are particularly (though not solely) interested in the equilibrium case, when magnetic braking of the star is balanced by the positive accretion torque. When $\varpi_{\rm m} < \varpi_{\rm co}$, we expect the gas from the disrupted disc to flow down the field lines onto the star. It is unclear whether mass and angular momentum will be accreted when $\varpi_{\rm m} > \varpi_{\rm co}$, especially if there is an Alfvén wave-driven wind emanating from the polar regions (Section 10.4). Note that except in the equilibrium case, the steady state integral (10.48) will have a non-zero constant of integration. From now on we suppose $\varpi_{\rm m} < \varpi_{\rm co}$.

If the flow of gas from the disc onto the star does not begin until near $\varpi_{\rm m}$, then the torque exerted on the star by the accreted matter is

$$T_{\rm a} = \dot{M}\Omega_K(\varpi_{\rm m})\varpi_{\rm m}^2. \tag{10.66}$$

(If significant accretion begins further out, this term will be modified, along with the values of B_ϕ^+ already noted.) The magnetic torque acting on the star is

$$T_{\rm m} = -\int_{\varpi_{\rm m}}^{\varpi_{\rm d}} \varpi^2 B_z B_\phi^+ \,{\rm d}\varpi, \tag{10.67}$$

where $\varpi_{\rm d}$ is the disc radius, and contributions from both surfaces $\pm z_+$ are included. Campbell shows that when the diffusivity is due to buoyancy,

$$T_{\rm m} = -\frac{4}{9}|\varpi^3 B_\phi^+ B_z|_{\varpi_{\rm m}} \left[\frac{\xi^{3/2} - (1-\xi)^{3/2}}{(1-\xi)^{1/2}}\right]. \tag{10.68}$$

The condition (10.64) for $\varpi_{\rm m}$ relates $T_{\rm a}$ and $T_{\rm m}$, yielding for the total torque on the star

$$T = \dot{M}\varpi_{\rm m}^2\Omega_K(\varpi_{\rm m})\left[1 - \frac{4}{9}\frac{\xi^{3/2} - (1-\xi)^{3/2}}{(1-\xi)^{1/2}}\right]. \tag{10.69}$$

which has the equilibrium solution $T = 0$ when $\xi_{\rm eq} = 0.88$. With turbulent diffusion, the net torque computed similarly is

$$T = \dot{M}\varpi_{\rm m}^2\Omega_K(\varpi_{\rm m})\left[1+\frac{8}{25}\frac{(1-25\xi/13)}{(1-\xi)}\right], \qquad \xi_{\rm eq} = 0.82. \tag{10.70}$$

The conclusion that $\xi_{\rm eq} < 1$ but close to unity (rather than a factor 2 or 3 smaller, as in the Ghosh–Lamb models) was derived first in Wang (1987), and subsequently is confirmed by Yi (1995) and Wang (1995).

A primary aim of the previous lengthy analysis has been achieved by our arriving at formulae (10.69) and (10.70) that can be used to follow the rotational evolution of the star as it contracts towards the main sequence. We emphasize once more that the correct formulae may turn out to be sensitive to the gross deficiencies in our understanding of turbulence, for example to deviations from the simple assumption of a magnetic Prandtl number near unity. Equally important in the present problem is the assumption that the poloidal field linking star and disc remains essentially a curl-free dipolar field. As long as calculations such as those of the next section never yield $|B_\phi|/B_{\rm p}$ other than small, then they are at least qualitatively acceptable. Equally, the work summarized in Section 3.4 shows that equations which allow this ratio to approach unity would need to be replaced by others which allow for the transition to a force-free field, with a much reduced flux linkage between star and disc and so also a reduced mutual torque.

10.3 Pre-main sequence rotational evolution

With these caveats in mind, we now report first on recent applications of the previous formulae by Collier Cameron and Campbell (1993). They have studied the rotational evolution of a pre-main sequence solar-type star, contracting down the Hayashi track and exchanging angular momentum with a surrounding accretion disc, according to the equation

$$\frac{\rm d}{{\rm d}t}(I\Omega_s) = I\frac{{\rm d}\Omega_s}{{\rm d}t} + \Omega_s\frac{{\rm d}I}{{\rm d}t} = T, \tag{10.71}$$

where the stellar moment of inertia $I = k^2MR^2$ and T is given either by (10.69) or (10.70). The fully convective star is described by an $n = 3/2$ polytrope, with $k^2 = 0.2$ (Rucinski 1988), gravitational energy $\mathcal{V} = -6GM^2/7R$ (Chandrasekhar 1939), thermal energy $U = -\mathcal{V}/2$ and total energy $U+\mathcal{V} = \mathcal{V}/2$ for a $\gamma = 5/3$ gas. (In a nearly uniformly rotating star, the rotational kinetic energy $Mk^2\Omega^2R^2/2 \approx 0.17(\Omega^2R^3/GM)(-\mathcal{V})$ and so yields a modest correction to the total energy even for the most rapidly rotating stars, with $\Omega^2R^3/GM \simeq 0.25$.) The contraction rate is given by

$$-\frac{\rm d}{{\rm d}t}\left(-\frac{3GM^2}{7R}\right) = \frac{6GM}{7R}\frac{{\rm d}M}{{\rm d}t} - \frac{3GM^2}{7R^2}\frac{{\rm d}R}{{\rm d}t} = L = 4\pi\sigma R^2T_e^4. \tag{10.72}$$

In the Hayashi phase the star moves almost vertically in the Hertzsprung–Russell diagram, with $T_e \approx$ constant and so with $L \propto R^2$.

From (10.71) and (10.72),

$$\frac{d\Omega_s}{dt} = \frac{T}{k^2MR^2} - \Omega_s\left(\frac{\dot{M}}{M} + \frac{2\dot{R}}{R}\right)$$
$$= \frac{T}{k^2MR^2} - \Omega_s\left(\frac{5\dot{M}}{M} - \frac{56\pi\sigma R^3T_e^4}{3GM^2}\right). \quad (10.73)$$

This shows the three processes affecting the rotation. Spin-up due to contraction has a characteristic time-scale which increases from about $\simeq 10^4$ yr at the top of the Hayashi track to a few times 10^6 yr at the bottom. The mass accretion rate is tied to the magnetic braking rate through (10.69) and (10.70). It does not stay constant during the Hayashi phase, since accretion seems to have come to a halt by the time stars reach the main sequence. The first term on the right is the effect of the torque with its competing magnetic and accretion terms, with time-scales depending on the strength of the stellar field and on the accretion rate. The system will tend to evolve through a seqence of equilibrium states with $T = 0$, provided the magnetic coupling time-scale is shorter than the Hayashi time-scale.

The results of this paper are summarized as follows.

(1) For most of the models, the inner disc radius ϖ_m is a few times R.

(2) For stellar fields of a few hundred gauss and accretion rates of a few times $10^{-8}M_\odot$ yr^{-1}, the star can indeed evolve in quasi-equilibrium during the Hayashi phase, with rotation periods of between 2 and 10 days, in good agreement with the observations of T Tauri stars (Bouvier 1991). The inner disc radius is just within the corotation radius.

(3) The field strengths required to bring the T Tauri stars to their equilibrium rotation rates within the Hayashi time are comparable with the ambient field strengths needed in the Alfvén wave-driven models of Lago (1984) and Hartmann *et al.* (1982) (cf. Section 10.4). The estimated values of the inner disc radii are able to explain why the Hα profiles in the classical T Tauri stars have symmetric wings, showing that the inner parts of the stellar wind are not obscured by the disc, whereas red-shifted optical forbidden line emission from the outer wind is often absent.

(4) At slightly lower field strengths, the star–disc coupling is more complicated. At high enough accretion rates there may be a limit–cycle type of behaviour: the star is braked until the inner edge of the disc can establish itself within the corotation radius, whereupon the positive net accretion torque spins the star up until again the disc is disrupted beyond the corotation radius.

(5) The crucial parameter in the theory so far is the accretion rate, determined by the supply of gas in the disc. As the supply of gas declines, the ratio of magnetic to accretion torques will rise, but both will in fact become small; braking then occurs only via the stellar wind.

(6) It is suggested that the theory offers a clue to explain the spread in the rotation rates of the majority of G and K dwarfs in young clusters, already alluded to and discussed in Chapter 8. The minority of very rapidly rotating solar-type stars could be those with less massive discs and which therefore suffered less efficient braking.

In a later paper (1995), Collier Cameron *et al.* have extended the theory for solar-type pre-main sequence stars to include the approach to the main sequence via the Henyey track. The stellar magnetic field is now taken explicitly as dynamo-generated, as in the work of Section 8.3. The aim specifically is to elicit the dependence of the ZAMS stellar rotation rates on the various input parameters: the initial rotation rate, the scaling factor giving the strength of the dynamo-built field, and the total mass accreted through the whole Hayashi and Henyey phases. The new work uses more sophisticated stellar models, in particular, the evolutionary code developed by Eggleton (1971, 1972). Gilliland's (1986) computations of moments of inertia and convective turnover times are updated, by use of the equation of state of Eggleton *et al.* (1973) but with more modern opacities (Rogers and Iglesias (1992) and Weiss *et al.* (1990) for temperatures below 10^4 K). In convective regions, the mixing length was taken as $2H_{\text{p}}$. The initial state was again a proto-star of $1\text{M}_\odot$ with radius $6\text{R}_\odot$, at the top of the Hayashi track. The evolution was followed for 10^8 yr, to the final settling down on the main sequence. Both I and $\dot{I}$ and the inverse Rossby number at the base of the convective zone were computed at each time-step.

A crucial quantity is $\dot{M}/\ddot{M}$, the time-scale on which the accretion rate from the disc is diminishing. $\dot{M}$ is taken to decline exponentially in a time-scale of 2×10^6 yr, based on estimates of the total disc masses derived by Beckwith *et al.* (1990) and André and Montmerle (1994) from infrared, sub-millimetre and millimetre-band fluxes. And as in the studies of main sequence rotation in Section 8.3, the simple dynamo law

$$B_0 = (B_0)_\odot \frac{\tau_c \Omega}{(\tau_c \Omega)_\odot} \tag{10.74}$$

is adopted, with τ_c again the turnover time at the base of the convective zone. Dynamo saturation was imposed by the imposition of an upper limit $B_{0,\max} =$ 2000 G. (This did not in fact affect the results significantly – in most cases, magnetic coupling with the disc kept Ω_s low enough for this limit never to be achieved.) Again two disc models were studied, with buoyancy and turbulent dissipation respectively.

The outstanding conclusion of the computations is that the stellar rotation rate on the ZAMS is a strong function of the mass of the proto-stellar accretion disc at low disc masses, but becomes nearly independent of disc mass for more massive discs. Consistent with this are the infrared observations of Edwards *et al.* (1993), which show that the minority of T Tauri stars that rotate above the norm are those which appear to have little or no surrounding disc material

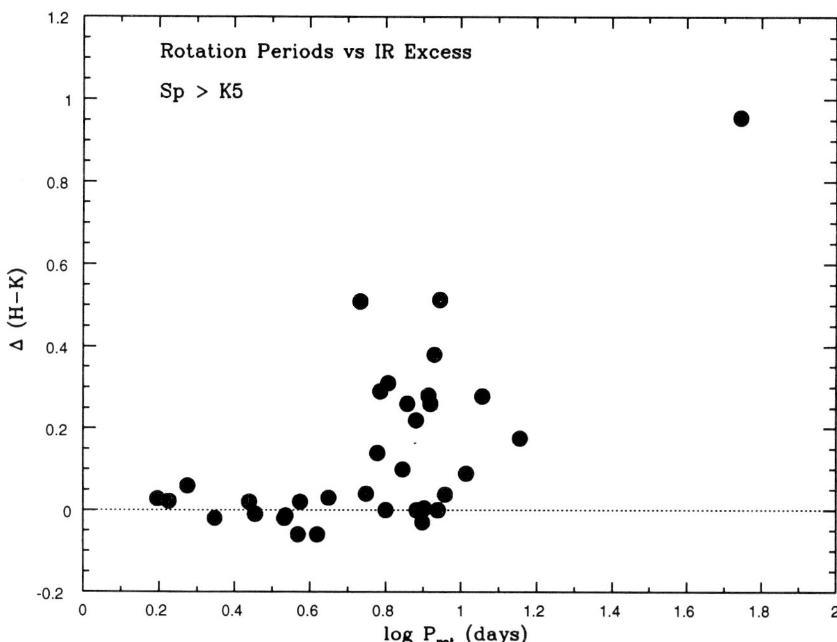

FIG. 10.3. The plot of H–K infrared excess against rotation period, showing the correlation between high T Tauri rotation with the presumed absence of a disc that would yield both high infrared excess and braking of the star. (After Edwards *et al.* 1993.)

(Fig. 10.3). Further, Fig. 10.4 shows that the ultimate distribution of rotation rates has a low velocity peak and extended high velocity tail, qualitatively similar to the observed distribution of solar-type stars in α-Persei and the Pleiades (Section 8.3).

There are many other interesting details; in particular, the authors speculate on the relevance of their models to the formation of the Solar System. Stauffer and Soderblom (1991) have argued that the slow rotators may have formed planets but not the rapid rotators. The disc-braking picture suggests once more a possible physical link between stellar rotation rates on the ZAMS and the formation of planetary systems. In this picture, the Sun would have evolved though its Hayashi main sequence phase in rotational equilibrium with a massive disc, transferring most of its angular momentum into the part of the disc that eventually formed the giant planets, and arriving on the ZAMS with a rotation rate typical of the cluster stars in the low-velocity peak. A rapid rotator, on the other hand, would retain most of its initial angular momentum until it arrived on the ZAMS, subsequently losing this angular momentum via a hot, magnetically-controlled stellar wind. If the Sun's behaviour is typical of stars in

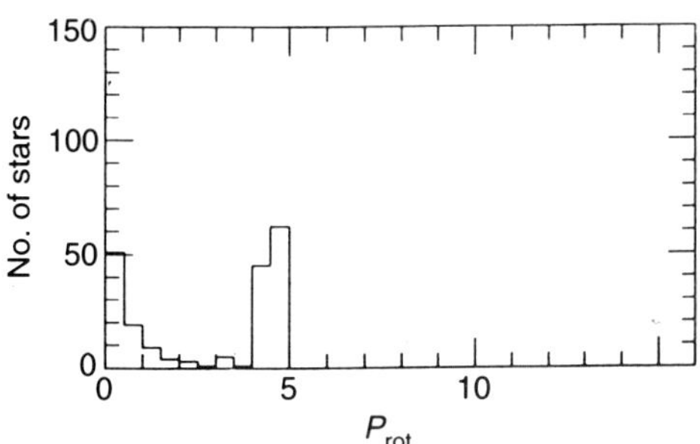

FIG. 10.4. Histogram of the ZAMS period distribution for an ensemble of 200 models with $B_{0,\odot} = 30$ G, the initial Ω at 15 per cent of the break-up value, and a log-normal distribution of initial disc masses. (After Collier Cameron *et al.* 1995.)

the low velocity peak, this suggests in turn that the fraction of G stars with disc remnants containing Jupiter-like planets would be similar to the fraction of G stars in the low velocity peak in young clusters, i.e. about 30 to 50 per cent.

More recently, Armitage and Clarke (1996) have performed similar computations. Their procedure differs in two ways. First, they adopt the Livio–Pringle prescription (10.14) for B_ϕ^+. As noted above, this is acceptable for $\varpi < \varpi_{\rm co}$, but beyond corotation it has to be replaced, for example by Wang's formula (10.15). It turns out (Armitage, personal communication) that this correction does not much affect the results during the braking phase, which is sensitive mainly to the values of B_ϕ/B_z near corotation, though it does prolong this phase somewhat. Second, and probably more importantly, rather than prescribing the accretion rate $\dot{M}$ as a function of time, the authors use the time-dependent equation for the evolution of the disc area density $\Sigma(\varpi, t)$, again defined as in (10.45). Thus the steady mass transfer equation (10.46) is replaced by

$$\frac{\partial \Sigma}{\partial t} + \frac{1}{\varpi}\frac{\partial}{\partial \varpi}(\varpi \Sigma v_\varpi) = 0; \tag{10.75}$$

and likewise (10.47) – the z-integrated form of (10.41) – is replaced by

$$\frac{\partial}{\partial t}(\Sigma \Omega \varpi^2) + \frac{1}{\varpi}\frac{\partial}{\partial \varpi}(\varpi \Sigma v_\varpi \Omega \varpi^2) = \frac{1}{\varpi}\frac{\partial}{\partial \varpi}\left(\nu \Sigma \varpi^3 \frac{\partial \Omega}{\partial \varpi}\right) + \frac{\varpi}{2\pi} B_z B_\phi^+. \tag{10.76}$$

Elimination of $\partial\Sigma/\partial t$ between (10.75) and (10.76) yields an expression for $(\Sigma \varpi v_\varpi)$ to be substituted back into (10.75); with $\Omega = \Omega_K = (GM/\varpi^3)^{1/2}$, and with B_ϕ^+ given by (10.14), we arrive at (e.g. Pringle 1981; Livio and Pringle 1992)

$$\frac{\partial \Sigma}{\partial t} = \frac{3}{\varpi}\frac{\partial}{\partial \varpi}\left[\varpi^{1/2}\frac{\partial}{\partial \varpi}(\nu \Sigma \varpi^{1/2})\right] + \frac{1}{\varpi}\frac{\partial}{\partial \varpi}\left[\frac{\Omega - \Omega^*}{\Omega}\frac{B_z^2 \varpi^{5/2}}{\pi (GM)^{1/2}}\right]. \tag{10.77}$$

With B_z given by (10.25), the total magnetic torque on the disc is the integral, taken between $\varpi_{\rm m}$ and ∞

$$\int \left(B_\phi^+ B_z(\varpi,\, z^+)\frac{\varpi}{2\pi}\right) 2\pi\varpi\, {\rm d}\varpi = \int \frac{B_0^2 R^6}{4}\left(1 - \frac{\Omega^*}{\Omega}\right)\frac{{\rm d}\varpi}{\varpi^4}$$
$$= -\frac{B_0^2 R^6}{12\varpi_{\rm m}^{3/2}}\left[\varpi_{\rm m}^{-3/2} - 2\varpi_{\rm co}^{-3/2}\right], \quad (10.78)$$

where $\varpi_{\rm m}$ is the inner radius, notionally defined by magnetic disruption, for example as discussed in Section 10.2.4. The total torque on the star is the sum of (10.78) with sign reversed, plus the accretion torque $\dot{M}(\varpi_{\rm m})\Omega_K(\varpi_{\rm m})\varpi_{\rm m}^2$.

The main conclusions of Armitage and Clarke are close to those of Collier Cameron *et al.* They confirm that it is the initial disc mass that controls the evolution of the star-disc system. For sufficiently massive discs – of order $0.1{\rm M}_\odot$ – a stellar field of about 1 kG is able to regulate the spin rate to near the observed values for the classical T Tauri phase. In particular, they show that models with a wide range of initial spins evolve, typically within 10^5 yr, to a state of quasi-equilibrium in which the net torque on the star is small. Again, they agree with Collier Cameron *et al.* (1995) that the final stellar rotation period is independent of the disc mass, provided it is high enough for magnetic braking to operate: the systems with the highest disc masses begin with the star spinning up more through accretion, but this is compensated by a longer disc lifetime and so a more prolonged time over which magnetic braking can occur (Fig. 10.5). Once sufficient mass has been accreted, the inner radius of the disc will move out beyond corotation, frictional transport of angular momentum becomes weak, and the magnetic field acts to expel the remnant of the disc. In systems with lower disc masses, this expulsion occurs earlier, before any significant magnetic braking has occurred.

Our discussion has been of evolution over long time-scales of stars that are quasi-steady. However, T Tauri stars were in fact discovered because of their marked photometric and spectrophotometric variability. As noted, there is a broad distinction between the CTTs and the WTTs, explicable in terms of the presence or absence respectively of an accretion disc that extends fairly close to the photosphere. In some cases, the variability manifests itself through the system's switching itself between strong- and weak-line states in a time-scale of decades (Herbig and Bell 1988), much shorter than can plausibly be associated with dynamical changes in the disc at large distances. In Clarke *et al.* (1995) and (in more detail) in Armitage (1995), it is argued convincingly that variation in TT stars is (at least in part) a consequence of modulation of the accretion flow by spontaneous variations in the stellar magnetic field, occurring on time-scales $\ll 10^5$yr. The discussion is again formulated in terms of the Livio–Pringle (1992) argument from reconnection for determining the limitations in B_ϕ^+, yielding again (10.77) for the time variation of Σ, but it is equally applicable to the Collier Cameron–Campbell models. The crucial feature is again the magnetic disruption of the disc at the inner radius $\varpi_{\rm m}$, predicted to occur when the magnetic torques

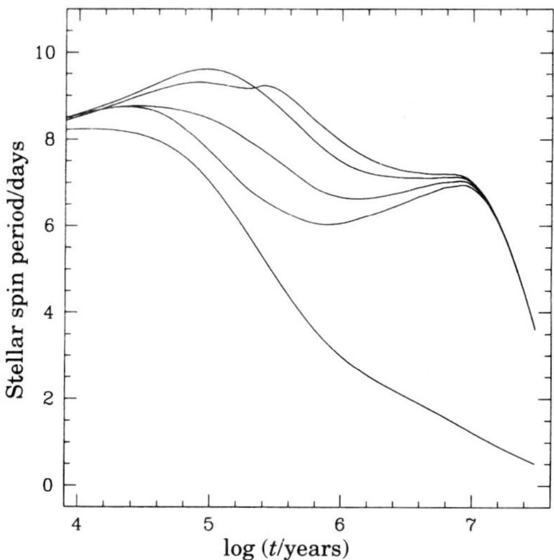

FIG. 10.5. Time-dependence of the stellar spin period for models with different initial disc masses. From top downwards, at $t = 10^6$ yr, the curves represent disc masses of 3×10^{-3}, 10^{-2}, 3×10^{-2}, $10^{-1}M_\odot$ and $10^{-3}M_\odot$. For the first four cases, the final stellar rotation is independent of the disc mass. In the case with a very low disc mass the ZAMS star will rotate much more rapidly. (After Armitage and Clarke 1996.)

start to exceed the viscous torques, with consequent accretion of gas onto the star. An increase in B_0 by a factor of order 2 can force ϖ_m into the domain beyond ϖ_{co}, where the magnetic torques give rather than remove angular momentum and so inhibit the accretion of gas (Fig. 10.6). The authors justly suggest that archival data relating to the spectrophotometric variability of T Tauri stars may provide an indirect record of magnetic activity cycles in low-mass pre-main sequence stars and therefore also a clue to the dynamo cycles in these stars.

10.4 Winds driven by Alfvén waves

A standard picture of the formation of the hot corona surrounding late-type stars appeals at least in part to the dissipation of magnetosonic waves, generated in the subphotospheric convective zone (e.g. Osterbrock 1961; Priest 1982). If the balance of heat input and radiative and conductive losses fixes the coronal temperature at near 10^6 K, then the thermal pressure far from the star is much higher than the combined pressures of the interstellar medium, and the corona expands as a stellar wind. As discussed in Section 7.5, a cooler corona surrounding a rotating star and brought into near corotation through magnetic coupling can then be forced into expansion by centrifugal driving, but both the mass loss and the associated angular momentum loss will be weak unless the

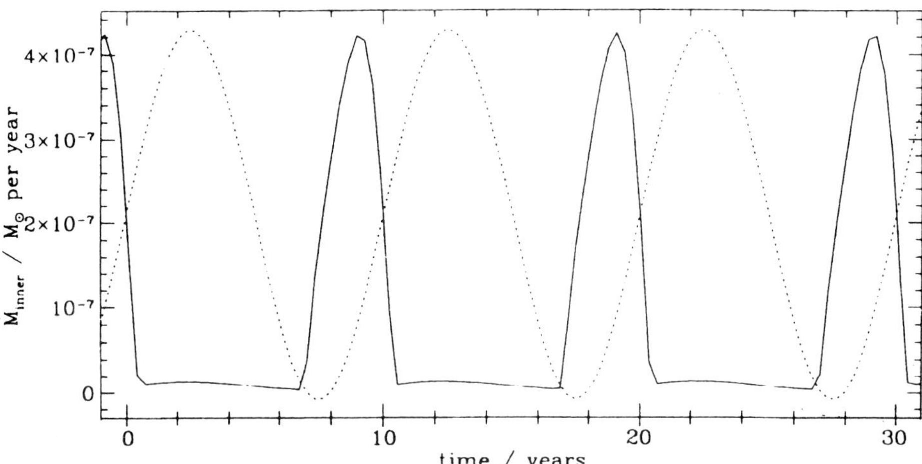

FIG. 10.6. Modulation of accretion rate and disc flux for a cycle with ϖ_m crossing ϖ_{co}. The solid line shows accretion rate, the dotted line the stellar field. (From Armitage 1995.)

point of centrifugo-gravitational balance is near the coronal base. In particular, the estimated coronal temperatures for T Tauri stars seem to be too low to drive a thermal wind, even with the lower gravitational field of a pre-main sequence star; while a centrifugally-driven wind can keep a contracting star just below the photospheric centrifugal limit, but cannot slow up the rotation to the modest values that are required for many zero-age main-sequence stars.

In fact, undamped Alfvén waves exert an additional pressure on the corona, analogous to a radiation pressure, which accelerates the plasma. The kinetic energy thus fed into the wind is at the expense of the wave energy flux, which therefore decreases with distance from the star, but this is quite distinct from normal damping which converts wave energy directly into heat. This Alfvén wave-driven wind model for T Tauri stars is treated in this section. In Section 12.7.1 a similar formalism is applied to describe support from Alfvén waves against gravity in molecular clouds.

The treatment follows that given by Parker (1965) and Belcher (1971). The study is restricted to the simplest case: a steady, spherically symmetric outflow along a radial zero-order field, and perturbations that are just Alfvén wave fields, polarized in the ϕ-direction. (Omission of the θ-polarization allows a simple construction of a global solution.) We look for solutions of the form

$$\rho = \rho(r), \qquad \mathbf{v} = [v(r),\, 0,\, \delta v(r,t)]\,, \qquad \mathbf{B} = [B(r),\, 0,\, \delta B(r,t)]\,. \tag{10.79}$$

If B and ρ were uniform, then the implicit assumption of divergence-free motions with ρ consequently time-independent would be valid even for large amplitudes (cf. Section 3.1). The same result can be taken over into the small wavelength,

JWKB treatment adopted here.

The time-dependent, dissipation-free equation of motion, the continuity equation and the field-freezing equation are (2.35) (with $\mathbf{F}_\nu$ dropped), (2.38) and (2.88). Substitution of (10.79) yields from respectively the r, θ, ϕ-components of (2.35)

$$v\frac{\partial v}{\partial r} - \frac{(\delta v)^2}{r} + \frac{1}{\rho}\frac{\partial p}{\partial r} + \frac{1}{8\pi\rho r^2}\frac{\partial}{\partial r}(r\delta B)^2 + \frac{\partial V}{\partial r} = 0, \tag{10.80}$$

$$\left((\delta v)^2 - \frac{(\delta B)^2}{4\pi\rho}\right)\frac{\cot\theta}{r} = 0, \tag{10.81}$$

$$\frac{\partial}{\partial t}(\delta v) + \frac{v}{r}\frac{\partial}{\partial r}(r\delta v) = \frac{B}{4\pi\rho r}\frac{\partial}{\partial r}(r\delta B). \tag{10.82}$$

The continuity equation is simply

$$\frac{1}{r^2}\frac{\partial}{\partial r}(r^2\rho v) = 0, \tag{10.83}$$

the divergence condition becomes

$$\frac{1}{r^2}\frac{\partial}{\partial r}(r^2 B) = 0, \tag{10.84}$$

and the one surviving (ϕ)-component of (2.88) yields

$$\begin{aligned}\frac{\partial}{\partial t}(\delta B) &= \frac{1}{r}\frac{\partial}{\partial r}[r(B\delta v - v\delta B)] \\ &= \left(B\frac{\partial}{\partial r}\delta v - v\frac{\partial}{\partial r}\delta B\right) + \frac{1}{r}(v\delta B - B\delta v) + \frac{v\delta B}{\rho}\frac{\partial\rho}{\partial r},\end{aligned} \tag{10.85}$$

on use of (10.83) and (10.84).

We now use the JWKB small wavelength technique, writing

$$\begin{aligned}\delta B(r,t) &= [\delta B_1(r) + \mu\delta B_2(r) + ...]\exp\mathrm{i}(S(r) - \omega t), \\ \delta v(r,t) &= [\delta v_1(r) + \mu\delta v_2(r) + ...]\exp\mathrm{i}(S(r) - \omega t),\end{aligned} \tag{10.86}$$

where

$$\mu = 2\pi/kh, \qquad k = \partial S/\partial r. \tag{10.87}$$

Here ω is the constant wave-frequency, and h is the characteristic scale-length for variations in ρ, v and B, assumed to be less than but comparable with r. Thus μ is the small ratio of the Alfvén wavelength to h. For consistency, the quantities δB_1, δv_1 etc. should turn out to have scale-lengths of the order of h. Insertion of (10.86) into (10.80) and (10.85) yields the terms of zero order in μ:

$$(\mathrm{i}kv - \mathrm{i}\omega)\,\delta v_1 - \frac{\mathrm{i}kB}{4\pi\rho}\delta B_1 = 0, \tag{10.88}$$

$$-\mathrm{i}kB\,\delta v_1 + (\mathrm{i}kv - \mathrm{i}\omega)\,\delta B_1 = 0, \tag{10.89}$$

whence

$$\omega = k\,(v \pm v_{\mathrm{A}}), \qquad \delta v_1 = \mp\frac{\delta B_1}{(4\pi\rho)^{1/2}}, \qquad v_{\mathrm{A}} = \frac{B}{(4\pi\rho)^{1/2}}. \tag{10.90}$$

The upper and lower signs in (10.90) correspond to waves propagating respectively outwards and inwards. To construct the first order approximation, the zero-order solutions (10.90) are inserted and terms of first order in μ are retained, yielding for outward-propagating waves:

$$\mu\left(\mathrm{i}kv_{\mathrm{A}}\,\delta v_2 + \mathrm{i}k\frac{B}{4\pi\rho}\,\delta B_2\right) + \frac{1}{r}\left[\frac{B}{4\pi\rho} + \frac{v}{(4\pi\rho)^{1/2}}\right]\delta B_1 + \left[\frac{B}{4\pi\rho}\frac{\partial}{\partial r}(\delta B_1) + v\frac{\partial}{\partial r}\left(\frac{\delta B_1}{(4\pi\rho)^{1/2}}\right)\right] = 0, \tag{10.91}$$

$$\mu\,(\mathrm{i}kv_{\mathrm{A}}\,\delta B_2 + \mathrm{i}kB\,\delta v_2) + \left[\frac{1}{r}\,(v + v_{\mathrm{A}}) + \frac{v}{\rho}\frac{\partial\rho}{\partial r}\right]\delta B_1 - \left[B\frac{\partial}{\partial r}\left(\frac{\delta B_1}{(4\pi\rho)^{1/2}}\right) + v\frac{\partial}{\partial r}(\delta B_1)\right] = 0. \tag{10.92}$$

Elimination of δv_2, δB_2 leaves a differential equation for δB_1

$$\frac{1}{\delta B_1}\frac{\partial}{\partial r}(\delta B_1) = \frac{1}{4\rho}\frac{\partial\rho}{\partial r}\left(\frac{3v + v_{\mathrm{A}}}{v + v_{\mathrm{A}}}\right). \tag{10.93}$$

Equations (10.83) and (10.84) yield

$$r^2\rho v = r_0^2\rho_0 v_0, \qquad r^2 B = r_0^2 B_0, \tag{10.94}$$

where the suffix zero defines a convenient reference level. The Alfvénic Mach number is given by

$$M_{\mathrm{A}} = \frac{v}{v_{\mathrm{A}}} = \frac{v_0}{(v_{\mathrm{A}})_0}\left(\frac{\rho_0}{\rho}\right)^{1/2}. \tag{10.95}$$

Insertion of (10.95) into (10.93) and integration yields

$$\delta B_1(r) = \delta B_0\left(\frac{\rho}{\rho_0}\right)^{3/4}\left(\frac{1 + (v_{\mathrm{A}}/v)_0}{1 + [(v_{\mathrm{A}}/v)_0](\rho/\rho_0)^{1/2}}\right) \tag{10.96}$$

with the associated $\delta v_1(r)$ given by (10.90) (Parker 1965).

Substitution of δB_1 and δv_1 into (10.80) yields the extra driving term

$$\frac{1}{8\pi\rho}\frac{\partial}{\partial r}\left(\frac{(\delta B_0)^2}{2}\left(\frac{\rho}{\rho_0}\right)^{3/2}\frac{[1 + (v_{\mathrm{A}}/v)_0]^2}{\{1 + [(v_{\mathrm{A}}/v)_0](\rho/\rho_0)^{1/2}\}^2}\right). \tag{10.97}$$

The equation now yields the wave-modified Bernoulli integral

$$\frac{1}{2}v^2 + a^2 \log\left(\frac{\rho}{\rho_0}\right) - \frac{GM}{r} + \frac{(\delta B_0)^2}{8\pi\rho_0}\left(\frac{\rho}{\rho_0}\right)^{1/2}\left(1+\frac{(v_{\rm A})_0}{v_0}\right)^2 \times \frac{3/2 + (v_{\rm A}/v)_0(\rho/\rho_0)^{1/2}}{\{1+[(v_{\rm A}/v)_0](\rho/\rho_0)^{1/2}\}^2} = \text{constant.} \tag{10.98}$$

(In the case of a rapidly rotating star, one would need to include the extra sink and source terms $\Omega^2\varpi^2/2$ and $\alpha\Omega\varpi^2$ – cf. Section 7.5.) Multiplication of (10.98) by the constant mass flux $(4\pi\rho v r^2)$ leads after some manipulation to a modified expression for the constant total energy flux F, given by

$$\frac{F}{4\pi r^2} = v\left[\frac{\rho v^2}{2} + a^2\rho\log\left(\frac{\rho}{\rho_0}\right) - \rho\frac{GM}{r}\right] + \frac{\rho(\delta v_1)^2}{2}\frac{v}{2} + \frac{(\delta B_1)^2}{4\pi}\frac{v+v_{\rm A}}{2}. \tag{10.99}$$

In this expression, the energy flux due to the Alfvén waves has been written as the sum of two terms: the mean rate of convection by the bulk flow of the wave kinetic energy density, and the radial component of the Poynting vector. From the form (10.99), it is seen that the wave energy flux across a sphere r decreases monotonically to zero at infinity. As the waves propagate and are convected outwards they do work on the wind; in this dissipation-free system the energy they lose all goes into accelerating the wind.

For details of the modification of the basic Parker solution by this direct input of energy and momentum from the waves, the reader is referred to Belcher (1971). The model has been applied convincingly by Lago and Penston (1982) and Lago (1984), especially to the T Tauri star RU Lupi. A complete treatment of the rotational evolution of T Tauri stars in general, on the lines of the work summarized in Section 10.3, should include a contribution from Alfvén wave-driven winds flowing along open field lines.

10.5 Instability in a magnetic rotating disc

We return now to the physics of magnetic accretion discs. It was early recognized that microviscosity is far too small to enforce significant angular momentum redistribution. The original disc theory therefore contained the turbulent viscosity as a crucial phenomenological parameter, estimated essentially on dimensional grounds, as in (10.56), but without a convincing physical explanation of the origin of the turbulence. For a recent succinct account with references of the different approaches to the problem in the literature, the reader is referred to Papaloizou and Lin (1995). In this section we consider the instabilities to be expected in a magnetic disc, with an eye on the possible spontaneous generation and maintenance of turbulence. For the moment the magnetic field is supposed prescribed, but some of the results will turn out to be relevant to dynamos operating in discs (Sections 10.6 and 10.7).

The powerful interaction between magnetism and rotation – a recurrent theme in this volume – has long been recognized. Inside a star, pressure and gravity are in approximate balance, with centrifugal force at most a significant perturbing force, so that changes in the Ω-field brought about by the passage of torsional Alfvén waves lead to modest changes in the pressure–density–temperature field. By contrast, already in the study of roughly spherical stellar winds (Chapter 7) it was seen that in the case of a rapid rotator, the enforcement of near uniform rotation by magnetic stresses yields a centrifugal force that exceeds the restraining gravitational field and so either assists the thermal pressure in driving the wind or becomes the dominant driver in a 'centrifugal wind'.

Early work on disc-like centrifugo-gravitational equilibrium, such as protostellar discs (Mestel 1960) or disc-like galaxies (Hoyle and Ireland 1960), noted that the presence of a large-scale magnetic field with a poloidal component would normally lead to ongoing secular evolution of the angular momentum distribution with respect to mass, and so also of the mass distribution in space. The initial shear will again change the toroidal magnetic field and so generate new magnetic torques that redistribute angular momentum; but the gas simultaneously moves radially, so as to remain in close centrifugal balance with the gravitational field. The only disc-like bodies which exert a gravitational force $\propto \varpi$ and which could therefore be kept in equilibrium by the centrifugal force of *uniform* rotation are the flattened Maclaurin spheroids (which are in any case gravitationally unstable to both large-scale and small-scale modes (Hunter 1963)). In all realistic cases, the gravitational field enforces a non-uniform rotation field, which will interact with the radial component of a magnetic field to drive an ongoing angular momentum redistribution; but the growth of the toroidal field and so of the magnetic torques is linear in time, leading to comparatively slow evolution.

10.5.1 *Axisymmetric torsional instabilities*

For decades, there has been in the literature a prediction of an exponentiating instability that could lead to a significant macroviscosity. Velikhov (1959) and independently Chandrasekhar (1960; 1961, p. 384) had studied the hydromagnetic instability of inviscid Couette flow – the circular flow of conducting liquid between two coaxial cylinders, in the presence of a uniform magnetic field $B_0\hat{\mathbf{z}}$ parallel to the axis. The treatment is global, with the condition $v_\varpi = 0$ applied at the two bounding cylinders $\varpi = \varpi_1, \varpi_2$. It is readily shown (cf. Chandrasekhar 1961, p. 386, equation (51)) that an axisymmetric solution having a Lagrangian displacement ξ with the (z, t)-dependence of the form $\exp \mathrm{i}(\omega t + k_z z)$, has ξ_ϖ satisfying

$$(\omega^2 - k_z^2 v_{\mathrm{A}z}^2)^2 \left[\frac{\mathrm{d}}{\mathrm{d}\varpi}\left(\frac{\mathrm{d}(\xi_\varpi \varpi)}{\varpi\,\mathrm{d}\varpi}\right) - k_z^2 \xi_\varpi\right] = -k_z^2\left[\kappa^2(\omega^2 - k_z^2 v_{\mathrm{A}z}^2) + 4\Omega^2 k_z^2 v_{\mathrm{A}z}^2\right]\xi_\varpi, \tag{10.100}$$

where $v_{\mathrm{A}z} = B_0/\sqrt{4\pi\rho}$, and κ is the epicyclic frequency, defined by

$$\kappa^2 = \frac{1}{\varpi^3}\frac{\mathrm{d}(\Omega\varpi^2)^2}{\mathrm{d}\varpi}. \tag{10.101}$$

Multiplication of (10.100) by $\varpi\xi_\varpi^*$ and integration over $\varpi_1 < \varpi < \varpi_2$ shows that ω^2 must be real, so that the modes are either stable oscillations ($\omega^2 > 0$) or exponentially growing ($\omega^2 < 0$). The condition for stability is then found to be

$$\frac{I_1 B_0^2}{4\pi\rho} > -\int_{\varpi_1}^{\varpi_2} \left(\frac{\mathrm{d}\Omega^2}{\mathrm{d}\varpi}\right) \varpi^2 \xi_\varpi^2 \,\mathrm{d}\varpi, \tag{10.102}$$

where

$$I_1 = \int_{\varpi_1}^{\varpi_2} \varpi \left[\left(\frac{1}{\varpi}\frac{\mathrm{d}}{\mathrm{d}\varpi}(\varpi\xi_\varpi)\right)^2 + k_z^2 \xi_\varpi^2\right] \mathrm{d}\varpi. \tag{10.103}$$

Since $I_1 > 0$ and is bounded, we require

$$\frac{\mathrm{d}\Omega^2}{\mathrm{d}\varpi} \geq 0 \tag{10.104}$$

to ensure stability for an arbitrarily weak field B_0; while a negative Ω^2-gradient can be stabilized by taking B_0 sufficiently great. In the analogous problem in a non-conducting fluid, there is no flux-freezing constraint $\partial\mathbf{B}/\partial t = \nabla \times (\mathbf{v} \times \mathbf{B})$, and axisymmetric modes are stable provided that κ^2 defined by (10.101) satisfies the familiar Rayleigh criterion

$$\kappa^2 > 0. \tag{10.105}$$

Chandrasekhar noted that (10.105) is not recovered as $B_0 \to 0$: (10.104) remains valid because flux-freezing persists for arbitrarily small B_0.

It should be noted that the instability occurs even in systems in which the unperturbed magnetic field is too weak to affect the zero-order equilibrium. The cause of the instability to axisymmetric modes is easily understood. A z-dependent, outwardly directed radial displacement field yields a non-uniform rotation along the magnetic field, which in the familiar way generates a z-dependent B_ϕ-component. The consequent magnetic torques attempt to establish isorotation, and the excess centrifugal forces act to drive the displaced gas further out. The outward motion is resisted by the poloidal forces exerted by the distorted poloidal field; for sufficiently large wavelengths the centrifugal force wins and the instability grows.

An analogous hydromagnetic problem was subsequently treated by Fricke (1969) for a Boussinesq fluid, and in the 'local', small wavelength approximation, which implicitly assumes the effects of boundaries are small. Fricke's analysis was redeveloped by Balbus and Hawley (1991), and has since been discussed by the same and other authors in a number of papers. It is convenient to reproduce here the original Balbus–Hawley analysis (with some slight modifications). In the zero-order state, $\Omega = \Omega(\varpi)$, but the other time-independent variables may depend on both z and ϖ, as allowed by the equations. Again the special case with $B_\phi = 0$ is adopted; also, $B_\varpi = 0$ is assumed, so that in the unperturbed state (and also if the perturbations are independent of z), there is no angular momentum evolution. Local Eulerian disturbances of the form $\exp \mathrm{i}(k_\varpi\varpi + k_z z - \omega t)$

are studied in the short wavelength approximation, i.e. with $k_\varpi \varpi \gg 1$, $k_z z \gg 1$. The Boussinesq approximation is adopted, so that magneto-acoustic waves are filtered out. The unperturbed equations and the associated perturbed equations (with only dominant terms retained) are then as follows.

Continuity:

$$\rho \nabla \cdot \mathbf{v} + \frac{\mathrm{d}\rho}{\mathrm{d}t} = 0 : \tag{10.106}$$

$$k_\varpi \, \delta v_\varpi + k_z \, \delta v_z = 0. \tag{10.107}$$

Field-freezing:

$$\frac{\partial \mathbf{B}}{\partial t} = \nabla \times (\mathbf{v} \times \mathbf{B}) : \tag{10.108}$$

ϖ-component

$$-\mathrm{i}\omega \, \delta B_\varpi = \mathrm{i}k_z B_z \, \delta v_\varpi ; \tag{10.109}$$

z-component

$$-\mathrm{i}\omega \, \delta B_z = -\mathrm{i}k_\varpi B_z \, \delta v_\varpi = \mathrm{i}k_z B_z \, \delta v_z \tag{10.110}$$

by (10.107); ϕ-component

$$-\mathrm{i}\omega \, \delta B_\phi = (\mathrm{i}k_z B_z \, \delta v_\phi + \Omega' \varpi \, \delta B_\varpi), \tag{10.111}$$

where the perturbed divergence condition

$$0 = \varpi \nabla \cdot \delta \mathbf{B} = \Omega \big[(1 + \mathrm{i}\varpi k_\varpi) \, \delta B_\varpi + \mathrm{i}k_z \varpi \, \delta B_z \big] \tag{10.112}$$

has been used in (10.111). (Note that even though $\varpi k_\varpi \gg 1$ by assumption, the term $1 \times \delta B_\varpi$ in (10.112) must be retained as it is of the same order as $(\Omega' \varpi) \delta B_\varpi$; its neglect would lead to an erroneous form for δB_ϕ, with $(\Omega \varpi)'$ replacing $\varpi \Omega'$.)

Equation of motion:

$$\frac{\mathrm{d}\mathbf{v}}{\mathrm{d}t} + \frac{1}{\rho} \nabla \left(p + \frac{\mathbf{B}^2}{8\pi} \right) - \frac{1}{4\pi\rho} (\mathbf{B} \cdot \nabla) \mathbf{B} - \nabla V = 0; \tag{10.113}$$

ϖ-component

$$-\mathrm{i}\omega \, \delta v_\varpi + \frac{\mathrm{i}k_\varpi}{\rho} \, \delta p - 2\Omega \, \delta v_\phi - \frac{\delta \rho}{\rho^2} \frac{\partial p}{\partial \varpi} + \frac{\mathrm{i}k_\varpi}{4\pi\rho} (B_z \delta B_z) - \frac{\mathrm{i}k_z}{4\pi\rho} B_z \, \delta B_\varpi = 0; \tag{10.114}$$

z-component

$$-\mathrm{i}\omega \, \delta v_z + \frac{\mathrm{i}k_z \delta p}{\rho} - \frac{\delta \rho}{\rho^2} v \frac{\partial p}{\partial z} = 0; \tag{10.115}$$

ϕ-component

$$-\mathrm{i}\omega \, \delta v_\phi + \frac{\kappa^2}{2\Omega} \, \delta v_\varpi - \mathrm{i}k_z B_z \, \delta B_\phi = 0, \tag{10.116}$$

where κ is again defined by (10.101). In deriving these perturbed forms, terms such as $\delta B (\partial (\mathbf{B}) / \partial \varpi)$ have been dropped compared with $(k_\varpi, \, k_z) B \, \delta B$, etc.

The disturbances are assumed isentropic, with $\gamma = 5/3$. Because of the Boussinesq approximation, δp is put equal to 0 except in the equation of motion, where it gives rise to the buoyancy term; hence

$$0 = \frac{\mathrm{d}}{\mathrm{d}t}\ln(p\rho^{-5/3}) = \mathrm{i}\omega\frac{5}{3}\frac{\delta\rho}{\rho} + \delta v_z \frac{\partial \ln(p\rho^{-5/3})}{\partial z} + \delta v_\varpi \frac{\partial \ln(p\rho^{-5/3})}{\partial \varpi}. \quad (10.117)$$

Balbus and Hawley use (10.107) to eliminate δv_ϖ, and then (10.109)–(10.112), (10.115) and (10.116) to express all perturbed quantities in terms of δv_z, deriving finally the dispersion relation by substitution into the ϖ-component (10.114). Some simplification of the rather tedious algebra comes from noting that in the unperturbed steady state, with $\Omega = \Omega(\varpi)$ and magnetic forces making no contribution, the curl of the equation of motion (10.113) requires that the isobaric and isochoric surfaces adjust so as to coincide, so that $p = p(\rho)$. The final form of the dispersion relation then becomes:

$$\frac{k^2}{k_z^2}\tilde{\omega}^4 - \left[\kappa^2 + \left(\frac{k_\varpi}{k_z}N_z - N_\varpi\right)^2\right]\tilde{\omega}^2 - 4\Omega^2 k_z^2 v_{\mathrm{A}z}^2 = 0, \quad (10.118)$$

where

$$k^2 \equiv k_\varpi^2 + k_z^2, \quad (10.119)$$

$$\tilde{\omega}^2 \equiv \omega^2 - k_z^2 v_{\mathrm{A}z}^2, \quad (10.120)$$

and N_z, N_ϖ jointly make up the Brunt–Väisälä (B–V) frequency (cf. (3.126)):

$$N^2 \equiv N_z^2 + N_\varpi^2 = -\frac{3}{5\rho}\frac{\partial p}{\partial z}\frac{\partial \ln(p\rho^{-5/3})}{\partial z} - \frac{3}{5\rho}\frac{\partial p}{\partial \varpi}\frac{\partial \ln(p\rho^{-5/3})}{\partial \varpi} \quad (10.121)$$

The dispersion relation (10.118) is a special case of that found by Fricke (1969). Note that B_z enters in the combination $k_z B_z$. When N_ϖ, N_z are negligible, as is often the case, for example in an accretion disc, then the Chandrasekhar equation (10.100) does indeed reduce to (10.118) on insertion of the local approximation $\partial/\partial\varpi \simeq ik_\varpi$ with $k_\varpi \varpi \gg 1$.

It is assumed that both the epicyclic frequency κ and the B–V frequency N and the associated frequencies N_ϖ, N_z are real, so that the *non-magnetic* disc would be stable against inviscid, adiabatic, axisymmetric perturbations. The instabilities of the magnetic disc emerge from the dispersion relation (10.118). The transition from stability to instability occurs at $\omega^2 = 0$, or $\tilde{\omega}^2 = -k_z^2 v_{\mathrm{A}z}^2$, for which (10.118) reduces to

$$k_\varpi^2(k_z^2 v_{\mathrm{A}z}^2 + N_z^2) - 2k_\varpi k_z N_\varpi N_z + k_z^2\left(\frac{\mathrm{d}\Omega^2}{\mathrm{d}\ln\varpi} + N_\varpi^2 + k_z^2 v_{\mathrm{A}z}^2\right) = 0. \quad (10.122)$$

If this quadratic in k_ϖ allows real solutions, then unstable modes exist as ω passes through zero; thus to ensure stability, the discriminant must be negative:

$$k_z^4 v_{\mathrm{A}z}^4 + k_z^2 v_{\mathrm{A}z}^2 \left(N^2 + \frac{\mathrm{d}\Omega^2}{\mathrm{d}\ln\varpi} \right) + N_z^2 \frac{\mathrm{d}\Omega^2}{\mathrm{d}\ln\varpi} > 0. \tag{10.123}$$

Since in general $N_z^2 > 0$ by assumption, condition (10.123) can hold for arbitrarily small k_z if and only if

$$\frac{\mathrm{d}\Omega^2}{\mathrm{d}\varpi} \geq 0. \tag{10.124}$$

Thus the local analysis again replaces the familiar non-magnetic Rayleigh criterion $\mathrm{d}(\Omega\varpi^2)^2/\mathrm{d}\varpi > 0$ by the same stringent stability criterion (10.104), which will normally be violated.

The theory predicts instability for $k_z < (k_z)_{\mathrm{crit}}$, given by the vanishing of the left-hand side of (10.123). The most interesting cases are with the B–V frequency small (rotation velocities supersonic), for which the critical wavenumber is

$$(k_z)_{\mathrm{crit}} = \frac{1}{v_{\mathrm{A}z}} \left| \frac{\mathrm{d}\Omega^2}{\mathrm{d}\ln\varpi} \right|^{1/2} \tag{10.125}$$

The stability criterion (10.124) does not apply in the disc mid-plane, where $N_z^2 = 0$, but is replaced there by

$$N_\varpi^2 + \frac{\mathrm{d}\Omega^2}{\mathrm{d}\varpi} \geq 0, \tag{10.126}$$

which, however, for supersonic velocities differs little from (10.124).

The analysis is greatly simplified by use of the generalized Boussinesq approximation, which not only ignores compressibility in the continuity equation but also assumes $|\delta p|/p \ll |\delta\rho|/\rho$ in the energy equation (10.117). If the zero-order $B_\varpi \neq 0$, then as noted it will cause a slow redistribution of mass and angular momentum, so the zero-order state is not strictly steady. However, as long as $B_\varpi^2 \ll 4\pi\rho\Omega^2\varpi^2$, the time-dependence that will now be present in the 'unperturbed' quantities will be weak compared with the exponential growth of the fastest disturbances. The essential change is the replacement of $k_z^2 v_{\mathrm{A}z}^2$ by $(\mathbf{k} \cdot \mathbf{v}_{\mathrm{A}})^2$ in (10.120) and (10.118), and generalizations of the above arguments show that the stability criterion is unaffected.

The instability is possible because of the energy available in the shear, compared with that in a uniformly rotating fluid with the same angular momentum. However, in the case of an inviscid, non-magnetic fluid, *conservation of vorticity* ensures that this energy cannot be released unless there is a point of inflection in the velocity profile (e.g. Lin 1955; Drazin and Reid 1981). As emphasized earlier by Lynden-Bell (1966), the role of the Lorentz force in this type of problem is to break the conservation of vorticity, so making available states of lower energy, and this effect persists however weak the field.

The Chandrasekhar analysis is for a *cylindrical* zero-order state with no finite characteristic scale in the z-direction, so the $\exp ik_z z$ assumption for small perturbations involves no further approximation. The F–B–H local treatment requires that $k_\varpi \varpi \gg 1$ – ϖ-wavelength small compared with ϖ – and also z-wavelengths small compared with the disc thickness. As an example, consider again the thin, isothermal Keplerian disc, for which the epicyclic frequency $\kappa = \Omega$, and the density

$$\rho = \rho_0(\varpi)\exp(-z^2/H^2), \qquad H = 2^{1/2}c_s/\Omega(\varpi), \tag{10.127}$$

satisfies the z-hydrostatic equation

$$c_s^2 \frac{\partial \ln \rho}{\partial z} = -\frac{GMz}{\varpi^3} = -\Omega^2 z. \tag{10.128}$$

The largest growth rates are typically $\approx 0.75\Omega$. From (10.125), the ratio $(\lambda_{\rm crit}/2H)$ of the critical wavelength to the disc thickness is $(\pi/6^{1/2})(v_{{\rm A}z}/c_s)$, or

$$\left(\frac{\lambda_{\rm crit}}{2H}\right)^2 = \frac{\pi^2}{3\beta} \tag{10.129}$$

in terms of a plasma $\beta = 8\pi\rho c_s^2/B_z^2$. Thus we have the striking characteristic properties of the instability: it occurs provided the magnetic energy is well below the thermal (high β), and then with a growth rate dependent on the local rotation frequency but independent of the field strength.

As the thickness $2H$ is clearly a crude upper limit to the domain of validity of the local theory, it is tempting to extrapolate and estimate the field-strength capable of stabilizing the disc by putting the ratio (10.129) equal to unity. This requires that $\beta \approx 1$ – the magnetic and thermal energy densities comparable; or equivalently,

$$v_{{\rm A}z} \equiv \frac{B_z}{(4\pi\rho)^{1/2}} \approx (\Omega\varpi)\frac{H}{\varpi}. \tag{10.130}$$

However, for an accurate estimate one needs again a global treatment, as in Papaloizou and Szuszkiewicz (1992), who study the stability to axisymmetric disturbances of a disc again with zero toroidal but with a more general curl-free poloidal field, and without use of the Boussinesq approximation. The field-strength *sufficient* to stabilize is estimated now as

$$v_{{\rm A}z} \approx (\Omega\varpi)\left(\frac{z}{\varpi}\right)^{1/2}. \tag{10.131}$$

The square-root appearing in (10.131) represents the effect of the magnetic energy in the quasi-vacuum region outside the disc. However, the condition has not been shown to be necessary, so it is possible that stabilization against axisymmetric modes may occur at a value somewhere between (10.130) and (10.131).

The axisymmetric F–B–H local analysis can be performed with the zero-order B_ϕ-component non-zero, yielding apparently the same dispersion relation

(10.118), i.e. without B_ϕ appearing explicitly. If the local analysis were strictly applicable, this would be understandable, for an axisymmetric perturbation will not generate an extra poloidal component from B_ϕ and so will not increase further the interaction with the rotation field that drives the instability. However, as emphasized by Knobloch (1992) and others, when shear flows are under study, a local analysis can yield results that are not strictly correct. Knobloch formulated the eigenvalue problem analogous to the Velikhov–Chandrasekhar problem, but with a zero-order B_ϕ-component as well, showing that the pure exponentially-growing mode predicted by the local analysis cannot then exist. Kumar *et al.* (1994), following earlier work by Tayler (1973) and Chanmugam (1979) (cf. Section 3.6.5), use global analysis to derive sufficient conditions for stability against axisymmetric modes. They find no conflict between local and global analysis when $B_\phi = 0$ and provided B_z is weak enough for instability to be possible for wavelengths much less than the thickness of the disc. But when $B_\phi \neq 0$, they confirm Knobloch's conclusion that an unstable mode necessarily shows *overstability* – exponentially growing oscillations. In an illustrative example, with the domain an annulus with $\delta\varpi \ll \varpi$, they find the most unstable mode behaves like $\exp -(\mathrm{i}\omega t)$, with

$$\omega = \frac{3\mathrm{i}\Omega}{4} - \frac{\sqrt{15}}{4}\frac{B_\phi}{(4\pi\rho)^{1/2}}. \tag{10.132}$$

Other workers who have taken up the linear problem include Blaes and Balbus (1994), who introduce ambipolar diffusion into the instability analysis. Curry *et al.* (1994) have studied the global problem, but with free boundaries replacing rigid. They find the growth rates of unstable modes are less than but comparable with the corresponding local growth rates.

10.5.2 *Non-axisymmetric instabilities*

Dubrulle and Knobloch (1993) study both axisymmetric and non-axisymmetric modes, with an assumed zero-order toroidal field. They derive the necessary condition for instability (equivalently, with the inequality reversed, a sufficient condition for stability), and valid for both axisymmetric and non-axisymmetric perturbations:

$$\frac{1}{\varpi^2}\frac{\mathrm{d}}{\mathrm{d}\varpi}\left[\frac{(\varpi^2 B_\phi)^2}{4\pi\rho}\right] - \varpi^2\frac{\mathrm{d}\Omega^2}{\mathrm{d}\varpi} > 0. \tag{10.133}$$

This illustrates how a strong enough B_ϕ can suppress the V–Ch–F–B–H instability.

In recent studies, Ogilvie and Pringle (1996) and Terquem and Papaloizou (1996) consider the non-axisymmetric instability of a differentially rotating ideal fluid containing an azimuthal (toroidal) magnetic field. A primary motivation is the search for possible dynamo action in a disc; again one is interested in any process that may complete the poloidal → toroidal → poloidal cycle (Chapter 6; Sections 10.6 and 10.7). Ogilvie and Pringle specialize to a cylindrical model, with $\partial/\partial z = 0$ in the unperturbed state, and with the fluid incompressible. Their analysis is global, and like that of Coleman *et al.* (1995) and Matsumoto and Tajima (1995), it is performed using non-shearing coordinates (cf. Section 10.7). It is known that in general, a purely toroidal field is itself spontaneously unstable; however, they start with the field $B_\phi \propto 1/\varpi$, which corresponds to the minimum magnetic energy for a given toroidal magnetic flux (cf. Tayler 1973). Like Curry *et al.* they treat the global eigenvalue problem, but the results are not greatly different for the much simpler local analysis. The same basic result emerges: a magnetic field, however weak, can destabilize a sheared rotating flow if $\mathrm{d}\Omega^2/\mathrm{d}\varpi < 0$. In the absence of diffusion, the instability grows preferentially at arbitrarily small scales. Suppression of the instability occurs only for fields so strong that the mode of longest possible wavelength allowed by the geometry requires through field line bending an energy input too great for the actual energy release from the shear. But in striking contrast to the basic Balbus–Hawley result – stabilization reached when the magnetic energy reaches the thermal – suppression of the instability now does not occur until much higher field strengths, with the magnetic energy in approximate equipartition with the *rotational* energy.

Terquem and Papaloizou (like Foglizzo and Tagger 1995) study the stability of a stratified, compressible, disc-like equilibrium. Discs containing a purely toroidal field are always unstable. Using local analysis, they find again a class of unstable modes driven by shear, and not dependent on the equilibrium stratification (Balbus–Hawley-type), and a class driven by buoyancy (Parker-type). As expected, the first class of instability predominates when the field is weak and the underlying medium is strongly stable against convection; the second class predominates if the field is strong.

10.5.3 *Interchange instability*

Following earlier work by Spruit and Taam (1990) and Lubow and Spruit (1995), Spruit *et al.* (1995) have studied interchange instabilities in a disc permeated by a poloidal field, analogues of the modes discussed in Section 3.6.5. Their treatment is in the thin disc approximation, using equations similar to (10.75, 10.76, 10.77) that result from integration through the disc thickness. The microresistivity and microviscosity are both assumed negligibly small. As by their nature, interchange motions do not bend field lines, the motions are nearly parallel to the disc plane: the total velocity field $\mathbf{u}$ can be taken as two-dimensional and independent of height z within the disc. The equation of motion is then

$$\Sigma\frac{\mathrm{d}\mathbf{u}}{\mathrm{d}t} = \Sigma\left[\frac{\partial \mathbf{u}}{\partial t} + (\mathbf{u}\cdot\nabla)\mathbf{u}\right] = \frac{B_z}{4\pi}[\mathbf{B}] - \frac{GM}{\varpi^2}\Sigma\hat{\varpi}, \tag{10.134}$$

where $[\mathbf{B}]$ is the jump in $\mathbf{B}$ across the disc. The continuity condition is

$$\frac{\partial\Sigma}{\partial t} + \nabla\cdot(\Sigma\mathbf{u}) = 0, \tag{10.135}$$

and the perfect conductivity condition yields

$$\frac{\partial B_z}{\partial t} + \nabla \cdot (B_z \mathbf{u}) = 0. \tag{10.136}$$

Equations (10.135) and (10.136) combine into

$$\mathrm{d}(\Sigma/B_z)/\mathrm{d}t = 0. \tag{10.137}$$

During the perturbation the disc is assumed to remain flat and perpendicular to the z-axis – i.e. no corrugation occurs. Outside the disc the density is low enough for the field to be force-free. It is assumed that in practice the curl-free approximation is adequate; hence for $z > 0$,

$$\mathbf{B} = \nabla\Phi, \qquad \nabla^2\Phi = 0, \tag{10.138}$$

which with the boundary conditions

$$\frac{\partial \Phi}{\partial z} = B_z^+ \qquad \text{at } z = 0 \qquad \Phi \to 0 \text{ as } z \to \infty \tag{10.139}$$

fix Φ and so also B_ϖ^+ and B_ϕ^+ at $z = 0$. Because no corrugation occurs, the field is antisymmetric about the midplane, with B_z preserving sign but B_ϖ, B_ϕ changing sign; hence $[\mathbf{B}] = [2B_\varpi^+, 2B_\phi^+, 0]$ both in the unperturbed state (as in the model of Section 10.2), and in the perturbed state.

In the unperturbed state, $B_\phi^+ = 0$, $\mathbf{u} = [0,\, \Omega(\varpi)\varpi]$. Because of the magnetic support, Ω is non-Keplerian, satisfying

$$\Omega^2\varpi = \frac{GM}{\varpi^2} - g_{\mathrm{m}}, \qquad g_{\mathrm{m}} \equiv \frac{B_\varpi^+ B_z}{2\pi\Sigma}, \tag{10.140}$$

with the associated shear

$$S \equiv \varpi\frac{\mathrm{d}\Omega}{\mathrm{d}\varpi} = -\frac{3}{2}\Omega - \frac{g_{\mathrm{m}}}{2\Omega}\frac{\mathrm{d}}{\mathrm{d}\varpi}\ln(\varpi^2 g_{\mathrm{m}}). \tag{10.141}$$

The time-dependent equations are linearized about this state, and then treated in the shearing sheet model of Goldreich and Lynden-Bell (1965). In this model, local Cartesian coordinates $(x,\, y)$ in the radial and azimuthal directions are used around a point $\varpi = \varpi_0$, in a frame rotating at the rate $\Omega_0 = \Omega(\varpi_0)$, and the shear is approximated as constant:

$$\varpi = \varpi_0 + x, \qquad \frac{1}{\varpi_0}\frac{\partial}{\partial\phi} = \frac{\partial}{\partial y}, \qquad \Omega - \Omega_0 = \frac{Sx}{\varpi_0}. \tag{10.142}$$

The external magnetic scalar potential Φ is constructed in the short wavelength approximation, yielding

$$\mathbf{B}'(x, y, z, t) = \mathbf{b}(t)\exp(\mathrm{i}k_x x + \mathrm{i}k_y y - k|z|), \tag{10.143}$$

where

$$k = (k_x^2 + k_y^2)^{1/2}, \tag{10.144}$$

$$b_x = -\mathrm{i}(k_x/k) b_z \operatorname{sgn}(z), \tag{10.145}$$

$$b_y = -\mathrm{i}(k_y/k) b_z \operatorname{sgn}(z). \tag{10.146}$$

In a shearing coordinate frame, $x' = x$, $y' = y - Sxt$, $z' = z$, $t' = t$, the wave vector is now time-dependent:

$$\mathbf{k}(t) = (-Stk_y,\, k_y). \tag{10.147}$$

By the introduction of the shearing coordinates, the linear dependence of the shear flow on x is removed, in exchange for the introduction of an explicit time dependence, which leads to (10.147). The resulting system of equations for the perturbed quantities depends on the three parameters s, K_y, a:

$$s = \frac{S}{\Omega}, \tag{10.148}$$

$$K_y = k_y L, \qquad L = \frac{B_z^2}{2\pi\Sigma\Omega^2}, \tag{10.149}$$

$$a = \frac{N_{\mathrm{m}}^2}{\Omega^2}, \qquad N_{\mathrm{m}}^2 = -\frac{g_{\mathrm{m}}}{\Omega^2}\frac{\mathrm{d}}{\mathrm{d}\varpi}\ln\frac{\Sigma}{B_z}, \tag{10.150}$$

where S is the shear rate (10.141), and L is the characteristic length-scale of the system. The newly-defined quantity N_{m} is the *magnetic buoyancy frequency*; for in the case of a stable magnetic gradient, it is the frequency of interchange displacements, analogous to the Brunt–Väisälä frequency in a convectively stable stratification (cf. Section 3.6.2). The short wavelength model is applicable if $K \geq |a|$.

The case of uniform rotation had been studied earlier by Spruit and Taam (1990) and Lubow and Spruit (1995). Because of the absence of shear, the wave number does not depend on time, and the solutions are $\propto \exp(\sigma t)$. The interchange modes are stable when $a > 0$, behaving then like an internal gravity wave (Section 3.6.2). They become unstable if $a < 0$, i.e. if $\mathrm{d}[\ln(\Sigma/B_z)]/\mathrm{d}\varpi > 0$. The local growth rate – $\approx (\varpi/g_{\mathrm{m}})^{1/2}$ – is rather short. The general solutions show that shear has a dramatic stabilizing effect: significant growth is possible only if

$$g_{\mathrm{m}}\frac{\mathrm{d}}{\mathrm{d}\varpi}\ln\frac{\Sigma}{B_z} > 2S^2 = 2\left(\varpi\frac{\mathrm{d}\Omega}{\mathrm{d}\varpi}\right)^2. \tag{10.151}$$

The instability occurs on a short length-scale, and with a power law rather than an exponential law of growth. If $S \simeq \Omega$, and L is the characteristic length-scale of Σ/B_z, then (10.151) can be written

$$\frac{B^2}{2\pi\Sigma} > \Omega^2 L; \tag{10.152}$$

or in terms of the disc semi-thickness $H = \Sigma/2\rho$,

$$\frac{B^2}{4\pi H} > \rho(\Omega^2\varpi)\frac{L}{\varpi} = \rho\Omega^2\varpi \tag{10.153}$$

if $L \simeq \varpi$. The requirements for instability are then not only that the gradient of Σ/B_z be positive, but that the magnetic force contribute as much as the centrifugal force to support against gravity. Such strongly magnetic (though non-self-gravitating) discs could arise during the process of star formation in the presence of the galactic magnetic field, as discussed in detail in Chapters 11 and 12. The work of Spruit *et al.* suggests strongly that the evolution of such discs may sometimes be affected as much by interchange instability as by ambipolar diffusion.

10.5.4 *Resumé*

One may also rewrite (10.153) for a thermally supported disc by substituting from (10.127) for $\Omega^2 = 2c_s^2/H^2$, yielding

$$\frac{v_\mathrm{A}^2}{c_s^2} > 2\frac{\varpi}{H}. \tag{10.154}$$

Recall that the original V–Ch–F–B–H instability, discussed in Section 10.5.1, is suppressed when $v_\mathrm{A}^2/c_s^2 \approx 1$, whereas the analogous non-axisymmetric instabilities of a toroidal field, discussed in Section 10.5.2, persist until $v_\mathrm{A}^2 \simeq \Omega^2\varpi^2$, or $v_\mathrm{A}^2/c_s^2 \simeq (\varpi/H)^2$. Thus when $\mathrm{d}(\Sigma/B_z)/\mathrm{d}z > 0$, the interchange instability will start up after the V–Ch–F–B–H instabilities have subsided but before the toroidal field instabilities have been suppressed.

Once linear analysis has shown that a system is unstable to various classes of perturbation, interest immediately focuses both on their mutual interaction and on the non-linear development of the system. It was seen in Section 3.6.2 that magnetic fields exerting a Lorentz force opposing gravity are liable to a Rayleigh–Taylor-type instability. Shu (1974) pioneered the effect of rotation with shear on the Parker instability, showing that the onset of instability is unaffected by a weak shear. The problem was taken up again by Foglizzo and Tagger (1994, 1995), who study in particular the transition between the Parker and Balbus–Hawley instabilities. They show that the strength of the differential rotation, as measured by the Oort constant $A = \varpi(\partial\Omega/2\,\partial\varpi)$ is crucial. If the rotation curve decreases more slowly than the flat rotation curve ($-A/\Omega < 1/2$), then the two instability domains are separated by a stable domain; but if $-A/\Omega > 1/2$, the two modes may occur simultaneously.

A series of papers has tackled the extension of the stability problem into the non-linear domain (Hawley and Balbus 1991, 1992; Brandenburg *et al.* 1995; Hawley *et al.* 1995; Matsumoto and Tajima 1995). The most recent simulations do show the system apparently reaching a saturated state of magneto-turbulence. The application to possible dynamo action is dicussed in Section 10.7.

The attraction of the V–Ch–F–B–H modes is that they show a possible, simply understood physical process for the spontaneous development of accretion disc turbulence. However, it should be noted that the interpretation of these simulations is not unambiguous; and in fact the description of this work as 'non-linear Balbus–Hawley' instability is considered by some to be somewhat misleading. The linear work shows that once a strong toroidal component has been generated, then the buoyancy instabilities can take over as a major means of supply of turbulent energy, with the ultimate source as always the energy in the shear. Nevertheless, though more clarificatory work remains to be done, it is clear that the presence of a weak magnetic field changes the whole stability problem for a disc, suggesting a plausible means by which the the kinetic energy of the differential rotation may be tapped so as to maintain the turbulence that had previously been postulated, so accounting for the secular evolution of the disc.

10.6 Disc dynamos: mean field theory

The work of Sections 10.2–10.4 assumes dynamo action occurs just in the convective envelope of the central star, generating the field which then interacts with the disc material. The astronomical consequences drawn are certainly encouraging; but it is inevitable that one should now ask whether dynamo action occurs also in the turbulent rotating disc, and if so, what qualitatively new effects emerge.

In Chapter 6, the standard mean-field dynamo equations were formulated and studied in spherical geometry, for application to stellar convective zones. It was recognized early, however, that dynamo action could occur efficiently in systems forced by high angular momentum into a disc-like structure, with pressure–gravity balance holding just in one dimension, but with centrifugal acceleration nearly balancing gravity in the other two. As before, we are interested initially in exploring the consequences of the simple mean-field equations, while recognizing that they are essentially the simplest possible parametrizations of a much more complicated hydromagnetic process. Some important recent work that goes beyond mean-field theory by incorporating the results of Section 10.5 is reported in the next section.

Application of the standard equations to disc-like galaxies, with their extensive domains of highly non-uniform rotation, was pioneered by Parker (1971) and Stix (1975), and later by Rosner and DeLuca (1989), by Ruzmaikin et al. (1988) and by Rüdiger *et al.* (1993). Some more recent work has introduced coupling with density wave theory of spiral structure (Chiba and Tosa 1990 ; Mestel and Subramanian 1991; Subramanian and Mestel 1993; Moss *et al.* 1990). Pioneering analytical studies of dynamos in accretion discs were made by Takahara (1979) and Pudritz (1981). More extensive numerical calculations were made by Stepinski and Levy (1988, 1990), by Camenzind (1991), and later by Torkelsson and Brandenburg (TB 1994), whose work is now summarized.

As before, the induction equation takes the form

$$\frac{\partial \mathbf{B}}{\partial t} = \nabla \times [\mathbf{v} \times \mathbf{B} + \alpha \mathbf{B} - \eta(\nabla \times \mathbf{B})], \tag{10.155}$$

The slow radial motion consequent on angular momentum redistribution within the disc is neglected, so the only macroscopic velocity to be inserted in (10.155) is that of the non-uniform rotation:

$$\mathbf{v} = \Omega \varpi \mathbf{t}. \tag{10.156}$$

Both the α-parameter and the turbulent resistivity η are assumed isotropic. Appropriate dynamo numbers are defined:

$$C_\alpha = \frac{\alpha_0 R}{\eta_\mathrm{d}} \tag{10.157}$$

and

$$C_\Omega = \frac{\Omega_0 R^2}{\eta_\mathrm{d}}, \tag{10.158}$$

where α_0 is a typical value for α, R is the outer radius of the disc, η_d is the disc diffusivity at $\varpi = R$, and Ω_0 the maximum of the angular velocity in the disc.

Axisymmetric solutions are sought, with poloidal and toroidal parts

$$\mathbf{B}_\mathrm{p} = \nabla \times (a\mathbf{t}), \qquad \mathbf{B}_\mathrm{t} = b\mathbf{t}, \tag{10.159}$$

where the potential a and the flux function P are as before related by $a = -P/\varpi$. Equation (10.155) yields

$$\left(\frac{\partial}{\partial t} - \eta D^2\right) a = \alpha b \tag{10.160}$$

and

$$\left(\frac{\partial}{\partial t} - \eta D^2\right) b = -\alpha(D^2 a) + \varpi \mathbf{B}_\mathrm{p} \cdot \nabla\Omega + \frac{1}{\varpi}\left[\nabla(\varpi b) \cdot \nabla\eta - \nabla(\varpi a) \cdot \nabla\alpha\right], \tag{10.161}$$

where D^2 is again the Stokes operator, satisfying $D^2 a = -\mathbf{t} \cdot \nabla\times(\nabla\times(a\mathbf{t}))$ (cf. (5.75)).

The medium outside the disc is assumed by TB to be of low effective conductivity, implying that $\nabla\times\mathbf{B} \approx 0$. It is not clear that this is a good approximation for the pre-main sequence star problem, for which ionization, for example by X-rays from the stellar corona, may be significant. As noted above and in Section 3.4, a better assumption for a conducting low-density domain is the force-free approximation (e.g. Aly and Kuijpers 1990; van Ballegooijen 1994): if there is magnetic interchange of angular momentum between star and disc, as discussed in the earlier sections, there must be a poloidal current flow linking the two domains, and the contribution $\mathbf{j}_\mathrm{p} \times \mathbf{B}_\mathrm{t}/c$ to the poloidal Lorentz force must be balanced by $\mathbf{j}_\mathrm{t} \times \mathbf{B}_\mathrm{p}/c$, so forcing $\mathbf{B}_\mathrm{p}$ to deviate from the strictly curl-free structure (cf. (10.12)). We shall quote the results of TB but note that there is scope for generalization. In their computations, TB find it convenient to follow Stepinski and Levy and apply vacuum boundary conditions at the surface of a sphere of low conductivity in which the disc is embedded.

A standard Shakura–Sunyaev (1973) thin accretion disc model is adopted. With the notation slightly modified from that in (10.56), the phenomenological turbulent viscosity is written

$$\nu_t = \alpha_{\mathrm{SS}} z_0 c_s, \tag{10.162}$$

where c_s and z_0 are respectively the local generalized sound speed and disc semi-thickness. The S–S parameter α_{SS} is given the perhaps optimistic value of 0.1. By balancing the component of gravity $g(\varpi)(z_0/\varpi)$ normal to the equator against the pressure gradient per gram c_s^2/z_0, one derives again the simple standard result

$$c_s \approx \left(g(\varpi)\frac{z_0^2}{\varpi}\right)^{1/2} \approx \Omega(\varpi) z_0. \tag{10.163}$$

If it is assumed that the turbulent correlation length cannot exceed z_0, then an upper limit to the magnitude of the dynamo α-parameter is $\alpha_0 \simeq z_0\Omega_0$, where Ω_0 is the maximum angular velocity in the disc. A further restriction again takes the turbulent magnetic Prandtl number as of order unity, so that $\eta_t \simeq \nu_t$. The other parameters are the degree of flatness of the disc, measured by z_0/R, and the plasma $\beta = 8\pi p/B^2$.

To construct dynamo models, the angular velocity, the α-effect and the diffusivity must be specified as functions of position. The rotation law is taken as Keplerian in the outer parts of the disc, but as the theory is to be applied all the way to the axis, the law goes over to rigid rotation in the innermost parts, to avoid a singularity at $\varpi = 0$. The law adopted is

$$\Omega(\varpi) = \Omega_0 \frac{R}{\varpi_0}\left(\frac{3}{2^{2/3}}\right)^{1/n}\left[1 + \left(\frac{\varpi}{\varpi_0}\right)^{3n/2}\right]^{-1/n}. \tag{10.164}$$

A large value for $n \simeq 10$ ensures a rapid transition (near $\varpi = \varpi_0$) from Keplerian to rigid rotation. The α-effect function adopted is as in Stepinski and Levy (1988, 1990):

$$\alpha = \alpha_0 z \frac{\Omega}{\Omega_0} F(z), \tag{10.165}$$

where $F(z) = 1$ on the equator and drops rapidly to zero beyond $|z| = z_0$. The limiting example is

$$F(z) = [\Theta(z + z_0) - \Theta(z - z_0)] \tag{10.166}$$

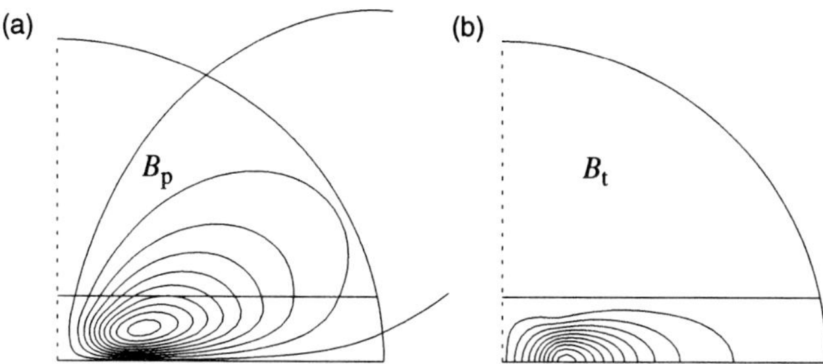

FIG. 10.7. An example of the most easily excited (steady S0) mode in a flat disc: (a) the $\mathbf{B}_p$-lines; (b) the constant B_t-contours.

with the Heaviside step function Θ in practice approximated by the continuous, rapidly varying function $[1 + \tanh(z/d)]/2$, where $d = 0.2$. The magnetic diffusivity adopted is

$$\eta = \eta_{\text{out}} + (\eta_{\text{disc}} - \eta_{\text{out}})F(z) \tag{10.167}$$

with $\eta_{\text{out}}/\eta_{\text{disc}} \gg 1$.

As in studies of dynamos in spherical geometry, the first (linear) models constructed are for critical dynamo numbers $C_\alpha C_\Omega$, yielding pure parity models. These marginal modes are either symmetric (S0 or 'quadrupolar') or antisymmetric (A0 or 'dipolar') about the equatorial plane. Although in a realistic disc, C_α is expected to be positive, for completeness, calculations are performed also for negative values.

In the first set of models, the disc is thick and 'flared', with $z_0 = 0.25(\varpi + R)$, $\varpi_0 = 0.25$, $C_\Omega = 100\eta_{\text{out}}/\eta_{\text{disc}}$, and with $\eta_{\text{out}}/\eta_{\text{disc}} = 20$ or 50. It is found that the most easily excited mode is dipolar when $C_\alpha > 0$ and quadrupolar when $C_\alpha < 0$. In general, for $C_\alpha > 0$, oscillatory solutions are found, with field migration towards the equator, accompanied by an outward radial motion; for $C_\alpha < 0$, both z- and ϖ-motions of the field are reversed. In some regions this dynamo-generated field exerts a negative torque, greater in magnitude than the positive torque exerted elsewhere, suggesting that the magnetic torque can contribute to accretion through transport of angular momentum.

Results for constant thickness discs with $z_0/R = 0.25$ and again with $\varpi_0 = 0.25$ show some qualitative differences. The most easily excited modes are steady rather than oscillatory and of S0 form (Fig. 10.7). Large variations in C_Ω yield only modest changes in $C_\alpha C_\Omega$ for the marginal modes, suggesting that these modes are in the $\alpha\Omega$ rather than $\alpha^2\Omega$ regime. Quadrupolar fields seem to be more efficient than dipolar fields in promoting accretion, in partial agreement with recent results by Rüdiger *et al.* (1993) for galactic dynamos.

The limitations of linear theory have already been emphasized. TB make a preliminary study, again in the spirit of earlier work in spherical geometry. The familiar 'α-quenching'

$$\alpha = \frac{\alpha_0}{1 + \alpha_B B^2} \tag{10.168}$$

is introduced, with $\alpha_B = 1/\mathbf{B}_{\rm eq}^2$ where the equipartition field satisfies $\mathbf{B}_{\rm eq}^2 = 4\pi\rho v_t^2$. No parity is imposed; instead, following Brandenburg *et al.* (1989), the continuous measure P of the field parity is computed from the solutions:

$$P = \frac{E^S - E^A}{E^S + E^A}, \tag{10.169}$$

where E^S, E^A are the energies of the respective parts.

It is found that when $C_\alpha > 0$ – the plausible case – the solutions settle into a state with a definite value of P; but with $C_\alpha < 0$, the solutions keep oscillating irregularly. In general, with positive C_α, the field structure is independent of the disc model, and is roughly the same as in the linear theory, except for being somewhat more evenly distributed in the disc. The parity of the field is mixed, with a small dipolar component as well as the dominant quadrupolar. The authors note that all other non-oscillatory mixed parity solutions known show a slow change in P, and they cannot rule out the same effect in their work.

When $C_\alpha < 0$, the field shows chaotic behaviour, varying on short time-scales (Fig. 10.8). The details of the solutions are questionable, as they predict variations on scales less than the disc semi-thickness, which is too small for mean-field theory to be applicable; but the prediction of chaos appears to be genuine.

The other non-linear effect introduced is again magnetic buoyancy, simulated by the new velocity term

$$\mathbf{v}_b = \gamma_b \mathbf{B}^2 \hat{\mathbf{z}}, \tag{10.170}$$

where γ_b is a strength parameter (Moss *et al.* 1990). The exact geometrical shape of the disc appears to be insignificant in the non-linear domain. It is found that for positive C_α, the behaviour is similar to the linear case – a steady quadrupolar solution – but the buoyancy effect leads to the presence of a considerable amount of toroidal flux in the surroundings of the disc. An interesting detail is that the magnetic field growth rate accelerates for a short time before the dynamo saturates. This is again due to buoyancy, which transports the field away from the equator into regions with a stronger α-effect.

More striking changes are found for negative C_α. The solution is oscillatory in both energy and parity, but the P–oscillations are damped, tending towards $P = 1$. The dynamo action appears also to be much more restricted, being concentrated into two regions at the inner edge and close to the upper and lower surfaces.

As emphasized by the authors, the aim of this paper is not to model the magnetic field of a real accretion disc, but to demonstrate that the standard

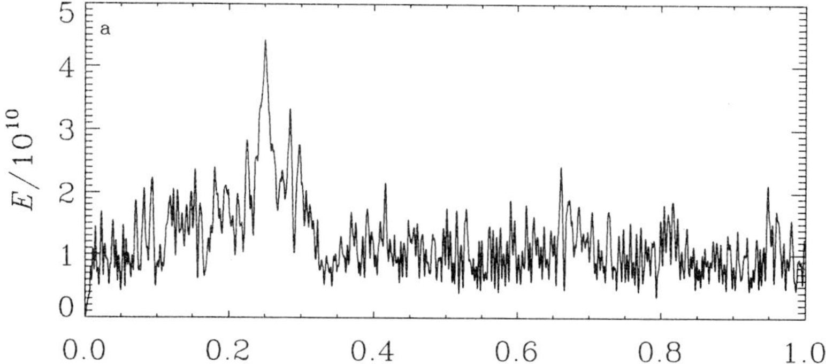

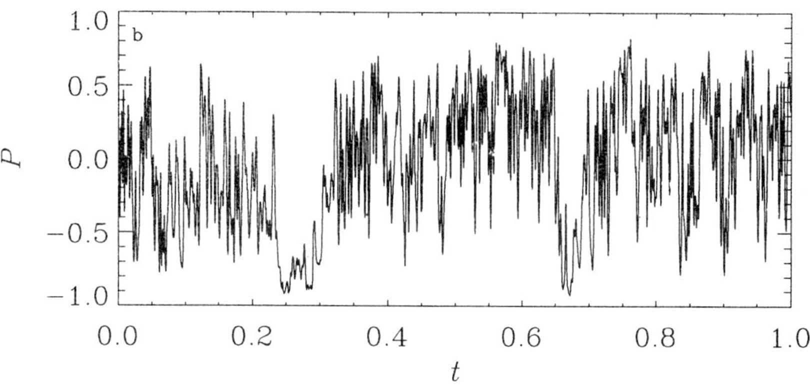

FIG. 10.8. An example of a dynamo with α-quenching given by (10.168), in a disc of uniform thickness. Chaotic behaviour is shown for (a) the magnetic energy and (b) the parity P given by (10.169).

dynamo equations can act in an accretion disc, and to elicit some guidelines to the qualitative behaviour of a disc dynamo. Thus it is confirmed that the strongest fields are built up where the gradient of the angular velocity is strongest, which is always in the innermost parts of the disc, and this part of the field is expected to be only weakly affected by the outer boundary conditions. This suggests that these preliminary models are reasonably reliable in the regions where the field attains its maximum, but that they cannot be used to predict field strength and structure further out, where it is believed that the carbonaceous chondrites formed in the solar nebula (Levy 1978).

10.7 Disc dynamos: recent developments

The discussion of the last section is in the spirit of what may now be called 'classical dynamo theory', with both the enhanced resistivity and the α-effect introduced as parametrizations of the interaction between the turbulence and the magnetic field. The work of Section 10.5 encourages the search for disc dynamos which do not depend on a postulated non-proven hydrodynamic turbulence. Tout and Pringle (1992) outlined such a model. The starting point is the familiar generation by shear of B_ϕ from a pre-existing B_ϖ, with the Parker magnetic buoyancy instability limiting the growth (cf. Sections 4.7 and 10.5.4). The Balbus–Hawley instability generates B_ϖ from an existing B_z, with buoyancy again limiting its growth. The Parker instability converts horizontal to vertical field (Horiuchi *et al.* 1988), but with alternating signs across the disc. The random juxtaposition of oppositely-signed B_z-components leads to reconnection which is postulated as the principal process of flux-destruction, offsetting the generation of B_z by the buoyancy instability (primarily from the dominant B_ϕ-component). This coupling of the three field components suggests the possibility of dynamo action, with the Keplerian gravitational field – which maintains the shear – again the energy source. The Tout–Pringle treatment is still semi-quantitative, but is justly claimed as 'a step in the right direction'.

In a new detailed simulation of a disc dynamo by Brandenburg *et al.* (1995, 1996), turbulent viscosity and resistivity are assumed but no phenomenological α-effect is imposed. The equations spontaneously generate through the non-linear V–Ch–F–B–H instability turbulence with the degree of anisotropy necessary to yield dynamo action: the system acts like a dynamo that generates its own turbulence, with the energy source as before the gravitational energy released as the gas responds to the ongoing redistribution of angular momentum by the magnetic and Reynolds stresses. In fact one motivation of the work is to show that the Balbus–Hawley instability does not depend on the magnetic field's being externally maintained, for example in the central star or in the galactic background, but can be generated locally, without imposing any extra conditions on the turbulence.

Another significant advance is that in some of the models, allowance is made for partial ionization through the inclusion of ambipolar diffusion. This not only increases the effective magnetic diffusion but makes it anisotropic, since the diffusion occurs in the direction of $\mathbf{j}_\perp$ – cf. Section 2.7, especially (2.130).

The problem is tackled in a local Cartesian reference frame (x, y, z), referring respectively to the radial, toroidal and vertical directions. The systematic variation of the variables in the x-direction is suppressed by solving only for deviations $\mathbf{u}$ from the Keplerian shear flow $u_y^0 = -3\Omega_0 x/2$ – following Wisdom and Tremaine (1988) – with Ω_0 the angular velocity at the origin of coordinates, located at ϖ_0 (cf. Section 10.5.3).

The equations to the vector potential $\mathbf{A}$, $\mathbf{u}$, ρ and the thermal energy per gram (written as e rather than u, to avoid confusion with $|\mathbf{u}|$) are:

$$\frac{D\mathbf{A}}{Dt} = \mathbf{u} \times \mathbf{B} + (3/2)\Omega_0 A_y \hat{\mathbf{x}} - (4\pi\eta/c)\mathbf{j}, \tag{10.171}$$

$$\frac{D\mathbf{u}}{Dt} = -(\mathbf{u}\cdot\nabla)\mathbf{u} + \mathbf{g} - \frac{1}{\rho}\nabla p + \mathbf{f}(\mathbf{u}) + \frac{1}{\rho}\frac{\mathbf{j}\times\mathbf{B}}{c} + \frac{1}{\rho}\nabla\cdot(2\nu\rho\mathbf{S}), \tag{10.172}$$

$$\frac{D\ln\rho}{Dt} = -(\mathbf{u}\cdot\nabla)\ln\rho - \nabla\cdot\mathbf{u}, \tag{10.173}$$

$$\frac{De}{Dt} = -(\mathbf{u}\cdot\nabla)e - \frac{p}{\rho}\nabla\cdot\mathbf{u} + \frac{1}{\rho}\nabla(\chi\rho\nabla e) + 2\nu\mathbf{S}^2 + \frac{4\pi\eta}{\rho c^2}\mathbf{j}^2 + Q. \tag{10.174}$$

Here, the time-derivative $D/Dt \equiv \partial/\partial t + u_y^0\,\partial/\partial y$ follows the Keplerian shear flow u_y^0 which occurs explicitly only in D; elsewhere only its constant x-gradient appears. The term $\mathbf{f}(\mathbf{u}) = \Omega_0(2u_y, -u_x/2, 0)$, describing epicyclic deviations from circular motion, arises from the Coriolis force and the part $= -3/2\Omega_0 u_x$ of the inertial force. In the z-direction, again gravity $\mathbf{g} = -\Omega_0^2\mathbf{z}$. The gas has the equation of state $p = 2\rho e/3$, and the tensor $S_{ij} = (u_{i,j} + u_{j,i})/2 - \delta_{ij}u_{k,k}/3$. The cooling term per gram Q must be included to prevent secular heating. In contrast to the optically thick model of Sections 10.2–10.4, the law $Q = -\sigma_{\text{cool}}(e - e_0)$ is adopted with the suffix 0 refering to the initially uniform value.

An initial isothermal state is assumed with again $\ln\rho = \ln\rho_0 - z^2/H_0^2$, where the sound speed $c_s = (2e_0/3)^{1/2} = H_0\Omega_0/2^{1/2}$. Stress-free, insulating boundary conditions are applied at top and bottom – i.e. all of $\partial u_x/\partial z$, $\partial u_y/\partial z$, u_z, $\partial e/\partial z$ vanish. Periodic boundary conditions are assumed for the (toroidal) y-direction. In the (radial) x-direction, 'quasi-periodic' boundary conditions that allow for the Keplerian shear in the toroidal direction are imposed on the dependent variables F (Wisdom and Tremaine 1988):

$$F(L_x, y, z) = F(0, y + 3\Omega_0 L_x t/2, z), \tag{10.175}$$

where L_x is the radial extent of the box. In any such local simulation, one has to watch that the problem studied is a genuine dynamo and not an amplifier (cf. Section 6.1): i.e. the boundary conditions must not be such as to prevent the field from decaying to zero, even in the absence of fluid motions. In the present problem, the magnetic field is assumed vertical at top and bottom, but the periodic and quasi-periodic boundary conditions ensure that the total flux vanishes, so allowing the field to decay to zero if the motions become too weak.

The numerical results show some interesting contrasts with those for a solar-type dynamo. The mean magnetic energy density is 6 or 7 times the kinetic – a marked super-equipartition. The ratio changes relatively slightly even when the total activity level changes by a factor 10, illustrating the close coupling between the turbulence and the magnetic field. The thermal/magnetic energy ratio is always large, varying between 10 and 100.

Energy goes from the Keplerian motion into both magnetic and turbulent energies, roughly in the ratio 6:1. The characteristic feature of this dynamo is that the Lorentz force does work on the gas, unlike a convectively-driven dynamo where work is done against the Lorentz force. Two-thirds of the energy going

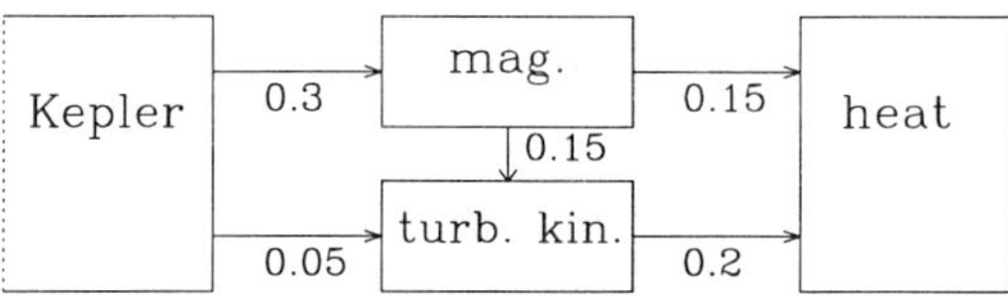

FIG. 10.9. Sketch of the energy budget. Energy tapped from the Keplerian motion goes into magnetic and kinetic energy, and is finally converted into heat. The numbers give the approximate energy fluxes in units of $\overline{\mathbf{B}^2\Omega/2}$.

from the Keplerian motion into the turbulence first passes through a phase of magnetic energy. Even though most of the rotational energy is converted first into magnetic energy, most of the heating of the disc is viscous rather than Ohmic. This is a consequence of the turbulence being driven by the Lorentz force. Adapting slightly the authors' words: the role of the kinetic and magnetic energy reservoirs is that of a catalyst that allows us to tap the energy in the Keplerian flow, but to perform this conversion, the turbulence and the dynamo have to act together. The energy budget is sketched in Fig. 10.9.

Consistent with this model being a genuine dynamo is the fact that the field is built up not just on small scales but also on the largest scales (defined by the imposed periodicity). It is important to bring out both the similarities and the contrasts of the present model with the now traditional α-effect dynamos discussed in Chapters 6 and 8 and in the last section. The most striking contrast is that the magnetic energy is found to dominate over the kinetic on all scales. The mean field generated again performs a cyclic migration, reminiscent of the dynamo waves found in many $\alpha\Omega$ models. Analogously to the regeneration of the poloidal field from the toroidal by the α-effect, in the present model $\overline{B}_x$ must be regenerated from $\overline{B}_y$ by an electromotive force $\overline{\mathcal{E}}_y = \overline{(\mathbf{u} \times \mathbf{B}/c)}_y \equiv \alpha_{\mathrm{dyn}}\overline{B}_y$. Recall from Section 6.4 that in α-effect dynamos – in which the turbulence is *not* magnetically-driven – α_{dyn} is predicted to be effectively a negative multiple of the helicity $(\mathbf{u} \cdot \nabla \times \mathbf{u})$. The magnitude of α_{dyn} is found to be in agreement with the α-effect models, but not the sign. Thus analytic description of the process yielding the mean emf along the mean magnetic field lies hidden in the computations.

In a follow-up paper (Brandenburg *et al.* 1996), the set (10.171) et seq., modified to allow for curvature effects, leads to a finite net accretion flow, enabling computation of the parameter α_{SS} appearing in the standard form (10.162) for the phenomenological turbulent viscosity. The values found lie between 0.004 and 0.008. In the simulations done so far, α_{SS} increases with higher numerical resolution: further work is awaited to see whether this trend continues or whether the values converge satisfactorily.

10.8 Centrifugal winds from discs

Angular momentum transport from stellar polar caps by the Alfvén wave-driven winds of Section 10.4 are essentially another example of the laminar transport discussed in Chapter 7. As pointed out by Blandford and Payne (1982) and Blandford (1989), a laminar magnetic field, threading a disc and of the appropriate structure, can yield a centrifugally-driven wind which will also remove angular momentum and so drive the same secular evolution as is most often attributed to some form of viscosity, phenomenological or otherwise.

Consider again a Keplerian disc, with magnetic field lines leaving the disc at an acute angle to the outward-drawn radius. With the field maintaining isorotation, the gravitational-plus-centrifugal field is given by $\mathbf{F} = -\nabla\Phi$, with the equipotential passing through the point $(\varpi_0, 0)$ having the form

$$\Phi = -\frac{GM}{\varpi_0}\left(\frac{\varpi_0}{(\varpi^2+z^2)^{1/2}} + \frac{\varpi^2}{2\varpi_0^2}\right). \tag{10.176}$$

Displacement of a particle from the equilibrium position $(\varpi_0,\ 0)$ along a field line with the local shape $z = \lambda(\varpi - \varpi_0)$ yields $(\partial^2\Phi/\partial\varpi^2)_{\varpi_0} < 0$ if $\lambda < 3^{1/2}$. This corresponds to the field line's leaving the disc surface at an angle greater than $\pi/6$ to the z-direction (Blandford and Payne 1982). A disc particle at this point is then in an unstable position: when disturbed, it will be shot off centrifugally. When the angle is less than $\pi/6$ there is a potential barrier that must be overcome by giving the gas a finite temperature, just as in the standard stellar wind theory, whereas when the angle exceeds $\pi/6$ a pure centrifugal wind is possible, with the mass flux along the field line fixed by the supply available in the disc and the strength of the disturbances feeding gas into the favourable field line.

As will be seen below, it is uncertain whether a centrifugal wind emanating from a disc is in fact an important feature; however, this extension of the theory of Chapter 7 is at least of pedagogic value. We study the simplest case of a cold steady wind along field lines that are nearly parallel to the disc, but which retain the $B_\varpi \propto 1/\varpi^2$ structure of a split monopole field (Mestel and Spiegel 1968 (unpublished); Sakurai 1987; Spruit 1994, 1996). The formalism of Chapter 7 is applicable, with the differences that now the sound speed is zero, and the angular velocity $\alpha(P)$ of a field line is not the uniform value assumed for the star but that of the 'footpoint' where the line intercepts the disc. For a Keplerian disc, we replace α by

$$\Omega_0 = (GM/\varpi_0^3)^{1/2}, \tag{10.177}$$

where ϖ_0 is the radial coordinate of the footpoint. The poloidal field has just the ϖ-component

$$B_\varpi = B_0(\varpi_0/\varpi)^2. \tag{10.178}$$

Bernoulli's equation (7.19) becomes

$$\frac{1}{2}v^2 + \frac{1}{2}\Omega^2\varpi^2 - \Omega_0\Omega\varpi^2 - \frac{GM}{\varpi} = E(\varpi_0), \tag{10.179}$$

where for brevity the suffix p on v is to be understood, and ϖ_0 is a convenient label for a field line. Following Sakurai (1985), for each line we define the parameters

$$\omega = \frac{\Omega_0^2 \varpi_{\rm A}^3}{GM}, \qquad \tilde{\beta} = \frac{v_{\rm A}^2 \varpi_{\rm A}}{GM}, \tag{10.180}$$

where the suffix A again refers to the Alfvénic point as defined by the poloidal field, and the tilde on $\tilde{\beta}$ is to distinguish this usage of beta from that in Chapter 7. Then

$$\varpi_{\rm A}/\varpi_0 = \omega^{1/3}, \qquad v_{\rm A}/v_0 = \tilde{\beta}^{1/2}\omega^{-1/6}, \tag{10.181}$$

where $v_0 \equiv \Omega_0 \varpi_0$ is the orbital velocity at the footpoint. As before, the singularity-free solution for Ω is given by (7.15):

$$\frac{\Omega}{\Omega_0} = \frac{(1 - \rho_{\rm A}\varpi_{\rm A}^2/\rho\varpi^2)}{(1 - \rho_{\rm A}/\rho)} = \frac{(1 - v/v_{\rm A})}{(1 - v\varpi^2/v_{\rm A}\varpi_{\rm A}^2)}, \tag{10.182}$$

where the last step makes use of the continuity equation (7.7)

$$\rho v/B = (\rho v/B)_{\rm A} = \eta(\varpi_0) \tag{10.183}$$

in the form appropriate for radial flow

$$\rho v \varpi^2 = (\rho v \varpi^2)_{\rm A} = \eta B_0 \varpi_0^2. \tag{10.184}$$

As in Section 7.5, (10.179) is converted into an equation of the form $H(\rho, \varpi) = E$ by substitution from (10.183) and (10.182), and the critical points determined from $\partial H/\partial \rho = 0$, $\partial H/\partial \varpi = 0$. Because of the assumed vanishing of the sound speed, the fast magnetosonic point is at infinity, while the slow point is on the disc with $\varpi = \varpi_0$, $\Omega = \Omega_0$ and $v = 0$. (From (10.184), a finite specific mass flux η then implies infinite $\rho(\varpi_0)$. In reality, the small but finite sound speed will keep the density finite, just as the cold, zero-thickness disc with finite area density is a good approximation to a thin disc with a small but finite sound speed.) Equation (10.179) applied at ϖ_0 then yields $E = -3\Omega_0^2\varpi_0^2/2$. At infinity, from (10.182)

$$(\Omega\varpi^2)_\infty = \Omega_0 \varpi_{\rm A}^2 (1 - v_{\rm A}/v_\infty), \tag{10.185}$$

and so from (10.179) and (10.181),

$$\left(\frac{v_\infty}{v_0}\right)^2 = \left[2\omega^{2/3} - 3 - \frac{2(\tilde{\beta}\omega)^{1/2}}{v_\infty/v_0}\right]. \tag{10.186}$$

The toroidal field component B_ϕ is again given by (7.4) and (7.7); in the present notation,

$$B_\phi = -(\Omega_0 - \Omega)\varpi B_\varpi / v. \tag{10.187}$$

In the cold gas limit, the fast magnetosonic speed, reached at infinity, is identical with the Alfvén speed defined by the total field. From (10.187) and (10.185),

$$-(\varpi B_\phi)_\infty/4\pi = \eta(v_{\rm A}/v_\infty)\Omega_0\varpi_{\rm A}^2, \tag{10.188}$$

whence the speed at the fast point is given by

$$v_f^2 = v_\infty^2 = \left([B_\varpi^2 + B_\phi^2]/4\pi\rho\right)_\infty = \Omega_0^2\varpi_{\rm A}^2(v_{\rm A}/v_\infty), \tag{10.189}$$

or

$$(v_\infty/v_0) = (\tilde{\beta}\omega)^{1/6}. \tag{10.190}$$

Substituting (10.190) into (10.186) completes the solution of the problem by yielding a relation between $\tilde{\beta}$ and ω:

$$\tilde{\beta} = \omega\left(\frac{2}{3} - \omega^{-2/3}\right)^3. \tag{10.191}$$

At $\varpi_{\rm A}$, from (10.179) and (10.181)

$$\Omega_A/\Omega_0 = 1 - \left[(1 - (\tilde{\beta} - 2)/\omega - 3/\omega^{2/3}\right]^{1/2}, \tag{10.192}$$

and from (10.187), the pitch angle

$$-(B_\phi/B_\varpi)_{\rm A} = \tilde{\beta}^{-1/2}\left[2 - 3\omega^{1/3} - \tilde{\beta} + \omega\right]^{1/2}. \tag{10.193}$$

A simple way of checking one's algebra is to use, for example, (10.188) and (10.185), or (10.192) and (10.193), to confirm that the total flow of angular momentum is indeed given again by 'effective corotation' as implied from (7.9) and (7.14).

The mass flux η per unit poloidal flux tube is

$$\eta = \left(\frac{\rho v}{B}\right)_{\rm A} = \frac{B_{\rm A}}{(4\pi v_{\rm A}/v_0)v_0} = \frac{B_0(\varpi_0/\varpi_{\rm A})^2}{4\pi v_0(v_{\rm A}/v_0)} = \frac{B_0}{4\pi v_0}\frac{1}{(\tilde{\beta}\omega)^{1/2}} = \frac{\eta^*}{(\tilde{\beta}\omega)^{1/2}}, \tag{10.194}$$

where

$$\eta^* \equiv \frac{B_0}{4\pi v_0} = \frac{B_0}{\Omega_0\varpi_0} \tag{10.195}$$

is the natural unit of mass flux in this model (Spruit 1996). In terms of the dimensionless quantity $\mu \equiv \eta/\eta^*$, the parameters given in (10.180) become

$$\omega = [(3/2)(1 + \mu^{-2/3})]^{3/2}, \qquad \tilde{\beta} = (\mu^2\omega)^{-1}. \tag{10.196}$$

As μ increases, the Alfvénic point $\varpi_{\rm A} = \varpi_0\omega^{1/3}$ decreases monotonically to the limiting value $\varpi_0(3/2)^{1/2}$. The angular momentum flux per unit flux tube is

$$\eta\Omega_0\varpi_{\rm A}^2 = \eta^*\Omega_0\varpi_0^2(3\mu/2)(1 + \mu^{-2/3}), \tag{10.197}$$

and the terminal flow speed is $v_\infty = v_0\mu^{-1/3}$. This last formula brings out strikingly the contrast with the thermo-centrifugal expansion of a hot stellar corona

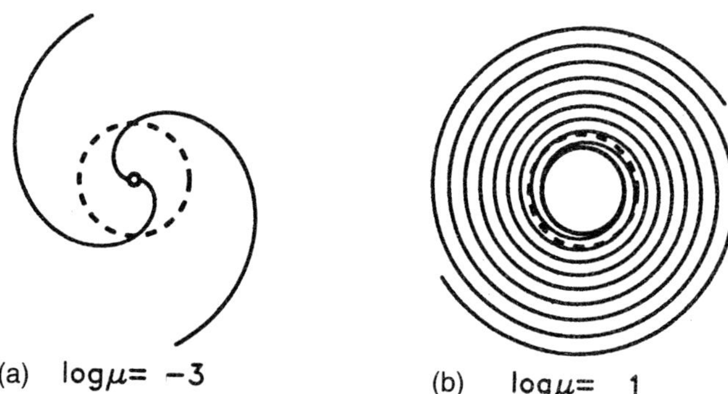

FIG. 10.10. Shape of the field lines in a cold, centrifugally-driven wind: (a) low mass-loss, modestly wound; (b) high mass-loss, tightly wound. Dashed line is the Alfvénic radius.

– the standard stellar wind as discussed in Chapter 7, for which v always exceeds the high sound speed, and the speed at infinity always exceeds the velocity of escape; whereas in the present problem, when $\mu \gg 1$, the mathematics yields dense, slowly moving winds.

There is a similar contrast in the pitch angle, given by (10.193), which for $\mu \ll 1$ has the modest limiting value $(19/8)^{1/2}$, with B_ϕ and B_p comparable, but for $\mu \gg 1$ tends to 0.55μ, corresponding to a tightly wound spiral with B_ϕ dominant (Fig. 10.10). Likewise, when $\mu \ll 1$, approximate corotation again holds within the Alfvén radius, whereas for $\mu \gg 1$ corotation fails hopelessly almost from the start. In one sense, these winds can reasonably be given the title 'centrifugal' for all μ-values, because with zero sound speed, the only driver present in Bernoulli's equation (10.179) is the term $\Omega_0\Omega\varpi^2$, arising from the work done on the gas by the same torque that forces Ω up above the value $(GM/\varpi^3)^{1/2}$. However, if one writes down the radial component of the equation of motion (which in (10.179) is combined with the ϕ-component), it is found that for $\mu \gg 1$ it is the gradient of $B_\phi^2/8\pi$ which continually accelerates the gas radially, through and beyond ϖ_A, ensuring that it reaches infinity, even though at ϖ_A its speed is below the escape speed. Figure 10.11 shows how some of the predictions of the cold wind model vary with mass loss.

This picture of a wind driven centrifugally from a centrifugally supported disc is very appealing, but there are difficulties in the way of a fully self-consistent model. In a quasi-steady state, the inward dragging of field lines by the accretion flow must be balanced by resistive diffusion (Königl 1989), so that as in (10.21)

$$-\frac{v_\varpi B_z}{\eta} = (\nabla\times\mathbf{B})_\phi \approx \frac{B_\varpi^+}{H}, \tag{10.198}$$

with η again the magnetic diffusivity. The complete solution for the field is given

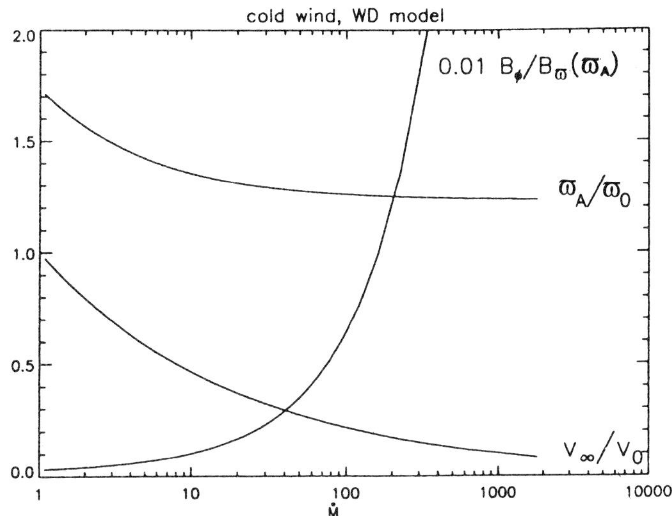

FIG. 10.11. Cold, centrifugally-driven wind. Dependence on mass-loss rate $-\dot{M}$ of the Alfvén radius ϖ_A, the asymptotic wind speed v_∞ and the pitch angle B_ϕ/B_ϖ. (After Spruit 1994.)

in Lubow *et al.* (1994*a*); for the present, the important point is to note that the field line emanating from the disc is inclined to the disc plane at the angle

$$B_z/B_\varpi^+ \approx \eta/(-v_\varpi)H. \tag{10.199}$$

In the standard disc model with purely viscous angular momentum transport, $v_\varpi \simeq -\nu/\varpi$ by (10.20), and so

$$\frac{B_z}{B_\varpi^+} \simeq \left(\frac{\eta}{\nu}\right)\left(\frac{\varpi}{H}\right). \tag{10.200}$$

Thus if the Prandtl number $\eta/\nu \simeq 1$, $B_z/B_\varpi^+ \gg 1$ – the angle of inclination is closer to $\pi/2$ than $\pi/3$.

One possible way out is to have the wind itself responsible for the angular momentum transport and so also for the velocity v_ϖ in at least the outer parts of the disc. This has been studied by Lubow *et al.* (1994*b*) in the limit of zero viscosity. The self-consistency requirement is that the mass loss carry off angular momentum according to the familiar effective corotation result, so giving rise to an accretion velocity in the disc v_ϖ; and the magnetic field structure is again fixed by the balance of inward field dragging against macro-diffusion. It is found that there do exist solutions with $-v_\varpi \gg \eta/\varpi$, yielding inclination angles below $\pi/3$. But Lubow *et al.* point out that such solutions may be globally unstable: a slight increase in $|v_\varpi|$ decreases the inclination and so acts to increase the mass loss rate, which in turn increases $|v_\varpi|$; and likewise a slight decrease in $|v_\varpi|$ is not spontaneously reversed.

There is therefore this question mark over the likely appearance and persistence of steady centrifugo-magnetic flows from cold discs. We shall nevertheless provisionally keep them as an open option, in view of our very imperfect understanding of much of the physics, especially of magneto-turbulence in a disc.

10.9 Wind collimation

We now return to the collimation problem, touched on in Section 7.4. This is part of the general problem of the equilibrium and stability of the magnetosphere, especially the wind domain. The wind associated with a CTT-star may be in part emitted from the turbulent subphotospheric regions of the star itself, perhaps driven initially by Alfvén wave-pressure (cf. Section 10.4), and in part may emanate centrifugally-driven from a cool surrounding disc, flowing along general field lines that are not inclined by too large an angle to the disc (Section 10.8). Only the low mass-loss centrifugal winds are considered here, so that within the Alfvénic surface the field remains predominantly poloidal. However, as in the high-speed, hot stellar winds discussed in Chapter 7, well beyond the Alfvénic surface the steady-state equations yield the result (7.41), showing that once the gas is effectively coasting, if the flow and field were to remain essentially radial the B_ϕ-component would dominate over B_p; the consequent Lorentz force would pinch near the axis and so act to modify both the poloidal field and the wind flow.

In any self-consistent model, the magnetospheric poloidal field must adjust its structure so that the balance of poloidal forces is maintained. The toroidal field $B_\phi \mathbf{t}$ is maintained by the current density $\mathbf{j}_\mathrm{p} = (c/4\pi)\nabla \times (B_\phi \mathbf{t}) = (c/4\pi)\nabla(\varpi B_\phi) \times (\mathbf{t}/\varpi)$, so the (poloidal) Lorentz force due to $\mathbf{B}_\mathrm{t}$ is

$$\mathbf{F}^\phi = \frac{\mathbf{j}_\mathrm{p} \times B_\phi \mathbf{t}}{c} = -\frac{1}{8\pi\varpi^2}\nabla(\varpi B_\phi)^2. \tag{10.201}$$

On the axis $B_\phi = 0$, and so in the neighbourhood of the axis, F^ϕ_ϖ is negative – a pinching force. Thus it is tempting to associate a predicted steady growth in $|B_\phi|$ with the observed formation of collimated jets, especially in winds emitted by young stellar objects.

A fully convincing picture must fit such a local effect into a global picture. Well within the Alfvénic surface the toroidal field is nearly torque-free, satisfying $\varpi B_\phi = \beta(P) =$ constant on the field line P, so that the toroidal field continues to exert a pinching force – with the component of $\mathbf{F}^\phi$ normal to $\mathbf{B}_\mathrm{p}$ acting towards the axis – as long as $|\beta|$ continues to increase as $|P|$ increases from its zero value on the axis. In the hot wind problem of Chapter 7, there is no surrounding disc, and the wind zone is partly separated from the equator by the dead zone (cf. Fig. 7.1) within which $B_\phi = 0$. On the separatrix between the wind and dead zones, by Ampère's law there flows a surface current with a poloidal component $J_{\rm p} = (c/4\pi)(B_\phi)_{\rm w}$ and a toroidal component that maintains the discontinuity in $\mathbf{B}_{\rm p}$, required so that the pressure balance condition

$$p_{\rm d} + \frac{\mathbf{B}_{\rm d}^2}{8\pi} = p_{\rm w} + \frac{\mathbf{B}_{\rm p}^2}{8\pi} + \frac{\mathbf{B}_{\rm t}^2}{8\pi} \tag{10.202}$$

holds. Near and beyond the Alfvénic surface, the dead zone has disappeared: the hot stellar wind problem and the CTT wind problem are now similar in having the wind zone comprising the whole outer magnetosphere. Construction of both the B_ϕ-component and the distorted $\mathbf{B}_{\rm p}$-field (cf. Section 7.4) is subject again to a boundary condition at the equator analogous to (10.202): as B_ϕ and B_ϖ are of opposite sign in the northern and southern hemispheres, the thermal pressure at the equatorial plane of the disc will adjust to balance the pinching effect of the total $\mathbf{B}$ just outside the disc.

Lynden-Bell (1996) illustrates the role of the external pressures in the dynamics of the collimation problem through some detailed magnetostatic modelling of force-free fields. He emphasizes that 'left to itself a toroidal magnetic field has no net tendency to make its configuration expand or contract in radius overall. Thus the presence of an external pressure is essential to an equilibrium pinch.' The 'true role' of the toroidal field is as a 'concentrator of pressure'. These statements are illustrated by the various integral relations discussed in Section 3.4. However, one can distinguish between 'active' and 'passive' external pressure. In the picture as just outlined, the pressures are 'passive'. The B_ϕ-field is generated by twisting on the plane $z = 0$, the twist is propagated along field lines, and the poloidal field and the frozen-in gas adjust, yielding both pinching at the axis and build-up of a compression in the surroundings. Sakurai's model (1985) (which has no dead zone) is an example of this. Beyond $\rm S_A$, the predominantly toroidal field causes asymptotic collimation; but as noted by Spruit (1996), because of the absence of an *externally generated* pressure, the flow does not reach full collimation until well beyond the predicted distance at which a stable wind would reach its interstellar termination shock.

At the other extreme, in a cylindrically-symmetric structure with just a toroidal field, by (10.201) the ϖ-component of equilibrium requires that $\varpi B_\phi = \beta =$ constant. An external pressure p_e imposed on the external radius ϖ_2 is balanced by the toroidal field pressure of $B_\phi(\varpi_2) = (8\pi p_e)^{1/2}$; and at a radius ϖ_1 the equilibrium condition requires the magnetic pressure $B_\phi^2(\varpi_1)/8\pi = (\varpi_2/\varpi_1)^2 p_e$ which can be much greater than p_e. This example is of course highly artificial, not least because the angular momentum flow function β is not fixed by twisting at some z-level but by assuming no contribution of a poloidal field to poloidal equilibrium. In reality, even in a structure that is independent of both ϕ and z, there will be a poloidal component $B_z(\varpi)$ which will exert a ϖ-component of magnetic pressure gradient, responding to the force (10.201), while the surface requirement that magnetic pressure balance the imposed p_e will involve both components of $\mathbf{B}$. Nevertheless, Lynden-Bell is able to construct an illustrative

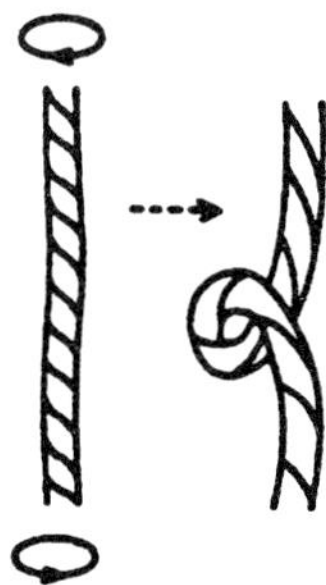

FIG. 10.12. The kink instability.

example of a 'magnetic tower', with the axial field (necessarily poloidal) exerting a pressure $= 41.1p_e$.

The prime difficulty with appealing for collimation to the 'hoop stresses' associated with the toroidal field is the tendency towards the dynamical kink instability, as discussed in Section 3.4, and emphasized especially by Eichler (1993). The conditions assumed in idealized instability studies are seldom exactly realized, either in the laboratory or in the cosmos, and probably more work – analytical and numerical, and extending into the non-linear domain – is required for definitive answers. Spruit *et al.* (1997) argue that it is certainly plausible that the kink instability sets in when field lines make more than one turn about the axis over the jet domain (an extrapolation of the Kruskal–Shafranov condition – cf. Section 3.6.3). They also quote experimental support for Taylor's conjecture of near conservation of magnetic helicity $K = \int \mathbf{B} \cdot \mathbf{A}\, d\tau$ even if there is some diffusion (Section 3.5). They show that this helps one to understand how the essentially non-axisymmetric instability works, since K is a measure of the amount of twist in the field. When a cylinder develops a kink, one field line twist about the axis is exchanged for one turn due to the looping path of the axis (Fig. 10.12). By reducing the strength of the mean toroidal field, measured by turns about the axis, this reduces the hoop stresses. Simultaneously, the longitudinal field is increased, so increasing the magnetic pressure which is destabilizing. Their time-scale estimates suggest strongly that as long as the instability growth rate is close to the Alfvén rate $B_\phi/[(4\pi\rho)^{1/2}\varpi]$, then the high degrees of hoop stress collimation predicted by the unperturbed, axisymmetric theory are impossible.

Because of this, Spruit *et al.* (1997) develop further the suggestion by Blandford (1994) and Spruit (1994, 1996) that the collimation observed in pre-main sequence stars and other bodies (such as active galactic nuclei) is again magnetic, but due essentially to a *poloidal* field anchored into an accretion disc that extends many decades in radius compared with the stellar and inner disc radii. The field can again be either dynamo-generated in the disc, for example according to one of the models discussed in Sections 10.6 or 10.7, or a compressed part of the galactic magnetic field present in the parent gas cloud (cf. Chapter 11). In both cases the field is likely to increase in strength towards the disc centre: a dynamo-generated field will tend to saturate when its energy density is at some fraction of the gas pressure (e.g. Brandenburg *et al.* 1995; Hawley *et al.* 1995), while a compressed fossil field will increase inwards because of the non-homologous nature of the accretion flow. Even so, the total flux in the central regions can still be small compared with the total flux emerging from the disc. Spruit gives an example with the field at the disc having the normal component

$$B_z \propto \left(\frac{\varpi^2}{\varpi_i^2} + 1\right)^{-\nu/2} \tag{10.203}$$

where ϖ_i is the inner edge of the disc. The parameter ν lies between 0 and 2, so that B_z increases towards the centre, while the flux $\Phi(\varpi)$ crossing the disc within $\varpi \propto \varpi^{2-\nu}$ and so increases with ϖ. The self-similar disc model of Blandford and Payne (1982) is of this class with $\nu = 5/4$.

Just as in the theory of winds emitted from stars, with reasonable choice of the parameters, the Alfvénic surface S_A is far above the disc. Well within S_A, if the mass loss rates are moderate, the energy density of $\mathbf{B}_p$ dominates over the thermal, gravitational and kinetic energy densities and so also of the toroidal field B_ϕ generated by the wind. The local field structure is therefore well approximated by the curl-free condition, determined by the boundary condition (10.202). Figure 10.13 illustrates the case $\nu = 1$ (Cao and Spruit 1994). Along the disc surface, because of the high field strength near the centre, the field lines fan out away from the axis, but the concentration of most of the flux in the outer parts of the disc ensures that above the disc the field lines emanating from the central regions bend towards the axis.

The field lines have the approximate equation $\varpi \propto z^{\nu/2}$, so that $d\varpi/dz \propto z^{\nu/2-1}$ which goes asymptotically to zero for $\nu < 2$. However, the disc (10.203) is of infinite size. In more realistic models, with the disc of finite outer radius ϖ_e, a curl-free field will look like Fig. 10.13 in the inner regions, but on a large scale it will look dipolar, with all the field lines emanating from the disc returning to the disc plane between ϖ_e and infinity. Thus if the flow were to follow the field lines everywhere, that starting from the inner regions would be collimated out to large distances, but all the flow would ultimately be decollimated. However, beyond the Alfvénic surface S_A, the poloidal kinetic energy exceeds the poloidal magnetic energy; and with the tendency of the non-corotating wind to build up a toroidal field severely limited by the kink instability, the argument for the force-free or curl-free structure ceases to hold; and equally the flow becomes ballistic, with the field being distorted so as to follow the flow rather than the flow being channelled by the field. The ballistic flow will preserve an initially small angle between field line and axis; hence the predicted collimation will be better, the higher S_A is above the disc, and the larger the ratio ϖ_e/ϖ_i of the outer to the inner disc radii.

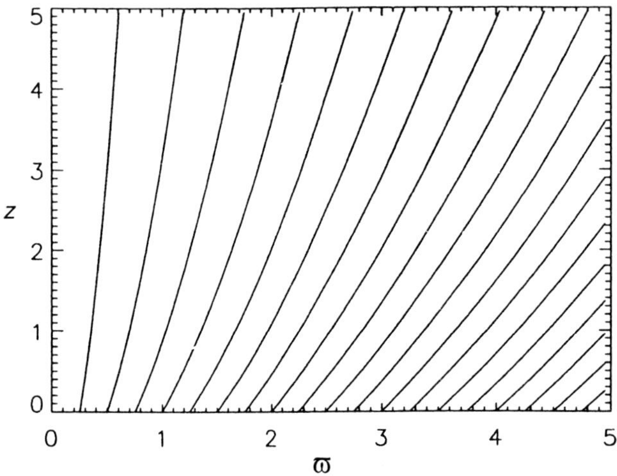

FIG. 10.13. Field lines of a curl-free field produced by a disc field with the normal component varying as $(\varpi^2/\varpi_i^2+1)^{-1/2}$. The collimating shape of the field lines is due to the magnetic flux in the outer parts of the disc. (After Spruit *et al.* 1997.)

10.10 Conclusion

As was recognized in the eighteenth century by Laplace and Kant, cosmogony subject to angular momentum conservation leads inevitably to a disc-like scenario. As will emerge in Chapter 11, the problem is to see how systems with low angular momentum can form. In his 1928 treatise *Astronomy and Cosmogony*, Sir James Jeans wrote: 'If one put all the angular momentum of the Solar System into the Sun (modelled as a liquid), the Sun would rotate 28 times as fast as it does now, but would still be far from centrifugal break-up'. And again: 'If the Sun once assumed the lenticular shape necessary for the shedding of matter by rotation, it is difficult to see how it could abandon it and become as spherical as it is now.' It was natural for a master of his craft like Jeans to think along those lines in 1928; equally, these quotations illustrate strikingly the radical changes in our thinking brought about by the introduction of elecromagnetic coupling into the picture.

How close we are – both in this chapter on the later phases of star formation and the next two chapters on the earlier phases – to a permanent picture remains to be seen. A chastening quotation – even more appropriate because of its generality – is from Chandrasekhar's 1949 lecture 'Turbulence: a physical theory of astrophysical interest'. Apropos of the von Weizsäcker cosmogony, he writes: 'It is the usual fate of cosmogonical theories not to survive'. No doubt many of our efforts to build theories essentially in terms of magnetohydrodynamics will look naive after some decades or less. Nevertheless, I would hazard a guess that just as some form of turbulence has remained an ingredient in cosmogony, so cosmical magnetic fields will continue to play a central role in future theories, both through the direct reaction of the large-scale Lorentz force on the mean flow, and through the spontaneous generation of turbulence, as discussed above. In particular, the magnetohydrodynamics of discs and the associated winds and jets will continue to be a thriving industry. Very recent work includes a review by Livio (1997), giving a unified treatment that attempts to cover all classes of jet-emitting bodies – not only young stellar objects but active galactic nuclei, high-mass and low-mass X-ray binaries, black hole X-ray transients, symbiotic stars, etc. Ogilvie and Livio (1998) return to the problem of launching an outflow from an accretion disc, with an eye especially on the cataclysmic variables.

References

Adams, F.C., Lada, C.J. and Shu, F.H. (1987). *Astrophysical Journal*, **312**, 788.

Adams, F.C., Lada, C.J. and Shu, F.H. (1988). *Astrophysical Journal*, **326**, 865.

Aly, J.J. (1980). *Astronomy and Astrophysics*, **86**, 192.

Aly, J.J. (1991). *Astrophysical Journal*, **375**, L61.

Aly, J.J. and Kuijpers, J. (1990). *Astronomy and Astrophysics*, **227**, 473.

André, P. (1987). In *Protostars and molecular clouds* (eds T. Montmerle and C. Bertout), p. 143. University of Arizona Press, Tucson.

André, P. and Montmerle, T. (1994). *Astrophysical Journal*, **420**, 837.

Anzer, U. and Börner, G. (1980). *Astronomy and Astrophysics*, **83**, 133.

Appenzeller, I., Jankovics, I. and Ostreicher, R. (1984). *Astronomy and Astrophysics*, **101**, 108.

Armitage, P.J. (1995). *Monthly Notices Royal Astronomical Society*, **274**, 1242.

Armitage, P.J. and Clarke, C.J. (1996). *Monthly Notices Royal Astronomical Society*, **280**, 458.

Balbus, S.A. and Hawley, J.F. (1991). *Astrophysical Journal*, **376**, 214.

Basri, G., Marcy, G.W. and Valenti, J.A. (1992). *Astrophysical Journal*, **390**, 622.

Bath, G.T., Evans, W.D. and Pringle, J.E. (1974). *Monthly Notices Royal Astronomical Society*, **166**, 113.

Beckwith, S.V.W., Sargent, A.I., Chini, R.S. and Güsten, R. (1990). *Astronomical Journal*, **99**, 924.

Belcher, J.W. (1971). *Astrophysical Journal*, **168**, 509.

Bertout, C. (1989). *Annual Review Astronomy Astrophysics*, **27**, 351.

Blaes, O.M. and Balbus, S.A. (1994). *Astrophysical Journal*, **421**, 163.

Blandford, R.D. (1989). In *Theory of accretion discs* (eds F. Meyer, W.J. Duschl, J. Frank and E. Meyer-Hofmeister), p. 35. Kluwer, Dordrecht.

Blandford, R.D. (1994). In *Cosmical magnetism* (ed. D. Lynden-Bell), p. 171. Kluwer, Dordrecht.

Blandford, R.D. and Payne, D.G. (1982). *Monthly Notices Royal Astronomical Society*, **199**, 883.

Bouvier, J. (1990). *Astronomical Journal*, **99**, 946.

Bouvier, J. (1991). In *Angular momentum evolution of young stars* (eds S. Catalano and J.R. Stauffer), p. 41. Kluwer, Dordrecht.

Bouvier, J. (1994). In *Cosmical magnetism, contributed papers* (ed. D. Lynden-Bell), p. 40. Institute of Astronomy, Cambridge.

Bouvier, J., Bertout, C., Benz, W. and Mayor, M. (1986). *Astronomy and Astrophysics*, **165**, 110.

Brandenburg, A., Krause, F., Meinel, R., Moss, D. and Tuominen, I. (1989). *Astronomy and Astrophysics*, **213**, 411.

Brandenburg, A. and nine others (1993). *Astronomical Journal*, **106**, 1, 372.

Brandenburg, A., Nordlund, Aa., Stein, R.F. and Torkelsson, U. (1995). *Astrophysical Journal*, **446**, 741.

Brandenburg, A., Nordlund, Aa., Stein, R.F. and Torkelsson, U. (1996). *Astrophysical Journal*, **458**, L45.

Camenzind, M. (1991). *Reviews Modern Astronomy*, **3**, 234.

Campbell, C.G. (1992*a*). *Geophysical Astrophysical Fluid Dynamics*, **63**, 179.

Campbell, C.G. (1992*b*). *Geophysical Astrophysical Fluid Dynamics*, **66**, 243.

Campbell, C.G. (1997). *Magnetohydrodynamics in binary stars.* Kluwer, Dordrecht.

Campbell, C.G. and Heptinstall, P.M. (1998). *Monthly Notices Royal Astronomical Society*, **299**, 31.

Cao, X.-W. and Spruit, H.C. (1994). *Astronomy and Astrophysics*, **287**, 80.

Chandrasekhar, S. (1939). *An introduction to the study of stellar structure.* Chicago University Press. (Dover edition 1957).

Chandrasekhar, S. (1949). *Astrophysical Journal*, **110**, 329.

Chandrasekhar, S. (1960). *Proceedings National Academy Sciences*, **46**, 253.

Chandrasekhar, S. (1961). *Hydrodynamic and hydromagnetic stability.* Clarendon Press, Oxford.

Chanmugam, G. (1979). *Monthly Notices Royal Astronomical Society*, **187**, 769.

Chiba, M. and Tosa, M. (1990). *Monthly Notices Royal Astronomical Society*, **244**, 714.

Clarke, C.J., Armitage, P.J., Smith, K.W. and Pringle, J.E. (1995). *Monthly Notices Royal Astronomical Society*, **273**, 639.

Coleman, C.S., Kley, W. and Kumar, S. (1995). *Monthly Notices Royal Astronomical Society*, **274**, 171.

Collier Cameron, A. (1995). *New Scientist*, 29 July, p. 36.

Collier Cameron, A. and Campbell, C.G. (1993). *Astronomy and Astrophysics*, **274**, 309.

Collier Cameron, A., Campbell, C.G. and Quaintrell, H. (1995). *Astronomy and Astrophysics*, **298**, 133.

Curry, C., Pudritz, R.E. and Sutherland, P. (1994). *Astrophysical Journal*, **434**, 206.

Davidson, K. and Ostriker, J.P. (1973). *Astrophysical Journal*, **179**, 585.

Drazin, P.G. and Reid, W.H. (1981). *Hydrodynamic stability.* Cambridge University Press.

Dubrulle, B. and Knobloch, E. (1993). *Astronomy and Astrophysics*, **274**, 667.

Edwards, S., Strom, S.E., Hartigan, P. *et al.* (1993). *Astronomical Journal*, **106**, 1, 372.
Eggleton, P.P. (1971). *Monthly Notices Royal Astronomical Society*, **151**, 351.
Eggleton, P.P. (1972). *Monthly Notices Royal Astronomical Society*, **156**, 361.
Eggleton, P.P., Faulkner, J. and Flannery, B.P. (1973). *Astronomy and Astrophysics*, **23**, 325.
Eichler, D. (1993). *Astrophysical Journal*, **419**, 111.
Foglizzo, T. and Tagger, M. (1994). *Astronomy and Astrophysics*, **287**, 297.
Foglizzo, T. and Tagger, M. (1995). *Astronomy and Astrophysics*, **301**, 293.
Frank, J., King, A.R. and Raine, D.J. (1992). *Accretion power in astrophysics*, (2nd edn). Cambridge University Press.
Fricke, K. (1969). *Astronomy and Astrophysics*, **1**, 388.
Ghosh, P. and Lamb, F.K. (1978). *Astrophysical Journal*, **223**, L83.
Ghosh, P. and Lamb, F.K. (1979). *Astrophysical Journal*, **234**, 296.
Gilliland, R.L. (1986). *Astrophysical Journal*, **300**, 339.
Goldreich, P. and Lynden-Bell, D. (1965). *Monthly Notices Royal Astronomical Society*, **130**, 125.
Hartmann, L., Edwards, S. and Avrett, E. (1982). *Astrophysical Journal*, **261**, 279.
Hawley, J.F. and Balbus, S.A. (1991). *Astrophysical Journal*, **376**, 223.
Hawley, J.F. and Balbus, S.A. (1992). *Astrophysical Journal*, **400**, 595.
Hawley, J.F., Gammie, C.F. and Balbus, S.A. (1995). *Astrophysical Journal*, **440**, 742.
Herbig, G.H. (1952). *Journal Royal Astronomical Society Canada*, **46**, 222.
Herbig, G.H. (1957). *Astrophysical Journal*, **125**, 612.
Herbig, G.H. (1962). *Advances Astronomy Astrophysics*, **1**, 47, 76.
Herbig, G.H. and Bell, K.R. (1988). *Lick Observatory Bulletin*, **1111**.
Horiuchi, T., Matsumoto, R., Hanawa, T. and Shibata, K. (1988). *Publications Astronomical Society Japan*, **40**, 147.
Hunter, C. (1963). *Monthly Notices Royal Astronomical Society*, **126**, 299.
Hoyle, F. and Ireland, J. (1960). *Monthly Notices Royal Astronomical Society*, **121**, 253.
Jeans, J.H. (1928). *Astronomy and cosmogony.* Cambridge University Press.
Jeffreys, H. (1924). *The Earth*, p. 55 et seq. Cambridge University Press.
Kenyon, S. and Hartmann, L. (1987). *Astrophysical Journal*, **323**, 714.
Knobloch, E. (1992). *Monthly Notices Royal Astronomical Society*, **255**, 25p.
Königl, A. (1989). *Astrophysical Journal*, **342**, 208.
Königl, A. (1991). *Astrophysical Journal*, **370**, L39.
Kumar, S., Coleman, C.S. and Kley, W. (1994). *Monthly Notices Royal Astronomical Society*, **266**, 379.
Lago, M.T.V. (1984). *Monthly Notices Royal Astronomical Society.* **210**, 323.
Lago, M.T.V. and Penston, M.V. (1982). *Monthly Notices Royal Astronomical Society*, **198**, 429.
Lamb, F.K., Pethick, C.J. and Pines, D. (1973). *Astrophysical Journal*, **184**, 271.
Levy, E.H. (1978). *Nature*, **276**, 481.

Li, J. (1996). *Astrophysical Journal*, **456**, 696.

Li, J., Wickramasinghe, D.T. and Rüdiger, G. (1996). *Astrophysical Journal*, **469**, 765.

Lin, C.C. (1955). *The theory of hydrodynamic stability.* Cambridge University Press.

Livio, M. (1997). In *Accretion phenomena and related outflows* (eds D.T. Wickramasinghe, G.V. Bicknell and L. Ferrario), p. 845. ASP Conference Series, **121**, San Francisco.

Livio, M. and Pringle, J.E. (1992). *Monthly Notices Royal Astronomical Society*, **259**, 23P.

Lovelace, R.V.E., Romanova, M.M. and Bisnovatyi-Kogan, G.S. (1995). *Monthly Notices Royal Astronomical Society*, **275**, 244.

Lubow, S.H. and Spruit, H.C. (1995). *Astrophysical Journal*, **445**, 337.

Lubow, S.H., Papaloizou, J.C.B. and Pringle, J.E. (1994*a*). *Monthly Notices Royal Astronomical Society*, **267**, 235.

Lubow, S.H., Papaloizou, J.C.B. and Pringle, J.E. (1994*b*). *Monthly Notices Royal Astronomical Society*, **268**, 1010.

Lüst, R. (1952). *Zeitschrift Naturforschung*, **7a**, 87.

Lynden-Bell, D. (1966). *Observatory*, **86**, 57.

Lynden-Bell, D. (1969). *Nature*, **223**, 690.

Lynden-Bell, D. (1996). *Monthly Notices Royal Astronomical Society*, **279**, 389.

Lynden-Bell, D. and Pringle, J.E. (1974). *Monthly Notices Royal Astronomical Society*, **168**, 603.

Matsumoto, R. and Tajima, T. (1995). *Astrophysical Journal*, **445**, 767.

Mestel, L. (1960). In *Vistas in astronomy*, Vol. 3 (ed. A. Beer), p. 296. Pergamon Press, London.

Mestel, L. and Subramanian, K. (1991). *Monthly Notices Royal Astronomical Society*, **248**, 677.

Montmerle, T., André, R., Casanova, S. and Feigelson, E.D. (1994). In *Cosmical magnetism, contributed papers* (ed. D. Lynden-Bell), p. 33. Institute of Astronomy, Cambridge.

Moss, D., Tuominen, I. and Brandenburg, A. (1990). *Astronomy and Astrophysics*, **228**, 284.

Najita, J.R. and Shu, F.H. (1994). *Astrophysical Journal*, **429**, 808.

Ogilvie, G.I. and Livio, M. (1998). *Astrophysical Journal*, **499**, 329.

Ogilvie, G.I. and Pringle, J.E. (1996). *Monthly Notices Royal Astronomical Society*, **279**, 152.

Osterbrock, D.E. (1961). *Astrophysical Journal*, **134**, 347.

Ostriker, E. and Shu, F.H. (1995). *Astrophysical Journal*, **447**, 813.

Papaloizou, J.C.B. and Lin, D.N.C. (1995). *Annual Review Astronomy Astrophysics*, **33**, 505.

Papaloizou, J. and Szuszkiewicz, E. (1992). *Geophysical Astrophysical Fluid Dynamics*, **66**, 223.

Parker, E.N. (1965). *Space Science Reviews*, **4**, 666.

Parker, E.N. (1971). *Astrophysical Journal*, **163**, 255.

Peek, B.M. (1942). *Journal British Astronomical Association*, **53**, 23.
Prendergast, K.H. and Burbidge, G.R. (1968). *Astrophysical Journal*, **151**, L83.
Priest, E.R. (1982). *Solar magnetohydrodynamics*. Reidel, Dordrecht.
Pringle, J.E. (1981). *Annual Review Astronomy Astrophysics*, **19**, 137.
Pringle, J.E. and Rees, M.J. (1972). *Astronomy and Astrophysics*, **21**, 1.
Pudritz, R.E. (1981). *Monthly Notices Royal Astronomical Society*, **195**, 881 and 897.
Rogers, F.J. and Iglesias, C.A. (1992). *Astrophysical Journal Supplement*, **79**, 507.
Rosner, R. and DeLuca, E. (1989). In *The center of the galaxy* (ed. M. Morris), p. 319. Kluwer, Dordrecht.
Rucinski, S.M. (1988). *Astronomical Journal*, **95**, 1895.
Rüdiger, G., Elstner, D. and Schultz, M. (1993). *Astronomy and Astrophysics*, **270**, 53.
Ruzmaikin, A.A., Shukurov, A.M. and Sokoloff, D.D. (1988). *Magnetic fields of galaxies*. Kluwer, Dordrecht.
Sakurai, T. (1985). *Astronomy and Astrophysics*, **152**, 121.
Sakurai, T. (1987). *Publications Astronomical Society Japan*, **39**, 821.
Shakura, N.I. and Sunyaev, R.A. (1973). *Astronomy and Astrophysics*, **24**, 337.
Shu, F.H. (1974). *Astronomy and Astrophysics*, **33**, 55.
Shu, F.H., Najita, J., Ruden, S.P. and Lizano, S. (1994*a*). *Astrophysical Journal*, **429**, 797.
Shu, F.H., Najita, J., Ostriker, E., Wilkin, F., Ruden, S.P. and Lizano, S. (1994*b*). *Astrophysical Journal*, **429**, 781.
Spruit, H.C. (1994). In *Cosmical magnetism* (ed. D. Lynden-Bell), p. 33. Kluwer, Dordrecht.
Spruit, H.C. (1996). *Magnetohydrodynamic jets and winds from accretion discs*. In *Physical processes in binary stars* (eds R.A.M.J. Wijers, M.B. Davies and C.A. Tout), p. 249. Kluwer, Dordrecht.
Spruit, H.C. and Taam, R.E. (1990). *Astronomy and Astrophysics*, **229**, 475.
Spruit, H.C., Stehle, R. and Papaloizou, J.C.B. (1995). *Monthly Notices Royal Astronomical Society*, **275**, 1223.
Spruit, H.C., Foglizzo, T. and Stehle, R. (1997). *Monthly Notices Royal Astronomical Society*, **288**, 333.
Stauffer, J.R. and Soderblom, D.R. (1991). *The evolution of angular momentum in solar mass stars*. In *The Sun in time* (eds C.P. Sonnett, M.S. Giampapa and M.S. Matthews), p. 832. University of Arizona Press, Tucson.
Stepinski, T.F. and Levy, E.H. (1988). *Astrophysical Journal*, **331**, 416.
Stepinski, T.F. and Levy, E.H. (1990). *Astrophysical Journal*, **362**, 318.
Stix, M. (1975). *Astronomy and Astrophysics*, **42**, 85.
Subramanian, K. and Mestel, L. (1993). *Monthly Notices Royal Astronomical Society*, **265**, 649.
Takahara, F. (1979). *Progress Theoretical Physics*, **62**, 629.
Tayler, R.J. (1973). *Monthly Notices Royal Astronomical Society*, **161**, 365.
Tayler, R.J. (1987). *Monthly Notices Royal Astronomical Society*, **227**, 553.

Terquem, C. and Papaloizou, J.C.B. (1996). *Monthly Notices Royal Astronomical Society*, **279**, 767.

Torkelsson, U. and Brandenburg, A. (1994). *Astronomy and Astrophysics*, **283**, 677.

Tout, C.A. and Pringle, J.E. (1992). *Monthly Notices Royal Astronomical Society*, **259**, 604.

van Ballegooijen, A.A. (1989). In *Accretion discs and magnetic fields in astrophysics* (ed. G. Belvedere), p. 99. Kluwer, Dordrecht.

van Ballegooijen, A.A. (1994). *Space Science Reviews*, **68**, 299.

Velikhov, E.P. (1959). *Soviet Physics JETP*, **36**, 995.

Vogel, S.N. and Kuhi, L.V. (1981). *Astrophysical Journal*, **245**, 960.

von Weizsäcker, C.F. (1943). *Zeitschrift Astrophysik*, **22**, 319.

von Weizsäcker, C.F. (1948). *Zeitschrift Naturforschung*, **3a**, 524.

Wang, Y.-M. (1987). *Astronomy and Astrophysics*, **183**, 257.

Wang, Y.-M. (1995). *Astrophysical Journal*, **449**, L153.

Weiss, A., Keady, J.J. and Magee, N.H. (1990). *Atomic data and nuclear data tables*, **45**, 209.

Wisdom, J. and Tremaine, S. (1988). *Astronomical Journal*, **95**, 925.

Yi, I. (1995). *Astrophysical Journal*, **442**, 768.

11

MAGNETISM AND STAR FORMATION I

11.1 Introduction

The simplest discussions of the dynamics of star formation involve just the mutual interaction of self-gravitation and thermal pressure. It is convenient to begin by studying how a cool interstellar gas cloud initially in hydrostatic equilibrium is brought to the brink of gravitational collapse. Application of the scalar virial theorem (3.64) to a spherical cloud of mass M, temperature T and mean molecular weight μ, subject to the pressure p_e of the warm interstellar medium, yields an equation for the radius R of the cloud (McCrea 1957):

$$(4\pi R^3)p_e = 3\int(4\pi\rho a^2 r^2)\,\mathrm{d}r + \mathcal{V} = 3Ma^2 - AGM^2/R \qquad (11.1)$$

where $a = (\mathcal{R}T/\mu)^{1/2}$ is the isothermal sound speed and $\mathcal{V}$ the gravitational energy, with A a number of order unity depending on the details of the density distribution, and increasing slowly with stronger central condensation. When $R \gg AGM/3a^2$, self-gravitation is weak, the density through the cloud does not vary much from the mean density $\bar{\rho} = M/(4\pi R^3/3)$, A has the value $\simeq 3/5$ appropriate to a uniform sphere, and (11.1) reduces effectively to Boyle's law $p_e \simeq \bar{\rho}a^2$. If p_e is supposed to increase slowly, then R decreases and the self-gravitation term becomes steadily more important. As a good first approximation one may take A to be constant, so that (11.1) predicts a turning value for p_e at $R = \tilde{R} = 4AGM/9a^2$, with the corresponding value $\tilde{p}_e = 3Ma^2/16\pi\tilde{R}^3 = \bar{\rho}a^2/4$. Thus provided

$$\Pi \equiv (p_e/a^8)G^3M^2 < 3^7/4^5A^3\pi, \qquad (11.2)$$

the relation (11.1) with A taken as constant yields two equilibria, but that with $R < \tilde{R}$ is manifestly unstable against changes in p_e or a. The state with $R > \tilde{R}$ is stable, and a steady increase in p_e or decrease in a leads to adjustment of the cloud to a new equilibrium, until the limit (11.2) is reached. Any further increase in Π leads to gravitational free-fall in a characteristic time $t_f \simeq (G\bar{\rho})^{-1/2}$.

An exact treatment of the problem (Ebert 1955; Bonnor 1956) uses the Emden solution (e.g. Chandrasekhar 1939) for the density within an isothermal sphere in hydrostatic equilibrium and so takes implicit account of the slow variation of A. The turning value for p_e is found to occur when $A = 0.733$, with $\tilde{R} = 0.41GM/a^2$, $\tilde{p}_e = 1.2Ma^2/4\pi\tilde{R}^3 = 0.4\bar{\rho}a^2$. It is not surprising that there are differences in the numerical factors; rather, one is gratified that a judicious use of the integral virial theorem yields parameter values that are so close to the exact results, so encouraging one to apply the theorem and obtain quick results for problems where exact solutions are difficult to obtain.

These formulae can be inverted to yield $\tilde{R}$ and M in terms of the mean density $\bar{\rho}$:

$$\tilde{R} = 0.76(\mathcal{R}T/\mu G\bar{\rho})^{1/2} \simeq \lambda_J, \tag{11.3}$$

$$M = 1.9(\mathcal{R}T/\mu G)^{3/2}(1/\bar{\rho})^{1/2} \equiv M_J. \tag{11.4}$$

Apart from the numerical factor, $\tilde{R}$ is the same as the Jeans length λ_J that emerges from the classical study of gravitational instability in a uniform, unbounded, isothermal medium of density $\bar{\rho}$ (Section 3.3). Likewise M is close to the mass of a sphere of radius λ_J. As noted in Chapter 3, there are well-known objections to the original Jeans treatment, deriving from its inadequate attention to the zero-order state that is being perturbed; it is nevertheless appropriate to refer to the critical length and mass as the 'Jeans length' and the 'Jeans mass'.

The microphysics of the gas enters crucially through the different heating and cooling processes that determine the temperature. In a typical HI cloud with $T \simeq 100$ K and $\mu \simeq 1$, $M_J \simeq 3 \times 10^3 M_\odot/(\rho/10^2 m_H)^{1/2} \simeq 3 \times 10^3 M_\odot$ if $n \equiv \rho/m_H = 10^2$. When n reaches 10^4 or more the atomic hydrogen is converted largely into molecular hydrogen H_2, with a simultaneous formation of other molecules such as CO which act as efficient coolants. Below $n \simeq 10^{10}$, the optical depth τ_J over a Jeans length in the infrared and millimetre bands is small, and a balance between heating by the galactic cosmic ray and photon fields and cooling by dust grains and molecules yields a temperature $\simeq$ 15–20 K, so that with $\mu = 2$, M_J now becomes $\simeq 7 \times 10^{-3}[(T/15)^{3/2}/(n/10^{10})^{1/2}]M_\odot$. Beyond $n = 10^{10}$, τ_J exceeds unity, and the compressional heat generated at the rate $\simeq (p/\rho t_f)$ gm^{-1}s^{-1} is effectively trapped, so that the temperature of a freely-falling mass begins to rise adiabatically; at this transition point $M_J \simeq 7 \times 10^{-3} M_\odot$.

The value (11.4) is the maximum mass of a cloud of given $a = (\mathcal{R}T/\mu)^{1/2}$ that can be maintained in stable equilibrium at density $\bar{\rho}$. Since M_J for typical HI cloud densities and temperatures is so large, while also M_J decreases systematically with increasing $\bar{\rho}$ and decreasing T, one is led to a tentative picture in which a gravitationally collapsing cloud spontaneously converts itself into a star cluster by systematic fragmentation into subcondensations of mass bounded below by the instantaneous value of M_J (Hoyle 1953). The dynamical problem is subtler than the Jeans-type problem, since we are concerned now with the spontaneous growth of instabilities within a cloud that is itself in a state of collapse. Analysis by Hunter (1962, 1964) and Lynden-Bell (1973) has elucidated how a local density perturbation in a cold, homogeneous, spherically symmetric collapsing cloud can indeed begin to separate out. The essential point is that although the time of collapse $t_f \simeq (G\rho_0)^{-1/2}$ of a cloud from the initial density ρ_0 is only slightly longer than the corresponding time $[G\rho_0(1+\delta\rho_0/\rho_0)]^{-1/2}$ for the self-amplification of the blob of density $(\rho_0 + \delta\rho_0)$, this simply means that it is only after a time $t \simeq t_f$ that the excess self-gravitation of the blob manifests itself to yield $(\rho + \delta\rho)_{\rm t}/\rho_{\rm t} \gg 1$ (Mestel 1965). In a cloud with a finite internal pressure and with its temperature T either staying constant or (more plausibly) decreasing, self-amplification begins only when the blob mass exceeds the instantaneous Jeans mass M_J, defined by the mean density. This suggests a hierarchical fragmentation picture similar to Hoyle's, terminating when the last fragments become opaque, for isothermality is then replaced by the adiabatic condition $T \propto \rho^{\gamma-1}$, and from (11.4), M_J will now begin to increase with ρ (since $\gamma > 4/3$ in this domain). Detailed studies (Low and Lynden-Bell 1976; Rees 1976; Silk 1977) confirm Hoyle's proposal that fragmentation according to this model will not proceed beyond the Jeans mass at the density $n \simeq 10^{10}\ \mathrm{cm}^{-3}$ for which $\tau_J \simeq 1$, but the predicted minimum masses remain embarrassingly low – $M \simeq (0.007–0.1)M_{\odot}$. An alternative way of applying the opacity criterion (Mestel and Spitzer 1956) has the final fragment not isothermal, but rather with the internal temperature gradient necessary to maintain hydrostatic equilibrium as a proto-star, and contracting at the rate fixed by heat flow against the opacity. For self-consistency, this should yield a time-scale longer than the instantaneous free-fall time: the consequent upper limit on the mass was estimated by Gaustad (1963) as $0.2M_{\odot}$ – still uncomfortably low.

It can be argued that the opacity criterion sets a lower limit to the final mass, but that it would be a realistic estimate for a typical proto-stellar mass only if fragmentation proceeds with maximum efficiency. In fact, some of the difficulties facing Hoyle's model were emphasized early by Layzer (1963). An essential feature of any successful fragmentation process is the conversion of much of the initial gravitational energy of the cloud into conserved macroscopic kinetic energy of self-gravitating blobs, which must therefore have acquired geometrical cross-sections sufficiently small for dissipative collisions to be comparatively infrequent. There are a number of studies in the literature which attempt to follow the evolution of a fragmenting cloud into the non-linear domain, taking explicit account of collisions, which can lead sometimes to coalescence of fragments and sometimes to disruption (see Silk 1985 for a survey). The final stellar mass spectrum is determined by the details of the dynamics, with the opacity criterion setting a lower limit.

We regard the picture emerging from these studies as a paradigm against which to contrast the problems that arise in a more realistic scenario. As already noted in Chapter 1, the evidence for global magnetic fields in the Milky Way and other galaxies is now overwhelming (Sofue *et al.* 1986; Beck *et al.* 1996). Particularly impressive is the overall agreement between the field direction in M51 as inferred respectively from the polarization of synchrotron radio emission and from optical polarization induced by magnetically aligned dust grains (Scarrott *et al.* 1987). Faraday rotation observations combined with estimates from pulsar dispersion of the mean electron density yield a typical background field strength of 3–4 μG for the Milky Way.

Although some workers (e.g. Kulsrud and Anderson 1992) argue for a cosmological origin for the galactic magnetic flux, most workers think in terms of galactic dynamos acting on a weak cosmological seed field (e.g. Rees 1987, 1994). The special problems of dynamo action on a galactic scale are reviewed elsewhere (e.g. Beck *et al.* 1996). Growth of the field is likely to be limited by spontaneous MHD instabilities. The Rayleigh–Taylor–Parker instability (e.g. Parker 1966, 1970), a generalization of the type of problem discussed in Section 3.6.2(b), arises in a plane-parallel stratified system in which the Lorentz force contributes significantly – along with turbulent and cosmic ray pressure – to support against the galactic gravitational field. The consequent accumulation of gas by flow down the field lines is one possible scenario for the formation of massive gas clouds. Once the self-gravitation of the cloud gas dominates over the galactic field, it will cause further contraction in all three dimensions. The trans-**B** motion will amplify and distort the frozen-in field, so generating local Lorentz forces which act against the trans-field component of gravity.

It was noted early (Chandrasekhar and Fermi 1953; Mestel and Spitzer 1956) that studies of cloud dynamics should include magnetic forces. In particular, it was predicted (Mestel 1965, 1969; Strittmatter 1966; Field 1970) that there should exist such massive dense clouds containing enhanced large-scale fields, forming by compression of the magnetized galactic gas, and maintained in equilibrium by self-gravitation. The general picture was confirmed by the first measurements of the Zeeman effect on the 21-cm line (Verschuur 1969). Subsequent Zeeman measurements in molecular clouds from up to a decade ago gave steadily accumulating support, summarized by Myers and Goodman (1988): HI and OH absorption and emission lines yielded field strengths of 10–100 μG in at least eight molecular clouds, while fields of $(3\text{–}8) \times 10^3$ μG were inferred from the Zeeman effect in OH maser emission lines from dense cores associated with compact HII regions. Unpredicted and therefore more surprising was the discovery that typical line widths are (0.89–9.2) $\mathrm{km\,s^{-1}}$, as compared with a sound speed of $\simeq 0.25(T/15)^{1/2}$ $\mathrm{km\,s^{-1}}$ (Zuckerman and Palmer 1974). The full understanding of these highly supersonic motions is perhaps the most immediately challenging problem in this area (cf. Section 12.7). However, with the line-widths interpreted in terms of a roughly isotropic turbulent pressure, Myers and Goodman found no obvious contradiction with simple theory (cf. Section 11.3). More recent observations (Crutcher *et al.* 1993, 1996) appear to yield mixed results (Section 12.1). It seems likely that improved observational techniques will again be subjecting theory to more stringent tests.

We proceed deductively, studying first the effect of the trapped magnetic flux and of any associated turbulent pressure field on the overall problem of gravitational collapse and fragmentation. Subsequently, we endow the cloud with the angular momentum to be expected from both the galactic rotation and the interstellar turbulence, and study the mutual interaction of the pressure, gravitational, magnetic and rotation fields. Characteristic plasma problems – hydromagnetic stability, deviation from flux freezing, reconnection – will be seen to emerge during the systematic development of the theory.

A complete theory, covering both the early and late phases of star formation, as well as the pre-main sequence phase studied in Chapter 10, will not only elucidate the role of the galactic magnetic field in star formation, including its possible influence on the initial mass function, but will also answer the related but distinct astronomical problem already underlined in Chapters 5–10: why are the stars all magnetically 'weak', in contrast to the gas clouds in which they are born; and at what epoch is most of the primeval flux lost?

11.2 Magneto-thermo-gravitational equilibrium

We adopt an idealized model of the warm interstellar medium, with uniform density ρ_0, temperature T_0, mean molecular weight μ_0, sound speed a_0 and pressure $p_0 = \mathcal{R}\rho_0 T_0/\mu_0 = \rho_0 a_0^2$. The medium is threaded by a magnetic field $\mathbf{B}_0$ which may also be taken as uniform, since the maintaining currents $\mathbf{j}_0$ flow in circuits large compared with the scale of the gas clouds to be studied. Now imagine that a cloud forms, for example by local cooling of a mass M to a temperature well below T_0, so that the external pressure p_0 initiates contraction in all three dimensions. Since flux-freezing is initially a good approximation (Section 11.7), the essentially non-homologous motions both compress and distort the field, generating a local force density $(\nabla \times \mathbf{B})\times\mathbf{B}/4\pi$ opposing the self-gravitation of the cloud. It is intuitively clear that a sufficiently strong trapped magnetic flux will be able to prevent indefinite gravitational collapse: after dissipation and radiation of the macroscopic kinetic energy released, a cool cloud of low angular momentum and weak turbulent pressure will instead settle into a state with magnetic and gravitational forces in approximate balance in the two dimensions perpendicular to $\mathbf{B}_0$. Discussion of precise models, with force balance holding at each point, is postponed until Sections 12.2–12.3. In the spirit of McCrea's paper (1957) cited above, we begin by adopting simple forms for the matter and magnetic fields in order to derive global results from the virial theorems.

11.2.1 *A spherical cloud model*

To isolate the significant non-dimensional parameter, consider first the simplest example (Fig. 11.1(b)). The cloud is modelled as a uniform sphere of radius R and density $\bar{\rho}$, with the internal field $\bar{\mathbf{B}}$ arising from the uniform compression of $\mathbf{B}_0$ from the radius R_0 under strict flux-freezing, so that $\bar{\mathbf{B}} = (R_0/R)^2\mathbf{B}_0$. Beyond R_0 the field is supposed unaltered, while between R and R_0, the (divergence-free) radial field with $B_r = B_0 R_0^2 \cos\theta/r^2$ is continuous at R and R_0 with the normal components of $\bar{\mathbf{B}}$ and $\mathbf{B}_0$ respectively. This structure is itself a simplification of Fig. 11.1(a) which pictures a cloud forming within our idealized interstellar medium by a continuous, non-homologous, spherically symmetric collapse (Mestel 1966). Now apply the scalar virial theorem (3.64) to the volume within R_0. The total magnetic energy is

$$\begin{aligned}\mathcal{M} &= (4\pi R^3/3)(\bar{B}^2/8\pi) + (1/8\pi)\int_0^{\pi}\int_R^{R_0} 2\pi r^2 \sin\theta (B_0^2 R_0^4 \cos^2\theta/r^4)\, dr\, d\theta \\ &= \frac{1}{6}\bar{B}^2 R^3 + \frac{1}{6}B_0^2 R_0^4\left(\frac{1}{R} - \frac{1}{R_0}\right). \qquad (11.5)\end{aligned}$$

The surface integral over R_0 can be computed directly, or more simply by noting that when $R = R_0$, the sum of the volume and surface terms must vanish, since the field is then uniform, exerts no forces, and so cannot make any net contribution to the virial theorem. This surface term is therefore $-B_0^2R_0^3/6$, and the total magnetic contribution at radius R is

$$\frac{1}{3}B_0^2R_0^4\left(\frac{1}{R}-\frac{1}{R_0}\right)=\frac{F^2}{3\pi^2}\left(\frac{1}{R}-\frac{1}{R_0}\right) \tag{11.6}$$

where

$$F=\pi B_0R_0^2=\pi\bar{B}R^2 \tag{11.7}$$

is the flux within the cloud. For the moment we ignore deviations from uniform density, so taking the gravitational energy of the cloud as $-3GM^2/5R$. If the internal and external pressure terms are small, the scalar virial theorem then predicts a balance between the magnetic and gravitational terms at the radius $R<R_0$, satisfying

$$1-\frac{R}{R_0}=\frac{9\pi^2GM^2}{5F^2}. \tag{11.8}$$

Thus according to this simple model, the magnetic forces alone are capable of holding up the cloud provided the flux/mass parameter

$$f\equiv\frac{F}{\pi^2G^{1/2}M} \tag{11.9}$$

exceeds the critical value $f_c=3/5^{1/2}\pi$. Further, if f is not close to f_c, R/R_0 will not be small, and the increase in B/B_0 from (11.7) will be modest; but as $f\to f_c$, the model yields large values for B/B_0.

Since the magnetic forces are essentially anisotropic, acting primarily in the two directions perpendicular to $\mathbf{B}_0$, it is more reasonable to derive a first criterion for equilibrium from the two trans-field diagonal components of the tensorial virial theorem (3.62). Retention also of the internal and external pressure terms then yields the generalization of the McCrea equation (11.1) for a cloud of constant sound speed a:

$$p_e=\frac{3Ma^2}{4\pi R^3}-\frac{3GM^2}{20\pi R^4}+\frac{F^2}{5\pi^3R^4}\left(1-\frac{R}{R_0}\right) \tag{11.10}$$

$$=\frac{3Ma^2}{4\pi R^3}(1-\epsilon)-\frac{3GM^2}{20\pi R^4}\left(1-\frac{4\pi^2f^2}{3}\right), \tag{11.11}$$

where

$$\epsilon=\frac{4F^2}{15\pi^2Ma^2R_0}=\frac{4\pi^2f^2}{15}\left(\frac{GM}{a^2R_0}\right). \tag{11.12}$$

(From now on, the external pressure p_e is kept a free variable, distinct from the pressure $\rho_0a_0^2$ of the idealized interstellar medium introduced above.)

Without the terms in p_e and a, (11.10) differs from (11.8) only in the modified critical value $f_c=3^{1/2}/2\pi$, lower by a factor $\simeq 0.65$. With the thermal pressure terms retained, the same critical value $f_c=3^{1/2}/2\pi$ emerges, defined by the

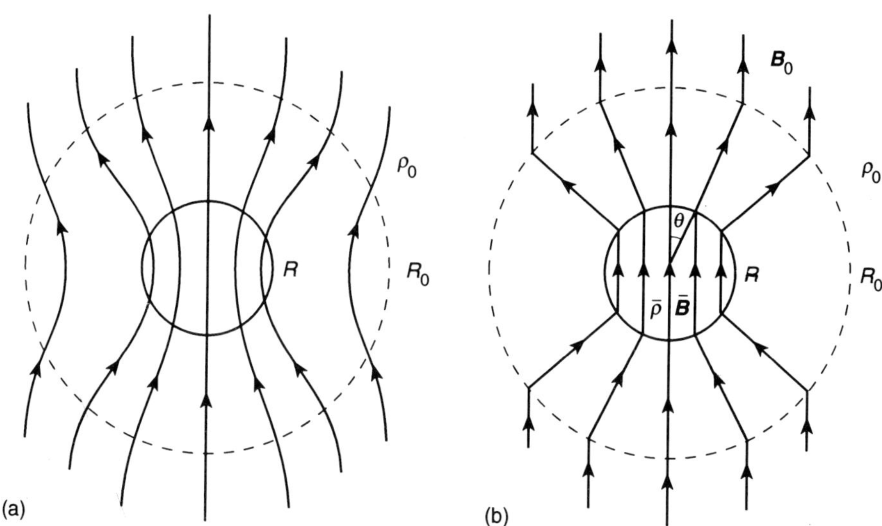

FIG. 11.1. (a) A field arising by non-homologous spherically symmetric compression within the idealized interstellar medium of uniform density ρ_0 and permeated by a uniform field $\mathbf{B}_0$. (b) Simulation of (a) by the spherical cloud model, of uniform density $\bar{\rho}$, and with a uniform magnetic field $\bar{\mathbf{B}}$ arising by compression of the background field $\mathbf{B}_0$.

vanishing of the coefficient of $1/R^4$ in (11.11). With F prescribed, then (11.11) predicts that a cloud of mass $M < M_c$, with

$$M_c = F/\pi^2 G^{1/2} f_c, \tag{11.13}$$

has equilibria with R monotonically decreasing with increasing p_e, whatever the value of ϵ. If, however, $f < f_c = 3^{1/2}/2\pi$, then (11.11) without the thermal terms, like (11.8), predicts that no equilibria exist; while with the thermal terms retained, equilibria can exist only if GM/a^2R_0 is low enough to make $\epsilon < 1$. Conclusions can then be drawn analogous to those found from (11.1). There is a turning value $\tilde{p}_e$ of p_e at $\tilde{R} = 4GM(1 - f^2/f_c^2)/15a^2(1 - \epsilon)$, with $\tilde{p}_e = (3M/4\pi\tilde{R}^3)a^2(1 - \epsilon)/4 \equiv \bar{\rho}a^2(1 - \epsilon)/4$. For equilibria with $f < f_c$ to exist, the condition analogous to (11.2) (with $A = 3/5$) must hold: the quantity $\Pi \equiv (p_e/a^8)G^3M^2(1 - f^2/f_c^2)^3/(1 - \epsilon)^4$ must be less than $(5^3 3^4/4^5\pi) \simeq 3.1$. If this is satisfied, then this treatment predicts two classes of equilibria with R respectively greater than and less than $\tilde{R}$, but again only those with $R > \tilde{R}$ are grossly stable. Increase of p_e, decrease in a, or increase in M through accretion will lead to indefinite collapse, once Π exceeds the above limit.

Substitution of $\tilde{R}$ and $\tilde{p}_e$ into (11.10) yields an equation for the maximum mass $M_{\rm cr}$ that can be held in equilibrium by the joint effect of magnetic force and thermal pressure, when F and a are prescribed:

$$M_{\rm cr} = \left[\frac{3}{4\pi\bar{\rho}}\left(\frac{15a^2}{4G}\right)^3\right]^{1/2}\left[1-\left(\frac{M_c}{M_{\rm cr}}\right)^2\right]^{-3/2} \equiv c_J M_J \left[1-\left(\frac{M_c}{M_{\rm cr}}\right)\right]^{-3/2}, \tag{11.14}$$

where M_c is again the magnetic critical mass defined by (11.13), and the term in ϵ (which comes from the contribution of the background field $\mathbf{B}_0$ to the virial surface term) has been neglected. The constant c_J depends on the precise definition of the thermal Jeans mass M_J; if one again adopts the McCrea definition, given by the same turning point condition but with $f = 0$ also, then $c_J = 1$. As expected, in a cool cloud the correction to the estimate $M_{\rm cr} = M_c$ will be small. A result analogous to (11.14) applies to models which allow for spontaneous flattening along the field (Section 11.2.2; Mouschovias and Spitzer 1976; Tomisaka *et al.* 1988). McKee (1989) shows that in the cases of interest, (11.14) can be written $M_{\rm cr} = M_J + M_c$ to within about 5% accuracy.

11.2.2 *A spheroidal model*

An obvious blemish of the above model is the assumption that the cloud stays spherical, whereas in reality the cloud will spontaneously flatten parallel to $\mathbf{B}_0$. The increase in $\bar{\rho}$ increases the gravitational force density, but the consequent further distortion of $\mathbf{B}$ increases the magnetic force; provided f is sufficiently large, after dissipation of the kinetic energy supplied from the gravitational field a non-turbulent cloud will settle into a state of magneto-thermo-gravitational equilibrium, essentially with gravity balanced by thermal pressure in the $\mathbf{B}_0$ direction and by magnetic force in the other two dimensions. If the cloud is weakly turbulent with motions that are at most sonic, their effect on the gross dynamics of the cloud can be incorporated through a factor of up to 2 in the thermal pressure, so that the equilibrium models for cool massive molecular clouds would be sheet-like (cf. the analysis in Section 12.2). There is some evidence for flattened structures within molecular clouds such as the Taurus complex (Moneti *et al.* 1984), but apparently not to the extent implied by a temperature of 15 K, while as already noted, in many clouds observed line broadening implies highly supersonic motions. Analysis of this 'Alfvénic turbulence' is postponed until Section 12.7; for the moment we adopt the simplest parametrization, supposing the turbulence to be isotropic and effectively increasing the mean 'sound speed' a over the cloud to anything between the real sound speed and the mean value of the fast magnetohydrodynamic wave speed. The simplest generalization of Fig. 11.1(b) is then the model of Fig. 11.2, with the equilibrium cloud approximated as a uniform, oblate spheroid $\bar{S}$, symmetric about $\mathbf{B}_0$, of density $\bar{\rho}$, semi-axis R across $\mathbf{B}_0$, semi-axis Z along $\mathbf{B}_0$, and with the field $\bar{\mathbf{B}}$ within $\bar{S}$ uniform and parallel to $\mathbf{B}_0$. The eccentricity $e = (1 - Z^2/R^2)^{1/2}$ is then clearly a measure of the strength of the turbulence. As before, the flux F within the cloud enables us to define a radius R_0 by (11.7). The cloud mass M can likewise be written either in terms of its density $\bar{\rho}$:

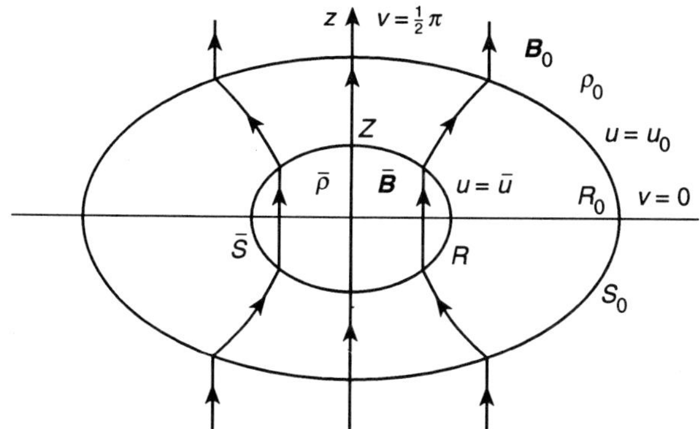

FIG. 11.2. Cloud modelled as an oblate spheroid $\bar{S}$ of eccentricity e, again with uniform density $\bar{\rho}$ and a uniform field $\bar{\mathbf{B}}$ arising by compression of the background field $\mathbf{B}_0$. The confocal spheroid passing through $(R_0, 0)$ is the analogue of the sphere R_0 in Fig. 11.1(b). Between $\bar{S}$ and S_0 the field lines lie on the orthogonal hyperboloids.

$$M = \frac{4\pi}{3}\bar{\rho}ZR^2 = \frac{4\pi}{3}(1-e^2)^{1/2}\bar{\rho}R^3 = \frac{2\pi}{3}\bar{\sigma}R^2, \tag{11.15}$$

where the projected area density

$$\bar{\sigma} = 2\bar{\rho}Z, \tag{11.16}$$

or in terms of the background density ρ_0

$$M = K\left[\frac{4\pi}{3}\rho_0 R_0^3\right] = \frac{2\pi}{3}\rho_0 R_0^2 L_0, \qquad L_0 = 2KR_0. \tag{11.17}$$

If the cloud has formed subject to strict flux-freezing, then R_0 is the radius and $L_0 = 2KR_0$ is the axial length of a spheroidal 'parent cloud' which cooled and condensed to form $\bar{S}$; but even if significant trans-field flow has taken place, the study of cloud *equilibrium* is unaltered, with just the interpretation of the parameters requiring modification (see below).

The appropriate structure of the external field in this model is a straightforward generalization of Fig. 11.1(b). Of the set of oblate spheroids confocal with $\bar{S}$, the one passing through the points $\varpi = R_0$, $z = 0$ (labelled S_0) is the analogue of the sphere of radius R_0. Beyond S_0, the field $\mathbf{B} = \mathbf{B}_0$; within $\bar{S}$, $\mathbf{B} = \bar{\mathbf{B}}$; and between $\bar{S}$ and S_0 the field lines lie in the orthogonal hyperboloids, with the normal components of $\mathbf{B}$ continuous on $\bar{S}$ and S_0. The discontinuities in the tangential components of $\mathbf{B}$ again imply toroidal surface currents and so finite forces per unit area on $\bar{S}$ and S_0; within $\bar{S}$ and beyond S_0 the uniform fields $\bar{\mathbf{B}}$

and $\mathbf{B}_0$ exert no volume forces, but in the domain between $\bar{S}$ and S_0, $\nabla \times \mathbf{B}$ and so also $(\nabla \times \mathbf{B}) \times \mathbf{B}/4\pi$ are non-zero. There is of course no claim that this is even an approximate representation of the *detailed* magnetic force distribution, but again, substitution of these simple forms for the $\mathbf{B}$ and ρ fields into the virial integrals yields quick global results, to be compared later with those found from accurately constructed models.

Analytical details are given in Appendix A. From (11.146), the 11-component of the virial theorem yields a condition for the equilibrium of the spheroid of eccentricity e and semi-major axis R:

$$p_e(1-e^2)^{1/2} = \frac{3Ma^2}{4\pi R^3} - \frac{3GM^2}{20\pi R^4}\left[Q(e) - \frac{5\pi^2}{6}f^2 m(e,\eta)\right], \tag{11.18}$$

where η is the non-dimensional semi-major axis

$$\eta = \frac{R}{R_0}, \tag{11.19}$$

again $f = F/\pi^2 G^{1/2} M$, the magnetic contribution $m(e,\eta)$ is defined by (11.141)–(11.145), and the gravitational function $Q(e)$ by (11.140).

In terms of the limit $m(e,0)$ of $m(e,\eta)$ as $\eta \to 0$ at fixed e (given by (11.148)), (11.18) can be written

$$p_e(1-e^2)^{1/2} = \frac{3Ma^2}{4\pi R^3}\left[1 - \frac{\pi^2 f^2}{6}\frac{GM}{a^2 R_0}\frac{m(e,0)-m(e,\eta)}{\eta}\right] - \frac{3GM^2}{20\pi R^4}\left[Q(e) - \frac{5\pi^2 f^2}{6}m(e,0)\right]. \tag{11.20}$$

This is the generalization of (11.10) for $e \neq 0$, for when $e = 0$, $Q = 1$ and $m = 8(1-\eta)/5$ and we recover (11.10). For $e > 0$, $[m(e,0) - m(e,\eta)]/\eta$ increases monotonically to 1.6 at $\eta = 0$ for all e, from a value at $\eta = 1$ between 1.6 (for $e = 0$) and 1.184 (for $e = 1$). If for the moment e is treated as a prescribable parameter, one can discuss (11.18) and (11.20) analogously to (11.10) et seq. Thus if a^2 is assumed constant, then the critical parameter $f_c(e)$ and the associated critical mass $M_c(e)$ are again defined by the vanishing of the coefficient of $1/R^4$ in (11.20):

$$f_c(e) = [1.2Q(e)/\pi^2 m(e,0)]^{1/2}, \tag{11.21}$$

$$M_c(e) = F/\pi^2 G^{1/2} f_c(e), \tag{11.22}$$

where F is the prescribed flux threading the spheroid. As in the discussion following (11.12), when $f > f_c(e)$, (i.e. $M < M_c(e)$ for prescribed F), there always exists a radius R satisfying (11.20), whereas if $f < f_c(e)$, provided $p_e G^3 M^2/a^8$ is sufficiently small, there are two equilibria, with the lower density model the stable one. The parameter $f_c(e)$ increases monotonically from the value $f_c(0) = 3^{1/2}/2\pi$

to $f_c(1) = 2(0.6)^{1/2}/\pi$, with a corresponding decrease in $M_c(e)$ for a given flux by the factor $0.8(5)^{1/2} \simeq 1.8$ (Strittmatter 1966).

When $f > f_c(e)$, one can again illustrate the mutual interaction of magnetism and self-gravitation by temporarily dropping both pressure terms in (11.18). With e prescribed, the equilibrium value of η is then given by

$$m(e, \eta) = (1.2Q(e)/\pi^2)(1/f^2) = m(e, 0)f_c^2(e)/f^2 \tag{11.23}$$

from (11.21). By (11.19) and (11.22), this gives an estimate of the semi-major axis R of the cloud model of mass M and eccentricity e, expressed in terms of the critical mass $M_c(e)$ for the prescribed flux F, with $R_0 = (F/\pi B_0)^{1/2}$ again the radius at which frozen-in flux F would yield $\bar{B} = B_0$, the background field-strength. Again it is found that unless f is close to $f_c(e)$, (11.23) predicts η-values that are not small, with correspondingly modest increases in $\bar{B}/B_0$ over unity. But when f is close enough to $f_c(e)$, so that $\eta \ll 1$, then (11.147) and (11.23) yield

$$R/R_0 = \eta = (5m(e, 0)/8)(1 - f_c^2(e)/f^2) \tag{11.24}$$

$$= (5m(e, 0)/8)(1 - M^2/M_c^2(e)). \tag{11.25}$$

The coefficient in (11.24) is unity when $e = 0$, falling monotonically to $15\pi/64$ at $e = 1$.

The obvious next step is to restore the pressure terms and to treat e not as an independently prescribable parameter, but rather to estimate it from the 33-virial component. The apparently most straightforward procedure applies the theorem again to the volume within the surface surrounding S_0, yielding (11.153) with the magnetic term given by (11.150). However, this has the objection that as $e \to 1$, although by (11.152), $H(e) \to 0$, $n(1, \eta)$ stays positive and finite and in fact is of order unity once $\eta \neq 1$, so that (11.153) demands a high effective sound speed even in a flattened body. This unacceptable feature is an artefact of our model field, which exerts z-force components at high z-values even when the cloud has flattened into a thin sheet near $z = 0$. When weighted by z these forces make spurious large contributions to the 33-virial component. If an accurately constructed field – for example one of the models of Section 12.2, with $\mathbf{j} = 0$ outside the disc – were inserted into the virial theorem, the volume integrals would be much smaller, equivalent to the integral $\int x_3 T_{3j} n_j \, \mathrm{d}S$ over a surface just outside the cloud spheroid $\bar{S}$. When η is not close to 1, a far better simulation of the 33-magnetic contribution is given by (11.154), which is the sum of the $\bar{S}$ volume and surface integrals with our model field substituted.[3] We therefore use (11.153) but with $n(e, \eta)$ replaced by $q(e)$ given by (11.154), yielding

[3]The 11-magnetic contribution to the $\bar{S}$ volume and surface integrals has the form $(F^2/6\pi^2 R)(1 - e^2)^{1/2}$ as $e \to 1$, larger than (11.154) but still vanishing at $e = 1$. This is again a consequence of our particular model which has the field lines leaving $\bar{S}$ normally: a realistic estimate of the lateral effect of the magnetic forces must include the integrals over the volume betwen $\bar{S}$ and S, as in the above treatment.

$$p_e(1-e^2)^{1/2} = \frac{3Ma^2}{4\pi R^3} - \frac{3GM^2}{20\pi R^4}\left[H(e) - \frac{5f^2}{6}q(e)\right]. \tag{11.26}$$

Equations (11.18) and (11.26) then yield a relation for f in terms of e and η:

$$Q(e) - H(e) = \frac{5\pi^2 f^2}{6}[m(e,\eta) + q(e)] \tag{11.27}$$

with the effective sound speed given by

$$\frac{a^2 R}{GM}\left[1 - \frac{4\pi p_e R_0^3 \eta^3 (1-e^2)^{1/2}}{3Ma^2}\right] = \frac{[q(e)Q(e) + m(e,\eta)H(e)]}{5[q(e) + m(e,\eta)]}. \tag{11.28}$$

Since m increases monotonically as η decreases at constant e, one can define a new critical parameter $\tilde{f}(e)$ and associated critical mass $\tilde{M}(e)$ – to replace $f_c(e)$ and $M_c(e)$, given respectively by (11.21) and (11.22) – by putting $\eta = 0$ in (11.27):

$$\tilde{f}(e) = \left\{1.2[Q(e) - H(e)]\Big/[q(e) + m(e,0)]\right\}^{1/2}\Big/\pi \tag{11.29}$$

and

$$\tilde{M}(e) = F/\pi^2 G^{1/2}\tilde{f}(e). \tag{11.30}$$

For example, when $e = 0.866$, corresponding to an axis ratio $Z/R = 0.5$, $m(e,0) = 0.8833$, $q(e) = 0.0969$ and (11.29) yields $\pi\tilde{f}(e) = 0.8765$, whereas (11.21) yields $\pi f_c(e) = 1.388$, whence $\tilde{M}(e)/M_c(e) = 1.6$. From (11.28), a^2R/GM is then $\simeq 0.1760$ at small η. This effective sound speed a both maintains the finite eccentricity and contributes to lateral support, so increasing by the factor $\simeq 1.6$ the critical mass required for indefinite collapse. As $e \to 1, H(e)$ and $q(e) \to 0$; by (11.28), $a \to 0$, and so $\tilde{f}(e)$ and $f_c(e)$ become identical. At small η, (11.27), (11.29) and (11.147) yield as a substitute for (11.24) at small η

$$\begin{aligned} R/R_0 &= (5/8)[m(e,0) + q(e)][1 - \tilde{f}^2(e)/f^2] \\ &= (5/8)[m(e,0) + q(e)][1 - M^2/\tilde{M}^2(e)]. \end{aligned} \tag{11.31}$$

This is a tolerable approximation even when M is not close to $\tilde{M}(e)$, so that η is not small.

11.3 Applications

11.3.1 *The accumulation length*

Consider first a sequence of models with the *same eccentricity*, and with f close to but greater than $\tilde{f}(e)$, defined by (11.29), so that M is less than but close to $\tilde{M}(e)$ for the flux F, and $\eta \equiv R/R_0 \ll 1$. Now suppose f approaches $\tilde{f}(e)$, either by accretion of gas or by flux leakage; then R decreases according to (11.31), so $\bar{\rho} \simeq \tilde{M}(e)/(4\pi(1-\ e^2)^{1/2}R^3/3)$, and (11.30) yields

$$\bar{B} \simeq \pi \tilde{f}(e) G^{1/2} \tilde{M}(e)/R^2 \propto \bar{\rho}^{2/3}, \tag{11.32}$$

with the required effective sound speed given by (11.28):

$$a^2 \propto GM/R \propto \bar{\rho}^{1/3}. \tag{11.33}$$

Relations (11.32) and (11.33) describe the *isotropic* contraction of a body with nearly constant flux and mass, but with F/M decreasing slightly towards the limit given by (11.29).

If a^2 does not increase according to the '$\gamma = 4/3$' law (11.33) but, for example, stays constant (effective isothermality), then a^2R/GM decreases, and (11.28), (11.29) and (11.30) yield e increasing towards unity, $\tilde{f}(e)$ increasing and $\tilde{M}(e)$ decreasing.

If $M \ll \tilde{M}(e)$ – or $f \gg \tilde{f}(e)$ – then by (11.31), $\eta \equiv R/R_0$ is no longer small, and the approximation (11.147) for $m(e,\eta)$ is inappropriate. It is easy to check numerically from (11.141)–(11.145) that the magnetic term $5\pi^2 f^2 m(e,\eta)/6$ in (11.18) will far outweigh the gravitational term $Q(e)$ unless $\eta \simeq 1$, with $\bar{B} \simeq B_0$: masses much below the critical value achieve equilibrium at R less than but very close to R_0, with a very slight distortion of the field sufficient to generate the magnetic force required to balance gravity.

It is instructive to consider further the case of a cool, weakly turbulent cloud, which flattens to form a spheroid with $e \approx 1$. (For brevity, from now on the word 'cool' is used to imply both a low thermal pressure and weak turbulence.) The virial treatment then predicts indefinite collapse with the field lines frozen in provided $M > M_c(e = 1)$. We use (11.22), with $f_c(1) = 2(0.6)^{1/2}/\pi$ given by (11.21), and the definitions (11.7) and (11.15) to write the collapse condition in terms of $\bar{B}$ and $\bar{\sigma}$:

$$\bar{\sigma} > \bar{\sigma}_c \simeq (1/4\pi)(15/G)^{1/2}\bar{B} \tag{11.34}$$

(Field 1970; Nakano 1981, 1984). Note that no length-scale perpendicular to the field appears: the condition for indefinite gravitational collapse in all three dimensions of a cold body, permeated by a uniform field, relates the mass per unit area normal to the field direction to the magnetic flux threading the same area. The length L_0 defined in (11.17) when combined with the two expressions (11.7) for the invariant flux F defines an *accumulation length* along $\mathbf{B}_0$:

$$L_0 = \frac{B_0}{\rho_0}\left(\frac{\bar{\sigma}}{\bar{B}}\right) \tag{11.35}$$

and the collapse condition (11.34) for a disc-like body imposes a lower limit

$$(L_0)_c \simeq \frac{1}{4\pi}\left(\frac{15}{G}\right)^{1/2}\frac{B_0}{\rho_0} \simeq 700\frac{(B_0/3\times 10^{-6})}{(n_0/1)}\ \text{pc}, \tag{11.36}$$

with standard galactic background values inserted.

The same exercise can be performed for a cloud maintained at a small eccentricity by an effective sound speed. As noted above, the modifications to both the magnetic and gravitational terms plus the direct contribution of the sound speed to lateral support yield a critical mass $\tilde{M}(e) > M_c(e)$ (cf. (11.29) and (11.30)). The critical accumulation length is a factor 2 or so longer than (11.36).

These basic results should be compared with the predictions of the linear Jeans-type analysis (Section 3.3), which yields unstable trans-field motions of wavelengths exceeding the magnetic Jeans length λ_B, whereas along the field the minimum wavelength is the ordinary Jeans length λ_J, which is much less than λ_B in a cool cloud. The virial treatment shows clearly that the linear analysis is a misleading guide to the non-linear development: the collapse of a body of mass $4\pi\lambda_B^2\lambda_J/3$ would in fact be halted by the magnetic curvature force. Further, in the cold limit *the scale across the field is arbitrary.* It fixes the flux through the incipient cloud and therefore the mass necessary for collapse to occur; under field-freezing, it is the accumulation length L *along* the field which determines whether the cloud in fact acquires enough mass to undergo indefinite collapse $(L > L_c)$, or whether it will settle into an equilibrium state.

The value of $(L_0)_c$ is of the order of the magnetic Jeans length λ_B in a cold medium. It is also of the order of the minimum wavelength for the Rayleigh–Taylor–Parker instability, and so may indicate how self-gravitating masses may form spontaneously from the interstellar medium. Even so, one wonders whether the magnitude of $(L_0)_c$ given by (11.36) is a hint that there has been a significant departure from flux-freezing during cloud formation. If clouds are sometimes formed following violent input of energy into the interstellar medium, then an effective turbulent resistivity may correspond to substantial trans-field diffusion (cf. Section 4.3). The distance L_0 then becomes notional: one may still conveniently write the cloud mass as in (11.17) in terms of the background density ρ_0, the radius R_0 defined by (11.7) and a length L_0 along $\mathbf{B}_0$, but the requirement that L_0 must exceed 700 pc in order that the resulting flattened cloud may undergo gravitational collapse – or even cause a significant gravitational amplification of the local magnetic field – will no longer imply physical accumulation along $\mathbf{B}_0$ of background gas extending for this distance.

11.3.2 *The B–ρ relations in a cool cloud*

A cool cloud with $f > f_c(1)$ should reach equilibrium as a highly flattened structure, for which the relation (11.25) between $M/M_c(1)$ and R/R_0 is an adequate approximation. It is then of interest to derive the variation of $\bar{B}$ with $\bar{\rho}$ in a sequence of such clouds, all with the same value of a but with f steadily decreasing towards the limit $f_c(1)$.

The approach to equilibrium is again pictured as occurring in two stages. First, the gas in the prolate spheroidal parent cloud (real or notional) with parameters (ρ_0, R_0, L_0) defined in (11.17) is supposed to have cooled to the sound speed $a \ll a_0$ and consequently to flow down the field lines until hydrostatic equilibrium along $\mathbf{B}_0$ is reached. This stage is again idealized as a highly oblate $(e_0 \simeq 1)$ spheroid of semi-minor axis Z_0, uniform density $\bar{\rho}_0 = \rho_0 L_0/2Z_0$, and semi-major axis R_0 related to the flux F threading the cloud by $F = \pi B_0 R_0^2$. The 33-component (11.153) of the virial theorem can be written as

$$\frac{3Z_0}{R_0} \simeq H(e_0) = \frac{5R_0 a^2}{GM} - \frac{10R_0^2}{GM}\frac{p_e}{\rho_0 L_0}\frac{Z_0}{R_0} \tag{11.37}$$

where use has been made of the definitions (11.15) (the magnetic term vanishes as the field is still uniform). The ratio of the gravitational term to the external pressure term reduces with the help of (11.15) to $(\pi G\rho_0^2 L_0^2/5p_e)$, which is conveniently written in terms of the critical accumulation length (11.36) as

$$\frac{3}{2}\frac{B_0^2}{8\pi p_e}\left(\frac{L_0}{(L_0)_c}\right)^2 \simeq 0.2\frac{(B_0/3\times 10^{-6})^2 2\mu_0}{(n_0/1)(T_0/10^4)}\left(\frac{L_0}{(L_0)_c}\right)^2 \tag{11.38}$$

with standard values inserted. When $L_0 \ll (L_0)_c$ this ratio is small, and the cloud density $\bar{\rho}_0$ at this interim phase is determined essentially by balance of the two pressure terms in (11.37):

$$\begin{aligned} \bar{\rho}_0 &= p_e/a^2 = \rho_0(T_0/\mu_0)/(T/\mu) \simeq 10^3(\mu/2\mu_0)[(T_0/10^4)/(T/20)]\rho_0, \\ Z_0 &= \rho_0 L_0/2\bar{\rho}_0 = [5\times 10^{-4}(2\mu_0)\{(T/20)/(T_0/10^4)\}]L_0. \end{aligned} \tag{11.39}$$

The cloud subsequently adjusts laterally to achieve magneto-gravitational equilibrium in the direction perpendicular to $\mathbf{B}_0$ while maintaining equilibrium along $\mathbf{B}_0$. The internal frozen-in field $\bar{\mathbf{B}}$ is then given by (11.7). As already noted, when $M/M_c(1) = L_0/(L_0)_c \ll 1$, (11.18) yields $\eta \equiv R/R_0$ less than but close to unity, and so also $\bar{B}/B_0 = 1/\eta^2$ is only slightly greater than unity: the cloud has so little mass that only a very slight distortion of the galactic field lines is sufficient to generate the magnetic force able to balance gravity. As $M/M_c(1)$ increases, so does the distortion of the field necessary for equilibrium. When $M/M_c(1) \to 1$, by (11.25) R/R_0 becomes small, and $\bar{B}/B_0 = (R_0/R)^2$ becomes large. Since $H(e) \simeq 3Z/R$ and $\bar{\rho} = 3M/4\pi ZR^2$, the ratio of the gravitational term to the external pressure term in (11.26) now exceeds the estimate (11.38) essentially by the large factor $(R_0/R)^4$. Since also $q(e)$ becomes small as $e \to 1$, the 33-equation (11.26) now approximates to

$$4\pi G\bar{\rho}Z^2 = 5a^2, \tag{11.40}$$

with

$$4\pi\bar{\rho}ZR^2/3 = M = 4\pi\bar{\rho}_0 Z_0 R_0^2/3 \tag{11.41}$$

by mass conservation.

Equations (11.40), (11.41), (11.17) and (11.36) jointly yield

$$\bar{\rho} = \frac{9}{20\pi}\frac{GM^2}{a^2R^4} = \left(\frac{\pi G\rho_0^2 L_0^2}{5a^2}\right)\left(\frac{R_0}{R}\right)^4 = \frac{3B_0^2}{16\pi a^2}\left(\frac{L_0}{(L_0)_c}\right)^2\left(\frac{R_0}{R}\right)^4$$

$$= \frac{3B_0^2}{16\pi a^2}\left(\frac{M}{M_c(1)}\right)^2\left(\frac{R_0}{R}\right)^4, \tag{11.42}$$

whence from (11.7) follow the equivalent forms

$$\frac{\bar{B}}{\bar{\rho}^{1/2}} = \left(\frac{20}{9\pi G}\right)^{1/2}\left(\frac{F}{M}\right)a = \left(\frac{16\pi}{3}\right)^{1/2}\left(\frac{M_c(1)}{M}\right)a$$
$$= \left(\frac{16\pi}{3}\right)^{1/2}\left(\frac{f}{f_c(1)}\right)a. \tag{11.43}$$

In order that R/R_0 be small enough for (11.40) to replace $\bar{\rho} = \bar{\rho}_0$, given by (11.39), f must already be close to $f_c(1)$ with R/R_0 given approximately by (11.24) or (11.25) with $e \approx 1$, and $5m(1,0)/8 = 15\pi/64$ by (11.149). Now suppose $f/f_c(1)$ decreases further, either by accretion of further cooled gas, or by slow flux leakage (cf. Section 11.7), but with the sound speed a still staying constant. Then by (11.25) and (11.42)

$$\left(\frac{M}{M_c(1)}\right)^2 = \left(1 - \frac{64}{15\pi}\frac{R}{R_0}\right) = \left[1 - \left(\frac{\bar{\rho}^*}{\bar{\rho}}\right)^{1/4}\left(\frac{M}{M_c}\right)^{1/2}\right] \tag{11.44}$$

where a convenient standard density is defined by

$$\bar{\rho}^* = \left(\frac{64}{15\pi}\right)^4\left(\frac{3B_0^2}{16\pi a^2}\right) = \frac{665 m_H \mu (B_0/3\times 10^{-6})^2}{(T/20)}. \tag{11.45}$$

Note that if T were $\simeq 10^4$ K, then with $B_0 \simeq 3\times 10^{-6}$, (11.45) would yield a density similar to that of the warm interstellar medium; but with $T \simeq 20$ K, typical of a molecular cloud, $\bar{\rho}^*$ is about three orders higher. When $\bar{\rho} = \bar{\rho}^*$, (11.44) yields $M/M_c = 0.52$ and $R/R_0 = 0.53$, so one can define a corresponding B^* by $\bar{B}^*/B_0 = (R_0/R)^2 = 3.6$.

A steady decrease in $f/f_c(1)$ implies a corresponding increase in $M/M_c(1)$ towards unity, with $\bar{\rho}/\bar{\rho}^*$ increasing according to (11.44), while $\bar{B}/\bar{\rho}^{1/2}$ given by (11.43) decreases slightly. As $M/M_c(1)$ approaches unity, this model predicts that small changes in $M/M_c(1)$ produce large changes in R/R_0 and so in $\bar{B}$ and $\bar{\rho}$, but in this isothermal model $\bar{B}/\bar{\rho}^{1/2}$ is almost constant. For illustration, we may relate simultaneous values of $\bar{B}$ and $\bar{\rho}$ to the standard values $\bar{\rho}^*$ and $\bar{B}^*$ by writing $\bar{B}/\bar{B}^* = (\bar{\rho}/\bar{\rho}^*)^n$ and then tracing the change in the index n (see Table 11.1).

As the approximation (11.41) does not hold for the first three lines, the corresponding figures are notional. However, as $\bar{\rho}/\bar{\rho}^*$ gets large, corresponding to $M/M_c(1) \to 1$, then by (11.43) the changes in $\bar{B}$ and $\bar{\rho}$ approximate to $\bar{B} \propto \bar{\rho}^{1/2}$; for example, the last two rows in Table 11.1 are related by $\bar{B} \propto \bar{\rho}^{0.48}$ (see Fig. 11.3).

These elementary results demonstrate clearly that one cannot expect a universal relation between observed values of $\bar{B}$ and $\bar{\rho}$; even if the effective sound

Table 11.1 The variation in density, radius, field strength and index n as the mass M approaches the critical value.

$M/M_c(1)$	$\bar{\rho}/\bar{\rho}^*$	$\bar{R}/R_0$	$\bar{B}/B_0$	n
0.30	0.13	0.67	2.2	0.24
0.40	0.32	0.62	2.6	0.285
0.52	1.0	0.53	3.6	0.30
0.80	38	0.27	14.0	0.38
0.85	122	0.20	25.0	0.395
0.90	622	0.14	51.0	0.41
0.95	10^4	0.072	194.0	0.43

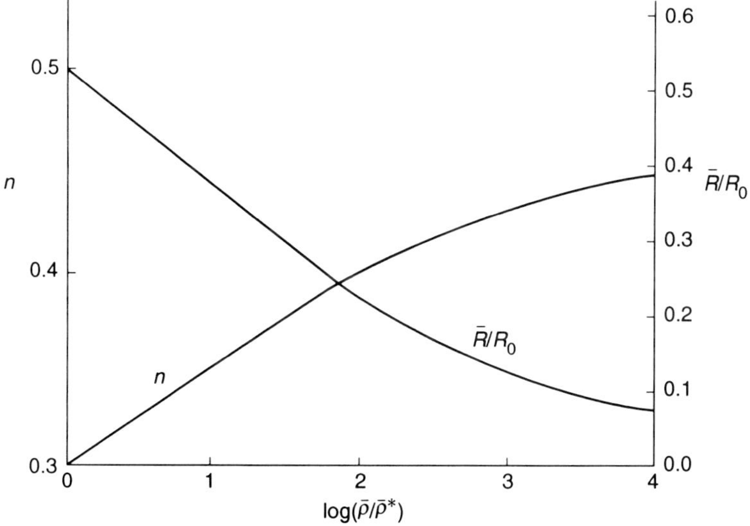

FIG. 11.3. The variation of the index n in the relation $\bar{B}/\bar{B}^* = (\bar{\rho}/\bar{\rho}^*)^n$ and of $\bar{R}/R_0$ with $\bar{\rho}/\bar{\rho}^*$ and $\bar{\rho}/\rho_0$ for standard values $B_0 = 3 \times 10^{-6}$, $n_0 = 1$, $\bar{T} = 20$ and $\mu = 1.75$ ($\simeq$ 25% He, 75% H).

speed were identical in all clouds, the value of the ratio M/M_c is crucial. As noted above, when $M/M_c \ll 1$, $\bar{B}/B_0$ has hardly changed but (11.39) predicts a large increase in density associated with a low cloud temperature: e.g. if $T \simeq 100$ K (typical of an HI cloud) and $T_0 \simeq 10^4$ K (the 'warm interstellar medium'), then $\bar{\rho}_0 \simeq 200\rho_0$, a high HI cloud density. But when M/M_c approaches unity, by (11.24) R/R_0 becomes small and $\bar{B} \propto (R_0/R)^2 = K\bar{\rho}^{1/2}$ approximately if a stays constant, with $K \propto a$ but independent of the cloud mass. What is quite wrong (though it is implicit in some published work) is to adopt the law $B = K\rho^{1/2}$, determine K by inserting values on B and ρ appropriate to the warm background, and then apply it to a cool isothermal medium. As is clear from the derivation, the $B \propto \rho^{1/2}$ law is a good approximation relating equilibrium states with the *same* temperature, and with varying but near critical mass/flux ratios.

11.3.3 *Strongly turbulent clouds*

The highly flattened cloud model may be appropriate if the rate of energy supply falls below the spontaneous decay rate of the turbulence. A cloud with the highly superthermal velocities v_t indicated by the measured line-widths in typical molecular clouds and with f close to the critical value f_c is modelled better as a moderately oblate spheroid. For definiteness, take again the case with $Z/R = 1/2$ and f close to $\tilde{f} = 0.88/\pi$, so that by (11.30), (11.40) and (11.41) (with a now replaced by v_t)

$$\bar{B} \simeq 0.88G^{1/2}M/R^2 \simeq 0.15(4\pi\bar{\rho})G^{1/2}R = 0.3(4\pi\bar{\rho})G^{1/2}Z \qquad (11.46)$$

$$\simeq 0.66(4\pi\bar{\rho})^{1/2}v_t, \qquad (11.47)$$

with

$$v_t \simeq 0.2(4\pi\bar{\rho})^{1/2}G^{1/2}R, \qquad (11.48)$$

equivalent to (11.33). The numerical factor in (11.47) is insensitive to variations in Z/R (cf. (11.43)). The result (11.47) does not involve the cloud mass: when $f \simeq f_c$, the equilibrium achieved by the lateral adjustment of $\bar{B}$ and $\bar{\rho}$ and the longitudinal adjustment of $\bar{\rho}$ yields inevitably an Alfvén speed $v_A = \bar{B}/(4\pi\bar{\rho})^{1/2}$ which by (11.48) is close to the typical speed of the turbulence that is responsible for the longitudinal support against gravity, whether the cloud is moderately or highly oblate about $\mathbf{B}_0$ (Shu *et al.* 1987; Myers and Goodman 1988). Thus at this stage of the discussion the term 'Alfvénic turbulence' is almost tautological. Equivalently, (11.46) yields

$$\bar{B} \simeq (1.5v_t^2/G^{1/2}Z). \qquad (11.49)$$

As noted in Section 11.1, Myers and Goodman (1988) presented encouraging observational support for this picture of combined magnetic and turbulent support of molecular clouds. They summarized data from 14 clouds where Zeeman splitting and line-widths can be accurately measured. As the observed field strengths are greater than the standard value of 3×10^{-6} G for the background galactic field B_0, it is plausible to take the clouds to have $f \simeq f_c$, so that the theoretical estimate $\bar{B} \simeq G^{1/2}M/R^2$ is appropriate. The clouds are assumed to be kept moderately oblate by kinetic support along the field, and as the observed widths Δv (yielding v_t) are easier to measure than the column density $\bar{\rho}Z$, the theoretically predicted field was estimated from (11.49). The comparison with the observed $\bar{B}$ values was found to be satisfactory (to a factor 2) over a range from 10^{-5} to 10^{-2} G both for different clouds and for three subregions within the Orion cloud. The kinetic condition (11.48) was used to estimate $\bar{\rho}$. Comparison of (11.47) with observation seemed to show clearly the dependence on $\bar{\rho}$, but the effect of $v_{\rm t}$ was within the observational scatter. It remains to be seen whether this tentative observational support will persist (cf. Section 12.1).

11.4 Gravitational collapse and fragmentation under flux-freezing

Consider again the spherical model of Fig. 11.1(b), but now suppose the cloud so massive that the forces exerted by the distorted magnetic field are never strong enough to prevent gravitational collapse. The trans-$\mathbf{B}_0$ components of the tensorial virial theorem yield instead of (11.10) the inequality

$$\frac{GM^2}{5R} > \frac{4F^2}{15\pi^2 R}\left(1 - \frac{R}{R_0}\right) + Ma^2\left(1 - \frac{p_e}{\bar{\rho}a^2}\right) \tag{11.50}$$

with $\bar{\rho} = M/(4\pi R^3/3)$. For the moment the artificial problem is studied with the cloud supposed remaining spherical during the collapse, so that the numerical coefficients of the terms in (11.50) are unaltered. Then when $R \ll R_0$, (11.50) can be written again in terms of $f \equiv F/\pi^2 G^{1/2} M$:

$$\left(1 - \frac{f^2}{f_c^2(0)}\right) > \frac{5a^2 R}{GM}\left(1 - \frac{p_e}{\bar{\rho}a^2}\right) \tag{11.51}$$

where $f_c(0) = 3^{1/2}/2\pi$ (the value of expression (11.21) when $e = 0$). The external pressure p_e assists the contraction, but its effect is progressively less important as $\bar{\rho}$ increases. Provided $a^2 R \to 0$ with R, then we recover the condition $f < f_c(0)$ for the indefinite collapse of a cold spherical body with prescribed flux F. This condition can be written alternatively as a lower limit on R for given $\bar{B}$, $\bar{\rho}$:

$$R > \lambda_B = \frac{3^{1/2}}{2\pi G^{1/2}}\left(\frac{\bar{B}}{\bar{\rho}}\right) \tag{11.52}$$

with the associated minimum mass

$$M = 4\pi\bar{\rho}R^3/3 > \frac{3^{1/2}}{2\pi^2 G^{3/2}}\left(\frac{\bar{B}}{\bar{\rho}^{2/3}}\right)^3. \tag{11.53}$$

The length λ_B is again of the order of the 'magnetic Jeans length' for small-amplitude waves in a cold medium, propagating across the field $\bar{\mathbf{B}}$. As already noted, the condition for indefinite collapse sets λ_B as a lower limit in all three dimensions.

If the collapsing cloud were to remain spherical, then with f less than but close to $f_c(0)$, no subcondensations could separate out at any epoch, as long as flux-freezing holds. This follows because a subsphere of radius $R' = \lambda R$ $(\lambda < 1)$ would contain mass $M' = \lambda^3 M$ and flux $F' = \lambda^2 F$, so yielding the parameter $f' = f/\lambda > f \simeq f_c(0)$, whereas for this subsphere to be able to separate out of the cold spherical cloud, the relation (11.51) would have to hold with $M = M'$ and the radius $\ll R'$, requiring $f' < f_c(0)$. This result is an example of the

'$\gamma = 4/3$' behaviour of magnetic forces under isotropic contraction: if all length-scales change by the same factor k, mass conservation yields $\bar{\rho} \propto 1/k^3$ and flux-freezing $\bar{B} \propto 1/k^2$, whence $\bar{B}^2 \propto \bar{\rho}^{4/3}$, and so the minimum mass that can separate out of the medium – given by (11.53) – remains invariant. The arguments apply equally (with slight variations in the numerical factors) to a presumed isotropic contraction of the more realistic oblate model of Fig. 11.2.

We have already noted that isotropic contraction of a sphere or an oblate spheroid requires an effective sound/turbulent speed a increasing like $\bar{\rho}^{1/6}$, implying a turbulent pressure that like the magnetic pressure $\bar{B}^2/8\pi$ increases like $\bar{\rho}^{4/3}$. In order to bring out the crucial importance for the fragmentation problem of the collapse geometry, we now study instead the case of cool, isothermal clouds which are weakly turbulent, with the effective sound speed of the same order as the real speed. Again a subcritical value for the flux/mass parameter $f < f_c(0)$ is assumed, so that the magnetic forces are unable to prevent gravitational collapse. Even if initially spherical, the cloud will spontaneously flatten, at first just because of the essential anisotropy of the magnetic forces, subsequently because the gravitational field in a spheroidal body acts always to increase the eccentricity, attempting to turn an oblate spheroid into a 'pancake' and a prolate spheroid into a 'cigar' (Lin *et al.* 1965). Following on from the last section and Appendix A, we again model the now contracting isothermal cloud as an oblate spheroid $\bar{S}$ of semi-axes (R, Z) and uniform density $\bar{\rho}$, which steadily adjusts to the approximate hydrostatic condition (11.40) along the field as R decreases. The discussion of Section 10.2 implies that the conditions for indefinite collapse will be satisfied a fortiori after flattening: from (11.18), (11.21), (11.22) and (11.140), the analogue of (11.51) for $\eta \equiv R/R_0 \ll 1$, $e \simeq 1$ and p_e negligible is

$$\left(1 - \frac{f^2}{f_c^2}\right) > \frac{5a^2R}{Q(1)GM} = \frac{20a^2R}{3\pi GM} \tag{11.54}$$

with $f_c^2(1) = 3.2 f_c^2(0)$, which indeed increases the upper limit on f. More significantly, from (11.43) $\bar{B} \propto \bar{\rho}^{1/2}$, as the contraction continues. With this weaker $\bar{B}$-$\bar{\rho}$ dependence – which follows from preferential flow down the field lines – formula (11.53) yields a minimum spherical mass now decreasing like $\bar{\rho}^{-1/2}$ as $\bar{\rho}$ increases, suggesting that even though flux-freezing persists, the magnetic forces, acting alone, do not necessarily prevent fragmentation.

This process of 'magnetic fragmentation' is illustrated in Fig. 11.4. Suppose the contraction of the cloud as a whole to be temporarily frozen at radius R and semi-minor axis Z, and consider a subspheroid S' of semi-axes (R', Z) centred on the midpoint of S, of mass $M' = 4\pi\bar{\rho}ZR'^2/3 = M(R'/R)^2$ and containing magnetic flux $F' = \pi\bar{B}R'^2 = F(R'/R)^2$, so that its flux/mass parameter $f' = F'/\pi^2G^{1/2}M' = f$. Imagine this spheroid attempting to condense out of $\bar{S}$, first by contracting laterally to form a sphere of radius $r' = Z$, and then by isotropic contraction, so that Fig. 11.1(b) is again an appropriate representation of the field local to S'. At radius $r' = Z$, a necessary condition for the condensation to be able to continue contracting gravitationally against the forces exerted by the

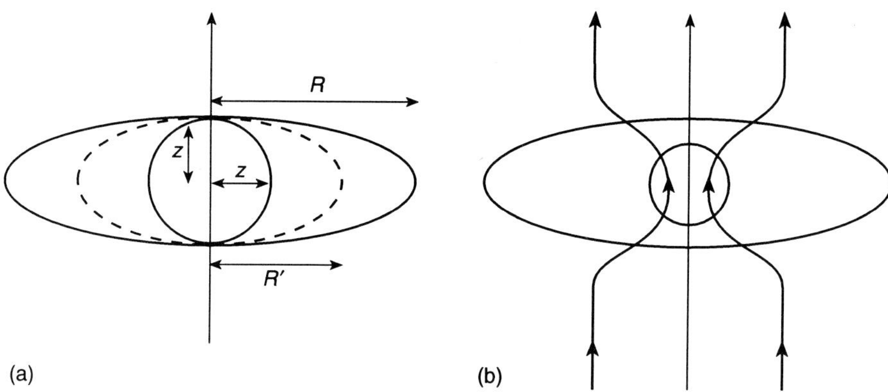

FIG. 11.4. Illustration of possible fragmentation of a cloud with subcritical f.
(a) Subspheroid (R', Z) contracting into sphere of radius $r' = Z$.
(b) Subsequent isotropic contraction, generating magnetic field as in Fig. 11.1(b).

locally distorted field is the analogue of (11.51) (with the p_e term dropped):

$$\left(1 - \frac{f'^2}{f_c^2(0)}\right) = \left(1 - \frac{4\pi^2 f'^2}{3}\right) > \frac{5a^2 Z}{GM'} = 3\left(\frac{Z}{R'}\right)^2 \qquad (11.55)$$

on substitution of $M' = 4\pi\bar{\rho}ZR'^2/3$ and of a^2 from (11.40). This sets a lower limit on R' and hence on M' in terms of Z, which in turn depends on the effective sound speed a. Thus with f given, the minimum spherical mass M' allowed by the virial theorem to separate out of the oblate cloud $\bar{S}$ of semi-major axis R is

$$\begin{aligned} M_{\min} &= (4\pi\bar{\rho}Z^3/3)(R'/Z)^2 = 3(4\pi\bar{\rho}ZR^2/3)(Z/R)^2/(1 - 4\pi^2 f^2/3) \qquad (11.56) \\ &= 3(Z/R)^2 M/(1 - 4\pi^2 f^2/3) = 3(5a^2 R/3GM)^2 M/(1 - 4\pi^2 f^2/3) \end{aligned}$$

by (11.40) and (11.41). As an example, adopt parameters appropriate to a typical cool massive molecular cloud: $M \simeq 10^4 M_\odot$, $T = 15$ K, $a^2 \simeq 6.2 \times 10^8$, $R \simeq 10^{19}$ cm, whence $Z \simeq 3 \times 10^{16} = (3 \times 10^{-3})R$, $\bar{\rho} = 10^6 m_H$, and $M_{\min} \simeq 0.23 M_\odot/(1 - 4\pi^2 f^2/3)$.

The crucial point in the argument is that after flattening, the ratio f' for the subspheroid has the same value f as for the spherical cloud. As discussed earlier, the condition that magnetic forces alone are unable to prevent indefinite contraction of a spherical (or oblate spheroidal) mass is that f is not too large, and this can be written as imposing a lower limit on the necessary accumulation length, *along* the magnetic field, of gas with a given volume density, but with no constraint perpendicular to the field (cf. (11.35) and (11.36)). Thus if the spherical cloud has a flux/mass parameter f less than the critical value

$f_c(0) = 3^{1/2}/2\pi$, the process of flattening along the field necessarily accumulates into a subsphere or subspheroid enough mass for $f' = f$ also to be subcritical. The condition (11.56) shows that in a spherical mass with f subcritical, the gross effect of the magnetic forces is to dilute gravity by the positive factor $1 - f^2/f_c^2(0) = 1 - 4\pi^2 f^2/3 < 1$. The finite thermal energy limits the spontaneous flattening of the cloud to the thickness Z, of the order of the Jeans length (11.3) for a medium of density $\bar{\rho}$ and temperature T. If $f \ll f_c(0)$, then the same analysis predicts $M_{\min}$ close to the Jeans mass (11.4) (cf. Ledoux 1951). If $f/f_c(0)$ is less than unity but not negligible, $M_{\min}$ is increased by the factor $(1-f^2/f_c^2(0))^{-1}$, so remaining of the same order as in the non-magnetic problem, unless $f/f_c(0)$ is very close to unity.

As the hypothetical spherical subcondensation contracts further, increasing its density ρ_c but with its temperature staying constant (or increasing less strongly than $\rho_c^{1/3}$), then we can picture a continuing process of spontaneous flattening and further fragmentation. It must be emphasized that as with the non-magnetic problem outlined in Section 11.1, the virial condition is no more than a censor: it can indicate when collapse and subsequent fragmentation are allowed, but a more subtle dynamical discussion is required to show that it will occur. Nevertheless, one can conclude that a large-scale magnetic field is in itself an insuperable barrier *neither to the collapse nor to the fragmentation of a slowly rotating cloud.* Whether the cloud is maintained in magneto-gravitational equilibrium or whether it can be brought into indefinite collapse under magnetically-diluted gravity is determined by the parameter $f = F/\pi^2 G^{1/2} M$; and when f is subcritical, and also if spontaneous flattening occurs along the field, gravitationally bound subcondensations can form.

The argument must not be overstated. It is not claimed that a disc-like body *in magneto-gravitational equilibrium* must be unstable to the gravitational separation-out of subspheres or subspheroids. In fact, the virial arguments have shown that the critical value f_c is larger by a factor ≈ 2 for a flattened body than for a sphere, implying that M_c corresponding to a given flux F decreases under flattening by the same factor. Inverting the argument: for a subsphere to condense out of a disc in equilibrium while conserving its flux, it would need to have the mass/flux ratio appropriate to a *sphere.* After a modest degree of contraction of the subsphere, the forces exerted by the locally distorted field would exceed gravity – the sphere would find that it has too little mass for indefinite collapse, in agreement with the conclusions of Shu and Li (1997) and Li and Shu (1997). What is being argued is that if a spherical cloud is supercritical in mass/flux, implying that the real or notional accumulation length exceeds the critical value for a *sphere*, then the same criterion will hold for a subsphere forming by further accumulation of gas down the field. With the parent sphere supercritical, a fortiori the disc is supercritical and would undergo two-dimensional collapse. To show that the subsphere does actually separate out – i.e. that its projected density runs systematically ahead of that of the disc as a whole – one would need to follow the dynamical evolution of the system into the non-linear domain, analogously to the non-magnetic problem outlined in Section 11.1.

The picture with the whole cloud flattening into an oblate spheroid is clearly highly idealized but is in fact not necessary for the argument. A significant point is that as in the problem without magnetism, at a given density $\bar{\rho}$, the minimum masses predicted in this geometry are normally fixed by the temperature of the gas, which remains low until the gas becomes opaque. If fragmentation occurs with maximum efficiency, it appears that the magnetic field, *acting alone*, is not going to resolve the difficulty noted in Section 11.1 that the resulting masses are embarrassingly small. In subsequent sections we discuss how this conclusion could be altered radically if a significant turbulent field is maintained, and if the cloud has a realistic angular momentum.

It should be noted that condensations forming in this way will necessarily be 'strongly magnetic'. The condition $f \simeq f_c$ implies magnetic and gravitational energies comparable, and the process just outlined yields $f' \simeq f_c$ for the successive fragments forming by gravitational instability following flattening. As emphasized in Section 5.1, all observed stellar fields are of energy much below the virial upper limit. It is quite unacceptable to explain this by appealing once more just to anisotropic condensation from the warm intercloud medium. The minimum accumulation length $(L_0)_c$ – given by (11.36) – for gravitational collapse to be possible is already uncomfortably (though not outrageously) long; to produce 'weakly magnetic' bodies under strict flux-freezing it would need to be far longer. Thus suppose the gas that ultimately forms a star of density ρ^*, radius R^* and mean field strength B^* originally formed a prolate spheroid along $\mathbf{B}_0$ of major axis L_0 and semi-minor axis r_0. Under strict flux-freezing, $R^*/r_0 = (B_0/B^*)^{1/2}$, and from mass conservation $2\rho^* R^{*3} = \rho_0 L_0 r_0^2$, yielding

$$\begin{aligned} L_0 &= 2(\rho^*/\rho_0)(B_0/B^*)R^* \qquad (11.57) \\ &= (1.2\times10^{11}/B^*)(\rho^*/1)(B_0/3\times10^{-6})(R^*/10^{11})/(n_0/1) \quad \text{pc.} \end{aligned}$$

Then even if $B^* = 3\times10^6$ G – about the strongest internal field suggested by some recent studies of magnetic stars (cf. Chapter 9) – L_0 would need to be $\simeq 40$ kpc. One can invert the argument by inserting a tolerable value for L_0 so as to deduce R^* and M^* in terms of B^*: for example, with $L_0 \simeq 1200$ pc, then again the choice of $\rho^* \simeq 1$, a typical stellar value, and $B^* \simeq 3\times10^6$ G in (11.57) fixes R^* at 3×10^9 cm and the mass M^* at $6\times10^{-5}M_\odot$ – again a clear *reductio ad absurdum*. It is in fact advantageous if the magnetic flux within successive condensing fragments remains near the Jeans virial limit well into the opaque phase (cf. Sections 11.5 and 11.6), but there is no doubt that most of the primeval flux must be lost at some epoch before arrival on the main sequence.

These points have to be emphasized because a spurious argument has appeared in the literature, claiming that one does not need deviation from flux-freezing to account for the weakness of the fields present in main sequence stars: the application of a law like $B \propto \rho^{1/2}$ will yield modest B-values correlated with ρ-values of main-sequence order. It should now be clear that the argument is fallacious, through its failure to consider the masses of the roughly spherical bodies that are supposed to form by preferential flow of gas down the field lines, without any accompanying flux leakage. We have seen that if the ratio of magnetic to gravitational energy is to be as small as inferred from observation, the accumulation length must be ludicrously long. Equally, if instead a roughly spherical body forms from a parent cylindrical cloud, with the accumulation length again greater than but of the order of the critical value (11.36), then the gravitational energy must necessarily be of the same order as the magnetic, whatever the mass of the sphere; hence if the sphere is supposed to have simultaneous (B, ρ) values equal to those in the Sun, its mass would need to be far below a stellar value, implying a correspondingly small trans-field radius of the accumulating cylindrical mass.

11.5 The angular momentum problem

The problems for star formation set by a realistic angular momentum are more severe than those due to the galactic magnetic field. Consider the simplest possible example – a cool, spherical cloud of mass M, radius R, uniform density ρ, and rotating with uniform angular velocity Ω . The angular momentum $H = I\Omega = 2MR^2\Omega/5$, the rotational kinetic energy $T = I\Omega^2/2 = H^2/2I = 5H^2/4MR^2$, and the gravitational energy $\mathcal{V} = -3GM^2/5R$. If, initially, the ratio

$$\epsilon \equiv 2T/|\mathcal{V}| = 2\Omega^2R^3/3GM = \Omega^2/2\pi G\rho = 25H^2/6GM^3R \tag{11.58}$$

has the small value ϵ_0, then the initial collapse of the cloud will be nearly spherically symmetric, with ρ increasing like $1/R^3$. If also no torques are exerted on the cloud, H stays constant, $\Omega \propto 1/R^2 \propto \rho^{2/3}$ and so $\epsilon \propto \rho^{1/3}$. Since under adiabatic compression, the thermal energy per gram of a perfect gas increases like $\rho^{\gamma-1}$, one can thus say that under isotropic contraction, angular momentum conservation implies a 'centrifugal γ' = 5/3.

The steady increase in ϵ implies that contraction in two dimensions must ultimately be slowed up by the centrifugal forces, so that the cloud spontaneously flattens parallel to the rotation axis. As an example, take as initial values $\Omega_0 = 10^{-15}\ \mathrm{s}^{-1}$ – of the order of the local orbital angular velocity about the galactic centre – and $\rho_0 \simeq 100m_H\ \mathrm{cm}^{-3}$; then $\epsilon_0 \simeq 1/40$, and isotropic contraction by a factor 40 would yield $\epsilon = 1$ when $\rho \simeq 6 \times 10^6 m_H\ \mathrm{cm}^{-3}$. Because of the interstellar turbulence, a cloud brought to the verge of gravitational collapse could have an angular momentum some ten times higher, so that the departure from initially spherical collapse would occur at densities still further below protostellar densities.

Suppose for simplicity that after flattening and dissipation of the gravitational energy released, the cloud achieves equilibrium as a spheroid of uniform angular velocity $\bar{\Omega}$, uniform density $\bar{\rho}$, eccentricity $\bar{e} \simeq 1$, and semi-axes (R, Z). On the equator at $(\varpi, 0)$ the gravitational field is $\simeq (3\pi/4)GM\varpi/R^3$, so that centrifugal balance is achieved with $\bar{\Omega} = (3\pi/4)^{1/2}(GM/R^3)^{1/2} = \pi(G\bar{\rho}^{1/2}(Z/R)^{1/2}$, and with R less by the factor 4/3 than the radius at which centrifugal balance holds in a sphere with the same mass and angular momentum. The semi-minor axis Z is from (11.40) given approximately in terms of the sound speed a by $a/(4\pi G\bar{\rho}/5)^{1/2}$, so that

$$\frac{Z}{R} = \frac{5Ra^2}{3GM} \ll 1 \tag{11.59}$$

in a cool cloud. In terms of the density ρ_s of a sphere of the same mass and of radius R,

$$\bar{\rho} = \rho_s(R/Z) = (3GM/5Ra^2)(3M/4\pi R^3). \tag{11.60}$$

Such a highly flattened, disc-like structures will be Jeans unstable. Infinite, plane-stratified isothermal models (Spitzer 1942; Ledoux 1951) were shown by Ledoux to be unstable against plane or cylindrical perturbations, the minimum unstable wavelength being the Jeans length corresponding to one-half the mid-plane density, which is somewhat greater than the disc thickness. Provided the ratio $\bar{\Omega}^2/(4\pi G\bar{\rho}/3) = (3\pi/4)(Z/R) = (\pi/2)(Ra^2/GM)$ is small, rotation does not interfere with the instability (Toomre 1964), but equally there is nothing in the Jeans-type analysis to violate detailed angular momentum conservation by axisymmetric modes. As the simplest example, consider as a potential fragment a subsphere located at the cloud centre with radius of the order of the semi-thickness Z, so that its mass $M_f = 4\pi\bar{\rho}Z^3/3 = M(2Ra^2/3GM)^2$. The ratio of centrifugal force of *spin* to the sphere's self-gravitation is

$$\frac{\bar{\Omega}^2 Z}{(4\pi G\bar{\rho}Z^3/3)/Z^2} = \frac{\bar{\Omega}^2}{4\pi G\bar{\rho}/3} = \left(\frac{3\pi}{4}\right)\left(\frac{Z}{R}\right) \equiv \bar{\epsilon} \ll 1. \tag{11.61}$$

If this sphere is able to break free of the gravitational field of the rest of the cloud and contract, conservation of its angular momentum of spin will again cause centrifugal force to increase more rapidly than self-gravitation. If the fragment remains spherical, centrifugal balance will be reached at radius $r = \bar{\epsilon}Z = 3\pi Z^2/4R$, with the spin $\Omega = \bar{\Omega}Z^2/r^2 = \bar{\Omega}/\bar{\epsilon}^2 = (16/9\pi^2)(R/Z)^2\bar{\Omega} = \bar{\Omega}(2GM/Ra^2)^2$, and the density given by $\bar{\rho}(Z/r)^3 = \bar{\rho}(4/3\pi)^3(R/Z)^3$. Again, if the fragment is sufficiently cool it will flatten, and smaller subcondensations may separate out, but they in turn will run into the same difficulty: conservation of angular momentum sets an automatic limit to the density that can be reached by the contraction of a given mass, while flattening followed by fragmentation systematically reduces the masses of the fragments that reach stellar densities.

As a *reductio ad absurdum*, consider a main sequence star of mean density ρ rotating near the centrifugal limit, so that $\Omega \simeq (G\rho)^{1/2}$, and the angular momentum per unit mass $\simeq \Omega R^2 \simeq (G\rho)^{1/2}(M/\rho)^{2/3} = G^{1/2}M^{2/3}\rho^{-1/6}$. If the gas that formed the star has accumulated from the interstellar medium of density ρ_0, angular velocity Ω_0, along a cylinder of radius d and length l parallel to the rotation axis, then $M \simeq \rho_0 l d^2$, the angular momentum per unit mass $\simeq \Omega_0 d^2$, and $\rho_0 l = M/d^2 = (M\Omega_0)/G^{1/2}M^{2/3}\rho^{-1/6} = \Omega_0\rho^{1/6}M^{1/3}/G^{1/2}$. With $\rho_0 \simeq 1$ gm cm^{-3}, $\Omega_0 \simeq 10^{-15}$, $\rho_0 = 1m_H$ cm^{-3} and $M = M_\odot$, then $l = 10^2$ kpc (Hoyle 1945). An increase in Ω_0 would make l even longer, while accumulation from matter with a larger 'lever arm' d would reduce M below stellar order.

The contrast with the problems set by frozen-in magnetic flux is striking. As discussed in Sections 11.2 and 11.3, once a supercritical cloud (of any mass) has formed, for example by flow down the galactic field lines of a length of gas exceeding the minimum accumulation length (11.36), then during further isotropic contraction the mean force exerted by the distorted magnetic field increases at the *same rate* as the gravitational – the 'magnetic γ' = 4/3. Again, spontaneous flattening of a cool magnetic cloud will normally occur, with likely consequent fragmentation, but there is no limit set by the magnetic field to the indefinite isotropic contraction of a subcondensation, whatever its mass, provided always that $G^{1/2}M/F$ is supercritical. And even if the length (11.35) were to be taken literally rather than notionally (i.e. if turbulent diffusion across the field during cloud formation is ignored – cf. Section 11.3), the value $(L_0)_c = 1$ kpc, though long, is not absurdly so, as is the 10^2 kpc just deduced.

The discussion naturally bifurcates at this point. One is interested in accurate models of non-magnetic clouds in equilibrium under self-gravitation, centrifugal acceleration, and internal and external thermal pressure. Such isothermal analogues of the Ebert–Bonnor–McCrea models (discussed in Section 11.1) have been constructed by Kiguchi *et al.* (1987), for four different distributions of angular momentum. Results of particular interest are: if the models are to be stable against either global contraction and expansion or ring formation, the mass cannot exceed 31 times the E–B–M maximum for a non-rotating cloud; the mean density cannot exceed six times the surface density (fixed by pressure balance with the warm external medium); and the mean rotation velocity cannot exceed $2.7a$. Steady increase in the external pressure (or decrease in a) causes either non-homologous gravitational collapse or axisymmetric ring formation. As long as there are no departures from axisymmetry, and if there is no internal friction, these instabilities will develop retaining the essential properties of the approximate discussion above: each element of gas will conserve its angular momentum, and the 'angular momentum problem' will persist. Non-axisymmetric instabilities such as bar formation in a strictly friction-free system may cause some interchange of angular momentum via gravitational torques, but the conservation of vorticity (Kelvin's theorem) remains a severe constraint, which is however relaxed in models involving spiral shocks (Spruit 1987).

One can argue that in a non-magnetic cloud, the whole picture in which stars form by accumulation of local matter within a cloud is misconceived. For example, in McCrea's model (1960) of a turbulent cloud, the very process by which the turbulence decays – mutual collision and adherence of 'floccules' moving with random supersonic speeds – causes efficient redistribution of gas; there is an effective turbulent friction which tends to accumulate gas in the centre, with most of the angular momentum carried by only a fraction of the mass. More conventional disc-like models appeal to an explicit turbulent friction with a variety of physical origins. It is indeed important to try and follow the evolution of weakly magnetic rotating clouds, if only because the galactic field may have taken a few times 10^8 yr to be built up by dynamo action (Zel'dovich *et al.* 1983), so that star formation in old Population II and hypothetical Population III systems may have occurred without the intervention of magnetic fields (Rees 1987, 1994). It is certainly premature to claim that star formation cannot occur unless a magnetic field is present, capable of dealing with the angular momentum problem; however, once a field of the observed strength has been built up, its interaction with the rotation of the gas should be as significant as its role in the equilibrium and contraction of gas clouds, as already discussed.

As an illustration, make the most optimistic assumption and suppose that magnetic coupling between a cool, highly oblate cloud or fragment and its surroundings is so efficient that contraction from an initial radius R_0 to a smaller radius R occurs not under conservation of angular momentum, but instead with constant angular velocity Ω_0. The ratio of centrifugal acceleration to the opposing component of gravity at a radius $\bar{R}$ is then

$$\alpha(\bar{R}) \equiv \frac{4\Omega_0^2\bar{R}^3}{3\pi GM} = \left(\frac{\bar{R}}{R_0}\right)^3 \frac{4\Omega_0^2 R_0^3}{3\pi GM} \equiv \left(\frac{\bar{R}}{R_0}\right)^3 \alpha_0. \tag{11.62}$$

For $R < \bar{R}$, magnetic coupling is supposed much weaker, for example because of significant flux leakage, or detachment of the cloud field from the background, so further contraction occurs with angular momentum conserved, yielding a steadily increasing ratio

$$\begin{aligned}\alpha(R) &= \frac{4\Omega^2 R^3}{3\pi GM} = \frac{4}{3\pi}\left(\frac{\Omega^2 R^4}{GM}\right)\frac{1}{R} = \frac{4}{3\pi}\left(\frac{\Omega_0^2 \bar{R}^4}{GM}\right)\frac{1}{R} \\ &= \left(\frac{\bar{R}}{R_0}\right)\alpha(\bar{R}) = \left(\frac{\bar{R}}{R_0}\right)^4 \alpha_0 \left(\frac{R_0}{R}\right)\end{aligned} \tag{11.63}$$

by (11.62), so that centrifugal balance is reached at the radius R given by

$$\frac{R}{R_0} = \left(\frac{\bar{R}}{R_0}\right)^4 \alpha_0. \tag{11.64}$$

If the disc-like structure is maintained at fixed a, then by (11.42), the associated density $\bar{\rho}$ is given by $\bar{\rho}/\rho_0 = (R_0/\bar{R})^4 = (R_0/\bar{R})^{16}/\alpha_0^4$; for example, if $\alpha_0 \simeq 1$, $\rho_0 = 10m_H$, and $\bar{R}/R_0 = 0.1$, then $\bar{\rho} = 10^{17}m_H$. This should not be taken literally, for before such densities are reached the gas will have become opaque; the trapped thermal energy would force the cloud to become roughly spherical, with a density $\bar{\rho}/\rho_0 = (R_0/\bar{R})^3 = (R_0/\bar{R})^{12}/\alpha_0^3$, yielding for the same initial parameters the value $\bar{\rho} \simeq 10^{13}m_H$ – still well into the opaque domain, though below main-sequence densities.

These numbers reflect that the assumption of contraction down to $\bar{R}$ with $\Omega = \Omega_0$ implies a high efficiency of angular momentum transport which may be exaggerated; nevertheless, they are a strong incentive to study magnetic braking in depth.

11.6 Magnetic braking by Alfvén waves

11.6.1 *An axisymmetric cylindrical model*

To minimize mathematical difficulties, we begin by imposing axial symmetry, making both the cloud and its surroundings rotate about an axis parallel to the local galactic field $\mathbf{B}_0$ (although this is probably not the most plausible geometry). As in Section 11.2, the cloud is idealized as an oblate spheroid of density $\bar{\rho}$, semi-axes R and Z, and mass $M = 4\pi\bar{\rho}R^2Z/3$; the cloud surface is $(\varpi, \tilde{z})$, with $\tilde{z} = \pm Z(1 - \varpi^2/R^2)^{1/2}$. In the simplest example (Fig. 11.5) the poloidal field is identical with the uniform background field $\mathbf{B}_0$, both within the cloud and in the surrounding medium of density ρ_0, and the braking problem can be discussed in terms of simple torsional Alfvén waves (Ebert 1960; Section 5.2.1).

At any time, in the medium ρ_0 the angular velocity Ω and the toroidal component B_ϕ generated by the shear are related by

$$\rho_0\varpi\frac{\partial\Omega}{\partial t} = B_0\frac{\partial}{\partial z}\left(\frac{B_\phi}{4\pi}\right), \tag{11.65}$$

$$\frac{\partial B_\phi}{\partial t} = \varpi B_0\frac{\partial\Omega}{\partial z}, \tag{11.66}$$

yielding the standard one-dimensional wave equation

$$\frac{\partial^2\Omega}{\partial t^2} = v_\mathrm{A}^2\frac{\partial^2\Omega}{\partial z^2}, \qquad v_\mathrm{A} = \frac{B_0}{(4\pi\rho_0)^{1/2}}, \tag{11.67}$$

valid outside the cloud within the cylinder $\varpi = R$. Inside the cloud, identical equations hold with $\bar{\rho}$ replacing ρ_0 and $\bar{v}_\mathrm{A} = B_0/(4\pi\bar{\rho})^{1/2}$ replacing v_A. To illustrate spin-down of the cloud, we adopt the simplest initial conditions: outside the cloud $\Omega = \Omega_0$; inside the cloud $\Omega = \bar{\Omega}$; and everywhere the field is untwisted, i.e. $B_\phi = 0$. The shear at the surface $\tilde{z}(\varpi)$ generates torsional Alfvén waves along each field line, propagating both away from and into the cloud. The ongoing adjustment of the density and the poloidal field to the changing forces exerted by the toroidal field is ignored; and likewise each poloidal field line is treated independently of the others. Because of the equatorial symmetry we need consider only $z \geq 0$. The boundary condition at $z = \infty$ is clearly that there be no incoming wave, so the appropriate solution of (11.67) for $z > \tilde{z}$ is

$$\Omega = f(z - v_\mathrm{A}t) \tag{11.68}$$

with

$$B_\phi = -\varpi(4\pi\rho_0)^{1/2}f(z - v_\mathrm{A}t) + g(\varpi). \tag{11.69}$$

From the initial conditions, $f(u) = \Omega_0$ for $u > \tilde{z}$, $f(\tilde{z}) = \bar{\Omega}$, and

$$g(\varpi) = \varpi(4\pi\rho_0)^{1/2}f(z) = \varpi(4\pi\rho_0)^{1/2}\Omega_0, \tag{11.70}$$

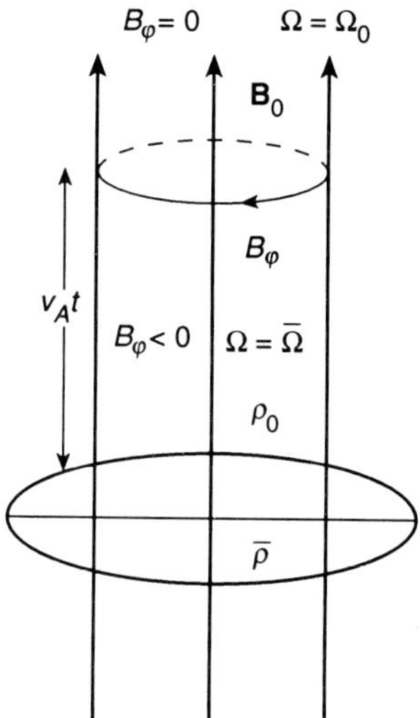

FIG. 11.5. Illustrative model of magnetic braking. A uniform field $\mathbf{B}_0$ permeates the rotating oblate cool cloud of density $\bar{\rho}$ and the surrounding warm medium of density ρ_0. Torsional Alfvén waves generated by the shear at the cloud surface transport angular momentum to infinity. (After Ebert 1960.)

so

$$B_\phi = -\varpi(4\pi\rho_0)^{1/2}[f(z - v_A t) - \Omega_0]. \tag{11.71}$$

Hence at any point z, by (11.68), (11.70) and (11.71), $\Omega = \Omega_0$, $B_\phi = 0$ as long as $(z - \tilde{z}) > v_A t$ – i.e. as long as the wave has not had time to travel from $\tilde{z}$ to the observation point z; but when $z = \tilde{z} + v_A t$, Ω jumps to $\bar{\Omega}$ and B_ϕ to $-(4\pi\rho_0)^{1/2}(\bar{\Omega} - \Omega_0)$.

A complete study involves following also the propagation of waves into the cloud and their repeated partial reflections at $z = \tilde{z}$. However, the time in which a wave crosses the cloud is

$$2\tilde{z}\big/v_A = 3\pi^{1/2}(M/F)(\tilde{z}/Z)(\rho_0/\bar{\rho})^{1/2}\big/\rho_0^{1/2}, \tag{11.72}$$

where again $F = \pi B_0 R_0^2$ is the flux through the cloud. We may anticipate that with M prescribed, when $\rho_0/\bar{\rho}$ is small this time will be smaller than the characteristic braking time. It is then reasonable to treat the cloud as having along each internal field line a uniform angular velocity $\bar{\Omega}(t)$ which decreases steadily with time from the initial value $\bar{\Omega}(0) \equiv \bar{\Omega}$. The boundary condition to be applied is then

$$\bar{\Omega}(t) = \Omega(\tilde{z}, t) = f(\tilde{z} - v_A t) \qquad t > 0 \tag{11.73}$$

so that condition (11.70) is supplemented by

$$f(u) = \bar{\Omega}(t) \qquad u < \tilde{z}. \tag{11.74}$$

The instantaneous value of $\bar{\Omega}(t)$ and the associated value (11.71)) of B_ϕ are thus propagated from the cloud with the Alfvén speed; at time t, $\Omega(z,t) = f(z - v_{\mathrm{A}}t) = f(\tilde{z} - v_{\mathrm{A}}t') = \bar{\Omega}(t')$ and $B_\phi(z,t) = -\varpi(4\pi\rho)^{1/2}(\bar{\Omega}(t') - \Omega_0)$, where $t' \equiv t - (z - \tilde{z})/v_{\mathrm{A}} > 0$. By (2.48), the magnetic torque transports per second across unit area normal to $\mathbf{B}_0$ the angular momentum $-\varpi B_\phi(\mathbf{k}\cdot\mathbf{B}_0)/4\pi = \varpi^2(4\pi\rho_0)^{1/2}B_0(\bar{\Omega}(t') - \Omega_0)/4\pi$.

In this strictly axisymmetric problem as formulated there is no coupling between different field lines, so the braking process would set up shears between field lines with differing values of $\tilde{z}$. We prefer to accept that in reality there would be sufficient internal coupling to keep the cloud rotating more or less uniformly; we therefore determine $\bar{\Omega}(t)$ for the whole cloud by writing

$$\begin{aligned} -\frac{\mathrm{d}}{\mathrm{d}t}(2MR^2\bar{\Omega}/5) &= \text{efflux of angular momentum per second} \\ &= 2B_0(4\pi\rho_0)^{1/2}(\bar{\Omega}(t) - \Omega_0)\int_0^R \varpi^2(2\pi\varpi\,\mathrm{d}\varpi)/4\pi \\ &= FR^2\rho_0^{1/2}(\bar{\Omega}(t) - \Omega_0)/2\pi^{1/2}, \end{aligned} \tag{11.75}$$

whence

$$(\bar{\Omega}(t) - \Omega_0) = (\bar{\Omega} - \Omega_0)\exp(-t/t_b) \tag{11.76}$$

with

$$t_b = 4\pi^{1/2}M/5F\rho_0^{1/2}. \tag{11.77}$$

In the time t_b the waves travel a distance $v_{\mathrm{A}}t_b = (16Z/15)(\bar{\rho}/\rho_0)$, defining a volume with about the same moment of inertia as the cloud. The ratio of t_b to the travel time of waves along the semi-minor axis of the cloud is $(4/15)(\bar{\rho}/\rho_0)^{1/2}$, showing that only if $\bar{\rho}/\rho_0 \gg 1$ will the treatment that implicitly averages over the internal wave propagation be very accurate; however, detailed treatments (Mestel and Paris 1980; Mouschovias and Paleologou 1980) confirm that the braking time is of the physically reasonable form (11.77) for all values of $\bar{\rho}/\rho_0$.

11.6.2 *Braking by a radially distorted field*

The simple example of Section 11.6.1 is in fact not a realistic model of the early phases of star formation. The adoption of an internal field $\mathbf{B}$ identical with the background field $\mathbf{B}_0$ implies that the cloud – formed by accumulation of gas down $\mathbf{B}_0$ – has insufficient mass to cause any noticeable distortion of the field: the parameter f is well above the critical value f_c for the particular cloud geometry, so that not only is the cloud unable to contract significantly in the two trans-field dimensions, but neither can any subcondensations separate out (cf. Section 11.4). If instead f is close to (though still greater than) f_c, then as discussed above, equilibrium is achieved by a lateral contraction which causes the necessary strong distortion of the poloidal field. Simultaneously, the cloud will be spun up, but again the shear generates torsional Alfvén waves which transport angular momentum to infinity along the field lines, and the cloud is brought into corotation with the surroundings in the characteristic time t_b appropriate to the new magnetic field geometry. The parameter f will steadily decrease, due either to further mass accretion or to flux leakage, but provided the characteristic time $-f/\dot{f}$ is much longer than the free-fall time t_f, the cloud will respond by contracting slowly through a series of magneto-gravitational equilibria. The magnetic torques will maintain approximate corotation provided $t_b < -f/\dot{f}$. The problem is different if f is less than f_c, for then the cloud will collapse at the magnetically-diluted free-fall rate unless held in equilibrium either by centrifugal forces or by a strong turbulent pressure. To make a first prediction of the rotational evolution of the cloud we again need an estimate of the braking time t_b, to compare now with the free-fall time t_f.

Consider then the spheroidal cloud model as in Fig. 11.2, with the uniform internal field $\bar{B} = F/\pi R^2$, but with the external field out to S_0 approximated by a roughly radial field with $B_r \simeq \bar{B}R^2/r^2$. The time of travel of waves out to a distance $\tilde{r}$ is

$$t = \int_R^{\tilde{r}} [r^2(4\pi\rho_0)^{1/2}/\bar{B}R^2]\,\mathrm{d}r \simeq \pi(4\pi\rho_0)^{1/2}\tilde{r}^3/3F. \tag{11.78}$$

An estimate for t_b is given again by equating the moment of inertia $(8\pi\rho_0\tilde{r}^5/15)$ of the sphere of radius $\tilde{r}(\gg R)$ to that of the cloud of mass $M = 4\pi\rho_0 R_0^3 K/3 = 4\pi\bar{\rho}ZR^2/3$. This yields $\tilde{r}/R \simeq K^{1/5}(R/R_0)^{2/5}$, whence from (11.78)

$$\begin{aligned} t_b &\simeq \frac{\pi^{1/2}}{2}\left(\frac{M}{\rho_0^{1/2}F}\right)\left(\frac{R}{R_0}\right)^{6/5}\frac{1}{K^{2/5}} \\ &= \frac{2\pi^{1/2}}{3}\left(\frac{R_0\rho_0^{1/2}}{B_0}\right)\left(\frac{R}{R_0}\right)^{6/5}K^{3/5}. \end{aligned} \tag{11.79}$$

Note that with $K > 1$, as is plausible, the time (11.79) is less than (11.77) by the factor $5(R/R_0)^{6/5}/8K^{2/5}$, due to the combined effect in this geometry of the increased field-strength near the cloud and the increased moment of inertia of the gas set into rotation. The time of travel along the semi-minor axis of the cloud is now

$$(3\pi^{1/2}/2)(M/F\bar{\rho}^{1/2}) \simeq 3(Z/R)^{1/2}(R/R_0)^{3/10}K^{-1/10}t_b, \tag{11.80}$$

which will be less than t_b for quite modest values of Z/R and R/R_0, suggesting again that the time-scale of braking is fixed essentially by the external wave propagation.

The expression (11.79) can be written in various other forms. Substitution from (11.7) and (11.15) yields

$$t_b = (2\pi^{1/2}/3)(\bar{\rho}/\rho_0)^{1/2}(\bar{\rho}^{1/2}Z/\bar{B})(R/R_0)^{6/5}/K^{2/5} \tag{11.81}$$

$$= (5^{1/2}/3)(\bar{\rho}/\rho_0)^{1/2}(a/\bar{B}G^{1/2})(R/R_0)^{6/5}/K^{2/5} \qquad (11.82)$$

on use of (11.40) which is valid over the bulk of the cloud. Note that (11.82) should not be read as implying that $t_b \to 0$ as $a \to 0$: a reduction in a for a cloud of given mass implies a compensating increase in $\bar{\rho}$, but with t_b essentially unchanged, as is clear from (11.79). The travel time (11.80) along the cloud semi-axis becomes $2^{1/2}a/\bar{B}G^{1/2}$ and this does decrease with a, because the Alfvén speed $\propto \bar{\rho}^{-1/2}$, but the distance of travel $Z \propto \bar{\rho}^{-1}$, yielding a travel time $\propto \bar{\rho}^{-1/2} \propto a$. The form (11.82) may be useful to apply to clouds for which $a, \bar{B}$ and $\bar{\rho}$ may be observationally inferred. For theoretical studies the original form (11.79) is preferable, as it displays the two global quantities F and M and the density ρ_0 of the medium that absorbs the angular momentum.

Another form for (11.79) results on use of (11.22) to write F in terms of the critical mass $M_c(e)$:

$$t_b = (1/2\pi^{3/2})(1/f_c(e))(1/G\rho_0)^{1/2}(M/M_c(e))(R/R_0)^{6/5}(1/K^{2/5}). \qquad (11.83)$$

For the limiting case $e = 1$, the same expression results from (11.82) on substitution for $\bar{B}/\bar{\rho}^{1/2}$ from (11.43), as it clearly must. (Recall that the conditions under which the relation $\bar{B}/\bar{\rho}^{1/2}$ = constant holds and the value of the constant have been discussed in Section 11.3. At least one spurious result has appeared in the literature, based on the quite unjustified assumption $\bar{B}/\bar{\rho}^{1/2} = B_0/\rho_0^{1/2}$, which when substituted into (11.81) yields $t_b \simeq (a/B_0G^{1/2})(R/R_0)^{6/5}/K^{2/5}$. This expression is manifestly wrong, for since it goes to zero with a, it would imply that the braking time of a cloud of given mass becomes small as the cloud 'cools' and so flattens.)

The characteristic time of gravitational collapse along its major axis of the oblate spheroid of eccentricity e is

$$t_f \simeq (4\pi G\bar{\rho}(1-e^2)Q(e)/3)^{-1/2} = (GMQ(e)/R^3)^{-1/2}, \qquad (11.84)$$

with $Q(e)$ given by (11.140). The form (11.83) for t_b is convenient for constructing the ratio t_b/t_f: on substitution for ρ_0 from (11.17),

$$\frac{t_b}{t_f} = \left[\frac{(Q(e))^{1/2}}{3^{1/2}\pi f_c(e)}\right](M/M_c(e))(R_0/R)^{3/10}K^{1/10}$$
$$\equiv \lambda(e)(M/M_c(e))(R_0/R)^{3/10}K^{1/10}, \qquad (11.85)$$

where $\lambda(e)$ is of order unity. Equivalently, by (11.15) and (11.17),

$$t_b/t_f = \lambda(e)(M/M_c(e))(\bar{\rho}/\rho_0)^{1/10}(1-e^2)^{1/20} \qquad (11.86)$$

(Nakano 1989; McKee *et al.* 1993), showing that deviation from $t_b/t_f \propto M/M_c$ occurs only after $\bar{\rho}$ has increased by many powers of 10. The important point that emerges is that *the ratio* (M/M_c) *plays a double role in the theory*: it not only determines whether or not the cloud can contract indefinitely against the forces exerted by the frozen-in field, but it is also the primary factor determining whether or not the braking time is longer or shorter than the free-fall time. A cloud that begins as strongly supercritical, collapsing at the magnetically-diluted free-fall rate, will suffer little angular momentum loss, and so will spin up to reach centrifugal balance at moderate densities. Subsequent contraction will occur at the rate fixed by the slower magnetic-braking process (which is still faster than the likely decrease in f – cf. Section 11.7). By contrast, for a strongly subcritical cloud ($M \ll M_c$), (11.85) and (11.86) predict $t_b \ll t_f$. As noted, contraction of a subcritical cloud occurs if M/M_c steadily increases through flux-leakage, in a time $-f/\dot{f}$ that is certainly longer than t_f in the molecular cloud phase – a necessary condition if magneto-gravitational equilibrium is to be a good approximation. Thus $t_b \ll -f/\dot{f}$ a fortiori, and the magnetic stresses should indeed maintain corotation as the cloud contracts, implying highly efficient transfer of angular momentum. At some density, enough flux will have leaked out for M/M_c to pass through unity. The now supercritical cloud will then go over into magnetically-diluted free-fall; and even if no further significant braking were to occur, it will not reach a state of magneto-thermo-centrifugo-gravitational equilibrium until high densities are reached (cf. Section 11.5).

The tentative conclusion that the rotational history of a cloud is likely to be sensitive to the mass/flux ratio is confirmed by the work summarized in Appendix B. The spherical and spheroidal cloud models of Section 11.2 and Appendix A are adopted; the associated poloidal field structures have all their lines infinite, so maximizing the efficiency of outward angular momentum transport by Alfvén waves. As already noted, as long as changes in the parameter f occur in a time long compared with t_b, then a detailed study of the braking of a subcritical cloud may reasonably adopt a time-independent structure for ρ and $\mathbf{B}_\mathrm{p}$. In supercritical cases, the steady loss of angular momentum ensures continuing contraction with simultaneous changes in the ρ- and $\mathbf{B}_\mathrm{p}$-fields as well as in the Ω-field, so that a fully rigorous treatment of the combined braking–contraction problem is difficult even with the adoption of our comparatively simple ρ and $\mathbf{B}_\mathrm{p}$ models. However, the conclusions drawn from the ratio (11.85) suggest an approximate step-function approach. At each epoch of the contraction the cloud is supposed rotating with uniform angular velocity Ω_c and to be in near mechanical equilibrium. The instantaneous rate of braking is studied via a subsidiary problem (called – perhaps unfortunately – a 'pseudo-problem' in the several papers: Gillis, Mestel and Paris 1974, 1979 (GMPI,II); Mestel and Paris 1979, 1984 (MPI,II)), in which ρ and $\mathbf{B}_\mathrm{p}$ are again time-independent. At the initial epoch of the subsidiary problem the cloud has a uniform angular velocity $\bar{\Omega}$ and the surroundings have $\Omega = \Omega_0$. At epoch t in the subsidiary problem the total angular momentum transported from the cloud is written in the form

$$(2MR^2/5)(\bar{\Omega} - \Omega_0)\tilde{F}(t) \tag{11.87}$$

where the transport function $\tilde{F}(t)$ depends also on the field parameters and on the density field outside the cloud. From $\tilde{F}(t)$ we construct an effective derivative which yields the instantaneous rate of magnetic braking at the contraction epoch considered. We then identify $\bar{\Omega}$ with Ω_c and write down (11.178) for the loss of angular momentum, to be coupled in its non-dimensional form (11.180) with the virial equations for the contraction of the cloud.

For all cases, the actual variation of the cloud eccentricity during the contraction depends on the degree of turbulent support. The most detailed computations (MPII) assume the cloud contracts isothermally, with any turbulence at most sonic, so the eccentricity steadily increases. This is estimated to have only a modest effect on the rate of braking (in (11.83), by (11.22), $M_c(e)f_c(e)$ is constant), but it does increase the lateral gravitational field by up to a factor 1.5. Some earlier computations (MPI) treated the contraction of spherical clouds, which can be thought of as the limiting case with strong Alfvénic turbulent support along the field (cf. Section 12.7); the main results are qualitatively similar.

The overall conclusions of the numerical work are as stated. For a wide choice of initial conditions it is found that clouds with markedly *supercritical* mass/flux contract in near equilibrium, with centrifugal forces remaining comparable with self-gravitation and magnetic force: as long as flux leakage is slow it is magnetic braking and not free-fall that determines the rate of contraction. *Subcritical* clouds are held in approximate magneto-gravitational equilibrium, assisted by centrifugal force and pressure. As discussed in Section 11.7, under slow flux leakage the cloud contracts steadily; the magnetic stresses keep the cloud in near corotation with the surroundings, so that centrifugal force is a decreasing fraction of gravity. In the computations in MPII, the assumed isothermality implies a steady increase in eccentricity, so that $M_c(e)$ decreases and $f_c(e)$ correspondingly increases by up to Strittmatter's factor ≈ 1.8 as e goes from 0 to 1. However, the simultaneous increase in $(Q(e))^{1/2}$ is by the factor $[Q(1)/Q(0)]^{1/2} = (3\pi/4)^{1/2} \approx 1.5$, so the reduction in the coefficient λ in (11.85) is at most ≈ 0.83. The most important effects follow when the crucial ratio $M/M_c(e)$ goes through unity, so that the actual contraction time of a slowly rotating body goes over into that of magnetically diluted free-fall. The limitations of the virial theorem treatment of the poloidal dynamics will inevitably show up when one asks for an accurate estimate of the density which such an initially subcritical cloud or fragment can reach before centrifugal force is able to balance the now positive excess of gravity over magnetic force. It could turn out that the most important limitation on the braking efficiency in this phase derives instead from the changing topology of the field through successive detachment of cloud field lines from the background field (cf. Section 12.6).

11.6.3 *A perpendicular magnetic rotator*

The discussion so far has assumed symmetry about the common direction of the angular momentum vector and the magnetic axis. Angular momentum transport along a poloidal field $\mathbf{B}_{\mathrm{p}}$ of prescribed structure is then due just to the toroidal magnetic tensions generated by the shear, which yield complicated Alfvén-like torsional waves propagating along mutually independent poloidal field lines. In a non-axisymmetric system, for example when the angular momentum and magnetic axes are mutually orthogonal, shearing will in addition generate magnetic pressure gradients about the rotation axis, causing interchange of angular momentum between different flux tubes (cf. Section 9.3). Intuitively, one expects that the total rate of loss of angular momentum is again fixed by the Alfvén travel time, with the magnetic pressures acting to prevent any marked relative shearing of individual field lines. The simplest illustrative example of the perpendicular rotator (Mouschovias and Paleologou 1979) solves exactly for the braking of a cylindrical cloud of density $\bar{\rho}$, radius R and height L, immersed in a medium of density ρ_0, rotating about its axis and with the basic field $\mathbf{B}_{\rm p}$ directed everywhere along cylindrical radii and of magnitude $\bar{B}R/\varpi = F/\pi L\varpi$. The resulting time-scale is shorter by the factor $(\rho_0/\bar{\rho})^{1/2}$ than the result (11.77) for a cloud with a cylindrical field aligned parallel to the rotation axis. The model is clearly somewhat artificial through its requiring a line of magnetic poles on the cylinder axis. However, replacing $\mathbf{B}_{\rm p}$ by a split cylindrical monopole field (with B_ϖ reversing sign across an azimuthal plane) should yield essentially the same result, which is easily understood from the assumed field geometry, for the cylinder of density ρ_0 with the same moment of inertia as the cloud has radius $\varpi_b = (\bar{\rho}/\rho_0)^{1/2}R$, so that the time of travel of an Alfvén wave out to ϖ_b is

$$\int_R^{\varpi_b} \mathrm{d}\varpi\,(4\pi\rho_0)^{1/2}\varpi/\bar{B}R = (\pi^{1/2}M/\rho_0^{1/2}F)(\rho_0/\bar{\rho})^{1/2}, \tag{11.88}$$

which is indeed essentially expression (11.77) reduced by the factor $(\rho_0/\bar{\rho})^{1/2}$. By contrast, a similar argument applied to a field structure with quasi-radial external field lines, as in Fig. 11.1 or 11.2, should not yield markedly different results whether the rotation and magnetic axes are parallel or perpendicular. This is confirmed in MPII, Appendix IV, where again the same field structure as in Appendix A is adopted, and the transport function $\tilde{F}_\perp$ for the perpendicular case – defined as in (11.172) – is constructed and shown to exceed $\tilde{F}_\parallel$ by a factor less than 2. Following Mouschovias and Paleologou, one can argue that for a general obliquity angle χ, the somewhat greater transport of the perpendicular component leads to a slow but systematic decline in χ: the cloud experiences a braking torque and also a precessional torque, tending to make the cloud angular momentum vector parallel to the local galactic magnetic field, frozen into the surrounding medium with effectively infinite moment of inertia.

11.6.4 *Fragmentation of a rotating magnetic cloud*

The prima facie simplest case is of a cloud with a supercritical mass–flux ratio, and with the magnetic and rotation axes aligned. Continuous magnetic transport of angular momentum will allow steady contraction in the two dimensions perpendicular to the common axis; and if the gas stays isothermal (or a fortiori if it cools), the lateral contraction is accompanied by flattening along the axis. The arguments of Sections 11.4 and 11.5, when combined, suggest that further fragmentation of the cloud is again not forbidden. Because the mass–flux ratio is supercritical, the spontaneous flow down the field yields submasses which are likewise supercritical. Further, the same flow reduces the ratio $\Omega^2/(4\pi G\rho/3)$, so that any fragment that separates out following the flattening will begin its contraction with a slow rotation about its mass-centre. If the angular momentum of spin were to be conserved, then again the consequent spin-up of the fragment would limit the allowed contraction; but further magnetic braking allows further contraction and possible fragmentation. There is again the danger that if fragmentation were to occur with maximum efficiency, the typical masses produced would be embarrassingly small.

The evolution may be different in the case of a supercritical cloud but with the magnetic and rotation axes mutually orthogonal (Mestel 1965). If the cloud is markedly prolate about the magnetic axis, the centrifugal forces will then inhibit flow along the field. As long as the field is nearly frozen in, contraction of the cloud following loss of angular momentum will now be nearly isotropic, so that fragmentation cannot occur. Relaxation of strict flux-freezing may allow fragmentation, but (admittedly very rough) estimates suggest that the final masses should be larger than for the aligned case, perhaps embarrassingly large (Campbell and Mestel 1987). A fully convincing discussion requires study in depth not only of the flux leakage problem (Section 11.6) and the related question of magnetic field topology, but also of the macroscopic stability of rotating magnetic clouds.

11.7 Flux leakage

11.7.1 *General discussion*

The problem of flux leakage during the early stages of star formation is a direct application of the MHD of a lightly ionized gas, described by the three-fluid approximation. In the notation of Section 2.7, the equations to the motion of the electron, ion and neutral gases are respectively

$$-en_e\left[\mathbf{E}+\frac{\mathbf{v}_e\times\mathbf{B}}{c}\right]+n_e\mathbf{F}_{ei}+n_e\mathbf{F}_{en}-\nabla p_e=0, \tag{11.89}$$

$$Zen_i\left[\mathbf{E}+\frac{\mathbf{v}_i\times\mathbf{B}}{c}\right]+n_i\mathbf{F}_{in}-n_e\mathbf{F}_{ei}-\nabla p_i+n_im_i\,\nabla V=n_im_i\frac{\mathrm{d}\mathbf{v}_i}{\mathrm{d}t}, \tag{11.90}$$

$$-(n_i\mathbf{F}_{in}+n_e\mathbf{F}_{en})-\nabla p_n+n_nm_n\,\nabla V=n_nm_n\frac{\mathrm{d}\mathbf{v}_n}{\mathrm{d}t}, \tag{11.91}$$

where, for example, $\mathbf{F}_{in}$ is the mean force on an ion due to collisions with neutrals, etc. Adding (11.89) and (11.90) yields the equation to the mean motion of the gas of electrons-plus-ions:

$$\frac{\mathbf{j}\times\mathbf{B}}{c}+(n_i\mathbf{F}_{in}+n_e\mathbf{F}_{en})-\nabla p_e-\nabla p_i+n_im_i\,\nabla V=n_im_i\frac{\mathrm{d}\mathbf{v}_i}{\mathrm{d}t}. \tag{11.92}$$

In a strongly magnetic cloud, the magnetic force density is comparable with the gravitational force on the neutral bulk of the gas; hence since in an HI cloud, and *a fortiori* in a molecular cloud, $n_i/n \simeq n_e/n \ll 1$, it is a good approximation to drop the pressure, gravitational and inertial terms in (11.92), which then becomes

$$\mathbf{j} \times \mathbf{B}/c = (\nabla \times \mathbf{B}) \times \mathbf{B}/4\pi = -(n_i \mathbf{F}_{in} + n_e \mathbf{F}_{en}). \tag{11.93}$$

Equation (11.93) implies that the magnetic force, acting directly on the charged particles, is balanced by the frictional force exerted by the neutral particles; its effect on the neutral particles is mediated by the same friction, appearing in (11.91) with reversed sign. The sum of (11.91) and (11.92) yields the equation to the bulk motion of the whole gas

$$(\nabla \times \mathbf{B}) \times \mathbf{B}/4\pi - \nabla p + \rho \nabla V = \rho\, \mathrm{d}\mathbf{v}/\mathrm{d}t \tag{11.94}$$

where $p \simeq p_n, \rho \simeq n_n m_n, \mathbf{v} \simeq \mathbf{v}_n$ in a lightly ionized gas.

The dominant frictional term in (11.93) is $n_i \mathbf{F}_{in}$, the 'ambipolar diffusion' term (cf. Section 2.7). The neutral particles are primarily hydrogen atoms in an HI cloud or H_2 in a molecular cloud, and so are considerably less massive than either metallic or molecular ions. The ion–neutral friction may therefore be calculated in the mean-free-path approximation by supposing that at each 'collision' the relative velocity of the lighter particle as seen by the heavier is randomized, so that the momentum gained by the ion at each collision is $m_n(\mathbf{v}_n - \mathbf{v}_i)$. The number of collisions per second felt by an ion is

$$n_n < \sigma_{in}(v_\mathrm{t})_n > \equiv \frac{1}{\tau_{in}} \tag{11.95}$$

where σ_{in} is the momentum transfer collision cross-section, $< \sigma_{in}(v_\mathrm{t})_n >$ is the collision rate, averaged over the velocity distribution of the neutral particles, and so τ_{in} is an appropriately defined collision time. Thus

$$\mathbf{F}_{in} = n_n < \sigma_{in}(v_\mathrm{t})_n > m_n(\mathbf{v}_n - \mathbf{v}_i), \tag{11.96}$$

and (11.93) yields for the drift velocity $\mathbf{v}_d \equiv (\mathbf{v}_i - \mathbf{v}_n)$

$$(\nabla \times \mathbf{B}) \times \mathbf{B}/4\pi = \alpha (n_i/n_n) \rho^2 \mathbf{v}_d, \tag{11.97}$$

with

$$\alpha = \frac{< \sigma_{in}(v_\mathrm{t})_n >}{m_n}. \tag{11.98}$$

The neglect of the inertial term in (11.93) is fully justified, for if the magnetic and friction terms are not in close balance, the change in $\mathbf{v}_i$ required to achieve balance occurs in the time $\simeq (m_i/m_n)\tau_{in}$, which is normally well below the macroscopic time-scale of the problem under study.

Osterbrock (1961) has computed σ_{in} for the low energy particles expected in HI regions or molecular clouds. The interaction occurs through the polarization of the neutral particle by the monopole electric field of the ion, yielding at large distances the potential $-e^2p/2R^4$ where p is the polarizability of the neutral particle and R is the separation of the centres of charge. The momentum transfer cross-section is

$$\sigma_{in} = 2\pi \int_0^\infty (1 - \cos\chi)\, b\, db \tag{11.99}$$

where χ is the angle of scattering suffered by a neutral particle with impact parameter b. If all the orbits are calculated using the $1/R^4$ approximation, then for a particle of speed v, $\sigma_{in} = 2.21\pi(e^2p/m_nv^2)^{1/2}$; if some allowance is made for the stronger interaction when $b < (4e^2p/m_nv^2)^{1/4}$, the coefficient is increased to 2.41π. The polarizability is known experimentally for He and H_2, and theoretically for H. At $T = 10^2$ K, σ_{in} is typically $\simeq 10^{-14}$ cm^2. Since $\sigma_{in} \propto 1/v$, $< \sigma_{in}v >$ is independent of the temperature, being about 1.5×10^{-9} cm^3 sec^{-1} for a cloud with standard cosmical H_2 and He abundances (Nakano 1984), so yielding a collision time $\tau_{in} = 0.67 \times 10^8 n_n$ s, and a mean value $\alpha \simeq 4 \times 10^{14}$ cm^3 s^{-1} g^{-1} for the coefficient in (11.97).

Equation (11.97) describes the drift of the ions relative to the neutral particles under the magnetic force. As in a fully ionized gas, the electrons are strongly coupled to the ions via the Coulomb force: the degree of charge separation $|n_iZ - n_e|/n_e = |\nabla \cdot \mathbf{E}|/(4\pi n_e e)$ is easily estimated from (11.89) to be much below unity (cf. Section 2.1). Equation (11.89) is well approximated by

$$c\mathbf{E} = -\mathbf{v}_e \times \mathbf{B}. \tag{11.100}$$

As in (2.55) for a fully ionized gas, the terms ∇p_e, $n_e\mathbf{F}_{ei}$ are normally small. The term $n_e|\mathbf{F}_{en}|$ may exceed $n_e|\mathbf{F}_{ei}|$ because there are many more neutral particles with which electrons can collide; however, $n_eF_{en}/n_iF_{in} \simeq (m_e/m_i)^{1/2}$, and by (11.93),

$$\frac{n_iF_{in}}{n_eev_eB/c} \simeq \frac{|\mathbf{j} \times \mathbf{B}/c|}{n_eev_eB/c} \simeq \frac{|\mathbf{v}_i - \mathbf{v}_e|}{v_e} \ll 1 \tag{11.101}$$

even for the low ion–electron densities in molecular clouds. Thus (11.90) approximates to

$$c\mathbf{E} = -\mathbf{v}_i \times \mathbf{B}, \tag{11.102}$$

consistent with (11.100) since $\mathbf{v}_i \sim \mathbf{v}_e$. Together with Faraday's law, (11.102) yields

$$\frac{\partial \mathbf{B}}{\partial t} = \nabla \times (\mathbf{v}_i \times \mathbf{B}) : \tag{11.103}$$

the field is inductively coupled to the ionized component of the gas, which moves relative to the neutral bulk according to (11.97), driven by the Lorentz force against the ion–neutral frictional resistance. The process is often referred to as 'ambipolar diffusion', but the term 'plasma drift' is to be preferred.

The energy equation to the problem is well approximated by

$$\mathbf{j} \cdot \mathbf{E} = \frac{\mathbf{j} \times \mathbf{B}}{c} \cdot \mathbf{v}_i = -n_i\mathbf{F}_{in} \cdot (\mathbf{v}_i - \mathbf{v}_n) - n_i\mathbf{F}_{in} \cdot \mathbf{v}_n \tag{11.104}$$

$$= \alpha \frac{n_i}{n_n} \rho^2 (\mathbf{v}_n - \mathbf{v}_i)^2 + \frac{\mathbf{j} \times \mathbf{B}}{c} \cdot \mathbf{v}_n \tag{11.105}$$

on use of (11.97) and (11.100) with $\mathbf{v}_i \approx \mathbf{v}_e$. The first term on the right in (11.104) and (11.105) is again the dominant dissipative term, due to ion–neutral collisions; the second term is the work done by the Lorentz force – acting through the mutual friction – in decelerating the neutrals.

The time which plasma-plus-field drift, acting alone, would need to reduce significantly the flux through an oblate cloud of axes (R, z) is given by

$$t_d \simeq \frac{R}{v_d} \approx \frac{(4\pi\alpha)(n_i/n)\rho^2}{|(\nabla \times \mathbf{B}) \times \mathbf{B}|} \approx \alpha \left(\frac{n_i}{n}\right) \frac{9\pi}{4} \frac{M^2}{F^2} \frac{R}{z}, \tag{11.106}$$

where $|\nabla \times \mathbf{B}|$ is again estimated by B/z, $M = 4\pi R^2 z/3$ and $F = \pi B R^2$. For a cloud in which the initial flux is less than but close to the critical value which would stop gravitational collapse, i.e. with $f = F/\pi^2 G^{1/2} M \lesssim f_c$, (11.106) yields

$$t_d = \left(\frac{R}{z}\right) \frac{\alpha(n_i/n)}{G} \tag{11.107}$$

if, for example, the value $f_c = 3^{1/2}/2\pi$ inferred from (11.10) is inserted. The value of the coefficient in (11.107) clearly depends on the particular cloud model, but the combination (11.107) will inevitably appear.

Most of this general analysis was given in Mestel and Spitzer (1956). In any particular application, the crucial question is how the drift velocity $\mathbf{v}_d$ of the ions relative to the neutrals, given by (11.97), compares with the bulk velocity of the gas as a whole – essentially $\mathbf{v}_n$; and this in turn depends strongly on n_i/n. Suppose, for example, that the cloud as a whole is gravitationally contracting. If the ratio n_i/n_n is not too low, so that (11.97) yields $v_d \ll v_n$, to a good approximation the field is frozen into the gas as a whole. In (11.104) the dissipation term, though much greater than the Ohmic dissipation due to electron–baryon collisions (cf. Section 2.7), is then much less than the modulus of the work term $|(\mathbf{j} \times \mathbf{B}) \cdot \mathbf{v}_n|/c$: most of the gravitational energy released becomes magnetic energy through the positive $-\mathbf{j} \cdot \mathbf{E}$ term, with the field strength increasing with ρ according to the discussion of Sections 11.2 and 11.3. At the other extreme, if $v_d \gg v_n$, the ionized component and the inductively coupled field are driven outwards by the Lorentz force, and most of the excess energy in the magnetic field is dissipated through the ion–neutral friction. As the field lines straighten, the Lorentz force declines, and the system approaches a state with $\mathbf{v}_i \approx 0$: the neutral gas contracts across the field, with the work being done on the field all being dissipated, and the flux/mass ratio of the collapsing gas becomes small.

The case studied by Mestel and Spitzer did in fact assume the cool, non-rotating cloud to be initially supercritical in $G^{1/2}M/F$, so that gravitational collapse of the neutral bulk continues at a rate close to free-fall. The crucial parameter n_i/n fixing the ion–neutral friction and so also the drift velocity v_d is determined by the relevant ionization and neutralization processes for the system under study. If v_d were to become larger than or comparable with the free-fall velocity – 'rapid diffusion' – the cloud would lose most of its magnetic flux in the early phases. This would explain why all observed stars are magnetically 'weak' (cf. Section 5.1), but at the cost of having the magnetic field become dynamically unimportant early in the star formation process. Equally, if the relative velocity v_d remains small compared with free-fall – i.e. if the diffusion is 'slow' – the magnetic field of the collapsing cloud remains strong and is, for example, able to transport excess angular momentum. As was recognized subsequently (Mestel 1965), even with the field nearly frozen in, fragmentation is not ruled out but may occur following spontaneous flattening along the field (Section 11.4); but the dissipation of most of the magnetic flux, as required by observation of newly born stars, must occur later, and perhaps by a different physical process.

Mestel and Spitzer estimated that in dusty HI clouds, shielded from ionizing galactic ultraviolet radiation, the ion density would fall spontaneously within a free-fall time to a value low enough to allow rapid diffusion. The most efficient process of plasma decay would be by attachment of ions and electrons to dust grains, at the rate $\mathrm{d}n_i/\mathrm{d}t = -n_i n_g \sigma_g \langle v_i \rangle_T$ in an obvious notation, leading to an exponential decline in the plasma density. Other processes such as radiative recombination of atomic ions or dissociative recombination of molecular ions occur at a rate proportional to $n_e n_i \propto n_i^2$ and so yield a slower, algebraic time-dependence; but even so, the cross-section for dissociative recombination seemed large enough for n_i to fall to the level at which rapid flux loss could occur, even without attachment to dust grains.

However, as pointed out by Hayakawa *et al.* (1961) and Cameron (1962), and discussed later by Spitzer and Tomasko (1968) and Pikel'ner (1968), low energy cosmic rays maintain a density of ions (mainly protons) in dense HI clouds long after the galactic ultraviolet radiation has been effectively extinguished. Galactic X-rays also contribute to the level of ionization. In addition, the cited calculations by Osterbrock increased the value of σ_{in} from the value 10^{-16} adopted by Mestel and Spitzer to 10^{-14}. The later papers by Draine and colleagues (1983, 1986) give a thorough discussion of the various couplings and rate coefficients. The upshot is that in HI clouds, the typical time t_d of flux loss by ambipolar diffusion is much longer than the free-fall time (and is normally longer than the Hubble time), so that a slowly rotating cloud, supercritical in $G^{1/2}M/F$, would collapse with effective conservation of flux. However, in dense molecular clouds n_i/n is significantly lower, and the above simple model yields t_d shorter than the Hubble time, so that the ambipolar diffusion process can be astronomically important. For the supercritical case, again the question is whether the speed of flux leakage is less than or greater than the free-fall speed.

Starting with Spitzer (1963), a number of authors have looked more carefully at the role of dust grains of different sizes in both the physics and the dynamics of the problem. Not only are grain surface reactions important in helping to fix the ion and electron densities, but also the charged grains contribute to the coupling between the neutral gas and the magnetic field (Elmegreen 1979; Nakano and

Umebayashi 1986*a,b*; Nishi *et al.* 1991 and references therein). Nishi *et al.* find that already at densities greater than 2×10^4 cm^{-3}, the smaller charged grains are more efficient than ions at coupling neutral gas to the magnetic field. However, as originally pointed out by Spitzer (1963), at high enough densities the ion density gets so low that it is the charged grains that carry the current, but they suffer Ohmic-type dissipation through collisions with the neutrals. Although there has been controversy (some of it for semantic reasons) on the issue, a series of detailed treatments (Norman and Heyvaerts 1985; Nakano and Umebayashi 1986*a,b*; Nakano 1988; Nishi *et al.* 1991) concur in concluding that t_d still remains longer than the free-fall time t_f by a factor 10 or more throughout the whole diffuse molecular cloud phase, so that a supercritical cloud will again retain the bulk of its flux during its free-fall.

Consider now a cloud with the mass/flux ratio subcritical. As discussed in Section 11.6 and Appendix B, the magnetic stresses may provisionally be assumed to keep the cloud in corotation with the surroundings, so that the centrifugal forces are small and the cloud is kept in approximate magneto-gravitational equilibrium, as described by the dominant terms in (11.18). It is at once clear that *the same forces exerted by the gravitationally distorted magnetic field which maintain the cloud in equilibrium are also responsible for its evolution*, through causing both redistribution of flux within the cloud and leakage from the cloud into the surrounding medium. Simultaneously, the cloud density field adjusts, attempting to reach a new state of equilibrium. As long as the characteristic time t_d of plasma-plus-field drift is longer than the free-fall time t_f, which we have seen by (11.85) to be close to the magnetic braking time, evolution will be through a series of such quasi-steady states.

11.7.2 *Quasi-steady contraction of an oblate spheroidal model*

As a first illustration we adopt again the model of Fig. 11.2, with the cloud a highly oblate uniform spheroid of semi-axes (Z, R), and with the magnetic field at all epochs retaining the structure analysed in Appendix A. To maintain the contracting spheroid at uniform density and with a uniform internal field $\bar{\mathbf{B}}$, the velocities of both the neutral bulk and of the plasma-plus-field must be taken as homologous:

$$(v_n)_\varpi = (\dot{R}/R)\varpi, \qquad (v_i)_\varpi = (\dot{R}_i/R)\varpi. \tag{11.108}$$

At time t the spheroid is threaded by flux $F(t)$ with the associated internal field $\bar{B} = F/\pi R^2$. The outward diffusion of flux is simulated by the introduction of a notional spheroid S_i containing the flux $F(t)$, and of semi-major axis R_i with the value R at time t and $R + \delta R_i$ at $t + \delta t$, so yielding for the instantaneous rate of change of $\bar{B}$

$$\dot{\bar{B}}/\bar{B} = -2\dot{R}_i/R. \tag{11.109}$$

Once field lines have leaked out of the dense spheroid they are assumed to relax to the structure of Fig. 11.2. The changing flux $F(t)$ within the cloud is then given by

$$\dot{F}/F = \dot{\bar{B}}/\bar{B} + 2\dot{R}/R = -2(\dot{R}_i - \dot{R})/R. \tag{11.110}$$

The relative velocity $(\dot{R}_i - \dot{R})$ is given by first substituting (11.108) into (11.97) and then constructing the 11-virial component, analogously to (11.139). With the help of (11.145) we find

$$\alpha(n_i/n)\{(\dot{R}_i - \dot{R})/R\} \int (\rho^2\varpi^2/2)\,\mathrm{d}\tau = (3/20\pi)\alpha(n_i/n)(M^2/zR)(\dot{R}_i - \dot{R})$$
$$= (F^2/6\pi^2 R)m(e,\eta) \qquad (11.111)$$

with e the instantaneous eccentricity of the cloud; whence from (11.110) and (11.9)

$$\dot{f}/f^3 = -(20\pi^3/9)(1-e^2)^{1/2}Gm(e,\eta)/\alpha(n_i/n). \qquad (11.112)$$

Since $\pi\tilde{f}$ or πf_c are close to unity, (11.112) yields

$$\alpha(n_i/n)/7G \qquad (11.113)$$

as a rough estimate for the characteristic time t_d in which f changes by a factor 2. If $n_i/n \simeq 5 \times 10^{-4}$, typical of an HI cloud, then with α at least 4×10^{14} cgs, t_d exceeds the Hubble time, so plasma drift is ignorable. However, as already noted, in a molecular cloud n_i/n is much lower, being determined essentially by a balance of the rate of ionization ζn_n against the rate of destruction of ions, which is normally proportional to n_i^2, so that

$$n_i/n_n = K_i/n_n^{1/2} \qquad (11.114)$$

with K_i a constant. In dense molecular clouds, ionization appears to be due primarily to galactic cosmic rays, for which $\zeta \simeq 10^{-17}$ (e.g. Spitzer 1978). Dust grains act as a locus for recombination, reducing considerably the ion–electron density. Elmegreen (1979) found $K_i \simeq 10^{-5}$ cm$^{-3/2}$, a value that appears to be a fair estimate for dense molecular clouds (McKee *et al.* 1993). However, as estimated by McKee (1989), photoionization by far-ultraviolet radiation keeps much of the molecular gas at a higher degree of ionization, with K_i values that can be as much as a factor 30 greater. Variations of n_i/n_n through an individual cloud are important (see below), but in a first discussion, both the cloud density and K_i are taken as constant, and the neutral gas as pure molecular hydrogen, so that (11.112) then becomes

$$\frac{\dot{f}}{f^3} = -\frac{20\pi^3}{9}\frac{Gm(e,\eta)(1-e^2)^{1/4}M^{1/2}}{\alpha K_i(2m_{\mathrm{H}})^{1/2}(4\pi/3)^{1/2}R^{3/2}}. \qquad (11.115)$$

As flux leaks out, the cloud contracts; from Appendix B, the 11-virial equation (with centrifugal forces negligible) yields

$$\ddot{R} = -\frac{GM}{R^2}\left[Q(e) - \frac{5\pi^2}{6}f^2 m(e,\eta)\right]. \qquad (11.116)$$

The equations are non-dimensionalized again by writing $R = \eta R_0$, $M = 4\pi\rho_0 R_0^3 K/3$, and by introducing a free-fall time t_f:

$$t = Tt_f, \qquad t_f = (4\pi G\rho_0 K/3)^{-1/2}. \tag{11.117}$$

To follow the contraction of the cloud model, one needs strictly to know how the turbulent support and hence the eccentricity e vary. For simplicity, e is assumed constant; it will be seen that the most important qualitative conclusions are unaffected. It is then convenient to define

$$\kappa = \frac{f^2}{f_c^2} \tag{11.118}$$

with $f_c(e)$ for a prescribed eccentricity again given by (11.21). The flux leakage equation (11.115) now becomes

$$\frac{1}{\kappa^2}\frac{\mathrm{d}\kappa}{\mathrm{d}T} \simeq -24 f_c^2 (1-e^2)^{1/4}\epsilon\frac{m(e,\eta)}{\eta^{3/2}}, \qquad \epsilon \equiv \frac{(G/m_H)^{1/2}}{\alpha K_i}. \tag{11.119}$$

The non-dimensional parameter $\epsilon \simeq 0.05$ when K_i is given the standard value 10^{-5}. The contraction equation (11.116) becomes

$$\frac{\mathrm{d}^2\eta}{\mathrm{d}T^2} = -\frac{Q(e)}{\eta^2}\left[1 - \frac{m(e,\eta)}{m(e,0)}\kappa\right]. \tag{11.120}$$

Under strict magneto-gravitational equilibrium, (11.120) reduces to (11.23).

As discussed in Sections 11.2 and 11.3, as f/f_c approaches unity, η becomes small. By (11.147), $m(e,\eta) \simeq [m(e,0) - 8\eta/5]$, and (11.120) and (11.119) become (with use of (11.21))

$$\ddot{\eta} = -\frac{Q(e)}{\eta^2}\left[(1-\kappa) + \frac{8}{5m(e,0)}\eta\kappa\right] \tag{11.121}$$

and

$$\frac{\dot{\kappa}}{\kappa^2} = -\frac{2.9\epsilon Q(e)(1-e^2)^{1/4}}{\eta^{3/2}}\left(1 - \frac{8\eta}{5m(e,0)}\right). \tag{11.122}$$

As long as the contraction is through states in near equilibrium, $m(e,\eta)\kappa \simeq m(e,0)$, or

$$\eta \simeq \frac{5}{8}m(e,0)\left(1 - \frac{1}{\kappa}\right), \tag{11.123}$$

and (11.122) becomes

$$\frac{\dot{\kappa}}{\kappa^2} = -\frac{2.9\epsilon Q(e)(1-e^2)^{1/4}}{\kappa\eta^{3/2}}. \tag{11.124}$$

The inertial term in (11.120) – neglected in (11.123) – is in modulus

$$|\ddot{\eta}| = (5m/8)\left|\frac{\ddot{\kappa}}{\kappa^2} - \frac{2\dot{\kappa}^2}{\kappa^3}\right| = (5m/8)[2.9\epsilon Q(1-e^2)^{1/4}]^2 \frac{\kappa + 1/2}{\kappa(\kappa-1)\eta^3}, \qquad (11.125)$$

and this becomes comparable with $Q(e)/\eta^2$ when

$$(2.9)^2\epsilon^2 Q(1-e^2)^{1/2} \simeq (\kappa-1)^2/(\kappa+1/2). \qquad (11.126)$$

With the standard value $\epsilon = 0.05$ inserted, for $e = 0$, (11.126) yields $\kappa \simeq 1.19$, and (11.123) the corresponding value $\eta \simeq 0.15$. When $e \simeq 1$, the numbers depend on $Z/R = (1-e^2)^{1/2}$: when $Z/R \simeq 0.1$, $\kappa \simeq 1.089$, $\eta \simeq 0.06$.

These numbers illustrate (cf. Nakano 1990) how as f approaches and passes through the critical value f_c, the appropriate description of the contracting cloud changes from quasi-magneto-gravitational balance to a balance of inertia and the small difference between gravity and magnetic force, represented by the last term in (11.121). As κ falls below unity, (11.121) shows that the cloud – now with supercritical M/F – will go over to collapse in magnetically-diluted free-fall, described by

$$\ddot{\eta} = -Q(e)\frac{(1-\kappa)}{\eta^2}; \qquad (11.127)$$

and since η is typically small, (11.122) becomes

$$\dot{\kappa} = -2.9\epsilon Q(e)(1-e^2)^{1/4}\kappa^2/\eta^{3/2} = -0.145 Q(e)(1-e^2)^{1/4}\kappa^2/\eta^{3/2} \qquad (11.128)$$

with $\epsilon = 0.05$. One can then verify that $\kappa = (f/f_c)^2$ will not fall to a small value during the subsequent collapse to the end of the molecular cloud phase. A lower limit to the rate of collapse of the cloud from an initial state η_a, with associated κ_a, is given by putting κ_a into (11.127). With e also constant, (11.127) is integrated by means of the substitution $\eta = \eta_a \cos^2\theta$, yielding

$$\theta + \frac{\sin 2\theta}{2} = \left[2Q(e)(1-\kappa_a)/\eta_a^3\right] T, \qquad \dot{\theta} = \left[Q(e)(1-\kappa_a)/2\eta_a^3\right]^{1/2} \sec^2\theta, \qquad (11.129)$$

where the small initial velocity at η_a has been ignored. Equation (11.128) can then be integrated to yield

$$\frac{1}{\kappa} - \frac{1}{\kappa_a} = 0.2\frac{(Q(e))^{1/2}(1-e^2)^{1/4}}{(1-\kappa_a)^{1/2}} \log(\sec\theta + \tan\theta). \qquad (11.130)$$

A value $\cos\theta = 0.1$ yields $\eta/\eta_a = 10^{-2}$, with $\rho/\rho_a \simeq 10^6$, corresponds to densities of 10^{10}–10^{11}, near the end of the molecular cloud phase. Then even with e put equal to 0, a value $\kappa_a = 0.25$ ($f/f_c = 0.5$) inserted into the right-hand side of (11.130) will yield a final value $\kappa = 0.21$. If further one inserts this value rather than 0.25 into just the right-hand side of (11.130) (to allow for the higher contraction rate at lower κ), then (11.130) predicts a final value only slightly less than 0.21. Similar results hold a fortiori for a non-zero eccentricity. (For a full discussion see Nakano 1990.)

The overall conclusion is that with standard values for the degree of ionization, ambipolar diffusion in a molecular cloud is fast enough to convert a cloud from a subcritical to a supercritical M/F ratio; but since the typical time of gravitational collapse is significantly shorter than the diffusion time, a collapsing supercritical body will retain the bulk of the remnant flux through the molecular cloud phase. The results are not strongly model-dependent; it will be found in Section 12.5 that the inevitably approximate virial treatment exaggerates somewhat the dynamical effect of a given amount of magnetic flux, so that the fraction diffusing out during the collapse phase should be even less than the above estimates.

The above treatment is merely illustrative. The assumption of a uniform cloud is now especially restrictive, for with $n_i/n_n \propto n_n^{-1/2}$, the time for plasma-plus-flux to drift out of a region decreases with increasing mean density. As discussed by Nakano (1976, 1983) and Mouschovias (1987, 1994, 1996), an appropriate generalization is to a molecular cloud again with subcritical M/F, but with local regions of enhanced density, and with the 'background field' B_0 now the mean field within the cloud. Over the time of interest the cloud may effectively retain its total flux, but differential flux leakage within the cloud can reduce the flux F' threading a dense core of mass M', so that the local ratio F'/M' decreases from a supercritical to a subcritical value. This suggests a new mode of fragmentation (Nakano 1976, 1983), in which the bulk of the cloud remains magnetically supported, but diffusion leads to the gravitational collapse of cloud cores (cf. Section 12.4). The process is in some ways closer to that in the original Jeans picture, in that the core separates out from a non-collapsing cloud rather than having to collapse more rapidly than the cloud as a whole.

APPENDIX A

The model of Fig. 11.2

The cloud in Fig. 11.2 is an oblate spheroid $\bar{S}$ symmetric about $\mathbf{B}_0$, of semi-axis $\bar{R}$ across $\mathbf{B}_0$, semi-axis Z along $\mathbf{B}_0$, and eccentricity e. It is convenient to use $R_0 = (F/\pi B_0)^{1/2}$ (cf. (11.7)) as a length-scale, with non-dimensional lengths η and ξ defined by

$$\bar{R} = \eta R_0, \qquad Z = \xi R_0, \qquad e = (1 - \xi^2/\eta^2)^{1/2}. \tag{11.131}$$

We introduce the set of oblate spheroidal coordinates based on $\bar{S}$. Cylindrical polar coordinates (ϖ, ϕ, z) are written as

$$\varpi = c\cosh u \cos v, \qquad z = c\sinh u \sin v, \qquad \phi = w, \tag{11.132}$$

so that each confocal spheroid

$$\frac{\varpi^2}{c^2\cosh^2 u} + \frac{z^2}{c^2\sinh^2 u} = 1 \tag{11.133}$$

has the same two focal points $\pm c$ and eccentricity $e = \operatorname{sech} u$. The displacement vector $\mathbf{dr}$ has components $(h_1\, du, h_2\, dv, h_3\, dw)$, where

$$h_1 = h_2 = c(\cosh^2 u - \cos^2 v)^{1/2}, \qquad h_3 = \varpi = c\cosh u \cos v. \tag{11.134}$$

In terms of parameters of the cloud (defined by $u = \bar{u}$, $e = \bar{e}$),

$$\begin{gathered} c = \bar{R}\bar{e} = R_0\eta\bar{e} = R_0\eta(1 - \xi^2/\eta^2)^{1/2}, \\ \bar{R} = c\cosh\bar{u} = \eta R_0, \qquad \cosh\bar{u} = 1/\bar{e} = (1 - \xi^2/\eta^2)^{-1/2}. \end{gathered} \tag{11.135}$$

The member S_0 of the family of spheroids passing through the circle ($\varpi = R_0, z = 0$) has parameter $u = u_0$, where

$$\cosh u_0 = R_0/c = 1/\eta(1 - \xi^2/\eta^2)^{1/2} = 1/\eta\bar{e}; \tag{11.136}$$

its semi-minor axis $z_0 = c\sinh u_0 = R_0(1 - \eta^2 + \xi^2)^{1/2}$ and eccentricity $e_0 = (\eta^2 - \xi^2)^{1/2} = \operatorname{sech} u_0 = \eta\bar{e}$.

The model magnetic field is based on this coordinate system (Mestel and Paris 1984). Within $\bar{S}$, the field is parallel to $\mathbf{B}_0$ and of strength $\bar{B}$. In the domain between the spheroids $\bar{S}$ and S_0, the magnetic field lines are defined by $\phi \equiv w =$ constant and the hyperboloids $v =$ constant, so that $\mathbf{B} = (B_u, 0, 0)$, with the divergence condition $\nabla \cdot \mathbf{B} = 0$ yielding $\partial(h_2 h_3 B_u)/\partial u = 0$, whence from (11.134),

$$\mathbf{B} = (B_u, 0, 0), \qquad B_u = A(v)/\cosh u(\cosh^2 u - \cos^2 v)^{1/2}. \tag{11.137}$$

Continuity of the normal component B_u of $\mathbf{B}$ at the surface of the spheroid $\bar{u}$ with the normal component of the uniform internal field $\bar{\mathbf{B}}$ fixes $A(v)$, yielding

$$B_u = \frac{\bar{B}\cosh^2\bar{u}\sin v}{\cosh u(\cosh^2 u - \cos^2 v)^{1/2}}. \tag{11.138}$$

The definition (11.7) of R_0 becomes $\bar{B}\cosh^2\bar{u} = B_0\cosh^2 u_0$; the same argument applied to the spheroid S_0 then shows that B_u is continuous with the component normal to S_0 of the uniform background field $\mathbf{B}_0$ beyond S_0. The model field of Fig. 11.2 thus satisfies the minimum requirements imposed by Maxwell's equations.

From now on it is convenient to drop the bar over R and e. The 11-component of the tensorial virial theorem (3.63) for a non-rotating axisymmetric magnetic gas gloud in equilibrium is

$$\mathcal{M}_{11} - 2\mathcal{M}_{11} + \int p\,\mathrm{d}\tau + \int x_1\,(T_{1j} - p_e\delta_{1j})\,n_j\,\mathrm{d}S = 0 \tag{11.139}$$

with p_e again the pressure of the warm interstellar medium. We apply the theorem to a surface just surrounding the spheroid S_0 – i.e. to the whole domain with a distorted magnetic field. The volume within S_0 is called V_0, that within $\bar{S}, \bar{V}$; the interstellar pressure p_e is supposed to extend down to $\bar{S}$, where there is a sharp increase in density to $\bar{\rho}$ and a corresponding drop in the effective sound speed to a. For a spheroid of eccentricity e and radius R, the gravitational term

$$\mathcal{V}_{11} = -\frac{GM^2}{5R}Q(e), \qquad Q(e) = \frac{3}{2}\left(\frac{\arcsin e}{e^3} - \frac{(1-e^2)^{1/2}}{e^2}\right). \tag{11.140}$$

$Q(e)$ increases monotonically from the value 1 for $e = 0$ to $3\pi/4$ for $e = 1$. (Note that when K in (11.17) is large – implying a highly anisotropic cloud formation process – then the mass of the cloud is far above the mass $4\pi\rho_0 R_0^3/3$ of the sphere R_0 with the background density ρ_0, so justifying the neglect of the self-gravitation of the mass within S_0 lying outside the cloud.) The thermal pressure surface term reduces to $-p_e\int \partial x_1/\partial x_1\,\mathrm{d}\tau = -p_e V_0$, combining with the thermal volume terms to yield $\bar{\rho}a^2\bar{V} + p_e(V_0 - \bar{V}) - p_e V_0 = Ma^2(1 - p_e/\bar{\rho}a^2)$.

The magnetic volume terms reduce to $-2\mathcal{M}_{11}+2\mathcal{M}_{11}+\mathcal{M}_{33} = \int_{V_0}(B_z^2/8\pi)\,\mathrm{d}\tau$. The surface integral over the surface just outside S_0 can be computed directly, but is again most easily found by noting that if the uniform field were to extend through the whole region within S_0, then the magnetic forces would be zero everywhere, so that there would be zero net magnetic contribution to the virial equation; this surface contribution is therefore $-(B_0^2/8\pi)(4\pi R_0^3(1 - e_0^2)^{1/2}/3)$. The magnetic volume contribution from $\bar{S}$ is $(\bar{B}^2/8\pi)(4\pi R^3(1 - e^2)^{1/2}/3)$. It is convenient to combine these two contributions, to yield with the help of (11.7) and (11.131)

$$\frac{F^2}{6\pi^2 R}\left((1-e^2)^{1/2}-\eta(1-e_0^2)^{1/2}\right) \equiv \frac{F^2}{6\pi^2 R}L, \tag{11.141}$$

say. The remaining magnetic term is the volume integral over the hyperboloidal domain, where $B_z = B_u(\partial z/h_1\partial u)_v$; use of (11.132), (11.134) and (11.135) yields for this

$$\begin{aligned}\frac{1}{8\pi}\int\int\int \frac{B_0^2\cosh^4 u_0 \sin^2 v \cosh^2 u \sin^2 v}{\cosh^2 u(\cosh^2 u-\cos^2 v)^2}\,d\tau \\ = \frac{c^3 B_0^2 \cosh^4 u_0}{2}\int_0^{\pi/2}\int_{\bar{u}}^{u_0}\frac{\sin^4 v\cosh u\cos v}{(\cosh^2 u-\cos^2 v)}\,du\,dv \\ = (F^2/6\pi^2 R)P\end{aligned} \tag{11.142}$$

where

$$P = (3/4e)[\tilde{P}(u_0)-\tilde{P}(\bar{u})] \tag{11.143}$$

with

$$\tilde{P}(u) = (\sinh u)/3 - \sinh^3 u + \arctan(\sinh u) + \sinh^4 u \arctan(\mathrm{cosech} u). \tag{11.144}$$

The total magnetic contribution is thus

$$\left(\frac{F^2}{6\pi^2 R}\right)(L+P) \equiv \left(\frac{F^2}{6\pi^2 R}\right)m(e,\eta) \tag{11.145}$$

and the 11-component of the virial theorem for a cloud in equilibrium can now be written:

$$p_e(M/\bar{\rho}) = Ma^2 - \frac{GM^2}{5R}Q(e) + \frac{F^2}{6\pi^2 R}m(e,\eta). \tag{11.146}$$

When $e \to 0$, $m(e,\eta)$ reduces to $8(1-\eta)/5$ for all η between 0 and unity, in agreement with (11.10) and (11.12). When $e \to 1$, $\bar{u}$ and $\tilde{P}(\bar{u}) \to 0$; and if also $\eta \to 0$, $e_0 \to \eta \to 0$, $u_0 \to \mathrm{arccosh}(1/\eta)$, $L \to -\eta \to 0$, $m \to P \to 3\pi/8$. For all eccentricities, at small η

$$m(e,\eta) \simeq m(e,0) - 8\eta/5 \tag{11.147}$$

with

$$\begin{aligned}m(e,0) = \frac{3}{4e}\left[\frac{\pi}{2} - \frac{(1-e^2)^{1/2}}{3e} + \frac{(1-e^2)^{3/2}}{e^3} - \arctan\frac{(1-e^2)^{1/2}}{e}\right] \\ +(1-e^2)^{1/2} - \frac{3(1-e^2)^2}{4e^5}\arctan\frac{e}{(1-e^2)^{1/2}},\end{aligned} \tag{11.148}$$

and

$$1.6 = m(0,0) > m(e,0) > m(1,0) = 3\pi/8. \tag{11.149}$$

When $e \to 0$, $m(e,0) \to 1.6$; as e increases to unity, $m(e,0)$ decreases to $3\pi/8$.

By similar analysis (see Mestel and Paris 1984 for details), the magnetic terms in the 33-virial component, applied again to the volume within the surface just outside S_0, combine into

$$-\frac{F^2}{6\pi^2 R}n(e,\eta) \equiv -\frac{F^2}{6\pi^2 R}\left[P + m - \{\arctan(\sinh u)\}_{\bar{u}}^{u_0}/e\right]. \tag{11.150}$$

The gravitational term is

$$\mathcal{V}_{33} = -(GM^2/5R)H(e), \tag{11.151}$$

with

$$H(e) = 3(1-e^2)^{1/2}[e - (1-e^2)^{1/2}\arcsin e]/e^3 \tag{11.152}$$

decreasing monotonically from unity at $e = 0$ to $\simeq 3Z/R$ as $e \to 1$. The 33-component of equilibrium becomes

$$p_e(M/\bar{\rho}) = Ma^2 - \frac{GM^2}{5R}H(e) - \frac{F^2}{6\pi^2 R}n(e,\eta). \tag{11.153}$$

However, for reasons noted in Section 11.2, a more realistic estimate for the gross magnetic contribution to z-equilibrium is in fact given by application of the virial theorem to a surface just outside the spheroid $\bar{S}$ rather than S_0. The surface term $\int_{\bar{S}} x_3 T_{3j} n_j \, \mathrm{d}S$ then combines with the volume term within $\bar{S}$ to yield

$$\begin{aligned}\frac{F^2}{6\pi^2 R}q(e) \equiv &- \frac{F^2}{6\pi^2 R}(1-e^2)^{3/2} \\ &\times \left[\frac{(3-e^2)}{e^4} - \frac{3(1-e^2)^{1/2}}{e^5}\arctan\frac{e}{(1-e^2)^{1/2}}\right]\end{aligned} \tag{11.154}$$

instead of (11.150), so that (11.153) holds with $n(e,\eta)$ replaced by $q(e)$.

APPENDIX B

Magnetic braking by Alfvén waves – detailed treatment

As discussed in Section 11.6, the evolution of a cloud subject to magnetic braking of its rotation is treated by studying first a series of subsidiary problems in which the structure of the cloud and of its associated poloidal field is assumed time-independent. The simplest generalization of the cylindrical field model (described by (11.65)–(11.67)) which simulates the distorted structure to be expected from gravitational contraction is that of Fig. 11.1(b), with the cloud spherical and with radial field lines between R and R_0. Within R and beyond R_0, Ω and B_ϕ will again satisfy (11.67) with $v_A = \bar{B}/(4\pi\bar{\rho})^{1/2}$ and $B_0/(4\pi\rho_0)^{1/2})$ respectively. In any domain where $\mathbf{B}_p$ is non-uniform, Ω and B_ϕ now satisfy (5.14) and (5.15):

$$\frac{\partial^2\Omega}{\partial t^2} = \frac{\mathbf{B}_p \cdot \nabla\left(\varpi^2\mathbf{B}_p \cdot \nabla\Omega\right)}{4\pi\varpi^2\rho} \tag{11.155}$$

and

$$\frac{\partial^2 B_\phi}{\partial t^2} = \varpi\mathbf{B}_p \cdot \nabla\left(\frac{\mathbf{B}_p \cdot \nabla(\varpi B_\phi)}{4\pi\varpi^2\rho}\right). \tag{11.156}$$

Substitution of $\mathbf{B}_p = (\bar{B}R^2\cos\theta/r^3)\mathbf{r}$ yields

$$\frac{\partial^2\Omega}{\partial t^2} = v_A^2\frac{\partial^2\Omega}{\partial r^2} \tag{11.157}$$

and

$$\frac{\partial^2(rB_\phi)}{\partial t^2} = r^2 B_r\frac{\partial}{\partial r}\left(\frac{B_r}{4\pi\rho r^2}\frac{\partial(rB_\phi)}{\partial r}\right) \tag{11.158}$$

with

$$v_A = B_r/(4\pi\rho)^{1/2} = (\bar{B}R^2\cos\theta/r^2)/(4\pi\rho)^{1/2}. \tag{11.159}$$

The standard one-dimensional wave equation for both Ω and (rB_ϕ) – with v_A constant on each field line – requires the ad hoc and implausible assumption $\rho \propto 1/r^4$. Thus even in this simplest possible simulation of the distorted field, in general the waves in the radial field domain do not travel with constant profiles; and in all cases they will again suffer repeated reflections at the field discontinuities at R and R_0.

For a model in which the cloud is a cool local condensation in a hot medium with a long thermal scale-height, the most appropriate assumption takes for ρ the background density ρ_0. When the cloud is meant to represent a subcondensation forming within a more or less isothermal cool massive cloud, the surrounding gas may flow in with the accretion density field $\rho \propto 1/r^{3/2}$ (Bondi 1952). The law $\rho \propto 1/r^3$ has ρ continuous at both R and R_0 if the cloud mass $M = 4\pi\rho_0 R_0^3/3$, i.e. if $K = 1$ in (11.15), implying no preferential flow of gas down $\mathbf{B}_0$ during the cloud's formation. In GMPII, the generalization of the Ebert cylindrical problem of Section 11.6 is constructed for $\rho = \rho_0(R_0/r)^n$ in the radial field domain, with numerical results worked out for $n = 0, 3/2, 3$ and with $K = 1$. Because of the assumptions of axisymmetry and of a fixed structure for $\mathbf{B}_\mathrm{p}$, there is again no coupling between individual field lines. Along each field line – parametrized conveniently by the constant angle θ in the radial domain – the subsequent propagation of the Alfvén waves is studied by taking the Laplace transforms $L_\Omega = \int_0^\infty \mathrm{e}^{-st}\,\Omega\,\mathrm{d}t, L_B = \int_0^\infty \mathrm{e}^{-st}\,B_\phi\,\mathrm{d}t$. The first problem solved is the analogue of the Ebert problem, with initial conditions $B_\phi = 0$ everywhere, and $\Omega = \Omega_0$ for $r > R, \Omega = \bar{\Omega}$ for $r \le R$. In the three zones of Fig. 11.1(b), the solutions then have the forms

$$\begin{aligned}\text{Zone I}: L_\Omega &= A\cosh(sz/\bar{V}) + \bar{\Omega}/s\\ \text{Zone II}: L_\Omega &= B_1 L_1(r) + B_2 L_2(r) + \Omega_0/s\\ \text{Zone III}: L_\Omega &= C\exp(-sz/V_0) + \Omega_0/s,\end{aligned} \tag{11.160}$$

where

$$\bar{V} = \bar{B}/(4\pi\bar{\rho})^{1/2}, \qquad V_0 = B_0/(4\pi\rho_0)^{1/2}, \tag{11.161}$$

$$L_1(r) = 2^{-\nu}\Gamma(1-\nu)D^\nu \mathrm{I}_{-\nu}(D), \qquad L_2(r) = 2^\nu\Gamma(1+\nu)D^{-\nu}\mathrm{I}_\nu(D), \tag{11.162}$$

with

$$D(s,r) = (2\nu R_0/V_0\cos\theta)s(r/R_0)^{1/2\nu}, \qquad \nu = 1/(6-n), \tag{11.163}$$

and $\mathrm{I}_\nu, \mathrm{I}_{-\nu}$ are modified Bessel functions in standard notation (e.g. Abramowitz and Stegun 1965). The continuity of Ω and B_ϕ on the surfaces R, R_0 for $t > 0$ implies continuity of L_Ω and L_B, so fixing the coefficients A, B_1, B_2, C as complicated bilinear expressions in $L_{1,2}(R_0)$ and $L_{1,2}(R)$.

Construction of the complete solution is laborious, requiring the location of two sequences of poles with rather subtle properties, given by the vanishing of the common denominator of the coefficients. The solution shows the expected qualitative behaviour: propagation of the discontinuity in Ω with the local Alfvén speed, successive partial reflections at the field discontinuities at R and R_0, and asymptotic approach everywhere of Ω to Ω_0 and B_ϕ to zero as the initial excess angular momentum of the cloud is propagated to infinity. The total angular momentum transported from the cloud at time t, given by application of (2.48), allows definition of the *transport function* $F(\nu,\eta,t)$ with $\eta \equiv R/R_0$:

$$-\int_0^t \mathrm{d}t\left(\int(\varpi B_\phi/4\pi)\mathbf{B}_\mathrm{p}\cdot\mathbf{n}\,\mathrm{d}S\right) \equiv (2MR^2/5)(\bar{\Omega}-\Omega_0)F(\nu,\eta,t). \tag{11.164}$$

Typical examples of F are shown in Fig. 11.6.

As already emphasized, this solution simulates the behaviour of Ω in a cloud of *sub*critical mass, held in equilibrium essentially by the balance of magnetic and gravitational forces. For a supercritical cloud the problem as formulated is a subsidiary problem: one is interested in behaviour at small rather than large t (see the discussion below). An earlier paper (GMPI) studied a simplified

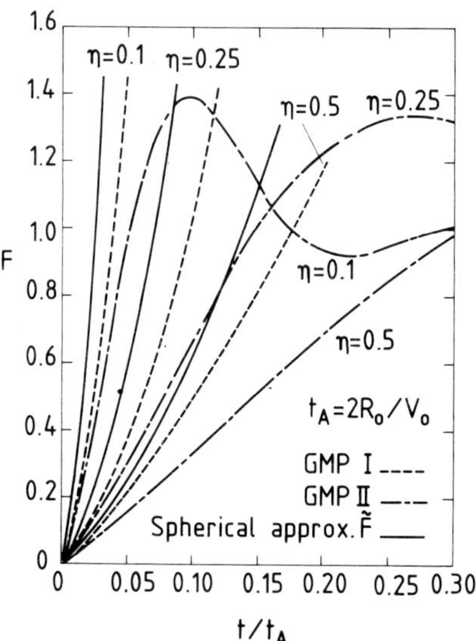

FIG. 11.6. The transport functions, defined in (11.164), for different cases.

subsidiary problem, with the wave propagation into the cloud suppressed through imposition of the extra boundary condition $\Omega(R) = \bar{\Omega}$ for $t \geq 0$. The solution is then of form (11.160) with $A = 0$; the detailed analysis – especially the search for zeros of the common denominator of B_1, B_2 and C in the complex plane – is much simplified. Because of the absence of the successive reflections and transmissions at R, the profile of Ω is also simpler: the discontinuity again propagates from the cloud surface with the local Alfvén speed, and Ω is always *monotonic decreasing* along a field line. (This is a consequence of the variation of v_A – (11.157) does not predict propagation of a given profile without distortion.) Intuitively, it is clear that the maintenance of a fixed $\bar{\Omega}$ at R will systematically exaggerate the rate of angular momentum transport, and the new transport function F' – again defined by (11.164) – is found always to be larger than F (Fig. 11.6). A still higher upper limit emerges when the outward decrease in Ω is ignored, so that the propagation of the wave is supposed to cause an instantaneous jump from Ω_0 to $\bar{\Omega}$ (essentially what is assumed in deriving the rough estimate (11.79)). At time t, a wave travelling along the radial field line with parameter θ reaches the radius $r(t)$ ($< R_0$) given by

$$t = \int_R^{r(t,\theta)} \frac{\mathrm{d}r}{v_A} = \frac{(4\pi\rho_0)^{1/2}}{3\bar{B}R^2\cos\theta}[r^3(t,\theta) - R^3]. \tag{11.165}$$

According to this latest approximation, the angular momentum deposited at time t is then

$$2(\bar{\Omega}-\Omega_0)\int_0^{\pi/2}\sin\theta\,\mathrm{d}\theta\int_R^{r(t,\theta)}2\pi\rho_0 r^2\sin^2\theta\,\mathrm{d}r \equiv \frac{2MR^2}{5K}(\bar{\Omega}-\Omega_0)\tilde{F}(\eta,t) \tag{11.166}$$

where the new transport function $\tilde{F}$ is

$$\begin{aligned}\tilde{F}(\eta,t) &= -\eta^3 \\ &+ \frac{3\eta^3}{2}\left[\frac{9(1+x)^{8/3}}{7x}\left(\frac{1}{8}+\frac{1}{11x}-\frac{3}{88x^2}\right)-\frac{3}{8x}\left(1-\frac{9}{77x^2}\right)\right]\end{aligned} \tag{11.167}$$

with

$$\eta = R/R_0, \qquad x = (6/\eta^3)(t/t_{\mathrm{A}}), \qquad t_{\mathrm{A}} = 2R_0(4\pi\rho_0)^{1/2}/B_0. \tag{11.168}$$

The characteristic time t_{A} is the time of travel of an Alfvén wave across the sphere R_0, for which the cloud field is just the undistorted local galactic field $\mathbf{B}_0$. The factor K in (11.166) again comes from the mass (11.15), which takes explicit account of accumulation of gas by preferential flow down the field lines. The plots of $\tilde{F}$ against t for these values of η confirm that $\tilde{F} > F' > F$. When $x \gg 1$,

$$F = (4.78/\eta^2)(t/t_{\mathrm{A}})^{5/3}. \tag{11.169}$$

The models of Fig. 11.2 are clearly more appropriate than those of Fig. 11.1(b) for a cool, mildly turbulent cloud, because of their taking some cognizance of the spontaneous flattening; but a rigorous treatment of the analogous subsidiary problem would be still more laborious than the work in GMPI and II, because of the complicated algebraic form of the non-radial field in the domain between $\bar{S}$ and S_0 as discussed in Appendix A. However, the experience gained from study of the spherical cloud model enables *upper limits* to the braking rate to be estimated with minimal effort (MPII). Again, one ignores the outward decrease of Ω, so that the gas behind the wave-front propagating along a field line is assumed set into corotation with the cloud. In the notation of Appendix A, a wave-front from $u = \bar{u}$ along the field line $v =$ constant reaches the point $u < u_0$ at time

$$t = \int_{\bar{u}}^{u} h_1\,\mathrm{d}u\frac{(4\pi\rho_0)^{1/2}}{B_u} = \frac{(4\pi\rho_0)^{1/2}}{B_0\cosh^2 u_0}\frac{c}{\sin v}\int_{\bar{u}}^{u}(\sinh^2 u+\sin^2 v)\cosh u\,du. \tag{11.170}$$

This defines for a chosen value of t (sufficiently small) a value $u(t,v)$; in normalized form

$$\frac{2\sin v}{e^3\eta^3}\left(\frac{t}{t_{\mathrm{A}}}\right) = \left(\frac{1}{3}\sinh^3 u+\sinh u\sin^2 v\right)_{\bar{u}}^{u(t/t_{\mathrm{A}},v)} \tag{11.171}$$

where η and t_{A} are as in (11.168). The approximation of corotation behind the wave-front then yields as the analogue of (11.166) and (11.167)

$$\frac{2MR_0^2\eta^2}{5K}(\bar{\Omega}-\Omega_0)\tilde{F}(\eta,t/t_{\rm A}) \tag{11.172}$$

with

$$\tilde{F}(\eta,t/t_{\rm A}) = (15e^2/4)\,(t/t_{\rm A}) + (15\eta^3 e^5/2)\int_0^{\pi/2}\cos^3 v\left(\sinh^5 u/5+\sin^2 v\sinh^3 u/3\right)_{\bar{u}}^{u}\,{\rm d}v \tag{11.173}$$

for the angular momentum transported in time t from the whole cloud. Numerical solution of (11.171) for $0 < \theta < \pi/2$ and given $t/t_{\rm A}$ enables calculation of the integrand in (11.171) and hence of the new transport function $\tilde{F}(\eta,t/t_{\rm A})$. It is found that for small $t/t_{\rm A}, \tilde{F} \simeq (15/4)(t/t_{\rm A}) + {\rm O}(t/t_{\rm A})^2$ irrespective of e. In fact the simple function (11.167) computed for spherical clouds is a good approximation to $\tilde{F}$ in (11.172) for all e: the angular momentum transport is not sensitive to the shape of the cloud (as long as no field line detachment occurs, as is implied by the adoption of the structure of Fig. 11.2.)

The computations of MPII follow approximately the contraction of a rotating supercritical isothermal magnetic cloud, again modelled as a uniform oblate spheroid with instantaneous semi-axes $R = \eta R_0, Z = \xi R_0$, eccentricity $e = (1-\xi^2/\eta^2)^{1/2}$, and uniform angular velocity Ω_c. The 11- and 33-virial equations of Section 3.3 are again used, but now with the kinetic and inertia terms retained. The adoption of a uniform spheroid at all epochs implies that the contraction is assumed to be homologous, with the instantaneous poloidal velocity at the point (ϖ, z) given by $[(\dot{R}/R)\varpi,(\dot{Z}/Z)z] = [(\dot{\eta}/\eta)\varpi,(\dot{\xi}/\xi)z]$. The two kinetic terms are $(M\dot{R}^2/5+M\Omega_c^2R^2/5)$, and the inertia term $\ddot{I}_{11}/2 = M(R\ddot{R}+\dot{R}^2)/5$. A convenient centrifugal parameter is $\alpha = (\Omega_c^2R^3/GM)/Q(e)$, so that centrifugal balance corresponds to $\alpha = 1$. The generalization of (11.18) then becomes with the help of (11.15)

$$\frac{{\rm d}^2\eta}{{\rm d}T^2} = -\frac{1}{\eta^2}\left[Q(e)(1-\alpha)-\frac{5\pi^2}{6}f^2 m(e,\eta)-\eta l\left(1-\frac{\Pi\xi\eta^2}{K}\right)\right] \tag{11.174}$$

where $\Pi = p_e/\rho_0 a^2$, $l = 5a^2R_0/GM$, and the non-dimensional time T is defined in terms of an appropriate free-fall time t_f by

$$T = t/t_f, \qquad t_f = (4\pi G\rho_0 K/3)^{-1/2}. \tag{11.175}$$

Likewise the generalization of (11.26) is

$$\frac{{\rm d}^2\xi}{{\rm d}T^2} = -\frac{1}{\xi\eta}\left[H(e)+\frac{5\pi^2}{6}f^2 q(e)-\eta l\left(1-\frac{\Pi\xi\eta^2}{K}\right)\right]. \tag{11.176}$$

Finally, the pseudo-problem solution (11.172) is used to estimate the instantaneous rate of transport of angular momentum. The angular velocity $\bar{\Omega}$ is identified with the instantaneous value of Ω_c at the epoch T when the cloud has non-dimensional axes η, ξ. At epoch $\bar{t}/t_\mathrm{A}$ in the pseudo-problem (with η, ξ, e *fixed*), the magnetic stresses are supposed to have transported a small fraction ϵ of $(2MR_0^2/5)(\Omega_c - \Omega_0)$, the excess angular momentum of the cloud over that associated with the background rotation Ω_0: i.e. the epoch $\bar{t}/t_\mathrm{A}$ is defined by

$$\tilde{F}(\eta, \bar{t}/t_\mathrm{A}) = \epsilon K \tag{11.177}$$

with $\tilde{F}$ given by (11.172). The corresponding instantaneous rate of braking can then be used in the real problem:

$$-\frac{\mathrm{d}}{\mathrm{d}t}\left(\frac{2}{5}MR_0^2\eta^2\frac{\Omega_c}{\Omega_0}\right) = \frac{\epsilon}{\bar{t}}\left(\frac{2}{5}MR_0^2\eta^2\right)\left(\frac{\Omega_c}{\Omega_0} - 1\right). \tag{11.178}$$

A non-dimensional angular momentum is conveniently defined by

$$h = \frac{\Omega_c \bar{R}^2}{\Omega_0 R_0^2} = \frac{\Omega_c}{\Omega_0}\eta^2; \tag{11.179}$$

(11.178) then transforms into

$$-\frac{\mathrm{d}h}{\mathrm{d}t} = \frac{\epsilon}{\bar{t}/t_\mathrm{A}}\frac{\pi 6^{1/2} f\eta^2}{6}\left(\frac{h}{\eta^2} - 1\right). \tag{11.180}$$

Recall that the function $\tilde{F}$ defined by (11.172) is not very different from the analogous function (11.167) for the spherical cloud problem, and that if ϵ is not too small this in turn can often be well approximated by (11.169), whence from (11.177), $t/t_\mathrm{A} \simeq (\epsilon K\eta^2/4.78)^{3/5}$, and (11.180) becomes

$$-\frac{\mathrm{d}h}{\mathrm{d}T} \simeq 1.64\epsilon^{2/5}K^{-3/5}f\eta^{4/5}\left(\frac{h}{\eta^2} - 1\right). \tag{11.181}$$

Once $\Omega_c/\Omega_0 = h/\eta^2 \gg 1$, (11.181) predicts a characteristic non-dimensional braking time $\simeq \eta^{6/5}/fK^{2/5}$, of the same order as that given by the rough estimate (11.79).

The centrifugal parameter α appearing in (11.174) can be written as

$$\alpha = \frac{\Omega_c^2\bar{R}^3}{GMQ(\bar{e})} = \frac{h^2}{\eta}\frac{1}{Q(\bar{e})}\left(\frac{\Omega_0^2 R_0^3}{GM}\right). \tag{11.182}$$

Equations (11.174), (11.176) and (11.181) can then be numerically integrated from given initial conditions. The results are not sensitive to the choice of ϵ in (11.177). A typical example is given in Fig. 11.7. The overall conclusion (MPII) is that as predicted: the contraction of a supercritical cloud tends to hug the line of magneto-centrifugo-gravitational balance, and is far from the line of corotation. The approximations used in deriving (11.181) are in the direction of exaggerating

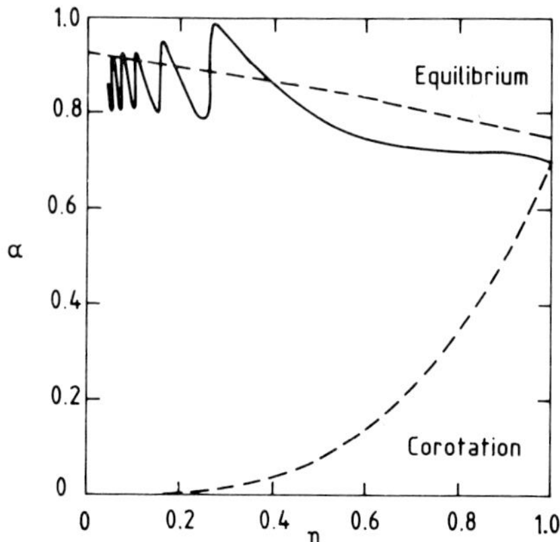

FIG. 11.7. An example of the contraction of a cloud, supercritical in $G^{1/2}M/F$, subject to magnetic braking.

the rate of braking, not least the use of a pseudo-problem which always has an infinite shear at the cloud surface. One can therefore feel confident that an exact numerical treatment will yield a fortiori the same general conclusion.

The earlier study (MPI) followed the evolution of both supercritical and subcritical *spherical* clouds, by implication supposed prevented from flattening by Alfvénic turbulence – cf. Section 12.7. The treatment was a little more rigorous, in that the subsidiary problem solved for each epoch of the contraction allows for an initial non-zero value of $\dot{\bar{\Omega}}$, so adding the term $\bar{\Omega}/s^2$ to the Zone I solution in (11.160). The corresponding transport function is constructed in GMPII, Section 3, and included in the analogue of the braking equation (11.178) employed in MPI, Section 4. The inclusion of this term increases the braking efficiency by up to a factor 2, especially when $R/R_0 \equiv \eta$ is still near unity, but the asymptotic behaviour and the overall qualitative conclusions are unchanged.

In MPI Section 5, solutions are constructed for subcritical clouds subject to simultaneous braking and flux leakage; the results confirm the expectation (cf. Section 11.6) that near corotation is maintained as long as $f/\dot{f}$ > the free-fall time t_f (Fig. 11.8).

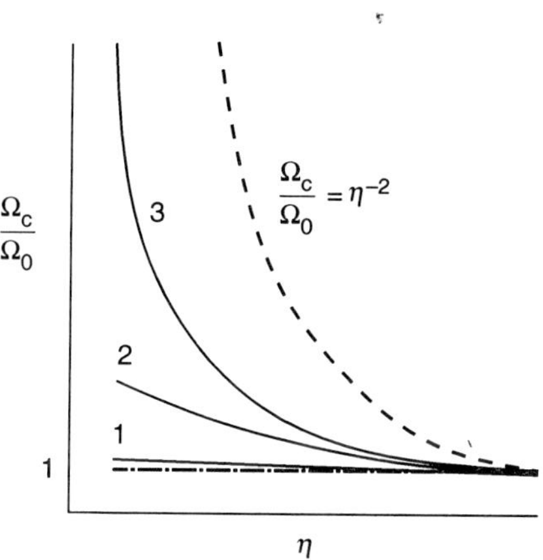

FIG. 11.8. An example of the contraction of a cloud subcritical in $G^{1/2}M/F$, subject to simultaneous magnetic braking and flux leakage. The dashed curve $\Omega_c/\Omega_0 = \eta^{-2}$ is that of angular momentum conservation; the dot-dashed curve $\Omega_c/\Omega_0 = 1$ is that of corotation. The curve for $\Omega_c(\eta)/\Omega_0$ with the label 1 is with the ratio of the flux-leakage time $-f/\dot{f}$ to the free-fall time t_f nearly unity. For all smaller values of this ratio, the curve is still closer to the corotation line. The curves 2 and 3 that deviate strongly from corotation have this ratio ≈ 5 and 10 respectively. (Adapted from Mestel and Paris 1979.)

12

MAGNETISM AND STAR FORMATION II

12.1 Resumé

It is convenient at this point to pause and summarize the essential qualitative results that have emerged. An adequate theory will elucidate how the galactic magnetic field may sometimes inhibit the formation of stars, and otherwise may modify profoundly the details of the physics and dynamics of the problem, in particular affecting the form of the ultimate mass spectrum. The persistence of cool massive molecular clouds can plausibly be put down to the Lorentz forces exerted by a gravitationally distorted magnetic field, which can hold a cloud in equilibrium provided the flux/mass ratio is supercritical: i.e. provided the parameter $f \equiv F/\pi^2 G^{1/2} M$ is greater than f_c, where both f_c and so also the associated critical mass $M_c(F)$ depend only modestly on the degree of flattening from a spherical structure (Section 11.2). If the cloud is only weakly turbulent, with speeds at most of sonic order, then one would expect the cloud to adopt a highly flattened structure; but with supersonic turbulent motions with an energy density of the order of the magnetic, the cloud can be either moderately or highly oblate, depending on the level of the energy input that maintains the turbulence against dissipation.

The mass–flux relation can be interpreted as requiring a minimum accumulation length *along* the large-scale magnetic field if a gravitationally collapsing body is to form (Section 11.3.1). Once f has become subcritical,[4] then collapse with flux conservation cannot be halted by magnetic forces alone Reduction of f can occur either by accretion of further cooled gas down the galactic field lines, or by lateral diffusion of the neutral gaseous component across the magnetic field, which remains inductively coupled to the ionized component (Section 11.7). This 'ambipolar diffusion' is inevitable, as it is caused by the same Lorentz force that maintains the cloud in near equilibrium. The diffusion is limited by friction due to ion–neutral and grain–neutral collisions, and in fact its time-scale does not become short enough to be of interest until n_i/n_n falls to values of the order inferred for molecular clouds.

The interaction between angular momentum and magnetism yields different results for subcritical and supercritical masses. A contracting cloud with a finite angular momentum will spin up, generating torsional Alfvén waves which transport angular momentum to infinity along undetached field lines. The ratio

[4]The terms 'supercritical' and 'subcritical' are used sometimes as in, for example,'a cloud supercritical in f', and sometimes as in the equivalent phrase 'a cloud subcritical in $G^{1/2}M/F$', or in obvious variations such as 'subcritical in mass'. No confusion should arise.

of the magnetic braking time t_b to the free-fall time t_f is roughly proportional to the ratio $M/M_c(F)$. Hence once a supercritical, weakly turbulent cloud or fragment has formed – i.e. once $M/M_c > 1$ – its initial contraction will be well approximated by magnetically diluted free-fall, with angular momentum conservation, until centrifugal forces have grown to values close to self-gravitation. The subsequent contraction is in quasi-equilibrium, with centrifugal force increasing steadily so as to take up the bulk of the difference between the magnetic and gravitational forces, and in a time determined by the rate of braking (which for near-critical clouds will not exceed by much the free-fall time). By contrast, a subcritical cloud or fragment is easily brought into approximate corotation with the surroundings. Slow flux leakage (either from the cloud as a whole, or from a cloud core into the rest of the cloud) enables further contraction to occur, again with near-corotation maintained, and so with a *de*creasing ratio of centrifugal force to gravity. This can lead ultimately to the formation of supercritical fragments with much less specific angular momentum than in a mass which was already supercritical at the lowest molecular cloud densities. The subsequent magnetically-diluted free-fall will not be halted by the growing centrifugal forces until much higher densities are reached. Thus the rotational history of a cloud appears to depend sensitively on whether it begins its life as subcritical or supercritical in $F/G^{1/2}M$ (Section 11.6 and Appendix 11B). There is no difficulty in generalizing these conclusions to allow for a turbulent contribution to support that can be of the same order as the magnetic.

The treatment has been mainly for simple cloud models of uniform density, both contraction and flux diffusion being assumed to occur homologously. In reality, one can expect differential flux diffusion, with proportionately higher leakage out of cloud cores with a significantly lower n_i/n_n ratio. The conclusions can be taken over, with 'cloud core' replacing 'cloud'. Thus differential leakage can lead to a new form of fragmentation, with the subcritical cloud as a whole retaining the bulk of its flux and so remaining essentially in magneto-gravitational-turbulent-equilibrium, while a cloud core, initially subcritical in M/F, becomes supercritical and goes over ultimately into gravitational collapse.

Once a mildly supercritical cloud or fragment has formed, its subsequent history may depend rather sensitively on the energy supply that feeds the turbulence. The very existence of a minimum accumulation length *along* the field – necessary for the formation of an oblate body, supercritical in mass – but with no length across the field entering, shows that flattening along the field allows the possibility of fragmentation (Section 11.4). If, however, the turbulence is able to keep the contraction of the mass as more or less isotropic, then fragmentation of a cloud that is subject to strict flux-freezing is prevented.

We have seen that an initially supercritical cloud is likely to retain a strong centrifugal field during its contraction. If the rotation and magnetic axes are more or less parallel, then if the turbulence is weak, flow down the field lines is unaffected by centrifugal force, and fragmentation is possible. However, if the axes are more nearly mutually perpendicular, contraction can occur following angular momentum transport but it will be quasi-isotropic. Fragmentation then depends on flux-leakage. Very rough estimates suggest that the final masses reaching the opaque phase would be significantly larger than for cases with the axes parallel. However, again the persistence or decay of Alfvénic turbulence is equally important.

It is by no means obvious that the final spectrum of bodies reaching the main sequence is fixed in the early phases of star formation that are being discussed in these two chapters. Nevertheless, it is important to note what effects the presence of the galactic field may have on the masses that enter the opaque phases, at densities $\simeq 10^{10}$ cm^{-3} or so. Consider again a cloud or fragment that has become supercritical and is collapsing. Again it is the strength of the Alfvénic turbulence that appears to be crucial. In Section 11.4 it was pointed out that if flow down the field lines is uninhibited, and if the analogue of Hoyle's opacity criterion is what determines the end of fragmentation, then we may still be faced with the embarrassment of final fragment masses that are much too small (cf. Section 11.1). However, suppose instead that the supercritical fragment is kept quasi-spherical by the persisting turbulence, so that further fragmentation does not occur. The contraction is approximately isotropic, with $\bar{B} \propto \bar{\rho}^{2/3}$, and the fragment mass is given roughly by (11.53) with the coefficient increased to $6^{1/2}/\pi^2$ to allow for the effect of the turbulence (supposed to double the dilatory effect of the magnetic field – cf. Section 12.7):

$$M \simeq (6^{1/2}/\pi^2)(\bar{B}/\bar{\rho}^{2/3})^3(1/G^{2/3}). \tag{12.1}$$

Provided flux remains frozen-in all the way to the opaque phase, and that enough angular momentum has been lost for centrifugal forces to remain small up to the opaque phase, then it is indeed the magnetic properties of the cloud which have determined the masses of the 'proto-stars' – defined as opaque, self-gravitating masses. How these masses are related to the ultimate stellar masses depends on the as yet ill-understood later phases. We can anticipate that the fragment will normally have retained enough angular momentum for centrifugal forces to become important well before the main sequence is reached.

These results have made extensive use of the virial theorems. The method is vindicated by its quick yield of approximate quantitative results, from which emerge qualitative ideas that serve as a valuable guide: the parameter $f = F/\pi^2 G^{1/2} M$, defining magnetically supercritical and subcritical bodies; the accumulation length; the B–ρ relations expected in different geometries; fragmentation following anisotropic flow; the sensitivity of rotational evolution to small variations in f; and contraction controlled either by braking or by flux leakage. But equally, the discussion has shown up some of the limitations of a method that inevitably involves judicious guesswork. For detailed comparison with observation, one wants to replace models with $\bar{\rho}$ and $\bar{B}$ assumed uniform by $(\rho, \mathbf{B})$ fields that satisfy the equations at each point rather than in the mean. A reliable answer to some theoretical questions – for example the actual amount of angular momentum lost by a cloud as it contracts (because of flux leakage) from M/M_c subcritical to supercritical – requires more precise numerical knowledge than the virial method can give (cf. Sections 11.6 and 11.7). The important question of field line detachment requires a more sophisticated treatment.

The rest of the chapter is devoted largely to a more detailed theoretical treatment of some of these issues. But the reader will justly ask: how much encouragement is the theorist receiving from observation? As stated in Section 11.3.3, comparison with observation summarized in Myers and Goodman (1988) appeared to yield no obvious discrepancy from the picture developed in Sections 11.2 and 11.3. More recently, Crutcher *et al.* (1993) again concluded that OH Zeeman observations of dark clouds yielded data consistent with the hypotheses that the clouds are in approximate virial equilibrium between magnetic and gravitational energy, with the central regions of density $\simeq 10^3\ \mathrm{cm}^{-3}$ approximately critical in flux/mass, and with the supersonic line-widths observed a manifestation of MHD motions such as Alfvén waves. The data were also claimed to be consistent with the detailed modelling of the evolution of cloud cores driven by ambipolar diffusion. However, more recent preliminary CN Zeeman observations of molecular cloud cores of higher density ($n(\mathrm{H}_2) \simeq 1.6 \times 10^6\ \mathrm{cm}^{-3}$) may be harbingers of a forced reappraisal of the theoretical picture. Crutcher *et al.* (1996) report upper limits for the line-of-sight B significantly less than the values expected from either a critical $G^{1/2}M/F$ or from internal motions at the Alfvén speed. They concede that the very small sample of clouds and also uncertainties in the theoretical estimates preclude definite conclusions at this time. Meanwhile, Padoan and Nordlund (1997) argue that the observed supersonic motions in dark clouds are not Alfvén waves: rather, their numerical experiments in the regime with the ratio β of gas to magnetic pressure above unity are consistent with observation, while the model with $\beta \ll 1$ has features that conflict with observation.

As in the theory of convection, discussed in Chapter 4, it may be that the supercomputers are again going to force a reappraisal. If it should turn out that ambipolar diffusion, though as noted, far larger than Ohmic, gives nevertheless an underestimate for flux leakage, then one hazards the guess that macroscopic instabilities are again playing a major role, especially since in diffuse clouds, heating and cooling times are much shorter than dynamical times, so that there is no likelihood of adiabatic stabilization.

12.2 Magneto-gravitational equilibrium: exact disc-like models

In this section we return to the problem of the structure of self-gravitating magnetic clouds in mechanical equilibrium. The initial motivation is to give a more rigorous justification of the inferences drawn from the virial theorem treatment. We begin with the mathematically simplest, disc-like models, which are an appropriate idealization for cool, weakly turbulent, slowly rotating clouds that have flattened along the field (Parker 1974; Mestel and Ray 1985; Barker and Mestel 1990, 1996; Ciolek and Mouschovias 1993; Basu and Mouschovias 1994; Li and Shu 1997; Shu and Li 1997; Zweibel and Lovelace 1997). The cold cloud is thus approximated as a zero-thickness, axisymmetric disc of finite *area* density $\sigma(\varpi)$. The field component B_z is continuous across the disc, but B_ϖ

changes sign; by Ampère's law, the discontinuity requires a sheet current density $J_\theta(\varpi) = (c/4\pi)2B_\varpi(\varpi)$, where $B_\varpi(\varpi) \equiv B_\varpi(\varpi, 0+)$. The ϖ-component of the magnetic force per unit disc area must then balance the area density of the gravitational force:

$$J_\theta B_z/c = B_\varpi B_z/2\pi = \sigma g \tag{12.2}$$

where $g(\varpi)$ is the gravitational acceleration in the $-\varpi$-direction, generated by the mass distribution $\sigma(\varpi)$.

12.2.1 *Finite disc models*

A faithful simulation of the clouds or fragments discussed in Chapter 11 adopts a disc model of finite radius R, immersed in a warm, low-density medium permeated by a uniform background field $\mathbf{B}_0$ parallel to the disc axis. Beyond $\varpi = R$ there is no cold gas condensed into the $z = 0$ plane, so $\sigma(\varpi)$, $J_\theta(\varpi)$ and so also $B_\varpi(\varpi, 0)$ all vanish for $\varpi > R$. The vector potential $\mathbf{A} = (0, A, 0)$ due to given disc currents J_θ can be written down as an integral and the associated component $B_z(\varpi, 0) = \partial[\varpi A(\varpi, 0)]/\varpi\,\partial\varpi$ constructed. The total equatorial z-component $B_z(\varpi)$ to be used in the balance equation (12.2) includes the contribution of the background field:

$$B_z(\varpi) = B_0 + \frac{\partial[\varpi A(\varpi, 0)]}{\varpi\,\partial\varpi}. \tag{12.3}$$

Dimensionless quantities are defined by

$$\varpi = xR, \qquad z = ZR,$$
$$A = C(B_0 R)a, \qquad J = cCB_0 b_x/2\pi, \qquad (B_\varpi, B_z) = CB_0(b_x, b_z), \tag{12.4}$$

where C is a numerical parameter pertaining to the particular model under study; whence (12.3) and (12.2) become

$$b_z = \frac{1}{C} + \frac{\partial(xa)}{x\,\partial x}, \tag{12.5}$$

$$b_x b_z = \frac{2\pi\,\sigma(\varpi)\,g(\varpi)}{C^2 B_0^2} \tag{12.6}$$

respectively.

We are explicitly separating the problem of cloud equilibrium from that of cloud formation: in particular, in general we make no a priori prescription as to the distribution of $\sigma(\varpi)$ with respect to the magnetic flux density $B_z(\varpi)$. Rather, we proceed by adopting different $\sigma(\varpi)$ functions with their associated $g(\varpi)$ fields, and then attempt to construct the $\mathbf{B}$-fields which satisfy equilibrium at each point. A constructed model contains implicitly a precise distribution of $\sigma(B_z)$ (equivalently, of the distribution of mass $M(\varpi)$ with respect to flux $F(\varpi)$). The reverse procedure, pioneered by Mouschovias (1976*a,b*) (for models with a finite sound speed), prescribes this distribution *ab initio*, for example, as that defined by a sphere of uniform density, permeated by a uniform field $\mathbf{B}_0$, and then supposes the system to relax to a state of magnetostatic equilibrium, subject to strict flux-freezing and so with conservation of the $M(F)$-relation. Either procedure prescribes – directly or indirectly – the initial $M(F)$-relation and so has a degree of arbitrariness: at this stage, the choice of technique is largely a matter of mathematical convenience.

We are interested in cloud models with flux/mass ratio close to the critical values estimated from the virial models, and which therefore have a marked degree of central condensation. In Appendix A, two sequences are considered.

(a) The 'moderately centrally condensed' sequence is defined by

$$g(\varpi) = V^2\varpi/(\varpi^2 + l^2R^2) = (V^2/R)[x/(x^2 + l^2)],$$
$$\sigma_a(\varpi) = (V^2l^2/\pi^2 GR)\tilde{\sigma}_a(x) \qquad 0 \leq l \leq 1 \tag{12.7}$$

with $\tilde{\sigma}_a(x)$ given by (12.126). With the convenient normalization

$$C^2B_0^2 = V^4/GR^2, \tag{12.8}$$

the balance equation (12.6) becomes

$$l^2x\tilde{\sigma}_a(x)/(x^2 + l^2) = (\pi/2)b_xb_z. \tag{12.9}$$

This class includes the special case of the point-singularity model with $l = 0$, for which $g(\varpi) = V^2/\varpi$ is the gravitational field that can be balanced by the centrifugal acceleration of the uniform velocity V, and the associated $\sigma_a(\varpi)$ reduces to

$$\frac{V^2}{2\pi G\varpi}\left[1 - \frac{2}{\pi}\arcsin\frac{\varpi}{R}\right] \tag{12.10}$$

(Mestel 1963). When l is finite, $g(x) \propto x$ for $x \ll l$ and goes over into a form roughly $\propto 1/x$ for $x > l$.

(b) The 'strongly centrally condensed' sequence is defined by

$$g(\varpi) = V^2lR\varpi/(\varpi^3 + l^3R^3) = (V^2/R)[lx/(x^3 + l^3)],$$
$$\tilde{\sigma}_b(\varpi) = (V^2l/\pi^2GR)\tilde{\sigma}_b(x) \qquad 0 \leq l \leq 1 \tag{12.11}$$

with $\tilde{\sigma}_b(x)$ given by (12.128). With the same normalization (12.8), (12.6) now becomes

$$l^2x\tilde{\sigma}_b(x)/(x^3 + l^3) = (\pi/2)b_xb_z. \tag{12.12}$$

Again $g(x) \propto x$ for $x \ll l$, but now $g \propto 1/x^2$ roughly for $x > l$.

The method of construction of the magnetic field is described in Appendix A. For either of these sequences the choice of l leaves C as the only free parameter. Of the two terms in (12.5), $1/C$ represents the background field and the other the field due to the disc currents. Some examples of the run of B_ϖ, B_z with ϖ

in the disc are given in Figs 12.6 and 12.7, and some illustrative field lines in Fig. 12.9.

The total mass of a disc model defined by (12.7) or (12.11) is

$$M = (2V^2R/\pi G)\tilde{M} \tag{12.13}$$

where

$$\tilde{M}_a = l^2 \int_0^1 \tilde{\sigma}_a(x) x \, \mathrm{d}x, \qquad \tilde{M}_b = l \int_0^1 \tilde{\sigma}_b(x) x \, \mathrm{d}x \tag{12.14}$$

for sequences (a) and (b) respectively. Once an equilibrium distribution $b_z(x)$ has been constructed for a given mass model and a choice of C, the total flux F threading the disc can be found:

$$F = 2\pi C B_0 R^2 \int_0^1 x b_z(x) \, \mathrm{d}x = (2\pi V^2 R/G^{1/2}) \int_0^1 x b_z \, \mathrm{d}x. \tag{12.15}$$

by (12.8). As expected from the virial method discussed in Section 11.2, it is found that for each choice of $\tilde{\sigma}(x)$ and the associated $g(x)$, converging models in magneto-gravitational equilibrium exist only for values of $1/C$ larger than a limiting value, which yield (cf. (11.9)):

$$f \equiv \frac{F}{\pi^2 G^{1/2} M} = \frac{\int_0^1 b_z x \, \mathrm{d}x}{\tilde{M}} > f_c. \tag{12.16}$$

The precise value f_c of the critical flux/mass parameter depends to some extent on the particular mass model under study. As $1/C$ decreases towards the limiting value, f decreases monotonically, and simultaneously the flux becomes more centrally condensed. In sequence (a), f_c decreases slowly but monotonically with decreasing l to the value $\simeq 0.74$ for the point-singularity ($l = 0$) model. In sequence (b), however, there is a striking difference, as shown in Fig. 12.1. On branch A, f_c again decreases as l decreases towards $\simeq 0.4$, but for smaller l, on branch B, f_c increases, reaching $\simeq 1.29$ at $l = 0.2$ and $\simeq 4.28$ at $l = 0.1$. The region above the curve is populated by points representing discs with supercritical f values for given l. As noted, the corresponding curve for sequence (a) models has just the monotonic branch A all the way to $l = 0$. (It is remarkable that the value $f_c = (2/\pi)(3/5)^{1/2} \simeq 0.49$, found in Section 11.2 by the virial method applied to a uniformly dense flattened spheroid, is so close to the values $f_c \simeq 0.74$ found on the branch A, a further tribute to the usefulness of the virial theorem.)

There is a further qualitative difference associated with the contrasting curves of f_c against l. For all of sequence (a), and for sequence (b) with $l \gtrsim 0.4$, the limiting equilibrium model for each l has a ring of neutral points at the edge of the disc (cf. Fig. 12.7); whereas in the most centrally condensed models – sequence (b) with $l < 0.4$ – the limiting model has B_z reaching its minimum – necessarily

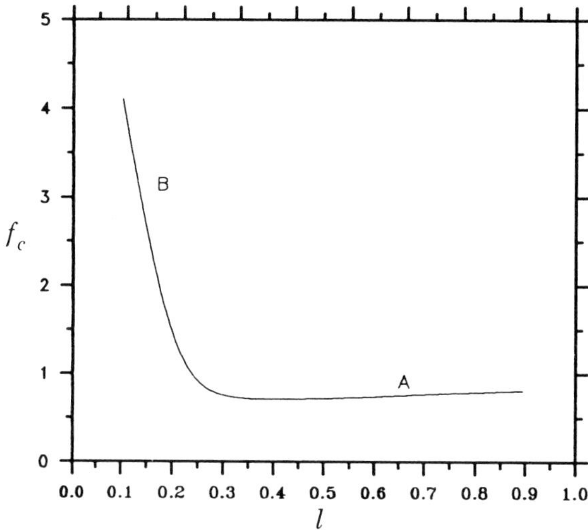

FIG. 12.1. The limiting value f_c of the flux/mass parameter f plotted against the mass distribution parameter l for sequence (b) – the strongly condensed discs with $g(\varpi)$ and the associated $\sigma(\varpi)$ given by (12.11) and (12.128).

positive – within $x = 1$. When f is close to f_c, all centrally condensed mass distributions (of either sequence) require for mechanical equilibrium a corresponding centrally condensed current distribution. Each current loop $J_\theta(\varpi')\mathrm{d}\varpi'$ generates a field at the disc with $B_z < 0$ or $B_z > 0$ according as $\varpi < \varpi'$ or $\varpi > \varpi'$. Outside the domain where the bulk of the current flows, the sum of the fields B_z due to all the loops will clearly be negative, so reducing the value of B_z below the background value B_0. However, mechanical equilibrium clearly requires that B_z stays positive within the disc (where $\sigma(\varpi)$ is finite). The numerical work then shows that when the mass central condensation is not too strong (sequence (a), and sequence (b) with $l > 0.4$), the limiting models have B_z reaching zero at the disc edge. The attempt to construct models with $f < f_c$ fails to converge, simply because the still more centrally concentrated current distribution – required for equilibrium in the central regions – would yield a total B_z with a zero somewhere in the outer parts of the disc, so that the balance condition (12.2) would locally fail. For stronger mass concentrations (sequence (b) with $l < 0.4$), the associated centrally condensed currents produce a minimum in the total B_z that is within the disc and so must be positive if equilibrium is to hold. Failure of convergence for $f < f_c(l)$ occurs when there is no way of distributing the reduced flux in such a way that $B_z B_x/2\pi$ is able to balance σg at all points.

As already noted, with $g(\varpi)$ and so also $\sigma(\varpi)$ prescribed, each converged solution for $B_z(\varpi)$ contains implicitly the distribution of b_z with $\tilde{\sigma}$, or equivalently the run of the dimensionless distribution of flux $\int_0^x b_z x\,\mathrm{d}x$ against mass

$\int_0^x \sigma x \, \mathrm{d}x$ between the origin and radius x. A convenient way of illustrating this distribution is to perform the 'unscrambling exercise', by which the disc is imagined converted – under the constraint of strict flux-freezing – into a 'parent cloud' of uniform volume density ρ_0 and permeated by the uniform background field $\mathbf{B}_0$. The transformation $\varpi \to \varpi', R \to R', x \equiv \varpi/R \to x' \equiv \varpi'/R'$, subject to detailed flux and mass conservation, yields the dimensionless coordinate $Z'(x') \equiv \sigma'/2\rho_0 R'$, defining the shape of the parent cloud (cf. Fig. 12.8 in Appendix A). If the model clouds do form under the constraint of strict flux-freezing from a hypothetical uniform background, then $Z'(x')$ does indeed represent the parent cloud; but if not (cf. Section 11.3.1), $Z'(x')$ is merely a convenient way of picturing the mass/flux distribution.

The evolution of the detailed flux/mass distribution within a molecular cloud is prima facie determined by slow ambipolar diffusion. As noted in Section 11.7, differential leakage within a cloud, for example more rapid diffusion within a central core, will lead to greater central concentration of mass and ultimate local breakdown in equilibrium – a form of fragmentation. The results summarized in Fig. 12.1 can be used to illustrate this. Consider a disc-like cloud built according to one of the sequence (b) models, with a given initial value of l, and with $f > f_c(l)$, so that it has a representative point above the curve AB. Suppose that flux leakage from the cloud as a whole is negligible, so that f stays constant, but that the shorter time-scale of diffusion in the core causes evolution towards greater central condensation. A rigorous treatment will again attempt to follow the actual drift of the field-plus-plasma relative to the neutral bulk, and the consequent associated adjustment of the density and so of the gravitational field (cf. Section 12.4). We can simulate this with the simplifying approximation that assumes the evolving gravitational field always conforms to one of our sequence (b) models, but with l monotonically decreasing, so that our representative point moves horizontally to the left. It is at once clear that models above but in the neighbourhood of branch A are stable: in fact a point that is just on branch A will move into the region above the curve. But once l has decreased to 0.4, further differential flux leakage and consequent reduction in l at constant f will cause the points ultimately to cross the curve: a point that is on branch B finds that to maintain equilibrium with a reduced l, the disc would need to have *more* flux than it actually has, implying breakdown in equilibrium, as predicted by Nakano (1976).

12.2.2 *Infinite disc models*

The elliptic integral technique summarized above can be somewhat cumbersome in practice, and it is sometimes more convenient to consider models in which σ, J_θ and B_ϖ are allowed to extend to infinity. An alternative method develops the Fourier–Bessel technique, used by Toomre (1963) for the study of disc-like galaxies. Outside the disc, $\mathbf{B}$ is both curl-free and divergence-free, so we may introduce a scalar potential, writing $\mathbf{B} = -\nabla\Phi$ with $\nabla^2\Phi = 0$. The solution satisfying finiteness conditions at $\varpi = \infty, z = \pm\infty$ is then written as a Hankel transform:

$$\Phi = (2\pi/c)\int_0^\infty S(k)\mathrm{J}_0(k\varpi)\exp(-k|z|)\,\mathrm{d}k \tag{12.17}$$

in standard notation. The surface current in the disc is given from Ampère's law as

$$J_\theta = (c/2\pi)(B_\varpi)_{0+} = \int_0^\infty S(k)k\mathrm{J}_1(k\varpi)\,\mathrm{d}k \tag{12.18}$$

so that

$$S(k) = \int_0^\infty J_\theta\varpi\mathrm{J}_1(k\varpi)\,\mathrm{d}\varpi = (c/2\pi)\int_0^\infty (B_\varpi)_{0+}\varpi\mathrm{J}_1(k\varpi)\,\mathrm{d}\varpi. \tag{12.19}$$

At the disc the total normal component is

$$B_z = B_0 + (2\pi/c)\int_0^\infty S(k)k\mathrm{J}_0(k\varpi)\,\mathrm{d}k. \tag{12.20}$$

The simplest case illustrating the method adopts the point-singularity disc model (Mestel 1963), in which the gravitational field is of the form (12.7) with $l = 0$:

$$g(\varpi) = V^2/\varpi, \tag{12.21}$$

generated by the analogue of (12.10) that extends to infinity:

$$\sigma(\varpi) = \frac{V^2}{2\pi G\varpi}. \tag{12.22}$$

The background field B_0 in (12.20) is now assumed to vanish, but it is convenient to retain the symbol B_0.

We normalize by writing

$$\begin{aligned} \varpi &= Rx, \qquad z = RZ, \qquad k = \kappa/R, \\ \mathbf{B} &= CB_0(b_x, b_z), \qquad \Phi = CB_0R\phi, \qquad S(k) = (cCB_0R^2/2\pi)s(\kappa), \end{aligned} \tag{12.23}$$

so that

$$\phi = \int_0^\infty s(\kappa)\mathrm{J}_0(\kappa x)\exp(-\kappa|Z|)\,\mathrm{d}\kappa, \tag{12.24}$$

$$(b_x)_{0+} = \int_0^\infty s(\kappa)\kappa\mathrm{J}_1(\kappa x)\,\mathrm{d}\kappa, \tag{12.25}$$

$$s(\kappa) = \int_0^\infty x(b_x)_{0+}\mathrm{J}_1(\kappa x)\,\mathrm{d}\kappa. \tag{12.26}$$

In general, the dimensionless z-component is given by

$$b_z = \frac{1}{C} + \int_0^\infty s(\kappa)\kappa\mathrm{J}_0(\kappa x)\,\mathrm{d}\kappa; \tag{12.27}$$

however, in this first example, the background field term $1/C$ is absent. The condition of equilibrium now takes on the simple form:

$$b_x b_z = \frac{1}{x^2}. \tag{12.28}$$

From the elementary properties of the Bessel function, it is easy to verify that the system has the solution

$$s(\kappa) = \frac{1}{\kappa}, \qquad b_x = b_z = \frac{1}{x}. \tag{12.29}$$

This solution has the same ϖ-dependence for σ and B_z, with the relation

$$B_z(\varpi) = 2\pi G^{1/2}\sigma(\varpi) \tag{12.30}$$

holding exactly from the origin to infinity. Thus in this model, the magnetic flux $F(\varpi)$ crossing the circle ϖ is proportional to the mass $M(\varpi)$, so yielding the value $2/\pi$ for the dimensionless flux/mass parameter f (Mestel 1984, unpublished; Zweibel and Lovelace 1997). Not surprisingly, the same relation (12.30) is found to hold very approximately over much of the range of x in the small l members of the 'moderately centrally condensed' sequence (a) of Section 12.2.1 (cf. Mestel and Ray 1985).

Shu and Li (1997) give the generic name 'isopedic' to discs which have the ratio B_z/σ independent of ϖ (cf. Appendix B). Note, however, that if such a relation were to hold strictly everywhere in a *finite* disc, then B_z would necessarily vanish at the edge $\varpi = R$, and so could not describe a system with a finite external z-component $\mathbf{B}_0$. But equally the equilibrium condition (12.2) would then require that B_ϖ remains finite at $\varpi = R$, although $\sigma(\varpi)$ vanishes there. The extension of the solution into the domain $\varpi > R$ cannot then be field-free. In particular, the non-vanishing equatorial $\pm B_\varpi$-components require the presence of equatorial sheet currents not only in $0 < \varpi < R$ but also between ϖ and ∞: even though by hypothesis there is no matter dense enough to feel a significant gravitational force, again the oppositely directed B_ϖ-components exert a pinching force that must be opposed by a finite thermal pressure. This raises once more the question of reconnection.

Return now to the picture with a background field $\mathbf{B}_0$. Again we do not impose any B_z–σ relation a priori, but rather select a mass distribution with its associated gravitational field and then construct the class of magnetic fields that satisfy mechanical equilibrium. Another simple case with a formally infinite extension is:

$$\sigma(\varpi) = \sigma_0 \exp(-\varpi^2/R^2) \tag{12.31}$$

with

$$g(\varpi) = \frac{\pi^2 G \sigma_0^{3/2}}{M^{1/2}} \varpi \mathrm{F}(1.5;\ 2;\ -\varpi^2/R^2) \tag{12.32}$$

where $M = \pi\sigma_0 R^2$ is the disc mass and F is a confluent hypergeometric function (Toomre 1963). With the same normalizations, the condition of equilibrium now becomes

$$\pi b_x b_z/2 = x\exp(-x^2)\mathrm{F}(1.5;\ 2;\ -x^2); \tag{12.33}$$

b_z now satisfies (12.27) with $1/C$ finite, and

$$C^2 B_0^2 = \pi^{7/2} G\sigma_0^2 = \pi^{3/2} GM^2/R^4 \tag{12.34}$$

is the analogue of (12.8)

Equations (12.24)–(12.33) form a closed set which can again be solved by iteration for values of $1/C$ that are not too small (Barker and Mestel 1990). Once a convergent solution has been found, the field lines over all space can again be constructed from the normalized flux function, which has the form

$$-p = x^2/2 + Cx\int_0^\infty s(\kappa)\mathrm{J}_1(\kappa x)\exp(-\kappa|Z|)\,\mathrm{d}\kappa, \qquad p = P/\pi^{3/4}G^{1/2}M, \tag{12.35}$$

with P as usual defined by (12.61).The limiting value found for $1/C$ is $\simeq 0.17$; when $1/C$ is smaller the iterative process fails to converge. Just as for the 'strongly centrally condensed' sequence (b), defined by (12.11), in the near-critical models, Cb_z has a pronounced minimum located in the domain where the density begins to exponentiate down, subsequently climbing to its asymptotic value of unity. Again, the finite disc density in these infinitely extended models forbids the appearance of a neutral point, which would violate the equilibrium condition (12.6).

A useful generalization of the mass distribution (12.31) adopts

$$\sigma(\varpi) = \frac{M_e}{\pi R^2}\exp(-\varpi^2/R^2) + \frac{M_s}{\pi\eta^2 R^2}\exp(-\varpi^2/\eta^2 R^2) \equiv \left(\frac{M_e + M_s}{\pi R^2}\right)\sigma_d(x) \tag{12.36}$$

where

$$\sigma_d(x) = \frac{\mathrm{e}^{-x^2} + (M_s/\eta^2 M_e)\,\mathrm{e}^{-x^2/\eta^2}}{1 + M_s/M_e} \tag{12.37}$$

is the dimensionless density and again $x = \varpi/R$. With $\eta < 1$, this model simulates a centrally condensed core Gaussian of mass M_s and length-scale R, superposed on an envelope Gaussian of mass M_e and scale R. The mass within the radius ϖ is

$$M(\varpi) = M_e(1 - \mathrm{e}^{-x^2}) + M_s(1 - \mathrm{e}^{-x^2/\eta^2}) \equiv (M_e + M_s)\tilde{M}(x). \tag{12.38}$$

When $\eta = 1$ the model reduces to the single Gaussian of the same total mass $M = (M_e + M_s)$. The gravitational field has the form

$$g(\varpi) = (\pi^{1/2} GM/R^2) g_d(x), \tag{12.39}$$

where

$$g_d(x) = \frac{x}{1 + M_s/M_e}\left[\mathrm{F}(1.5;\ 2;\ -x^2) + \frac{M_s/M_e}{\eta^3}\mathrm{F}(1.5;\ 2;\ -x^2/\eta^2)\right]. \quad (12.40)$$

The equilibrium equation (12.6) takes the form

$$\pi b_x b_z/2 = \sigma_d(x) g_d(x) \quad (12.41)$$

and solutions can again be constructed for a sequence of $1/C$ down to a limiting value.

12.2.3 *Collapsed core models*

These double Gaussian models can again be used to illustrate how differential diffusion within a cloud of constant mass and flux can lead to breakdown in equilibrium, with the consequent separation out and collapse of a core, once it has become supercritical in $G^{1/2}M/F$ (Barker and Mestel 1996). The core may ultimately form either a single star or a multiple system; its self-gravitation will then be balanced primarily by thermal pressure in the case of a single star, and by random or rotatory motions in the case of a multiple system. The uncondensed gas will adjust to a new state of magneto-gravitational equilibrium, feeling both its self-generated gravitational field and also that of the collapsed gas. An illustrative disc-like model with a collapsed core, of total mass M, adopts for the uncollapsed envelope the area density distribution (Barker and Mestel 1996)

$$\sigma(\varpi) = \frac{M}{\pi R^2}\sigma_{\mathrm{coll}}(x), \qquad \sigma_{\mathrm{coll}} = \left[\exp(-x^2/2) - \exp(-x^2/2\epsilon^2)\right]^2, \quad (12.42)$$

with the parameter $\epsilon < 1$. The mass of the envelope $M_e = M(1-\epsilon^2)^2/(1+\epsilon^2)$; that of the core $M_s = M - M_e = M\epsilon^2(3-\epsilon^2)/(1+\epsilon^2)$. Within the radius x, the dimensionless mass $\tilde{M}(x) = M(\varpi)/M$ is

$$\tilde{M}(x) = 1 - \exp(-x^2) - \epsilon^2\exp(-x^2/\epsilon^2) + \frac{4\epsilon^2}{1+\epsilon^2}\exp[-x^2(1+\epsilon^2)/2\epsilon^2]. \quad (12.43)$$

The gravitational field g_{coll} felt by the gas is the sum of the Coulomb field due to the collapsed mass M_{c} and that due to the density (12.42). With the same normalization as above, the field is conveniently written

$$g(\varpi) = \frac{\pi^{1/2}GM}{R^2}\tilde{g}_{\mathrm{coll}}(x) \quad (12.44)$$

where

$$\tilde{g}_{\mathrm{coll}}(x) = \frac{1}{\pi^{1/2}x^2}\frac{M_{\mathrm{c}}}{M} + x\mathrm{F}(1.5;\ 2;\ -x^2) + \frac{x}{\epsilon}\mathrm{F}(1.5;\ 2;\ -x^2/\epsilon^2) - \frac{2^{1/2}(1+\epsilon^2)^{1/2}x}{\epsilon}\mathrm{F}(1.5;\ 2;\ -(1+\epsilon^2)x^2/2\epsilon^2). \quad (12.45)$$

The dimensionless balance equation is again (12.41) with σ_d, g_d replaced by σ_{coll}, g_{coll}. The crucial difference from the non-collapsed double Gaussian models is through the presence of the collapsed mass at $x = 0$, which yields $g \propto 1/x^2$ near $x = 0$ instead of $g \propto x$. This is the reason for the choice (12.42) for σ, which yields $M_{\text{coll}} \propto x^4$, so that $b_x b_z \propto x^2$ as $x \to 0$.

With a particular value ϵ chosen, collapsed core solutions can again be found by iteration for values of $1/C$ down to a limiting value in general between 0.1 and 0.2. For each converged solution – whether a double Gaussian or a collapsed core model – one can construct dimensional and dimensionless flux functions $F(\varpi)$ and $\tilde{F}(x)$, related by

$$F(\varpi) = 2\pi \int_0^{\varpi} B_z \varpi \, \mathrm{d}\varpi = 2\pi C B_0 R^2 \int_0^x b_z x \, \mathrm{d}x \equiv 2\pi^{7/4} G^{1/2} M \tilde{F}(x) \quad (12.46)$$

on use of the normalization (12.34). Figure 12.2 contains plots of $\tilde{F}(x)$ against $\tilde{M}(x)$ for some double-Gaussian models and the collapsed core model, all for the same $M_s/M_e = 0.644$ and with $1/C = 1.0$. Note that over about one-half of the range in mass, the relation between $\tilde{F}$ and $\tilde{M}$ is nearly linear, reflecting a linear relation between B_z and $2\pi G^{1/2}\sigma$, as in the 'isopedic' models cited in Section 12.2.1. By (12.34), with B_0 and M fixed, a constant value for C implies a fixed radial scale R. The different double-Gaussian models show a decrease in $\tilde{F}(x)$ for given $\tilde{M}(x)$, correlated with a reduction in η, and so again illustrate the systematic central condensation of the cloud as flux leaks outwards. By contrast, the collapsed core model illustrates what happens when the leakage leads to a breakdown in magneto-gravitational equilibrium in the central regions. The 'missing flux' for $\tilde{M} < 0.4$ begins by collapsing with the core, thus playing no part in the equilibrium of the envelope. Following reconnection, most of the flux through the core will consist of finite field lines, while the disconnected field with infinite lines would spring back and again contribute to magneto-gravitational balance of the envelope. The new equilibrium structure will remain qualitatively similar to that described by (12.42).

12.2.4 *Disc models with partial turbulent support*

The construction of disc-like, zero-pressure equilibria precludes the presence of neutral points in any domain where there is finite mass density, requiring magnetic support against gravity according to (12.2). All the field lines emanating from the cloud must be 'infinite', and the condition $\nabla \cdot \mathbf{B} = 0$ puts a severe constraint on the strength of the mean field $\bar{B}$. As noted by McKee *et al.* (1993), although models constructed with undetached field lines can yield B_c/B_0 large, the mean value $\bar{B}/B_0$ over the cloud does not reach the value ≈ 10 that observers tentatively report for some clouds. Virial theorem treatments (Strittmatter 1966; Zweibel 1990) suggest that field line detachment will allow high enough $\bar{B}/B_0$-values, but at the cost of having magnetic neutral points. We can illustrate this further by studying finite disc-like models, with the area density and its associated gravitational field again prescribed, but now modifying (12.2) to include a

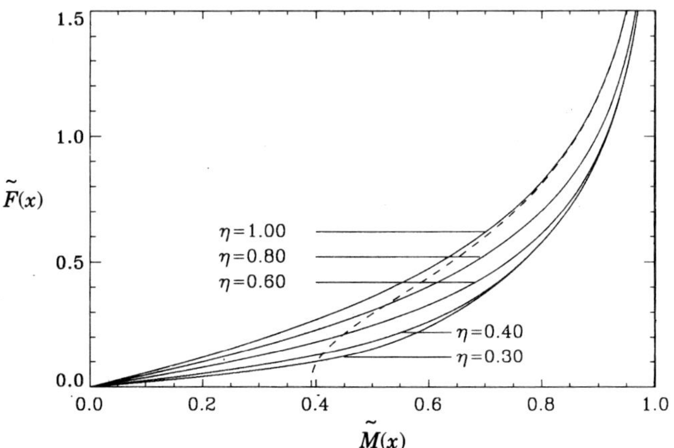

FIG. 12.2. Plots of $\tilde{F}(x)$ against $\tilde{M}(x)$ for double-Gaussian models (solid) and one collapsed core (dashed) model, all with $1/C = 1.0$ and $M_s/M_e = 0.644$.

two-dimensional turbulent pressure term:

$$\frac{\mathrm{d}}{\mathrm{d}\varpi}(\sigma v_t^2) + \sigma g = \frac{B_\varpi B_z}{2\pi}. \tag{12.47}$$

We consider just the limiting case of the moderately centrally condensed sequence together with the standard normalizations (12.4):

$$\begin{aligned} \sigma &= \frac{2\sigma_0}{\pi}\left(\frac{\arccos x}{x}\right), \qquad g = \frac{2\pi G\sigma_0}{x}, \\ M &= 4\sigma_0 R^2, \qquad C^2 B_0^2 = \frac{4\pi^2 GM^2}{16R^4}, \end{aligned} \tag{12.48}$$

so that (12.47) becomes

$$\frac{\mathrm{d}}{\mathrm{d}x}\left(\frac{\Lambda \arccos x}{x}\right) + \frac{\arccos x}{x^2} = \frac{\pi}{2} b_x b_z, \tag{12.49}$$

where

$$\Lambda \equiv 2Rv_t^2/\pi GM. \tag{12.50}$$

The simplest procedure is to prescribe a mutually self-consistent pair of components (b_x, b_z) and then use (12.49) and (12.50) to construct the necessary v_t^2 field. The example chosen has

$$b_x = b\frac{(1-x^2)^{1/2}}{x} \qquad 0 < x < 1 \tag{12.51}$$

with $b_x = 0$ for $x > 1$. Equation (12.136) then yields the auxiliary function

$$s(x) = \frac{b}{2\pi}\int_x^1 \frac{(1-t^2)^{1/2}}{t(t^2-x^2)^{1/2}}\,\mathrm{d}t = \frac{b}{4}\left(\frac{1}{x}-1\right), \tag{12.52}$$

whence from (12.135)

$$a(x) = b\left(x - \frac{\pi x^2}{4}\right), \tag{12.53}$$

leading to

$$\frac{b_z(x)}{b} = \left(\frac{1}{Cb} - \frac{\pi}{2}\right) + \frac{1}{x} \tag{12.54}$$

from (12.5). Beyond $x = 1$, from (12.139),

$$\frac{xa(x)}{b} = x - \frac{1}{2}(x^2-1)^{1/2} - \frac{x^2}{2}\arcsin\frac{1}{x}, \tag{12.55}$$

and

$$\frac{b_z}{b} = \frac{1}{Cb} + \left(\frac{1}{x} - \arcsin\frac{1}{x}\right). \tag{12.56}$$

When $1/Cb < \pi/2 - 1$, there is an O-type neutral point in the disc at $x_n = 1/\left(\pi/2 - 1/Cb\right)$ and a corresponding external X-type neutral point at x_0 defined by the vanishing of (12.56). The mean value of $B_z(\varpi)/B_0$ over the interval $(0,\varpi)$ within the disc is

$$\frac{\bar{B}_z(x)}{B_0} = \frac{2C}{x^2}\int_0^x b_z x\,\mathrm{d}x = 1 + Cb\left(\frac{2}{x} - \frac{\pi}{2}\right). \tag{12.57}$$

For example, if $b = 1$, $1/C = 0.1$, then $x_0 \approx 0.68$, containing 70 per cent of the cloud mass. Also, $\bar{B}_z(1)/B_0 \approx 4.3$; and $\bar{B}_z(x_0)/B_0 \approx 15$. The fraction of the total cloud flux undetached is given by $P(x_0)/P(x_n)$. Integration of the balance equation (12.49) with the boundary condition that Λ is finite at $x = 1$ then yields

$$\Lambda = \left(1 + \frac{\pi b^2}{2}x\right) + \frac{1}{\arccos x}\left(\frac{\pi b^2}{2}(1-x^2)^{1/2}\left[x\left(\frac{1}{Cb} - \frac{\pi}{2}\right) - 1\right]\right)$$
$$-\frac{x}{\arccos x}\log\left[\frac{1+(1-x^2)^{1/2}}{x}\right]\left(1 + \frac{\pi b^2}{2}\left(\frac{1}{Cb} - \frac{\pi}{2}\right)\right). \tag{12.58}$$

A necessary condition is that Λ is positive over all of $0 < x < 1$, requiring that $b < 1$. A typical case is plotted in Fig. 12.3.

The models illustrate the point made above: through the introduction of a significant turbulent contribution to lateral support of the disc-like cloud, the severe constraint on the field structure is lifted, allowing neutral points to form within the cloud (Fig. 12.4), and yielding a higher mean field over the bulk of

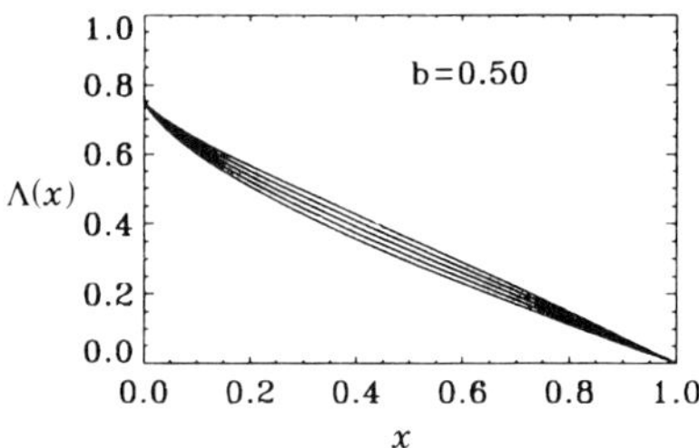

FIG. 12.3. The dimensionless mean squared turbulent velocity Λ for the Mestel disc model with $b = 0.5$ and a set of C-values.

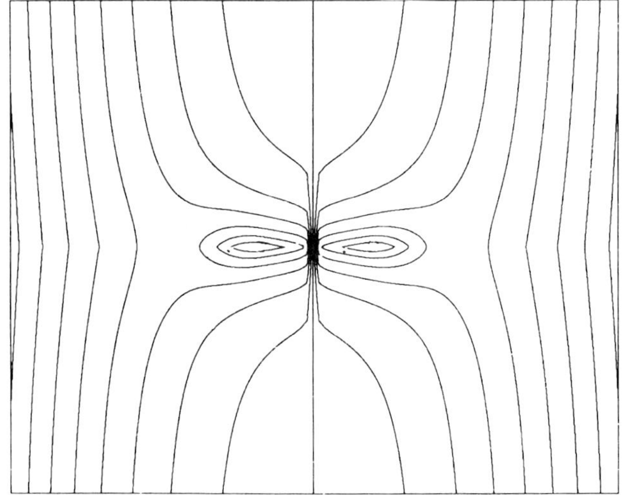

FIG. 12.4. A disc-like model with turbulent-plus-magnetic support against gravity, showing a domain with detached field lines.

the cloud. The obvious objection is that the most plausible turbulent motions are torsional oscillations of the field (Section 12.7); but this 'Alfvénic turbulence' will be weakest near the neutral ring, just where the turbulent support is most needed. A more acceptable model may have toroidal flux linking the poloidal (cf. Section 5.3), still allowing field line detachment, with the poloidal component still having a neutral ring, but with the total B staying finite.

12.3 Magneto-turbulent cloud models

We now turn to models in which a strong turbulent pressure is comparable in importance with gravity and the mean magnetic force. The picture is again similar to that of Fig. 11.2, with the cloud a local condensation from a warm, low-density medium, and symmetric about the direction of a uniform background field $\mathbf{B}_0$. The self-gravitation of the cloud is balanced by the pressure and the Lorentz force exerted by the distorted magnetic field:

$$-\frac{\nabla p}{\rho} + \nabla V + \frac{(\nabla \times \mathbf{B}) \times \mathbf{B}}{4\pi\rho} = 0, \tag{12.59}$$

with the gravitational potential V satisfying Poisson's equation

$$\nabla^2 V = -4\pi G\rho. \tag{12.60}$$

Again $\mathbf{B}$ is conveniently written in terms of the flux function P:

$$\mathbf{B} = -\frac{\nabla P \times \mathbf{t}}{\varpi} = \left(\frac{1}{\varpi}\frac{\partial P}{\partial z},\ 0,\ -\frac{1}{\varpi}\frac{\partial P}{\partial \varpi}\right) \tag{12.61}$$

in cylindrical polar coordinates based on the mass-centre. For reasons that will emerge, a general barotropic equation of state $p = f(\rho)$ is adopted for the combined thermal-plus-turbulent pressure. From the component along $\mathbf{B}$ of (12.59), $\mathbf{B} \cdot \nabla(V - \int \mathrm{d}p/\rho) = 0$, so that

$$V - \int \frac{f'(\rho)}{\rho}\,\mathrm{d}\rho = -h(P) \tag{12.62}$$

relates the variation of ρ with V along a field line in terms of the enthalpy function h. The component of (12.59) perpendicular to $\mathbf{B}$ becomes

$$(\nabla \times \mathbf{B}) \times \mathbf{B} = 4\pi\rho h'(P)\,\nabla P, \tag{12.63}$$

or by use of (12.61)

$$\nabla \times \mathbf{B} = -4\pi\varpi\rho h'(P)\mathbf{t}. \tag{12.64}$$

Expression (12.62) is an implicit relation through $h(P)$ between the density ρ, the associated potential V and the flux function P. In differential form, the mass within the volume defined by neighbouring flux tubes $P, P + \mathrm{d}P$ satisfies:

$$\frac{\mathrm{d}M(P)}{\mathrm{d}P} = 2\int_0^{z_s(P)} \mathrm{d}z\,\frac{2\pi\varpi\rho(\varpi, z)}{(\partial P/\partial\varpi)_z}, \tag{12.65}$$

where $z_s(P)$ is the z-coordinate of the intersection of the cloud surface with the flux surface P. With a given (barotropic) equation of state, and a prescribed mass–flux relation $M(P)$, self-consistent solutions of (12.60) and (12.64) may be sought by iteration, subject to (12.65) and to appropriate boundary conditions at the cloud surface.

The first studies (Parker 1973; Mouschovias 1976*a*,*b*; Tomisaka *et al.* 1988) were of models with an isothermal pressure $p = \rho a^2$, for which (12.62) yields

$$\rho = \frac{q(P)}{a^2} \exp\left(\frac{V}{a^2}\right), \tag{12.66}$$

where

$$q(P) = a^2 \exp[h(P)/a^2]. \tag{12.67}$$

Equation (12.64) then becomes in cylindrical polars

$$\frac{\partial^2 P}{\partial \varpi^2} + \frac{\partial^2 P}{\partial z^2} - \frac{1}{\varpi}\frac{\partial P}{\partial \varpi} = -4\pi\varpi^2 \frac{\mathrm{d}q}{\mathrm{d}P} \exp\left(\frac{V}{a^2}\right). \tag{12.68}$$

Likewise, (12.60) and (12.66) combine into

$$\frac{\partial^2 V}{\partial \varpi^2} + \frac{\partial^2 V}{\partial z^2} + \frac{1}{\varpi}\frac{\partial V}{\partial \varpi} = -\frac{4\pi G}{a^2} q(P) \exp\left(\frac{V}{a^2}\right). \tag{12.69}$$

Equation (12.66) relates ρ along a particular field line to the value $\rho_e(P)$ at the point where the field line P crosses the equator. Thus ρ cannot vanish at any point: as for a non-magnetic cloud, equilibrium requires a finite external pressure. The integral of (12.63) across a surface layer of finite volume density but infinitesimal thickness yields as the boundary condition the continuity of the total stress

$$(p\delta_{ij} + T_{ij})n_j = (p + \mathbf{B}^2/8\pi)n_i - (\mathbf{B}\cdot\mathbf{n})B_i/4\pi, \tag{12.70}$$

where $\mathbf{n}$ is the unit normal vector. Provided $(\mathbf{B}\cdot\mathbf{n})$ is non-zero, its continuity across S implies by (12.70) continuity also of the tangential component of $\mathbf{B}$ and so also of p. In the warm external medium p is nearly constant on each field line (the term V/a^2 drops to a small value across the cloud surface, so $p = \rho a^2 \simeq q(P)$). In the simplest case, with $p(\infty)$ having the uniform value p_0, the condition $q(P) = p_0$ is propagated along the field lines from infinity all the way to the surface of the cloud, and (12.64) and (12.69) reduce outside the cloud to $\nabla\times\mathbf{B} = 0$ and Laplace's equation $\nabla^2 V = 0$ respectively. The flux function $P(\varpi, z) \to -B_0\varpi^2/2$ as $(\varpi, z) \to \infty$, to yield the uniform background field $\mathbf{B}_0$.

Insertion of (12.66) into (12.65) yields

$$q(P) = \frac{a^2}{2\pi}\frac{\mathrm{d}M}{\mathrm{d}P} \Big/ \int_0^{z_s(P)} \frac{\partial}{\partial P}[\varpi^2(z,P)] \exp\{V[\varpi(z,P),z]/a^2\}\,\mathrm{d}z \tag{12.71}$$

to be substituted into (12.68) and (12.69), valid within the cloud. With a prescribed $M(P)$ relation, the cloud structure is determined by iterative solution of these equations and of the 'vacuum' equations valid outside the cloud, with the cloud surface $z = z_s(P)$ emerging as part of the solution from the continuity conditions on $\mathbf{B}$ and ρ. Again it is worth noting that in our notation, the most recent calculations (Tomisaka *et al.* 1988) yield f_c between 0.5 and 1.0, to be compared with the virial theorem estimate of 0.49 for a flattened body, and the typical value 0.75 found for the exact disc-like models. The smallness of the deviations from the virial estimates is gratifying, but as noted in Section 12.1, in some applications – for example to the magnetic braking problem – one may need the greater accuracy which only the detailed numerical treatment can yield.

If the value adopted for a were that corresponding to the thermal temperature of an actual molecular cloud, then as remarked, the cloud would be quite well described by one of the disc-like models of Section 12.2. A more realistic approximation for the equation of state is that used by Lizano and Shu (1988): p is written as the sum of the isothermal term and an isotropic turbulent part

$$p_t = K \ln(\rho/\rho_0) \tag{12.72}$$

where K is a constant and ρ_0 a reference density. This particular barotropic law yields an effective wave speed

$$v_t \equiv (\mathrm{d}p_t/\mathrm{d}\rho)^{1/2} = (K/\rho)^{1/2}, \tag{12.73}$$

consistent with the observed correlation $\Delta v \propto \rho^{-1/2}$ found for the non-thermal part of molecular line-widths (e.g. McKee 1989), with K estimated as $(1.2\text{–}7.2) \times 10^{-11}$ dyn cm^{-2}. A tentative theoretical justification will emerge later.

In a cloud or cloud core which is supercritical in F/M, the turbulence makes no essential qualitative change but just reduces the distortion in the field lines necessary for support against gravity; whereas if F/M is markedly subcritical, a strong turbulence can take up the difference between gravity and Lorentz force and so prevent collapse. With the recognition that in observed molecular clouds, Alfvénic turbulence and the mean magnetic forces contribute comparably to support against gravity, it is helpful to follow Tomisaka *et al.* and Lizano and Shu and make explicit what is implicit, for example, in Section 11.4, defining M_{cr} as the minimum mass that can collapse against the Lorentz force exerted by the mean magnetic field and the associated Alfvénic turbulence.

12.4 Evolution through flux diffusion

As discussed in Section 11.7.1, the magnetic forces that hold the cloud in equilibrium also cause it to evolve by driving the outward diffusion of field-plus-plasma relative to the neutral bulk. The field diffuses at the speed $\mathbf{v}_d$ given by (11.97), so that the instantaneous mass $M(P,t)$ within the axisymmetric volume defined by the value of P changes according to

$$\frac{\partial M(P,t)}{\partial t} = \int_P \rho \mathbf{v}_d \cdot \mathbf{n}\,\mathrm{d}S = \frac{1}{\alpha}\int \frac{[(\nabla\times\mathbf{B})\times\mathbf{B}]\cdot\mathbf{n}\,\mathrm{d}S}{4\pi(n_i/n_n)\rho} \tag{12.74}$$

$$= \frac{4\pi h'(P)}{\alpha}\int_0^{z_s(P,t)} \frac{\varpi(z;P,t)[(\partial P/\partial\varpi)^2 + (\partial P/\partial z)^2]\,\mathrm{d}z}{(n_i/n_n)\,\partial P/\partial\varpi} \tag{12.75}$$

on use of (12.61) and (12.63). Given a relation for n_i/n_n, for example (11.114), and starting at time t from an equilibrium cloud model with a known equation of state and a definite mass–flux distribution $M(P,t)$, (12.75) yields $M(P,t+\delta t)$, which is then used for the construction of a new equilibrium model. If the cloud is isothermal, then (12.71) yields at once the new function $q(P)$ to be fed into (12.68) and (12.69). The process is continued until at some epoch, failure of convergence will indicate that equilibrium is no longer possible: as illustrated by the work of Section 11.7, flux leakage has converted a region initially supercritical in flux/mass to one that is so subcritical that the pressure is unable to preserve equilibrium, and collapse of at least a cloud core ensues.

Calculations along these lines have been performed by Nakano (1979, 1984), by Lizano and Shu (1989), and by Mouschovias and colleagues (1987, 1993, 1994, 1996), as already noted. The results depend to some extent on the assumptions made about the equation of state, which in turn determines whether the cloud is closer to respectively a quasi-spherical or a disc-like structure.

Nakano studied the evolution of a class with different mass but with the same value for B_0, the density $n_s = p_0/m_{\rm H}a^2$ = the density in nucleons cm^{-3} on the inner side of the cloud surface, and for the initial value of $G^{1/2}M/F$. He found that his numerical results are well represented by the relation

$$\frac{M_s}{M_{\rm J}} = 0.62\left(\frac{M}{M_{\rm J}}\right)^{0.4}, \tag{12.76}$$

where M_s is the mass of the final core that separates out, and $M_{\rm J}$ the Jeans mass in the initial state.

Similar computations have been performed by Lizano and Shu (1989) for clouds with an equation of state that is the superposition of an isothermal part $p = \rho a^2$ and the turbulent part (12.72). It is instructive to return at this point temporarily to the virial treatment of Section 11.2. The turbulence is supposed strong enough to justify ignoring departures from sphericity, so that (11.10) may be used, with the sound speed replaced by $(a^2 + v_t^2)^{1/2}$, where v_t is given by (12.73):

$$p_0 = \frac{3Ma^2}{4\pi R^3} + K - \frac{3GM^2}{20\pi R^4}\left[1 - \frac{f^2}{f_c^2}\left(1 - \frac{R}{R_0}\right)\right], \tag{12.77}$$

with p_0 again the pressure exerted on the surface by the external gas, and $f_c^2 = 3/4\pi^2$. Now specialize further by taking the turbulence to be Alfvénic, with $K = \bar{B}^2/4\pi$, whence $f^2 = F^2/\pi^4 GM^2 = 4KR^4/\pi GM^2$, and (12.77) can be written

$$\left(\frac{4\pi p_0 R_0^4}{3GM^2}\right)\left(\frac{R}{R_0}\right)^4 = \left(\frac{a^2 R_0}{GM}\right)\left(\frac{R}{R_0}\right) - \frac{1}{5} + \left(\frac{9}{20} - \frac{1}{5}\frac{R}{R_0}\right)\left(\frac{f}{f_c}\right)^2. \tag{12.78}$$

At small R the surface term may be ignored. A reduction in f due to flux leakage yields

$$\mathrm{d}\left(\frac{R}{R_0}\right)\left(\frac{a^2R_0}{GM}-\frac{f^2}{5f_c^2}\right)=-\mathrm{d}\left(\frac{f^2}{f_c^2}\right)\left(\frac{9}{20}-\frac{R}{5R_0}\right)>0. \tag{12.79}$$

Thus if $f^2/f_c^2 > 5a^2R_0/GM$, then R/R_0 decreases as flux leaks out, by (12.78) reaching zero when $f/f_c = 2/3$ rather than unity, because of the support due to the Alfvénic turbulence term. If, instead, $f^2/f_c^2 < 5a^2R_0/GM$, then the mass considered is small enough for stability of equilibrium to depend on the finite thermal pressure. The field adjusts itself until it is straight, with $R = R_0$, and so exerts no forces, but the non-zero field strength implies again a finite turbulent contribution to equilibrium at the radius R_0, satisfying

$$4\pi p_0 R_0^3 = 3Ma^2 - \frac{3GM^2}{5R_0} + 4\pi K R_0^3. \tag{12.80}$$

Now return to the exact treatment. The adopted turbulent pressure (12.72) yields from (12.62) the enthalpy function

$$h(P) = -V - K/\rho + a^2\log\rho + \text{constant}. \tag{12.81}$$

There is now no simple way of inverting (12.62) into an analogue of (12.66); the function $h(P)$ has to be determined by iteration, as part of the simultaneous solution of (12.60), (12.64) and (12.65). Lizano and Shu confirm that there are three possible outcomes. If the cloud mass M is close to but exceeds M_{cr} (defined in Section 12.3), then the cloud as a whole collapses with the flux nearly frozen-in; its subsequent history is discussed in Section 12.5. If initially $M_s < M < M_{\mathrm{cr}}$, then it is the pressure (dominantly turbulent) that prevents gravity from overwhelming the magnetic force, and evolution is through ambipolar diffusion. In condensations of small mass (in Lizano and Shu's terminology, 'subumbral' masses), the magnetic field is able steadily to straighten itself so that ultimately it exerts zero force, and gravity is balanced by pressure alone. By contrast, in superumbral masses, flux leakage is at first automatically compensated by further contraction in magneto-gravitational-turbulent equilibrium. These latter cases are of particular interest, since they lead ultimately to the collapse of a low mass proto-star.

The assumption $v_t^2 \propto 1/\rho$ has theoretical justification *ex post facto*. Even in cases where flux leakage from a cloud core causes further contraction, with increase in ρ, the calculations of Lizano and Shu show that during the approach to the supercritical state, B does not in fact increase very much; hence the Alfvén speed $v_{\mathrm{A}} \propto 1/\rho^{1/2}$ approximately, so that Alfvénic turbulence will indeed satisfy $v_t \propto 1/\rho^{1/2}$. Like Nakano (1979), Lizano and Shu find that their clouds evolve towards a singular structure reminiscent of that found earlier for a non-magnetic, isothermal sphere.

More recently, there has been a series of studies of the analogous problem for weakly turbulent, disc-like equilibrium models (Mouschovias 1996 and references therein). Especially noteworthy are those by Basu and Mouschovias (1994), who include angular momentum, confirming in detail that during evolution in quasi-equilibrium through slow flux leakage, centrifugal forces remain small. Lizano and Shu had found that the initially plausible assumption of quasi-equilibrium breaks down when the central density concentration exceeds a factor $\simeq 10^2$. Fiedler and Mouschovias (1993) and later Basu and Mouschovias (1994) retained the inertial terms and followed the evolution further. They found that the motions remain submagnetosonic, so that use of the quasi-magnetostatic equations is a tolerable approximation up to close to the predicted singular core formation. In their treatment of ambipolar diffusion, Mouschovias and colleagues follow Draine *et al.* (1983, 1986) in including the effect of neutral–grain collisions, which can reduce somewhat the rate of ambipolar diffusion.

Recall that in the treatment of magneto-gravitational equilibrium in Sections 12.2 and 12.3, there was no a priori choice of the distribution of mass with *flux*; instead, various illustrative mass distributions in space were selected, and examples constructed of the associated possible magnetic fields. The 'unscrambling' exercise was then performed to *infer* the $\sigma(B_z)$ relation implicit in each viable model. The reverse procedure effectively prescribes $\sigma(B_z)$ (or equivalently, $M(F)$) and then constructs the joint $\sigma(\varpi)$ and $B_z(\varpi)$ fields that yield equilibrium. As a particular constructed model evolves through flux diffusion, it will clearly begin by retaining some memory of the original $\sigma(B_z)$. However, according to Li and Shu (1997), it appears that in fact there is a singular isothermal state that (in the authors' words) emerges as an 'attractor' from the phase of quasi-magnetostatic evolution by ambipolar diffusion. As the cloud models approach the point of onset of collapse – referred to by these authors as the 'gravomagneto catastrophe' – the numerical simulations suggest that the volume density behaves approximately like r^{-2}, and the mass-to-flux ratio σ/B_z has an approximately constant value. Some properties of such 'isopedic' states are summarized in Appendix B.

12.5 Gravitational collapse

Parallel to the extensive detailed computations of the slow contraction of clouds and cloud cores through ambipolar diffusion, there have been a number of studies that follow the collapse of supercritical masses. As noted in Section 11.7.1, during the molecular cloud phase, the typical ambipolar diffusion time remains markedly longer than the free-fall time, so that even magnetically-diluted gravitational collapse will be with near flux-freezing.

Galli and Shu (1993) have studied the collapse of weakly magnetic, initially quasi-spherical clouds. The bulk of the mass collapses approximately radially, and is identified tentatively with an incipient proto-star. As discussed in Section 11.3, the possibility still exists that such a supercritical mass may fragment following flattening, but it is at least plausible that the final opaque bodies forming will be towards the upper end of the mass spectrum.

The distorted external field structure is qualitatively as expected (Mestel 1966), with a dominant radial component except near the equator. The deflection of the infalling gas by the field causes the formation of a surrounding magnetically pinched equatorial disc which is not in radial equilibrium, and is in fact many times larger than any expected centrifugally-supported disc. Some of the computations show the beginning of field line detachment through the formation of neutral points in the pinched equatorial zone (cf. Section 3.7.1). Although this is not strictly consistent with the field-freezing equation adopted, and is in fact due to numerical diffusion, nevertheless it is argued plausibly that the calculations mimic the realistic behaviour (cf. Section 12.6).

More recently, Nakamura *et al.* (1995) and then Li and Shu (1997) have studied collapsing models at the other extreme – weakly turbulent, strongly magnetic disc-like cloud cores, the stage following the development of the limiting attractor referred to above. The collapse studied is assumed to be self-similar, so the model remains approximately isopedic: the area density distribution $\sigma(\varpi)$ is again $\propto 1/\varpi$, and so the radial gravitational force $g \propto 1/\varpi$ (cf. (12.21)), while the radial magnetic force stays approximately a constant fraction of g.

The model also possesses a distribution of angular momentum per unit mass $\propto \varpi$, corresponding to a flat rotation curve, and so yielding a centrifugal field again $\propto 1/\varpi$, like the gravitational field in the disc (12.22). As noted, the computations of Basu and Mouschovias (1994), which incorporated magnetic braking into the study of cloud cores forming by ambipolar diffusion, found such a flat rotation curve emerging. The solution shows close similarity to the self-similar collapse of a non-magnetic, isothermal sphere (Shu 1977).

As long as flux remains frozen in, the magnetic tensions prevent fragmentation of the disc. As remarked in Section 11.4, this is consistent with the virial theorem work, which finds that a disc-like body in limiting magneto-gravitational equilibrium has significantly less mass than a less oblate body containing the same magnetic flux: a fragment attempting to separate out would find its collapse halted by the growing magnetic forces. This, however, can be changed by relaxation of strict flux-freezing (Zweibel 1998).

One looks forward to the exploitation of the newer supercomputers to follow further the collapse of both supercritical and marginally critical clouds and cloud fragments towards the opaque, pre-main sequence phases.

12.6 Field line detachment

The question of magnetic field topology has arisen in several contexts. The model of Section 12.2.4 shows strikingly how relaxation of the topological constraint on the field does allow a markedly stronger mean field within the bulk of the cloud. Also, it is possible that a reduction in the amount of flux linking a cloud with distant gas could significantly lower the efficiency of magnetic braking.

In any cloud model with strongly distorted field lines, besides the radial magnetic force component opposing gravity, there is also a strong magnetic pinching force acting towards the equator, which can lead to reconnection of field lines (Mestel 1966; Mestel and Strittmatter 1967). This is particularly marked in the disc-like models of Section 12.2, for which the sharp jump in B_ϖ from zero in the midplane of the disc to the values $\pm B_\varpi(\varpi, 0+)$ respectively yields a magnetic pressure $B_\varpi^2/8\pi$ acting on each face. Together with the pressure of the warm interstellar medium, it is balanced by the pressure of the cool cloud gas, less the effect of the z-component of gravitation, so fixing the local semi-thickness of the disc in terms of the local area density $\sigma(\varpi)$. However, again the pinching magnetic force acts directly only on the changed particles, and causes drift towards the equator, at a rate determined as before by balance with the ion–neutral friction. Alternatively, one can say that the partial pressure gradient of the neutral particles exceeds the gravitational force density by the value of the magnetic force density: the neutrals are therefore driven across the magnetic field away from the equator by their (gravitationally reduced) partial pressure gradient, opposed by the friction density. A similar development of sharp structures in the ambipolar drift time has been discussed more recently and in much greater detail in Brandenburg and Zweibel (1994, 1995) and Zweibel and Brandenburg (1997).

If one were dealing with a pure pinch model, with the field lines parallel to the equatorial plane, then this squeezing out of the neutrals would lead to a new equilibrium in which the initial ionized fraction of the gas is compressed to a new density $n_i = B_\varpi^2/16\pi kT$, T being the temperature and the factor 2 taking account of the electrons (only singly ionized species being assumed present). Since in the original molecular cloud n_i/n_n is very small – 10^{-6} or lower – the charged particles would now be compressed into a layer thinner by a factor 10^6 or more. This would inevitably shift the ionization equilibrium, so that most of the ions and electrons present at normal cloud densities would recombine. The newly neutralized gas would itself be squeezed out across the field, and the remaining charges would be recompressed, etc. The process would continue until Ohmic dissipation of the electron current (due for example to the development of the two-stream instability – cf. Section 2.8), leads to mutual annihilation of the oppositely directed field lines.

Applied to one of our models, this combination of neutralization and Ohmic diffusion leads to reconnection, so that some of the field lines detach from the background field. An example illustrating how normal Ohmic diffusion in a stationary medium can change the field topology in this way is treated in Mestel and Strittmatter (1967) – cf. Section 3.7.1. In reality, the reconnection will be greatly accelerated, as the local radial magnetic force per unit area remains at $B_z B_\varpi/2\pi$ but is now opposed by a gravitational force per unit area that is reduced by at least the thickness factor 10^6 already quoted. Thus the squeezing out in the z-directions of the original charged component is assisted by radial magnetic forces, and in a dynamical time-scale rather than in the comparatively longer recombination time. The process is reminiscent of the Petschek reconnection mechanism (Section 3.7.2). The longest time-scale in the sequence is that of the initial ambipolar diffusion in the z-directions of the hydrogen, helium and other neutral comonents. The analogue of (11.97) for the speed v_z of z-diffusion is

$$\frac{B_\varpi^2}{8\pi\bar{z}} \simeq \alpha(n_i/n_n)\bar{\rho}^2 v_z \tag{12.82}$$

where $\bar{\rho}$ is the local volume density and $\bar{z}$ the disc thickness. Equation (12.82) yields the characteristic time

$$\tau_z = (\bar{z}/v_z) = \alpha(n_i/n_n)(\bar{\rho}\bar{z})^2 8\pi/B_\varpi^2 = \alpha(n_i/n_n)2\pi\sigma^2/B_\varpi^2 \tag{12.83}$$

for the z-magnetic force effectively to separate the charged particles from the neutrals. In the ϖ-direction, (11.97) and (12.2) yield

$$\tau_\varpi = \varpi/v_d = \alpha(n_i/n_n)\pi\sigma^2(\varpi/z)/B_\varpi B_z, \tag{12.84}$$

whence

$$\tau_\varpi/\tau_z = (B_\varpi/B_z)(\varpi/2z). \tag{12.85}$$

The different disc models each have a central region where the density is nearly uniform and the field is essentially a homologously compressed background field $\mathbf{B}_0$, with $B_\varpi/B_z \ll 1$. The ratio (12.85) will be less than unity over this region in models that are moderately flattened, with $a^2R/GM \simeq 0.1$. But the discussion suggests that in bodies without strong turbulent support, the evolution via flux leakage discussed above tends to be accompanied by detachment of some field lines.

As already noted, the changed field topology in the outer parts of the disc is clearly inconsistent with the original picture of magneto-gravitational equilibrium, for near the neutral point the magnetic force changes sign. Equilibrium can be maintained by an ad hoc two-dimensional turbulent pressure, as in the illustrative example in the last section; however, in reality (cf. Section 12.7), the turbulence would exert a strong pressure in the z-direction, and the disc approximation would be implausible, at least in the outer regions.

A disc-like body could go over into a state of quasi-steady accretion, in which recurrent reconnection acts as a macroresistivity, enabling cooled equatorial gas to flow across field lines inwards towards the bulk of the cloud. The important point that emerges is that field lines that are distorted gravitationally into a hairpin-type structure exert strong pinching forces which try spontaneously to relax by reconnection. A cloud or fragment that has become supercritical in M/F will normally collapse non-homologously, and develop such a distorted field structure (Mestel 1966); the assumption for such a collapsing body of 'infinite', undetached field lines becomes more and more questionable.

12.7 Alfvénic turbulence

12.7.1 *Non-dissipative theory*

Finally, we return to the earlier phases, in particular to make a preliminary study of the turbulent motions inferred to be present inside the giant molecular clouds that are the locale of contemporary star formation. As shown by (11.47) and (11.48), the description of the motions within a near-critical molecular cloud as 'Alfvénic turbulence' is almost tautological. When $f \simeq f_c$, lateral magnetic support against gravity yields $\bar{B}^2 \simeq GM^2/R^4$, and simultaneous longitudinal support *requires* $v_t^2 \simeq \pi G\bar{\rho}Z^2 \simeq k^2(\bar{B}^2/4\pi\bar{\rho})$, with k typically close to unity: with v_t^2 prescribed, the cloud's adjustment of its semi-thickness Z and therefore of its density ensures that the Alfvén speed is close to v_t.

Highly supersonic motions normally cause large density fluctuations, and consequent rapid dissipation by shock formation. This was a primary reason why theorists tended to assume that any turbulence would be at most sonic, requiring just the replacement in the virial theorem of a^2 by $2a^2$, until the observation of line-widths an order of magnitude greater than the thermal widths (Zuckerman and Palmer 1974) forced a radical rethinking. With the simultaneous recognition of the crucial role of magnetic fields, it is tempting to argue that the observed motions are associated with Alfvén waves along the strong large-scale field, which in the linear approximation are divergence-free and so yield only small density fluctuations (Arons and Max 1975; Zweibel and Josafatsson 1983). Dewar (1970) had earlier pointed out that in the JWKB (small wavelength) approximation, the net effect of Alfvén waves is to yield an *isotropic* wave pressure. When combined with the continuity equation, the dissipation-free equation of motion is

$$\frac{\partial}{\partial t}(\rho v_i) + \frac{\partial}{\partial x_j}(\rho v_j v_i) = -\frac{\partial p}{\partial x_i} + \rho\frac{\partial V}{\partial x_i} - \frac{\partial}{\partial x_j}T_{ij} \qquad (12.86)$$

where the Maxwell stress tensor $T_{ij} = (B^2/8\pi)\delta_{ij} - B_iB_j/4\pi$. In a uniform medium threaded by a uniform magnetic field, an Alfvén wave has fluctuating fields δB_i, δv_i related by

$$\delta v_i = \pm\frac{\delta B_i}{(4\pi\rho)^{1/2}} \qquad (12.87)$$

(Sections 2.6, 3.1). When averaged over a period, the magnetic and Reynolds stress terms in (12.86) combine into

$$(p_w)_{ij} \equiv \frac{(\delta B)^2}{8\pi}\delta_{ij} - \frac{\delta B_i\,\delta B_j}{4\pi} + \rho\,\delta v_i\,\delta v_j = \frac{(\delta B)^2}{8\pi}\delta_{ij}. \qquad (12.88)$$

The Alfvén waves thus yield the term

$$-\partial(p_w)_{ij}/\partial x_j = -\nabla[(\delta B)^2/8\pi)] \qquad (12.89)$$

on the right-hand side of (12.86) – the force due to an isotropic pressure of strength

$$p_w = \frac{(\delta B)^2}{8\pi}. \qquad (12.90)$$

(See Zweibel 1994, McKee and Zweibel 1995 and Zweibel and McKee 1995 for a fuller discussion.)

An illustrative dynamical model of a wave-supported cloud adapts from Section 10.4 the theory of dissipation-free, torsional Alfvén waves of short wavelength, propagating in a medium with a radial and therefore non-uniform magnetic field. In that problem, because of their systematic deposit of momentum, the waves force the expansion as a steady, spherically-symmetric wind of a stellar envelope that is too cool for thermal pressure to initiate the outflow. To describe instead the support by wave-pressure of a cloud in a steady state, one can take

(10.96) and write $v = 0$ for the mean radial velocity, so that the fluctuations in the field behave like

$$\delta B \propto \rho^{1/4}. \tag{12.91}$$

The relation (12.91) represents the condition of constant wave-energy flux

$$4\pi r^2 \frac{(\delta B)^2}{8\pi} \frac{B_\mathrm{p}}{(4\pi\rho)^{1/2}} \propto \left(\frac{\delta B}{\rho^{1/4}}\right)^2 = \text{constant}. \tag{12.92}$$

This model assumes implicitly that the waves propagate out of the cloud into the surrounding low-density medium. More realistically, outgoing waves will be largely reflected at the cloud surface, but an inward travelling wave will gain momentum in the direction of increasing ρ, so that (12.91) again holds, and each wave exerts the mean pressure (12.90).

A complete treatment would need to include study of the region deep in the cloud which acts both as emitter, absorber and reflector of wave energy. In this exploratory study of a spherically symmetric model, the cloud is divided into two regions: a core $r < \tilde{R}$ of mass $\tilde{M}$ with a uniform magnetic field $\bar{B}$, where outgoing waves are generated and incoming waves are reflected; and a massive 'envelope' domain $r > \tilde{R}$ with a radial magnetic field $B_r = \pm\tilde{B}\tilde{R}^2/r^2$ and maintained in equilibrium just by the pressure exerted by the field of standing waves. By (12.90) and (12.91) the wave pressure obeys the $n = -2$ polytropic relation (Mestel 1991, unpublished; Holliman and McKee, in preparation)

$$p_w = C\rho^{1/2} = \left((\delta\tilde{B}_+)^2/8\pi\tilde{\rho}_+^{1/2}\right)\rho^{1/2}, \tag{12.93}$$

where $\delta\tilde{B}_+$, $\delta\tilde{\rho}_+$ refer to conditions at the base of the envelope. (From now on the +-suffix will be retained only where necessary for non-ambiguity.) Thus for the envelope,

$$\frac{\mathrm{d}}{\mathrm{d}r}(C\rho^{1/2}) = -\frac{G\rho(\tilde{M} + \int_{\tilde{R}}^{r} 4\pi\rho r^2\,\mathrm{d}r)}{r^2}. \tag{12.94}$$

We select the solution of this Emden problem that is singular if it is extrapolated to the origin (cf. Chandrasekhar 1939):

$$\rho = \left(\frac{5}{(12\pi)^2 G}\right)^{2/3} \frac{(\delta\tilde{B})^{4/3}}{\tilde{\rho}^{1/3}} r^{-4/3} \equiv K r^{-4/3}, \tag{12.95}$$

and

$$\tilde{M} = (12\pi K/5)\tilde{R}^{5/3} = \frac{1}{(60\pi G^2)^{1/3}} \frac{(\delta\tilde{B})^{4/3}}{\tilde{\rho}^{1/3}} \tilde{R}^{5/3}. \tag{12.96}$$

At $r = \tilde{R}_+$, just outside the core,

$$\tilde{\rho}_+ = \frac{1}{12\pi}\left(\frac{5}{G}\right)^{1/2} \frac{\delta\tilde{B}}{\tilde{R}}. \tag{12.97}$$

Integration of (12.95) over the volume, followed by insertion of (12.97) yields for the envelope mass

$$
\begin{aligned}
M_{\rm env} &= \frac{1}{(60\pi G^2)^{1/3}} \frac{(\delta\tilde{B})^{4/3}}{\tilde{\rho}^{1/3}} (R_s^{5/3} - \tilde{R}^{5/3}) \\
&= \frac{5^{1/6}}{(25)^{1/3} G^{1/2}} (\delta\tilde{B}) \tilde{R}^{1/3} (R_s^{5/3} - \tilde{R}^{5/3}),
\end{aligned} \tag{12.98}
$$

whence

$$
M_{\rm env}/\tilde{M} = (R_s/\tilde{R})^{5/3} - 1. \tag{12.99}
$$

With the magnetic field assumed continuous at the radius R_s of the cloud, the boundary condition is then continuity of the wave pressure p_w with the total pressure $p_{\rm ext}$ of the surrounding medium – thermal, turbulent, cosmic ray, but excluding the magnetic.

For the equilibrium of the uniformly dense core, we apply the scalar virial theorem (3.64). Continuity of the uniform field $\bar{B} \equiv F/\pi\tilde{R}^2$, for $r < \tilde{R}$, with a strictly radial field B_r outside, yields $B_r(\tilde{R}) = \bar{B}\cos\theta$. The virial surface and volume terms for the sphere $\tilde{R}$ are then equal, yielding jointly $\bar{B}^2\tilde{R}^3/3 = F^2/3\pi^2\tilde{R}$ (cf. the discussion leading up to (11.5) and (11.6)). However, in the treatment of the envelope, the radial field is assumed to have adjusted laterally to a θ-independent structure $B_r = \tilde{B}\tilde{R}^2/r^2$. Flux conservation then yields

$$
\tilde{B} = \bar{B}/2 = F/2\pi\tilde{R}^2 \tag{12.100}
$$

The virial equation for the core is

$$
\frac{3G\tilde{M}^2}{5\tilde{R}} - \frac{F^2}{3\pi^2\tilde{R}} + 4\pi C\tilde{R}^3 \left(\tilde{\rho}_+^{1/2} - \tilde{\rho}_-^{1/2}\right) = 0. \tag{12.101}
$$

The first term is the familiar gravitational energy of a uniform sphere, the second is again the sum of the mean magnetic volume and surface terms, while the last two terms make a rough allowance for a discontinuity in density, so that the volume and surface contributions from the wave pressure do not mutually cancel. With the definition analogous to (11.9)

$$
f = \frac{F}{\pi^2 G^{1/2}\tilde{M}}, \tag{12.102}
$$

and insertion of C from (12.93), $\tilde{\rho}_+$ from (12.97) and

$$
\tilde{\rho}_- = 3\tilde{M}/4\pi\tilde{R}^3, \tag{12.103}
$$

(12.101) reduces to

$$
\left(\frac{9}{5\pi^2 f^2} - 1\right) + \frac{3}{8}\left(\frac{\delta\tilde{B}}{\tilde{B}}\right)^2 \left(1 - \frac{3\sqrt{2}}{5^{1/4}\pi^{1/2} f^{1/2} (\delta\tilde{B}/\tilde{B})^{1/2}}\right) = 0. \tag{12.104}
$$

(From now on the suffix '+' on $\tilde{B}$ and $\delta\tilde{B}$ is dropped.) Also, from (12.93), (12.95) and the surface boundary condition $p_w(R_s) = p_{\mathrm{ext}}$,

$$p_{\mathrm{ext}} = \frac{(\delta\tilde{B})^2}{8\pi}\left(\frac{\tilde{R}}{R_s}\right)^{2/3} = \left(\frac{\delta\tilde{B}}{\tilde{B}}\right)^2\left(\frac{\tilde{B}}{B_s}\right)^2\frac{B_s^2}{8\pi}\left(\frac{\tilde{R}}{R_s}\right)^{2/3}, \tag{12.105}$$

so that

$$\frac{R_s}{\tilde{R}} = \left(\frac{\delta\tilde{B}}{\tilde{B}}\right)^{-3/5}\left(\frac{8\pi p_{\mathrm{ext}}}{B_s^2}\right)^{3/10}, \tag{12.106}$$

and by (12.99)

$$\frac{M_{\mathrm{env}}}{\tilde{M}} = \left(\frac{8\pi p_{\mathrm{ext}}}{B_s^2}\right)^{1/2}\left(\frac{\delta\tilde{B}}{\tilde{B}}\right)^{-1} - 1. \tag{12.107}$$

Equation (12.104) relates the flux/mass parameter f for the core to the relative amplitude $\delta\tilde{B}/\tilde{B}$ at the envelope base, required for the equilibrium of the core. If $f < 0.427$, then the first bracket in (12.104) is positive, i.e. the flux/mass ratio of the core is subcritical, so that the turbulent pressure is necessary to prevent core collapse. If for example $f = 0.3$, then $\delta\tilde{B}/\tilde{B} = 6.3$. For larger values of f, $\delta\tilde{B}/\tilde{B}$ must remain finite so as to support the envelope, but it is progressively smaller: for $f = 0.7$, it is 4.4, for $f = 1$, it is 3.8, and it reaches the limit 1.633 at $f = \infty$. Not surprisingly, the required values of $\delta\tilde{B}/\tilde{B}$ in this admittedly very special model are not small. In addition, since for consistency $R_s/\tilde{R} > 1$ and correspondingly $M_{\mathrm{env}} > 0$, (12.106) and (12.107) impose the upper limit

$$\left(\frac{\delta\tilde{B}}{\tilde{B}}\right)^2 < \frac{8\pi p_{\mathrm{ext}}}{B_s^2}. \tag{12.108}$$

12.7.2 *The effect of dissipation*

The waves will in fact be subject to dissipation. Again, the same process of plasma drift or ambipolar diffusion which allows slow evolution of the cloud (Section 11.7) also damps the waves. The simplest example has wavelengths short enough for the local density ρ to be taken as uniform, and likewise the mean field $\bar{B} = (4\pi\rho)^{1/2}v_{\mathrm{A}}$ to be locally uniform and in the z-direction $\hat{\mathbf{z}}$ of a local Cartesian set of axes. The fluctuating field $\delta\mathbf{B}$ is in the (x, y) plane and propagates along $\hat{\mathbf{z}}$ with phase-velocity ω/k_R and characteristic damping length k_I^{-1}:

$$\delta\mathbf{B} = \mathbf{b}\sin(k_R z - \omega t)\exp(-k_I z), \qquad \mathbf{b}\cdot\hat{\mathbf{z}} = 0, \tag{12.109}$$

which is the imaginary part of

$$\mathbf{b}\exp\mathrm{i}(kz - \omega t), \qquad k = k_R + \mathrm{i}k_I. \tag{12.110}$$

If in the wave motion the relative velocity $|\mathbf{v}_i - \mathbf{v}_n|$ of ions and neutrals is small compared with v_n, then in the linear approximation the bulk equation of motion is again

$$\rho\frac{\mathrm{d}\mathbf{v}}{\mathrm{d}t} = -\mathrm{i}\omega\rho\mathbf{v} = \frac{(\nabla\times\delta\mathbf{B})\times\bar{B}\hat{\mathbf{z}}}{4\pi} = \frac{\mathrm{i}k\bar{B}}{4\pi}\mathbf{b}, \tag{12.111}$$

where from now on $\mathbf{v}_n$ is replaced by the bulk velocity $\mathbf{v}$. As discussed in Sections 11.7 and 12.4 (cf. also Section 2.7), the variation of $\mathbf{B}$ is well described by assuming the field frozen into the ionized component, as in (11.103), which when combined with the equation of motion (11.97) yields

$$\begin{aligned}\frac{\partial\mathbf{B}}{\partial t} &= \nabla\times(\mathbf{v}\times\mathbf{B}) + \nabla\times[(\mathbf{v}_i - \mathbf{v})\times\mathbf{B}] \\ &= \nabla\times(\mathbf{v}\times\mathbf{B}) + \nabla\times\left[\frac{[(\nabla\times\mathbf{B})\times\mathbf{B}]\times\mathbf{B}}{4\pi\alpha(n_i/n_n)\rho^2}\right].\end{aligned} \tag{12.112}$$

In the present problem this reduces to

$$-\mathrm{i}\omega\mathbf{b} = \bar{B}\mathrm{i}k\mathbf{v} - \frac{v_{\mathrm{A}}^2k^2}{4\pi\alpha(n_i/n_n)\rho}\mathbf{b}. \tag{12.113}$$

Equations (12.111) and (12.113) combine into the dispersion relation

$$k^2 = \frac{(\omega^2/v_{\mathrm{A}}^2)}{1-\mathrm{i}D\omega}, \qquad D = 1/4\pi\alpha(n_i/n_n)\rho, \tag{12.114}$$

whence

$$k_R^2 = \frac{(\omega/v_{\mathrm{A}})^2}{2}\frac{[1+(1+\omega^2D^2)^{1/2}]}{1+\omega^2D^2}, \tag{12.115}$$

$$k_I^2 = \frac{(\omega/v_{\mathrm{A}})^2}{2}\frac{[-1+(1+\omega^2D^2)^{1/2}]}{1+\omega^2D^2}. \tag{12.116}$$

For weak or moderate damping these reduce respectively to the usual Alfvén wave relation

$$k_R = \omega/v_{\mathrm{A}}, \tag{12.117}$$

and the damping coefficient

$$k_I = \omega^2\big/(8\pi\alpha(n_i/n_n)\rho v_{\mathrm{A}}). \tag{12.118}$$

The weak damping condition $\omega < 4\pi\alpha(n_i/n_n)\rho$ puts a constraint on the wavelengths (Zweibel and Josafatsson 1983; Carlberg and Pudritz 1990);

$$\lambda = \frac{2\pi}{k_R} > \frac{2\pi\bar{B}}{(4\pi\rho)^{1/2}}\frac{1}{4\pi\alpha(n_i/n_n)\rho} = \frac{4\pi k_I}{k_R^2}. \tag{12.119}$$

(When this is violated, the second term in (12.112) is no longer just a perturbation, so that $|\mathbf{v}_i - \mathbf{v}|$ is no longer small compared with $|\mathbf{v}|$. The damping then prevents propagation of the waves, until the frequencies ω are so high that ions and neutrals are no longer coupled – clearly of no interest in the present problem (Kulsrud and Pearce 1969).)

As a first example, consider again the moderately oblate, nearly critical model of Section 11.2.2 with $Z/R = 1/2$, but without specifying how the uniformly dense cloud is supported along the field. Insertion again of the Elmegreen relations $n_i/n = K_i/n^{1/2}$ with $(G/m_H)^{1/2}/\alpha K_i = 0.05$ yields $\lambda > 2 \times 10^{-2} Z$, so allowing a fairly wide range of propagating wavelengths. However, when applied to the above wave-supported model, using (12.95), (12.97) and (12.103), one finds for the minimum wavelength at a radius r within the envelope

$$\lambda/r \approx 0.84\Big/(r/\tilde{R})^{5/3}, \tag{12.120}$$

showing that the small wavelength treatment needs to be replaced.

The weakly damped waves in the above two-zone model yield support by virtue of non-uniformity in the mean radial field and in the density. At the other extreme, one can picture propagating waves that are damped and so give support against gravity through the loss of momentum associated with their loss of energy. In the simplest example (Dewar 1970; Shu *et al.* 1987), the mean field $\bar{B}\hat{\mathbf{z}}$ is taken as uniform and the fluctuating wave field $\delta\mathbf{B}$ is as given by (12.109). The consequent magnetic force per unit volume, averaged over a period $2\pi/\omega$, is

$$\overline{(\nabla\times\mathbf{B}) \times \delta\mathbf{B}/4\pi} = -\overline{\nabla(\delta\mathbf{B})^2/8\pi} = |\mathbf{b}\exp(-k_I z)|^2 k_I \hat{\mathbf{z}}/8\pi \tag{12.121}$$

along the direction of $\hat{\mathbf{z}}$, – an example of the general result (12.89). Note that with the mean field uniform, damping of the wave is essential if there is to be deposition of z-momentum to provide support: if $k_I = 0$, the mean component $\bar{T}_{zz}$ of the Maxwell stress tensor would be independent of z, corresponding to a *constant* flow of z-momentum per unit area at the rate $\mathbf{b}^2/8\pi$. The picture is of waves continuously generated near $z = 0$ and propagating in the $\pm z$ directions, with those contributing most to the support being damped in the distance k_I^{-1}, of the order of the cloud semi-thickness Z.

To estimate the amplitude $|\mathbf{b}|$ necessary for the momentum input from the waves to balance gravity, consider again the spheroidal models of Section 11.2.2 with semi-major axis R and semi-minor axis $Z \approx k_I^{-1}$. The new term (12.121) contributes approximately $0.25\pi R^2 Z^2 \mathbf{b}^2 k_I/8\pi \approx ZR^2\mathbf{b}^2/32$, and the gravitational term $(GM^2/5R)H(e)$ as in (11.151), so that we require

$$\mathbf{b}^2 \approx 6.4(GM^2/R^4)(RH(e)/Z) = 6.4[\bar{B}^2/(\pi f)^2](RH(e)/Z) \tag{12.122}$$

with f defined again by (11.9). Not surprisingly, one finds that for near-critical clouds, $b/\bar{B}$ is not small but ≈ 2.5, and is insensitive to the value of the eccentricity e. The crucial point is that the cloud semi-thickness Z is fixed by the damping distance k_I. The larger k_I is (subject to (12.119)), the higher $\bar{\rho}$ and the smaller v_{A} and v are, but the wave-amplitude is forced to remain at least comparable with $\bar{B}$.

To sum up: in each attempt to apply wave support realistically, one finds that the fluctuation amplitudes required are not small compared with the local mean field. Linear analysis can therefore give one valuable qualitative and semi-quantitative insights, but its limitations are manifest. In particular, wave steepening, ignorable in the linear approximation, will occur, increasing the damping rate and so also the heat input into the cloud. Even if the energy sources are available to maintain the waves at the required ampitudes, one must also check that heat input is consistent with the radiative loss processes and the inferred temperatures (Zweibel and Josafatsson 1983).

12.8 Concluding remarks

How do the essential features of the star formation process, emerging tentatively from the approximate treatment of Chapter 11, appear now, especially following the reporting in the present chapter of a somewhat more rigorous treatment of some of the problems? And what can one say about prospects for elucidating the ultimate problem in a theory of star formation – the prediction of the Initial Mass Function?

(1) Magneto-gravitational equilibrium

The non-turbulent, disc-like models of Section 12.2 confirm the existence of a critical flux–mass parameter f_c whose value is of the same order as that found by applying the virial theorem to fields that are the simplest topologically similar simulations of the constructed fields. However, the degree of contraction that the cloud has undergone when the f_c-equilibrium state is reached is limited by the requirement that equilibrium should hold at each point, including those in the outer equatorial zone where the density is comparatively low, so putting a bound on the local value of the curl of the field. The results from the different groups that have computed equilibria with a finite effective sound speed appear to point to the same conclusion: it is not clear that any model with all the field lines 'infinite' will yield a relative mean field $\bar{B}/B_0 = F/\pi B_0 R^2$ which can approach the value of 10 or more inferred for the strongest observed cloud fields (McKee *et al.* 1993). The introduction of a turbulent pressure – as is implicit in the three-dimensional models described in Section 12.3 – does not alter this conclusion. For high enough contraction of the bulk of the cloud to be possible, it seems one requires some of the field lines to be detached from the background, yielding a structure with O- and X-type neutral points (Fig. 12.4). Within O the local value of $|\nabla \times \mathbf{B}|$ is reduced, so allowing further contraction with increase of B_z, but at O the magnetic force vanishes and beyond O it changes sign, so that the turbulent pressure has to balance both the gravitational and magnetic forces which are acting inwards together.

If this is correct, and if the turbulence is associated with the local presence of a magnetic field, then one is forced to explore the possibility that at least in the detached region, the cloud field is of the complex topology discussed for stellar fields in Section 5.3, with toroidal flux linking the poloidal. We have seen that such a structure seems to be necessary to keep the detached poloidal field stable. (In fact, since in a diffuse cloud there is no adiabatic stabilization – i.e. $\gamma = 1$ effectively – there is no need to restrict the stability discussion of Section 5.3 to disturbances that are perpendicular to gravity, so a detached, purely poloidal cloud field would be more liable to instability than would a stellar field.) Near the neutral points of the poloidal field, the turbulence would then consist of waves essentially along the toroidal field, but still exerting their isotropic pressure. Toroidal fields again arise naturally through interaction with non-uniform rotation, and the quasi-permanent linked toroidal flux could develop by a relaxation of flux freezing. It should not be beyond the capabilities of either existing or future computers to construct such cloud models, but it is not clear whether unambiguous observational confirmation would be available.

(2) Magnetic braking

Both the approximate discussion in Section 11.6 and the detailed discussions of the toroidal dynamics in Appendix 11B and in the references cited all simplify the analysis of the torsional Alfvén waves by the adoption of simple forms for the poloidal field, but this appears to have only a modest effect on the angular momentum transport along each individual 'infinite' field line. The sharp distinction found earlier between the rotational histories of respectively subcritical and supercritical clouds seems fairly robust as long as most field lines remain infinite, but it could become blurred if some field lines detach and the outer regions of a cloud have difficulty in disposing of excess angular momentum. However, it should be remembered that a domain that is kept rotating near the centrifugal limit will tend to drive a centrifugal wind, which will certainly brake (Draine 1983; Campbell and Mestel 1987). Further, it appears that in the models that have been constructed of subcritical cloud cores, contracting through flux leakage, the field lines stay fairly straight. The strong distortion of field lines that presages detachment occurs instead in supercritical, freely-falling clouds, for which magnetic braking is in any case rather ineffective.

(3) Gravitational collapse

Once a weakly turbulent cloud or fragment has become subcritical in the flux/mass parameter f, then gravitational contraction or collapse must ensue. Some of the qualitative features emerging from the study of equilibria persist in collapsing clouds, and were to some extent anticipated. In an early study (Mestel 1966), a uniform field was subject first to the distortion due to a spherical but non-homologous velocity field, simulating crudely that due to the free-fall of a cool, centrally condensed sphere. The resulting field exerts a Lorentz force density with both r- and θ-components. The radial component must be everywhere less than the gravitational force density, yielding a condition equivalent to $f < f_c$, with f_c not too different from the estimates given in Section 11.2. However, the θ-component is unbalanced and so compresses the gas towards the equator, increasing markedly the curvature of the field lines in the low-density outer regions of the cloud. The modified radial component now locally exceeds gravity: the gas holding apart the field lines in the northern and southern hemispheres is dragged outwards, so the field has no option but to reconnect in the outer regions. The reconnection will again be assisted by ambipolar diffusion, but it is the spontaneous dynamical effect of the distorted magnetic field that again appears crucial (cf. the discussion at the end of Section 12.6).

(4) Flux leakage

From the start, a clear distinction is drawn between two questions: (a) how do self-gravitating proto-stellar masses form in the presence of the galactic flux; and (b) why is any of the galactic flux retained by the observed main sequence and pre-main sequence stars clearly so much less than the virial limit?

We recall the conclusion of the detailed discussion by Nakano and Umebayashi (1986*a,b*), confirmed by later workers, that up to densities of $n_n \simeq 10^{11}$ cm^{-3}, leakage from molecular clouds by ambipolar diffusion remains slow – by at least one order of magnitude – compared with free-fall. This justifies treating the evolution of a cloud supercritical in f as through a sequence of quasi-steady states, in the time-scale set by diffusion, and with the magnetic field remaining dynamically significant. A fortiori, a cloud initially subcritical in f will collapse effectively retaining its flux. Further, as discussed in Sections 11.7 and 12.4, differential flux leakage enables a cloud in quasi-equilibrium to undergo an initial fragmentation that is to some extent analogous to the classical Jeans process. Subsequent further fragmentation of the gravitationally collapsing cloudlets can in principle occur following spontaneous preferential flow of gas down field lines (Section 11.4). So question (a) above is given a tentative answer. As just noted, changes in field topology, so far from being precluded, are assisted by the strong equatorial pinching forces exerted by gravitationally-distorted fields, but as long as flux leakage remains slow, the fields remain dynamically strong, so question (b) persists.

Beyond $n_n \simeq 10^{10}$–10^{11}, the optical depth through a Jeans mass exceeds unity, so that during subsequent contraction the compressional heat generated is largely trapped, and the temperature rises adiabatically. Beyond $T = 10^3$ K, thermal ionization increases the crucial ratio n_i/n_n to the level at which flux leakage through ambipolar diffusion is negligible. A molecular cloud may have a temperature of no more than 10 K at $n \simeq 10^{10}$, so it will increase to 10^3 as n goes up by 10^3 or so. In this interval, following earlier work by Spitzer (1963), Nakano and Umebayashi (1986*a,b*) and later Nishi *et al.* (1991) find that if the ionization is determined primarily by galactic cosmic rays, the field is no longer tied to the ions: the electrons are attached largely to the dust grains, and the current is carried by the ions which suffer a large Ohmic dissipation due to ion–neutral collisions. They conclude that the bulk of the primeval flux is destroyed in this epoch. Interestingly enough, they estimate that the remnant flux at $T = 10^3$ K is less than that in the strongest magnetic Ap stars.

Without questioning at all these careful calculations, one wonders whether the magnetic flux problem is always resolved by enhanced micro-dissipative processes in a stationary proto-stellar medium – by a generalized Cowling-type process. First, one should note that ionization could be kept at a much higher level, for example by X-rays from nearby newly formed stars in the same cloud complex (Silk and Norman 1983), or by a subcosmic ray flux generated locally in magnetic reconnection processes, for example as the field of a fragment with subcritical F/M detaches itself from the cloud field. If so, both ambipolar diffusion and Ohmic diffusion will remain slow. And although most of the discussion has been of cold molecular clouds giving birth to Population I stars, there is the possibility that in the earliest epochs, Population II and (maybe) Population III stars have been formed from magnetized H and He gas at $T \simeq 10^4$ K (Hoyle 1953): collisional ionization will then have kept flux loss by large-scale ambipolar diffusion to a negligible rate at all epochs. These various possibilities suggest that – at least sometimes – the bulk of the primeval flux is lost in the pre-main sequence phase, for example through magnetic buoyancy, in at most a thermal time-scale. It has been suggested (Hoyle personal communication; Lynden-Bell 1977) that such a reservoir of magnetic energy could be an alternative power source for the T Tauri phenomenon.

(5) Turbulence and instability

It is clear that an outstanding problem for the future is an improved analysis of the turbulence. In the small amplitude limit, a central issue is: for what wavelength range do there exist – for non-special geometries – analogues of the torsional Alfvén waves, with $\delta\mathbf{B}$ perpendicular to the local mean field $\bar{\mathbf{B}}$ and with $\delta\rho = 0$? And even so, to describe realistically the motions within molecular clouds, the analysis of such modes must be extended into the non-linear domain. An important question is: does the presence of the magnetic field still modify the motions sufficiently for the consequent dissipation to be markedly reduced below what would be expected for supersonic turbulence in a non-magnetic cloud? With a plentiful energy supply available, is the magnetic field inevitably accompanied by a strong turbulent field? Promising recent studies of non-linear MHD turbulence in the present context are by Goldreich and Sridhar (1995) and Gammie and Ostriker (1996).

A related question is that of the stability of the various cloud models. The energy source of the turbulence is usually spoken of as being 'external' to the cloud, for example from buffeting by local HII regions, or from the release of gravitational-plus-rotational energy of collapsing fragments already forming within the cloud (cf. Appendix 11B). If these sources were to dry up, would the steady dissipation of the existing turbulence cause the cloud to flatten into a structure as in one of the disc-like models of Section 12.2 and Appendix A or B; or are these models themselves unstable to non-axisymmetric modes that convert gravitational energy into turbulent energy, so causing further evolution of the cloud? Studies by Garlick (1979), Li and Shu (1997), Zweibel and Lovelace

(1997) and Lovelace and Zweibel (1997) have not yielded such instabilities in plausible disc models, as long as strict flux-freezing holds; but relaxation of this constraint through ambipolar diffusion may allow incipient fragmentation and also significant energy release, especially in clouds close to the critical flux/mass ration (Zweibel 1998).

(6) The mass spectrum

Herbig (1962*a,b*) made the tentative inference from observation that the formation process might differ for the more massive and less massive stars respectively. The idea was developed by Eggen (1976), who introduced the term 'bimodal star formation'. On the theoretical side, Shu *et al.* (1987) and Lizano and Shu (1989) made a link with the clear distinction that had emerged between clouds that form respectively as supercritical or subcritical in M/F. In their picture, O and B stars form by the fragmentation of supercritical clouds or clumps. However, within molecular clouds that are held up against self-gravitation by the joint effect of magnetic force and the associated turbulent pressure, slow differential flux leakage enables the precursors of low mass stars to separate out as supercritical fragments.

In both modes, one would expect the combined effect of the magnetic field and the supersonic Alfvénic turbulence to shift upwards the masses of the initial fragments that can form, as compared with those expected in the classical fragmentation picture. As a fragment contracts, its overall shape is again fixed by the strength of the turbulence. If the fragment flattens markedly, a roughly spherical subcondensation may in turn separate out. The minimum mass is given by the approximate formula (11.56) with the sound speed a now replaced by a turbulent speed v_t

$$M_{\rm min} = \frac{3(5v_t^2 R/3GM)^2 M}{(1 - f^2/f_c^2)}. \qquad (12.123)$$

The magnetic field enters directly into (12.123) through the denominator; but as emphasized, unless f is very close to f_c, it is the likely persistence of strongly supersonic v_t-values that is crucial in keeping $M_{\rm min}$ high, by limiting flow down the field lines. If, instead, v_t were to fall to just the sound speed a, then fragmentation could in principle continue down to the opacity limit (cf. Section 11.1), again yielding embarrassingly small final masses. So it appears that if fragmentation is assumed to occur with maximum efficiency, but of course subject to basic dynamical constraints, then the mass spectrum of proto-stars entering the opaque phase can indeed be shifted to markedly higher values by the magnetic field, but as much by its indirect effect in maintaining the turbulence as through its direct contribution to support against gravity (measured by f).

It can be argued plausibly that the actual mass function of stars reaching the main sequence is determined by more subtle dynamical processes than are envisioned in the fragmentation picture, even with the sound speed replaced by a superthermal turbulent speed. It should be clear that the potentialities of the theory of the early phases of star formation, on the lines outlined in this chapter, with magnetic fields playing a crucial role, are far from being exhausted. An adequate theory will need to cover also the later phases, linking up with some of the topics of Chapter 10, such as the theory of bipolar outflows from proto-stellar objects, especially from magnetic accretion discs. A theory able to predict a near universal function presumably contains a self-regulatory process, with an omnipresent magnetic field possibly playing an important role.

In such a difficult area, one expects progress in our understanding to depend even more than usual in astrophysics on close liaison between theory and observation. One can hardly doubt that in the earliest phases, the galactic magnetic field cannot be ignored. But on the ultimate question – the physical explanation of the mass spectrum – some authors give magnetic fields a secondary role, appealing just to the statistical properties of random supersonic flows in a self-gravitating medium. Silk's approach (1995) is semi-empirical, leaning on the observed superthermal line-widths in molecular cloud cores. He argues that the motions are generated by proto-stellar outflows, which play a regulatory role by limiting the mass accretion by a forming star. The model of Padoan *et al.* (1997), though based rather on sophisticated numerical experiments, has in the authors' words a 'strong observational counterpart', being linked with the claim (quoted in Section 12.1) that the motions in molecular clouds are markedly super-Alfvénic. As remarked above, if also direct Zeeman observations point unambiguously to a more rapid flux leakage than currently estimated, this will give a fillip to further studies on the stability of magnetic fields in gas clouds.

(7) Galactic spiral arms

If the mass spectrum of the stars finally forming in a particular galactic region is shifted upwards by a modest amount through an increase in the local galactic field strength, then this should yield a marked increase in the local integrated stellar luminosity, because of the sensitivity of the $L(M)$ dependence. The galactic spiral arm phenomenon is often plausibly linked with the presence of a spiral structure in the background gravitational field, described by some version of the rotating density wave theory (e.g. Lin 1971). The accompanying spiral shock (Roberts 1969) compresses the gas and its frozen-in field: the spiral density wave then manifests itself by the generation of a corresponding bright star spiral.

When the evidence emerged for a large-scale galactic magnetic field, there were many attempts to try and account for the spiral arms as an essentially MHD phenomenon (e.g. Chandrasekhar and Fermi 1953, p. 113). There is now a consensus (cf. Chapter 1) that the field is too weak by at least a factor 5, yielding Lorentz forces that are correspondingly at least 25 times too small, and current studies (e.g. Binney and Tremaine 1987) concentrate on purely dynamical theories of spiral structure. However, according to the present line of thought, the compressed field can act as a tracer of the wave, and also as a catalyst that facilitates the release of nuclear energy through massive star formation.

The large-scale galactic field is presumed generated and maintained by dynamo action, provisionally described by the 'standard' dynamo equations of Chapter 6. In the applications to disc-like geometry in Section 10.6, the generated fields are assumed to be axisymmetric. However, at least some disc-like galaxies (e.g. M81) seem able to generate 'bisymmetric' ($\exp \mathrm{i}\phi$) modes rather than axisymmetric (ϕ-independent) modes. It may be that a non-axisymmetric α-effect term is required, to offset the tendency of the strong shear to generate axisymmetric modes (Mestel and Subramanian 1991; Subramanian and Mestel 1993). The compressive effect of the density wave will probably tend to generate a non-axisymmetric turbulent field. Perhaps a stronger deviation from axisymmetry will arise from the spiral stellar distribution, again especially if the magnetic field has shifted the mass spectrum upwards, for it is the small fraction of massive stars that is primarily responsible for the input of energy into the interstellar turbulence, through the development of expanding HII regions, planetary nebulae, novae and supernovae. This 'bootstrap' picture appeals to a spiral **B**-field to produce by its catalytic effect a non-axisymmetric turbulent field that has the degree of anisotropy needed to keep the dynamo operating.

APPENDIX A
Exact disc-like models

Finite disc models are conveniently studied using the spheroidal shell technique to describe the density and gravitational fields, familiar from studies of disc-like galaxies (e.g. Mestel 1963). Two sequences are considered.

(a) The *'moderately centrally condensed'* sequence is defined by

$$g(\varpi) = V^2\varpi/(\varpi^2 + l^2R^2) = (V^2/R)[x/(x^2 + l^2)], \qquad 0 \le l \le 1, \qquad (12.124)$$

with the associated density distribution

$$\sigma_a(\varpi) = (V^2 l^2/\pi^2 GR)\tilde{\sigma}_a(x) \qquad (12.125)$$

where

$$\begin{aligned} \tilde{\sigma}_a(x) &= 0 && \text{for } x > 1, \\ &= \int_x^1 \tilde{F}(u)\,\mathrm{d}u/(u^2 - x^2)^{1/2} && \text{for } x \le 1, \\ \tilde{F}(u) &= \{u(u^2 + l^2)^{1/2}/l^2 \\ &\quad + \log[\{u + (u^2 + l^2)^{1/2}\}/l]\}/(u^2 + l^2)^{3/2}. \end{aligned} \qquad (12.126)$$

(b) The *'strongly centrally condensed'* sequence has

$$g(\varpi) = V^2\varpi l R/(\varpi^3 + l^3R^3) = (V^2/R)[lx/(x^3 + l^3)],$$

$$\tilde{\sigma}_b(\varpi) = (V^2 l/\pi^2 GR)\tilde{\sigma}_b(x), \qquad 0 \le l \le 1, \tag{12.127}$$

where again $\tilde{\sigma}_b(x) = 0$ for $x > 1$, and for $x \le 1$

$$\tilde{\sigma}_b(x) = (1 - x^2)^{1/2} G(1) - \int_x^1 (u^2 - x^2)^{1/2} G'(u)\, \mathrm{d}u,$$

$$G(u) = (1/u) \int_0^{u^2} (l^3 - a^{3/2}/2)\, \mathrm{d}a \Big/ (u^2 - a^2)^{1/2} (a^{3/2} + l^3)^2. \tag{12.128}$$

Magneto-gravitational equilibrium under these prescribed density-gravitational fields are described respectively by (12.9) and (12.12). The choice of l leaves C as the only free parameter. To construct the magnetic field one needs formulae relating b_x and b_z.

Cartesian components of the Coulomb gauge vector potential $\mathbf{A}$ and the volume current field $\mathbf{j}$ are related by the Poisson integral

$$c\mathbf{A}(\mathbf{r}) = \int \mathbf{j}'\, \mathrm{d}\tau / |\mathbf{r} - \mathbf{r}'|. \tag{12.129}$$

As the present problem is axisymmetric, the observation point P may be taken as $\mathbf{r} = (\varpi, 0, z)$ in cylindrical polars. Now introduce temporarily a set of Cartesian axes $O(X, Y, Z)$ with $OZ \equiv Oz$ and with P having Cartesian coordinates $(X, 0, Z) \equiv (\varpi, 0, z)$ (see Fig. 12.5). At P, $\mathbf{A}$ has Cartesian components $(0, A, 0)$.

At a general source point $P'(\varpi', \theta', 0) = (\varpi' \cos\theta', \varpi' \sin\theta', 0)$ in Cartesians, the surface current $J'_\theta(\varpi')$ flowing in the azimuthal direction has the Cartesian components $J'_\theta(-\sin\theta', \cos\theta', 0)$; hence the Poisson integral yields as the only non-zero (dimensionless) component

$$a(P) = \frac{1}{2\pi} \int\int \frac{b_x(x') x' \cos\theta'\, \mathrm{d}x'\, \mathrm{d}\theta'}{[(x - x'\cos\theta')^2 + (x'\sin\theta')^2 + Z^2]^{1/2}} = a(x, Z) \tag{12.130}$$

by axisymmetry. On the equator

$$a(x, 0) = \frac{1}{2\pi} \int_0^1 b_x(x') x'\, \mathrm{d}x' \int_0^{2\pi} \frac{\cos\theta'\, \mathrm{d}\theta'}{(x^2 + x'^2 - 2xx'\cos\theta')^{1/2}}. \tag{12.131}$$

The elliptic integral over θ' becomes (Copson 1947; Sneddon 1957)

$$(4/xx') \int_0^{\mathrm{Min}(x,x')} t^2\, \mathrm{d}t \Big/ [(x^2 - t^2)(x'^2 - t^2)]^{1/2}, \tag{12.132}$$

so that when $x > 1$ (outside the disc),

$$a(x, 0) = (2/\pi x) \int_0^1 b_x(u)\, \mathrm{d}u \int_0^u t^2\, \mathrm{d}t \Big/ [(x^2 - t^2)(u^2 - t^2)]^{1/2}, \tag{12.133}$$

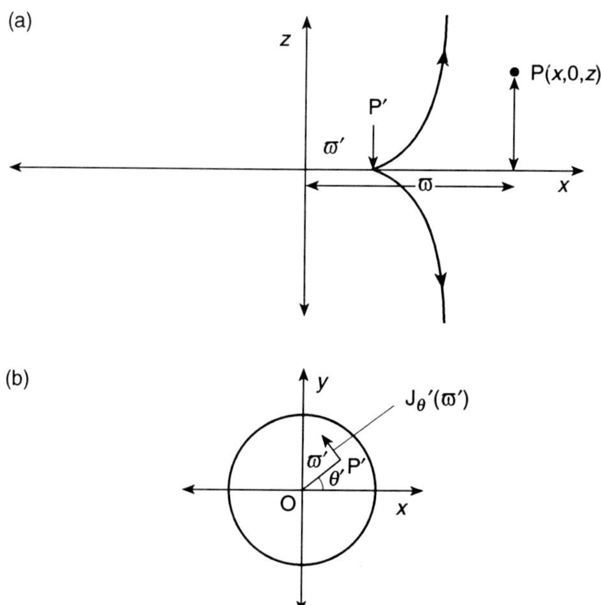

FIG. 12.5. The coordinate systems appropriate to the magneto-gravitational disc problem. (a) The observation point P and the source point P'. (b) The source point P' and the local current density.

and when $x < 1$ (within the disc)

$$a(x,0) = \frac{2}{\pi x}\int_0^x b_x(u)\,\mathrm{d}u \int_0^u t^2\,\mathrm{d}t \Big/ [(x^2-t^2)(u^2-t^2)]^{1/2} + \frac{2}{\pi x}\int_x^1 b_x(u)\,\mathrm{d}u \int_0^x t^2\,\mathrm{d}t \Big/ [(x^2-t^2)(u^2-t^2)]^{1/2}. \quad (12.134)$$

Inversion of the order of integration in (12.134) yields

$$xa(x) = 4\int_0^x t^2 s(t)/(x^2-t^2)^{1/2}\,\mathrm{d}t \quad (12.135)$$

where

$$s(x) = (1/2\pi)\int_x^1 b_x(t)/(t^2-x^2)^{1/2}\,\mathrm{d}t. \quad (12.136)$$

From the theory of Abel's integral equation (e.g. Sneddon 1957, again familiar from discussions of disc-like galaxies), (12.135) and (12.136) invert into

$$s(x) = \frac{1}{2\pi x^2}\frac{\mathrm{d}}{\mathrm{d}x}\int_0^x t^2 a(t)\,\mathrm{d}t/(x^2-t^2)^{1/2} \quad (12.137)$$

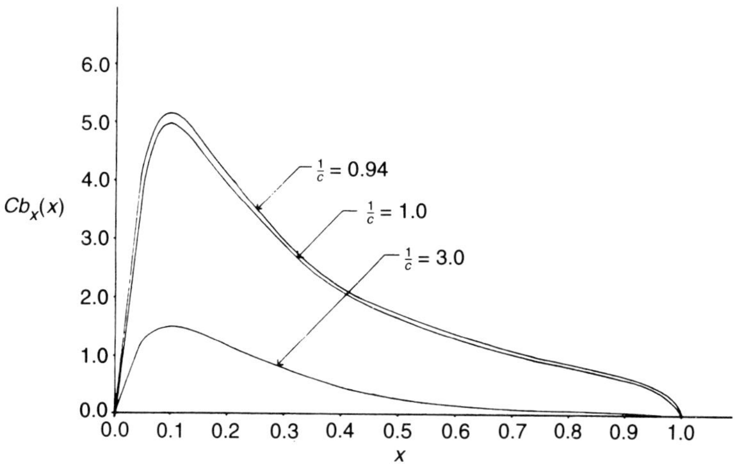

FIG. 12.6. Examples of sequence (a) models with $l = 0.1$: the run of $B_\varpi/B_0 = Cb_x$ with x for some values of C.

and

$$b_x(x) = -4\frac{\mathrm{d}}{\mathrm{d}x}\int_x^1 ts(t)\,\mathrm{d}t/(t^2 - x^2)^{1/2}. \tag{12.138}$$

Beyond $x = 1$, (12.133) becomes

$$xa(x) = 4\int_0^1 t^2 s(t)\,\mathrm{d}t/(x^2 - t^2)^{1/2}. \tag{12.139}$$

The problem can now be solved by iteration. Some examples of the run of $B_\varpi/B_0 = Cb_x$, $B_z/B_0 = Cb_z$ are given in Figs 12.6 and 12.7.

With $\tilde{\sigma}$ prescribed, each converged solution contains the run of the flux distribution against the mass distribution, each constructed for the area out from the origin to the radius x. This can be illustrated by the 'unscrambling' exercise described in Section 12.2. The transformation $\varpi \to \varpi'$, $R \to R'$, $x \to x'$, with the convention $x = 1$ implying $x' = 1$, defined by the term 'unscrambling'

$$B_z(\varpi)\varpi\,\mathrm{d}\varpi = B_0\varpi'\,\mathrm{d}\varpi', \qquad F = \pi B_0 R'^2 \tag{12.140}$$

yields

$$x' = \left[2C\int_0^x b_z x\,\mathrm{d}x\right]^{1/2}(R/R') = \left[\int_0^x b_z x\,\mathrm{d}x \Big/ \int_0^1 b_z x\,\mathrm{d}x\right]^{1/2} \tag{12.141}$$

with

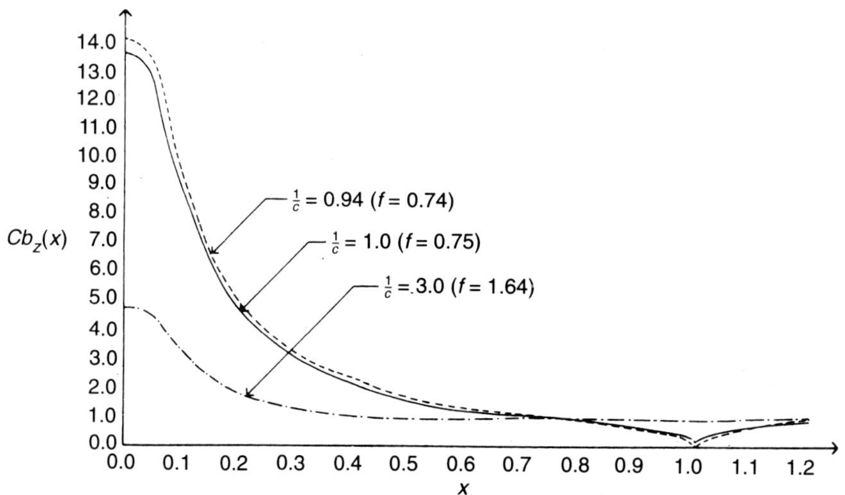

FIG. 12.7. As for Fig. 12.6: the run of $B_z/B_0 = Cb_z$.

$$R' = R\left[2C\int_0^1 b_z x\,\mathrm{d}x\right]^{1/2} = R(2C\tilde{M}f)^{1/2} \tag{12.142}$$

by (12.16). The flux-freezing constraint then yields a new area density $\sigma'(\varpi')$ given by

$$\sigma'(\varpi')\varpi'\,\mathrm{d}\varpi' = \sigma(\varpi)\varpi\,\mathrm{d}\varpi, \tag{12.143}$$

so that by (12.140) and (12.2)

$$\sigma'(\varpi')/B_0 = \sigma(\varpi)/B_z(\varpi) = B_\varpi(\varpi)/2\pi g(\varpi). \tag{12.144}$$

The dimensionless coordinate $Z'(x') = \sigma'(\varpi')/2\rho_0 R'$ of the parent cloud becomes

$$Z' = (B_0^3/16\pi G\rho_0^2 F)^{1/2}[G^{1/2}B_\varpi/g(\varpi)], \tag{12.145}$$

where

$$G^{1/2}B_\varpi/g(\varpi) = b_x(x^2+l^2)/x \qquad \text{for case (a)} \tag{12.146}$$

$$= b_x(x^3+l^3)/xl \qquad \text{for case (b).} \tag{12.147}$$

A selection of unscrambled models is shown in Fig. 12.8.

Solution of the equations for any model fixes $J_\theta(\varpi)$, so allowing construction of the external, curl-free field. The vector potential $\mathbf{A} = A\mathbf{t}$ of an axisymmetric poloidal field defines a flux function $P = -\varpi A$ which is constant on individual

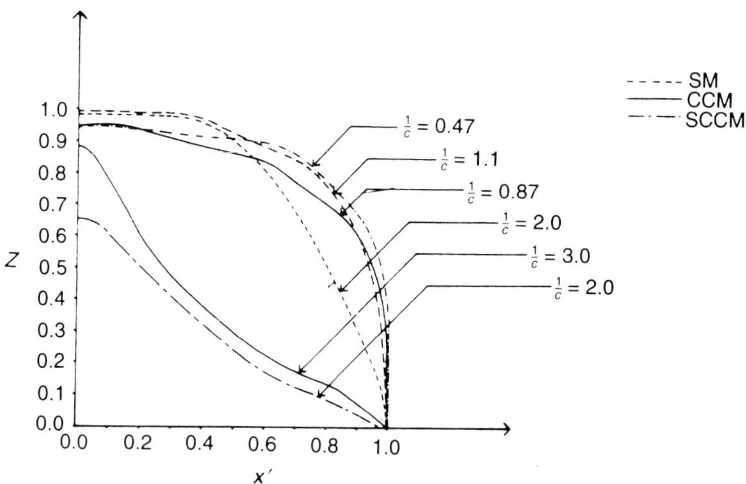

FIG. 12.8. Normalized shapes of the parent clouds of different disc models and with different C-parameter values.

field lines. The uniform background field $\mathbf{B}_0$ is derived from the potential $B_0\varpi/2$, whence from (12.61) the total field has the flux function in dimensionless form

$$p \equiv P/B_0R^2 = -\left\{\frac{x^2}{2} + \frac{C}{2\pi}G(x, Z)\right\} \tag{12.148}$$

where

$$G(x, Z) = x\int_0^1 b_z(x')x'\,\mathrm{d}x' \int_0^{2\pi} \frac{\cos\theta'\,\mathrm{d}\theta'}{(x^2 + x'^2 - 2xx'\cos\theta' + Z^2)^{1/2}} \tag{12.149}$$

can be expressed in terms of standard elliptic integrals (Mestel and Ray 1985). A typical set of field lines is shown in Fig. 12.9.

APPENDIX B

Isopedic disc-like models

As noted, recent work on the evolution of cloud cores through ambipolar diffusion has focused attention on 'isopedic discs'. We follow closely the treatment by Shu and Li (1997). In the current-free domain outside the disc, the magnetic field $\mathbf{B} = -\nabla\Phi$, where the scalar potential Φ satisfies Laplace's equation, with $B_z = -\partial\Phi/\partial z$ continuous with a prescribed value on the disc, and with $|\nabla\Phi|$ vanishing at infinity. (In the more general cases, a non-vanishing background z-component $\mathbf{B}_0$ is subtracted out and then superimposed at the end of the calculation.) The gravitational potential V likewise satisfies Laplace's equation,

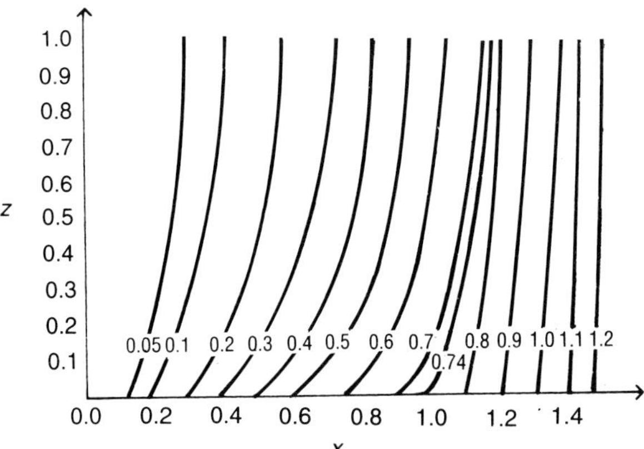

FIG. 12.9. Field lines for the sequence (a) model with $l = 0.2$ and $1/C = 0.89$.

with proper behaviour at infinity and with $(\partial V/\partial z)_+ = -2\pi G\sigma$. Thus with the isopedic condition

$$\sigma = \Lambda B_z, \qquad \lambda \equiv 2\pi G^{1/2}\Lambda, \tag{12.150}$$

supposed to hold,

$$\Phi = \frac{V}{2\pi G\Lambda} = \frac{1}{G^{1/2}\lambda}V. \tag{12.151}$$

The horizontal field component just north of the disc is then

$$B_\varpi^+ = -\frac{\partial\Phi}{\partial\varpi} = -\frac{1}{G^{1/2}\lambda}(g_{\mathrm{grav}})_\varpi = \frac{1}{G^{1/2}\lambda}g \tag{12.152}$$

with g as usual the inward-directed scalar gravitational component. Thus by (12.150) and (12.152), the magnetic tension force (cf. 12.2) is

$$\frac{B_z B_\varpi}{2\pi} = \frac{1}{2\pi}\left(\frac{2\pi G^{1/2}\sigma}{\lambda}\right)\left(\frac{g}{G^{1/2}\lambda}\right) = \frac{1}{\lambda^2}\sigma g, \tag{12.153}$$

By (12.2), the magneto-gravitational equilibrium of an infinitely thin, isopedic disc requires that $\lambda = 1$, e.g. as in (12.30).

Shu and Li consider also the z-equilibrium of an isopedic disc of finite thickness $z_0(\varpi)$. (Even if the disc has gone over into gravitational collapse in the ϖ-direction, magnetohydrostatic equilibrium under the disc's self-gravitation is likely to remain a good approximation in the z-direction. The following is a slight modification of their treatment.) From the one-dimensional Poisson equation,

$$g_z = -2\pi G\sigma(z), \qquad \sigma = 2\int_0^z \rho\,\mathrm{d}z, \tag{12.154}$$

We now use the symbol Σ for $\sigma(z_0)$, the total area density previously given the symbol σ. From (12.154), $-\pi G\sigma\,\mathrm{d}\sigma/\mathrm{d}z$ is the z-component of the gravitational force per unit volume. It is convenient to introduce the ratio η of the horizontal inward gravitational field to the downward value at z_0:

$$\eta = \frac{g}{2\pi G\Sigma}. \tag{12.155}$$

The magnetic tension term $(B_z/4\pi)\,\partial B_\varpi/\partial z$ remains the dominant contribution to the ϖ-component of the Lorentz force, but the z-component is dominated by the contribution to the magnetic pressure gradient of the rapidly varying B_ϖ-component. The gas pressure $p_g(z)$ then satisfies

$$p_g + \frac{B_\varpi^2}{8\pi} + \frac{\pi}{2}G\sigma^2 = p_g(0), \tag{12.156}$$

where $p_g(0)$ is the value in the mid-plane, varying slowly with ϖ. If the gas pressure outside the disc is negligible, then from (12.152), (12.156) and (12.155),

$$p_g(0) = \frac{\pi G\Sigma^2}{2\lambda^2}(\eta^2 + \lambda^2); \tag{12.157}$$

and from the isopedic assumption (12.150) for B_z, the total magnetic pressure achieves at z_0 its maximum value

$$p_m^+ = p_g(0) + \frac{\pi}{2}G\Sigma^2\left(\frac{1}{\lambda^2} - 1\right) = \frac{\pi G\Sigma^2}{2\lambda^2}\left(\eta^2 + 1\right) = \left(\frac{\eta^2+1}{\eta^2+\lambda^2}\right)p_g(0). \tag{12.158}$$

Thus the maximum magnetic and gas pressures are of the same order: even for the case $\lambda = 1$, with purely magnetic support against the radial component of gravity, the bracket in (12.158) is just unity. This illustrates that it is not the magnitude of the magnetic pressure or energy density that is large but the divergence of both the magnetic pressure and tension terms (a conclusion valid equally for the more general, non-isopedic models).

References for Chapters 11 and 12

Abramowitz, M. and Stegun, I.A. (1965). *Handbook of mathematical functions.* Dover, New York.

Arons, J. and Max, C.E. (1975). *Astrophysical Journal*, **196**, L77.

Barker, D.M. and Mestel, L. (1990). *Monthly Notices Royal Astronomical Society*, **245**, 147.

Barker, D.M. and Mestel, L. (1996). *Monthly Notices Royal Astronomical Society*, **282**, 317.

Basu, S. and Mouschovias, T.Ch. (1994). *Astrophysical Journal*, **432**, 720.

Beck, R., Brandenburg, A., Moss, D., Shukurov, A. and Sokoloff, D.D. (1996). *Annual Review Astronomy Astrophysics*, **34**, 155.

Binney, J.J. and Tremaine, S. (1987). *Galactic dynamics.* Princeton University Press.
Bondi, H. (1952). *Monthly Notices Royal Astronomical Society*, **112**, 195.
Bonnor, W.B. (1956). *Monthly Notices Royal Astronomical Society*, **116**, 351.
Brandenburg, A. and Zweibel E.G. (1994). *Astrophysical Journal*, **427**, L91.
Brandenburg, A. and Zweibel, E.G. (1995). *Astrophysical Journal*, **448**, 734.
Cameron, A.G.W. (1962). *Icarus*, **1**, 13.
Campbell, C.G. and Mestel, L. (1987). *Monthly Notices Royal Astronomical Society*, **229**, 549.
Carlberg, R.G. and Pudritz, R.E. (1990). *Monthly Notices Royal Astronomical Society*, **247**, 353.
Chandrasekhar, S. (1939). *An introduction to the study of stellar structure.* Chicago University Press. (Dover edition 1957.)
Chandrasekhar, S. and Fermi, E. (1953). *Astrophysical Journal*, **118**, 113, 116.
Ciolek, G.E. and Mouschovias, T.Ch. (1993). *Astrophysical Journal*, **418**, 774.
Crutcher, R.M., Troland, T.H., Goodman, A.A., Heiles, C., Kazès, I. and Myers, P.C. (1993). *Astrophysical Journal*, **407**, 175.
Crutcher, R.M., Troland, T.H., Lazareff, B. and Kazès, I. (1996). *Astrophysical Journal*, **456**, 217.
Dewar, R.L. (1970). *Physics Fluids*, **13**, 2710.
Draine, B.T. (1983). *Astrophysical Journal*, **270**, 519.
Draine, B.T. (1986). *Monthly Notices Royal Astronomical Society*, **220**, 133.
Draine, B.T., Roberge, W.G. and Dalgarno, A. (1983). *Astrophysical Journal*, **264**, 485.
Ebert, R. (1955). *Zeitschrift Astrophysik*, **37**, 217.
Ebert, R, (1960). In *Die Entstehung von Sternen*, p. 311. Springer, Berlin.
Eggen, O.J. (1976). *Quarterly Journal Royal Astronomical Society*, **17**, 472.
Elmegreen, B.G. (1979). *Astrophysical Journal*, **232**, 729.
Fiedler, R.A. and Mouschovias, T.Ch. (1993). *Astrophysical Journal*, **415**, 680.
Field, G.B. (1970). *Mémoires Société Royale Liège*, **19**, 29.
Galli, D. and Shu, F.H. (1993). *Astrophysical Journal*, **417**, 220, 243.
Gammie, C. and Ostriker, E. (1996). *Astrophysical Journal*, **466**, 814.
Garlick, A.R. (1979). *Astronomy and Astrophysics*, **73**, 337.
Gaustad, J.E. (1963). *Astrophysical Journal*, **138**, 1050.
Gillis, J., Mestel, L. and Paris, R.B. (1974). *Astrophysics Space Science*, **27**, 167.
Gillis, J., Mestel, L. and Paris, R.B. (1979). *Monthly Notices Royal Astronomical Society*, **187**, 311.
Goldreich, P. and Sridhar, S. (1995). *Astrophysical Journal*, **438**, 763.
Hayakawa, S., Nishimura, S. and Takayanagi, K. (1961). *Publications Astronomical Society Japan*, **13**, 184.
Herbig, G.F. (1962*a*). *Astrophysical Journal*, **135**, 736.
Herbig, G.F. (1962*b*). *Advances in Astronomy Astrophysics*, **1**, 47, 76.
Hoyle, F. (1945). *Monthly Notices Royal Astronomical Society*, **105**, 302.
Hoyle, F. (1953). *Astrophysical Journal*, **118**, 513.
Hunter, C. (1962). *Astrophysical Journal*, **135**, 594.

Hunter, C. (1964). *Astrophysical Journal*, **139**, 570.
Kiguchi, M., Narita, S., Miyama, S.M. and Hayashi, C. (1987). *Astrophysical Journal*, **317**, 830.
Kulsrud, R.M. and Anderson, S.W. (1992). *Astrophysical Journal*, **396**, 606.
Kulsrud, R.M. and Pearce, W.P. (1969). *Astrophysical Journal*, **156**, 445.
Layzer, D. (1963). *Astrophysical Journal*, **137**, 351.
Ledoux, P. (1951). *Annales d'Astrophysique*, **14**, 438.
Li, Z.-Y. and Shu, F.H. (1997). *Astrophysical Journal*, **475**, 237.
Lin, C.C. (1971). *Highlights of Astronomy*, **2**, 88.
Lin, C.C., Mestel, L. and Shu, F.H. (1965). *Astrophysical Journal*, **142**, 1431.
Lizano, S. and Shu, F.H. (1989). *Astrophysical Journal*, **342**, 834.
Lovelace, R.V.E. and Zweibel, E.G. (1997). *Astrophysical Journal*, **487**, 285.
Low, C. and Lynden-Bell, D. (1976). *Monthly Notices Royal Astronomical Society*, **176**, 367.
Lynden-Bell, D. (1973). In *Dynamical structure and evolution of stellar systems*, p. 91. Saas-Fee Lectures, Geneva Observatory.
Lynden-Bell, D. (1977). In *Star formation* (eds T. de Jong and A. Maeder), p. 295. Reidel, Dordrecht.
McCrea, W.H. (1957). *Monthly Notices Royal Astronomical Society*, **117**, 562.
McCrea, W.H. (1960). *Proceedings Royal Society A* **256**, 245.
McKee, C.F. (1989). *Astrophysical Journal*, **345**, 782.
McKee, C.F. and Zweibel, E.G. (1995). *Astrophysical Journal*, **440**, 686.
McKee, C.F., Zweibel, E.G., Goodman, A.A. and Heiles, C. (1993). In *Protostars and planets III* (eds E.H. Levy and J.I. Lunine), pp. 279, 327. University of Arizona Press, Tucson.
Mestel, L. (1963). *Monthly Notices Royal Astronomical Society*, **126**, 553.
Mestel, L. (1965). *Quarterly Journal Royal Astronomical Society*, **6**, 161, 265.
Mestel, L. (1966). *Monthly Notices Royal Astronomical Society*, **133**, 265.
Mestel, L. (1969). In *Plasma instabilities in astrophysics* (eds D.A. Tidman and D.G. Wentzel), p. 329. Gordon and Breach, New York.
Mestel, L. and Paris, R.B. (1979). *Monthly Notices Royal Astronomical Society*, **187**, 337.
Mestel, L. and Paris, R.B. (1980). Report EUR-CEA FC 1049.
Mestel, L. and Paris, R.B. (1984). *Astronomy and Astrophysics*, **136**, 98.
Mestel, L. and Ray, T.P. (1985). *Monthly Notices Royal Astronomical Society*, **212**, 275.
Mestel, L. and Spitzer Jr., L. (1956). *Monthly Notices Royal Astronomical Society*, **116**, 583.
Mestel, L. and Strittmatter, P.A. (1967). *Monthly Notices Royal Astronomical Society*, **137**, 95
Mestel, L. and Subramanian, K. (1991). *Monthly Notices Royal Astronomical Society*, **248**, 677.
Moneti, A., Pipher, J.L., Helfer, H.L., McMillan, R.S. and Perry, M.L. (1984). *Astrophysical Journal*, **282**, 588.
Mouschovias, T.Ch. (1976*a*). *Astrophysical Journal*, **206**, 753.

Mouschovias, T.Ch. (1976*b*). *Astrophysical Journal*, **207**, 141.
Mouschovias, T.Ch. (1987). In *Physical processes in interstellar clouds* (eds G.E. Morfill and M. Scholer), p. 453. Reidel, Dordrecht.
Mouschovias, T.Ch. (1994). In *Cosmical magnetism, contributed papers* (ed. D. Lynden-Bell), p. 24. Institute of Astronomy, Cambridge.
Mouschovias, T.Ch. (1996). In *Solar and astrophysical magnetohydrodynamic flows* (ed. K.C. Tsinganos), p. 505. Kluwer, Dordrecht.
Mouschovias, T.Ch. and Paleologou, E.V. (1979). *Astrophysical Journal*, **230**, 204.
Mouschovias, T.Ch. and Paleologou, E.V. (1980). *Astrophysical Journal*, **237**, 877.
Mouschovias, T.Ch. and Spitzer Jr., L. (1976). *Astrophysical Journal*, **210**, 326.
Myers, P.C. and Goodman, A.A. (1988). *Astrophysical Journal*, **326**, L27.
Nakamura, F., Hanawa, T. and Nakano, T. (1995). *Astrophysical Journal*, **444**, 770.
Nakano, T. (1976). *Publications Astronomical Society Japan*, **28**, 355.
Nakano, T. (1979). *Publications Astronomical Society Japan*, **31**, 697.
Nakano, T. (1981). *Progress Theoretical Physics, Supplement*, **70**, 54.
Nakano, T. (1983). *Publications Astronomical Society Japan*, **35**, 209.
Nakano, T. (1984). *Fundamentals Cosmical Physics*, **9**, 139.
Nakano, T. (1988). In *Galactic and extragalactic star formation* (eds R. Pudritz and M. Fich), p. 111. Kluwer, Dordrecht.
Nakano, T. (1989). *Monthly Notices Royal Astronomical Society*, **241**, 495.
Nakano, T. (1990). *Monthly Notices Royal Astronomical Society*, **242**, 535.
Nakano, T. and Umebayashi, T. (1986*a*). *Monthly Notices Royal Astronomical Society*, **218**, 663.
Nakano, T. and Umebayashi, T. (1986*b*). *Monthly Notices Royal Astronomical Society*, **221**, 339.
Nishi, R., Nakano, T. and Umebayashi, T. (1991). *Astrophysical Journal*, **368**, 181.
Norman, C.A. and Heyvaerts, J. (1985). *Astronomy and Astrophysics*, **147**, 247.
Osterbrock, D.E., (1961). *Astrophysical Journal*, **134**, 270.
Padoan, P. and Nordlund, A. (1998). Submitted to *Astrophysical Journal.*
Padoan, P., Nordlund, A. and Jones, B.J.T. (1997). *Monthly Notices Royal Astronomical Society*, **288**, 145.
Parker, D.A. (1973). *Monthly Notices Royal Astronomical Society*, **163**, 41.
Parker, D.A. (1974). *Monthly Notices Royal Astronomical Society*, **168**, 331.
Parker, E.N. (1966). *Astrophysical Journal*, **145**, 811.
Parker, E.N. (1970). *Interstellar gas dynamics* (ed. H.J. Habing), p. 168. Reidel, Dordrecht.
Pikel'ner, S.B. (1968). *Soviet Astronomy*, **11**, 737.
Rees, M.J. (1976). *Monthly Notices Royal Astronomical Society*, **176**, 483.
Rees, M.J. (1987). *Quarterly Journal Royal Astronomical Society*, **28**, 197.
Rees, M.J. (1994). In *Cosmical magnetism* (ed. D. Lynden-Bell), p. 155. Kluwer, Dordrecht.

Roberts, W.W. (1969). *Astrophysical Journal*, **158**, 123.
Scarrott, S.M., Ward-Thompson, D. and Warren-Smith, R.F. (1987). *Monthly Notices Royal Astronomical Society*, **224**, 299.
Shu, F.H. (1977). *Astrophysical Journal*, **214**, 488.
Shu, F.H. and Li, Z.-Y. (1997). *Astrophysical Journal*, **475**, 251.
Shu, F.H., Adams, F.C. and Lizano, S. (1987). *Annual Review Astronomy Astrophysics*, **25**, 23.
Silk, J. (1977). *Astrophysical Journal*, **214**, 152.
Silk, J. (1985). In *Birth and infancy of stars* (eds R. Lucas, A. Omont and R. Stora), p. 349. North-Holland, New York.
Silk, J. (1995). *Astrophysical Journal*, **438**, L41.
Silk, J. and Norman, C.A. (1983). *Astrophysical Journal*, **272**, L49.
Sneddon, I.N. (1957). *Elements of partial differential equations*, p. 178. McGraw-Hill, New York.
Sofue, Y., Fujimoto, M. and Wielebinski, R. (1986). *Annual Review Astronomy Astrophysics*, **24**, 459.
Spitzer Jr., L. (1942). *Astrophysical Journal*, **95**, 329.
Spitzer Jr., L. (1963). In *Origin of the solar system* (eds R. Jastrow and A.G.W. Cameron), p. 39. Academic Press, New York.
Spitzer Jr., L. (1978). *Physical processes in the interstellar medium*. Wiley, New York.
Spitzer Jr., L. and Tomasko, M.G. (1968). *Astrophysical Journal*, **152**, 971.
Spruit, H.C. (1987). *Astronomy and Astrophysics*, **184**, 173.
Strittmatter, P.A. (1966). *Monthly Notices Royal Astronomical Society*, **132**, 359.
Subramanian, K. and Mestel, L. (1993). *Monthly Notices Royal Astronomical Society*, **265**, 649.
Tomisaka, K., Ikeuchi, S. and Nakamura, T. (1988). *Astrophysical Journal*, **335**, 239.
Toomre, A. (1963). *Astrophysical Journal*, **138**, 385.
Toomre, A. (1964). *Astrophysical Journal*, **139**, 1217.
Verschuur, G.L. (1969). *Astrophysical Journal*, **156**, 861.
Zel'dovich, Ya.B., Ruzmaikin, A.A. and Sokoloff, D.D. (1983). *Magnetic fields in astrophysics*. Gordon and Breach, New York.
Zuckerman, B. and Palmer, P.E. (1974). *Annual Review Astronomy Astrophysics*, **12**, 279.
Zweibel, E.G. (1990). *Astrophysical Journal*, **340**, 550.
Zweibel, E.G (1994). In *Cosmical magnetism* (ed. D. Lynden-Bell), p. 73. Kluwer, Dordrecht.
Zweibel, E.G. (1998). *Astrophysical Journal*, **499**, 746.
Zweibel, E.G. and Brandenburg, A. (1997). *Astrophysical Journal*, **478**, 563.
Zweibel, E.G. and Josafatsson, K. (1983). *Astrophysical Journal*, **270**, 511.
Zweibel, E.G. and Lovelace, R.V.E. (1997). *Astrophysical Journal*, **475**, 260.
Zweibel, E.G. and McKee, C.F. (1995). *Astrophysical Journal*, **439**, 779.

13

PULSAR ELECTRODYNAMICS I

A collapsed object such as a white dwarf or a neutron star may retain some of the magnetic flux present in the parent star during its main sequence phase; or it may generate new flux by a dynamo or battery process (e.g. Blandford *et al.* 1983). Magnetohydrodynamic effects will manifest themselves in the star's magnetosphere provided there is a plentiful supply of plasma, for example by overflow from the Roche lobe of an evolving companion in a binary system. Some of the work discussed in Chapter 10 on pre-main sequence stars – for example on accretion discs – was in fact originally motivated by the discovery of collapsed X-ray sources, and there is now an extensive literature on such models for cataclysmic variables, novae, X-ray pulsars, etc. (Frank *et al.* 1992; Campbell 1997) to which the reader is referred. In Chapters 13 and 14, attention is instead focused on the electrodynamics of single collapsed stars (principally neutron stars), for which the supply of gas is supposed to come just from the star. Most workers adopt such a model as appropriate to the radio pulsars. The work of Chapter 13 involves just classical physics, and may be applicable also to rotating magnetic white dwarfs; whereas that of Chapter 14, which explicitly includes electron–positron pair production, is restricted to neutron stars.

13.1 The charged magnetosphere

13.1.1 *Introduction*

The central object in a pulsar is an obliquely rotating magnetic neutron star, with a mass $\simeq M_\odot$, a radius $R \simeq 10^6$ cm $\simeq 1.4 \times 10^{-5} R_\odot$ and $GM/R \simeq 0.15c^2$. The typical rotation period $\mathrm{P}_1 \equiv 2\pi/\alpha \simeq 1$ s yields $\simeq 3\times 10^{-7}$ for the centrifugal parameter $\epsilon_c \equiv \alpha^2 R^3/GM$, and even the Crab pulsar with $\mathrm{P}_1 \equiv 1/30$ s is a 'slow rotator', with $\epsilon_c \simeq 3\times 10^{-4}$, and the light-cylinder ('l-c') radius $\tilde{\omega}_c \equiv c/\alpha \simeq 200R$. Only for the millisecond pulsars with $\mathrm{P}_1 \simeq 10^{-3}$ does $\tilde{\omega}_c$ come close to R and ϵ_c to unity. The same enormous gravitational field ensures that for any reasonable temperature, the scale-height is so small that a thermally-supported atmosphere would exponentiate down to virtually zero density within a few centimetres. This is the prima facie justification historically for the first model of the pulsar magnetosphere, which predated the actual discovery of pulsars: in a prophetic paper, Pacini (1967) adapted the solution constructed earlier by Deutsch for the electromagnetic field of a perfectly conducting oblique rotatorwithout any external sources ρ_e and $\mathbf{j}$. Within the star the electric field as measured in the inertial frame is again

$$\mathbf{E} = -\alpha\,(\mathbf{k}\times\mathbf{r})\times\mathbf{B}/c \tag{13.1}$$

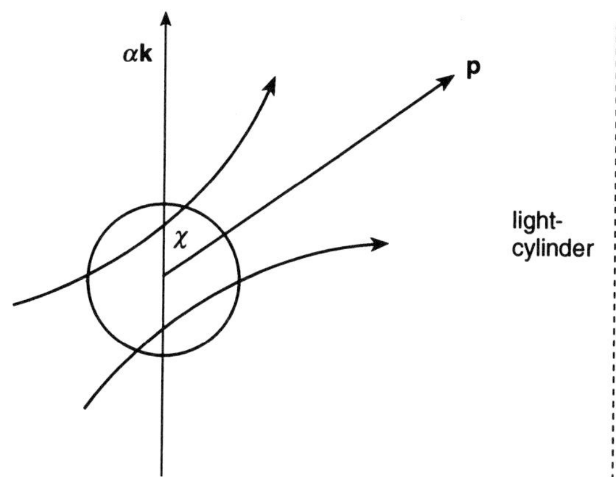

FIG. 13.1. Schematic diagram of the oblique rotator.

where **k** is the rotation axis, and the magnetic field over the stellar surface is taken as that of a dipole of moment **p** inclined to **k** by the angle χ (Fig. 13.1).

The external electromagnetic field is constructed from the vacuum Maxwell equations, subject to the field's remaining finite at infinity and to continuity at the star's surface of the tangential component of **E** and the normal component of **B**. Far from the star the dominant terms are those forming the wave field emitted by the rotating perpendicular component $p \sin \chi$ of the point dipole **p**, and transporting per second the energy

$$\left(2p^2\alpha^4/3c^3\right)\sin^2\chi = \left(B_s^2 R^6 \alpha^4/6c^3\right)\sin^2\chi, \tag{13.2}$$

where $B_s = 2p/R^3$ is the polar field strength. If the assumed obliquity χ is not very small, then the equating of (13.2) to the energy loss

$$-I\alpha\dot{\alpha} = 4\pi^2 I\dot{\mathrm{P}}_1/\mathrm{P}_1^3 \tag{13.3}$$

from a star of moment of inertia I, as inferred from the observed steady increase $\dot{\mathrm{P}}_1$, yields the canonical value of $B_s \simeq 10^{12}$ G (Gold 1969).

The value (13.3) is orders of magnitude greater than the energy carried off per second in the pulsed emission in radio waves, even after making generous allowance for beaming. Thus the radio noise is seen to be be a diagnostic of a much more powerful energy loss process. The Deutsch–Pacini model is a simple example of such a process – the energy and angular momentum are carried away by a classical electromagnetic wave of frequency equal to the rotation frequency of the star ($\simeq$ 1 Hz–1 kHz). Figure 13.2 illustrates the magnetic field lines of the rotating perpendicular dipole in vacuo.

The Deutsch–Pacini wave from a neutron star in vacuo is 'strong', in the

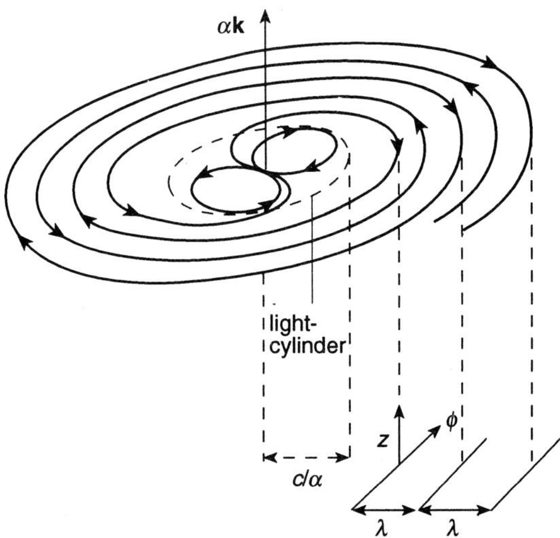

FIG. 13.2. The **perpendicular dipole p** in vacuo, rotating with angular velocity $\alpha\mathbf{k}$ where $\mathbf{k}\cdot\mathbf{p} = 0$. Sketch of the magnetic field lines in the equatorial plane, showing the transition near the light-cylinder radius $\varpi_c = c/\alpha$ to the 'far-field' radiation solution with wavelength $\lambda = \varpi_c$, $\mathbf{B}$ purely toroidal and $\mathbf{E}$ parallel to $\mathbf{k}$.

sense that a test particle would gyrate in the wave magnetic field with a frequency much greater than the wave frequency α. Thus an individual particle would see effectively a static field, and rapid acceleration would occur to energies of 10^{14}–10^{15} electonvolts (Ostriker and Gunn 1969; Gunn and Ostriker 1971). These particles would in turn emit 'synchro-Compton radiation', with a spectrum essentially that of synchrotron radiation, but with a polarization similar to that of the low-frequency strong wave (Rees 1971; Arons 1972; Blandford 1972). The non-appearance of a high degree of circular polarization in the Crab nebula is plausibly attributed to destruction of the low frequency wave by instabilities. Chapters 13 and 14 are devoted entirely to the 'pulsar magnetosphere' problem, and mainly (though not entirely) to the aligned 'non-pulsar'. A principal aim is to gain insight into the dominant energy loss processes, but with physics more realistic than in the vacuum models. For recent discussions on emission processes, especially on coherent mechanisms for the radio pulses, the reader is referred to Melrose (1996).

The Deutsch–Pacini model is like a dynamo on open circuit, with the rotation of the highly conducting neutron star generating by (13.1) enormous potential differences (of the order of $2\pi B_s R^2/c\mathrm{P}_1 \simeq 10^{17}/\mathrm{P}_1$ volts!) over the star's surface from pole to equator, but without any external current flow. When the obliquity

χ is finite, energy loss occurs through the non-vanishing displacement current; equally, the aligned dipolar model in vacuo, with $\chi = 0$, is axisymmetric and so is dead, because of the vanishing of the time-dependence.

The essence of the Goldreich–Julian (GJ) critique (1969) of the vacuum model is that the system will build up its own conducting leads – the magnetosphere spontaneously charges itself up by current flow from the star. In the simplest example the magnetic field at the star's surface is again assumed to be dipolar, and continuous in both components. Just within the star condition (13.1) yields

$$\mathbf{E} = (\alpha R B_s/3c)\left\{\left[1 - \mathrm{P}_2\left(\cos\theta\right)\right],\, \partial \mathrm{P}_2/\partial\theta,\, 0\right\}. \tag{13.4}$$

If the star has zero net charge, the external vacuum solution has the quadrupole scalar potential (continuous over the stellar surface)

$$\phi = -\left(\alpha R^5 B_s/3c\right)\mathrm{P}_2\left(\cos\theta\right)/r^3, \tag{13.5}$$

so there is a surface charge density

$$\sigma = \left[\left(E_r\right)_{\mathrm{ext}} - \left(E_r\right)_{\mathrm{int}}\right]/4\pi = -\left(\alpha R B_s/4\pi c\right)\cos^2\theta. \tag{13.6}$$

In the aligned case, this consists of a sheet of electrons; in the counteraligned, 'anti-pulsar' case (p and B_s negative), a sheet of ions. Whereas within the star, by (13.1) $\mathbf{E}$ is perpendicular to $\mathbf{B}$, just outside the star the solution (13.5) yields $\mathbf{E}\cdot\mathbf{B} = -\alpha R B_s^2 \cos^3\theta/c$ – the component $\mathbf{E}_\parallel$ is of the same order as $\mathbf{E}_\perp$. Thus the electric force $2\pi\sigma^2$ per unit area (normal to the star's surface) has a component along $\mathbf{B}$ that exceeds the restraining gravitational force component, by the factor $(eB_s/m_e c\alpha)\left(\alpha^2 R^3/2GM\right) \simeq 3\times 10^{18}\mathrm{P}_1\left(B_s/10^{12}\right)\left(\alpha^2 R^3/GM\right)$ in the aligned case, and so will far exceed unity for any observable pulsar. The same conclusion follows for an anti-pulsar, with m_e replaced by a heavy ion mass $\simeq 56 m_{\mathrm{H}}$. (It was at first thought that quantum mechanical binding would allow a significant $\mathbf{E}_\parallel$ at the star's surface for the anti-pulsar case, but it now appears that for both cases the appropriate dynamical boundary condition at the surface is $\mathbf{E}_\parallel \simeq 0$ (Flowers *et al.* 1977).) The aligned vacuum model is therefore not viable; charges are pulled out of the star into the surroundings, modifying profoundly both the external magnetic and electric fields. The GJ argument is immediately extensible to the oblique case: the vacuum Deutsch–Pacini model is modified by the inevitable presence of external charges (Cohen and Toton 1971; Mestel 1971).

The Deutsch–Pacini solution is perhaps the simplest example of a 'quasi-steady' model, in which the time-variation of any quantity, as seen in the non-rotating frame, is due just to the rigid rotation of the non-axisymmetric structure of the oblique rotator. This condition can be written as the equivalence in the inertial frame of two operators:

$$\partial/\partial t = -\alpha\,\partial/\partial\Phi \tag{13.7}$$

where Φ is the azimuthal angle defined by the rotation axis, and the operations are applied to scalars or to cylindrical or spherical polar components. (In the special case of axial symmetry, for example the aligned rotator, $\partial/\partial\Phi = 0$ implies $\partial/\partial t = 0$ – 'quasi-steady' becomes 'steady'.) With the constraint (13.7), Faraday's law of induction becomes

$$-c\nabla\times\mathbf{E} = -\alpha\,\partial\mathbf{B}/\partial\Phi = \nabla\times\left[(\alpha\mathbf{k}\times\mathbf{r})\times\mathbf{B}\right], \tag{13.8}$$

whence

$$\mathbf{E} = -(\alpha\mathbf{k}\times\mathbf{r})\times\mathbf{B}/c - \nabla\psi : \tag{13.9}$$

the quasi-steady electric field can always be written as the sum of the 'corotational' field $-(\alpha\mathbf{k}\times\mathbf{r})\times\mathbf{B}/c$ and the 'non-corotational' field $-\nabla\psi$. In each domain, $\nabla\psi$ is determined by the physics considered appropriate. Inside the rigidly rotating, perfectly conducting neutron star crust, $\nabla\psi \equiv 0$. The Deutsch–Pacini external vacuum model fixes ψ by imposing $\nabla\cdot\mathbf{E} = 0$, $\nabla\times\mathbf{B} = \partial\mathbf{E}/c\,\partial t = -\alpha\,\partial\mathbf{E}/c\,\partial\Phi$, subject to the usual boundary conditions. Following their critique of the aligned vacuum model, GJ postulated that $\nabla\cdot\mathbf{E} = 0$ should be replaced by the simple plasma condition

$$-\mathbf{E}\cdot\mathbf{B} = \mathbf{B}\cdot\nabla\psi = 0 : \tag{13.10}$$

charges distribute themselves so as to short out the component $\mathbf{E}_{\|}$ along $\mathbf{B}$. If (13.10) holds all the way along the field lines linking the star with the region considered – i.e. if there are no 'gaps' (cf. Section 13.2.5) – then the uniform value ψ_s of ψ within the star is propagated outwards, and the vacuum condition is replaced by the corotating plasma condition (13.1), with the associated charge density

$$\rho_e = \rho_{\mathrm{GJ}} = \frac{\nabla\cdot\mathbf{E}}{4\pi} = -\left(\frac{\alpha}{2\pi c}\right)\mathbf{k}\cdot\left[\mathbf{B} - \frac{1}{2}\mathbf{r}\times(\nabla\times\mathbf{B})\right]. \tag{13.11}$$

Within the l-c, condition (13.1) yields $E < B$, so that the individual charges are subject to the $c\,(\mathbf{E}\times\mathbf{B})/\mathbf{B}^2$ drift, which for a purely poloidal field $\mathbf{B}_{\mathrm{p}}$ reduces to the corotation velocity $\alpha\tilde{\omega}\mathbf{t}$. GJ argued further that since corotation of individual charges cannot hold beyond the l-c, there must be a continuous flow of charge along those field lines that cross the l-c. They postulated an electrically driven wind emanating from each polar cap: an electron stream in a channel containing the axis, surrounded by a collar of ions carrying off positive charge at the same rate, but flowing through a background sea of corotating electrons, sufficient to ensure that the net charge density continues to satisfy (13.11). These currents generate a toroidal field component $\mathbf{B}_{\mathrm{t}} = B_{\phi}\mathbf{t}$; the resulting $\mathbf{E}\times\mathbf{B}$ drift has the magnitude $\alpha\varpi B_{\mathrm{p}}/B$ which is subluminal within the l-c.

An electrodynamically equivalent system replaces the outflowing ion stream by an inflowing electron stream (Section 13.2). In general, a 'live' aligned model will have a poloidal current system as pictured in Fig. 13.3. The currents leave

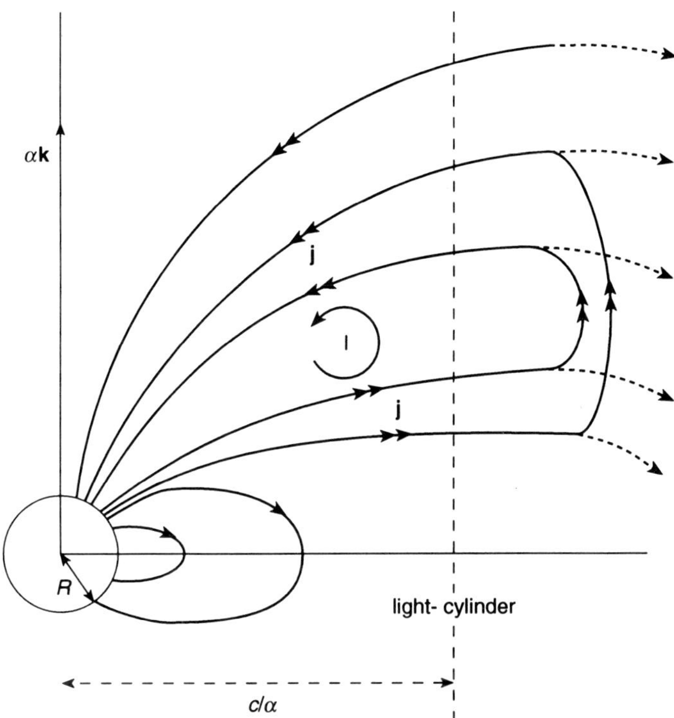

FIG. 13.3. Schematic current system set up in a 'live' aligned pulsar model. Shown are some typical poloidal field lines that close within the l-c and so can carry corotating charges, and some 'open' poloidal lines that cross the l-c. The magnetospheric poloidal currents cross field lines primarily beyond the l-c, giving up to the charges the angular momentum picked up by the equal and opposite magnetic torques exerted on the star.

and return to the star within two polar caps of angle $\theta_{\rm pc}$. By (2.48), the total (negative) torque L exerted on the star is given by

$$L = 2\int_0^{\theta_{\rm pc}} \frac{\varpi B_\phi}{4\pi} B_r 2\pi R^2 \sin\theta \, \mathrm{d}\theta. \tag{13.12}$$

For illustration, write the net total current from the star between the pole and the angle θ as $I(\theta) = K\theta(\theta_{\rm pc} - \theta)$, where K and so also I are negative in sign. The value of I with the maximum modulus is $I_{\rm m} = K\theta_{\rm pc}^2/4$, reached at $\theta = \theta_{\rm pc}/2$. By Ampère's law, at θ the associated B_ϕ is given by

$$(2\pi\varpi)B_\phi = 4\pi I(\theta)/c. \tag{13.13}$$

Over the small angle polar caps, $B_r \approx B_s$, and the torque (13.12) becomes

$$L \approx \frac{2B_s R^2}{c} \int_0^{\theta_{\rm pc}} K\theta(\theta_{\rm pc} - \theta)\theta \, d\theta = \frac{KB_s R^2}{6c} \theta_{\rm pc}^4 = \frac{2I_{\rm m} F_{\rm pc}}{3\pi c}, \tag{13.14}$$

where $F_{\rm pc} \simeq (\pi R^2 \theta_{\rm pc}^2) B_s$ is the poloidal flux over the northern cap. The associated energy loss rate $-\dot{\mathcal{E}}$ is $\alpha|L|$. The rotating magnetospheric charges will modify the poloidal field (cf. Section 13.1.3), but for a first estimate it may be taken as still roughly like a vacuum dipolar field out to the l-c. Assume also that $\theta_{\rm pc}$ is given by the last dipolar field line to reach the l-c: $\sin^2 \theta_{\rm pc}/R = \alpha/c$, or $\theta_{\rm pc}^2 = \alpha R/c$, with $B_{\rm p}$ at the l-c typically $(B_s/2)(\alpha R/c)^3$. In stellar wind theory (Chapter 7), $|B_\phi| \approx B_{\rm p}$ at the Alfvénic surface. From (13.13), if a similar relation holds at the l-c, $|I| \simeq cB_s(\alpha/c)^2 R^3/4$, yielding for the energy loss associated with the torque (13.14)

$$-\dot{\mathcal{E}} \approx B_s^2 R^6 \alpha^4 / 6c^3, \tag{13.15}$$

of the same order as the Deutsch–Pacini value (13.2).

This same estimate, found by GJ (1969) for their original model with charge-separated electron–ion winds, will emerge in the electron circulation model discussed below, and also in the the model with an electron–positron wind discussed in Chapter 14. On dimensional grounds one can hardly expect any other result for an active model, and it gives one extra confidence that the estimate for B_s from (13.2) is firmly based.

13.1.2 *Charge-separated domains*

Before beginning the long (and so far only partially successful) attempt on a more detailed and rigorous treatment of the magnetosphere problem, we follow up the original GJ argument and ask in particular how much of the semi-quantitative picture of a plasma that emerged in Section 2.1 will survive for the pulsar problem. Condition (13.1) also holds in both the dead zone (with closed field lines) and the wind zone of the corona surrounding a uniformly rotating, uncollapsed magnetic star (Chapters 7 and 8), but with a crucial difference: the hot stellar corona is thermally supported along $\mathbf{B}$ and so behaves like a normal plasma, with the charge density of both the ionic and electronic components far greater than the absolute value of the net charge density (13.11), which requires only a very small degree of charge separation. In the pulsar case, we have seen that the thermal pressure is negligible. If the charges were mass-less, then the GJ approximation (13.10) would hold exactly in the absence of dissipation (cf. Section 13.2.3 below), and there would be no restriction on the densities of electrons and ions other than that the difference in positive and negative densities should yield (13.11). But in reality there will be a net component along $\mathbf{B}$ of the non-electromagnetic forces – gravitational-plus-inertial – acting on both electrons and ions. As an example, consider a domain near the star, with $|\nabla \times \mathbf{B}|$ small and with $\mathbf{k}\cdot\mathbf{B} > 0$, so that by (13.11) $\rho_e < 0$ – a local excess of electrons over ions. Suppose also that gravity dominates over centrifugal force, and that in a steady state there is no steady flow into or out of the star in this domain, so that the electrons are held in equilibrium by a small component $\mathbf{E}_{\|}$ acting towards the star, with $(-e\mathbf{E} + m_e\mathbf{g})\cdot\mathbf{B} = 0$. An ion of mass $m_i = Am_{\mathrm{H}}$ and charge Ze then feels the net force component $(m_i\mathbf{g} + Ze\mathbf{E})\cdot\mathbf{B}/B = (Am_{\mathrm{H}} + Zm_e)\,\mathbf{g}\cdot\mathbf{B}/B$ and so will be forced back to the star; while if centrifugal forces dominate, the ion would be driven towards the equator. In a zone of steady electron outflow from the polar regions, the small but non-zero component $\mathbf{E}_{\|}$ that lifts the electrons from the gravitational well would again assist gravity to hold back the ions. Within the framework of classical theory one is then led at first to a tentative picture of a *charge-separated* magnetosphere: when $\rho_e = \nabla\cdot\mathbf{E}/4\pi < 0$, only electrons are present, with number-density $n_e = -\rho_e/e$; and likewise when $\rho_e > 0$, only ions are present, with $n_i = \rho_e/Ze$.

It will be seen below that even in a classical model, strict charge separation is unlikely to persist over the whole magnetosphere. However, it is also likely that unless efficient pair production or surface spallation occurs, conditions will still differ radically from a classical plasma, in that the density $|\rho_e|$ given by (13.11) will not be small compared with $n_e e$ or $n_i Ze$ but of the same order, and it is instructive to see how this affects the elementary order-of-magnitude discussion of basic plasma physics (Mestel 1971). From (13.11),

$$|\rho_e|/n_i Ze \simeq (\alpha B/2\pi c)\,/\,(\rho/Am_{\mathrm{H}})\,Ze = 4\left(B^2/8\pi\rho c^2\right)(\alpha/\omega_{ig}) \qquad (13.16)$$

with $\omega_{ig} = ZeB/Am_{\mathrm{H}}c$ = non-relativistic Larmor gyration frequency of an ion in the local field B. In a normal plasma, $B^2/8\pi\rho c^2 \ll 1$ usually, and also the ratio of macroscopic to microscopic frequencies $\alpha/\omega_{ig} \ll 1$, so justifying the conclusion that $|\rho_e| \ll n_i Ze$. But if a magnetospheric domain is such that the ratio $|\rho_e|/n_i Ze$ is O(1), then (13.16) must be read as fixing the mass density:

$$B^2/8\pi\rho c^2 \simeq (\omega_{ig}/4\alpha) \gg 1, \qquad (13.17)$$

showing that the Coulomb force is so strong that only a very small charge density is required to yield the corotational electric field (13.1), with $E_{\|} \ll E_{\perp}$. Again, in a non-relativistic plasma the convection current $\rho_e\mathbf{v}$ due to the bulk motion of a gas with the net charge density (13.11) usually makes a negligible contribution to the total current, which is due primarily to the relative motion of electrons and ions. Thus, for example, with $|\mathbf{v}| \simeq \alpha\tilde{\omega}$ = the velocity of corotation,

$$\begin{aligned} 4\pi\,|\rho_e\mathbf{v}|\,/c\,|\nabla\times\mathbf{B}| &= \left[(4\pi/c)\,(\alpha B/2\pi c)\,(\alpha\tilde{\omega})\right]\Big/\left[|\nabla\times\mathbf{B}|\,(\tilde{\omega}/B)\,(B/\tilde{\omega})\right] \\ &= 2\,(\alpha\tilde{\omega}/c)^2\Big/\,|\nabla\times\mathbf{B}|\,(\tilde{\omega}/B)\,, \end{aligned} \qquad (13.18)$$

and if $|\nabla\times\mathbf{B}| \simeq B/\tilde{\omega}$, this ratio is $\simeq (\alpha\tilde{\omega}/c)^2$ which is small in domains with non-relativistic speeds. In fact – cf. Section 2.1 – as long as terms of order $(v/c)^2$

are dropped (e.g. if the Galilean rather than the Lorentz transformation is used), then for consistency one is forced to drop terms such as the convection current, and likewise to ignore the electric force density $\rho_e\mathbf{E}$ compared with the magnetic force density $\mathbf{j}\times\mathbf{B}/c$ in the bulk equation of motion. But in a nearly charge-separated domain the convection current is of the order of the total current, and one must use (13.18) to deduce that

$$|\nabla\times\mathbf{B}|\,\tilde{\omega}/B \simeq \mathrm{O}\,(\alpha\tilde{\omega}/c)^2 \ll 1 \tag{13.19}$$

as long as $\alpha\tilde{\omega}/c \ll 1$. This justifies the neglect of the $\nabla\times\mathbf{B}$ term in (13.11); but near the l-c, the rotation of the GJ charge density will cause a serious deviation of the $\mathbf{B}$-field from the curl-free extension of the star's dipolar field (cf. Section 13.1.3). The same conclusion holds if $\rho_e\mathbf{v}$ is not the whole of the current but is still of the same order as the 'conduction current'. Equally, one can no longer argue that there is a near mutual cancellation of the electric force, due to $|\rho_e|$ being much less than $n_e e$ and $n_i Ze$; instead, the electric and magnetic force densities will be comparable.

Similar arguments apply to the non-aligned problem, except that there is also the displacement current term. In order of magnitude,

$$|\partial\mathbf{E}/c\,\partial t| \simeq (\alpha/c)\,(\alpha\tilde{\omega}B/c) \simeq (\alpha\tilde{\omega}/c)^2\,(B/\tilde{\omega}) \tag{13.20}$$

which is $\simeq 4\pi\,|\rho_e|\,v/c$: the displacement and material currents are of the same order, yielding comparable deviations from the $\nabla\times\mathbf{B} = 0$ approximation as the l-c is approached, so that the rotating charges modify the structure of the Deutsch–Pacini wave (cf. Section 13.3.3).

13.1.3 *The aligned model with strictly corotating charges*

Most of the discussion in this chapter will be on the axisymmetric magnetosphere problem. It is illuminating to begin with the simplest generalization of the vacuum model, following the GJ critique. The region within the l-c is supposed filled with the charge-density (13.11) which follows from the hypothesis (13.1). If the charges have no poloidal motion, the poloidal current $\mathbf{j}_\mathrm{p}$ and so also the toroidal field $\mathbf{B}_\mathrm{t}$ are zero, and the drift $c\,(\mathbf{E}\times\mathbf{B})\,/B^2$ reduces strictly to the corotation velocity $\alpha\tilde{\omega}\mathbf{t}$ all the way to the l-c. Corotation beyond the l-c is forbidden by special relativity, but as an illustration one can complete the specification of this electrodynamically dead model by postulating a vacuum beyond the l-c. The poloidal field $\mathbf{B}_\mathrm{p}$ is again described by the flux function $P\,(\tilde{\omega}, z)$:

$$\mathbf{B}_\mathrm{p} = -\nabla\times(P\mathbf{t}/\tilde{\omega}) = -\nabla P\times\mathbf{t}/\tilde{\omega} \tag{13.21}$$

$$= (\partial P/\tilde{\omega}\partial z,\; 0,\; -\partial P/\tilde{\omega}\partial\tilde{\omega})\,. \tag{13.22}$$

Within the l-c, the GJ electric field is then

$$\mathbf{E} = -\alpha\tilde{\omega}\mathbf{t}\times\mathbf{B}_\mathrm{p}/c = (\alpha/c)\,\nabla P, \tag{13.23}$$

and the charge-density (13.11) can be written in the alternative ways

$$\rho_e = \left(\frac{\alpha}{4\pi c}\right)\nabla^2 P = -\left(\frac{\alpha}{2\pi c}\right)\left\{B_z - \frac{1}{2}\tilde{\omega}\left(\nabla\times\mathbf{B}\right)_\phi\right\}. \tag{13.24}$$

The Ampère equation with the corotating charge-density

$$\left(\nabla\times\mathbf{B}\right)_\phi = 4\pi\rho_e\alpha\tilde{\omega}/c \tag{13.25}$$

converts (13.24) and (13.25) into

$$\rho_e\left[1-(\alpha\tilde{\omega}/c)^2\right] = -\alpha B_z/2\pi c = (\alpha/2\pi c)\,\partial P/\tilde{\omega}\,\partial\tilde{\omega} \tag{13.26}$$

and

$$\nabla^2 P\left[1-(\alpha\tilde{\omega}/c)^2\right] = 2\,\partial P/\tilde{\omega}\,\partial\tilde{\omega}, \tag{13.27}$$

equivalent to

$$\frac{\partial^2 P}{\partial\tilde{\omega}^2}+\frac{\partial^2 P}{\partial z^2}-\frac{1}{\tilde{\omega}}\frac{1+(\alpha\tilde{\omega}/c)^2}{1-(\alpha\tilde{\omega}/c)^2}\frac{\partial P}{\partial\tilde{\omega}} = 0. \tag{13.28}$$

The rotating charges manifest themselves through the terms in $(\alpha\tilde{\omega}/c)^2$. Well within the l-c these terms are small, and the field approximates to a curl-free structure (as anticipated in (13.19)), satisfying

$$\frac{\partial^2 P}{\partial\tilde{\omega}^2}+\frac{\partial^2 P}{\partial z^2}-\frac{1}{\tilde{\omega}}\frac{\partial P}{\partial\tilde{\omega}} = 0. \tag{13.29}$$

Beyond the l-c to infinity, the vacuum assumption replaces (13.28) by (13.29). Except for the millisecond pulsars, the ratio $\alpha R/c$ is small enough for the star to be treated as a point dipole, with

$$P \to -\frac{B_s R^3}{2}\frac{\tilde{\omega}^2}{(\tilde{\omega}^2+z^2)^{3/2}} \tag{13.30}$$

at the origin. An acceptable solution must by symmetry satisfy $B_{\tilde{\omega}} \propto \partial P/\partial z = 0$ at $z = 0$, for a discontinuity in $B_{\tilde{\omega}}$ at the equator would imply an equatorial current sheet, subject to unbalanced Maxwell stresses.

The variables are scaled by

$$\tilde{\omega} = \tilde{\omega}_c x, \qquad z = \tilde{\omega}_c Z, \qquad B = B_{\mathrm{lc}}\overline{\mathbf{B}}, \qquad P = \tilde{\omega}_c^2 B_{\mathrm{lc}}\overline{P},$$
$$\tilde{\omega}_c \equiv c/\alpha = \text{radius of the light cylinder}, \tag{13.31}$$

where B_{lc} is a convenient measure of field strength on the l-c. Equation (13.28) then reduces to

$$\overline{P}_{xx} - \frac{1}{x}\frac{1+x^2}{1-x^2}\overline{P}_x + \overline{P}_{ZZ} = 0 \tag{13.32}$$

and the form (13.30) to

$$\overline{P} \to -\frac{1}{2}A\frac{x^2}{(x^2+Z^2)^{3/2}}, \qquad A = (B_s/B_{\rm lc})\,(\alpha R/c)^3. \tag{13.33}$$

Equation (13.29) becomes

$$\overline{P}_{xx} - \frac{1}{x}\overline{P}_x + \overline{P}_{ZZ} = 0. \tag{13.34}$$

(From now on the bar on P is dropped, and Z is replaced by z.)

This problem is conveniently solved by writing $P(x, z)$ as a Fourier integral (Mestel and Wang 1979; Mestel and Pryce 1992):

$$P\,(x, z) = \int_0^\infty f\,(x, k)\cos\,(kz)\;\mathrm{d}k, \tag{13.35}$$

which automatically satisfies $\partial P/\partial z = 0$ at $z = 0$. Equation (13.32) then yields for f

$$f_{xx} - \frac{1}{x}\frac{1+x^2}{1-x^2}f_x - k^2 f = 0. \tag{13.36}$$

The l-c ($x = 1$) is a singularity of (13.28) and hence of (13.36); the solution for $0 \le x \le 1$ is found first and then extended beyond the l-c by using the transform of (13.29). The boundary condition (13.33) at $x = 0$ requires

$$\begin{aligned} f &= \frac{2}{\pi}\int_0^\infty P\cos\,(kz)\;\mathrm{d}z \to -\frac{A}{\pi}x^2\int_0^\infty \frac{\cos\,(kz)}{(x^2+z^2)^{3/2}}\,\mathrm{d}z \\ &= -\frac{A}{\pi}\,(xk)\,\mathrm{K}_1\,(kx) = \frac{A}{\pi}\,(xk)\,\mathrm{H}_1^{(1)}\,(\mathrm{i}kx) \end{aligned} \tag{13.37}$$

where K_1 is the modified Bessel function of the second kind and $\mathrm{H}_1^{(1)}$ is the first Hankel function (Erdelyi *et al.* 1954, Vol. 1, p. 11; Abramowitz and Stegun 1965, p. 375). The dominant terms near the origin are

$$f = -\frac{A}{\pi}\left[1 + \frac{1}{2}\,(kx)^2\log\,(kx) + \mathrm{O}\left(x^2\right)\ldots\right]. \tag{13.38}$$

The most general solution of (13.36) that reduces near the origin to (13.38) is

$$f = -\frac{A}{\pi}\left\{\left[1 + \frac{k^2}{2}x^2\log x + \left(\frac{4+k^2}{8}\right)x^2 + \ldots\right] + a_k\left[x^2 + \left(\frac{4+k^2}{8}\right)x^4 + \ldots\right]\right\}. \tag{13.39}$$

Near $x = 1$ there is an irregular solution $\propto \ln\,(1-x)$ which is rejected, as the field must stay finite at the l-c. The regular solution is

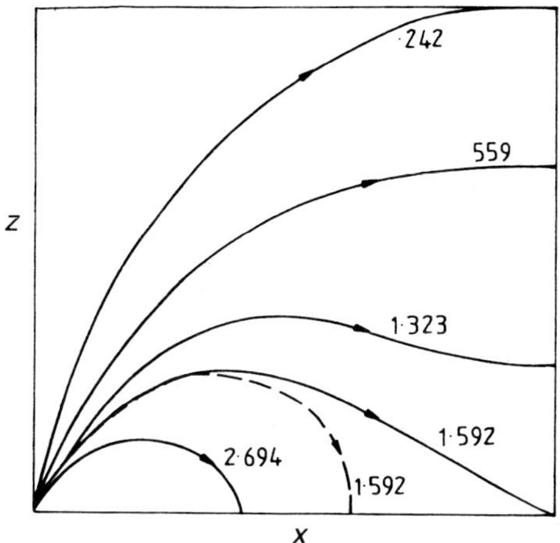

FIG. 13.4. The field within the l-c as modified by the corotating GJ charge density filling the l-c. The figures refer to the the value of $-2P/A$. The dashed line is that belonging to the vacuum dipolar field.

$$f = -(A/\pi)\, b_k \left[1 + \left(k^2/4\right)(1-x)^2 + \dots\right]. \tag{13.40}$$

It is convenient to define the function $u(x,k)$ for each value of k as that solution of (13.36) which is normalized to unity at $x = 1$, so that

$$f(x,k) = -(A/\pi)\, b_k u(x,k), \qquad u(1,k) = 1 \tag{13.41}$$

with $u(1,k) = 1 + \left(k^2/4\right)(1-x)^2 + \dots$ near $x = 1$. The function $u(1,k)$ is then extended inwards numerically to values of x small enough for the series expressions in (13.39) to be adequate (with a few extra terms if necessary). Thus at a point $\tilde{x}$ conveniently near the origin, (13.39) and its first derivative must be continuous with $-(A/\pi)\, b_k u(\tilde{x},\lambda)$ and $-(A/\pi)\, b_k u\,(\partial u/\partial \tilde{x})$ respectively, so that b_k and a_k are then fixed uniquely. Comparison of (13.39) and (13.41) shows that

$$b_k = 1/u(0,k). \tag{13.42}$$

To construct the Fourier integral (13.35), the functions $u(x,k)$ must be computed for a sufficient spread of k-values. Figure 13.4 shows the resultant field, with a neutral point where the l-c crosses the equator, and with all the other field lines approaching the l-c normally. The dotted line in the figure passes through the point where $B_z = 0$, by (13.26) the line separating the electron and ion domains. For $x \ll 1$, this line is defined by $\mathrm{P}_2(\cos\theta) = 0$.

The solution of this illustrative problem by the Fourier integral method is a paradigm for the construction of the **B** and **E** fields in more realistic models.

The standard JWKB method (e.g. Jeffreys and Jeffreys 1956) yields for large k the asymptotic results $a_k \simeq 0$, $b_k \simeq \pi k \exp(-k)/2^{1/2}$, so that

$$f \simeq -(A/\pi)\left[x/\left(1-x^2\right)\right]^{1/2}[\pi k/2]^{1/2}\,\mathrm{e}^{-kx} \tag{13.43}$$

for points not near $x=0$, $x=1$; near $x=0$

$$f \simeq -(A/\pi)(xk)\,\mathrm{K}_1(kx) \tag{13.44}$$

and near $x=1$

$$f \simeq -(A/\pi)\left(\pi k/2^{1/2}\right)\mathrm{e}^{-k}\,\mathrm{I}_0[k(1-x)]\,. \tag{13.45}$$

Convergence at $z \neq 0$ is assisted by the oscillatory nature of $\cos kz$, but it is the exponential k-dependence of $f(x,k)$ which ensures rapid convergence of the Fourier integral even at $z=0$.

By use of (13.41) and (13.42), the integral (13.35) may be written

$$P(x,z) = -(A/2\pi)\int_{-\infty}^{\infty}\frac{u(x,k)}{u(0,k)}\exp(\mathrm{i}kz)\,\mathrm{d}k. \tag{13.46}$$

For each value of z the range of integration may now be extended to form a closed contour in the complex k-plane by the addition of the infinite semicircle in the upper half-plane if $z>0$, in the lower if $z<0$. The integral then reduces to a sum over the residues at the poles of the integrand, which occur at the zeros of $u(0,k)$. Substitution of $k=\mathrm{i}\kappa$ reduces (13.36) to the standard Sturm–Liouville form, yielding the usual infinite sequence of interweaving eigensolutions

$$v_n(x) \equiv u(x,\kappa_n)\,, \qquad \begin{cases} u(0,\kappa_n)=0 \\ u(1,\kappa_n)=1 \end{cases} \tag{13.47}$$

with real eigenvalues κ_n and with $v_n(x)$ possessing n nodes (including the origin). The quantities α_n, h_n defined by

$$\alpha_n = v_n''(0)/2, \qquad h_n = \int_0^1\left[\left(1-x^2\right)v_n^2(x)/x\right]\mathrm{d}x \tag{13.48}$$

yield the residue of the integrand in (13.46) at the pole $\mathrm{i}\kappa_n$. For large n,

$$\kappa_n \simeq n\pi, \qquad \alpha_n \simeq (-1)^{n-1}n\pi/2^{1/2}, \qquad h_n \simeq 2/n\pi^2. \tag{13.49}$$

Expression (13.46) thus transforms into the dimensional form

$$P(\tilde{\omega},z) = -B_sR^3(\alpha/c)\sum_{n=1}^{\infty}\frac{\alpha_n v_n(\alpha\tilde{\omega}/c)\exp(-\kappa_n\alpha z/c)}{\kappa_n h_n} \tag{13.50}$$

(Michel 1973, 1991; Mestel and Pryce 1992). At points not close to $z=0$, expression (13.50) reduces to just the $n=1$ eigensolution. As z decreases, more and more terms have to be included. At $z=0$ the damping term $\exp(-n\pi\alpha z/c)$ becomes unity, and to arrive at the value 1.592 for $(-2P/A)$ on the limiting field line through the point (1,1), convergence of the series $-2\sum\alpha_n/\kappa_n h_n$ has to be induced by a judicious weighting of the partial sums (Michel 1991). In the corresponding series expressions for values near the equator of B_x and B_z the terms are even worse behaved. The obvious conclusion is that for points near the equator, nothing is gained by the transformation into (13.50). As the Fourier integral converges rapidly, especially with the help of fast Fourier-transform techniques, in this problem (and in the more realistic models discussed below) it is far easier to construct P directly from (13.35), having first performed integrations for $f(x,k)$ for a sufficient number of k-values.

Beyond the l-c, the vacuum equation (13.30) for P has the solution, behaving properly at infinity, given by the Fourier transform

$$f=-(A/\pi)\,E_k x\mathrm{K}_1(kx)\,. \tag{13.51}$$

From (13.41), continuity of B_x at $x=1$ yields

$$E_k\mathrm{K}_1(k)=b_k. \tag{13.52}$$

The component of B_z at $x=1+$ is then fixed:

$$\begin{aligned}(B_z)_{1+}&=-B_{\mathrm{lc}}\int_0^\infty\left(\frac{\partial f}{\partial x}\right)\cos(kz)\,\mathrm{d}k\\&=B_s(\alpha R/c)^3\,\frac{1}{\pi}\int_0^\infty\left[1+k\mathrm{K}_1'(k)/\mathrm{K}_1(k)\right]\cos(kz)\,\mathrm{d}k\end{aligned} \tag{13.53}$$

on use of (13.51), (13.52) and (13.31). The discontinuity between (13.53) just to the right of the l-c and the zero value just to the left implies the toroidal current sheet

$$J_\phi=-(c/4\pi)\,\Delta[B_z]=-c/4\pi\,(B_z)_+\,. \tag{13.54}$$

The electric field beyond the l-c satisfies the vacuum condition $\nabla\cdot\mathbf{E}=0$, which is conveniently written from (13.9) as an equation for the non-corotational potential ψ:

$$\nabla^2\psi=-\left(\frac{\alpha}{2\pi c}\right)\mathbf{k}\cdot\left\{\mathbf{B}-\frac{1}{2}\mathbf{r}\times(\nabla\times\mathbf{B})\right\}=-\frac{\alpha B_z}{2\pi c} \tag{13.55}$$

since $\nabla\times\mathbf{B}=0$ in this region. Within the l-c, ψ has the same constant value (conveniently taken as zero) as within the star. The same zero value is propagated to infinity along the axis, so the solution we require of (13.55) is that which vanishes on the l-c and at infinity. This is readily found by taking again the Fourier transform of the non-dimensional form of (13.55) with B_z written in terms of (13.51) and (13.52). Again there is a discontinuity in $E_{\tilde{\omega}}$ maintained by a surface charge density σ on the l-c:

$$4\pi\sigma=\Delta[E_{\tilde{\omega}}]=-(B_z)_+-(\partial\psi/\partial\tilde{\omega})_+ \tag{13.56}$$

(recall that $(E_x)_- = (B_z)_- = 0$).

This completes the solution of the electrodynamic problem as formulated, but again the question arises whether the model is dynamically viable. The net Maxwell stresses exerted on the l-c, due to the fields both within and without $x = 1$, have the $\tilde{\omega}$-component

$$-\frac{1}{8\pi}\Delta\left[\left(\mathbf{E}^2 - 2E_{\tilde{\omega}}^2\right) + \left(\mathbf{B}^2 - 2B_{\tilde{\omega}}^2\right)\right] = \Delta\left(E_{\tilde{\omega}}^2 - B_z^2\right)/8\pi$$
$$= 2\pi\sigma^2\left[1 - \left(J_\phi/\sigma c\right)^2\right] \quad (13.57)$$

where use is made of (13.54) and (13.56). Likewise the stresses have the z-component

$$(1/4\pi)\left(E_z\,\Delta E_{\tilde{\omega}} + B_z\,\Delta B_{\tilde{\omega}}\right) = \sigma B_{\tilde{\omega}}\left(1 - J_\phi/\sigma c\right). \quad (13.58)$$

If $(\partial\psi/\partial\tilde{\omega})_+$, fixed by (13.55) and the boundary conditions, had turned out to be zero all the way along the l-c, then by (13.54) and (13.56), $J_\phi/\sigma c$ would be unity and both (13.57) and (13.58) would vanish: the electric and magnetic stresses would mutually cancel, and the surface currents J_ϕ would be due to the corotation of the surface charges σ with the star. One could then try and see whether the finite rest mass of the charges (effectively put equal to zero in the GJ approximation) could be accommodated by the insertion of a thin vacuum gap separating the GJ zones from the l-c, modifying the electromagnetic field only slightly but keeping the centrifugal forces finite and small. However, not surprisingly, the actual solution of the equations (Mestel and Wang 1982) yields $(\partial\psi/\partial\tilde{\omega})_+$ in general non-zero, positive for $0 < z < 0.625$ and negative for $z > 0.625$; σ is found to be negative up to $z \simeq 0.7$ and positive beyond, whereas J_ϕ is negative for $z > 0.55$ and positive for $0 < z < 0.55$. Thus this proposal for the pulsar magnetosphere fails hopelessly. The original GJ argument for a charged magnetosphere depends on the unbalanced electric stresses on the surface charges that appear on a rotating magnetic neutron star in vacuo. The present model predicts both surface charges and currents on the l-c, but the associated Maxwell stresses do not cancel: the GJ problem is merely transferred from the star to the l-c.

It is no surprise that a model in which the boundary of the charge-bearing region is prescribed a priori should yield discontinuities and a consequent surface charge–current distribution that is subject to unbalanced stresses: for if dead models exist, the shape of the surface separating charged and vacuum regions should emerge as a consequence of the boundary conditions. Axisymmetric steady-state dead models have been constructed numerically by Krause-Polstorff and Michel (1985*a,b*). All the region beyond the l-c and large regions within it are vacuum. Within the charged regions again the GJ condition $\mathbf{E}\cdot\mathbf{B} = 0$ holds, and there are no field discontinuities and so no surface charges and currents. Contiguous with the polar regions there are corotating negatively charged regions. Around the equatorial region is a super-rotating positive region, linked with the star by field lines that pass through a vacuum gap along which $\mathbf{E}\cdot\mathbf{B} \neq 0$, so allowing deviation from corotation with the star. However, these models all have a large net positive charge, and it is questionable whether they could in practice survive conversion into live models, following either discharging by the interstellar medium, or electron–positron pair production under the strong **B**-aligned electric fields in the gaps.

The solution pictured in Fig. 13.4 illustrates the general considerations of Section 13.1.2. Some of its features – the zone with field lines closing within the l-c, the growing deviation from the curl-free structure as the l-c is approached – will persist in the more realistic models studied below, but the perpendicular crossing of the l-c is special to the corotation assumption.

As already emphasized, the corotating charge-separated magnetosphere cannot be extended beyond the l-c. For reasons that will become clear in Sections 14.3 and 14.4, it is nevertheless instructive to consider what happens if instead of the vacuum (curl-free) condition (13.29), the same relativistic force-free equation (13.28) is extrapolated to apply beyond the l-c, but now subject to the constraints that both B_x and B_z are continuous – i.e. without any current sheets on the l-c. Then in the solution of (13.28) by Fourier transformation, for $x > 1$ the function f_k satisfying (13.36) must behave near $x = 1$ like $b_k(-A/\pi)[1 + k^2(x-1)^2/4 + \ldots]$ – cf. (13.40). This selects $f_k \simeq b_k(-A/\pi)\mathrm{I}_0[k(x-1)]$ as the appropriate solution just to the right of $x = 1$. When $x > 1$, the JWKB method yields for $k \gg 1$ the solution $f_k \simeq (-A/\pi)(A_k\,\mathrm{e}^{kx} + B_k\,\mathrm{e}^{-kx})[x/(x^2-1)]^{1/2}$. Identification with the asymptotic form of $\mathrm{I}_0[k(x-1)]$ then yields the values $B_k = 0$, $A_k = (\pi k/2)^{1/2}\,\mathrm{e}^{-2k}$, so that beyond $x = 1$ and for large k, $f_k \simeq -(A/\pi)\,(\pi k/2)^{1/2}\exp k(x-2)$. Thus insistence on continuity at $x = 1$ of both f_k and f'_k yields a solution for P that is singular beyond $x = 2$. Equivalently, one can invert the argument: the solution of (13.36) which is non-singular at $x = \infty$ is $C_k\mathrm{K}_0(kx)$, and inward extrapolation of this to $x = 1$ allows one to fix the constant C_k through continuity of f_k (required for continuity of B_x and so of P), but will not yield $\partial P/\partial x = 0$, as required for continuity of B_z (zero at $x = 1$ in the example above).

The reason why one cannot satisfy both continuity at the l-c and finiteness at infinity is that the l-c is a singularity of the differential equation (13.28), so that when expanding about $x = 1$, only one of the two independent solutions may be retained. Similar behaviour may be expected if a realistic physical model for the domain beyond the l-c still yields a differential equation for P that like (13.28) has a singular surface (cf. Sections 14.3 and 14.4, and the analogous discussion of the non-relativistic but non-force-free problem in Section 7.4).

13.2 A classical model

13.2.1 *Outline of the model*

From now on we concentrate on 'live' models, subject to steady spin-down. The first model studied is in the first instance strictly classical, without electron–positron pair production. Its essentials are depicted in Fig. 13.5. The rotation and magnetic axes are taken as aligned. Electrons leave the star's polar re-

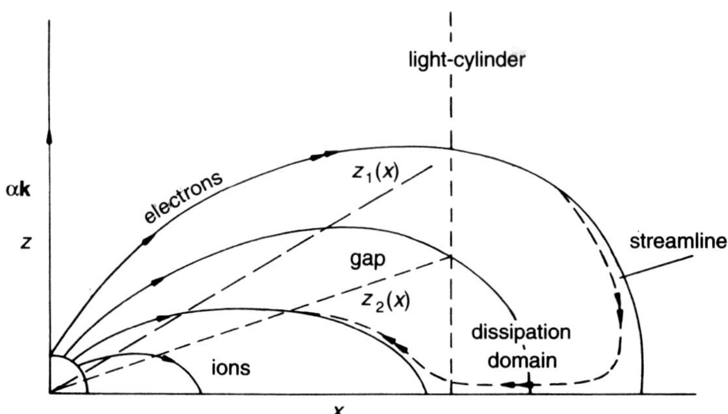

FIG. 13.5. Schematic diagram of the classical model.

gions as a sub-relativistic stream, picking up energy from the electric force and angular momentum from the magnetic torque. In the neighbourhood of the l-c, dissipation-free flow breaks down and the electrons radiate, incoherently or possibly coherently. It is the associated radiation of angular momentum which enables the electrons to drift across magnetic field lines under the essentially quadrupolar electric field, ultimately to complete their circuits back to the star at lower latitudes within the same polar cap, so resolving the 'return current problem'.

In order to maintain the required relation between the loss of energy and angular momentum (cf. 13.137), most of the angular momentum must be radiated beyond the l-c. For this reason, the work summarized studies the superficially simplest case, in which the flow is dissipation-free out to and beyond the l-c. Whether this is how a classical 'pulsar' comes to terms with its environment, or whether there must also be energy dissipation within the l-c, can be decided only after the attempted solution by iteration of the field and relativistic flow equations. The analysis appropriate to the different domains and the limited progress made towards an approximate synthesis are presented in the next few sections.

Study in depth of a largely charge-separated model is justified for intrinsic pedagogic reasons, but also to bring out both the similarities to and contrasts with the more realistic model discussed in Chapter 14, which incorporates pair production. (For further details see Mestel *et al.* (1985) and Fitzpatrick and Mestel (1988*a,b*), the latter papers being from now on referred to as FM*a,b*.) Meanwhile, it is instructive to derive a rough estimate of the power to be expected from an active classical model. It will be found in Section 13.2.3 that the electrons of GJ density $\rho_e \simeq -\alpha B_z/2\pi c$ leave the polar caps with speeds typically $\simeq c/2$, yielding a current density $\simeq -\alpha B_s/4\pi$. As before, the polar cap is defined by the last field line to reach the l-c. Recall that with the vacuum dipolar approximation for $\mathbf{B}_\mathrm{p}$ this yields a polar cap angle θ_pc and an associated cap radius R_pc

$$\theta_\mathrm{pc} \simeq (\alpha R/c)^{1/2} = (R/\varpi_c)^{1/2}; \qquad R_\mathrm{pc} \approx R^{3/2}/\varpi_\mathrm{c}^{1/2} \tag{13.59}$$

and so a polar cap area $\pi\alpha R^3/c$. The magnetospheric currents increase θ_pc by a factor of order unity – for the model of Section 13.1.2, by $(1.592)^{1/2}$. (Note that for a neutron star with a rotation period as short as 0.1 s and an aligned dipolar field, $\varpi_c/R \approx 500$, so that the polar angle θ_pc is no more than 2.6°.) From (13.5) the potential increase between 0 and θ_pc generated by the stellar rotation is

$$\simeq \left(\alpha B_s R^2 \theta_\mathrm{pc}^2/2c\right) R \simeq \alpha^2 R^3 B_s/2c^2. \tag{13.60}$$

Thus in completing their circuits (via a dissipation domain), the currents will yield a radiated power

$$\simeq 2\left(\pi\alpha R^3/c\right)\left(-\alpha B_s/4\pi\right)\left(-\alpha^2 R^3 B_s/2c^2\right) = \alpha^4 B_s^2 R^6/4c^3 \tag{13.61}$$

– again of the same order as the Deutsch–Pacini value (13.2).

13.2.2 *Steady dissipation-free axisymmetric systems*

We begin by developing a general formalism.

(a) The electromagnetic field

Live pulsar models will have regions with poloidal magnetospheric currents $\mathbf{j}_\mathrm{p}$ which generate a toroidal field, conveniently written as

$$\mathbf{B} \equiv B_\phi \mathbf{t} = -\left[4\pi S\left(\tilde{\omega}, z\right)/c\right]\left(\mathbf{t}/\tilde{\omega}\right). \tag{13.62}$$

The poloidal components of Ampère's law yield

$$\mathbf{j}_\mathrm{p} = c\nabla\times\mathbf{B}_\mathrm{t}/4\pi = -\nabla S\times\mathbf{t}/\tilde{\omega}, \tag{13.63}$$

so that S is the Stokes stream function describing the poloidal current field. The steady state electric field has the scalar potential ϕ which from (13.9) and (13.21) can be written in terms of the flux function P and the 'non-corotational potential' ψ

$$\phi = -\alpha P/c + \psi, \tag{13.64}$$

with the charge density ρ_e given by

$$-4\pi\rho_e = \nabla^2\phi = \nabla^2\psi - (\alpha/c)\,\nabla^2 P. \tag{13.65}$$

The toroidal component of Ampère's law is

$$4\pi j_\phi/c = \nabla\times\mathbf{B}_\mathrm{p} = \nabla^2 P/\tilde{\omega} - \left(2/\tilde{\omega}^2\right)\partial P/\partial\tilde{\omega}, \tag{13.66}$$

so that (13.65) and (13.66) yield jointly

$$-4\pi\rho_e = \nabla^2\psi + (2\alpha/c)\left[B_z - \tilde{\omega}\left(\nabla\times\mathbf{B}\right)_\phi/2\right], \tag{13.67}$$

which is just expression (13.11) with the non-corotational contribution included, instead of being suppressed as in the GJ approximation.

If now the charge separation condition is imposed, the current is a pure convection current: $\mathbf{j} = \rho_e\mathbf{v}$, or

$$\mathbf{j}_\mathrm{p} = \rho_e\mathbf{v}_\mathrm{p} = -\nabla S\times\mathbf{t}/\tilde{\omega} = c\nabla\times\mathbf{B}_\mathrm{t}/4\pi, \tag{13.68}$$

$$\mathbf{j}_\mathrm{t} = j_\phi\mathbf{t} = \rho_e\mathbf{v}_\mathrm{t} = \rho_e v_\phi\mathbf{t} = \rho_e\Omega\tilde{\omega}\mathbf{t} = c\nabla\times\mathbf{B}_\mathrm{p}/4\pi, \tag{13.69}$$

where now the local angular velocity $\Omega(\varpi, z)$ is not put equal to the stellar angular velocity α as in the model of Section 13.1.2. Equations (13.67) and (13.66) can be rewritten

$$\nabla^2\psi = -4\pi\rho_e\left(1 - \alpha\Omega\tilde{\omega}^2/c^2\right) + 2\left(\alpha/c\right)\partial P/\tilde{\omega}\,\partial\tilde{\omega} \tag{13.70}$$

and

$$\nabla^2 P\left(1 - \alpha\Omega\tilde{\omega}^2/c^2\right) - 2\,\partial P/\tilde{\omega}\,\partial\tilde{\omega} = -\left(\Omega\tilde{\omega}^2/c\right)\nabla^2\psi. \tag{13.71}$$

The deviation of the solution of (13.71) from the magnetic field model constructed for $\tilde{\omega} < c/\alpha$ from (13.27) will depend on the magnitude of departure of Ω from corotation, and on the contribution of $\nabla^2\psi$, which is itself related to P and Ω by (13.70).

The scalings (13.31) are supplemented by

$$\begin{aligned} &\mathbf{E} = B_\mathrm{lc}\overline{\mathbf{E}}, && S = \left(c\tilde{\omega}_c B_\mathrm{lc}/4\pi\right)\overline{S}, && (\phi,\psi) = B_\mathrm{lc}\tilde{\omega}_\mathrm{c}\left(\overline{\phi},\,\overline{\psi}\right),\\ &\rho_e = \left(B_\mathrm{lc}/4\pi\tilde{\omega}_\mathrm{c}\right)\overline{\rho}_e, && \mathbf{j} = \left(cB_\mathrm{lc}/4\pi\tilde{\omega}_c\right)\overline{\mathbf{j}}, &&\\ &\mathbf{v} = c\mathbf{V}, && \Omega = \alpha w. && \end{aligned} \tag{13.72}$$

For definiteness, the standard field strength B_lc is taken as

$$B_\mathrm{lc} \equiv B_s(\alpha R/c)^3/2, \tag{13.73}$$

the value for a vacuum dipolar field at the point where the l-c crosses the equator. The field equations now become (again with dropping of the bar)

$$\mathbf{B}_\mathrm{p} = -\nabla P\times\mathbf{t}/x, \qquad \mathbf{B}_\mathrm{t} = -\left(S/x\right)\mathbf{t}, \qquad \phi = -P + \psi, \tag{13.74}$$

$$\mathbf{E} = \nabla P - \nabla\psi = -x\mathbf{t}\times\mathbf{B}_\mathrm{p} - \nabla\psi, \tag{13.75}$$

$$\nabla^2 P = 2\,\partial P/x\,\partial x + j_\phi x, \tag{13.76}$$

$$\nabla^2\psi = \nabla^2 P - \rho_e. \tag{13.77}$$

The charge separation conditions (13.68) and (13.69) now become

$$\mathbf{j}_{\mathrm{p}} = -\nabla S \times \mathbf{t}/x = \rho_e \mathbf{V}_{\mathrm{p}}, \tag{13.78}$$
$$\mathbf{j}_{\mathrm{t}} = j_\phi \mathbf{t} = \rho_e \omega x \mathbf{t}, \tag{13.79}$$

whence the normalized forms of (13.70) and (13.71) are

$$\nabla^2 \psi = 2\,\partial P/x\,\partial x - \rho_e \left(1 - \omega x^2\right), \tag{13.80}$$

and

$$\nabla^2 P \left(1 - \omega x^2\right) - 2\,\partial P/x\,\partial x = -\omega x^2\,\nabla^2 \psi. \tag{13.81}$$

(b) The dynamics of a cold dissipation-free electron gas

From now on we deal specifically with an aligned rather than a counter-aligned pulsar, so that the magnetic flux function P defined in (13.21) and taken to be zero on the axis is negative elsewhere. The gas flowing out of the poles consists of electrons of mass m and charge $-e$ and negative charge density ρ_e. The equation of continuity is automatically satisfied by the introduction into (13.68) of the stream function S, which is again taken to be zero on the axis and is positive elsewhere. The equation of motion (with gravity ignorable)

$$-\left(e/m\right)\left(\mathbf{E} + \mathbf{v}\times\mathbf{B}/c\right) = \mathbf{v}\cdot\nabla\left(\gamma\mathbf{v}\right) = \nabla\left(\gamma c^2\right) - \mathbf{v}\times\left\{\nabla\times\left(\gamma\mathbf{v}\right)\right\},$$
$$\gamma = \left(1 - \mathrm{v}^2/c^2\right)^{-1/2} \tag{13.82}$$

transforms into

$$\mathbf{v}\times\mathbf{B}^*/c = \nabla\phi^*, \tag{13.83}$$

with

$$-e\phi^*\left(S\right) = -e\phi + \gamma m c^2 \tag{13.84}$$

and

$$\mathbf{B}^* \equiv \mathbf{B} - \left(mc/e\right)\nabla\times\left(\gamma\mathbf{v}\right). \tag{13.85}$$

Equation (13.84) is the energy integral – the constancy of ϕ^* on streamlines follows immediately from taking the scalar product of (13.83) with $\mathbf{v}$, which reduces to $\mathbf{v}_{\mathrm{p}}\cdot\nabla\phi^* = 0$ since ϕ^* is single-valued and axisymmetric. Likewise (13.83) implies

$$\mathbf{v}_{\mathrm{p}} = \kappa\mathbf{B}^*_{\mathrm{p}} = \kappa\left(-\nabla P^*\times\mathbf{t}/\tilde{\omega}\right) \tag{13.86}$$

where κ is a scalar function and the divergence-free vector $\mathbf{B}^*$ can be written in terms of the generalized flux function

$$eP^*\left(S\right)/c = eP/c + \gamma m\Omega\tilde{\omega}^2. \tag{13.87}$$

The lines P^* = constant are identical with the streamlines, which deviate, however, from the magnetic field lines through the presence of the inertial term in (13.87). In an axisymmetric system, an electron feels no toroidal electric force: to increase its angular momentum, it must have a component of poloidal velocity normal to $\mathbf{B}_{\rm p}$ for the magnetic field to exert the required torque.

From (13.68) and (13.86)

$$-\nabla S \times \mathbf{t}/\tilde{\omega} = \rho_e \kappa \left(-\nabla P^* \times \mathbf{t}/\tilde{\omega}\right), \tag{13.88}$$

so that

$$\rho_e \kappa = \mathrm{d}S/\mathrm{d}P^*. \tag{13.89}$$

Also, from (13.86),

$$\mathbf{v} \times \mathbf{B}^* = \mathbf{v}_{\rm p} \times \mathbf{B}_{\rm t}^* + \mathbf{v}_{\rm t} \times \mathbf{B}_{\rm p}^* = \left(\Omega \varpi \mathbf{t} - \kappa \mathbf{B}_{\rm t}^*\right) \times \mathbf{B}_{\rm p}^*, \tag{13.90}$$

so that (13.83) becomes

$$c \nabla \phi^* = -\left(\Omega - \kappa B_\phi^*/\tilde{\omega}\right) \nabla P^*; \tag{13.91}$$

hence

$$\Omega = \kappa B_\phi^*/\tilde{\omega} - c\, \mathrm{d}\phi^*/\mathrm{d}P^* \tag{13.92}$$

where

$$B_\phi^* = B_\phi - (mc/e)\, \mathbf{t} \cdot \nabla \left(\gamma \mathbf{v}_{\rm p}\right), \tag{13.93}$$

consistent with (13.85). Equations (13.86) and (13.91) can be written jointly as

$$\mathbf{v} = \kappa \mathbf{B}^* - c\left(\mathrm{d}\phi^*/\mathrm{d}P^*\right) \tilde{\omega} \mathbf{t} \equiv \kappa B^* + \alpha\left(S\right) \tilde{\omega} \mathbf{t}, \tag{13.94}$$

representing flow along a line of $\mathbf{B}^*$ together with an angular velocity $\alpha\left(S\right)$, uniform along the streamline considered. The formal similarity with the solution of the 'kinematic' MHD equation (cf. Chapters 5, 7 and 8) is manifest, but it must be recalled that $\mathbf{B}^*$ defined by (13.85) itself depends on $\mathbf{v}$.

The energy and angular momentum integrals (13.84) and (13.87) can be combined via (13.64) to give

$$\gamma\left(1 - \alpha \Omega \tilde{\omega}^2/c^2\right) - e\psi/mc^2 = -\left(e/mc^2\right)\left(\phi^* + \alpha P^*/c\right) \equiv \Gamma\left(S\right). \tag{13.95}$$

The magnetic force $-e\mathbf{v} \times \mathbf{B}/c$ does no direct work on a charge. However, the kinetic energy γmc^2 changes both through variations of the non-corotational potential ψ along a streamline, and also through the relativistic 'centrifugal slingshot' term $\gamma m \alpha \Omega \tilde{\omega}^2$. This arises from the 'inertial drift' which by (13.87) takes a particle from one field line to a neighbouring one, and the consequent gain in energy from the corotational potential $-\alpha P/c$ is just this slingshot term. From (13.87) and (13.22), the actual deviation from strict poloidal flow along $\mathbf{B}_{\rm p}$ is of order $\gamma \Omega/(eB/mc)$. Near the star, where $B \simeq B_s \simeq 10^{12}$ G, this ratio would approach unity only for ludicrously high values of γ (10^{14} or so). Even beyond the l-c the inertial drift is significant only for highly relativistic particles (see below), but once $\alpha\tilde{\omega}/c$ and $\Omega\tilde{\omega}/c$ are of order unity, the slingshot term can make a non-negligible contribution to the energetics even if the poloidal flow geometry is still nearly along $\mathbf{B}_{\rm p}$.

Inside the perfectly conducting stellar crust, $\mathbf{E}$ is given by (13.1), so the non-corotational potential ψ is a conveniently chosen constant ψ_s. Near the star, the relative deviation due to inertia from constancy of ϕ on field lines is by (13.84) $\simeq \gamma mc^2/e(\alpha P/c) \simeq (\gamma\alpha/(eB_s/mc))(c/\alpha R)^2$, greater by the factor $(\tilde{\omega}_c/R)^2$ than the corresponding ratio in (13.87), but still very small for any reasonable value of γ, even for pulsars of period $\mathrm{P}_1 = 1$. Thus near the star $\phi^* \simeq \phi$ as well as $P^* \simeq P$, and from (13.93), $\alpha(S) = -c\,\mathrm{d}\phi/\mathrm{d}P = \alpha$. However, it is worth noting that *as long as no dissipative process intervenes*, then however close to the star relativistic acceleration takes place, $\phi^*(S)$ is always given by ϕ at points where γmc^2 is negligible, and likewise $P^* = P$ at the same points. Hence $\alpha(S)$ is inevitably forced to be equal to α: with *steady, non-dissipative* flow along field lines emerging from the star, the rigid rotation of the stellar crust is propagated into the magnetosphere, yielding uniform rotation rather than just isorotation, with the flow $\mathbf{v} = \kappa\mathbf{B}^* \simeq \kappa\mathbf{B}$ superposed.

It is convenient to introduce the non-dimensional velocity $V_0(S)$ into (13.89) by writing

$$\rho_e v_{\rm p}/B_{\rm p}^* = \rho_e\kappa = \mathrm{d}S/\mathrm{d}P^* = -\alpha V_0\,(S)\,/2\pi. \tag{13.96}$$

With the scalings (13.31) and (13.72), then (13.96), (13.84), (13.87), (13.85), (13.93) and (13.94) become (again with dropping of the bar)

$$\rho_e V_{\rm p}/B_{\rm p}^* = \mathrm{d}S/\mathrm{d}P* = -2V_0\,(S)\,, \tag{13.97}$$

or

$$\mathbf{j}_{\rm p} = \rho_e\mathbf{V}_{\rm p} = \frac{\mathrm{d}S}{\mathrm{d}P*}\mathrm{B}_{\rm p}^*; \tag{13.98}$$

$$\phi - \epsilon_g\gamma = \phi^*\,(S)\,, \tag{13.99}$$

$$P + \epsilon_g\gamma\omega x^2 = P^*\,(S)\,, \tag{13.100}$$

$$\mathbf{B}_{\rm p}^* = \mathbf{B}_{\rm p} - \epsilon_g\nabla\left(\gamma\omega x^2\right)\times\mathbf{t}/x, \tag{13.101}$$

$$B_\phi^* = B_\phi - \epsilon_g\mathbf{t}\cdot\nabla\times\left(\gamma\mathbf{V}_{\rm p}\right), \tag{13.102}$$

and

$$\mathbf{V} = \left(-2V_0\,(S)\,/\rho_e\right)\mathbf{B}^* - x\left(\mathrm{d}\phi^*/\mathrm{d}P^*\right)\mathbf{t}, \tag{13.103}$$

which becomes

$$\mathbf{V} = \left(-2V_0\,(S)\,/\rho_e\right)\mathbf{B}^* + x\mathbf{t} \tag{13.104}$$

when the non-dissipative flow domain considered extends without a break all the way back to the star. The small parameter

$$\epsilon_g = \alpha/\omega_g \qquad \omega_g = eB_{\rm lc}/mc = (eB_s/mc)\,(\alpha R/c)^3\,/2 \tag{13.105}$$

is the ratio of the macroscopic frequency α to the microscopic (non-relativistic) gyration frequency ω_g of an electron in the field $B_{\rm lc}$ defined by (13.73). In terms of standard values,

$$\epsilon_g = 7.8 \times 10^{-8} \mathrm{P}_1^2 \left(B_s/10^{12}\right)^{-1} \left(R/10^6\right)^{-3}. \tag{13.106}$$

Finally, the energy–angular momentum integral (13.95) becomes

$$\gamma \left(1 - \omega x^2\right) = \psi/\epsilon_g + \Gamma\left(S\right) \tag{13.107}$$

with $\Gamma(S)$ fixed by the boundary conditions at the star.

13.2.3 *Dissipation-free flow penetrating the light-cylinder*

Field lines that leave the star sufficiently near to the poles will not close within the l-c, and so carry electron currents into the domain beyond. The analysis of Section 13.2.2(b) has shown that dissipation-free motion of a pure electron gas is well described by the familiar picture of flow along $\mathbf{B}$ plus isorotation, as long as $\gamma\Omega/\left(eB/mc\right)$ is small; and modest γ-values also imply by (13.107) that ψ remains small because of the smallness of the parameter ϵ_g, so that the GJ neglect of the $\nabla^2\psi$ contribution to ρ_e is justified. We begin by studying flow that satisfies GJ conditions till beyond the l-c (Goldreich and Julian 1969; Scharlemann and Wagoner 1973; Mestel *et al.* 1979; Burman 1984).

The *corotating* magnetosphere model of Section 13.1.3 was necessarily restricted to regions within the l-c. By analogy with stellar wind theory, one expects that outflowing gas described by the MHD-type equation (13.104) will have Ω lagging behind α, so that flow penetrating the l-c is not immediately ruled out by special relativity. And as we have seen that the small parameter ϵ_g in (13.107) ensures that ψ is small unless γ is large, one can begin by following GJ and dropping the $\nabla^2\psi$ term in (13.80), writing instead

$$\rho_e = -2B_z/\left(1 - \omega x^2\right). \tag{13.108}$$

The neglect of inertial drifts implies that (13.97) may be written

$$\rho_e V_{\rm p}/B_{\rm p} = {\rm d}S/{\rm d}P = -2V_0\left(P\right); \tag{13.109}$$

and likewise the toroidal component of (13.104) combines with (13.74) to yield

$$V_\phi \equiv \omega x = \left(2V_0 S/\rho_e x\right) + x. \tag{13.110}$$

In terms of V_0, S, and $\mathbf{B}_{\rm p}$ these then yield

$$\rho_e = -2\left(B_z - V_0 S\right)/\left(1 - x^2\right), \tag{13.111}$$

$$V_{\rm p} = V_0 B_{\rm p}\left(1 - x^2\right)/\left(B_z - V_0 S\right), \tag{13.112}$$

$$\omega = \left(B_z - V_0 S/x^2\right) / \left(B_z - V_0 S\right) = 1 - \left(V_0 S/x^2\right)\left(1 - x^2\right) / \left(B_z - V_0 S\right), \tag{13.113}$$

whence

$$\frac{1}{\gamma^2} = 1 - v_{\rm p}^2 - \omega^2 x^2 = C\left(1 - x^2\right)/A, \tag{13.114}$$

with

$$A \equiv \left(B_z - V_0 S\right)^2, \tag{13.115}$$

$$\begin{aligned} C &\equiv B_z^2 - V_0^2 \left\{B_{\rm p}^2\left(1 - x^2\right) + S^2/x^2\right\} \\ &= B_z^2 - V_0^2 \left\{B^2\left(1 - x^2\right) + S^2\right\}. \end{aligned} \tag{13.116}$$

The special case $V_0 = 0$ yields $\omega = 1$, $\mathbf{V}_{\rm p} = 0$, $1/\gamma^2 = 1 - x^2$ – zero outflow implying corotation, as in Section 13.1.3. In general, from (13.109), $S(P) = 2\langle V_0\rangle |P|$, where $\langle V_0\rangle$ is the average over the field lines between 0 and P. Well within the l-c the toroidal magnetospheric currents hardly affect the poloidal field structure (cf. (13.19)), the field being essentially the irrotational $P = -\sin^2\theta/r = -x^2/r^3$. Also, except for millisecond pulsars, the field lines leave the polar cap outflow regions at small values of $\sin\theta = x/r$, so that near the star $B_{\rm p} \simeq B_z = (2/r^3)\left[1 + {\rm O}(x^2/r^2)\right]$ and so $V_0 S/B_z \simeq V_0\langle V_0\rangle x^2$, whence from (13.112) $V_{\rm p} = V_0\left[1 + {\rm O}(x^2/r^2)\right]$: V_0 is essentially the emission velocity from the star. Note, however, that from (13.113), $\omega = 1 - V_0\langle V_0\rangle$: sub-relativistic outflow yields a significant deviation from corotation of the rotational velocity of the gas, right down to the star's surface.[5]

Equations (13.111)–(13.113) are reminiscent of those arising naturally in the magnetic stellar wind problem (Chapter 7), with the l-c playing the part of the Alfvénic surface. We are at the moment explicitly looking for GJ flows that pass smoothly through the l-c without any singularities developing, whence by (13.109) and (13.112)

$$\left(B_z\right)_{\rm lc} = V_0 S = -\frac{1}{2} S \frac{{\rm d}S}{{\rm d}P}. \tag{13.117}$$

Application of l'Hôpital's rule then yields expressions for ρ_e, $V_{\rm p}$ and ω at the l-c; in particular

$$\frac{1}{\gamma^2} = -\frac{2C'}{A''}\left[4V_0\left\{S\left(-B_z'\right)_{\rm lc} - V_0\left(B_{\rm p}^2 + S^2\right)_{\rm lc}\right\}\right] / \left(B_z'\right)_{\rm lc}^2 \tag{13.118}$$

where the derivative B_z' is along the field-streamline considered. One constraint immediately follows from (13.117): since both V_0 and S are positive in an outflow domain, $\left(B_z\right)_{\rm lc}$ must also be positive. This in fact follows from (13.108): with $\rho_e < 0$ and $\omega x < 1$, B_z must be > 0, a generalization of the result found for the

[5]It will be seen in Section 14.1.2 that the actual flow has small-scale spatial oscillations superposed on the mean GJ flow, described by (13.108)–(13.116).

corotating magnetosphere in Section 13.1.3. Thus in a charge-separated model, GJ flow from the polar caps through the l-c is restricted to field lines with $B_z > 0$ at all points within the l-c.

To make further progress, one has to proceed by iteration, though the choice of the best first guess for the form of $\mathbf{B}$ is not obvious. If the effect of the magnetospheric toroidal currents is ignored, then $\mathbf{B}_\mathrm{p}$ is the vacuum dipolar field with

$$P = -x^2/\left(x^2+z^2\right)^{3/2}, \qquad B_x = 3xz/\left(x^2+z^2\right)^{5/2}, \tag{13.119}$$
$$B_z = \left(2z^2-x^2\right)/\left(x^2+z^2\right)^{5/2}, \quad B_\mathrm{p} = \left(x^2+4z^2\right)^{1/2}/\left(x^2+z^2\right)^2.$$

Integration of (13.117) with $x = 1$ yields $S^2(1,z) = (8z^2-1)/2(1+z^2)^4$ which in turn yields

$$S = 2(-P)\left[1-9(-P)^{2/3}/8\right]^{1/2}, \tag{13.120}$$
$$V_0 = -\frac{1}{2}\,\mathrm{d}S/\mathrm{d}P = \left[1-3(-P)^{2/3}/2\right]\Big/\left[1-9(-P)^{2/3}/8\right]^{1/2}$$

to be propagated back to the star through the GJ domain, and into the region beyond the l-c. V_0 is typically $\simeq 1/2$, but it vanishes for $(-P)^{2/3} = 2/3$ (for which $B_z = 0$ at the l-c), and becomes unity on the axis ($P = 0$). The γ-values on the l-c found from (13.118) are modest for all but very large z, corresponding to small $-P$ (cf. the discussion in Section 14.2, leading up to (14.64)). Qualitatively similar results emerge with the adoption of a model field which takes some account of the magnetospheric currents. A fully self-consistent model for $\mathbf{B}_\mathrm{p}$ would be part of the final construction of a magnetosphere, with $\mathbf{B}_\mathrm{p}$ found from the function P satisfying (13.68). For the moment, the really important point is that once $\mathbf{B}_\mathrm{p}$ is chosen, and if GJ flow is to penetrate the l-c, then the $S(P)$ relation is fixed by (13.117), which derives from the requirement that there be no singularity on the l-c.

Beyond the l-c, B_z can be expected to decrease monotonically along a field-streamline, so that A is everywhere positive, whereas $-C$ will begin by being positive but can reach zero again – yielding γ infinite – at a point that depends rather sensitively on the precise structure of $\mathbf{B}_\mathrm{p}$. With the model field (13.158) adopted below, it is found that γ becomes formally infinite before the equator is reached along all field-streamlines. The curve defined by $\gamma \Rightarrow \infty$ is called S_l. Note that in this GJ flow

$$\mathbf{E}^2 - \mathbf{B}^2 = x^2B_\mathrm{p}^2 - \left(B_\mathrm{p}^2 + S^2/x^2\right) = -\left[(-C) + B_z^2\right]/V_0^2, \tag{13.121}$$

so that $E < B$ even when $C = 0$ unless B_z simultaneously vanishes, by (13.108) yielding also $\omega x^2 = 1$. The energy–angular momentum integral (13.107) shows that the growth in γ implies a simultaneous growth in the modulus of the non-corotational potential ψ, again expressed in terms of the small parameter ϵ_g, given by (13.105) and (13.106):

$$\psi \simeq \epsilon_g \gamma \left(1 - \omega x^2\right) = \epsilon_g \gamma B_z \frac{\left(x^2 - 1\right)}{\left(V_0 S - B_z\right)} = \epsilon_g B_z \frac{\left(x^2 - 1\right)^{1/2}}{\left(-C\right)^{1/2}}. \tag{13.122}$$

When B_z and so by (13.108) also $(1 - \omega x^2)$ are positive, ψ increases with decreasing $-C$, so that the centrifugal slingshot term $\gamma\omega x^2$ and ψ act together to increase γ, whereas if B_z and $\left(1 - \omega x^2\right)$ are negative, the non-corotational potential moderates the slingshot term, which we recall is due to the work done by the corotational electric field.

Except for the case when B_z and C vanish together, $\psi \Rightarrow \infty$ along with γ as $-C \Rightarrow 0$, so that the GJ approximation is clearly breaking down. From (13.122), (13.105) and the normalizations (13.31) and (13.72), near the l-c, ψ becomes comparable with ϕ when γ reaches the value

$$\gamma_m \simeq \frac{1}{\epsilon_g} = \frac{\omega_g}{\alpha} \gg 1; \tag{13.123}$$

and from (13.87) or (13.100), simultaneously the inertial drift term is no longer small. Deviation from flow along the field is prima facie to be welcomed as a possible way of yielding the return current to the star that is clearly required. However, as noted by Wang (personal communication), the inertial drifts in dissipation-free flow cannot be the sole means of trans-field motion. The flow is still subject to the continuity equation in the form (13.96), with $\rho_e\kappa$ constant on a current streamline S = constant, yielding the velocity field (13.103) in which the coefficient $-2V_0\left(S\right)/\rho_e$ multiplying $\mathbf{B}_{\mathrm{p}}^*$ necessarily retains its positive sign along a trajectory. By symmetry, the currents must return to the same polar cap from where they start; hence to get a return current from this formalism, $\mathbf{B}_{\mathrm{p}}^*$ given by (13.101) would need to have a sense opposite to $\mathbf{B}_{\mathrm{p}}$, requiring absurdly large γ-values in the inertial term in (13.101). A model without dissipation is in any case incomplete, as it fails to yield spin-down of the star. Introduction of dissipative loss of energy and angular momentum will, however, break the constraint on $\rho_e\kappa$, and the difficulty just noted will vanish. The magnetic torque L on the star is again due to the currents flowing in and out of both polar caps, which generate the B_ϕ-component. By successive use of (13.62), (13.72) and (13.73),

$$\begin{aligned} -L &= -2\int_0^{\theta_{\mathrm{pc}}} \frac{\varpi B_\phi}{4\pi}(\mathbf{B}\cdot\mathbf{n})2\pi R^2 \sin\theta\,\mathrm{d}\theta = B_s R^2 \int_0^{\mathrm{pc}} \frac{4\pi S}{c}\cos\theta\sin\theta\,\mathrm{d}\theta \\ &= \frac{1}{2}B_s^2 R^5 \left(\frac{\alpha}{c}\right)^2 \int_0^{\theta_{\mathrm{pc}}} S\cos\theta\sin\theta\,\mathrm{d}\theta, \end{aligned} \tag{13.124}$$

where again in the last integral the bar on S is dropped. The precise dependence of S on θ depends on the complete solution, including the trans-field motion of the electrons in the dissipation domain, and the structure of the poloidal field as modified by the magnetospheric currents. For illustration, again approximate by taking the field to that of a vacuum dipole all the way to the l-c, as in (13.119) and (13.120). Over the stellar surface, $-P = \sin^2\theta/(\alpha R/c)$, with θ increasing from 0 on the axis to $\tilde{\theta} = (2/3)^{3/2}(\alpha R/c)^{1/2}$ along the field line $-P = (2/3)^{3/2}$ for which $V_0 = 0$. The contribution from $0 < \theta < \tilde{\theta}$ to the integral in (13.124) is $0.13(\alpha R/c)$; assuming a similar contribution from the returning electron domain yields for the total torque

$$-L = 0.13 B_s^2 R^6 \alpha^3/c^3 \tag{13.125}$$

with the associated energy loss $0.13 B_s^2 R^6 \alpha^4/c^3$ – again remakably close to the Deutsch–Pacini value (13.2) for high obliquity.

13.2.4 *Flow with radiative dissipation*

By (13.106) and (13.107), inertial drifts will have become significant near and beyond the l-c when $\gamma \simeq 1/\epsilon_g \simeq 1.25 \times 10^7/\mathrm{P}_1^2$, but even for slow pulsars with $\mathrm{P}_1 \simeq 1$, particles with these γ-values will emit radiation beamed in the direction of their instantaneous motion and so will feel radiation drag. This acts analogously to Ohmic friction between electrons and ions in a normal plasma, enforcing in both cases departure from freezing of the field into the electron gas. We now develop the theory assuming provisionally that no other important dissipative process sets in other than single-particle, incoherent radiation damping. The power $\mathcal{P}$ emitted by a relativistic particle of momentum $\mathbf{p}$ and energy T is (e.g. Landau and Lifshitz 1975)

$$\mathcal{P} = \left(2e^2/3m^2c^3\right)\gamma^2\left\{(\mathrm{d}\mathbf{p}/\mathrm{d}t)^2 - (\mathrm{d}T/c\,\mathrm{d}t)^2\right\} \tag{13.126}$$

with the associated momentum loss yielding a drag force

$$\mathbf{f} = -\left(\mathcal{P}/c^2\right)\mathbf{v}. \tag{13.127}$$

The simplest way of deriving (13.127) is by direct use of the Lorentz transformation. In its instantaneous rest-frame S', the electron loses energy $\Delta T'$ in time $\Delta t'$. With the excellent approximation of ignoring the Larmor damping force (necessarily small compared with the external field – cf. Landau and Lifshitz 1975, Section 75), there is no momentum loss in this frame, so that the four-vector representing the change in the particle energy and momentum is just $(0, 0, 0, \Delta T'/c)$. Take the x-axis along the direction of the particle velocity $\mathbf{v}$ as seen in the 'laboratory frame' S (i.e. the frame in which the pulsar has angular velocity $\alpha\mathbf{k}$). Then the corresponding four-vector is $(\Delta p, 0, 0, \Delta T/c)$, where

$$\Delta p = \gamma v\,\Delta T'/c^2, \qquad \Delta T/c = \gamma\,\Delta T'/c, \qquad \Delta t = \gamma\,\Delta t'; \tag{13.128}$$

whence

$$\Delta T/\Delta t = \Delta T'/\Delta t' = \mathcal{P} \tag{13.129}$$

(power is Lorentz-invariant), and

$$\mathbf{f} = -\Delta\mathbf{p}/\Delta t = -\left(\mathcal{P}/c^2\right)\mathbf{v}. \tag{13.130}$$

The expression (13.130) is identical with the dominant term in the familiar relativistic generalization of the Lorentz–Larmor self-force. When radiation drag is supposed as just a small perturbation on the motion of a particle, the usual practice is to substitute into (13.126) the expressions for $\mathrm{d}\mathbf{p}/\mathrm{d}t$ and $\mathrm{d}T/\mathrm{d}t$ from the equation of motion with only the the external Lorentz force balancing the relativistic inertial term, and then to use the resulting expression for $\mathbf{f}$ in the next approximation. This procedure is not appropriate for the present problem, for which radiation damping is likely to be comparable with the Lorentz force. Instead, $\mathbf{p}$ and so $\mathbf{f}$ are left to be fixed by the geometry of the flow. In a steady state with $\mathrm{d}/\mathrm{d}t$ replaceable by $\mathbf{v}\cdot\nabla$, the dominant term in (13.126) reduces to

$$\mathcal{P} = \left(2e^2/3c^3\right)\gamma^4\left\{\mathbf{v}\times(\nabla\times\mathbf{v})\right\}^2. \tag{13.131}$$

The energy and angular momentum integrals (13.84) and (13.87) are now replaced by

$$\mathbf{v}\cdot\nabla\left(\gamma mc^2 - e\phi\right) = -\mathcal{P} \tag{13.132}$$

and

$$\mathbf{v}\cdot\nabla\left(\gamma m\Omega\tilde{\omega}^2 + eP/c\right) = -\left(\Omega\tilde{\omega}^2/c^2\right)\mathcal{P}, \tag{13.133}$$

and (13.95) by

$$\mathbf{v}\cdot\nabla\left[\gamma\left(1-\alpha\Omega\tilde{\omega}^2/c^2\right) - e\psi/mc^2\right] = -\left(\mathcal{P}/mc^2\right)\left(1-\alpha\Omega\tilde{\omega}^2/c^2\right). \tag{13.134}$$

Equation (13.133) shows how radiation of angular momentum enforces trans-field drift in the direction of decreasing P, i.e. equatorwards. As noted above, during any such motion the electrons acquire kinetic energy from the corotational part $-\alpha P/c$ of the potential. In (13.134), the first term on the right represents the direct loss of energy by radiation, while the second represents the gain in energy due to this 'dissipative drift', analogous to the inertial drift that yields the slingshot term $\gamma\alpha\Omega\tilde{\omega}^2/c^2$ in (13.95) and (13.134).

Returning to first principles: the energy source driving current through the magnetosphere is the potential variation $-\alpha P/c$ on the stellar surface due to the rotation of the magnetic star. In order to pick up this energy, particles must be able to cross field lines. Radiation of photons carrying angular momentum does indeed yield an equatorwards dissipative drift, enabling the (negatively charged) electrons to pick up the electrical energy available on field lines of higher potential. Clearly this requires that the gain of energy exceed that carried off by the photons. One can notionally integrate (13.134) along a trajectory followed by an electron, between emission from the star at time t_i and return at time t_f, to find the difference between the corresponding γ-values of a circulating electron:

$$(\gamma_f - \gamma_i) \simeq -\int_{t_i}^{t_f}\left(\mathcal{P}/mc^2\right)\left(1-(\Omega\tilde{\omega}/c)\,(\alpha\tilde{\omega}/c)\right)\mathrm{d}t \tag{13.135}$$

(recall that ψ is uniform on the star, and that $\alpha\Omega\tilde{\omega}^2/c^2 \ll 1$ well within the l-c). Since $\gamma_i \simeq 1$, and $\mathcal{P}$ is necessarily positive, it is clear that to ensure that γ_f exceeds unity, the dissipation must occur primarily beyond the l-c, with the weighting factor $[1-(\Omega\tilde{\omega}/c)(\alpha\tilde{\omega}/c)]$ negative, so ensuring that the energy released through the dissipation drift exceeds the associated energy loss. In this way, radiation of angular momentum – acquired from the star by magnetic torques – enables the electrons to pick up electrical energy supplied from the rotational energy of the star. Some of this energy is radiated along with the angular momentum, and the rest is retained by the returning electrons, as shown by (13.135), or by the similar equation found by integrating (13.132) around a trajectory, with use of (13.64):

$$(\gamma_f - \gamma_i) = \left(e\alpha/m^3\right)(P_i - P_f) - \int_{t_i}^{t_f} \left(\mathcal{P}/mc^2\right) \mathrm{d}t. \tag{13.136}$$

The result (13.135) links up with a comment made many times by Gold (e.g. 1980) and by Holloway (1977). There is a global constraint on steady or quasi-steady magnetospheric solutions. The kinetic energy $I\alpha^2/2$ and the angular momentum $I\alpha$ of the rotating star are the only sources of the energy and angular momentum being carried away, and their time-derivatives are related by

$$-\mathrm{d}\left(I\alpha^2/2\right)/\mathrm{d}t = \alpha\left[-\mathrm{d}\left(I\alpha\right)/\mathrm{d}t\right]. \tag{13.137}$$

An individual photon of energy $h\nu$, emitted at axial distance $\tilde{\omega}$ and beamed perfectly in the toroidal direction, carries off angular momentum $\tilde{\omega}h\nu/c$, yielding

$$h\nu/\tilde{\omega}\left(h\nu/c\right) = \alpha\left(c/\alpha\tilde{\omega}\right). \tag{13.138}$$

Thus the energy and angular momentum carried off by this photon have the ratio α only if emission is at the l-c; a photon emitted within the l-c carries off less angular momentum than $h\nu/\alpha$, while one emitted beyond the l-c carries off more or less than $h\nu/\alpha$, depending on the direction of emission. The *global* Gold–Holloway argument shows that the integrated emission in a self-consistently constructed model must satisfy (13.137). What the above discussion shows is that in the present model, there is a similar condition holding along each trajectory.

From (13.133), radiation drag is comparable with the Lorentz force if $\mathcal{P} \simeq ceB$, requiring at points near but beyond the l-c that $\gamma \simeq \gamma_d$, where

$$\gamma_d^4 \simeq \left(\omega_g/\alpha\right)\left[\left(c/\alpha\right)/\left(e^2/mc^2\right)\right] \simeq c^2 B_{\mathrm{lc}}/e\alpha^2. \tag{13.139}$$

The first expression in (13.139) is manifestly the product of two large numbers (the second bracket being approximately the ratio of the l-c radius to the classical electron radius). Thus the value of γ required to yield significant trans-field motion is

$$\gamma_d \simeq 1.4 \times 10^7 \left[\left(B_s/10^{12}\right)\left(R/10^6\right)^3/\mathrm{P}_1\right]^{1/4}. \tag{13.140}$$

With standard parameters, radiation damping keeps the emitting electrons at energies $\simeq \gamma_d mc^2 \simeq 7 \times 10^{12} \mathrm{P}_1^{1/4}$ eV, and the emitted radiation has typical frequency $\nu \simeq \gamma_d^3 \alpha$ by the familiar synchrotron argument – gamma-ray photons of energy $\simeq 5 \times 10^7 / \mathrm{P}_1^{7/4}$ eV. The maximum value γ_m that the circulating electrons can acquire is fixed by the potential difference (13.60) across the polar cap, defined as before by the last field line leaving the star to reach the l-c, which by (13.59) leaves the star at polar angle $\theta \simeq (R/\varpi_c)^{1/2}$. Thus by (13.73) and (13.105),

$$\gamma_m \simeq \frac{e\alpha^2 R^3 B_s}{2mc^4} \simeq \frac{eB_{\mathrm{lc}}}{\alpha mc} \simeq \frac{\omega_g}{\alpha} \simeq \frac{1}{\epsilon_g} \simeq 1.25 \times 10^7 \left(\frac{B_s}{10^{12}}\right)\left(\frac{R}{10^6}\right)^3 \mathrm{P}_1^{-2} \quad (13.141)$$

– identical with (13.123). Comparison with (13.139) shows that $\gamma_d \ll \gamma_m$ for short-period pulsars, but the two periods are uncomfortably close when $\mathrm{P}_1 \simeq 1$. It may be that a model in which dissipation is through incoherent gamma-ray emission alone is viable only for the more rapid pulsars, but instinctively one feels that the system will always find a way of emitting spin-down energy at a rate of order (13.1). In the present classical model, for example, one can imagine the high-γ electrons undergoing a shock transition; coherent radiation from the dense post-shock gas could occur, requiring values much lower than γ_d. A sophisticated treatment of the physics of the $e\pm$ plasma will almost certainly be required for the more realistic model of Chapter 14. For this illustrative classical model we concentrate on the gross electrodynamics and dynamics, postulating that even a slow rotator ($\mathrm{P}_1 \simeq 1$) would find a dissipative process occurring at the rate imposed by macroscopic considerations.

Equations (13.132) and (13.133) leave one component of the vector equation of motion. We anticipate that in the dissipation domain the flow can be adequately approximated by a balance between Lorentz force and radiation drag, i.e. that the inertial terms are small compared with either. Thus in this 'Stokes–Aristotle' approximation (cf. Toulmin and Goodfield 1961, p. 108)

$$-e\left(\mathbf{E} + \mathbf{v} \times \mathbf{B}/c\right) - \left(\mathcal{P}/c^2\right)\mathbf{v} = 0, \quad (13.142)$$

or in non-dimensional form

$$-\left(\mathbf{E} + \mathbf{V} \times \mathbf{B}\right) = K\mathbf{V}, \quad (13.143)$$

$$K = \mathcal{P}/ceB_{\mathrm{lc}}. \quad (13.144)$$

At each point, the scalar K and the velocity $\mathbf{V}$ are fixed by the local $\mathbf{E}$, $\mathbf{B}$ fields: manipulation of (13.143) with $\mathbf{V}^2 \simeq 1$ leads to

$$K^2 = \frac{1}{2}\left[\left(\mathbf{E}^2 - \mathbf{B}^2\right) + \left\{\left(\mathbf{E}^2 - \mathbf{B}^2\right)^2 + 4\left(\mathbf{E}\cdot\mathbf{B}\right)^2\right\}^{1/2}\right] \quad (13.145)$$

and

$$\mathbf{V}\left(K^2+\mathbf{B}^2\right)=\mathbf{E}\times\mathbf{B}-K\mathbf{E}-\mathbf{B}\left(\mathbf{E}\cdot\mathbf{B}\right)/K. \tag{13.146}$$

Equation (13.142) includes (13.132) and (13.133) without the inertial terms. With K given by (13.145), (13.131) and (13.144) may be used to predict the local value of γ_d when incoherent radiation damping is the dominant dissipative process. Neglect of inertia is again acceptable when $\gamma_d < \gamma_m \simeq 2/\epsilon_g$. There is an enormous saving of labour if particle trajectories can be calculated from (13.146) in terms of local quantities, instead of having to construct them through step-by-step integration.

Not surprisingly, the qualitative predictions of (13.145) and (13.146) depend on whether E is greater than or less than B. When $E \gg B$, $K^2 \simeq E^2$, and

$$\mathbf{E}^2\mathbf{V}\simeq -E\mathbf{E}+\mathbf{E}\times\mathbf{B}-\left(\mathbf{E}\cdot\mathbf{B}/E\right)\mathbf{B}. \tag{13.147}$$

Then $V_{\rm p}\simeq -\mathbf{E}/E+\mathrm{O}\left(B/E,\,B_{\rm p}^2/E^2\right)$, showing that the flow is predominantly poloidal and parallel to $-\mathbf{E}$. The toroidal velocity $\mathbf{V}_{\rm t}\simeq\left(\mathbf{E}\times\mathbf{B}_{\rm p}\right)/E^2$ unless $\mathbf{E}$ and $\mathbf{B}_{\rm p}$ are nearly parallel, in which case $\mathbf{V}_{\rm t}\simeq -\left(B_{\rm p}/E^2\right)\mathbf{B}_{\rm t}$.

Now suppose $E \ll B$, and also that $\mathbf{E}\cdot\mathbf{B}\neq 0$; then

$$K=\left[\left|\mathbf{E}\cdot\mathbf{B}\right|/B\right]\left\{1+\mathrm{O}\left(E^2/B^2\right)\right\}\ll B, \tag{13.148}$$

so that $\mathbf{E}+\mathbf{V}\times\mathbf{B}\simeq 0$, with $K\mathbf{V}$ a small correction – flow with high but finite magnetic Reynolds number. Equation (13.146) then yields

$$\mathbf{V}_{\rm p}\simeq -\mathrm{sgn}\left(\mathbf{E}\cdot\mathbf{B}\right)\mathbf{B}_{\rm p}/B+\left(\mathbf{E}\times\mathbf{B}_{\rm t}\right)/B^2, \tag{13.149}$$

$$\mathbf{V}_{\rm t}\simeq -\mathrm{sgn}\left(\mathbf{E}\cdot\mathbf{B}\right)\mathbf{B}_{\rm t}/B+\left(\mathbf{E}\times\mathbf{B}_{\rm p}\right)/B^2. \tag{13.150}$$

Unless $B_{\rm t}\gg B_{\rm p}$, $\mathbf{V}_{\rm p}$ is essentially parallel to $\mathbf{B}_{\rm p}$, driven by the component of $\mathbf{E}$ along $\mathbf{B}_{\rm p}$. $\mathbf{V}_{\rm p}$ is predicted to change sign across any line (other than a $\mathbf{B}_{\rm p}$-line) on which $\mathbf{E}\cdot\mathbf{B}$ changes sign. Such a sink of particles is unacceptable, and would imply a breakdown in the Stokes–Aristotle approximation, with inertial terms locally non-ignorable. With this caveat, one has qualitative rules of thumb: if $E>B$, then $\mathbf{V}_{\rm p}$ is primarily along $\mathbf{E}$; if $E<B$, then $\mathbf{V}_{\rm p}$ is primarily along $\mathbf{B}_{\rm p}$.

In GJ flow passing smoothly through the l-c, $E > B$ for $x > 1$: however, the non-zero $B_{\rm t}$ ensures that E stays below $(B_{\rm p}^2+B_{\rm t}^2)^{1/2}$ even when γ formally becomes infinite through the vanishing of C, unless B_z simultaneously vanishes (cf. (13.116) and (13.121)). In the computations of FM*a,b*, the electron flow mostly follows $\mathbf{B}_{\rm p}$ towards the equator, there to form an equatorial disc held in z-equilibrium primarily by a balance between electric disruptive and magnetic pinching stresses. Within the disc, $\mathbf{B}_{\rm t}$ vanishes, and $E>B$; the Stokes–Aristotle equations then reduce to

$$K^2=E_x^2-B_z^2, \tag{13.151}$$

$$V_\phi=-B_z/E_x>0, \tag{13.152}$$

$$V_x=-K/E_x=-\left(1-B_z^2/E_x^2\right)^{1/2}<0. \tag{13.153}$$

The principal trans-$\mathbf{B}_\mathrm{p}$ motion thus takes place in this equatorial zone. These equations make manifest the link between the radiation of angular momentum and the return electron current.

For a discussion of the necessary quantum corrections to this classical picture, with the deterministic trajectories becoming stochastic, see Kahn and da Costa (1994).

13.2.5 *Towards the construction of a global classical model*

The hypothetical classical model outlined in Section 13.2.1 is subject to some integral constraints. First of all, *the system should have zero net charge and so zero monopole electric field at infinity.* A net negative charge would be spontaneously lost, for electrons emitted close to the poles would feel at large distances a repulsive electric force that dominates over the magnetic deflecting force. If the system has a net positive charge, then again the electric potential at infinity dominates over the GJ potential $-P$, but now the (essentially non-corotational) electric field will attract distant electrons towards the star. Electrons flowing out of the poles along field lines close to the axis would therefore reach a surface on which E_z changes sign. Such a sink of particles in the magnetosphere is again clearly inconsistent with a steady state. The dead models of Krause-Polstorff and Michel (1985*a,b*) do all carry a net positive charge. It is possible that there exist live models, similar to the one under study, but with a net positive charge, and with the circulating electron currents emanating from the star only at polar angles θ_i large enough to ensure that $\mathbf{E}\cdot\mathbf{B}$ remains non-positive all the way round a field-streamline until the equatorial domain is reached. We restrict the discussion, however, to (hypothetical) models in which the whole of each polar cap emits electrons continuously, and so which are subject to the zero charge condition

$$\int_0^\infty \mathrm{d}z \int_0^\infty x\rho_e\left(x,z\right)\mathrm{d}x = 0. \tag{13.154}$$

As a corollary, the (equatorially symmetric) electric field far from the star must be dominantly quadrupolar, and therefore of strength at large distances less than that of any dipolar magnetic field. But this would violate the conditions for the equatorial return current (13.151–13.153), unless there is also a further constraint on the model: *the rotating magnetospheric charges must generate a magnetic dipole moment equal and opposite to that on the star.* In non-dimensional units:

$$0 = 1 + \frac{1}{2}\int_0^\infty \mathrm{d}z \int_0^\infty x^2 j_\phi\left(x,z\right)\mathrm{d}x \tag{13.155}$$

$$= 1 + \frac{1}{2}\int_0^\infty \mathrm{d}z \int_0^\infty x^3 \rho_e\left(x,z\right)\omega\left(x,z\right)\mathrm{d}x \tag{13.156}$$

where charge separation has been used in (13.156) by writing $j_\phi = \rho_e\omega x$. A further consistency condition is that the net electric quadrupole moment should be negative, in order to divert the electron flow from pole to equator at large distances:

$$\beta_2 \equiv \int_0^\infty \mathrm{d}z \int_0^\infty x\rho_e(x,z)\left(z^2 - \frac{1}{2}x^2\right)\mathrm{d}x < 0. \tag{13.157}$$

With the particle density–velocity field and the electromagnetic field mutually coupled so strongly, there is little hope of constructing a fully self-consistent solution to the problem, even with unlimited computer time: one must be content with an analytical-plus-numerical approach to an iterative scheme that looks promising, and which could in principle validate the model. Within the l-c, a zero-order description comprises besides the zone of outflowing electrons, a contiguous zone of corotating electrons and a positive charge zone encompassing the equator and separated from the electron zones by a vacuum gap. The gap must close at the stellar surface, and the inner part of the ion zone will either corotate with the star or super-rotate, as in the Krause-Polstorff and Michel dead models (1985*a,b*). Further out, the gap can be approximated as wedge-shaped, with the contiguous positive domain sub-rotating and in fact extending beyond the l-c (see Appendix). The returning electron current traverses the positive and vacuum zones, so that these descriptions are only approximate. The work in FM*a,b* begins by adopting a simple model for $\mathbf{B}_\mathrm{p}$ which is consistent with the above constraint. The model flux function adopted is

$$P^m(r,\mu) = -\frac{(1-\mu^2)}{r}\frac{\left[1 + (r/\bar{r})^2(5\mu^2 - 1)/\bar{s}\right]}{\left[1 + (r/\bar{r})^2\right]^2}. \tag{13.158}$$

When $r \ll \bar{r}$, this reduces to the field of a unit dipole, and when $r \gg \bar{r}$, to

$$P^m(r,\mu) = -\beta_3\left[3\left(1-\mu^2\right)\left(5\mu^2 - 1\right)/2r^3\right] + \mathrm{O}\left(1/r^5\right), \tag{13.159}$$

with the magnetic sextupole moment

$$\beta_3 = \frac{1}{2}\int_0^\infty \mathrm{d}z \int_0^\infty x^3\rho_e(x,z)\,\omega(x,z)\left(z^2 - x^2/4\right)\mathrm{d}x. \tag{13.160}$$

The procedure is then to use the field (13.158) to construct examples of the allowed charge-density and velocity fields, with the conditions (13.154), (13.156) and (13.157) setting limits on the parameters $\bar{r}$ and $\bar{s}$ in (13.158). In the next stage, the **B** and **E** fields are constructed from the Ampère–Maxwell and Poisson–Maxwell equations, to be compared with the initially assumed field (13.158) and the associated electric fields. These in turn can be used to determine anew the density–velocity fields. .

Details of a tentative classical model – with flow provisionally postulated to be dissipation-free out to and beyond the l-c – are given in FM*a,b*, and summarized in the Appendix to this chapter. The analytical-plus-numerical work to date has yielded encouraging results, but the effort involved in the construction of a plausibly self-consistent model remains formidable.

The essential physical point of the classical model is the link-up between the return current requirement and the spontaneous loss of energy and angular momentum through gamma-radiation from high-γ particles, energized from the potential differences on the stellar surface. In spite of some clear differences, noted above, the process of particle acceleration is similar to that in a classical 'centrifugal wind' problem (cf. Chapters 7 and 8), in that it is the perpendicular component $\mathbf{E}_{\perp}$ – responsible for the slingshot effect – that energizes the particles. As long as inertial and dissipative *drifts* are small, the gas velocity is well described by the familiar MHD integrals ((13.94) without the asterisks). The associated value of the non-corotational potential ψ is given by the Γ-integral (13.95) in the dissipation-free domain and by (13.134) when angular momentum dissipation is significant. The chain of causation is: $\mathbf{E}_{\perp} \rightarrow \gamma \rightarrow \mathbf{E}_{\|}$.

The classical model illustrates how dissipation-free flow spontaneously establishes domains in which acceleration to high γ-values occurs, leading inevitably to dissipation. The obvious objection to the model as a complete description of actual pulsars is that it proves too much; for although observations are now showing that gamma-ray emission from some pulsars is a significant fraction (up to 10 per cent) of the power (cf. Gunji *et al.* 1994), the bulk of the energy loss as inferred from the spin-down is in other modes. As already noted, if the circulating electrons were to radiate coherently, then the maximum γ-values would be kept much lower, and the emitted radiation would likewise be at much lower frequencies. However, with the overwhelming evidence that, for example, the Crab pulsar loses much of its spin-down energy in the form of a relativistic particle wind, it is clear that a purely classical model may be relevant only to the slower rotators (or perhaps only to stars with P_1 longer than the observed cut-off period).

The classical model may in fact contain the seeds of its own transformation. There is no reason to assume that the energy loss associated with the angular momentum loss in the dissipation domain should be equal to or even close to the total energy available from the potential difference between points of emission and return, and picked up via the inertial plus dissipative trans-field drift that enables the electrons to complete their circuits. High-γ electrons returning to the star with even a fraction of this energy are in principle able to generate by pair production and surface spallation a plasma of density far in excess of the GJ value, some of which may expand under its own one-dimensional pressure. The circulating primary electrons would then act as a trigger for the generation of a quasi-neutral relativistic wind. This is one of the features that could persist into the models, studied below, which differ in having relativistic acceleration occurring deep within the l-c.

13.3 Oblique rotators

In this final section, a start is made on applying some of the ideas emerging from the above study of axisymmetric 'non-pulsars' to obliquely rotating, essentially non-axisymmetric systems, which can describe the magnetospheres of observed pulsars.

13.3.1 *Quasi-steady, non-axisymmetric systems*

As defined in Section 13.1.1, a 'quasi-steady' system has the time-variation of any quantity as seen in the non-rotating frame as being due to an essentially non-axisymmetric structure being swung round at the rate α about the axis $\mathbf{k}$. The equivalence relation (13.7) applied to Faraday's law yields

$$\mathbf{E} = -\alpha(\mathbf{k}\times\mathbf{r})\times\mathbf{B}/c - \nabla\psi = -\alpha\varpi\mathbf{t}\times(\nabla\times\mathbf{A})/c - \nabla\psi. \tag{13.161}$$

In terms of the standard scalar and vector potentials,

$$\mathbf{E} = -\partial\mathbf{A}/c\,\partial t - \nabla\phi = \alpha\,\partial\mathbf{A}/c\,\partial\Phi - \nabla\phi, \tag{13.162}$$

where the upper case Φ is used for the azimuthal angle about the rotation axis $\mathbf{k}$. It is readily verified (e.g. by taking cylindrical polars) that for any vector $\mathbf{a}$,

$$\alpha\varpi\mathbf{t}\times(\nabla\times\mathbf{a}) + \alpha\,\partial\mathbf{a}/\partial\Phi = \nabla(\alpha\varpi a_\Phi), \tag{13.163}$$

so that (13.162) and (13.163) yield the gauge-invariant relation (Endean 1972*a*)

$$\psi = \phi - \alpha\varpi A_\Phi/c. \tag{13.164}$$

The equation of motion of a cold, friction-free gas of particles of charge q is

$$\partial\mathbf{p}/\partial t + (\mathbf{v}\cdot\nabla)\mathbf{p} = q(\mathbf{E} + \mathbf{v}\times\mathbf{B}/c), \qquad \mathbf{p} = \gamma m\mathbf{v}. \tag{13.165}$$

From the vector identity $(\mathbf{p}\cdot\nabla)\mathbf{p} = \nabla(\mathbf{p}^2/2) - \mathbf{p}\times(\nabla\times\mathbf{p})$, and with the use of $\mathbf{p}^2 = m^2c^2(\gamma^2-1)$,

$$\mathbf{v}\cdot\nabla\mathbf{p} = mc^2\nabla\gamma - \mathbf{v}\times(\nabla\times\mathbf{p}). \tag{13.166}$$

It is convenient to define $\mathbf{u}\equiv\mathbf{v}-\alpha\varpi\mathbf{t}$. Then with the help of (13.161), (13.163), (13.166) and the quasi-steady condition (13.7), (13.165) reduces to

$$q\,(\mathbf{u}\times\mathbf{B}/c - \nabla\psi) = mc^2\gamma - \mathbf{u}\times(\nabla\mathbf{p}) - \nabla(\alpha\varpi p_\Phi), \tag{13.167}$$

or

$$\mathbf{u}\times[\nabla\times(\mathbf{p}+q\mathbf{A}/c)] = \nabla\left[q\psi + mc^2\left(\gamma - \alpha\varpi v_\Phi/c^2\right)\right] \equiv mc^2\,\nabla\Gamma. \tag{13.168}$$

The quantity

$$\Gamma = q\psi/mc^2 + \gamma\left(1 - \frac{\Omega\varpi}{c}\frac{v_\Phi}{c}\right) \tag{13.169}$$

is thus seen to be constant on the lines of $\mathbf{u}$ and of $\nabla\times(\mathbf{p}+q\mathbf{A}/c)$ (Endean 1972*a,b*). This 'energy integral in the rotating frame' is identical in form with the integral that results from combining the individual energy and angular momentum integrals valid in axisymmetric systems (cf. (13.95) with $q = -e$ for electrons). It is the relativistic analogue of the Jacobi–Bernoulli integral (7.116), appropriate for the quasi-steady non-relativistic oblique rotator. Its derivation assumes no dissipation, either from mutual friction between different species or from radiation losses.

Particles leave the star with subrelativistic speeds; any subsequent acceleration is caused by the non-corotational potential ψ, with the integral (13.169) holding. Within the perfectly conducting stellar crust ψ is a constant, conveniently normalized to zero, and the density of the current-carrying electrons is so high that their velocities are non-relativistic with $\gamma = 1$; hence the value of Γ is effectively unity, the Endean integral becomes

$$1 - \frac{q\psi}{mc^2} = \gamma\left(1 - \frac{\Omega\varpi}{c}\frac{v_\phi}{c}\right), \tag{13.170}$$

and the equation of motion (13.167) reduces to the very simple form

$$\mathbf{u}\times\left[\nabla\times(\mathbf{p} + q\mathbf{A}/c)\right] = 0. \tag{13.171}$$

However, any particles that cross field lines as in Section 13.2.4 through a dissipative drift will in general have acquired a new constant Γ, and the right-hand side of (13.171) should be replaced by $\nabla\Gamma$. But as long as (13.171) holds, then one can immediately write down the generalization of the axisymmetric result

$$\mathbf{u} = \mathbf{v} - \alpha\varpi\mathbf{t} = (c/q)\kappa\nabla\times\left(\mathbf{p} + \frac{q\mathbf{A}}{c}\right) \equiv \kappa\tilde{\mathbf{B}} \tag{13.172}$$

with $\tilde{\mathbf{B}}$ divergence-free. Likewise, the equation of continuity $\partial n/\partial t = -\nabla\cdot(n\mathbf{v})$ reduces on use of the quasi-steady condition (13.7) to

$$\tilde{\mathbf{B}}\cdot\nabla(n\kappa) = 0. \tag{13.173}$$

(Burman and Mestel 1978, 1979).

As in the axisymmetric problem, at least for rapid rotators the inertial drifts remain small until γ becomes so large that neglect of radiative dissipation is no longer valid. Again, a principal aim is to discover when high γ-values arise inevitably in dissipation-free flow. For outflowing electrons ($q = -e$), the integral (13.169) can be written in normalized form

$$\gamma\left(1 - x^2 - xU_\Phi\right) = 1 + \tilde{\psi} \tag{13.174}$$

where $\alpha\varpi/c = x$, $e\psi/mc^2 = \tilde{\psi}$, $\mathbf{U} = \mathbf{u}/c$ and so $U_\Phi = u_\Phi/c = \Omega\varpi/c - x$. Substitution of (13.172) into the definition of γ yields

$$1/\gamma^2 = 1 - x^2 - 2xU_\Phi - (1+\eta)U_\Phi^2 \tag{13.175}$$

where

$$\eta = \frac{U_x^2 + U_z^2}{U_\Phi^2} \simeq \frac{B_x^2 + B_z^2}{B_\Phi^2} \tag{13.176}$$

as long as inertial drifts are small.

As noted by Kahn (1971) and da Costa and Kahn (1982), the behaviour of an outflowing particle as it approaches the l-c depends crucially on the sense about the rotation axis in which the field line points. In the axisymmetric problem, the B_Φ-component arises just from the poloidal motion of the particles, and outflowing gas gives a backwards twist about the axis, yielding $B_\Phi < 0$. In the non-aligned problem, by contrast, an undistorted dipolar field has some field lines that point forwards, and even with the field modified by both particle and displacement currents, if the obliquity is large, roughly half of the field lines should still point forwards. Outflowing electrons along such a line have $U_\Phi > 0$, but as they approach the l-c, U_Φ and so also U_ϖ, U_z must approach zero as the velocity of corotation approaches c. This is seen formally from (13.175): as $x \to 1$ from below, with $U_\Phi > 0$, $1/\gamma^2$ can stay non-negative only if $U_\Phi \to 0$, yielding $\gamma \to \infty$.

Along backward-pointing field lines, $U_\Phi < 0$, and so from (13.175) electrons can cross the l-c provided $|U_\Phi|$ is not too great. However, some of these particles face a difficulty beyond the l-c similar to that found in Section 13.2.3 for the axisymmetric problem. From (13.174), even with $\tilde{\psi}$ finite, γ become infinite when $xU_\Phi = 1-x^2$. Substitution into (13.175) then yields either $x = 1$ and $U_\Phi = 0$ (the case already noted), or $x^2 = 1 + 1/\eta$. In both cases, the breakdown in the flow – at the l-c for the forward-moving particles and on a surface somewhat further out for backward-moving – is again interpreted as a *reductio ad absurdum* of the dissipation-free assumption.

13.3.2 *Quasi-steady force-free domains*

As for the axisymmetric problem, when the inertial force density is small compared with the separate electric and magnetic force densities, the dynamical equation reduces to the force-free condition

$$\rho_e \mathbf{E} + \mathbf{j} \times \mathbf{B}/c = 0, \tag{13.177}$$

which now becomes with use of the quasi-steady condition (13.7)

$$(\nabla \cdot \mathbf{E})\mathbf{E} + (\nabla \times \mathbf{B} + (\alpha/c)\, \partial \mathbf{E}/\partial \Phi) \times \mathbf{B} = 0. \tag{13.178}$$

Now consider the analogue of the GJ hypothesis, with $\mathbf{E} \cdot \mathbf{B} = 0$ holding all the way from the star, so that

$$\mathbf{E} = -(\alpha\varpi/c)\mathbf{t} \times \mathbf{B} \tag{13.179}$$

$$= -(\alpha\varpi/c)(B_z,\ 0,\ -B_\varpi) \tag{13.180}$$

in cylindrical polars. Equation (13.178) then reduces to (Mestel 1973; Endean 1974; Mestel *et al.* 1976)

$$\nabla \times \tilde{\mathbf{B}} = \lambda \mathbf{B}, \tag{13.181}$$

where the vector $\tilde{\mathbf{B}}$ is defined by

$$\tilde{\mathbf{B}} = \left[B_x(1-x^2),\ B_\Phi,\ B_z(1-x^2)\right] \tag{13.182}$$

with $x = (\alpha\varpi/c)$ as usual. Taking the divergence of (13.181) yields the constraint $\mathbf{B}\cdot\nabla\lambda = 0$ – the scalar function λ is constant on magnetic field lines. The factor $(1-x^2)$ in (13.182) is due to the joint effect of the corotating GJ charge density and the displacement current associated with the rotating, non-axisymmetric electric field (13.179). The non-relativistic force-free equation $\nabla\times\mathbf{B} = \lambda\mathbf{B}$ of Section 3.3 is recovered when terms of order $(v/c)^2$ can be ignored, and its axisymmetric form (3.72) when $\partial/\partial\Phi \equiv \partial/\partial\phi = 0$, $\mathbf{B} = [\partial P/x\partial z,\ -S(P)/x,\ \partial P/x\partial x]$, and $\lambda = \alpha = \mathrm{d}S/\mathrm{d}P$.

Equations (13.181) and (13.182) may alternatively be derived by making a local Lorentz transformation from the frame S with Cartesian axes (x, y, z) parallel to the unit vectors $(\hat{\varpi}, \hat{\mathbf{\Phi}}, \hat{\mathbf{z}})$ to the frame S' with axes (x', y', z') instantaneously coinciding with (x, y, z) but moving with the local rotatory velocity $(0, \alpha\varpi, 0)$. In the frame S', the electric field $\mathbf{E}' = 0$, and the force-free condition reduces to $\mathbf{j}' = \lambda'\mathbf{B}'$. Use of the standard relations between electromagnetic quantities in S and S' then yields (13.181), with the scalars λ, λ' related by $\lambda = \lambda'(1-x^2)$ (Mestel 1973).

13.3.3 *The perpendicular corotating magnetosphere*

Equations (13.177) and (13.179) yield

$$\mathbf{j} = \rho_e \alpha\varpi\mathbf{t} + (c\lambda/4\pi)\mathbf{B} : \tag{13.183}$$

the total current is the sum of the corotating charge density ρ_e – by the assumption (13.179), the GJ charge density – together with the current parallel to $\mathbf{B}$. We consider here the analogue of the problem of Section 13.1.3 by putting $\lambda = 0$. Equations (13.181) and (13.182) then yield the non-dimensional form

$$\left[B_x(1-x^2), B_\Phi, B_z(1-x^2)\right] = -\left(\partial\chi/\partial x, \partial\chi/x\,\partial\Phi, \partial\chi/\partial z\right). \tag{13.184}$$

We specialize further to the perpendicular rotator, with B_x, B_Φ symmetric and B_z antisymmetric in the rotational equator. Near the origin, $\mathbf{B} = -\nabla\chi$, so χ reduces to the scalar potential generated by the stellar field. As before, the star is approximated as a point dipole, but now aligned along the Y-axis: the dipole moment $\mathbf{p} = (B_s R^3/2)\hat{\mathbf{Y}}$, generating the potential

$$\chi = p\frac{\varpi\sin\Phi}{(\varpi^2+z^2)^{3/2}}. \tag{13.185}$$

With the normalizations as in (13.31), the non-dimensional potential is the imaginary part of

$$\chi = \frac{x}{(x^2+z^2)^{3/2}} \exp(\mathrm{i}\Phi). \tag{13.186}$$

with the dipolar axis in the plane $\Phi = \pi/2$. We thus look for the function $\chi = \bar{\chi}(x,z)\exp(\mathrm{i}\Phi)$, reducing to (13.186) near the origin, and satisfying

$$x^2(1-x^2)\frac{\partial^2\chi}{\partial x^2} + x(1+x^2)\frac{\partial\chi}{\partial x} - (1-x^2)^2\chi + x^2(1-x^2)\frac{\partial^2\chi}{\partial z^2} = 0, \tag{13.187}$$

derived by imposing $\nabla\cdot\mathbf{B} = 0$. (From now on the bar over χ is dropped.)

The method of solution of (13.187) is closely parallel to that used in Section 13.1.3 (Mestel *et al.* 1999). The Fourier transform in z

$$g(x,k) = \frac{2}{\pi}\int_0^\infty \chi(x,z)\cos(kz)\,\mathrm{d}z \tag{13.188}$$

satisfies

$$x^2(1-x^2)g'' + x(1+x^2)g' - [(1-x^2)^2 + k^2x^2(1-x^2)]g = 0, \tag{13.189}$$

where the prime again signifies $\partial/\partial x$. The solution of (13.189) near the origin is conveniently written

$$g(x) = \frac{2}{\pi}k\mathrm{K}_1(kx) + \frac{2}{\pi}f(x) \tag{13.190}$$

where again K_1 is the modified Bessel function of the second kind (Erdelyi *et al.* 1954). Then $f(x)$ satisfies an equation with the same differential operator as in (13.189) but with the right-hand side now $-kx^2[2x\mathrm{K}_1' + (1-x^2)\mathrm{K}_1]$. The first term in (13.190) is the Fourier transform of (13.186) (without the $\exp(\mathrm{i}\Phi)$ factor), so the boundary condition at the origin is satisfied if $f(0)$ is finite. Near enough to the origin one can use series expansions for both K_1 and f, arriving at

$$g = \frac{2}{\pi}\left[\left(\frac{1}{x} + \frac{1}{2}(k^2+1)x\log x + 2x + \ldots\right) + a_k\left(x + (k^2-3)\frac{x^3}{8} + \ldots\right)\right]. \tag{13.191}$$

Near $x = 1$, there is the solution $g = 1 - k^2(1-x)^2 + \ldots$, which yields B_z singular and so is dropped, and the acceptable solution

$$g = \frac{2}{\pi}b_k\left[(1-x)^2 + \frac{1}{3}(1-x)^3 + \frac{1}{4}(1+k^2/2)(1-x)^4 + \ldots\right]. \tag{13.192}$$

Again, the numerical continuation inwards of the solution (13.192) must link up smoothly with (13.191) or its numerical continuation, so fixing b_k and a_k/b_k. For moderate values of k, this procedure works satisfactorily with the analytic form (13.191) adopted for the outgoing solution. For large k the link-up point is beyond the domain of validity of the series approximations. One then extends the outgoing solution numerically, but using values for K_1 and K_1' from the computer library. As in Section 13.1.3, for large k the JWKB method yields $a_k \approx 0$, $b_k \approx (\pi/\sqrt{2})k\,\mathrm{e}^{-k}$. Away from $x = 0$ and $x = 1$, $g \approx [(2k/\pi)(1 - x^2)/x]^{1/2}\exp(-kx)$. The whole solution may be constructed from (13.184) and the Fourier back-transform

$$\chi = \int_0^\infty g(x,k)\cos(kz)\,\mathrm{d}k. \qquad (13.193)$$

From (13.184) and (13.192), at the l-c, B_x is finite, while B_z, B_Φ vanish like $(1-x), (1-x)^2$ respectively. The x-component of the Poynting vector $\propto -B_x B_\Phi$ vanishes: as in the aligned case, transport of energy and angular momentum by a field satisfying $\mathbf{E}\cdot\mathbf{B} = 0$ requires current flow along the field lines. (Note the contrast with the vacuum rotator of Fig. 13.2.)

By (13.184), the field lines defined by $\chi(x,z)$ may be constructed from

$$\frac{(1-x^2)\,\mathrm{d}x}{(\partial\chi/\partial x)\sin\Phi} = \frac{x\,\mathrm{d}\Phi}{(\chi/x)\cos\Phi} = \frac{(1-x^2)\,\mathrm{d}z}{(\partial\chi/\partial z)\sin\Phi}. \qquad (13.194)$$

Near the origin χ reduces to (13.186) and $(1-x^2) \approx 1$, whence (13.194) integrate to

$$\frac{z^2}{(x^2+z^2)^{3/2}} = C, \qquad (13.195)$$

$$\cos\Phi = \frac{Dz^{1/3}}{[1-(Cz)^{2/3}]^{1/2}}, \qquad (13.196)$$

where C, D are constants of integration. Values of C close to unity pertain to field lines that emerge from near the dipolar equator and so remain in the near vacuum dipole domain for the whole of their length. Smaller values of C pertain to lines that emerge nearer to the poles. With C and D chosen, (13.195) and (13.196) fix z and Φ in terms of x a little way from the origin, after which one must integrate (13.194) with the computed function χ inserted. Some typical field lines are shown in Fig. 13.6.

If one tries to extend this model beyond the l-c by attaching a vacuum field, continuous at the l-c in B_x, E_z, E_Φ, and going to zero at infinity, the resulting discontinuities in B_Φ, B_z, E_x again imply a surface charge–current distribution with non-cancelling Maxwell stresses. There must therefore be a charge–current field beyond the l-c, and it is again instructive to make the illicit extrapolation of the corotating solution, imposing continuity of χ and its derivative. Analysis analogous to that of Section 13.1.3 yields $g \approx [(2k/\pi)(x^2-1)/x]^{1/2}\exp[k(x-2)]$, which by (13.193) yields a χ which blows up beyond $x = 2$: again, the absence of a second non-singular solution at the l-c prevents the construction of a smoothly continuous solution extending to infinity.

In a more realistic force-free model, there will be current flow along the field lines, described by a non-vanishing term $(c/4\pi)\lambda\mathbf{B}$ in (13.183), and yielding transport of energy and angular momentum across the l-c. If λ is small, then

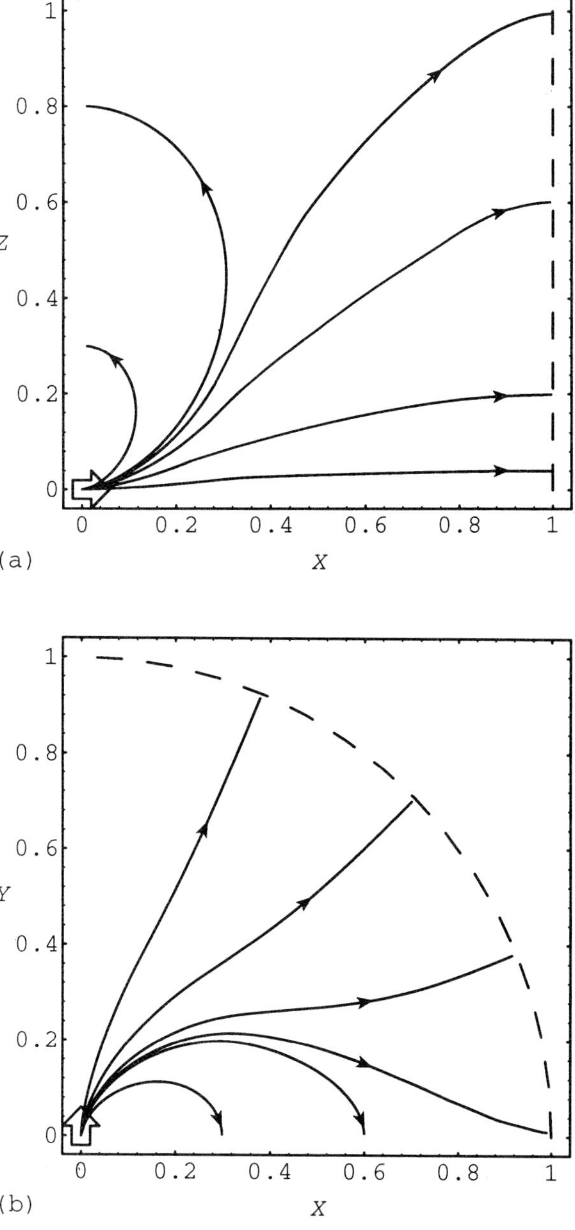

FIG. 13.6. The perpendicular model, with corotating GJ charges within the l-c but no field-aligned currents. (a) Field lines in the plane defined by the rotation and magnetic axes. (b) Field lines in the equatorial plane defined by the rotation axis. Note that field lines cross the l-c normally as in the analogous aligned problem (Fig. 13.4).

within the l-c, the correction $\mathbf{B}_1$ to the field may be found by writing the current system as $(c/4\pi)\lambda\mathbf{B}_0$, where $\mathbf{B}_0$ is the corotation field just constructed. For details, see Mestel *et al.* 1999.

APPENDIX

The classical magnetosphere model

The aligned classical model is as portrayed in Fig. 13.5. The initial model magnetic field is constructed from the flux function (13.158) To begin the iterative procedure, one needs zero-order expressions for the ρ_e and ω fields in the different domains. Consider first those within the l-c. The two curves $z = z_1(x)$, $z = z_2(x)$, the boundaries of the postulated vacuum gap, diverge from the point on the star where $\mathrm{P}_2(\cos\theta) = 0$. Above $z_1(x)$ the GJ condition $\mathbf{E}\cdot\mathbf{B} = 0$, holding all the way from the star, implies that the non-corotational potential ψ has the constant stellar value ψ_s. Below $z_2(x)$, again $\mathbf{E}\cdot\mathbf{B} = 0$, but now because of the gap, $\psi \neq \psi_s$ (except for those field lines which leave the star below the point where $\mathrm{P}_2 = 0$). We therefore write for the positively charged zone

$$\psi - \psi_s = \int^P \{1 - \tilde{\alpha}(P)\}\, \mathrm{d}P, \tag{13.197}$$

so that by (13.75)

$$\mathbf{E} = -\nabla\phi = \nabla P - \nabla\psi = \tilde{\alpha}(P)\nabla P. \tag{13.198}$$

Within the gap, by (13.80) ψ satisfies

$$\nabla^2\psi = -2B_z = 2\,\partial P/x\,\partial x, \tag{13.199}$$

while by (13.77) P satisfies

$$\nabla^2 P = 2\,\partial P/x\,\partial x. \tag{13.200}$$

The boundary conditions require that both $\nabla\psi$ and ∇P be continuous on the curves $z_1(x)$, $z_2(x)$.

Sufficiently far from the star the convergence of these curves on the stellar surface may be ignored. If now P is approximated as the vacuum dipolar field $-\sin^2\theta/r$, with $B_z = 2\mathrm{P}_2(\cos\theta)/r^3$ from (13.119), then (13.199) has the solution

$$\psi - \psi_s = \left[b + a\log(\tan\theta/2) - \sin^2\theta\right]/r \equiv g(\theta)/r, \tag{13.201}$$

with the curves z_1, z_2 reducing to $\theta = \theta_1$, $\theta = \theta_2$ – i.e. the gap is wedge-shaped. The boundary conditions $\nabla\psi = 0$ are satisfied when

$$a = 2\sin^2\theta_1\cos\theta_1, \qquad b = \sin^2\theta_1\{1 - 2\cos\theta_1\log(\tan\theta_1/2)\}, \tag{13.202}$$

$$\cos\theta_1 - \cos\theta_2 = 2\cos\theta_1\cos\theta_2\log\left\{\frac{\tan\theta_2/2}{\tan\theta_1/2}\right\}, \tag{13.203}$$

and

$$\tilde{\alpha}\,(P) = \sin^2\theta_1\cos\theta_1/\sin^2\theta_2\cos\theta_2 \equiv \tilde{\alpha}\ (\text{independent of P}), \tag{13.204}$$

so that in the positive domain,

$$\psi - \psi_s = (1-\tilde{\alpha})\,P = -\,(1-\tilde{\alpha})\sin^2\theta/r. \tag{13.205}$$

From (13.103) with $V_0 = 0$, and inertial effects ignorable, the toroidal velocity $V_t = -x\,\mathrm{d}\phi/\mathrm{d}P = -x(-1+\mathrm{d}\psi/\mathrm{d}P) = \tilde{\alpha}x$, showing that $\tilde{\alpha}$ is indeed the (non-dimensional) rotation of the positive domain. For $\theta_1 = \cos^{-1}(1/\sqrt{3}) = 55°$, $P_2(\cos\theta_1) = 0$, (13.203) yields $\theta_2 = \theta_1$ and so $\tilde{\alpha} = 1$ – the gap closes up. When $\theta_1 < 55°$, (13.203) yields $\theta_2 > 55°$, and (13.204) yields $\tilde{\alpha} < 1$ – *the positive domain subrotates.* Within the gap

$$\begin{aligned}\mathbf{E}\cdot\mathbf{B} &= \frac{2\sin^2\theta_1}{r^5}\left\{(\cos\theta - \cos\theta_1) + 2\cos\theta_1\cos\theta\log\left(\frac{\tan\theta/2}{\tan\theta_1/2}\right)\right\} \\ &\equiv f\,(\theta)\,/r^5\end{aligned} \tag{13.206}$$

with $f\,(\theta_1) = f\,(\theta_2) = 0$ by construction, and $f\,(\theta)$ having a maximum at a point in between. The direction of $\mathbf{E}\cdot\mathbf{B}$ is such as to maintain the gap, in the sense that a stray electron coming from the region $\theta < \theta_1$ would be driven back, and likewise a positive particle coming from $\theta > \theta_2$. The density of positive charge in the subrotating domain is from (13.198) $\rho_e = -\nabla^2\phi = \nabla^2 P - \nabla^2\psi = \tilde{\alpha}\,\nabla^2 P = -2\tilde{\alpha}B_z$, the non-corotational potential acting to reduce ρ_e below the value $-2B_z$ pertaining to the gap-free model.

With $\phi = P + \psi = -\sin^2\theta/r + \psi$, it is found from (13.205) and (13.201) that $\phi \propto 1/r$ in the positive domain and in the gap as well as in the electron domain, and since the gap boundaries are radial, it follows that within a sphere r the total charge $-\int(\partial\phi/\partial r)\,\mathrm{d}S$ is independent of r. There is a net positive charge within r, the maximum value being $8\pi/3$ when $\theta_1 = \theta_2 = \cos^{-1}(1/\sqrt{3})$ and the gap vanishes. (In this point-singularity model, the positive charges are located at the origin; in a realistic model with the gap closing up at $r = R$, they will be distributed over a volume which includes the domain where the magnetic field makes the transition from the vacuum dipolar form to the appropriate form within the neutron star crust.) Within a *cylindrical* region $\overline{x} =$ constant, the total charge is zero. This is most simply shown by integrating $E_x = -xB_z - \partial\psi/\partial x$ over the cylindrical surface. Since $B_z = -\partial V/\partial z$ with the scalar potential $V = z/r^3$, then $\int B_z\,\mathrm{d}z = 0$; thus $\int E_x\,\mathrm{d}z = -\int_0^{z_1(x)}(\partial\psi/\partial\overline{x})\,\mathrm{d}z$, and this is easily shown to vanish by use of (13.201) for the interval $z_2(x) < z < z_1(x)$ and (13.207) for $0 < z < z_2(x)$.

In a study focused on the large-scale properties of the magnetosphere, it is prima facie acceptable to treat the star as a point dipole and to ignore modifications that depend on the finite pulsar radius. Let us for the moment ignore also the modifications to these wedge-shaped gap models that must become significant as the l-c is approached, so that in particular the conclusion of zero charge within the l-c is preserved. It is then clear that the overall zero charge condition (13.154) cannot be satisfied unless the domain beyond the l-c contains not only circulating electrons but also positive charges which compensate for the negative charge of the electrons. This is possible with a finite gap, since the ion domain *subrotates* and so is not forbidden from extending beyond the l-c.

A fully self-consistent treatment would of course construct the appropriate generalization of the wedge–gap model, taking account of the deviation of $\mathbf{B}_\mathrm{p}$ from the vacuum dipole form, and of the effect on the electric field of the circulating electrons beyond the l-c, and simultaneously locating the compensating positive charges within a boundary on which $\mathbf{E}$ and $\mathbf{B}$ are continuous. In the first stage of an iterative scheme, stress is laid on the dynamics of the electron stream. The basic gap model just discussed is retained, but with some significant modifications. No attempt is made to determine the deviation of the gap boundaries from the wedge shape, and the relations (13.203) and (13.204) are retained, with θ_1 a new parameter. However, the model field (13.158) is used in the various formulae rather than the vacuum dipole field. Also, in constructing $\rho_e = \nabla \cdot \mathbf{E}$, the contribution of terms in $\nabla\times\mathbf{B}_\mathrm{p} = \rho_e \omega x \mathbf{t}$ are included (cf. e.g. (13.11)). Then in the subrotating ion domain within the l-c,

$$\rho_e = -2\tilde{\alpha} B_z / \left(1 - \tilde{\alpha} x^2\right); \tag{13.207}$$

in the corotating electron domain

$$\rho_e = -2B_z / \left(1 - x^2\right); \tag{13.208}$$

and in the electron outflow domain, ρ_e and ω are given by (13.111) and (13.113). These changes yield a non-zero total charge within the l-c, but compensating positive charges beyond are still required. They are located at this stage by fiat in a uniformly dense wedge-shaped extension of the $\theta > \theta_2$ domain, but with the uniform rotational *velocity* $\tilde{\alpha}$ that they would have if located on the l-c. There is no claim that in an accurately constructed model, the charges would distribute themselves in this simple way; this aspect of the overall model is purely illustrative. It will be seen below how the search for a fully self-consistent model forces careful scrutiny of the positive charge distribution.

Turn now to the electron domains beyond the l-c. Given the parameters $\overline{r}$ and $\overline{s}$, the condition (13.117) on the l-c fixes $S(P) \equiv S_1(P)$ and so $V_0 = -\mathrm{d}S/2\,\mathrm{d}P$; these in turn yield the curve S_1 beyond the l-c on which C defined by (13.116) vanishes, and the flow goes over from GJ type to dissipative, as already discussed. For most of calculations, it turns out that in the volume dissipation domain, the poloidal velocity $\mathbf{V}_\mathrm{p}$ may be taken as along $\mathbf{B}_\mathrm{p}$ (since $E < B$), with

$$\rho_e = -2V_0 B_\mathrm{p}/V_\mathrm{p}, \tag{13.209}$$

and

$$\omega = \left(1 - V_{\mathrm{p}}^2\right)^{1/2}/x. \tag{13.210}$$

In the equatorial sheet the area density $\sigma(x)$ and the trans-field flow V_x are related to the stream function $S(x, z)$ for the inflow into the sheet by

$$\sigma(x) = S(x, 0)/xV_x \tag{13.211}$$

from the continuity equation, with

$$\omega = \left(1 - V_x^2\right)^{1/2}/x. \tag{13.212}$$

To determine V_{p} and V_x, one needs to know the electric field, which should emerge in the next stage of the iterative scheme. For the first stage, simple analytical forms are adopted for the velocities, with large poloidal and small toroidal components at high latitudes, and small poloidal and large toroidal components near the equator. A further complication arises when the integrals (13.154), (13.146), (13.157) and (13.160) are computed; to ensure that the electric quadrupole moment (13.157) converges satisfactorily at infinity, explicit account must be taken of the small but non-zero component of flow across $\mathbf{B}_{\mathrm{p}}$.

Detailed numerical work shows that the various integral constraints put some preliminary limits on the parameters (FM*a*, p. 295): $\overline{s} \simeq 4.0$; $\overline{r}$ between 3.0 and 5.0, closer to 3.0; θ_1 between 30° and 40°, closer to 40°; and the limit of the extended positive domain between 1.8 and 3.0, closer to 1.8. It is found that the vanishing of the total charge is due basically to the negative electron disc region cancelling the positive domain extension; while the same two domains make the principal contributions to the magnetospheric dipole moment which cancels that due to the star. For both these constraints, the contributions of the other domains are negligible. The sextupole component β_3 given by (13.160) is dominated by the positive contribution from the equatorial disc, but the negative contribution from the volume charges beyond the l-c and the positive extension domain are also significant. The only integral to which the volume charges within the l-c contribute significantly is the quadrupole moment (13.157): the large positive contribution from the disc is counteracted and the total made just negative by contributions from all the volume domains.

The next stage in the iterative procedure is to use the charge-density and velocity fields to construct a *derived* flux function P^d for the new $\mathbf{B}_{\mathrm{p}}^d$ field that satisfies the Ampère–Maxwell equation (FM*b*). The contributions from the volume charge distribution are calculated by a Fourier transform (analogous to the method of Section 13.1.3), and from the equatorial sheet by an elliptic integral technique (similar to that discussed in Chapter 12, Appendix A). The derived field is found to be reasonably similar to the model field in its gross structure except near the equator, where the current sheet causes a strong local distortion of the field lines (equivalent to a thickening of the disc). Similar techniques are used for constructing a derived electric field $\mathbf{E}^d$ from the Poisson–Maxwell equation. Numerical experiments show that the possible values of the parameter $\overline{r}$ in (13.158) must now be restricted to a narrow range centred on $\overline{r} = 3.5$. The derived electric field is then qualitatively satisfactory, acting so as to drive electrons out of the poles towards the equatorial zone, everywhere except in a region near the extension beyond the l-c of the positively charged domain (see below).

It is mildly encouraging that with a narrowing of the allowed parameter range, the charge-current fields associated with the model field (13.158) yield both a derived **B**-field ($\mathbf{B}^d$) which retains at least the gross qualitative features of (13.158), and an associated **E**-field ($\mathbf{E}^d$) which is prima facie acceptable. Complications arise at the next stage, when the $\mathbf{E}^d$, $\mathbf{B}^d$ fields are used to reconstruct the flow. The obvious next step in the iterative procedure is to redetermine $S(P)$ from condition (13.117) for smooth GJ flow through the l-c, but with the derived function P^d inserted instead of the model field P^m given by (13.158). However, this new $S_2(P)$ relation, when combined with the derived field $\mathbf{B}_{\mathrm{p}}^d$, yields a new transition curve S_1 (found by the vanishing of C defined by (13.116)) which lies inside the l-c, contradicting the assumption behind (13.117). This need not be disastrous: the classical model as outlined in Section 13.2.1 could turn out to have its GJ domain terminating on a transition curve S_1 lying within the l-c, with relativistic acceleration and dissipation-limited flow beyond. More serious is that many of the particle trajectories found using the new $\mathbf{E}^d$, $\mathbf{B}^d$ fields are unacceptable. Further experimentation shows that, in general, small changes in the $S(P)$ relation can yield very large changes in the trajectories computed from (13.146) with given **E**, $\mathbf{B}_{\mathrm{p}}$ fields. This suggests that the requirement of GJ flow extending out to and through the l-c, leading to the relation (13.117), is just too severe: instead, one should abandon the constraint (13.117), and look for possible $S_3\,(P)$ relations yielding trajectories that are globally acceptable, i.e. with no particle sinks, and no poleward rather than equatorward drifts. This procedure will be self-consistent if the associated surface S_1 – the transition between GJ and dissipative flow – is located either within or on the l-c, so that (13.117) is no longer relevant. Some further calculations with such an $S_3(P)$ relation (Fitzpatrick, personal communication) yield the tentative conclusion that S_l is indeed likely to be within but close to the l-c (as surmised in Mestel *et al.* 1985). And again, since $B_\phi = -S/x$ will ensure that $E < B$ until close to the equator, the poloidal flow over the bulk of the dissipation zone should be nearly parallel to $\mathbf{B}_{\mathrm{p}}$, with the trans-field motion occurring primarily in the equatorial zone.

A fully convincing numerical demonstration remains a programme for the future. The difficulties in the way are quite subtle: local errors can react back on the whole model in the next iteration. It was noted above that the derived field $\mathbf{E}^d$ in FMb was qualitatively satisfactory except near to the positive domain extension: as that neighbourhood is approached, $\mathbf{E}\cdot\mathbf{B}$ changes from being negative to mildly *positive*, of magnitude $\simeq 0.05(|\mathbf{E}||\mathbf{B}|)$ – i.e. there is a surface within the dissipation domain on which $\mathbf{E}\cdot\mathbf{B} = 0$. Recall from (13.149) et seq. that on such a surface, E must exceed B to avoid sinks of particles. In the present problem this puts a severe constraint on the new $S_3\,(P) = x\,|B_{\mathrm{t}}|$ – it is forced to be markedly smaller than either $S_1(P)$ or $S_2(P)$, and this reacts sharply on both the shape and position of S_1. However, since $\mathbf{E}\cdot\mathbf{B}/|\mathbf{E}||\mathbf{B}|$ is small, it would apparently require only a modest rearrangement of the positive charges (which in the zero-order model were distributed by fiat) to yield both $\mathbf{E}\cdot\mathbf{B} = 0$ in the positive charge domain, and $\mathbf{E}\cdot\mathbf{B}$ negative in the *whole* of the volume dissipation domain. Without the surface in the dissipation domain on which $\mathbf{E}\cdot\mathbf{B} = 0$, the severe constraint on $S_3\,(P)$ disappears, and a self-consistent approximate model may be possible.

14

PULSAR ELECTRODYNAMICS II

14.1 Single-sign flow near the star

14.1.1 *Introduction*

The model of Chapter 13 was developed with guidance from the stellar wind theory of Chapter 7. In the first iteration, the strength of the current flowing out of the polar caps along each field line is determined by a singularity-free requirement at the light-cylinder, which plays a role analogous to the Alfvén surface. The parallel with a purely centrifugally-driven stellar wind is seen to be close, in that it is the component $\mathbf{E}_\perp$ that is responsible for energizing the outflowing electron gas, forced to drift across field lines by the effect first of the toroidal inertial force (i.e. by angular momentum conservation) and subsequently by radiation drag. Recall that in the pulsar problem, the only energy sources are the potential variations set up on the surface by the rotation of the star in its magnetic field. In all models, this energy must ultimately be picked up through the magnetospheric currents completing their circuits by crossing field lines; and the energy–angular momentum constraint discussed in Section 13.2.4 requires that this must occur beyond the l-c. A virtue of the classical model is that it puts these basic global requirements manifestly first: the current density along each circuit near the star is then not a free parameter of a local theory, but in principle emerges from the global constraints.

As against this, there is the prime objection to the classical model in its simplest form: its prediction that all the energy lost beyond the l-c is in the form of gamma rays rather than as a wind of particles. Most workers have concentrated instead on models in which electron–positron pairs are copiously produced near the star. We have seen that in fact the classical model predicts that the electrons return to the star with high energies, and they may generate a gas of pairs in the strong B regions near the star, as discussed below in Section 14.2.2. If such a gas were to expand and flow out as a wind, then the circulating primary electron current could conceivably lose much of its angular momentum not to the ether as gamma rays but to the wind through frictional interaction. Such a model would remain akin to the original 'classical' model only through having the acceleration of the primary particles take place near and beyond the l-c. It is, however, clearly more straightforward to have pair production occur during the initial outflow of the primary electrons from the star. From now on we concentrate on models in which a strong component $\mathbf{E}_\parallel$ is built up near the star, sufficient to accelerate to high γ-values electrons flowing out along the field $\mathbf{B}_\mathrm{p}$.

14.1.2 *Space–charge-limited flow*

We begin with a general discussion of electron flow out of the star. The treatment is essentially local, with the current density along each field line left as a free parameter. For both the aligned, axisymmetric, time-independent model and for the quasi-steady oblique rotator (a system 'steady in the rotating frame'), the electric field is again conveniently written as the sum of the corotational and non-corotational parts (cf. (13.9)):

$$\mathbf{E} = -\frac{(\alpha\mathbf{k}\times\mathbf{r})\times\mathbf{B}}{c} - \nabla\psi. \tag{14.1}$$

The presumed strong local acceleration will be caused by the parallel component $E_{\|} = -\mathbf{B}\cdot\nabla\psi/B = -\partial\psi/\partial s$, where s is length measured along the field. Note that all such models begin by 'living on borrowed capital'. The energy source remains the potential variation over the star's surface, to be tapped by currents that leave and return at different latitudes within the polar caps. However, suppose that ψ on the outflow field line with flux function P_1 increases from 0 in the star to ψ_1. When somewhere beyond the l-c an electron of charge $-e$ is forced to drift to the line $P_2 < P_1$, subsequently returning to the star, the further potential energy released is then $-e[(-\alpha P_1/c+\psi_1)-(-\alpha P_2/c)] = -e[(\alpha/c)(P_2-P_1)]-e\psi_1$ – with the initial gain $e\psi_1$ subtracted out.

Poisson's equation then becomes

$$\nabla^2\psi = -4\pi(\rho_e - \rho_{\mathrm{GJ}}), \tag{14.2}$$

where the corotation field contributes the Goldreich–Julian charge density (13.11)

$$\rho_{\mathrm{GJ}} \approx -\frac{\alpha}{2\pi c}(\mathbf{k}\cdot\mathbf{B}), \tag{14.3}$$

since by (13.19) the $\nabla\times\mathbf{B}$-contribution will be small near the star (see also the detailed illustrative model of Section 13.1.3). The energy–angular momentum integral (13.95) is well approximated by

$$\gamma = \gamma_s + \frac{e\psi}{mc^2} \tag{14.4}$$

where the gas emerging from the stellar surface has the Lorentz factor $\gamma_s \approx 1$ and ψ is given the value zero on the star. In a flow domain, density and velocity at different points are related by continuity :

$$\frac{\rho_e v}{B} = \frac{-nev}{B} = \frac{j}{B} = \text{constant}, \tag{14.5}$$

the current density j being still a pure convection current.

The simplest case is with the length-scale along the field-streamlines small compared with the transverse scale, so that (14.2) may be approximated by the one-dimensional model

$$\frac{\mathrm{d}^2\psi}{\mathrm{d}s^2} = -4\pi(\rho_e - \rho_{\mathrm{GJ}}). \tag{14.6}$$

The appropriate boundary condition at the star is the GJ condition $E_{\|} = -\partial\psi/\partial s = 0$. It is seen that when the electron number density exceeds the local GJ number density, i.e. when $(\rho_e - \rho_{\mathrm{GJ}}) < 0$, then $\partial\psi/\partial s$ and so also ψ increase outwards, causing by (14.4) spontaneous acceleration of the electrons.

For most pulsars, the flow domains are fairly narrow funnels emerging from the polar caps, bounded by domains with field lines that close within the l-c. When the length-scale along $\mathbf{B}$ approaches the polar cap width, then there is a significant transverse contribution $\nabla_{\perp}^2\psi$ to (14.2). Within the neighbouring dead zones, the GJ condition $\mathbf{E}\cdot\mathbf{B} = 0$ holds to a high approximation, so the extra boundary condition on ψ is that it vanish on the boundaries. This 'three-dimensional effect' is well approximated by writing $\nabla_{\perp}^2\psi = -\psi/D_{\perp}^2$ where $D_{\perp}$ is the transverse length-scale (Sturrock 1971; Fawley *et al.* 1977; Mestel 1981). Equation (14.2) now becomes

$$\frac{\mathrm{d}^2\psi}{\mathrm{d}s^2} = \frac{\psi}{D_{\perp}^2} - 4\pi(\rho_e - \rho_{\mathrm{GJ}}). \tag{14.7}$$

The new term acts like an extra negative charge, assisting the outward increase of ψ and hence of γ.

We introduce a standard current density j_0 at each point, defined by the GJ charge density supposed to move with speed c:

$$j_0 = \rho_{\mathrm{GJ}}c = -\alpha\mathbf{k}\cdot\mathbf{B}/2\pi = -\alpha B_z/2\pi, \tag{14.8}$$

where as before z is measured along the rotation axis $\mathbf{k}$. This will sometimes be referred to as the 'GJ current density', even though in a single-sign domain, it can be achieved only by a super-GJ density moving with a subluminal speed. It is convenient to normalize in terms of local quantities. Thus from (14.5) and (14.8) we can define the positive constant

$$J = j/(-\alpha B/2\pi) = (B_z/B)(j/j_0). \tag{14.9}$$

The dimensionless form of j_0 is similarly

$$J_0 = j_0/(-\alpha B/2\pi) = B_z/B, \tag{14.10}$$

a quantity that varies only because of the curvature of the field lines. Define also

$$\tilde{\psi} = e\psi/mc^2, \tag{14.11}$$

so that

$$\gamma = \gamma_s + \tilde{\psi}. \tag{14.12}$$

and

$$l = s/L,\quad L = c/\omega_{\mathrm{pl}},\quad \omega_{\mathrm{pl}} = (4\pi e^2 n_B/m)^{1/2},\quad n_B = (\alpha B^*/2\pi ce),$$

$$b = B/B^*, \tag{14.13}$$

where B^* is the field strength at the standard point $s = 0$. (When the polar cap angle $\theta_{\rm pc}$ is small, for all outflow field lines B^* is effectively the value B_s at the pole.)

The chosen length-scale L is close to the plasma wavelength for the GJ density at $s = 0$. It is also seen that L is related to the l-c radius $\varpi_c = c/\alpha$ by

$$L = \varpi_c \left(\frac{R}{\varpi_c}\right)^{3/2} \frac{1}{2\gamma_m^{1/2}} \ll 1 \tag{14.14}$$

where (cf. (13.141))

$$\gamma_m = \frac{eB_s R^3}{2mc^2}\left(\frac{\alpha}{c}\right)^2 = \frac{eB_{\rm lc}\varpi_c}{mc^2} \tag{14.15}$$

is approximately the maximum attainable value of γ, with $B_{\rm lc}$ again defined by

$$B_{\rm lc} = B_s(\alpha R/c)^3/2. \tag{14.16}$$

Along a dipolar field line $\varpi \propto r^{3/2}$, $B/B^* = b \approx (R/r)^3$; and by flux conservation, the lateral scale $D_\perp$ satisfies $2\pi\varpi D_\perp B =$ constant, so that $D_\perp^2 \propto r^3$. From (14.13), (14.14) and (13.59)

$$b\frac{D_\perp^2}{L^2} = \left(\frac{R^3}{r^3}\right)\left(R^2\theta_{\rm pc}^2\frac{r^3}{R^3}\right)\left(\frac{2\varpi_c\gamma_m}{R^3}\right) = 4\gamma_m. \tag{14.17}$$

Thus the dimensionless form of (14.7) is

$$\frac{1}{b}\frac{{\rm d}^2\psi}{{\rm d}l^2} = \frac{\psi}{\gamma_a} + \frac{J}{(1-1/\gamma^2)^{1/2}} - J_0, \tag{14.18}$$

where $\gamma_a \equiv 4\gamma_m$, the tilde on ψ has now been dropped, and the electron velocity v is written in terms of γ.

We consider now the different parameter domains, following a recent survey by Shibata (1996, 1997).

14.1.3 *Radial field lines*

With field line curvature for the moment ignored, by (14.10), $J_0 = B_z/B$ and so also J/J_0 are both constant. Consider first the one-dimensional problem, in dimensionless form described by (14.12) and by (14.18) without the three-dimensional term, and with b put equal to unity (so ignoring the slow variation of B with r):

$$\frac{{\rm d}^2\gamma}{{\rm d}l^2} = \frac{J}{(1-\gamma^{-2})^{1/2}} - J_0. \tag{14.19}$$

When $J < J_0$ – the current less than the GJ current density as defined in (14.8) – the simplest solution is then $\psi =$ constant and so also $\gamma =$ constant, with

$v/c = J/J_0$: 'GJ-flow', the current being due to the steady flow with *subluminal* velocity of a density equal to the local GJ density. In general, however, the boundary conditions at the star do not allow neglect of the second derivative of ψ and γ (Appendix B by Mestel and Pryce in Mestel *et al.* 1985; Shibata 1991). When J/J_0 is less than and not close to unity, then it is clear from the form of (14.19) that the actual solution for ρ_e and γ will consist of spatial oscillations about the local mean GJ flow, with $\rho_e/\rho_{\rm GJ}$ alternating between values respectively greater than unity (at low $(\gamma - 1)$), causing acceleration, and less than unity (at high $(\gamma - 1)$), causing deceleration.

Equation (14.19) integrates to yield

$$(\gamma')^2 = 2J(\gamma^2 - 1)^{1/2} - 2J_0(\gamma - 1), \tag{14.20}$$

with γ put equal to unity at the star's surface, where ψ' and so also γ' vanish. (The slight increase in γ over unity due to the finite but non-relativistic emission speeds from the star makes no essential difference to the results.) As the ratio J/J_0 is allowed to increase to unity, then (14.19) and (14.20) predict that both the maximum in γ and the oscillation wavelength approach infinity. Figure 14.1 shows solutions with $J_0 = 1$ ($B = B_z$) and three different values of $J = (j/j_0)J_0 = j/j_0$. It is seen that with $J = 0.9$, the wavelength is already well above the normalization scale L. When $J = 0.99$, Fig. 14.1(c) illustrates the growth in both the maximum γ and the wavelength. In general, when J/J_0 is close to unity, then near the star, where $v/c \ll 1$, $|\rho_e| \gg |\rho_{\rm GJ}|$, and the electron gas is strongly accelerated by this super-GJ negative charge density (cf. (14.6)).

In the one-dimensional treatment, monotonic increase in ψ and γ takes over from the oscillatory solution when $J/J_0 = 1$, with (14.20) yielding $(\mathrm{d}\psi/\mathrm{d}l)^2 \simeq 2J_0$ and so with γ increasing outwards indefinitely. In reality, there is the upper limit γ_m, given by (14.15), which will be built into a global model. In any case, even when one is still concerned just with local modelling, when the wavelength becomes comparable with the transverse length, one must take account of the three-dimensional term ψ/γ_a in (14.18), which at high γ by (14.12) is effectively γ/γ_a:

$$\frac{\mathrm{d}^2\gamma}{\mathrm{d}l^2} = \frac{\gamma}{\gamma_a} + \left[\frac{J}{(1-\gamma^{-2})^{1/2}} - J_0\right] \tag{14.21}$$

and the integrated form (14.20) has the extra term γ^2/γ_a on the right-hand side. Examples of the solution of (14.21) are given in Fig. 14.2. From (14.20), when $J < J_0 \simeq 1$, the one-dimensional model yields $\gamma \approx (1 - J/J_0)^{-1}$ for the maximum γ-value attained. For this value, the ratio of the three-dimensional term γ/γ_a to the modulus of the bracketed terms in (14.21) is $\approx (\gamma_a[1 - (J/J_0)]^2)^{-1}$; hence if $\gamma_a^{1/2}[1 - J/J_0] \gg 1$, the three-dimensional term merely modifies slightly the space–charge density wave (Fig. 14.2(a)). For J/J_0 larger, the new term assists the transition to solutions with an effectively monotonic increase in ψ and γ. A *unique* solution emerges if one now demands 'self-screening', i.e. if one imposes the condition $E_\parallel \propto \psi' = 0$ at infinity as well as on the star's surface (Fawley *et*

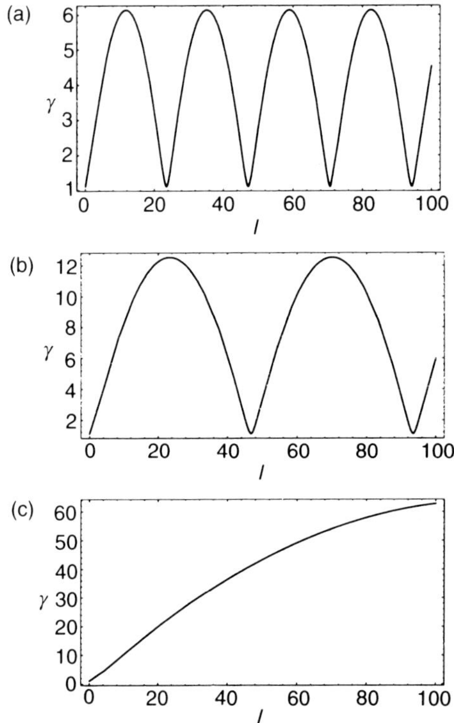

FIG. 14.1. Illustrative space–charge-limited flows in the one-dimensional approximation, ignoring field line curvature. For numerical convenience, the parameters chosen (differing somewhat from those appropriate to real pulsars) are: $\gamma_a = 10^4$, $(\varpi_c/L) = (c/\alpha L) = 10^5$, corresponding to $P_1 = 2.1 \times 10^{-2}$ s and $B_s = 8 \times 10^4$ G. The speed of emission from the star is for $\gamma = 1.1$. The solutions of (14.19) are for $J_0 = 1$ and $J = 0.9, 0.95, 0.99$ in (a), (b) and (c) respectively. (After Shibata 1997.)

al. 1977). The integral of (14.21) with $J = J_a = J_0(1-\epsilon)$ and with $\gamma \simeq 1$, $\gamma' = 0$ at the star's surface then yields

$$\frac{\gamma_\infty^2}{\gamma_a} + 2J_0(1 - \epsilon\gamma_\infty) - \frac{1}{\gamma_a} \approx 0, \tag{14.22}$$

while from (14.21), the simultaneous vanishing at infinity of γ'' yields $\gamma_\infty/\gamma_a \approx \epsilon J_0$; whence

$$\frac{J_a}{J_0} \approx 1 - \left(\frac{2}{J_0\gamma_a}\right)^{1/2}, \qquad \gamma_\infty \approx (2J_0\gamma_a)^{1/2}. \tag{14.23}$$

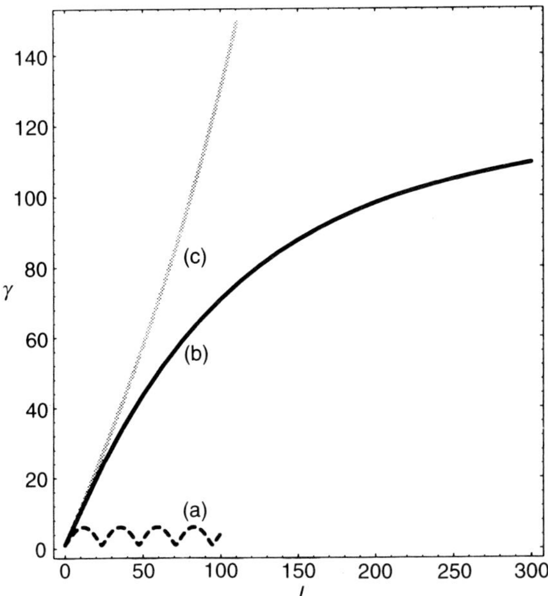

FIG. 14.2. Space–charge-limited flow, again ignoring field line curvature, but with the three-dimensional effect included, and with $\gamma_a = 10^4$, $J_0 = 1$. Case (a) is for J=0.9, so $\gamma_a^{1/2}(1 - J/J_0)$=10. Case (b) is for $J = 1 - 1.12\gamma_a^{-1/2} = 0.9888$, and so is just above the critical value J_a given by (14.23). Case (c) is with $J = 1$, showing exponential growth due to the three-dimensional term. (After Shibata 1997.)

It should be noted that in this critical solution, again it is the super-GJ density near the star that gives the gas its acceleration, later assisted by the three-dimensional effect. In order to satisfy the special boundary condition at infinity, the slightly sub-GJ current density must be chosen to ensure that once $v \approx c$, in (14.7) the decelerating term $-4\pi(\rho_e - \rho_{\rm GJ})$ exceeds the accelerating term $\psi/D_\perp^2$ by just the right value to yield a monotonic increasing solution for ψ and γ, with ψ' and so also γ' going asymptotically to zero.

Within the domain near the star where the one-dimensional equation (14.19) is valid, the modulus of the normalized excess Σ of the electron column density over the GJ column density density out to the height l is

$$\Sigma = \int_0^l (\rho_{\rm GJ} - \rho_e)\,\mathrm{d}l = \int_0^l \left(\frac{J}{(1-\gamma^{-2})^{1/2}} - J_0\right)\mathrm{d}l$$
$$= \int_0^l \gamma''\,\mathrm{d}l = \gamma'(l). \tag{14.24}$$

The complete (three-dimensional) critical (self-screening) solution has yielded the J/J_0 relation (14.23). The one-dimensional relations (14.19) and (14.20) predict that γ' has its maximum when $\gamma = [1-(J/J_0)^2]^{-1/2} \approx 2^{-3/4}(J_0\gamma_a)^{1/4}$ and $\gamma' = (2J_0)^{1/2}[1-(2J_0\gamma_a)^{-1/4}]^{1/2}$; hence

$$\Sigma_{\max} \approx (2J_0)^{1/2}, \tag{14.25}$$

achieved at a height of about one scaled unit of length. With this value of γ, it is seen that the fractional three-dimensional correction $(\gamma/\gamma_a J_0)$ in (14.21) $\approx (2J_0^{1/4}\gamma_a)^{-3/4}$ and so is indeed small.

Fawley *et al.* (1977) pointed out that the asymptotic value $\gamma_a^{1/2}$ given in (14.23) – which follows from imposition of their boundary condition $\psi' = 0$ at infinity – is in fact well below the value $\approx 10^7$ required to yield significant pair creation (cf. Section 14.2.2). If the local parameter $J/J_0 > 1$, (14.21) shows that both terms on the right accelerate the electrons. Once $v \approx c$, then the space–charge term yields

$$\gamma \approx (J-J_0)l^3/6. \tag{14.26}$$

Beyond $l \simeq \gamma_a^{1/3}$ the three-dimensional term dominates and this local treatment predicts that γ grows exponentially with the characteristic length $D_\perp = \gamma_a^{1/2}L$. Exponential growth predicted by a local model must be treated with caution, as it may quickly run into the limitations imposed by the global model, for example that there is the maximum attainable value (14.15) for γ. However, if this value $\gamma_m \gtrsim 10^7$, then prima facie it appears that when $J > J_0$, even in the straight field line case, γ may grow high enough for pairs to be produced in a distance of order $D_\perp$.

14.1.4 *The effect of field line curvature*

Any plausible field will have curved field lines, so that $J_0 = B_z/B$ is not a constant parameter but varies along a field-streamline. Along field lines that curve away from the rotation axis, including all the lines of an aligned dipolar field, J_0 decreases outwards. A hypothetical flow leaving the star with J/J_0 already greater than unity – i.e. with a super-GJ density – again suffers immediate relativistic acceleration. More plausibly, a flow which leaves the star with J/J_0 less than but comparable with unity will inevitably find J/J_0 reaching unity at a modest distance $l = l_0$ from the star: the GJ flow with a space–charge wave superposed must then go over into flow with monotonic acceleration (Mestel 1981; Mestel *et al.* 1985; Shibata 1991). An illustrative example is shown in Fig. 14.3.

It is convenient to write

$$J_0 = J + \left(\frac{\partial J_0}{\partial l}\right)_0 (l-l_0) \equiv J - \frac{l-l_0}{l_L}, \tag{14.27}$$

where the minus sign ensures that $l_L > 0$. If the field is essentially dipolar even at the star (i.e. higher multipoles are not important), then by (13.119), B_z/B changes over the scale ϖ_c, and so

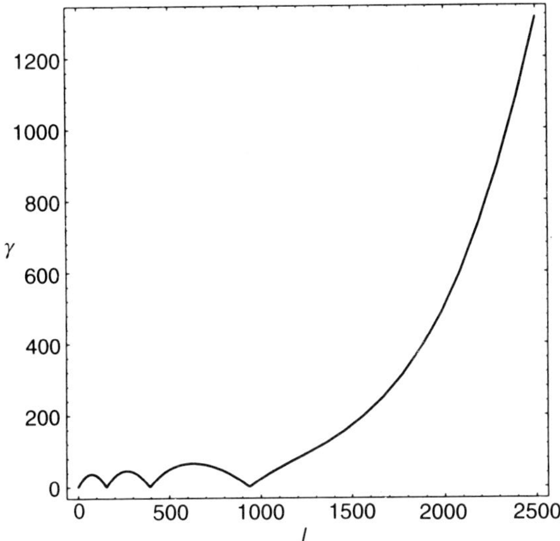

FIG. 14.3. Example of flow along field lines that curve away from the axis, showing how the space-charge wave goes over into monotonic acceleration. Equation (14.21) with $J_0 = 1 - l/l_L$ is solved numerically for the parameters $\gamma_a = 10^4$, $l_L = 10^5$ and $J = 0.98$. (After Shibata 1997.)

$$l_L \simeq \frac{\varpi_c}{L} \simeq \gamma_a^{1/2}\left(\frac{\varpi_c}{R}\right)^{3/2} \simeq 2.3\times 10^9\left(\frac{B_s}{10^6}\right)^{1/2}\left(\frac{R}{10^6}\right)^{3/2}\mathrm{P}_1^{1/2} \qquad (14.28)$$

from (14.13). Equation (14.21) then becomes

$$\frac{1}{b}\frac{\mathrm{d}^2\gamma}{\mathrm{d}l^2} = \frac{\gamma}{\gamma_a} + J\left(\frac{1}{(1-\gamma^{-2})^{1/2}} - 1\right) + \frac{l-l_0}{l_L} \approx \frac{\gamma}{\gamma_a} + \frac{l-l_0}{l_L} + \mathrm{O}(\gamma^{-2}). \quad (14.29)$$

For consistency, we retain the variation of $b \approx [(R+s)/R]^{-3} = [1 + l/(R/L)]^{-3}$, with $R/L \approx (\gamma_a l_L)^{1/3}$ by (14.28) and (14.14). Since $J = \mathrm{O}(1)$, the initial acceleration is by the one-dimensional term, yielding $\gamma \approx b(l-l_0)^3/6l_L$, the value γ being reached after a dimensionless distance $(\gamma/b)^{1/3} l_L^{1/3}$. The three-dimensional term dominates after $(l - l_0) \simeq (\gamma_a/b)^{1/2}$, corresponding to a length $\simeq$ the transverse scale; subsequent growth in γ is exponential.

It is seen that in this geometry, with field lines that curve away from the rotation axis, the change in the GJ term leads inevitably to further acceleration. There cannot be automatic screening of $E_{\|}$ within the classical domain: if screening is achieved, it must be through pair production and subsequent polarization of the $e^{\pm}$-plasma.

Now consider the field lines which curve towards the rotation axis – roughly half of the lines of a dipolar field with a significant obliquity between the magnetic and rotation axes. Since now J_0 increases outwards, the velocity of a GJ flow emanating from the star decreases outwards, so there is now no automatic breakdown: if the flow starts with J/J_0 somewhat below unity it can in principle persist, together with the superposed space–charge wave. However, just as in the case with J_0 constant, there is a unique monotonic solution with a critical current density J_b that is slightly less than the surface value of J_0, yielding $E_{\|} = 0$ at large distances – self-screening within the classical regime (Scharlemann *et al.* 1978). The behaviour is very similar to the straight field line case of Fawley *et al.* (1977) outlined above. Near the star, it is again the super-GJ electron density that causes rapid acceleration. (The integrated charge along the field can be thought of as the relic of the surface charge present in the Deutsch–Pacini vacuum solutions (cf. Section 13.1.1)). Once v is close to c, $(\rho_e - \rho_{\mathrm{GJ}})$ changes sign: in contrast to the case with curvature away from the rotation axis, in this 'starvation model' there is a self-screening effect before the onset of pair production. The three-dimensional term both assists the initial acceleration and moderates the screening effect.

Analogously to (14.27), one can approximate J_0 by

$$J_0 = J_{0*} + \frac{l}{l_L} \tag{14.30}$$

where J_{0*} is the value of J_0 at $l = 0$, and $(\partial J_0/\partial l)_{l=0} = 1/l_L > 0$. Equation (14.21) now becomes

$$\frac{1}{b}\frac{\mathrm{d}^2\gamma}{\mathrm{d}l^2} = \frac{\gamma}{\gamma_a} + J\frac{1}{(1-\gamma^{-2})^{1/2}} - J_{0*} - \frac{l}{l_L}. \tag{14.31}$$

Solutions for different values of J are shown in Fig. 14.4, including (d), the self-screening case $J = J_b$. As J approaches J_b from below, the wavelength of the space-charge density wave becomes steadily longer; when J reaches J_b, the first local maximum in γ (where $E_{\|} = 0$) is at infinity. This monotonically increasing classical solution begins with a linear phase

$$\gamma = \gamma_a(J_{0*} - J_b) + (\gamma_a/l_L)l, \tag{14.32}$$

predicting that $\gamma \to \gamma_a$ as $l \to l_L$. As illustrated in Fig. 14.4, if J slightly exceeds J_b, then the accelerating $E_{\|}$ persists, and the exponential growth due to the three-dimensional term dominates over the screening; while if J is slightly less than J_b, γ retains its wave-like structure.

14.1.5 *The effect of general relativity*

Space–time outside a spherically symmetric mass M with angular momentum $\mathbf{J}$ is described by the Kerr metric. This differs qualitatively from the Schwarzschild metric for a non-rotating body by having non-diagonal terms, which imply a local inertial frame to be rotating with respect to the inertial frame at infinity

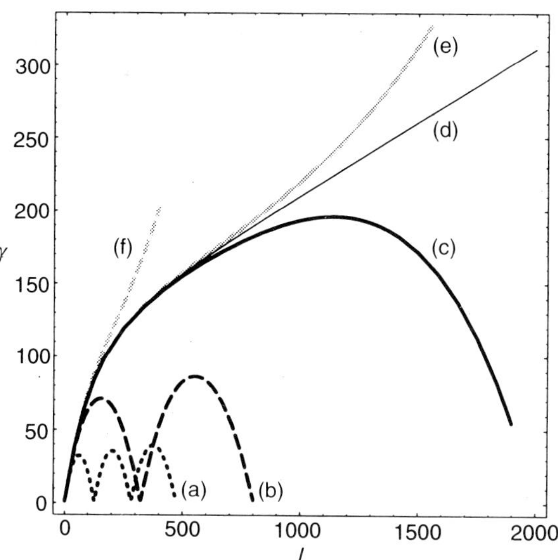

FIG. 14.4. Examples of flow along field lines that curve towards the axis, with $\gamma_a = 10^4$ and $l_L = 10^5$. The solutions of (14.31) are shown for $J_{0*} = 1$ and for (a) $J = 0.98$; (b) $J = 0.987$; (c) $J = 1 - 1.1\gamma_a^{-1/2}$; (d) $J = 1 - 1.09937\gamma_a^{-1/2}$; (e) $J = 1 - 1.099\gamma_a^{-1/2}$; (f) $J = 0.9895$. Curve (d) has $J \approx J_b$, the unique self-screening case. The steady outward increase in the peak Lorentz factor in the space–charge density waves in (a) and (b) is due to the outward decrease in $B/B^* = b$. (After Shibata 1997.)

(the Lense–Thirring effect or 'the dragging of inertial frames'). A star is rotating 'slowly' if the ratio $|\mathbf{J}|/Mcr_g \ll 1$, where $r_g = 2GM/c^2$ is the Schwarzschild radius. The spatial part of the Kerr metric can then be approximated by the corresponding part of the Schwarzschild metric:

$$\mathrm{d}l^2 = \frac{\mathrm{d}r^2}{\alpha^2} + r^2(\mathrm{d}\theta^2 + \sin^2\theta\,\mathrm{d}\phi^2) \tag{14.33}$$

where the red-shift factor

$$\alpha = (1 - r_g/r)^{1/2}, \tag{14.34}$$

the ratio of proper time at coordinate r to the Schwarzschild time t measured by a clock at infinity.[6] Macdonald and Thorne (1982) use a set of fundamental zero-angular-momentum observers (ZAMOs), each rotating with respect to the inertial frame at infinity with the local angular velocity

[6] In this section, 'Ω' is used for the rotation of the star and 'α' for the red-shift factor.

$$\omega \approx \frac{2G\mathbf{J}}{c^2 r^3} = j_s \left(\frac{r_g}{r}\right)\left(\frac{R^2}{r^2}\right)\boldsymbol{\Omega}, \tag{14.35}$$

where j_s is a number ≈ 0.4 for more or less uniformly dense stars. (Because of the slow rotation approximation, ω is independent of θ.) In this formalism, the effects of general relativity are contained in the two functions α and ω.

For analytic details the reader is referred to Muslimov and Tsygan (MT 1986, 1990, 1992*a,b*) and to Beskin (1990). In an axisymmetric steady state, Maxwell's equations as measured by ZAMOs become

$$\nabla \cdot \mathbf{B} = 0, \tag{14.36}$$

$$\nabla \times (\alpha \mathbf{B}) = \frac{4\pi}{c}\alpha \mathbf{j}, \tag{14.37}$$

$$\nabla \times (\alpha \mathbf{E}) = -\frac{1}{c}(\mathbf{B}\cdot\nabla\omega)(\varpi \mathbf{t}), \tag{14.38}$$

$$\rho_e = \frac{\nabla\cdot\mathbf{E}}{4\pi}, \tag{14.39}$$

where ρ_e is again the actual charge density, and the operator ∇ is that appropriate to the spatial metric (14.33). Near the star, the magnetospheric currents have little effect on the magnetic field structure (cf. Sections 13.1.2 and 13.1.3), so $\mathbf{j}$ may be dropped from the right-hand side of (14.37). With a dipolar distribution over the star's surface, (14.37) and (14.36) then yield

$$B_r = B_0 \frac{f(\eta)}{f(1)}\frac{1}{\eta^3}\cos\theta, \quad B_\theta = \frac{B_0 \alpha \sin\theta}{2\eta^3}\left[-\frac{2f(\eta)}{f(1)} + \frac{3}{(1-\varepsilon/\eta)f(1)}\right], \tag{14.40}$$

where

$$\eta = \frac{r}{R}, \qquad \varepsilon = \frac{r_g}{R} \approx 0.3\frac{M_0}{R_6},$$
$$f(\eta) = -3\left(\frac{\eta}{\varepsilon}\right)^3\left[\ln\left(1-\frac{\varepsilon}{\eta}\right) + \frac{\varepsilon}{\eta}\left(1+\frac{\varepsilon}{2\eta}\right)\right] \approx 1 + \frac{3\varepsilon}{4\eta} + \ldots \tag{14.41}$$

Equation (14.38) yields the generalization of the familiar (14.1):

$$\alpha\mathbf{E} = -\frac{1}{c}\left[(\boldsymbol{\Omega}-\omega)\times\mathbf{r}\right]\times\mathbf{B} - \nabla\psi : \tag{14.42}$$

there is the red-shift factor multiplying $\mathbf{E}$, and the corotational field $-(\boldsymbol{\Omega}\times\mathbf{r})\times\mathbf{B}/c$ is replaced by $-[(\boldsymbol{\Omega}-\omega)\times\mathbf{r}]\times\mathbf{B}/c$, so that the angular velocity entering is measured relative to the local frame rotating with ω.

From (14.39) and (14.42), the non-corotational potential ψ satisfies

$$\nabla\cdot\left(\frac{\nabla\psi}{\alpha}\right) = -4\pi(\rho_e - \rho_{\rm GJ}), \tag{14.43}$$

with the GJ charge density

$$\rho_{\rm GJ} = -\frac{1}{4\pi c}\,\nabla\cdot\left[\frac{1}{\alpha}(\Omega-\omega)r\sin\theta\,\mathbf{t}\times\mathbf{B}\right]. \tag{14.44}$$

With (14.40) substituted, in the pulsar polar caps this reduces to

$$\rho_{\rm GJ} = -\frac{\Omega B_s}{2\pi c}\,\frac{1}{\alpha\eta^3}\,\frac{f(\eta)}{f(1)}\left(1-\frac{\kappa}{\eta^3}\right)\cos\theta, \tag{14.45}$$

where $\kappa = \dfrac{2G|\mathbf{J}|}{\Omega R^3c^2} = j_s\left(\dfrac{r_g}{R}\right) \approx 0.4\varepsilon \approx 0.12\left(\dfrac{M_\odot}{R_6}\right)$.

In steady outflow of single sign charges along field lines, $\mathbf{j} = \rho_e\mathbf{v} = \rho_e k\mathbf{B}$, and by (14.37) and (14.36), the modified continuity condition is

$$\frac{\alpha\rho_e v}{B} = \text{constant on field-streamlines.} \tag{14.46}$$

From (14.40) and (14.45)

$$\alpha\frac{\rho_e}{\rho_{\rm GJ}}\frac{\rho_{\rm GJ}v}{B} = \frac{\alpha\left(\dfrac{\rho_e}{\rho_{\rm GJ}}\right)\left(\dfrac{v}{c}\right)c\left[-\dfrac{\Omega B_s}{2\pi c}\,\dfrac{1}{\alpha\eta^3}\,\dfrac{f(\eta)}{f(1)}\left(1-\dfrac{\kappa}{\eta^3}\right)\cos\theta\right]}{B_s\dfrac{f(\eta)}{f(1)}\dfrac{1}{\eta^3}\cos\theta} \tag{14.47}$$

$$= \left(\frac{\rho_e}{\rho_{\rm GJ}}\right)\left(\frac{v}{c}\right)\left(\frac{-\Omega}{2\pi}\right)\left(1-\frac{\kappa}{\eta^3}\right) = \text{constant on streamlines,} \tag{14.48}$$

where the difference between B and the component B_z along $\boldsymbol{\Omega}$ is for the moment ignored.

The importance of the modifications due to general relativity (the 'g–r effect') depends on the particular model considered for the flow near the star. From (14.48), 'GJ flow' – non-relativistic outflow with $\rho_e = \rho_{\rm GJ}$ and $\psi = 0$ in the mean – is possible, with the mean velocity

$$v/c \propto 1/(1-\kappa/\eta^3) \tag{14.49}$$

decreasing outwards initially by a factor $\approx (1/0.88) = 1.13$. Again, there will in general be a spatial density wave superposed. Further out, the effects of field line curvature summarized above need to be included. Thus for this type of solution the modification is quite modest.

By contrast, Muslimov and Tsygan (1992*a,b*) follow the approach of Arons and collaborators and look for the relativistic solution that is monotonic in ψ and γ. The boundary conditions are as before: $\psi = 0$ on the star and on the sides of the flow domain (bounded by the domain of field lines that close within the l-c), and $\mathbf{E}\cdot\mathbf{B} = 0$ on the star and also at 'infinity'. Again, it is this outer boundary condition – requiring screening of the aligned electric field *before* pair production occurs – that fixes the value of the current density j. They do not in fact discuss the initial acceleration by a super-GJ density from $v < c$ to $v \approx c$ – as seen, an essential feature of the Arons *et al.* solution – but instead assume $v \approx c$ everywhere outside the star. The continuity condition (14.46) then requires $\alpha\rho_e \propto B$ on a streamline. In their solution there remains a slight super-GJ density near $\eta = 1$;

$$\rho_e \approx -\frac{\Omega B_s}{2\pi c}\frac{1}{\alpha\eta^3}\frac{f(\eta)}{f(1)}\left[(1-\kappa) + \text{small positive term}\right] \tag{14.50}$$

which is close to but slightly greater in modulus than ρ_{GJ}. Further out, their solution is 'starved', like that of Arons *et al.*:

$$\frac{\rho_e}{\rho_{\mathrm{GJ}}} = (1-\kappa)\left(\frac{\rho_e}{\rho_{\mathrm{GJ}}}\right)_{\mathrm{star}} < 1. \tag{14.51}$$

Along all field lines within the polar cap, not too close to the last open field line (along which the $\psi = 0$ condition holds), the parallel electric field they construct has the value

$$E \sim \left(\frac{\Omega R}{c}\right)^2 B_s \left(\frac{GM}{Rc^2}\right)\cos\chi \tag{14.52}$$

at a height $\approx$ the polar cap width, where χ is again the obliquity angle. Compared with the value found by Arons *et al.* for their self-screening case, this contains the large factor $(\Omega R/c)^{-1/2}$. It is also smaller by the same factor compared with the field given by assuming all the potential variation across the polar cap is picked up at this height, by the effect of the three-dimensional term.

As noted, the value of the current is fixed by the far boundary condition. A larger current would yield $|E_{\|}|$ increasing outwards; a smaller current would yield a quasi-GJ system, with $v < c$ (no relativistic acceleration), and with spatial oscillations superposed. The curvature of the field lines does not come into the MT discussion. It will enter further out if the flow is described by the sub-relativistic GJ solution, as given by (14.49), rather than by the self-screening solution. Shibata has done some calculations which illustrate these conclusions (cf. Fig. 14.5). He solves the equations for ψ and so for γ, ignoring both field line curvature and the three-dimensional effect due to the finite width of the beam, but starting (like Arons *et al.*) with a non-relativistic flow and super-GJ density near the star. The equation for ψ is

$$\frac{d^2\psi}{dl^2} = \frac{J}{(1-1/\gamma^2)^{1/2}} - J_0, \tag{14.53}$$

where J and J_0 are defined analogously to (14.9) and (14.11). In the problem without general relativity, $J_0 = 1$; but with inclusion of the general relativistic modifications (14.45) and (14.46) to the GJ density and the continuity condition, $J_0 = (1-\kappa/\eta^3)$, where again $\eta = 1$ at the star's surface and $\kappa \approx 0.12$. At infinity,

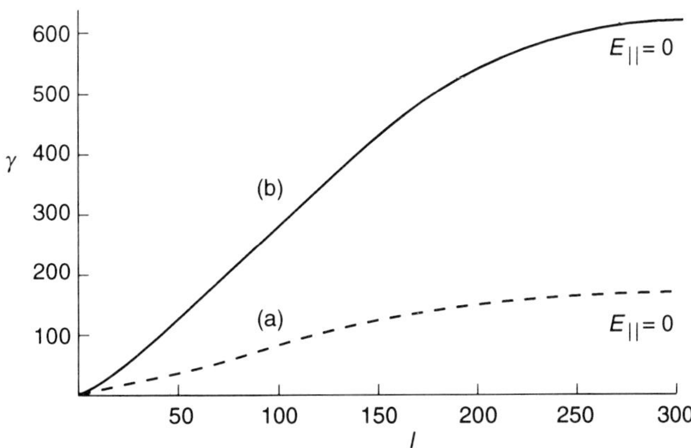

FIG. 14.5. Solutions of the one-dimensional equation ignoring field line curvature and the three-dimensional effect, with J chosen so as to satisfy the boundary condition $E_{\|} = 0$ both near and far from the star. The flow starts with non-relativistic speeds but with a super-GJ density near the star. Case (a) is without the g–r correction, with $J_0 = 1$ and $J = 1$. Case (b), for $J_0(\infty) = 1$, $J = 1$, is with the g–r correction included, showing a marked increase in the γ-values. (Calculations by Shibata, reported in Mestel 1996.)

(14.53) reduces to (14.19), for which, as seen, the Arons-type (self-screening) solution – *monotonic* in γ and ψ, and forced to satisfy $E_{\|} = 0$ not only at the star but also at infinity – must have $J = J_0(\infty) = 1$, which then yields $\gamma_\infty = \infty$. As the g–r effect reduces J_0 below unity near the star, the solution of (14.53) with the same condition $\psi' = 0$ at the star and infinity predicts ψ and so also γ everywhere larger. The same effect is found by MT who effectively insert the g–r correction into the three-dimensional equation (14.21); it will persist when their solutions are emended to start with non-relativistic emission from the star. As in Section 14.1.2, inclusion of the three-dimensional term ensures that γ_∞ is large but finite.

Shibata also calculates a case with J markedly below $J_0(\infty) = 1$, finding as before the space–charge wave superposed on GJ flow, and confirming that the effect of the g–r correction is then quite modest (Fig. 14.6).

14.1.6 *Discussion*

The previous sections have summarized the different possibilities for the flow of electrons out of the polar caps. Inevitably, most of the results have been stated in terms of a local parameter – the normalized current density – for each field line. To fix these parameters, this local analysis must be incorporated into a global model for the whole magnetosphere, including in particular the trajectories of the return currents. The obvious first example is the classical model of Chapter 13, where it is initially hypothesized that no breakdown in

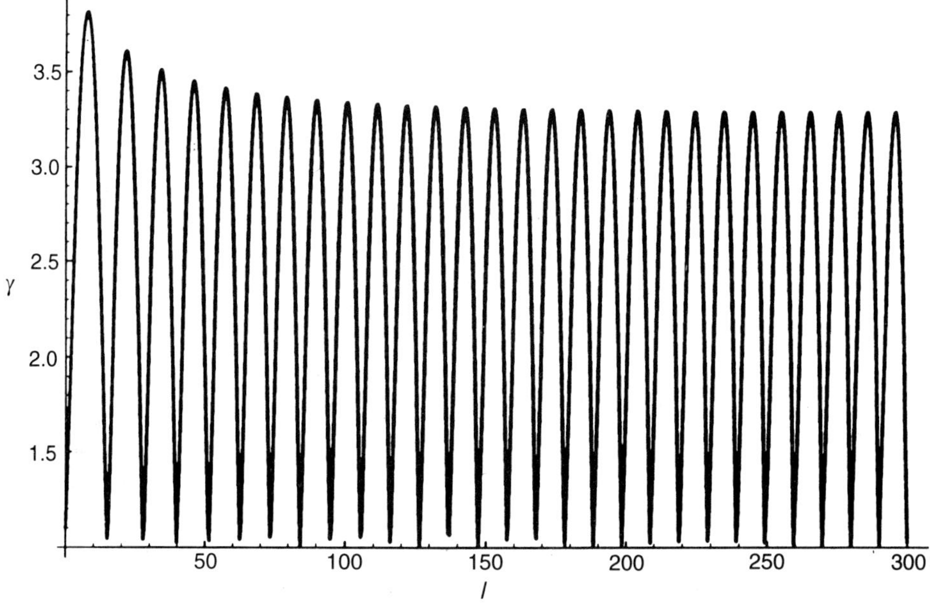

FIG. 14.6. Solution of the same equation as in Fig. 14.5, including the g–r effect, but with $J = 0.75$.

the GJ approximation occurs until the dissipation-free flow has passed through the l-c, with the necessary trans-field motion being linked with the radiation of angular momentum. For each field-streamline, this suffices to fix the emission velocity (and so also the current density) pertaining to the level where the ionic lattice structure of the pulsar crust ends. (Within this surface, the ion density and so also the compensating electron density increase rapidly so as to approach the mean value of $\approx 10^{36}\ \mathrm{cm}^{-3}$ for an iron lattice, much greater than the GJ density $\approx 6 \times 10^8/\mathrm{P}_1$, which is then as in a standard plasma just a charge-separation density.) Likewise, the departure due to finite inertia from strict GJ density in the mean is determined and found to be very small. Further constraints on the system include the vanishing of the net charge on the system, requiring that the electric field must have only a quadrupole component at infinity. The subsequent required modifications to this model, such as the likely location of the critical surface S_l somewhat within the l-c (cf. Section 13.2.5), do not introduce any essential change of principle: the free parameters in the local theory are to be fixed by global requirements.

Of the local models yielding acceleration to high γ-values near the star, that with flow along 'away'-curving field lines – for example, in the aligned rotator – has the advantage of predicting a spontaneous transition from GJ conditions to relativistic acceleration. Outflow subject to continuity of convection current, the Maxwell–Poisson equation and the GJ condition $\mathbf{E} \cdot \mathbf{B} = 0$ yields a monotonically increasing v/c, and when this approaches unity, the neglect of $E_{\|}$ is no longer possible. Note that there is no requirement of a large super-GJ density near the star. The initial excess of $|\rho_e|$ over $|\rho_{\rm GJ}|$ is just the very modest value required by the inertial forces acting on particles with γ close to unity, and the current density is essentially $\rho_{\rm GJ}v$, in modulus less than j_0. The subsequent overfilling of the flux-tubes is due purely to the geometry of the flow. As noted, the g–r correction is then of minor importance.

Within a strictly classical theory, the difficulty is that there is no obvious way of halting the growth in $|E_{\|}|$ and the associated build-up of negative charge until the constraint of flow parallel to $\mathbf{B}$ can be dropped, and we have seen that this must be primarily beyond the l-c. One cannot appeal to the variation of the GJ term in (14.7) to cause both acceleration and shielding. For this reason, Arons *et al.* have called 'away'-curvature 'unfavourable'. One can argue that without pair production, flow along 'away'-curving lines would have rather to conform to a classical model as in Chapter 13, with the current along each field line fixed on or near the l-c as part of the global solution, and with $E_{\|}/E_{\perp}$ and so also $(\gamma - 1)$ remaining small until beyond the l-c, after which strong inertial or radiative trans-field drift can occur, so resolving the return current problem.

By contrast, the original Arons *et al.* solutions (without the g–r term) – existing for the 'towards'-curvature case – stand out by having the current density fixed locally through the extra boundary condition $E_{\|} = 0$ at 'infinity', meaning at distances large compared with the polar cap radius. The condition reacts back strongly on the physical quantities at the star, for the postulated local acceleration requires a local super-GJ density. The required playing off against each other in (14.18) of the three-dimensional term and the effective charge density allows satisfaction of this extra boundary condition, but only for field lines with 'towards'-curvature, whence their description as being 'favourably' curved. Since the difference between the 'favourable' and 'unfavourable' cases derives not from their respective ability to accelerate particles, but rather to the possible *termination* of acceleration through the asymptotic *self-screening* of $E_{\|}$, it is preferable to use the more neutral terminology, 'towards'-curvature and 'away'-curvature respectively.

As noted, in a highly oblique rotator, about half of the field lines curve towards the axis. For definiteness, let us consider the perpendicular rotator, for which all lines (apart from those in the rotational equator) have 'towards' curvature, and for the moment ignore the possibility of vacuum gaps. A classical model alternative to that of Arons *et al.* would be an extension of that in Sections 13.3.3 and 13.3.4. Along field lines that close within the l-c, the GJ charge density again corotates with the star. Charges on field lines that reach the l-c will also flow out along field lines, but as seen in Section 14.1.4, because now B/B_z decreases with distance from the axis, insertion of $j = \rho_{\rm GJ}v$ into (14.9) shows that the geometry of the flow will now allow them to retain the GJ density, with the velocity v along B *decreasing* outwards like B/B_z. Again, it is only when the l-c is approached that the γ-values begin to grow, but now because of the growing importance of the *rotatory* motion. Inertial and in particular dissipative drifts associated with radiation losses can then allow particles to break away from field lines, either just within or just without the l-c (cf. Section 13.3.1), and to complete their circuits. A high-obliquity model would have ion as well as electron emission.

To validate this non-aligned classical model, one would need to attempt the self-consistent construction of the **E**, **B** fields, analogously to the work in FM*a, b* summarized in Chapter 13 (especially Section 13.2.5 and the Appendix). There is no suggestion that this is a high-priority or indeed a feasible enterprise. The point to be emphasized is that from the discussion so far, it is not clear why in the context of classical physics, the flow along individual field lines should opt for the self-screening solution, with the required strongly super-GJ density near the star, rather than the GJ-type solution, with high γ-values appearing near the l-c rather than near the star. This underlines the limitations of any local discussion, in which the current along a field line is a parameter, which one expects to be fixed as part of the global solution.

However, a prime motivation for studying relativistic acceleration near the star is that it can lead to efficient $e^{\pm}$-pair production, converting the primary-electron gas into a three-component plasma of high-γ primary electrons immersed in a sea of pairs (cf. Section 14.2.2). This can enable the spin-down of the star to occur primarily via a wind, and also allow scope for radio noise generation (e.g. Ruderman and Sutherland 1975; Melrose 1996). At the moment, the question is how this generation of an $e^{\pm}$-plasma reacts on the various flow models.

Recall that for each field line with 'towards'-curvature, there exists in the classical domain a critical current density which yields the Arons *et al.* self-screening, relativistic flow solution. If one is prepared to accept the large super-GJ density near the star, then one can clearly argue that retention of self-screening in the single-sign domain keeps the problem with pair production comparatively simple: the newly created electrons and positrons both continue their outward motion, and so a steady or quasi-steady state is possible, with the $E_{\|} = 0$ boundary condition now holding at a sharp pair formation front. This model was studied by Arons and Scharlemann (1979), and developed further by Arons (1981, 1983). If the current is supposed fixed, then the model is stable against fluctuations in the pair-creation rate; for a reduction in pair production leads to a higher $E_{\|}$ and so to a correcting increase in pair production, while an increase causes earlier screening of $E_{\|}$ and so a reduction in pair production.

More recently, Arons (1996) emphasizes that the MT results of Section 14.1.5, incorporating the general relativistic dragging of inertial frames, make self-screening more attractive; for as noted, the modified electric field at the polar cap height is larger than that predicted by Arons *et al.* for the 'towards'-curvature case by a factor $(\Omega R/c)^{-1/2}$, yielding correspondingly larger γ-values. Recall that the g–r modification to the GJ density is negligible far from the star, but in the self-screening solution, it is the boundary condition *at infinity* which fixes the actual current in terms of the GJ current J_0; whereas near the star, the g–r effect reduces by a modest but significant amount the strength of the GJ current, so that (as illustrated by the one-dimensional equation (14.53)), for a given J, the acceleration is more efficient, yielding markedly higher γ-values. And, significantly, the curvature of the field lines does not enter: the extra acceleration due to the g–r reduction in the GJ term occurs near enough to the star for the field lines to be taken as straight. But this still does not in itself answer why the system should opt for the solution with a super-GJ density near the star.

The problem is certainly more difficult for the more general solutions in which $E_{\|}$ is not classically self-screened, for example in flow without a super-GJ density at the star, but along away-curved field lines. It has been tacitly assumed in much of the literature that once efficient pair production has produced a dense $e^{\pm}$-plasma, screening quickly occurs through polarization. In the newly created plasma the electrons are then accelerated further and the positrons decelerated, and once the positrons have become non-relativistic, by continuity their density goes up and so polarization begins. The system may thus be able to create its own 'anode', in Arons's appropriate terminology, and so a steady or quasi-steady model would result without requiring a large super-GJ density near the star. The problem is that the very process of screening by the mutual separation of electrons and positrons can lead to positrons being driven back towards the star into the primary acceleration domain, reducing and perhaps changing the sign of the driving electric field. A recent study by Shibata *et al.* (1998) suggests strongly that pair production is unlikely to be efficient enough to prevent this. This would not necessarily be fatal but certainly complicating – it would imply that without self-screening the system would tend to settle into a non-steady, presumably irregularly oscillatory state. A steady state would perhaps be possible if the relatively streaming electrons and positrons were frictionally coupled via a plasma instability. It is possible that a non-linear relativistic extension of the two-stream instability of Section 2.8 will suffice: Usov (1987, 1988) has argued that although the growth rate is too low if the outflow of $e^{\pm}$-plasma is stationary, if the outflow is strongly non-stationary, conditions are much more favourable for development of the instability.

14.2 The aligned model with pair production

14.2.1 *Models with relativistic acceleration within the l-c*

We now return to the global problem, with most of the analysis confined to the aligned rotator. Recall that the discussion of the tentative classical model in Section 13.2.3 began by postulating a GJ domain extending out to and beyond the l-c. The electrons from the polar caps flow smoothly through the l-c, having acquired only moderate γ-values and so having suffered virtually no dissipation. The condition that there be no singularities at the l-c suffices to fix the (necessarily subluminal) velocities of emission from the star (admittedly in terms of the the **B**-field which is not known a priori but must be constructed by iteration). The merits and the limitations of the classical model have been noted. This chapter has focused attention again on the flow near the star. For the aligned rotator, again with the density close to the GJ-density in a domain surrounding the star, an increase in the emission speed along each field line – or equivalently an increase in the current density – ensures that GJ conditions will then break down before the l-c is reached. (As seen above, in this type of model the g–r correction has only a small local effect and will be ignored.)

It is convenient to restate the arguments in a formalism closer to that used in Chapter 13 and so more appropriate to a global study. Consider again the dimensionless equations (13.108), (13.109) and (13.110) – which incorporate the GJ condition $\nabla^2\psi = 0$ – at points well within the l-c, so that $\rho_e \simeq -2B_z$, $V_{\rm p} = -2V_0 B_{\rm p}/\rho_e \simeq V_0 B_{\rm p}/B_z$, and $\omega x \ll 1$. As before, the current stream function $S(P)$ is related to $V_0(P)$ by $V_0 = -{\rm d}S/2\,{\rm d}P$ by (13.120). Then because of the geometry of the dipolar field, with all field lines *curving away* from the axis, $V_{\rm p}$ *increases outwards* with $B_{\rm p}/B_z$. If $V_0 \lesssim 1$, $V_{\rm p}$ is predicted to reach the value 1 (i.e. $v_{\rm p} \to c$) on a surface S_l quite near to the star. This clearly means that the GJ approximation is breaking down. Beyond S_l, the continuity equation (13.109) with $V_{\rm p} \simeq 1$ now enforces $\rho_e \simeq -2V_0 B_{\rm p}$ to replace $\rho_e \simeq -2B_z$ given by (13.108). Thus the near shorting-out of $\mathbf{E}_\|$ – the essence of the GJ approximation – is no longer possible, because of the limitation $v < c$ imposed by special relativity. The Poisson–Maxwell equation (13.80) (with ωx^2 negligible) now becomes an equation to determine the growth of ψ beyond S_l:

$$\nabla^2\psi \simeq 2B_z\left[-1 + (B_{\rm p}/B_z)/(B_{\rm p}/B_z)_l\right], \tag{14.54}$$

where the suffix l refers to the above-defined surface S_l. The energy–angular momentum integral (13.107), likewise well approximated by $\gamma = \psi/\epsilon_g$, yields the corresponding increase in γ, rapid because of the smallness of the parameter

$$\epsilon_g = \alpha/\omega_g = 7.8 \times 10^{-8} \mathrm{P}_1^2 (B_s/10^{12})^{-1} (R/10^6)^{-3}. \tag{14.55}$$

More precisely, from (13.80), (13.78) and (13.107), with $V_{\rm p}$ written first in terms of γ, with terms in ω again neglected *but with* $\nabla^2\psi$ *retained,*

$$\nabla^2\psi = -2B_z - \frac{j_{\rm p}}{V_{\rm p}} = 2B_z\left[-1 + \frac{\overline{J}}{(1-1/\gamma^2)^{1/2}}\right] \tag{14.56}$$

$$= 2B_z\left[-1 + \overline{J}\frac{\psi/\epsilon_g + \Gamma}{\left\{(\psi/\epsilon_g + \Gamma)^2 - 1\right\}^{1/2}}\right]. \tag{14.57}$$

The variable $\overline{J}$, defined by

$$\overline{J} = -j_{\rm p}/2B_z = V_0 B_{\rm p}/B_z = -(B_{\rm p}/B_z)({\rm d}S/2\,{\rm d}P), \tag{14.58}$$

(differing slightly from J defined in (14.9)), is seen to be the actual current density $j_{\rm p} = \rho_e V_{\rm p} = -2V_0 B_{\rm p}$ (by the definition (13.97) of V_0), normalized in terms of a standard current density in which the local GJ charge-density $-2B_z$ is supposed to move with the speed of light. The surface S_l introduced above has $\overline{J} = 1$, and is notionally defined by putting $\gamma = \infty$ and $\nabla^2\psi = 0$ in (14.56). From the definition (14.58), $\overline{J}$ is seen to increase outwards with $B_{\rm p}/B_z$ along a dipolar field line. Thus within S_l, where $\overline{J} < 1$, there is always a value of γ and so of $V_{\rm p}$ for which the GJ condition $\nabla^2\psi = 0$ is allowed, whereas beyond S_l, $\overline{J} > 1$ and so $\nabla^2\psi > 0$ whatever the value of γ. Since γ must be finite, the transition to the non-GJ domain will actually start before S_l is reached.

We recall also from Section 14.1 that this analysis of the GJ-domain describes just the mean flow, on which there are superposed spatial oscillations of scale close to the plasma wavelength; whereas beyond S_l, (14.57) yields a monotonic increase in ψ and γ. The precise solution for ψ and γ between the star and S_l will depend on conditions at the stellar surface, where for example $\psi' = 0$ may be associated with an emission speed nearer to the (dimensionless) thermal speed than to $\overline{J}$, so that $|\rho_e| > |(\rho_e)_{GJ}|$. Whether these spatial oscillations in ρ_e and γ will turn out to be observationally important is unclear, but the exclusion *a priori* of any non-monotonic behaviour is unjustified, as it would enforce the adoption from *local considerations alone* of $\overline{J} \simeq 1$ just outside the star, with the surface S_l and so also the domain of relativistic acceleration *necessarily* starting at the stellar surface. Our view is that although this is certainly not ruled out, the issue will be resolved on global rather than on purely local grounds. The oscillatory behaviour is replaced by a monotonic increase in ψ and γ near and beyond S_l as $\overline{J}$ passes through unity. Again, the details will depend on the particular boundary conditions, but the asymptotic behaviour is universal (see for example Appendix B by Mestel and Pryce in Mestel *et al.* 1985; Shibata 1991).

We suppose provisionally that a one-dimensional treatment is adequate. We are interested in the attainment of γ-values of up to 10^6–10^7 or so, when again powerful incoherent radiation will occur. Already for values well below this, (14.56) is well approximated (in the appropriate one-dimensional form) by

$$\frac{\mathrm{d}^2\psi}{\mathrm{d}s^2} = 2B_z\left[-1 + \overline{J}\right], \tag{14.59}$$

and it is the increase of $\overline{J}$ above unity beyond S_l which is responsible for at least the initial acceleration to high γ-values (cf. the discussion in Section 14.1.4). Well within the l-c the field is nearly curl-free. For the moment a dipolar form over the stellar surface is still adopted. For all but the millisecond pulsars, dipolar field lines which extend to the l-c will emerge from the polar caps at small polar angles, i.e. with $x^2/(x^2+z^2) \ll 1$. From (13.119), at a point with coordinate x along a field line defined by P,

$$x^2/r^2 = x^2/\left(x^2+z^2\right) = x^{2/3}\left(-P\right)^{2/3} \equiv x^{2/3}Q \tag{14.60}$$

say, and

$$\begin{aligned}\overline{J} &= V_0 B_{\rm p}/B_z = V_0\left[1 - 3x^2/4\left(x^2+z^2\right)\right]^{1/2} / \left[1 - 3x^2/2\left(x^2+z^2\right)\right] \\ &\simeq V_0\left(1 + 9x^{2/3}Q/8\right). \end{aligned}\tag{14.61}$$

With $\overline{J} = 1$ at $x = x_l$, $V_0 \simeq 1 - 9x_l^{2/3}Q/8$, and

$$\frac{\mathrm{d}\overline{J}}{\mathrm{d}x} \simeq (3/4)\, V_0 Q x_l^{-1/3} \simeq (3/4) Q x_l^{-1/3} \tag{14.62}$$

at x_l. The element of length along a field line is $\mathrm{d}s = (2/3Q^{1/2}x^{1/3})\,\mathrm{d}x$, and since from (13.119), $B_z \simeq 2(-P)/x_l^2$, (14.59) transforms with the help of (14.61) into

$$\mathrm{d}^2\psi/\mathrm{d}s^2 = \left[9\,(-P)\,/2r_l^3\right]\left[(s - s_l) + \mathrm{O}\,(s - s_l)^2\right], \tag{14.63}$$

yielding

$$\psi - \psi_l \simeq \left[3\,(-P)\,/4r_l^3\right](s - s_l)^3\,. \tag{14.64}$$

This predicts that from (13.107), particles will be accelerated to the large value

$$\gamma \simeq \psi/\epsilon_g \tag{14.65}$$

after a distance

$$(s - s_l)\,/r_l \simeq \left[4\,(\epsilon_g\gamma)\,/3\,(-P)\right]^{1/3}. \tag{14.66}$$

For typical field lines that reach the l-c, $-P \simeq 1/3 - 1/10$; for definiteness, we take $-P = 1/3$. With substitution from (14.55), (14.66) predicts that only for a short-period pulsar like the Crab, with $\mathrm{P}_1 \simeq 1/30$, will γ reach 10^6–10^7 in a distance $(s - s_l) \ll r_l$, as is assumed in the approximate form (14.63) leading to (14.66).

However, we recall (cf. (14.7)) that if the dimensional $(s - s_l)$ becomes comparable with the lateral scale $D_\perp$, then the term $\nabla_\perp^2\psi$ – again approximated by $-\psi/D_\perp^2$ – becomes important. In the present normalization, by (14.60) the limiting dipolar field line linking the polar cap and the l-c and so passing through the point $(1, 0)$ has $-P = 1$ and so satisfies $x^2 = r^3$; hence near the polar cap, $D_\perp \approx x = r^{3/2}$, and (14.63) is replaced by

$$\frac{\mathrm{d}^2\psi}{\mathrm{d}s^2} = [9(-P)/2r_l^3](s - s_l) + \frac{\psi}{r^3}. \tag{14.67}$$

The solution (14.65) is approximately valid until the two terms on the right of (14.67) are comparable; for the $-P = 1/3$ field line, until $(s - s_l)^2 \approx 6s_l^3$, for which $\psi \equiv \tilde{\psi} = (9/2)(-P)6^{1/2}s_l^{3/2} = 3.7s_l^{3/2}$. After this, provided s stays close to s_l, (14.67) predicts $\psi \approx \tilde{\psi}\exp[(s - s_l)/s_l^{3/2}]$. For a slow pulsar, with $\mathrm{P}_1 = 1$, and with $r_l/R = 1$ (i.e. with the acceleration zone at the star's surface), along a typical field line γ-values $\simeq 10^7$ are reached after a distance $(s - s_l)/s_l \approx 0.14$.

The problem merits a more rigorous treatment; however, in the aligned or nearly aligned model, where virtually all field lines have 'away' curvature, then for pulsar periods down to $\mathrm{P}_1 = 1$, the provisional conclusions are summed up in Fig. 14.7. Outflow of gas with the GJ density that begins as slightly sub-luminal spontaneously converts itself at a surface S_l near the star into a flow

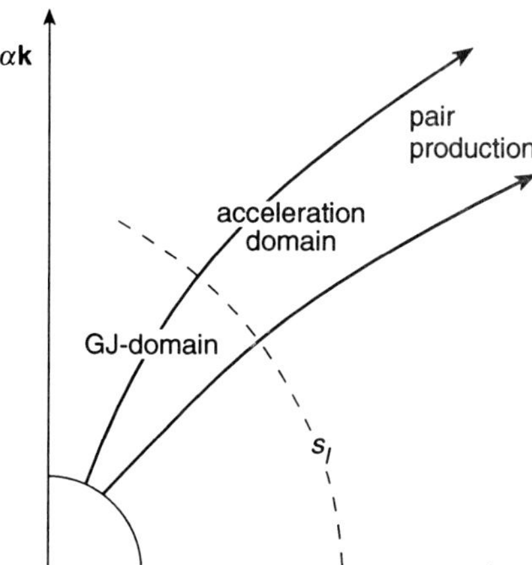

FIG. 14.7. Outflow along field lines curving away from the axis, with the initial speed less than but close to c. The surface S_l separates the GJ domain from the relativistic acceleration domain.

with a slightly super-GJ density moving relativistically. Within a distance that is a fraction of the radius of S_l, the accelerated particles have acquired γ-values sufficiently high for powerful emission of gamma-radiation.

The description is somewhat different on the axis, where $B_{\mathrm{p}}/B_z = 1$ and so the field line curvature effect vanishes. The quantity $\overline{J}$ defined in (14.58) now has the value unity not only on S_l but all the way through the dissipation-free zone; hence $V_0 = 1$, and a GJ domain extending back to the star would have electrons leaving the star with $V_{\mathrm{p}} = 1$ and $\gamma = \infty$. (Similarly, the GJ flows which pass smoothly through the l-c – studied in Section 13.2.3 – have $S = -2P$ as $P \to 0$, so $V_0 = -\mathrm{d}S/2\,\mathrm{d}P \to 1$.) The conclusion is that along the axis, $\nabla^2\psi$ cannot be dropped even when one is studying the mean flow within S_l. In a one-dimensional approximation, (14.56) reduces with the help of (13.98) and (14.58) to

$$\left(1 - 1/\gamma^2\right)^{1/2} \mathrm{d}^2\gamma/\mathrm{d}r^2 = \left(4/\epsilon_g r^3\right) \left[1 - \left(1 - 1/\gamma^2\right)^{1/2}\right], \tag{14.68}$$

showing clearly that the acceleration domain now extends all the way back to the star. The finite rest-mass of the electrons keeps ϵ_g finite (though small), ensuring that no infinite γ-values arise. It is readily shown that (14.68) predicts a rapid increase of γ from unity, with γ' stabilizing at $\simeq \left(\alpha R^3\right)^{-1/2}$. There is a narrow bundle of field lines about the axis, with $-P < \left(\epsilon_g r\right)^{3/4}$, within which the spatial oscillations discussed above have a wavelength comparable with the local value of r. Outside this bundle, the earlier picture survives: if the surface S_l is at a finite distance from the star, then within S_l the domain is one of smooth monotonic GJ flow, with small-scale stationary oscillations superposed (cf. FM*a*, Appendix A).

It should be noted that even if the surface S_l is taken at the star's surface, the model differs from those with a vacuum gap postulated at the star (Beskin *et al.* 1993), with the associated potential jump fixed by matching near the l-c. The Ruderman–Sutherland (1975) *counter*aligned pulsar model is explicit in its appeal to a vacuum gap at the surface, because earlier calculations suggested that *ions* are too tightly bound to be pulled out by the Goldreich–Julian $E_{\parallel}$-component; and it was the subsequent breakdown of this vacuum gap through pair-production that led to their model both for spin-down and for radio emission. There has never been a suggestion of such a strong work-function inhibiting the free flow of *electrons* from the surface; and in fact subsequent work (Flowers *et al.* 1977) has reduced the value of the ion work-function, so that the appropriate boundary condition – for both counteraligned and aligned pulsars – is the vanishing of $\mathbf{E}_{\parallel}$ at the star's surface. This is indeed assumed in the discussion in Section 14.1, with any local rapid increase in ψ related to the local current density.

As already emphasized, the curvature of the field lines away from the axis leads to 'overfilling' of the flux-tubes, in that the relativistic limitation on v forces the density up beyond the GJ value, and it is the consequent $E_{\parallel}$-component which causes monotonic acceleration of the electrons to high γ-values. This is again fully consistent with the result of Scharlemann *et al.* (1978), who show that no solutions exist for field lines curving way from the axis, if the boundary condition $\mathbf{B}\cdot\nabla\psi = 0$ is imposed both at the stellar surface and also far above it (but before pair-creation has set in).

The analysis can be extended to cases with S_l close enough to the l-c for terms in the angular velocity ω to be included, both in the expression for γ and in Poisson's equation (13.80), but with inertial drifts still negligible. Equations (13.104) and (13.97) then yield (Mestel *et al.* 1985)

$$-\frac{\rho_e}{V_0} = 2\left[\frac{\left\{B^2\left(1-x^2-1/\gamma^2\right)+S^2\right\}^{1/2}-S}{1-x^2-1/\gamma^2}\right], \tag{14.69}$$

$$\frac{V_{\mathrm{p}}}{B_{\mathrm{p}}} = \frac{\left\{B^2\left(1-x^2-1/\gamma^2\right)+S^2\right\}^{1/2}+S}{B^2}, \tag{14.70}$$

$$\omega = \frac{-\left(S/x^2\right)\left\{B^2\left(1-x^2-1/\gamma^2\right)+S^2\right\}^{1/2}+B_{\mathrm{p}}^2}{B^2}, \tag{14.71}$$

with

$$B^2 = B_{\mathrm{p}}^2 + B_{\mathrm{t}}^2 = B_{\mathrm{p}}^2 + S^2/x^2; \tag{14.72}$$

whence from (13.80)

$$\nabla^2\psi = -2B_z + 2V_0 SF\left(\gamma, x, \mathbf{B}\right), \tag{14.73}$$

where

$$F(\gamma, x, \mathbf{B}) = \frac{\{(1-x^2)/S\}\{B^2(1-x^2-1/\gamma^2)+S^2\}^{1/2} - 1/\gamma^2}{(1-x^2-1/\gamma^2)} \quad (14.74)$$

in which both V_0 and S are functions of the magnetic flux function P. Much of the earlier discussion can be taken over. Again, the surface S_l, with a GJ domain within and relativistic acceleration beyond, is defined by putting $\nabla^2\psi = 0$ and $\gamma = \infty$ in (14.73) and (14.74), i.e. by

$$\begin{aligned} 2B_z &= 2V_0 S F(\infty, x, \mathbf{B}) \\ &= -(\mathrm{d}S/\mathrm{d}P)\left\{B_{\mathrm{p}}^2(1-x^2) + S^2/x^2\right\}^{1/2} \end{aligned} \quad (14.75)$$

– equivalent to the condition $C = 0$ of Section 13.2.3, with C given by (13.116). When $x \ll 1$, $|S| = x|B_{\mathrm{t}}| \ll xB_{\mathrm{p}}$, and (14.75) reduces to $\overline{J}$ as given by (14.58) put equal to unity, as expected. If $S(P)$ is prescribed, (14.75) fixes the shape of S_l; alternatively, if this curve is prescribed, $S(P)$ follows by integration across field-streamlines.

We shall not be considering further models with S_l so far out that (14.73) (together with (14.74)) rather than (14.56) is the appropriate equation for ψ; for subsequent pair creation will occur only in strong field regions near the star, for which the correcting terms in (14.73) are negligible. Recall, however, that a fully self-consistent classical model as discussed in Section 13.2.5 may require the surface S_l to lie close to but within the l-c, with the flow penetrating the l-c being already dissipative. The constraint (13.117) imposed on GJ flow through the l-c then no longer holds; instead, the constraints imposed on $S(P)$ by the requirement that the trajectories beyond the l-c be acceptable may determine the shape of S_l, through the condition (14.75).

14.2.2 *Pair creation and the generation of a mixed plasma*

Energetic particles forced to move along a curved path – in the present case along the field lines – again suffer copious emission of gamma rays, beamed along $\mathbf{B}$. By the usual synchrotron-type formula, the typical energy of an emitted photon is

$$h\nu_c \simeq 3\gamma^3\hbar c/2R_c \quad (14.76)$$

where R_c is the radius of curvature of the field-streamline. The power emitted is given by (13.131) as

$$\mathcal{P}_c = \left(2e^2/3c^3\right)\gamma^4\left(c^2/R_c\right)^2. \quad (14.77)$$

The crucial new factor, first noted by Goldreich in a conference report and by Sturrock (1971), is that hard gamma rays moving across a very strong magnetic field convert spontaneously into electron–positron ($\pm e$) pairs. The mean-free-path for pair production by a photon with $h\nu > 2mc^2$ is (Erber 1966)

$$\ell \simeq \left[4/\left(e^2/\hbar c\right)\right](\hbar/mc)(B_q/B_\perp)\exp(4/3\chi) \quad (14.78)$$

where

$$B_q = mc^3/e\hbar = 4.4 \times 10^{13}\ \mathrm{G}, \tag{14.79}$$
$$\chi = \left(h\nu/2mc^2\right)\left(B_\perp/B_q\right) \tag{14.80}$$

and $B_\perp$ is the component of $\mathbf{B}$ perpendicular to the photon path.

This new physical effect – the generation of a mixed plasma from high-γ primary electrons – should occur whichever of the acceleration models discussed in Section 14.1 is the appropriate one. The basic picture has been applied to different models in many papers (Ruderman and Sutherland 1975; Arons and Scharlemann 1979; Jones 1980; Arons 1981, 1983, 1992; Daugherty and Harding 1982; Shibata 1991; Mestel and Shibata 1994; Usov 1996 and references therein). As already noted, if classical self-screening is already complete at the pair creation front, then neither the electrons nor the positrons will experience any further acceleration or deceleration by the macroscopic electric field, but will coast out together. But if self-screening is incomplete, then the electrons will be accelerated further and the positrons decelerated. If the secondary electrons are accelerated enough, they themselves will emit gamma rays and so lead to a pair production cascade. As noted, a common presumption is that spontaneous polarization of a sufficiently dense $e^{\pm}$-gas will ultimately screen out the accelerating electric field, so that the cascade must be self-limiting. In the simplest picture there is an outstreaming pair plasma accompanying the high-γ primary electron current.

From now on we return to the aligned or nearly aligned model, with all field lines curving away from the rotation axis. To illustrate the pair production process, we assume that the system has not opted for the g–r model, with a large super-GJ density at the star responsible for the initial acceleration, but instead starts near the star with a near-GJ density, becoming steadily overfull because of the continuity condition and the 'away'-curvature of the field lines, as discussed in Sections 14.1.4 and 14.2.1. However, the discussion of the global problem in the rest of the chapter merely assumes that a dense pair plasma has been created near the star, but will not depend sensitively on the details of the flow and electric fields near the star.

In the notation of Section 14.2.1, consider flow along a typical field line, with particles emitted from the star with speeds close enough to c for relativistic acceleration to begin at a surface S_l that is well within the l-c. For rapid rotators like the Crab pulsar, we can continue to use the one-dimensional approximation. Then from (14.64) in dimensional form,

$$\gamma \simeq \left(e/mc^2\right)\left(3\left(-\overline{P}\right)/8\right)\left(\alpha/c\right)^2 B_l \left(s - s_l\right)^3 \tag{14.81}$$
$$\simeq 3.2 \times 10^{-12}\left(B_s/10^{12}\right)\left(B_l/B_s\right)\left(s - s_l\right)^3/\mathrm{P}_1^2. \tag{14.82}$$

The dimensionless label $-\overline{P}$ in (14.81) for the particular field line considered has again been given in (14.82) the representative value 1/3.

Shibata (1991) and Mestel and Shibata (1994) (following the now classical treatment by Sturrock, Ruderman and Sutherland and others) give an approximate discussion of pair creation, appropriate for aligned or nearly aligned model rapid rotators. A photon emitted near but beyond S_l and moving initially along the local direction of $\mathbf{B}$ will see a perpendicular component $B_\perp$ after travelling the distance

$$L = (B_\perp/B_l)\,R_c \tag{14.83}$$

which on substitution from (14.76) and (14.80) becomes the typical pair creation distance

$$L = (4mc/3\hbar)\,(B_q\chi)\,R_c^2/B_l\gamma^3. \tag{14.84}$$

This distance is to be identified with the mean-free-path ℓ given by (14.78), which, however, depends exponentially on $1/\chi$ and hence on L through $B_\perp$. Thus since χ varies only logarithmically with ℓ, it may be given a standard value $\chi = 1/18$ (Ruderman and Sutherland 1975), whence (14.84) becomes

$$L \simeq 8.5\times 10^{22}\left(r_6^2/B_{12}\right)(R_c/R)^2\left(B_l\big/B_s\right)^{-1}\gamma^{-3} \tag{14.85}$$

(where $B_{12} = B_s/10^{12}$ and $r_6 = R/10^6$).

At least for rapid rotators, a provisional estimate for the upper limit γ_e attained by the primaries is therefore given by inserting L from (14.85) into (14.82):

$$\gamma_e \simeq 5.4\times 10^5\left(r_6^{3/5}/B_{12}^{1/5}\right)(R_c/R)^{3/5}\Big/\left(B_l/B_s\right)^{1/5}\mathrm{P}_1^{1/5}, \tag{14.86}$$

which when inserted back into (14.85) yields

$$L \simeq 5.55\times 10^5\left(r_6^{1/5}/B_{12}^{2/5}\right)(R_c/R)^{1/5}\,\mathrm{P}_1^{3/5}\Big/\left(B_l/B_s\right)^{2/5}. \tag{14.87}$$

The pair creation rate per primary particle is given by

$$\begin{aligned}\mathcal{P}_c/h\nu_c &\simeq \left(4e^2/9\hbar\right)(\gamma/R_c) && (14.88)\\ &= (97/r_6)\left(\gamma/\left(R_c/R\right)\right) \\ &= \left(5.2\times 10^7/r_6^{2/5}B_{12}^{1/5}\right)\Big/(R_c/R)^{2/5}\left(B_l/B_s\right)^{1/5}\mathrm{P}_1^{1/5}. && (14.89)\end{aligned}$$

from (14.86).

From (14.78)–(14.80), pair production is sensitive to changes in the field strength. A dipolar magnetic field falls off by a factor ≈ 2 over a distance $\approx r_l/3$. Provided $L < r_l/3$, a lower limit to the *multiplicity* M – the total number of pairs generated per primary electron – is given by neglecting possible further creation by accelerated secondary electrons:

$$M = \left(\mathcal{P}_c/h\nu_c\right)(r_l/3c) = \left(579r_6^{3/5}/B_{12}^{1/5}\right)(r_l/R)\left[(R_c/R)^{2/5}\left(B_l/B_s\right)^{1/5}\mathrm{P}_1^{1/5}\right]^{-1}. \tag{14.90}$$

Finally, the values γ_+, γ_- of the $e^{\pm}$ pairs are estimated roughly by

$$\gamma_{\pm} = h\nu_c/2mc^2 = (3\hbar/4mc)\,\gamma^3/R_c \tag{14.91}$$
$$= \left(4.5r_6^{4/5}/B_{12}^{3/5}\right)(R_c/R)^{4/5} \Big/ (B_l/B_s)^{3/5}\,\mathrm{P}_1^{3/5}. \tag{14.92}$$

All the formulae depend on the detailed field structure. The simplest case has the field at the star strictly dipolar, with $B_l/B_s \simeq (R/r_l)^3$, and

$$R_c/r \simeq (4/3\sin\theta) = (4/3)\left[(\alpha r/c)\left(-\overline{P}\right)\right]^{1/2}, \tag{14.93}$$

so that near r_l

$$R_c/R = (r_l/R)^{1/2}\left(4/3\left(-\overline{P}\right)^{1/2}\right) \Big/ (\alpha R/c)^{1/2}$$
$$= 69\mathrm{P}_1^{1/2}\,(r_l/R)^{1/2}\left(4/3\left(-\overline{P}\right)^{1/2}\right) \Big/ r_6^{1/2}. \tag{14.94}$$

Then from (14.86) et seq., with $-P = 1/3$ as a typical example,

$$\gamma_e \simeq 1.1\times10^7\left(r_6^{3/10}/B_{12}^{1/5}\right)\mathrm{P}_1^{1/10}\,(r_l/R)^{9/10}, \tag{14.95}$$
$$L \simeq 1.5\times10^6\left(r_6^{1/10}/B_{12}^{2/5}\right)\mathrm{P}_1^{7/10}\,(r_l/R)^{13/10}, \tag{14.96}$$
$$M \simeq 1.2\times10^2\left(r_6^{4/5}/B_{12}^{1/5}\right)(r_l/R)^{7/5}\,\mathrm{P}_1^{-2/5}, \tag{14.97}$$
$$\gamma_{\pm} \simeq 2.6\times10^2\left(r_6^{2/5}/B_{12}^{3/5}\right)(r_l/R)^{11/5}\,\mathrm{P}_1^{-1/5}. \tag{14.98}$$

As stated, the treatment is self-consistent when $L < r_l/3$. From (14.96), when $\mathrm{P}_1 = 1/30$, as for the Crab pulsar, the inequality holds for any relevant value of r_l/R. For longer period pulsars we have seen in any case that a more refined treatment is required.

The estimate (14.95) for γ_e must be compared with (14.15), the maximum possible value $\gamma_m = 1.3\times10^7 B_{12} r_6^3 \mathrm{P}_1^{-2}$ available from the potential difference across the polar cap. The ratio $\gamma_e/\gamma_m \simeq 0.8(1/r_6^{2.7}B_{12}^{1.2})\mathrm{P}_1^{2.1}(r_l/R)^{0.9}$, well below unity when $\mathrm{P}_1 \ll 1$. Although as noted, extrapolation of these formulae to longer P_1 is questionable, nevertheless the correct qualitative conclusions can be drawn. Thus when $\mathrm{P}_1 > 1.5$, the ratio will exceed unity even if S_l is at the star's surface. If the $e^{\pm}$ cascade is essential for the generation of coherent radio emission from pulsars, then as argued by Sturrock, Ruderman and Sutherland and others, there is a clear physical reason for pulsars' shutting off as their periods lengthen to beyond 1 s. A 'live' model with a longer period can spin down, perhaps according to some variation of the classical model discussed earlier, but the transition to a quantum magnetosphere via pair creation will not occur.

The existence of a cut-off is most strikingly illustrated through the familiar P–$\dot{P}$ diagram (Fig. 14.8). The simplest estimate for pulsar spin-down adopts the energy loss formula (13.2) for a perpendicular dipole in vacuo:

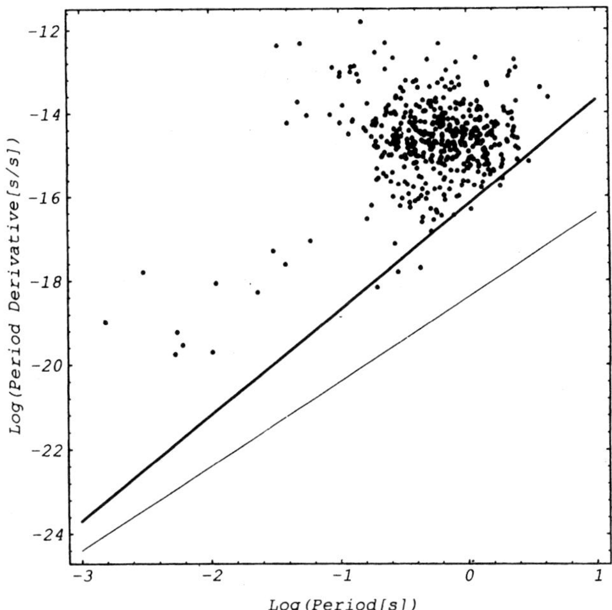

FIG. 14.8. Period versus period derivative diagram for radio pulsars with the theoretical death line. For the thick line, the field line curvature of the pure dipole field is assumed; for the thin line, the curvature radius is assumed to be of order of the stellar radius.

$$-I\alpha\dot{\alpha} = 4\pi^2 I\dot{\mathrm{P}}_1/\mathrm{P}_1^3 = (2\mu_0^2/3c^3)(2\pi/\mathrm{P}_1)^4 \tag{14.99}$$

where $\mu_0 = B_s R^3/2$ is the pulsar magnetic moment and I the moment of inertia. In spite of the restrictive assumptions under which (14.99) is derived, it is likely to be a reasonable estimate for both aligned and non-aligned models with plasma magnetospheres (GJ, FM$a,b$). The comparative insensitivity of $I \approx 10^{45}$ to variations in both the stellar mass M_s and the equation of state means that for a given star, the curve of $\log \dot{\mathrm{P}}_1$ against $\log \mathrm{P}_1$ has just μ_0 as a parameter. The cut-off point is given by putting $\gamma_e = \gamma_m$ on S_l, taken to coincide with the stellar surface. When $\mathbf{B}_\mathrm{p}$ is taken as purely dipolar all the way back to the star, the cut-off curve (with $r_6 = 1$) is then given by

$$\log \dot{\mathrm{P}}_1 \simeq 2.5 \log \mathrm{P}_1 - 16.2. \tag{14.100}$$

This curve is seen from Fig. 14.8 to lie slightly high in the $\mathrm{P}_1 - \dot{\mathrm{P}}_1$ plane. However, Ruderman and Sutherland (1975) argued that near to the star the field structure may very well have higher multipoles present, so that when S_l is near the surface, it is more realistic to take $R_c \simeq R$ rather than the value (14.93) appropriate for a pure dipolar field. This yields $\gamma_e = 5.4 \times 10^5 (r_6^{3/5}/B_{12}^{1/5})/\mathrm{P}_1^{1/5}$ when $r_l = R$. The value of γ_m is unaltered, as it is fixed by the field lines that reach the l-c, where only the dipolar part of the field should still be significant. Again equating γ_e to γ_m, we find for the cut-off line

$$\log \dot{\mathrm{P}}_1 \simeq -18.4 + 2 \log \mathrm{P}_1 \tag{14.101}$$

which lies satisfactorily below the observed points. The substitution of the smaller value $R_c = R$ rather than (14.94) yields somewhat smaller values of L, γ_e and $\gamma_\pm$ but a larger value for M, as compared with (14.95)–(14.98).

The whole problem clearly merits a much more precise treatment. It is a plausible conclusion that beyond the acceleration and pair-production domains, there is a plasma domain in which the outflowing primary electron stream – with $\gamma \simeq \gamma_e$ estimated at least for rapid rotators by (14.95) – is accompanied by a dense, moderate-γ $e^{\pm}$-gas which keeps $E_{\|}$ small. Associated with γ_e is the non-corotational potential

$$\psi_e(P) \simeq \epsilon\gamma_e \tag{14.102}$$

However, we still lack a detailed model of the transition from the mono-sign acceleration domain through the pair production domain into the plasma domain, with the decelerated positrons acting as the anode. As already hinted, the actual solution may be time-dependent. The relation $\psi_e(P)$ is in principle not fully determinable locally, but should emerge as part of a global solution.

The model with a dense $e^{\pm}$-plasma outflowing together with the primary electrons still has to resolve the return current problem. Unless there is a compensating outflow of positive charge, the excess of electrons must ultimately leave the outflow, cross field lines and return to the star. Within the l-c, the poloidal current will be described as in (13.78) by the stream function $S \simeq S(P)$, since inertial and dissipative drifts will be modest:

$$\mathbf{j}_{\mathrm{p}} = (\mathrm{d}S/\mathrm{d}P)\mathbf{B}_{\mathrm{p}} \tag{14.103}$$

with $\mathrm{d}S/\mathrm{d}P$ vanishing on the field line $\tilde{P}$ separating the outflowing electrons from the inflowing. The above discussion shows how electrons leaving the star with speeds near c may generate an $e^{\pm}$-stream accompanying the primary stream: the outflow part of the magnetosphere is converted into a plasma wind carrying a negative current. At first sight, the inflow domain would appear to consist of a pure electron gas. However, as remarked at the end of Chapter 13, the returning electrons are likely to be highly energetic, and may very well generate a plasma by spallation at the stellar surface and by pair production in the strong magnetic field near the star. The trans-field components of random motion will again be rapidly radiated away, but the plasma should expand under its effective one-dimensional pressure. For the rest of the discussion this is assumed to be the case. This new pulsar magnetosphere model (Fig. 14.10) in some ways looks like the magnetosphere of a normal star as discussed in Chapters 7 and 8, but there remain crucial differences. The field lines closing within the l-c form a dead zone separated into electron and ion (or positron) zones. Field lines crossing the l-c form a wind zone, with an electron current flowing out along field lines between the axis and the line $\tilde{P}$, and a return electron current between $\tilde{P}$ and the boundary P_c of the dead zone.

14.3 The outflowing plasma

The effect of the acceleration near and beyond S_l of the outflowing electrons, followed by copious pair production, is to convert the single-sign stream into a three-fluid plasma consisting of high-γ primary electrons and a dense, moderate-γ electron–positron gas (given respectively the suffixes e and +,−). The whole gas will begin by behaving rather like a normal, non-relativistic plasma, with the non-corotational potential $\psi(P)$ again nearly constant on each field-streamline, its value given by (14.102). As long as inertial drifts are small, the zero-order flow of all three components is described by the quasi-MHD approximation: in dimensionless form, e.g. for the high-γ electrons,

$$\mathbf{V}_e = \kappa_e \mathbf{B} + \tilde{\alpha} x \mathbf{t}, \tag{14.104}$$

where

$$\tilde{\alpha}(P) = 1 - \mathrm{d}\psi/\mathrm{d}P, \tag{14.105}$$

$$\begin{aligned} \mathbf{E} &= -x\,\mathbf{t}\times\mathbf{B} - \nabla\psi = \tilde{\alpha}\nabla P \\ &= -\tilde{\alpha}x\,\mathbf{t}\times\mathbf{B} = -\nabla\phi, \end{aligned} \tag{14.106}$$

$$\phi = -P + \psi. \tag{14.107}$$

Besides $\mathbf{B}_\mathrm{p}$, there is again the toroidal component $\mathbf{B}_\mathrm{t} = (-S/x)\mathbf{t}$, generated by the poloidal currents with the stream function $S \simeq S(P)$ as in the GJ domain within S_l and in the subsequent acceleration and pair production domains. An important change in the formalism from that developed in Section 13.2.2 is that ρ_e is now the sum of contributions from the three gases. In dimensionless form

$$\nabla \cdot \mathbf{E} = \rho_e = -\left(n_e + n_- - n_+\right); \tag{14.108}$$

and likewise $\mathbf{j}$ is no longer constrained to be the convection current $\rho_e\mathbf{V}$ but is the sum of three parts, each satisfying its own equation of continuity.

Superposed on the zero-order velocity (14.104) there will be small additional motions of each component. Perpendicular to $\mathbf{B}$ are the inertial drifts which yield a net Lorentz force density. As in the charge-separated theory of Section 13.2.2(b), in discussing the energy and angular momentum flow, account is taken implicitly of inertial drifts, even when they have little effect on the geometry of the flow. For the moment assume there is no frictional coupling between the three components, which therefore interact only through the large-scale electromagnetic fields. Then for each component there are the energy and angular momentum integrals (13.84) and (13.87). In dimensionless form the primary and secondary electrons satisfy (with the appropriate suffices)

$$\mathbf{V} \cdot \nabla (\gamma - \phi/\epsilon_g) = 0 \tag{14.109}$$

and

$$\mathbf{V} \cdot \nabla \left(P + \epsilon_g \gamma \omega x^2\right) = 0, \tag{14.110}$$

and the positrons the same equations with $-\epsilon_g$ substituted. For all three components, (14.109) and (14.110) combine with the ϕ, P, ψ and $\tilde{\alpha}$ relations in (14.105) and (14.107) to yield

$$\gamma \left(1 - \tilde{\alpha}\omega x^2\right) = \text{constant on a field-streamline.} \tag{14.111}$$

Thus each component increases in γ from its post-pair-production value according to the familiar relativistic slingshot formula. Inertial and dissipative drifts would however become large if $\tilde{\alpha}\omega x^2 \to 1$ (see below).

Both the continuity equation for each component and the Poisson–Maxwell equation must still be satisfied. This is brought about by small variations along $\mathbf{B}_{\rm p}$ of the non-corotational potential ψ, suppressed in the quasi-MHD approximation (14.104). Thus to next order, (14.111) for each component is replaced by the analogue of (13.107):

$$\begin{aligned} \epsilon_g \gamma_\pm \left(1 - \tilde{\alpha}\omega_\pm x^2\right) \pm \psi' &= \text{constant on a streamline,} \\ \epsilon \gamma_e \left(1 - \tilde{\alpha}\omega_e x^2\right) - \psi' &= \text{constant on a streamline} \end{aligned} \tag{14.112}$$

where ψ' is this extra contribution, over and above the ψ-contribution absorbed into the isorotational but non-corotational angular velocity $\tilde{\alpha}$. The discussion is simplest at points deep enough within the l-c for the slingshot terms in (14.112) to be small and likewise for the rotational velocity ωx to be small compared with $V_{\rm p}$. One can then show that when the multiplicity factor $M \gg \gamma_\pm^2$, then the adjustment of the densities of the three components to conform to Poisson's equation occurs with hardly any change in γ_e and only modest changes in $\gamma_\pm$. The associated variations in ψ' along a field-streamline remain small compared with the corotational potential $-P$.

Recall that the domain $P < \tilde{P}$, which carries the return current, is pictured also as having outflowing plasma, generated by the high-γ electrons returning to the regions near the stellar surface. The zero-order non-corotational potential is not given by (14.102) – which arises from the combined effects of outward acceleration and pair production in the domain $0 > P > \tilde{P}$ – but has just the uniform value ψ_s of the rigidly rotating star. The small deviations from ψ_s can be estimated as above.

The analysis is more complicated in regions near and beyond the l-c for which the terms in ω and B_ϕ cannot be neglected. This is most simply illustrated by ignoring the small differences in velocity between the three components and using (14.104) with $V_e^2 \simeq V^2 \simeq 1$ for relativistic flow, and (13.74) for B_ϕ:

$$\mathbf{V}^2 = \tilde{\alpha}^2 x^2 + \kappa^2 \mathbf{B}^2 + 2\kappa\tilde{\alpha} x B_\phi = \kappa^2 \mathbf{B}^2 - 2\kappa\tilde{\alpha} S + \tilde{\alpha}^2 x^2 = 1. \tag{14.113}$$

The quadratic (14.113) has the solution

$$\begin{aligned}\kappa &= \left\{\tilde{\alpha}S + \left[\tilde{\alpha}^2S^2 - \mathbf{B}^2\left(\tilde{\alpha}^2x^2 - 1\right)\right]^{1/2}\right\}\Big/B^2 \\ &= \left\{\tilde{\alpha}S + \left[B_{\rm p}^2\left(1 - \tilde{\alpha}^2x^2\right) + S^2/x^2\right]^{1/2}\right\}\Big/B^2.\end{aligned} \tag{14.114}$$

The positive sign before the radical is mandatory for continuous flow, as the negative sign would yield κ negative near the star, implying inflow rather than outflow. Breakdown occurs if the discriminant in (14.114) vanishes, i.e. if

$$B_{\rm p}^2\left(1 - \tilde{\alpha}^2x^2\right) + S^2/x^2 = 0. \tag{14.115}$$

From (14.104), $\omega x = \kappa B_\phi + \tilde{\alpha}x = -\kappa S/x + \tilde{\alpha}x$; hence

$$\kappa S \to \tilde{\alpha}x^2 - 1/\tilde{\alpha} \tag{14.116}$$

as $\tilde{\alpha}\omega x^2 \to 1$. Substitution of (14.116) into the quadratic (14.113) yields that simultaneously

$$\kappa = \left(\tilde{\alpha}^2x^2 - 1\right)^{1/2}/B_{\rm p}\tilde{\alpha}x, \tag{14.117}$$

so that from (14.116)

$$S = \left(\tilde{\alpha}^2x^2 - 1\right)^{1/2}xB_{\rm p} \tag{14.118}$$

and the discriminant in (14.114) vanishes. Also, from (14.106)

$$E^2/B^2 = \tilde{\alpha}^2x^2B_{\rm p}^2/\left(B_{\rm p}^2 + S^2/x^2\right) = 1. \tag{14.119}$$

Thus we have that the ultimate breakdown in the flow along a particular field line, with $\gamma \to \infty$, would occur at the point – necessarily beyond the l-c – when (14.116) holds and simultaneously $E \to B$. The breakdown is analogous to that found for charge-separated flow when C given by (13.116) vanishes, so that (13.114) yields infinite γ. A significant difference is shown by (13.121). In the charge-separated problem, the vanishing of C still leaves $E < B$ (unless B_z is simultaneously zero), and as a consequence there is still a dissipative domain in which the electron gas continues to move essentially along $\mathbf{B}_{\rm p}$ until the equatorial domain is reached with $B_{\rm t} \to 0$ and $E > B$. In the present problem, however, the limit to the flow (14.104), with $\gamma \to \infty$, is simultaneous with $E \to B$.

The position of the breakdown point along a given field-streamline is clearly to some extent model-dependent. So far the structure of $\mathbf{B}_{\rm p}$ has been taken as given. When γ is moderate the inertial terms are small, and (14.104) and (14.106) yield the appropriate form of the relativistic force-free equation: in dimensional form

$$\rho_e\mathbf{E} = e\left(-n_e - n_- + n_+\right)\left(-\tilde{\alpha}\alpha\tilde{\omega}\mathbf{t}\times\mathbf{B}/c\right)$$

$$= -\left[e\left(-n_e\right)\left(\mathbf{v}_e - \kappa_e\mathbf{B}\right)\times\mathbf{B} + \ldots +\right] = -\mathbf{j}\times\mathbf{B}/c. \qquad (14.120)$$

On use of Maxwell's equations, the form (14.106) for $\mathbf{E}$ and $\mathbf{B}_\mathrm{t} = -(S/x)\mathbf{t}$, this reduces to the dimensionless form

$$\nabla^2 P\left(1 - \tilde{\alpha}^2 x^2\right) - 2\,\partial P/x\,\partial x + S\,\mathrm{d}S/\mathrm{d}P - \tilde{\alpha}\left(\mathrm{d}\tilde{\alpha}/\mathrm{d}P\right)x^2\left(\nabla P\right)^2 = 0. \quad (14.121)$$

When $S = 0$ and $\tilde{\alpha} = 1$ – zero poloidal current and corotation with the star – (14.121) reduces to (13.27) discussed in Section 13.1.3:

$$\nabla^2 P\left(1 - x^2\right) - 2\,\partial P/x\,\partial x = 0. \qquad (14.122)$$

Recall that when (14.122) is extrapolated to apply beyond the l-c, with the constraints that both B_x and B_z are continuous – i.e. with no current sheets on the l-c – it was found that the solution cannot be extended to infinity, but instead it becomes singular at $x = 2$. The reason is the existence of a singular surface, defined by the vanishing of the factor multiplying the highest derivatives $\nabla^2 P$. Extrapolation of the corotating model beyond the l-c is clearly illicit, as already remarked. However, numerical experiments with (14.121), with $\tilde{\alpha}$ put equal to unity (for simplicity) and illustrative forms for the $S(P)$ relation show the same phenomenon: in solutions extended beyond the l-c, continuous in all three components of $\mathbf{B}$, the transform functions $f_k(x)$ and the Fourier back-transform begin to diverge typically at $x_{\mathrm{gr}} \simeq 1.4$ (cf. Fig. 14.9). The breakdown in the flow – indicated by $\gamma \to \infty$, because of $E/(B_\mathrm{p}^2 + B_\mathrm{t}^2)^{1/2}$ approaching unity – occurs on each field line very close to the point where the field begins to diverge, so that clearly one must consider next the appropriate replacement of the force-free condition (14.121).

These results appear to be qualitatively valid for a general choice of the poloidal current generating function $S(P)$. A recent paper (Contopoulos *et al.* 1998) inverts the problem, constructing numerically the special form for $S(P)$ which, when inserted into the force-free equation (14.121) (again with $\tilde{\alpha} = 1$), yields a function P that is both smoothly continuous at the l-c and well-behaved all the way to infinity. Within the l-c, the constructed field again has a wind zone, with field lines crossing the l-c, and a dead zone with closed field lines. Most of the return current flows along the boundary between the two zones, yielding a discontinuity in B_ϕ and so requiring a discontinuity in $\mathbf{B}_\mathrm{p}$ along the boundary to ensure continuity of the total electromagnetic pressure. Beyond the l-c the field is everywhere open, becoming quasi-radial at infinity, while at the equator there are discontinuities in both $\mathbf{B}_\mathrm{p}$ and B_ϕ, so requiring a plasma pressure to balance the total magnetic pressure. Thus this dissipationless solution is force-free everywhere outside the star except at this pinched equatorial sheet. Significantly, the authors also find that in their solutions the fields never satisfy the condition (14.118), so that γ stays finite, and the force-free assumption remains prima facie valid.

The Contopoulos *et al.* work is clearly an important extension of our knowledge of the allowed solutions of the perfectly conducting, relativistic force-free

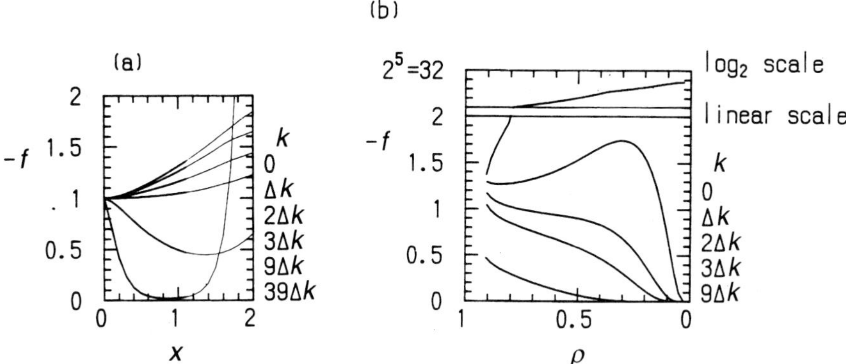

FIG. 14.9. Fourier cosine transform $f(x,k)$ of the flux function $P(x,z)$. In (a), $-f$ is plotted against x in the force-free domain ($0 \leq x \leq 1.1$) for different k-values (0, Δk, $2\Delta k$, $3\Delta k$, $9\Delta k$, $39\Delta k$, where $\Delta k = 0.196$). The poloidal current is given by $S(P) = -2P(1 - P/P_d)$. The thin curves indicate extrapolation of the force-free solution showing exponential growth beyond $x = x_{\text{gr}} \sim 1.2 - 1.4$. The associated flux function P shows violent spatial oscillations as x_{gr} is approached. In (b), $-f$ is plotted against $\rho = 1/x$ for the dissipation-free inertial wind domain (see Section 14.4). For each k, f is made continuous with the corresponding force-free solution, but to yield a non-singular open field structure to infinity, there must in general be a jump in $\partial f/\partial x$.

equation. However, one wonders whether in a magnetically dominant system – as is implied by the force-free assumption – the field would persist in a quasi-radial structure, which is associated rather with a field pulled out by a quasi-radial wind in a kinetically dominant system. It can be argued that near the l-c, the field will adopt a more general structure which does allow the γ-values of the centrifugally-driven, quasi-MHD flow to become large, pulling the field lines out and limiting the tendency of a force-free structure to reconnect across the equator, but possibly at the cost of requiring a domain in which dissipation enforces a local departure from the perfect conductivity condition.

It was seen in Section 13.2.3 that in a pure electron gas, for inertial effects to become important, γ must approach $\gamma_m = 1/\epsilon_g = \omega_g/\alpha$, and simultaneously $|\psi|$ becomes comparable with $\alpha |P| /c$. Even for slowly rotating pulsars, incoherent radiation of energy and angular momentum is already important for such γ-values, and indeed for the faster pulsars, the motion of high-γ particles is adequately described by (13.142) et seq., in which inertial terms are dropped compared with frictional drag. By contrast, in the present problem, with a dense $e^{\pm}$-plasma of multiplicity M as well as the primary high-γ electrons, the neglect of inertia ceases to be valid when $2Mn_e\gamma mc^2 \simeq B^2/8\pi$, i.e. for $\gamma \simeq (\omega_g/\alpha)/8M$, which because of the large factor M is much below the value γ_d given by (13.139) at which incoherent radiation would rapidly drain the particle energies. This does not mean that dissipation is negligible (cf. Section 14.5 and Appendix), but it lends support to the proposal that much – probably the bulk – of the energy and angular momentum loss from the star is carried by a quasi-neutral relativistic wind, with only a fraction – though an observable fraction – being lost as high-γ photons.

14.4 Dissipationless, non-force-free flow

The obvious next step is to derive the appropriate modification of (14.121) that incorporates inertia while still ignoring dissipative processes. In dimensional form (cf. (13.82)), the bulk poloidal equation of motion of the gas in a steady state is

$$\sum_{\mathrm{s}} \left[n_{\mathrm{s}} \nabla \left(\gamma_{\mathrm{s}} m_{\mathrm{s}} c^2 \right) - n_{\mathrm{s}} \mathbf{v}_{\mathrm{sp}} \times \left\{ \nabla \times \left(\gamma_{\mathrm{s}} m_{\mathrm{s}} \mathbf{v}_{\mathrm{sp}} \right) \right\} - n_{\mathrm{s}} \mathbf{v}_{\mathrm{st}} \times \left\{ \nabla \times \left(\gamma_{\mathrm{s}} m_{\mathrm{s}} \mathbf{v}_{\mathrm{st}} \right) \right\} \right]$$
$$-\mathbf{j}_{\mathrm{p}} \times \mathbf{B}_{\mathrm{t}}/c - \mathbf{j}_{\mathrm{t}} \times \mathbf{B}_{\mathrm{p}}/c - \rho_e \mathbf{E} = 0, \qquad (14.123)$$

where the suffix s covers the high-γ primary electrons and the secondary electron–positron pairs. As in the familiar two- or three-fluid models of non-relativistic plasmas (Chapter 2), the near cancellation of the positive and negative charge densities – which ensures that $\rho_e = \nabla \cdot \mathbf{E}/4\pi$ is satisfied – yields $|\rho_e|/n_{\pm} \simeq 1/M \ll 1$, where M is the large multiplicity factor. Thus in the bulk equation (14.123), the electric force density and also the part of the magnetic force density due to the convection current will be much reduced, so that for moderately high γ-values the inertial terms in (14.123) will be comparable with the electromagnetic, as surmised. The small trans-field inertial drifts are included implicitly in the magnetic force term, yielding a significant Lorentz body force.

As long as the inertial (and dissipative) terms are small compared with the Lorentz force per particle, the steady state equation for each species alone is again well approximated by

$$\mathbf{v}_{\mathrm{s}} = \kappa_{\mathrm{s}} \mathbf{B} + \beta \tilde{\omega} \mathbf{t}, \qquad \mathbf{E} = -\beta \tilde{\omega} \mathbf{t} \times \mathbf{B}/c. \qquad (14.124)$$

The function β satisfies the isorotation condition $\beta(P)$, as required by the vanishing of $\nabla \times \mathbf{E}$; in general, it differs from the uniform rotation α of the star if $\mathbf{E} \cdot \mathbf{B} \neq 0$ on any part of the field line linking the star with the point considered. Thus just beyond the acceleration and pair production domains, $\beta = \alpha\tilde{\alpha}$ with $\tilde{\alpha}$ fixed by (14.105) and (14.102). In the model developed below there is a further dissipation domain, with non-zero $\mathbf{E} \cdot \mathbf{B}$, beyond which the flow is again dissipationless but with a new $\beta(P)$.

The continuity equation allows the definitions

$$n_{\mathrm{s}} \kappa_{\mathrm{s}} = \eta_{\mathrm{s}} (P). \qquad (14.125)$$

In dimensional form, the energy–angular momentum integral (14.111) for each species when multiplied by (14.125) becomes

$$\eta_{\mathrm{s}} \gamma_{\mathrm{s}} m_{\mathrm{s}} c^2 \left(1 - \tilde{\alpha} \Omega_{\mathrm{s}} \tilde{\omega}^2/c^2\right) \equiv \epsilon_{\mathrm{s}} (P). \qquad (14.126)$$

Likewise, summation over the dimensional form of the angular momentum equation for each species yields (cf. (14.110))

$$\sum n_{\rm s}\mathbf{v}_{\rm s}\cdot\nabla\left(\gamma_{\rm s}m_{\rm s}\Omega_{\rm s}\tilde{\omega}^2\pm eP/c\right)=0 \tag{14.127}$$

(the plus sign referring to electrons, the minus to positrons). The second term in (14.127) is just $-\mathbf{j}_{\rm p}\cdot\nabla P/c=-\tilde{\omega}(\mathbf{j}_{\rm p}\times B_{\rm p}/c)\cdot\mathbf{t}=-\nabla\cdot(\tilde{\omega}B_\phi\mathbf{B}_{\rm p}/4\pi)$ as usual, so that (14.127) becomes with use of the continuity equation (14.125) for each species

$$\nabla\cdot\left(\sum n_{\rm s}\gamma_{\rm s}m_{\rm s}\Omega_{\rm s}\tilde{\omega}^2\mathbf{v}_{\rm s}-\tilde{\omega}B_\phi\mathbf{B}_{\rm p}/4\pi\right)=0. \tag{14.128}$$

With inertial drifts small, the quasi-MHD approximation (14.124) may again be substituted, yielding

$$\begin{aligned}-\tilde{\omega}B_\phi/4\pi&+\sum\gamma_{\rm s}m_{\rm s}\Omega_{\rm s}\tilde{\omega}^2\eta_{\rm s}\\&=\left(-\tilde{\omega}B_\phi/4\pi\right)\left(1-\sum 4\pi\gamma_{\rm s}m_{\rm s}\eta_{\rm s}\kappa_{\rm s}\right)+\beta\tilde{\omega}^2\sum\gamma_{\rm s}m_{\rm s}\eta_{\rm s}\\&=\ell\left(P\right)\equiv\left(\tilde{\omega}_{\rm c}B_{\rm lc}/4\pi\right)\bar{\ell}(P),\end{aligned} \tag{14.129}$$

where $\bar{\ell}(P)$ is the dimensionless form. Summation of (14.126) for the three species when combined with (14.129) then yields

$$-\beta\tilde{\omega}B_\phi/4\pi+\sum\gamma_{\rm s}m_{\rm s}\eta_{\rm s}c^2=\beta\ell\left(P\right)+\sum\varepsilon_{\rm s}\left(P\right) \tag{14.130}$$

$$=\epsilon(P)\equiv\left(\alpha\tilde{\omega}_{\rm c}B_{\rm lc}/4\pi\right)\bar{\epsilon}\left(P\right) \tag{14.131}$$

as the bulk energy integral. This is immediately recognizable as the joint flow along a unit poloidal flux tube of the total kinetic energy plus the Poynting flux of electromagnetic energy $(c/4\pi)(\mathbf{E}\times\mathbf{B})=(c/4\pi)(-(\beta\tilde{\omega}/c)\mathbf{t}\times\mathbf{B}_{\rm p})\times B_\phi\mathbf{t}=-\left(\beta\tilde{\omega}B_\phi/4\pi\right)\mathbf{B}_{\rm p}$. After some manipulation, (14.123) reduces to

$$\begin{aligned}&\Big[\nabla^2P\left(1-\beta^2\tilde{\omega}^2/c^2\right)-\beta\beta'\left(\nabla P\right)^2/4\pi c^2-2\partial P/\tilde{\omega}\partial\tilde{\omega}\\&-\sum\left(4\pi\gamma_{\rm s}m_{\rm s}\eta_{\rm s}\kappa_{\rm s}\right)\left(\nabla^2P-2\partial P/\tilde{\omega}\partial\tilde{\omega}\right)-\nabla P\cdot\nabla\left(\sum 4\pi\gamma_{\rm s}m_{\rm s}\eta_{\rm s}\kappa_{\rm s}\right)\\&+\sum\eta_{\rm s}\left(\log\eta_{\rm s}\right)'\left(4\pi\gamma_{\rm s}m_{\rm s}\kappa_{\rm s}\right)\left(\nabla P\right)^2\Big]/4\pi\tilde{\omega}^2-B_\phi\ell'/\tilde{\omega}\\&+\sum n_{\rm s}\left[\varepsilon_{\rm s}\left(\log\varepsilon_{\rm s}\right)'/\eta_{\rm s}+\gamma_{\rm s}m_{\rm s}v_{s\phi}\beta\tilde{\omega}\left(\log\beta\right)'\right.\\&\left.-\gamma_{\rm s}m_{\rm s}c^2\left(1-v_{s\phi}^2/c^2\right)\left(\log\eta_{\rm s}\right)'\right]=0,\end{aligned} \tag{14.132}$$

where primes are derivatives with respect to P, and where use has been made of (14.125), (14.126) and (14.129).

Consider now the flow beyond the pair production domain, with $\beta=\tilde{\alpha}\alpha$. Initially the flow is essentially force-free, with the Poynting flux the dominant mode of energy transport, so that $\varepsilon_{\rm s}(P)$ given by (14.126) is small (in fact zero in the zero-inertia limit). As inertia becomes important, $\varepsilon_s(P)$ stays constant on a field-streamline as long as the flow remains dissipationless, so the term $\sum n_s(d\varepsilon_s/dP)/\eta_s$ can be safely dropped from (14.132). And although η_s and $\tilde{\alpha}$ will also not be strictly the same constants on different streamlines, one may reasonably begin by ignoring their variation, so that (14.132) simplifies to

$$(1 - \tilde{\alpha}^2\varpi^2/c^2 - M_A^2)\nabla^2 P - 2(1 - M_A^2)(\partial P/\varpi\partial\varpi)$$
$$-\nabla P \cdot \nabla M_A^2 - 4\pi(\varpi B_\phi)dl/dP = 0, \qquad (14.133)$$

where the dimensionless quantity M_A – an Alfvénic Mach number – is defined by

$$M_A^2 \equiv \sum 4\pi\gamma_s m_s \eta_s \kappa_s. \qquad (14.134)$$

The dimensionless forms of (14.133) and (14.134) are

$$(1 - \tilde{\alpha}^2 x^2 - M_A^2)\nabla^2 P - 2(1 - M_A^2)\left(\frac{1}{x}\frac{\partial P}{\partial x}\right) - \nabla(M_A^2)\cdot\nabla P + S\frac{dl}{dP} = 0, \quad (14.135)$$

and

$$M_A^2 = \epsilon_g \sum(\gamma_s \eta_s^2/n_s). \qquad (14.136)$$

where as before $\epsilon_g = \alpha/\omega_g$. For each of the three species, the dimensionless $\overline{n}_s \equiv (4\pi\varpi_c e/B_{lc})n_s$. The dimensionless forms of (14.129) and (14.130) become

$$\bar{l} = S(1 - M_A^2) + \epsilon_g \tilde{\alpha} x^2 \sum \gamma_s \eta_s, \qquad (14.137)$$

and

$$\bar{\epsilon} = \tilde{\alpha} S + \epsilon_g \sum \gamma_s \eta_s$$
$$= \tilde{\alpha}\bar{l} + \tilde{\alpha} S M_A^2 + \epsilon_g(1 - \tilde{\alpha} x^2)\sum \gamma_s \eta_s. \qquad (14.138)$$

As already seen, the dissipationless, quasi-MHD flow described by (14.124) breaks down on a surface beyond the l-c when $\gamma \to \infty$. (From (14.111), this implies that $\tilde{\alpha}\omega x^2 \to 1$, and it is easy to check that this is equivalent to the limiting condition (14.116).) It was noted also that the *force-free* equation (14.121) (with $\tilde{\alpha}$ put equal to unity), when solved subject to continuity of P and P_x at the l-c, yields a flux function P which blows up on a surface close to that on which the breakdown occurs in the quasi-MHD flow. The essential mathematical reason is that the l-c $x = 1$ is a singular surface of the force-free equations, so that for each Fourier component only one of the two independent solutions near the l-c is non-singular. A solution that is everywhere finite beyond the l-c has enough freedom to satisfy continuity of $B_x \propto P_z$ but not also of $B_z \propto P_x$; enforcement of continuity at the l-c in general yields breakdown on some surface further out.

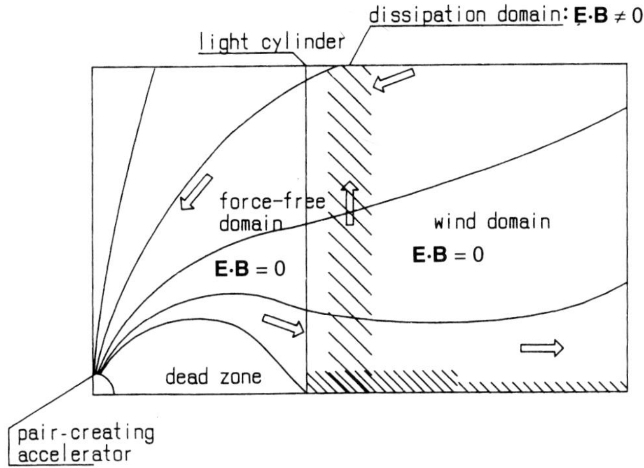

FIG. 14.10. Schematic magnetospheric model.

It is now seen that inclusion of inertia leads to a field equation (14.132) or (14.135) of similar form: there is again a singular surface, $(1 - \tilde{\alpha}^2 x^2 - M_{\mathrm{A}}^2) = 0$ instead of $(1 - \tilde{\alpha}^2 x^2) = 0$. It is possible that there exists a solution which is both smoothly continuous and everywhere regular – a high-γ generalization of the perfectly-conducting, force-free solution of Contopoulos *et al.* summarized in Section 14.3. The presence of the inertial term M_{A}^2 may assist by allowing the singular surface to adjust its shape. We shall instead discuss solutions for which the equations and boundary conditions require there to be a domain of dissipative flow. (The problem has a distant analogue in Riemann's classical work in gas dynamics, where the dissipation-free equations spontaneously predict breakdown in the flow, leading to shock formation.)

14.5 A magnetospheric model

The new magnetospheric model is pictured as in Fig. 14.10. There is the first domain of dissipation-free, quasi-MHD outflow of the wind consisting of the primary electrons plus the dense $e^{\pm}$-gas, that extends slightly beyond the l-c. Further out, there is a second such domain, but with a different set of parameters for the isorotation law and for the transport of energy and angular momentum. Separating them is a dissipation domain in which there is radiation of energy and angular momentum. This domain will in reality be of finite thickness, allowing for a strong local field-aligned component $\mathbf{E}_{\parallel}$ and for an associated deviation from isorotation. However, in this first phenomenological study, the dissipation domain is treated as a thin cylindrical layer $x = \tilde{x}$, but differing from an MHD shock in that there is a tangential discontinuity in $\mathbf{E}$. Thus ahead of the dissipation layer, $\mathbf{E}$ in dimensional form is given by $\mathbf{E} = -\tilde{\alpha}(P)\alpha\varpi\mathbf{t} \times \mathbf{B}/c$, where the isorotation function $\tilde{\alpha}(P)$ is again either fixed by the flow through the relativistic acceleration/pair production domains, or is unity, as discussed in Sections 14.2 and 14.3. After passing through the layer and entering the outer MHD domain, the flow feels the electric field $\mathbf{E} = -\tilde{\beta}(P)\alpha\varpi\mathbf{t} \times \mathbf{B}_{\mathrm{p}}/c$ (where $\beta = \tilde{\beta}\alpha$). The dimensionless isorotation *function* $\tilde{\beta}(P)$ should strictly emerge from the solution for the global structure; we, however, approximate by taking $\tilde{\beta}$ to be an adjustable *constant* less than unity. The energy and angular momentum integrals (14.130) and (14.129) hold with l and ϵ related to the corresponding quantities to the left of the dissipation layer by jump conditions, to be treated in the Appendix. A velocity field of the same form (14.124) holds in the outer domain. Ahead of the dissipation layer the field is approximately force-free; to the right, it satisfies the disspation-free but non-force-free equation (14.133). Clearly, there is a constraint on the value of β and on the current function S that not only must $\mathbf{B}_{\mathrm{p}}$ show no singularities, but that also there must be no similar breakdown in the flow as found in the absence of a dissipation layer.

Details of the first attempt to construct self-consistent models of this type are to be found in Mestel and Shibata (1994). Fig. 14.11 displays a typical constructed field. The Appendix below summarizes the procedure and lists predicted quantities of interest: the normalized wind angular velocity $\tilde{\beta}$ and the associated potential variation across the dissipation layer; the effect of the layer on the fractional Poynting flux; and the energy radiated from the dissipation layer. Other points of interest are the suggestion of some degree of flow collimation along the rotation axis, and a domain of closed magnetic field lines beyond the l-c. It is as yet unclear whether these latter are artefacts of some of the simplifying assumptions made, or whether they will persist in an accurately constructed model.

14.5.1 *Discussion*

The essential features of pulsar activity are the microwave emission, which is believed to be amplified in pair plasmas penetrated by primary particle beams near the star (e.g. Ruderman and Sutherland 1975), the high-energy pulsed emissions, which probably originate in the vicinity of the l-c, and the primary spin-down process, believed by most workers to be via a relativistic wind. The present tentative model incorporates pair production and a consequent relativistic wind. It also predicts radiation losses from the dissipation layer, which would appear as hard photon emission, of about a few per cent – possibly going up to about 10 per cent – of the total luminosity, depending on the model parameters. The relativistic slingshot effect will ensure that the primary electrons will again achieve γ-values large enough for significant radiative losses to occur: the strong γ-dependence of incoherent radiative damping enables the primary electrons to act as self-adjusting rheostats, with the radiation again in the gamma- and X-ray domains.

Recall that in resolving the return current problem by a dissipative drift across the field lines, the classical model of Chapter 13 requires that all the energy loss be in the form of hard photons. The present model predicts instead a modest but observable hard photon emission, with typically about 10 per cent of the return current occurring through the layer (the rest flowing back through the layer from the wind zone, as in a standard stellar wind model). It is mildly encouraging that some years ago, the Compton Gamma-Ray Observatory discovered three new 'gamma-ray pulsars' to add to the well-known Crab, Vela and Geminga pulsars, and that, more recently, two more have been added to the list. For these, the gamma-ray luminosity goes up to about 10 per cent of the spin-down luminosity L_Ω, varying roughly like the $L_\Omega^{1/2}$. Other observations are finding a higher proportion than earlier of spin-down energy going into hard photons (Gunji *et al.* 1994). The observational situation is reported by Thompson *et al.* 1997 and Thompson 1998. Saito *et al.* (1997) and Kuiper *et al.* (1998) report on two millisecond pulsars which show Crab-like double pulses in X-rays. It is too early to say whether the emission from the layer is indeed the primary source of gamma- and X-rays (cf. e.g. Cheng *et al.* 1986); but one can hazard a guess that the spontaneous development of a dissipative domain, with consequent radiative losses at some level, is a universal feature.

The model has some similarities with that of Beskin *et al.* (1993) but also some important differences. Their model has a large potential jump at the stellar surface, which they relate to the poloidal current function by applying a balance condition along the separatrix between the open and closed field line domains. As already remarked, in the absence of a significant quantum mechanical work function at the star's surface, it is questionable whether their 'gap' would in fact survive. It is more appropriate to impose the GJ condition $\mathbf{E}\cdot\mathbf{B} = 0$ at the star's surface, and construct the potential increase further out self-consistently from the equations for relativistic acceleration and pair production, as discussed in Section 14.1. Further, as pointed out by Lyubarskii (1992*a*), their balance condition implicitly suppresses any contribution to the Maxwell stresses of a discontinuity in the poloidal field strength across the separatrix field line, which could destroy their relation between the potential jump and poloidal current.

Again, as in Beskin *et al.* (and indeed as found by several other workers), the model finds a breakdown in MHD-type flow beyond the l-c, and also in an equatorial disc, but it is argued here that this is a *reductio ad absurdum* of the adoption of dissipation-free flow: the macroscopic constraints enforce a spontaneous development of a domain beyond the l-c where $\mathbf{E}\cdot\mathbf{B}$ is not small and where there are in general rapid changes in $\mathbf{B}$. A large multiplicity M ensures that the associated currents can be carried without requiring superluminal relative velocities of electrons and positrons. As in a normal plasma (cf. Chapter 2), the strong $E_\parallel$-component in the layer drives oppositely-directed charges in opposite senses (as contrasted with the same sense of drift enforced by the $E_\perp$-component). A large resistivity is needed to limit the consequent relative drift of the moderate-γ electrons and positrons, so that both species take part in the outflow. The ordinary microresistivity of the plasma is far too small, and the most obvious collective plasma two-stream instability, extrapolated into the non-linear domain (Section 2.8), also does not yield in the time available a powerful enough 'anomalous' resistivity when acting directly on the poloidal currents. However, the examples of force-free fields mentioned in Section 14.3 have toroidal components which become large compared with the poloidal as the surface of breakdown is approached. A likely spontaneous hydromagnetic instability of such a field would be equivalent to a large macroscopic resistivity (cf. Sections 3.6 and 10.9; Lyubarskii 1992*b*; Begelman 1998). Future work should focus on the macro- and microphysics of both the dissipative layer and the equatorial disc.

The suggestion came early that the spin-down of an aligned or nearly aligned pulsar should be by the relativistic analogue of a magnetic stellar wind. Michel (1969) and Goldreich and Julian (1970) pioneered the study of cold relativistic centrifugal winds (Section 7.9). Hot winds were studied by Kennel *et al.* (1983). Near the star the transport of energy and angular momentum is essentially all by the Poynting vector and the Maxwell stresses. As noted in Chapters 7 and 10, a strictly dissipation-free wind predicts a dominant toroidal field component at infinity and so also a Poynting flux that dominates over the particle energy flux: in the Michel/GJ notation (7.140), $\sigma \equiv \text{Poynting flux/kinetic energy flux} \gg 1$. This leads to a difficulty, pointed out by Rees and Gunn (1974). Their model for the energization of the Crab nebula by the obliquely rotating Crab pulsar combines a modified Deutsch–Pacini wave and a relativistic wind. The wind is decelerated by a strong shock at about 3×10^{17} cm from the pulsar. To make the post-shock speed decrease with radial distance, so as to match the observed 200 km s^{-1} expansion speed of the outer Crab nebula, one needs rather that σ be small upstream – a large σ upstream yields a large σ downstream. Kennel and Coroniti (1984) came to a similar conclusion for their $e^{\pm}$ wind. The Mestel–Shibata model discussed above is an improvement in that it has $\sigma \simeq 10$ after the dissipation layer, smaller but still too large. However, again the difficulty may disappear if it can be shown that the linear instabilities of a dominant toroidal field, when extrapolated into the non-linear domain, lead to rapid conversion of excess toroidal magnetic energy into kinetic energy (e.g. Bogovalov 1997; Begelman 1998). This should occur sufficiently far beyond the l-c to avoid synchrotron losses predicting too much X-ray and gamma-ray emission near the star.

The oblique rotator offers other opportunities. Coroniti (1990) points out that the wind emitted by an oblique rotator will be magnetically 'striped', with B_Φ of alternating sign. He argues for the systematic mutual annihilation of opposite polarity stripes as the wind flows out, an effect somewhat analogous to the Rädler-effect within an oblique rotator (Section 9.3). It was noted in Mestel *et al.* (1976) that in the oblique problem, the extrapolation of the GJ *ansatz* $\mathbf{E} \cdot \mathbf{B} = 0$ to infinity led to unacceptable conclusions. In a much more detailed study, Melatos and Melrose (1996) have recently argued that it is indeed the breakdown in the ability of the charges to short out the displacement current that enforces efficient dissipation of the large toroidal fields.

APPENDIX

The magnetosphere with pair production: an illustrative example

The model – discussed in detail in Mestel and Shibata (1994) – is portrayed schematically in Fig. 14.10. To the left of the dissipation layer the force-free approximation for the field is adopted. To the right, the field equation (14.135) contains the dimensionless inertial term $M_{\mathrm{A}}^2 \equiv \sum 4\pi\gamma_{\mathrm{s}} m_{\mathrm{s}} \eta_{\mathrm{s}} \kappa_{\mathrm{s}}$. As before, the primary electrons plus the $e^{\pm}$-gas are treated as a single plasma of density n and associated mean γ, so that in terms of dimensionless quantities and the parameter ϵ_g,

$$M_{\mathrm{A}}^2 = \epsilon_g \gamma \eta^2 / n = M \epsilon_g \gamma \eta^2 / n_e \tag{14.139}$$

where again n_e is the density of primaries and M the multiplicity factor. Clearly, some plausible assumption must be made about the analytical form of M_{A}^2. The relation

$$n \propto \tilde{n} \tilde{x}^2 / x^2 \tag{14.140}$$

is adopted, where $\tilde{n}$ is the density just to the right of $\tilde{x}$; its variation with z is ignored. In addition, γ is assumed constant. This is probably a fair approximation for an individual streamline. Most of the relativistic acceleration due to $\mathbf{E}_{\|}$ has occurred well within the l-c. The centrifugal slingshot term (due to $\mathbf{E}_{\perp}$) has a strong effect only when $\tilde{\beta}\omega x^2$ is near unity and the non-dissipative flow is approaching breakdown, and the model specifically restricts this behaviour to the dissipation layer. Thus the provisional approximation

$$M_{\mathrm{A}}^2 = \delta x^2 \tag{14.141}$$

is made, with the parameter δ, like $\tilde{\beta}$, taken as constant over the whole domain.

There will be a loss of both energy and angular momentum in the dissipation layer. As discussed in detail in Section 13.2.4, radiation of angular momentum yields a drift of particles to neighbouring field lines of different corotational potential. This spontaneously converts electric potential energy into kinetic energy, offsetting the inevitable kinetic energy loss accompanying the angular momentum loss. Thus although before the dissipation layer, $\varepsilon = (\epsilon - \tilde{\alpha} l)$ given by (14.126) is small, beyond the layer the analogous quantity $(\epsilon - \beta l)$ will not in general be negligible. Beyond the dissipation layer, the angular momentum and energy integrals (14.129) and (14.130) are written

$$m\gamma\eta\Omega\varpi^2 - \varpi B_\phi / 4\pi = l_{\mathrm{out}}(P), \tag{14.142}$$

$$m\gamma\eta c^2 - \tilde{\beta}\alpha\varpi B_\phi / 4\pi = \epsilon_{\mathrm{out}}(P), \tag{14.143}$$

where the suffix 'out' refers to the points $(\tilde{x} + 0, z)$, just to the right of the layer. There is now a contribution from the term $\Sigma(n_{\mathrm{s}}/\eta_{\mathrm{s}})\,\mathrm{d}\varepsilon_{\mathrm{s}}/\mathrm{d}P$ in (14.132). In the same phenomenological spirit, the ratio of energy loss to angular momentum loss is treated as a parameter to be adjusted: we write

$$\Lambda = \beta l/\epsilon \tag{14.144}$$

with Λ independent of P. The variation of η between different field lines is again neglected. With this series of draconian approximations applied to (14.132), we arrive at the field equation for the domain to the right of the layer:

$$(1-(\delta+\tilde{\beta}^2)x^2)\nabla^2 P - \frac{2}{x}\frac{\partial P}{\partial x} + \left(S + \frac{1-\Lambda}{\Lambda}\frac{\tilde{\beta}x^2}{\kappa}\right)\frac{\mathrm{d}l}{\mathrm{d}P} = 0, \tag{14.145}$$

where all quantities are in their dimensionless forms.

To the left of $\tilde{x}$, besides the dropping of the inertial terms, the force-free equation is approximated further: the small deviation of $\tilde{\alpha}$ from unity is ignored, so the equation to be solved has the form

$$(1-x^2)\nabla^2 P - \frac{2}{x}\frac{\partial P}{\partial x} + S\frac{\mathrm{d}S}{\mathrm{d}P} = 0. \tag{14.146}$$

As explained above, the expected dissipative process in the magnetosphere is assumed to occur within a thin cylindrical layer at $x = \tilde{x}$, at which the inner, force-free solution is linked up with the inertial wind solution by jump conditions. For $x < \tilde{x}$, the MHD kinematic condition and the associated electric field (in dimensional form) are

$$\mathbf{v} = \alpha\varpi\mathbf{t} + \kappa\mathbf{B}, \qquad \mathbf{E} = -(\alpha\varpi/c)\mathbf{t}\times\mathbf{B} = \nabla(\alpha P/c); \tag{14.147}$$

for $x > \tilde{x}$,

$$\mathbf{v} = \tilde{\beta}(\alpha\varpi)\mathbf{t} + \kappa\mathbf{B}, \qquad \mathbf{E} = \tilde{\beta}\nabla(\alpha P/c). \tag{14.148}$$

(From now on the tilde on β will be dropped, so β is a dimensionless quantity.) The flow is relativistic on both sides, and the condition $(v/c)^2 = 1$ fixes κ, analogously to (14.113). The mass flux η per unit flux tube is continuous, so a change in κ implies a corresponding change in n.

One of the parameters of the problem is the angular momentum loss in the dissipation domain, which is introduced by writing (cf. (14.142))

$$[-\varpi B_\phi/4\pi + \gamma m\Omega\varpi^2\eta]_{\text{out}} = l_{\text{out}} = l_0(P)(1-\Delta l), \tag{14.149}$$

where $l_0 = -\varpi B_\phi/4\pi$ is the inflow of angular momentum from the star into the layer along a unit flux tube, and Δl is taken as constant. The associated net energy loss is parametrized by Λ according to our assumption (14.144):

$$[-\beta\alpha\varpi B_\phi/4\pi + \gamma mc^2\eta]_{\text{out}} = \epsilon_{\text{out}}(P) = \beta\alpha l_0(P)(1-\Delta l)/\Lambda. \tag{14.150}$$

For the third dynamical condition we adopt the continuity of the (xz)-component of the tensor $(T_{ij} + nm\gamma v_i v_j)$, where T_{ij} the total Maxwell stress tensor $[(\mathbf{B}^2+\mathbf{E}^2)/8\pi]\delta_{ij}-(B_iB_j+E_iE_j)/4\pi$. This implicitly assumes that there is no radiation of z-momentum from the dissipation layer. The *integrated* loss of z-momentum from the whole system is certainly zero because of the equatorial symmetry of the problem. Further, relativistic beaming of radiation from particles moving along field lines will ensure much mutual cancellation within each hemisphere from regions with $B_z<0$ and $B_z>0$ respectively. However, the neglect of all local z-momentum loss is clearly a provisional approximation, to be checked subsequently. This third condition then reduces to the normalized form

$$\delta=\frac{(1-\tilde{x}^2)\,\Delta B_z+(1-\beta^2)\tilde{x}^2(B_z)_{\rm out}}{(B_z)_{\rm out}\tilde{x}^2}. \tag{14.151}$$

Some provisional choice must be made for the function $S(P)$ describing the toroidal field and so also the angular momentum flux in the force-free domain. From (14.58), the dimensionless quantities satisfy

$$\mathrm{d}S/\mathrm{d}P=j_{\rm p}/B_{\rm p}=-2(B_z/B_{\rm p})\bar{J}=-2V_0(P). \tag{14.152}$$

Near the poles, and near the star, $B_{\rm p}\simeq B_z$ and so $\bar{J}\simeq V_0\simeq 1$; hence $S\simeq -2P$ at small P. There must also be a value $\tilde{P}$ where $\mathrm{d}S/\mathrm{d}P=0$, with $j_{\rm p}=(\mathrm{d}S/\mathrm{d}P)B_{\rm p}$ negative for $P>\tilde{P}$ and positive for $P<\tilde{P}$. There will also be a field-streamline P_d on which S is again zero, so that there is no net flow of charge out of the star: P_d separates the dead zone from the active zone. The model has the return current density finite rather than compressed into a sheet on P_d (cf. Beskin *et al.* 1993). In this first example, we adopt the simple quadratic form

$$S(P)=-2P(1-P/P_d), \tag{14.153}$$

for which $\tilde{P}=P_d/2$.

Recall that the layer is a domain with a high (as yet unspecified) effective resistivity, so we require a local strong change in voltage, *of the right sign*, along a field line as it crosses the layer. From (14.148), it is seen that provided $\beta<1$ – isosubrotation to the right – then the potential difference across the layer is as required: $(\phi_{\rm out}-\phi_{\rm in})=\alpha(1-\beta)(P-\tilde{P})/c<0$ for $P<\tilde{P}$ and >0 for $P>\tilde{P}$ respectively.

The jump conditions can be solved to yield in dimensionless form

$$-(xB_\phi)_{\rm out}=\frac{1-\beta^2\tilde{x}^2/\Lambda}{1-(\delta+\beta^2)\tilde{x}^2}l_0(1-\Delta l) \tag{14.154}$$

and

$$\kappa_{\rm out}=\frac{(1-(\delta+\beta^2)\tilde{x}^2)}{(1-\Lambda-\delta\tilde{x}^2)}\frac{\delta\tilde{x}^2}{\beta}\frac{\Lambda}{4\pi l_0(1-\Delta l)}. \tag{14.155}$$

Details of the procedure adopted to obtain a numerical solution are given in Mestel and Shibata (1994). Once the position $\tilde{x}$ of the dissipative layer has

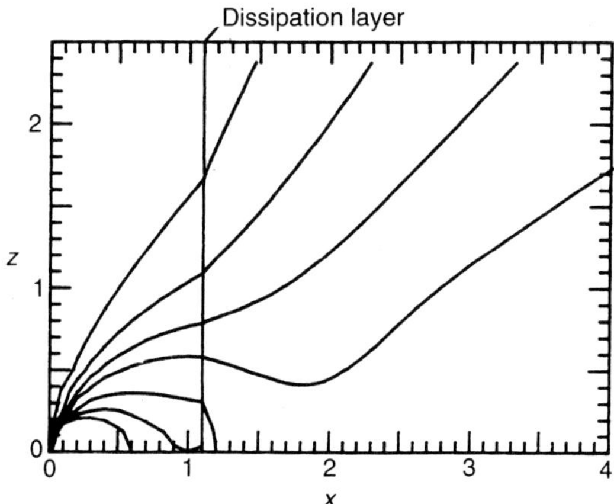

FIG. 14.11. The poloidal magnetic field structure for the global model with parameters $\beta = 0.877$ and $\hat{S}_{\text{out}} = 0.9$. The thin dissipative layer is located at $\tilde{x} = 1.1$, where the jump conditions are imposed. Within the layer the field is approximated to be force-free, while beyond the layer the dissipation-free wind with finite inertia is adopted. The poloidal current emerges at low latitudes and returns at high latitudes. Some of the current flows to infinity before returning to the star, but some returns via the dissipation layer. The layer loses energy and angular momentum via radiation. The layer has finite thickness with an electric field component along $\mathbf{B}_{\text{p}}$, so that the angular velocity of the wind is less than that of the inner, force-free domain.

been chosen, there remain four parameters: the subrotation constant β; the inertial parameter δ; the energy/angular momentum parameter Λ; and $\hat{S}_{\text{out}} = [-\varpi B_\phi(\tilde{x}+0)]/[-\varpi B_\phi(\tilde{x}-0)] = [-\varpi B_\phi(\tilde{x}+0)]/l_0$, the fractional angular momentum being carried by the Poynting vector just to the right of the layer. Beyond the thin, cylindrical dissipative layer at $x = \tilde{x}$, the poloidal component of the equation of motion perpendicular to the field has been reduced to the form (14.145). This has to be solved subject to the field-aligned equations – the conservation of mass, energy and angular momentum, and the isorotation law. If the poloidal field function P is known, then the field-aligned equations – subject to an appropriate boundary condition just outside the layer – have a uniquely determined solution, regular at the fast critical point (which is at infinity if thermal pressure is neglected); thus the quantities $l_{\text{out}}(P)$, S, $\bar{\kappa}$, etc. are obtained at points along each field-streamline. Equally, if the field-aligned flow is prescribed, the trans-field equation (14.145) can be solved for the poloidal field structure. Thus the construction of a complete, self-consistent solution requires iteration.

The various simplifying assumptions made – a cylindrical dissipative layer, constancy of β, δ, Δl and Λ – enable attack on (14.145) and (14.146) by Fourier transformation. Given a set of parameters, a trial function $P(x, z)$ is inserted into the non-linear term in (14.145), and the resulting linear equation is solved by the Fourier method (cf. Section 13.1.3), now applied for $\tilde{x} < x < \infty$. This new P is again fed into the non-linear term and the process continued until the input and output P functions are consistent. For $x < \tilde{x}$ a similar technique is applied to the much simpler force-free equation (14.146), with continuity at the l-c of P, P_x imposed. The solutions for P are linked at the layer $\tilde{x}$ by the jump conditions (14.151), (14.154) and (14.155).

To carry out this procedure, one needs to know the equatorial values $B_x(x)$. It is assumed that the field is 'open' beyond the equatorial radius $x_0 \equiv \delta^{-1/2}$ at which the poloidal flow speed reaches the poloidal Alfvén speed, i.e. $4\pi m\gamma n v_{\rm p}^2/B_{\rm p}^2 = x^2\delta = 1$. Within x_0 the field lines crossing the layer $\tilde{x}$ are closed, but the pair plasma accumulating in the equatorial disc $\tilde{x} < x < x_0$ either continuously or sporadically breaks out through 'turbulent diffusion'.

It turns out that quite stringent constraints can be placed on the parameter values. To keep δ finite, the jump ΔB_z at $\tilde{x}$ has to vanish at the point z_c where $B_z(z_c) = 0$. An arbitrarily chosen wind solution would not satisfy this condition: the required regularity of δ at the layer thus provides a boundary condition for the solution in the wind domain, fixing x_0 and so also the 'open' magnetic flux – the fraction crossing the layer that reaches infinity. Other constraints follow from the requirements that S and κ be positive; that the energy and angular momentum lost from the layer be positive; that the plasma beyond the layer does not rotate backwards; and that the angular momentum lost is less than the maximum allowed by the Gold–Holloway condition (cf. Section 13.2.4).

A typical constructed field is shown in Fig. 14.11. The results of the numerical calculations are summarized as follows:

1. The maximum allowed value of $\delta \simeq 0.2$.
2. For the solution to exist, the normalized angular velocity of the wind domain β is restricted to a narrow range, with $\beta \sim 0.9$. The associated potential variation in the dissipation layer is typically $(1-\beta)B_{\rm lc}\varpi_c = 6.6 \times 10^{11}(B_{12}r_6^3)[(1-\beta)/0.1]{\rm P}_1^{-2}$ V, where $\rm P_1$ is again the period in seconds. It is interesting that the light cylinder for the wind, $x = \beta^{-1}$ coincides roughly with the dissipation layer.
3. The fractional Poynting flux going out through the layer is given by $\beta\hat{S}_{\rm out}$ with $0.65 < \hat{S}_{\rm out} \leq 1$. The remaining fraction is converted partly into radiation from the layer and partly into wind plasma energy, which could be kinetic energy of bulk motion or random motion.
4. For the illustrative example with $\beta = 0.877$ and $\hat{S}_{\rm out} = 0.9$ (Fig. 14.11), the kinetic energy flux of the wind plasma is $\epsilon_{\rm K.E.} = 0.188$ of the total energy flux, and $\delta = 0.1$ implying $\gamma \simeq \delta(\gamma_m/M)$ ($\simeq 10^6$ for the Crab parameters). This is in striking contrast with the standard relativistic centrifugal wind theory with a radial field (Section 7.9), which predicts $\gamma_\infty = (\gamma_m/M)^{1/3}$: the reduction across the dissipation layer of the fractional Poynting flux is associated with a marked increase in the bulk kinetic energy. However, it must be stressed that the largest possible δ is $\simeq 0.2$, and therefore the wind emanating from the dissipation layer is still Poynting-flux dominated.

5. The energy radiated away from the dissipation layer is 2.3 per cent of the total energy flux for the model $\beta = 0.877$ and $\hat{S}_{\text{out}} = 0.9$. This value has uncertainty of $\sim$ 20 per cent because of the uncertainty in the fitting of $\kappa(\bar{x})$. In the several fitted models obtained, the radiative loss of the layer is at most 10 per cent of the total energy flux.
6. The poloidal field kinks at the layer, polewards at high latitudes and equatorwards at the low latitudes. The open field lines are more or less radial at around a few light-cylinder radii. At large distances, the poloidal magnetic field strength on the equator is found to decrease more rapidly than that for the radial geometry, suggesting collimation of the flow along the rotation axis (cf. Section 10.9). The collimation is stronger for smaller δ.
7. The amount of magnetic flux for $\beta = 0.877$ and $\hat{S}_{\text{out}} = 0.9$ that crosses the l-c is $0.666 P_{\text{lc}}$, where P_{lc} is the flux parameter of the last field line to reach the l-c. The remaining third of this 'open' flux forms a domain of closed field lines, outside the dead zone of field lines that close within the l-c. The plasma flowing along these field lines must form initially a dense equatorial disc. The model assumes implicitly that there is a further dissipative process occurring which enables the accumulating plasma and the associated energy and angular momentum to escape across the field. This may be triggered by a macroscopic instability following excess accumulation of gas.

References for Chapters 13 and 14

Abramowitz, M. and Stegun, I.A. (1970). *Handbook of mathematical functions* (9th edn). Dover, New York.

Arons, J. (1972). *Astrophysical Journal*, **177**, 395.

Arons, J. (1981). *Astrophysical Journal*, **248**, 1099.

Arons, J. (1983). *Astrophysical Journal*, **266**, 215.

Arons, J. (1992). In *Magnetospheric structure and emission mechanisms of radio pulsars* (eds J.A. Gil, T.M. Hankins and J.M. Rankin), IAU Colloquium **128**, p. 56. Pedagogical University Press, Zielona Gora, Poland.

Arons, J. (1996). In *Pulsars: problems and progress* (eds S. Johnston, M.A. Walker and M. Bailes), IAU Colloquium **160**, p. 177. *Astronomical Society Pacific Conference Series*, **105**.

Arons, J. and Scharlemann, E.T. (1979). *Astrophysical Journal*, **231**, 854.

Begelman, M.C. (1998). *Astrophysical Journal*, **493**, 291.

Beskin, V.S. (1990). *Soviet Astronomy Letters*, **16**, 286.

Beskin, V.S., Gurevich, A.V. and Istomin, Ya.N. (1993). *Physics of the pulsar magnetosphere*. Cambridge University Press.

Blandford, R.D. (1972). *Astronomy and Astrophysics*, **20**, 135.

Blandford, R.D., Applegate, J. and Hernquist, L. (1983). *Monthly Notices Royal Astronomical Society*, **204**, 1025.

Bogovalov, S.V. (1997). *Astronomy and Astrophysics*, **327**, 662.

Burman, R.R. (1984). *Proceedings Astronomical Society Australia*, **5**, 467.

Burman, R.R. and Mestel, L. (1978). *Australian Journal Physics*, **31**, 455.
Burman, R.R. and Mestel, L. (1979). *Australian Journal Physics*, **32**, 681.
Campbell, C.G. (1997). *Magnetohydrodynamics in binary stars*. Kluwer, Dordrecht.
Cheng, K.S., Ho, C. and Ruderman, M.A. (1986). *Astrophysical Journal*, **300**, 500, 522.
Cohen, J.M. and Toton, E.T. (1971). *Astrophysical Letters*, **7**, 213.
Contopoulos, I., Kazanas, D. and Fendt, C. (1998). *Astrophysical Journal*, in press.
Coroniti, F.V. (1990). *Astrophysical Journal*, **349**, 538.
da Costa, A.A. and Kahn, F.D. (1982). *Monthly Notices Royal Astronomical Society*, **199**, 211.
Daugherty, J.K. and Harding, A.K. (1982). *Astrophysical Journal*, **252**, 337.
Endean, V.G. (1972*a*). *Nature Physical Science*, **237**, 72.
Endean, V.G. (1972*b*). *Monthly Notices Royal Astronomical Society*, **174**, 125.
Endean, V.G. (1974). *Astrophysical Journal*, **187**, 359.
Erber, T. (1966). *Reviews Modern Physics*, **38**, 626.
Erdelyi, A., Magnus, W., Oberhettinger, F. and Tricomi, F.G. (1954). *Tables of integral transforms*. McGraw-Hill, New York.
Fawley, W.M., Arons, J. and Scharlemann, E.T. (1977). *Astrophysical Journal*, **217**, 227.
Fitzpatrick, R. and Mestel, L. (1988*a*). *Monthly Notices Royal Astronomical Society*, **232**, 277.
Fitzpatrick, R. and Mestel, L. (1988*b*). *Monthly Notices Royal Astronomical Society*, **232**, 303.
Flowers, E.G., Lee, J.-F., Ruderman, M.A., Sutherland, P.G., Hillebrandt, W. and Müller, E. (1977). *Astrophysical Journal*, **215**, 291.
Frank, J., King, A.R. and Raine, D.J. (1992). *Accretion power in astrophysics* (2nd edn). Cambridge University Press.
Gold, T. (1969). *Nature*, **221**, 25.
Gold, T.(1980). *Ninth Texas symposium on relativistic astrophysics* (eds J. Ehlers, J.J. Perry and M. Walker), p. 448. New York Academy of Sciences.
Goldreich, P. and Julian, W.H. (1969). *Astrophysical Journal*, **157**, 869.
Goldreich, P. and Julian, W.H. (1970). *Astrophysical Journal*, **160**, 971.
Gunji, S., Hirayama, M., Kamae, T., Miyazaki, S., Sekimoto, Y., Takahashi, T *et al.* (1994). *Astrophysical Journal*, **428**, 284.
Gunn, J.E. and Ostriker, J.P. (1971). *Astrophysical Journal*, **165**, 523.
Holloway, N.J. (1977). *Monthly Notices Royal Astronomical Society*, **181**, 9.
Jeffreys, H. and Jeffreys, B.S. (1956). *Methods of mathematical physics* (3rd edn). Cambridge University Press.
Jones, P.B. (1980). *Monthly Notices Royal Astronomical Society*, **192**, 847.
Kahn, F.D. (1971). Paper read at Royal Astronomical Society, unpublished.
Kahn, F.D. and da Costa, A.A. (1994). In *Cosmical magnetism* (ed. D. Lynden-Bell), p. 99. Kluwer, Dordrecht.
Kennel, C.F. and Coroniti, F.V. (1984). *Astrophysical Journal*, **283**, 694.

Kennel, C.F., Fujimura, F.S. and Okamoto, I. (1983). *Astrophysical Geophysical Fluid Dynamics*, **26**, 147.

Krause-Polstorff, J. and Michel, F.C. (1985*a*). *Astronomy and Astrophysics*, **144**, 72.

Krause-Polstorff, J. and Michel, F.C. (1985*b*). *Monthly Notices Royal Astronomical Society*, **213**, 43p.

Kuiper, L., Hermsen, W., Verbunt, F. and Belloni, T. (1998). Submitted to *Astronomy and Astrophysics.*

Landau, L.D. and Lifshitz, E.M. (1975). *The classical theory of fields*. Pergamon Press, Oxford.

Lyubarskii, Yu.E. (1992*a*). In *Magnetospheric structure and emission mechanisms of radio pulsars* (eds J.A. Gil, T.M. Hankins and J.M. Rankin), IAU Colloquium **128**, p. 112. Pedagogical University Press, Zielona Gora, Poland.

Lyubarskii, Yu.E. (1992*b*). *Soviet Astronomy Letters*, **18**, 5, 356.

Macdonald, D.A. and Thorne, K.S. (1982). *Monthly Notices Royal Astronomical Society*, **198**, 345.

Melatos, A. and Melrose, D.B. (1996). *Monthly Notices Royal Astronomical Society*, **279**, 1168.

Melrose, D.B. (1996). In *Pulsars: problems and progress* (eds S. Johnston, M.A. Walker and M. Bailes), IAU Colloquium **160**, p. 139. *Astronomical Society Pacific Conference Series*, **105**.

Mestel, L. (1971). *Nature Physical Science*, **233**, 149.

Mestel, L. (1973). *Astrophysics Space Science*, **24**, 289.

Mestel, L. (1981). In *Pulsars* (eds W. Sieber and R. Wielebinski), IAU Symposium **95**, p. 9. Reidel, Dordrecht.

Mestel, L. (1996). In *Pulsars: problems and progress* (eds S. Johnston, M.A. Walker and M. Bailes), IAU Colloquium **160**, p. 417. Astronomical Society Pacific Conference Series **105**.

Mestel, L. and Pryce, M.H.L. (1992). *Monthly Notices Royal Astronomical Society*, **254**, 355.

Mestel, L. and Selley, C.S. (1970). *Monthly Notices Royal Astronomical Society*, **149**, 197.

Mestel, L. and Shibata, S. (1994). *Monthly Notices Royal Astronomical Society*, **271**, 621.

Mestel, L. and Spruit, H.C. (1987). *Monthly Notices Royal Astronomical Society*, **226**, 57.

Mestel, L. and Wang, Y.-M. (1979). *Monthly Notices Royal Astronomical Society*, **188**, 799.

Mestel, L. and Wang, Y.-M. (1982). *Monthly Notices Royal Astronomical Society*, **198**, 405.

Mestel, L., Wright, G.A.E. and Westfold, K.C. (1976). *Monthly Notices Royal Astronomical Society*, **175**, 257.

Mestel, L., Phillips, P. and Wang, Y.-M. (1979). *Monthly Notices Royal Astronomical Society*, **188**, 385.

Mestel, L., Robertson, J.A., Wang, Y.-M. and Westfold, K.C. (1985). *Monthly Notices Royal Astronomical Society*, **217**, 443.

Mestel, L., Panagi, P. and Shibata, S. (1998). Submitted to *Monthly Notices Royal Astronomical Society.*

Michel, F.C. (1969). *Astrophysical Journal*, **158**, 727.

Michel, F.C. (1973). *Astrophysical Journal*, **180**, 207.

Michel, F.C. (1991). *Theory of neutron star magnetospheres.* University of Chicago Press.

Muslimov, A.G. and Tsygan, A.I. (1986). *Astronomicheskii Zhurnal*, **63**, 958.

Muslimov, A.G. and Tsygan, A.I. (1990). *Astronomicheskii Zhurnal*, **67**, 263.

Muslimov, A.G. and Tsygan, A.I. (1992*a*). In *Magnetospheric structure and emission mechanisms of radio pulsars* (eds J.A. Gil, T.M. Hankins and J.M. Rankin), IAU Colloquium **128**, p. 248. Pedagogical University Press, Zielona Gora, Poland.

Muslimov, A.G. and Tsygan, A.I. (1992*b*). *Monthly Notices Royal Astronomical Society*, **255**, 61.

Ostriker, J.P. and Gunn, J.E. (1969). *Astrophysical Journal*, **157**, 1395.

Pacini, F. (1967). *Nature*, **216**, 567.

Rees, M.J. (1971). *Nature Physical Science*, **230**, 55.

Rees, M.J. and Gunn, J.E. (1974). *Monthly Notices Royal Astronomical Society*, **167**, 1.

Ruderman, M.A. and Sutherland, P.G. (1975). *Astrophysical Journal*, **196**, 51.

Saito, Y., Kawai, N., Kamae, T., Shibata, S., Dotani, T. and Kulkarni, S.R. (1997). *Astrophysical Journal Letters*, **477**, L37.

Scharlemann, E.T. and Wagoner, R.V. (1973). *Astrophysical Journal*, **182**, 951.

Scharlemann, E.T., Arons, J. and Fawley, W.M. (1978). *Astrophysical Journal*, **222**, 297.

Shibata, S. (1991). *Astrophysical Journal*, **378**, 239.

Shibata, S. (1996). In *Pulsars: problems and progress* (eds S. Johnston, M.A. Walker and M. Bailes), IAU Colloquium **160**, p. 409. *Astronomical Society Pacific Conference Series*, **105**.

Shibata, S. (1997). *Monthly Notices Royal Astronomical Society*, **287**, 262.

Shibata, S., Miyazaki, J. and Takahara, F. (1998). *Monthly Notices Royal Astronomical Society*, **295**, L53.

Sturrock, P.A. (1971). *Astrophysical Journal*, **164**, 529.

Thompson, D.J. (1998). In *Neutron stars and pulsars* (eds N. Shibazaki, N. Kawai, S. Shibata and T. Kifune), p. 273. *Frontiers Science Series* **24**, Universal Academy Press, Tokyo.

Thompson, D.J., Harding, A.K., Hermsen, W. and Ulmer, M.P. (1997). In *Proceedings fourth Compton symposium* (eds C.D. Dermer, Strickman, J.D. and Kurfess, J.D.), p. 39. *American Institute Physics Conference Proceedings*, **410**.

Toulmin, S. and Goodfield, J. (1961). *The fabric of the heavens*. Penguin Books, London.

Usov, V.V. (1987). *Astrophysical Journal*, **320**, 333.

Usov, V.V. (1988). In *Plasma astrophysics*, p. 257. ESA SP-285 **1**.

Usov, V.V. (1996). In *Pulsars: problems and progress* (eds S. Johnston, M.A. Walker and M. Bailes), IAU Colloquium **160**, p. 323. *Astronomical Society Pacific Conference Series*, **105**.

INDEX

CARDIFF UNIVERSITY
PRIFYSGOL CAERDYDD